中国石油集团石油管工程技术研究院
"十三五"科技成果汇编
（2016—2020年）

中国石油集团石油管工程技术研究院
石油管材及装备材料服役行为与结构安全国家重点实验室　编

石油工业出版社

内 容 提 要

本书汇集了中国石油集团石油管工程技术研究院和石油管材及装备材料服役行为与结构安全国家重点实验室"十三五"重大科技成果、科技论文、发布标准和授权专利，从侧面反映了"十三五"科技工作的主要成果，涉及超高钢级管材断裂与变形控制技术、复杂工况油套管柱失效控制与完整性技术、地下储气库关键技术及应用，以及柔性复合管开发与应用等关键技术的研究及工程应用。

本书适于从事石油管研究的工程技术人员和管理人员阅读参考。

图书在版编目（CIP）数据

中国石油集团石油管工程技术研究院"十三五"科技成果汇编 . 2016—2020 年／中国石油集团石油管工程技术研究院等编 . —北京：石油工业出版社，2021.8
ISBN 978-7-5183-4795-7

Ⅰ. 中… Ⅱ. ①中… Ⅲ. ①石油管道-科技成果-汇编-中国-2016-2020 Ⅳ. ①TE9783

中国版本图书馆 CIP 数据核字（2021）第 156395 号

出版发行：石油工业出版社
　　　　　（北京安定门外安华里 2 区 1 号　100011）
　　网　　址：www.petropub.com
　　编辑部：（010）64523687　图书营销中心：（010）64523633
经　销：全国新华书店
印　刷：北京中石油彩色印刷有限责任公司

2021 年 8 月第 1 版　2021 年 8 月第 1 次印刷
787×1092 毫米　开本：1/16　印张：38.25
字数：976 千字

定价：220.00 元
（如出现印装质量问题，我社图书营销中心负责调换）
版权所有，翻印必究

《中国石油集团石油管工程技术研究院"十三五"科技成果汇编（2016—2020年）》编辑委员会

主　任：刘亚旭

副主任：霍春勇　赵新伟　秦长毅　殷　明

委　员：（按姓氏笔画排序）

马秋荣　尹成先　卢攀辉　池　强　张　华　张广利

杨专钊　李记科　杜迅风　佘伟军　陈宏达　李建民

陈娟利　张鸿博　林　凯　罗金恒　戚东涛　韩礼红

翟云萱　樊治海

主　编：刘亚旭

副主编：冯耀荣　马秋荣

编辑组：宫少涛　陈宏远　冯　春　付安庆　李厚补　马卫锋

王建军　张　丹　方　伟　杨　溪　谢文江　杨　放

刘心可

前　　言

　　石油管工程是从石油管的服役工况条件出发，重点围绕石油管的力学行为、环境行为、先进材料的成分/组织/性能/工艺相关性、失效控制及预测预防、安全评价与完整性等领域开展系统研究，揭示石油管的失效机理和规律，提出失效控制方法，发展全寿命周期的安全可靠性和完整性技术，确保石油管的长期服役安全。

　　中国石油集团石油管工程技术研究院是中国石油天然气集团有限公司的直属研究院，也是我国唯一专业从事石油管及装备材料应用基础和工程应用研究的专业机构，经过几十年的发展，已成为国内石油行业在石油管工程技术领域唯一集"科学研究、质量监督、工程技术服务"为一体的综合性技术中心与核心科研机构，是为中国石油集团石油管工程技术提供决策支持的"参谋部"，开展石油管工程技术创新的"研发中心"，保障石油管质量安全的"检测评价中心"，为重大管道工程和油气田勘探开发项目提供石油管技术支持与服务的"技术中心"。

　　石油管工程技术研究院以石油管材及装备材料服役行为与结构安全国家重点实验室、中国石油集团石油管工程重点实验室、陕西省石油管材及装备材料服役行为与结构安全重点实验室为创新平台，围绕我国石油天然气勘探开发、管道储运和炼油化工用石油管材及装备材料服役安全领域面临的挑战和难题，扎实开展原始创新和关键技术研究，发展石油管材及装备的应用关键技术，建立失效控制理论，形成石油管材及装备材料服役行为与结构安全核心技术体系，确保石油管材及装备服役安全，构建了石油管工程核心技术体系，取得了一大批有重要影响力的科技成果，技术创新能力持续提升，形成了人才、装备、技术、资质品牌四大优势，核心竞争力和综合实力显著增强，在我国油气工业科技进步中发挥了重要作用。

　　"十三五"期间，石油管工程技术研究院和石油管材及装备材料服役行为与结构安全国家重点实验室以国家、集团公司、陕西省和院级重大科技项目为抓手，围绕我国油气工业重大工程和生产需求推进研发创新，高强度管线钢应用关键技术取得新突破，石油管新产品新技术开发取得新进展，石油石化材料领域重大失效事故分析取得新认识，原始创新和超前储备技术研究取得新成果，学科体系和核心技术体系不断完善。"十三五"期间，主持或参与国家级项目/课题37项，其中作为项目长单位承担国家重点研发计划2项，牵头承担国家重点研发课题9项，承担国家自然科学基金项目11项，承担省部级课题81项。获国家级、省部级、集团公司级和各类社会级成果51项，其中国家技术发明奖二等奖1项；累计发表论文800余篇，其中SCI/EI收录论文126篇；出版专著15部；授权专利270件，其中发明专利167件；修订国际标准5项，发布国家标准15项、行业标准48项。取得的一系列技术创新成果在中俄东线、西气东输四线等重点建设天然气管道工程和塔里木油田、西南油气田等苛刻环境和页岩气项目开发中，提供了重要的技术支持。

本书收集了石油管工程技术研究院和石油管材及装备材料服役行为与结构安全国家重点实验室"十三五"重大科技成果28项，代表性科技论文52篇，收录了授权专利和发布标准目录，从侧面反映了"十三五"石油管科技工作的主要成绩。

面向"十四五"，石油管工程技术研究院将瞄准建设国际一流研究院的目标，大力实施人才发展战略、技术创新战略、对外合作战略，积极构建创新联合体，共同推动新材料、新技术在油气田开发、管道建设及运行、新能源利用等领域的创新发展，助力我国能源技术变革和升级，实现石油管和装备材料领域的科技自立自强和产业链供应链的自主可控。为我国石油工业的发展，为集团公司建设世界一流综合性国际能源公司作出新的更大的贡献。

编　者

2021年4月

目 录

第一篇 科技成果篇

大口径高压输气管道断裂控制技术及应用 …………………………………（ 3 ）
大口径高压输气管道变形控制技术及应用 …………………………………（ 6 ）
基于应变的稠油蒸汽热采井套管柱设计方法及工程应用 …………………（ 7 ）
地下储气库运行与安全保障技术研究 ………………………………………（ 9 ）
海底管道腐蚀控制技术研发及应用 …………………………………………（ 10 ）
特殊用途管线钢管应用技术研究 ……………………………………………（ 11 ）
输气管道提高强度设计系数工程应用研究 …………………………………（ 12 ）
100亿立方米调峰能力储气库重大关键技术及应用 ………………………（ 13 ）
西部油田非金属管规范化选材与应用研究 …………………………………（ 14 ）
管道全尺寸爆破试验场建设及试验关键技术 ………………………………（ 15 ）
新型气密封特殊螺纹套管研发及应用 ………………………………………（ 16 ）
高钢级大口径焊管残余应力及控制技术研究 ………………………………（ 18 ）
油气管道系统完整性关键技术与工业化应用 ………………………………（ 19 ）
石油管材及装备防腐涂镀层开发与应用关键技术 …………………………（ 20 ）
高强度铝合金钻杆关键技术研究及应用 ……………………………………（ 21 ）
枯竭气藏型储气库管柱设计、选材与评价技术 ……………………………（ 22 ）
X70大应变管线钢管及应用关键技术 ………………………………………（ 23 ）
复杂多场耦合工况环境中油井管柱腐蚀控制理论研究 ……………………（ 24 ）
石油钻采机械用材料技术体系研究 …………………………………………（ 25 ）
X80高强度管道服役安全关键技术研究及应用 ……………………………（ 26 ）
西部油气田地面管网腐蚀评价及治理技术 …………………………………（ 27 ）
稠油蒸汽热采井套管柱应变设计及工程应用配套技术 ……………………（ 28 ）
厚壁高钢级管线钢与钢管应用基础理论研究 ………………………………（ 29 ）
高钢级管道断裂控制技术 ……………………………………………………（ 30 ）
OD1422mm X80管线钢管研制及应用技术 …………………………………（ 31 ）
复杂工况油气井管柱腐蚀控制技术及工程应用 ……………………………（ 33 ）

西部油气田集输管线内腐蚀控制技术及工程应用 ……………………………………………（36）
石油工业用高性能膨胀管及其性能评价技术 ………………………………………………（38）

第二篇　论文篇

输送管产品开发与工程应用技术支撑体系 …………李鹤林　霍春勇　池　强　等（41）
石油科研院所推进高质量发展探索与实践 …………刘亚旭　翟云萱　谢文江　等（51）
石油管及装备材料科技工作的进展与展望 …………刘亚旭　李鹤林　杜　伟　等（57）
复杂工况油套管柱失效控制与完整性技术研究进展及展望
　………………………………………………………冯耀荣　付安庆　王建东　等（64）
石油管工程技术进展及展望 …………………………冯耀荣　张冠军　李鹤林（74）
X80钢级 ϕ1422mm 大口径管道断裂控制技术 ………霍春勇　李　鹤　张伟卫　等（86）
316L衬里复合管道主要失效形式及其完整性检测技术研究
　………………………………………………………赵新伟　魏　斌　杨专钊　等（94）
高性能钻杆研发及应用进展 …………………………秦长毅　蒋家华　蒋存民　等（103）
复合材料增强管线钢管研究进展 ……………………………………………马秋荣（110）
X90管线钢母材和焊缝在近中性模拟溶液中不同加载电位下的应力腐蚀行为
　………………………………………………………罗金恒　雒设计　李丽锋　等（115）
热影响区软化的X70管线环焊缝应变容量分析 ……陈宏远　张建勋　池　强　等（123）
油气输送管道用钢管标准的发展历程及趋势 ………李为卫　谢　萍　杨　明　等（130）
煤层气井用抽油杆腐蚀疲劳寿命的影响因素 ………李德君　王　伟　庞　斌　等（140）
地下储气库注采管柱气密封螺纹接头优选分析 ……王建军　孙建华　薛承文　等（148）
X90超高强度输气钢管材料本构关系及断裂准则 …杨锋平　罗金恒　李　鹤　等（154）
双金属机械复合管环焊工艺及强度匹配设计研究现状及趋势
　………………………………………………………杨专钊　王高峰　闫　凯　等（165）
柔性管内衬高密度聚乙烯的气体渗透行为研究 ……张冬娜　李厚补　戚东涛　等（180）
外径1422mm的X80钢级管材技术条件研究及产品开发
　………………………………………………………张伟卫　李　鹤　池　强　等（190）
扩散温度对TC4合金表面Cu/Ni复合镀层结构及耐蚀性能的影响
　………………………………………………………朱丽霞　罗金恒　武　刚　等（200）
Constitutive Equation for Describing True Stress-Strain Curves over
　a Large Range of Strains ………………… Cao Jun　Li Fuguo　Ma Weifeng　et al（208）
Failure Analysis of a Sucker Rod Fracture in an Oilfield
　………………………………………… Ding Han　Zhang Aibo　Qi Dongtao　et al（217）
Effect of Streaming Water Vapor on the Corrosion Behavior of Ti60 Alloy under a Solid
　NaCl Deposit in Water Vapor at 600℃ ………… Fan Lei　Liu Li　Cui Yu　et al（226）

Corrosion Behavior of Reduced-Graphene-Oxide-Modified Epoxy Coatings on N80
　　Steel in 10.0wt% NaCl Solution ……… Feng Chun　Cao Yaqiong　Zhu Lijuan　et al（247）
Downhole Corrosion Behavior of Ni-W Coated Carbon Steel in Spent Acid & Formation Water
　　and Its Application in Full-Scale Tubing … Fu Anqing　Feng Yaorong　Cai Rui　et al（258）
Fatigue and Corrosion Fatigue Behaviors of G105 and S135 High-Strength Drill Pipe Steels
　　in Air and H_2S Environment ……………… Han Lihong　Liu Ming　Luo Sheji　et al（272）
Analysis of Corrosion Behavior on External Surface of 110S Tubing
　　……………………………………… Han Yan　Li Chengzheng　Zhang Huali　et al（293）
The Microstructure Evolution of Dual-Phase Ferrite-Bainite X70 Pipeline Steel with Plastic
　　Deformation at Different Strain Rates …… Ji Lingkang　Xu Tong　Wang Haitao　et al（304）
Deformation Stability of a Low-Cost Titanium Alloy Used for Petroleum Drilling Pipe
　　………………………………………… Jiang Long　Feng Chun　Liu Huiqun　et al（316）
Experimental and Simulation Investigation on Failure Mechanism of a Polyethylene Elbow
　　Liner Used in an Oilfield Environment ……… Kong Lushi　Fan Xin　Ding Nan　et al（321）
Failure Analysis and Solution to Bimetallic Lined Pipe
　　……………………………………………… Li Fagen　Li Xunji　Li Weiwei　et al（331）
Development and Application of Sour Service Pipeline Steel with Low Manganese Content
　　………………………………………… Li Yanhua　Chi Qiang　Li Weiwei　et al（337）
Investigation on Leakage Cause of Oil Pipeline in the West Oilfield of China
　　………………………………… Liu Qiang　Yu Haoyu　Zhu Guochuan　et al（344）
Failure Analysis of the 13Cr Valve Cage of Tubing Pump Used in an Oilfield
　　…………………………………………… Long Yan　Wu Gang　Fu Anqing　et al（356）
Corrosion Behavior Research of Aluminized N80 Tubing in Water Injection Well
　　………………………………………… Lu Caihong　Feng Chun　Han Lihong　et al（368）
Tribological Properties of Ni/Cu/Ni Coating on the Ti-6Al-4V Alloy after Annealing at
　　Various Temperatures ……………………… Luo Jinheng　Wang Nan　Zhu Lixia　et al（374）
Effect of Type B Steel Sleeve Rehabilitate Girth Weld Defect on the Microstructure
　　and Property of X80 Pipeline ……… Ma Weifeng　Ren Junjie　Zhou Huiping　et al（387）
A Review of Dynamic Multiaxial Experimental Techniques
　　………………………………………… Nie Hailiang　Ma Weifeng　Wang Ke　et al（396）
Analysis of Cracks in Polyvinylidene Fluoride Lined Reinforced Thermoplastic Pipe
　　Used in Acidic Gas Fields …………… Qi Guoquan　Yan Hongxia　Qi Dongtao　et al（410）
Influence Factors of X80 Pipeline Steel Girth Welding with Self-Shielded Flux-Cored Wire
　　………………………………………… Qi Lihua　Ji Zuoliang　Zhang Jiming　et al（419）
Mechanical Properties of Girth Weld with Different Butt Materials Severed for
　　Natural Gas Station …………………… Ren Junjie　Ma Weifeng　He Xueliang　et al（431）

Research on the Steel for Oil and Gas Pipelines in Sour Environment ……… Shao Xiaodong (439)
Assessment of Hydrogen Embrittlement via Image-Based Techniques in Al-Zn-Mg-Cu
    Aluminum Alloys …………… Su Hang Hiroyuki Toda Kazuyuki Shimizu et al (447)
Failure Analysis of Casing Dropping in Shale Oil Well during Large Scale
    Volume Fracturing ……………… Wang Hang Zhao Wenlong Shu Zhenhui et al (468)
A Comprehensive Analysis on the Longitudinal Fracture in the Tool Joints of Drill Pipes
    ……………………………………… Wang Xinhu Li Fangpo Liu Yonggang et al (480)
Effects of Chloride Concentration on CO_2 Corrosion of Novel 3Cr2Al Steel
    in Simulated Oil and Gas Well Environments
    …………………………………… Tan Chengtong Xu Xiuqing Xu Lining et al (489)
Mechanical Performance of Casing in In-Situ Combustion Thermal Recovery
    …………………………………… Yang Shangyu Han Lihong Feng Chun et al (503)
Corrosion Inhibition Performance of 5-(2-Hydroxyethyl)-1,3,5-Triazine-2-Thione
    for 10# Carbon Steel in NH_4Cl Solution
    …………………………………… Yin Chengxian Ban Xiling Wang Yuan et al (515)
Investigation into the Failure Mechanism of Chromia Scale Thermally Grown on an Austenitic
    Stainless Steel in Pure Steam ……… Yuan Juntao Wang Wen Zhang Huihui et al (527)
NbC-TiN Co-Precipitation Behavior and Mechanical Properties of X90 Pipeline Steels
    by Critical-Temperature Rolling Process
    …………………………………… Zhang Jiming Huo Chunyong Ma Qiurong et al (539)
Stress Analysis of Large Crude Oil Storage Tank Subjected to Harmonic Settlement
    …………………………………… Zhang Shuxin Liu Xiaolong Luo Jinheng et al (547)
Electrochemical Corrosion Behavior of 15Cr-6Ni-2Mo Stainless Steel with/without Stress
    under the Coexistence of CO_2 and H_2S
    …………………………………… Zhao Xuehui Feng Yaorong Tang Shawei et al (553)
Fracture Failure Analysis of C110 Oil Tube in a Western Oil Field
    …………………………………… Zhu Lixia Kuang Xianren Xiong Maoxian et al (565)

第三篇　专利及标准

授权专利目录 ……………………………………………………………………… (575)
发布标准目录 ……………………………………………………………………… (589)

第四篇　"十三五"科技大事记

第一篇　科技成果篇

大口径高压输气管道断裂控制技术及应用

随着西气东输一线、二线、三线的建设运行，我国天然气管道逐步迈入"高钢级、大口径、高压力"时代，ϕ1422mmX80 钢管的研制应用使得管道年输量由二线、三线的 $300×10^8m^3$ 提高到中俄东线的 $450×10^8m^3$，为我国天然气资源的引进和输送提供了有力的保障。管道建设和服役中产生的损伤和缺陷，大大增加了裂纹萌生的可能性，高压大口径输气管道一旦发生裂纹起裂，就可能引发灾难性后果，因此，确定准确的管线钢止裂韧性指标成了管道延性断裂控制研究领域的热点和难点。ISO 3183 中推荐的 5 种止裂韧性计算方法均是基于全尺寸爆破试验数据库对 Battelle 双曲线模型的修正，而全尺寸实物爆破试验仍旧是世界上公认的解决高钢级管线断裂控制问题最为有效和可行的办法。国外仅三个国家掌握全尺寸爆破试验的关键技术，西二线全尺寸爆破试验是由意大利 CSM 公司完成。

"石油管材及装备材料服役行为与结构安全国家重点实验室管道断裂控制试验场"是国内唯一可进行以天然气为介质的管道断裂控制（全尺寸管道爆破）试验平台。其中包括爆破实验区和辅助生产区，设置 CNG 供气装置、储气管、试验管、初始裂纹引入装置、天然气引燃装置及数据采集系统等主要试验设施；通过止裂韧性计算确定天然气管道全尺寸爆破试验方案，在试验起裂管引入初始裂纹，在内压驱动下裂纹向两侧试验管段扩展，在此过程中通过对管道裂纹扩展速度、试验温度、压力以及减压波压力变化、管道局部残余塑性变形、局部应变场和管道形状变化等参数进行测试和分析，全面评估高压长输天然气管道的止裂韧性；通过对管道泄放的天然气进行点燃引爆后产生的冲击波、地震波、热辐射、土压力、噪声、飞溅物等内容进行测试和分析，全面分析和评估管道天然气爆破（爆炸）后的危害范围及程度。

重点实验室针对高压大口径天然气管道断裂控制技术，将消化吸收和自主创新相结合，进行了系统性的理论探索、技术开发和工程应用研究。结合国际上近年在天然气管道动态延性断裂研究领域最新的理论和积累的全尺寸试验数据，针对我国天然气管道服役环境及运行条件的特点开展研究，解决了天然气管道动态延性断裂控制的多项关键技术问题，包括天然气减压特性定量分析及天然气减压波计算模型开发、高压输气管道裂纹扩展阻力曲线计算模型开发、高压输气管道动态断裂数值模拟技术。在理论研究和模型开发的基础上，首次进行了 ϕ1422mm X80、ϕ1219mm X90 全尺寸气体爆破试验，对建立的理论模型和技术进行试验验证和完善，并运用该技术为西气东输二线等天然气管线提出了合理的止裂韧性指标，并制订了断裂控制方案。通过全面的研究和理论创新，系统地建立了 X80、X90 高压大口径天然气管道断裂控制技术，保证了管线的安全可靠性。

1 主要创新点

1.1 试验场设计建设

该试验场的建设属国内首创，并且相比其他国家，该试验场可进行全尺寸爆破试验的管道口径、压力及采用天然气试验介质的综合能力最强。

（1）进行总平面设计时，采用挪威 DNV 公司的事故后果分析软件（PHAST）进行多工况条件下气体爆炸模拟计算，定量评估泄放天然气爆炸后的热辐射、冲击波、噪声等影响范围,同时与相关实际试验数据曲线相对比,为试验场平面布置提供设计依据,保证了试验场的安全性。

(2) 开展储气管锚固墩大推力(最大约1440.6tf)计算和设计研究,并在国内首次设计实施了带斜锚杆的抗大推力管道锚固墩。

(3) 首次设计了上下游大压差及压力大幅度波动条件下的调压装置和供气工艺管道。

(4) 采用超高速数据采集系统,完成600余个试验数据高速、同步和连续采集,满足爆破试验海量数据安全、真实和准确采集的需要,保证有效数据获取率不低于90%。

(5) 设计了一种管道线性聚能切割器及天然气同步触发联锁引燃装置。

1.2 全尺寸爆破试验技术

开发形成了高压大口径天然气管道全尺寸气体爆破试验成套装置和技术,使我国具备了进行高压、大口径、高钢级天然气管道全尺寸断裂行为试验,以及管道爆炸危害试验的能力。包括:

(1) 天然气云团自动点燃装置和技术;
(2) 初始裂纹引入装置和技术;
(3) 断裂速度测试装置和技术;
(4) 减压波测试和分析技术;
(5) 天然气燃爆冲击波测试和分析技术;
(6) 天然气燃爆热辐射测试和分析技术;
(7) 地震波测试和分析技术。

1.3 高压大口径X80天然气管道断裂控制

成功进行了12MPa、ϕ1422mm×21.4mm X80直缝埋弧焊管天然气全尺寸爆破试验,13.3MPa、ϕ1422mm×21.4mm X80螺旋缝埋弧焊管天然气全尺寸爆破试验,其中13.3MPa、ϕ1422mm×21.4mm X80螺旋缝埋弧焊管天然气全尺寸爆破试验为世界首次,通过这些试验验证了ϕ1422mm X80管道的止裂韧性,证实了螺旋缝埋弧焊管的止裂能力高于直缝埋弧焊管,为中俄东线ϕ1422mm X80管道的应用提供了技术支撑。

(1) 成功开展一次12MPa,ϕ1422mm×21.4mm X80直缝埋弧焊管天然气全尺寸爆破试验。

(2) 成功开展世界首次13.3MPa,ϕ1422mm×21.4mm X80螺旋缝埋弧焊管天然气全尺寸爆破试验。

(3) 全尺寸爆破试验数据采集率在90%以上,数据采集精度在0.25级以上。

1.4 天然气管道爆破(爆炸)危害效应研究

结合两次ϕ1422mm X80(直缝、螺旋缝)及1次ϕ1219mm X90管道全尺寸爆破试验,测试得到了天然气管道爆炸冲击波、热辐射及地震波的传播规律,揭示了冲击波、热辐射及地震波危害效应的产生机理,形成了适用于全尺寸管道爆炸冲击波、热辐射和地震波危害效应的分析方法,为天然气管道爆破(爆炸)危害评价提供基础试验数据和技术支持。具体包括:

(1) 提出了开放空间大规模气云爆炸冲击波与热辐射效应的测试仪器、传感器与数据采集方案,建立了适用于天然气管道爆炸热辐射和冲击波的测试方法及试验场内测试系统的三维构建及布置,成功应用到了天然气管道泄漏气云爆炸冲击波和热辐射测试中,得到了有效的试验数据;

(2) 通过ϕ1422mm X80 12MPa直缝钢管,ϕ1422mm X80 13.3MPa螺旋钢管,ϕ1219mm X90 12MPa钢管全尺寸爆破实验得到的实验数据的拟合,建立了天然气管道爆破地震波、空气冲击波危害效应对建筑物危险作用范围，用于指导天然气管道设计时安全距离的确定。

2 与国内外同类技术对比

与国内外同类技术对比情况见表1。

表1 与国内外同类技术对比情况

序号	本项目研究获得的成果、技术	与国内/国际上同类成果、技术比较	水平评价
1	全尺寸爆破试验场及试验技术	全尺寸爆破试验场功能完善，采集信号种类涵盖了压力、温度、断裂速度、应力应变、冲击波、热辐射、地震波等，采集速率最高可达1MHz	国际先进
2	φ1422mm X80 直缝焊管全尺寸气体爆破试验研究	与俄罗斯φ1422mm X80全尺寸爆破试验相比，采用天然气作为试验介质，试验难度更大	国际先进
3	φ1422mm X80 螺旋焊管全尺寸气体爆破试验研究	在国际上首次进行φ1422mm X80螺旋焊管全尺寸爆破试验，验证了φ1422 21.4mm X80螺旋缝焊管的止裂能力	国际领先水平
4	同沟铺设管道全尺寸爆破试验研究	在国际上首次进行12MPa φ1219mm×22mm X80同沟铺设管道全尺寸爆破试验。通过试验测得了X80同沟铺设管道所受爆炸冲击载荷，以及在此载荷下的压扁和屈曲程度	国际先进
5	天然气管道爆炸危害研究	系统地测试了爆炸冲击波、热辐射及地震波危害，与国外相比，测试数据更加丰富完善，分析手段更加合理	国际先进

3 经济、社会效益和意义

管道全尺寸爆破试验场的建立，不仅可以节约试验费用，缩短项目建设周期，还可以按照国内基础研究的节奏适时开展爆破试验研究工作，掌控及保护第一手试验数据和资料，降低对国外试验机构依赖度，有效提高我国石油管道科技研发实力打下的强有力的基础，为进一步完善油气管道设计优化和运行安全研究工作提供了更加广阔的平台。同时，通过已开展的爆破试验，揭示天然气管道爆破冲击波、地震波、热辐射危害效应的产生机理、传播规律，形成爆破冲击波、地震波、热辐射等参数的工程计算和分析方法，为全面认识天然气管道爆破危害效应提供理论和技术支持，为国家有效防止管道事故危害，保证人民财产和生命安全作出贡献。

4 推广应用前景

目前已经开展的φ1422mm X80 螺旋焊缝管/直缝管全尺寸爆破试验形成的研究成果已应用于中俄东线天然气管道设计和建设。

本成果可在油气管道建设和运行过程中全面应用和推广。不仅可用于高钢级、大口径管道研究，以及新建油气管道的设计、管材采购、质量控制等多个环节，也可用于在役管线的安全评价和完整性管理。

相关成果"天然气管道全尺寸爆破试验技术"入选2016年中国石油天然气集团公司十大科技进展。

大口径高压输气管道变形控制技术及应用

实验室承担了中国石油天然气集团公司重大科技专项"西气东输二线管道断裂与变形控制关键技术研究",中国石油天然气集团公司"十二五"技术开发项目"特殊地区管道建设关键技术研究"和应用基础项目"高强度管道失效控制的应用基础研究"等研究课题。

目前建设的一些管道项目沿线地质条件十分复杂,如西气东输二线穿越22条活动断层,中亚D线途径大量的9度强震地区,中缅管道将通过9度强震区和5条活动断裂带,并途径多个采矿沉降地区,施工难度极大,被称为世界管道建设史上难度最大的工程之一。这些地质条件对管道的设计、材料、施工都提出了巨大的挑战。因此,为保证可能发生较大变形地段的X70钢级管道安全,在局部可能发生较大变形的地段采用了大变形钢管。针对西二线、西三线、中亚D线用X80大变形钢管,以及中缅用X70大变形钢管,通过全尺寸弯曲试验和变形行为研究、批量试制程序和质量控制及评估方法研究、大变形钢管冷弯管技术研究等相关的技术攻关,攻克了国产大变形管线钢的关键技术,形成国产大变形钢管变形行为预测、生产制造、性能控制等系列应用技术,实现大变形钢管的技术突破,推动国产X70、X80大变形钢管的生产应用水平,并最终保证了上述油气管道工程的顺利实施。主要创新点:(1)研究提出了抗大变形钢管的关键技术指标和检测评价方法。(2)通过单炉和小批量试制/千吨试制,确定了生产工艺的可行性及稳定性,提出了大变形钢管的小批量试制流程、质量控制方案。(3)通过钢管的屈曲变形仿真计算和全尺寸压缩弯曲试验,确定了基于应变地区用钢管屈曲应变指标。(4)在国内外首次提出了时效试验方法细则。(5)通过课题研究,撰写完成了X70、X80抗大变形钢管的工程技术标准,为高应变直缝埋弧焊管在基于应变设计地区许用应变确定提供了依据。

相关成果"中缅天然气管道设计施工及重大安全关键技术研究与应用"获2016年中国石油天然气集团公司科技进步奖特等奖。

基于应变的稠油蒸汽热采井套管柱设计方法及工程应用

项目围绕稠油热采井套损治理和失效预防，依托 2011 年中国石油天然气集团公司"油井管柱完整性技术研究（2011A-4208）"项目，深化 2005 年中国石油天然气集团公司国际合作项目"油气管道/柱风险评价（05D40101）"内热采井套损机理研究，与新疆油田、辽河油田及中国海油热采井套损治理相结合，联合渤海装备、衡阳钢管及新冶钢集团等制造单位联合攻关，形成本项技术成果，实现了套损机理、管柱设计、管材选用评价、工程配套技术及标准化一体化成果，为以蒸汽热采为主要方式的稠油工业高效、安全、经济开发提供了成套技术支撑。

通过套损普查与分析，掌握了套损主要模式及失效机理，指出套损主要来源于长期服役产生的热-弹塑性变形。井下超过 180℃ 的大温差将造成管材热屈服，过量塑性变形将导致管材缩径及断裂。塑性变形伴随着持久性包申格效应，产生循环硬化，由于螺纹连接部位应力/应变集中持续增加，螺纹接头将发生脱扣或者断裂。油田大量使用的 API 偏梯形螺纹和螺纹脂，在 200℃ 以上环境将失去密封能力，导致注入蒸汽泄漏，被井下泥岩层吸收，引发地层间大的横向载荷，导致管柱发生剪切变形及错断。

针对循环变温服役环境，明确了套管柱累积应变构成，建立了套管柱需求应变计算方法和试验测试中的管柱应变判断系统及方法。依据套管热弹塑性变形及规律，依据材料均匀变形能力，建立了许用应变计算方法。针对注汽、采油及动态热循环不同状态，建立了套管柱设计及安全评价三准则模型。针对管柱螺纹连接部位的应力/应变集中效应，突破 API Spec 5CT 套管标准约束，提出"管体—管端—螺纹连接"强度错配新方法，在确保管体发生塑性变形范围内，管端及螺纹连接应力不超过其屈服强度，预防脱扣及断裂失效。据此，提出管端镦粗、二次热处理工艺及管材局部加速冷却工艺，实现了管柱强度错配设计。依据上述理论指导和专利技术联合制造厂家开发了 80SH、90SH 及 110SH 三个钢级热采套管新产品，针对热采井仅需维持注汽阶段的气密封，考虑现场测井接箍定位需要，开发了台肩型和气密封特殊螺纹两种接头新产品。

在工程配套技术方面，从螺纹密封面检测方法、装备及实物模拟试验工装设计等方面入手，建立了蒸汽热采井套管柱工况模拟试验评价方法，完成了 80SH、90SH、110SH 及不同结构螺纹接头产品的试验评价。通过高温应力松弛效应分析，取消了传统的预应力固井，简化了工艺。针对技术套管与套管头焊接需求，建立了焊接接头适用性评价方法。针对注汽间隙管柱测井及频繁清洗需求，研发了同心管射流负压冲砂洗井方法。

本项技术成果获得国家授权发明专利 14 件，授权实用新型专利 16 件，开发套管新产品 3 种、螺纹接头新产品 2 种、形成管柱强度错配新工艺 2 项，制定并发布石油行业新标准 4 项，发表科技论文 17 篇。相关科技成果在 2015 年《中国石油报》进行了报道，项目组成员受邀在 2013 年世界石油大会和 2016 年 SPE 加拿大稠油国际会议上也对这些科技成果进行了学术报告，获得国内外专家高度评价。

在新疆油田完成了 8 口井现场试验，与同期常规井相比变形降低 42%，实现了 5 年零套损效果。在辽河油田蒸汽吞吐井组转火驱井方面开展了 10 余口井套损机理及承载能力分析，为火驱井点火工艺优化提供了技术支撑。在渤海湾热采井套损治理方面，完成了多井次变形及失效分析。在渤海装备、衡阳华菱及新冶钢等制造企业获得推广应用，在热采井套管新产品开发、现有管材工艺改进及常规套管质量提升等方面发挥了重要作用。油田工程应用配套技术在油田稠油开发方面获得规模应用，为油田高效、安全开发节约了大量资金。

相关成果获2016年中国石油天然气集团公司技术发明奖二等奖。

地下储气库运行与安全保障技术研究

"地下储气库运行与安全保障技术研究（2012B-3407）"为 2012 年中国石油天然气集团公司重点科技项目，主要针对储气库管柱完整性评价、在役溶腔稳定性评价、压缩机故障诊断、选材及安全评价标准体系等开展研究攻关，属于油气储运安全技术领域。

本项目以我国目前已建的枯竭油气藏型和盐穴型两类储气库为工程依托，将消化吸收和自主创新相结合，针对当前储气库面临的管柱评价依赖国外、在役溶腔稳定性评价缺乏工具手段、压缩机故障诊断预警难、选材及安全评价标准空白等运行安全保障技术需求，针对储气库运行安全保障技术开展了理论探索、技术开发和现场应用研究。通过项目研究攻关，在国内首次建立了地下储气库管柱完整性评价和储气库压缩机故障自适应诊断技术，建立了在役老腔稳定性评价方法，首次开发了专用评估软件，编制了管柱选材、安全评价、风险评估、声纳检测等标准规范，填补国内空白，支撑了储气库运行安全保障技术体系。

项目取得了系列创新性研究成果，包括基于三轴强度的含缺陷储气库管柱剩余强度评价方法、基于数值模拟和实物试验的储气库管柱螺纹密封性完整性评价方法、基于层次分析法的地下溶腔稳定性模糊综合评价方法、单腔和双腔稳定性数值分析方法、适用于变工况条件下的储气库注采压缩机自适应故障诊断技术、盐穴型储气库完整性评价软件、基于全井段测井数据的地下储气库老井套管柱强度分析软件、储气库注采压缩机自适应诊断系统，以及管柱选材及安全评价标准等。

通过本项目制定并发布企业标准 5 项，开发专用软件 3 套，申请国家专利 8 件，其中发明专利 7 件，授权发明专利 5 件，发表学术论文 16 篇，其中 SCI/EI 检索 3 篇，项目成果进一步完善和丰富了储气库的运行安全保障技术体系，对安全工程学科的发展具有重要推动作用。

本项目采用边开发、边应用的模式，不仅为地下储气库安全运行提供了技术保障，而且促进了我国地下储气库安全管理的技术进步。研究成果已成功应用于长庆油田、辽河油田、新疆油田、大港油田等储气库的管材选用和管柱评价，应用于大张坨储气库德莱塞蓝压缩机整机及关键部件，京 58 和永 22 储气库群的 4 台往复式压缩机故障诊断，以及金坛西 1 井和西 2 井的老腔稳定性分析评价，为储气库现场管柱密封性、耐蚀性及管材成本合理控制、管柱适用工况条件和运行检测周期确定、压缩机故障诊断、老腔稳定性提供了合理化建议和决策依据，为储气库安全运行提供了技术保障，产生直接经济效益 3000 多万元，节约投资约 3 亿元，经济与社会效益显著。同时，随着我国在建储气库大量投入注采运行和未来储气库建设的重大需求，项目成果具有广阔的应用前景。

相关成果获 2016 年中国石油天然气集团公司科技进步奖二等奖。

海底管道腐蚀控制技术研发及应用

本项目属于油气田开发领域应用基础研究类项目,主要为海底管道安全运行提供理论依据和解决措施。

本项目以我国东南沿海海洋油气开发为工程依托,紧密结合海底管道防腐设计、材质性能、施工工艺及安全运行的实际需求,开展了系统性的理论分析、技术开发和现场应用研究,取得了5项创新成果,总体技术达到国际先进水平。

(1)创新了基于全生命周期的海底管道防腐设计新方法:以案例分析为基础,全生命周期视角选取腐蚀环境、模拟实验和模型预测相结合定性管材耐蚀区间,辅以材质、工艺、药剂等多种防腐措施优化和完整性管理配套统筹考量进行防腐设计,打破了传统采用极端工况设计后技术富余经济不足的局限。研究成果为东海西湖凹陷区块八条海管设计了经济适用的防腐方案,确保了具有战略意义的管道按时建成并多年运行未出事故。

(2)拓展了海底管道内防腐产品及腐蚀评价装置体系:建立了海洋环境用双金属复合管性能评价方法和技术指标,编制了产品订货技术条件;开发了 TG520 高温 CO_2 缓蚀剂,解决了带压套管防腐问题;国内首次设计了多相流腐蚀模拟实物装置和顶部腐蚀评价系统,摆脱了复杂流态腐蚀评价难的困扰。研究成果丰富了海底管道防腐选材评价手段,保障了复合管订货、验收和安全使用,确保了崖城作业区套管诱喷作业安全施工。

(3)开发了海底管道外防腐补口技术及专用牺牲阳极产品:设计了干湿两用热收缩带的补口底漆和聚酰胺脂泡沫填充工艺,制定了补口工艺规程,形成了海管快速补口技术,解决了高温高湿高腐蚀环境补口施工质量控制问题,开发了海泥环境用牺牲阳极材料,提升了管道防腐质量。研究成果保障了西气东输二线香港支线补口施工安全,累计完成补口 1613 道,一次性报检合格率 100%。

(4)完善了海洋用双金属复合管施工工艺:搞清了 S 型铺管法对海洋环境用复合管材性能影响规律,给出安全铺管弯曲半径,开发了端部堆焊结构及现场焊接工艺,形成了焊接工艺评定技术,解决了海洋环境用复合管焊接问题,确保了铺管施工安全。研究成果促成了双金属复合管在崖城、番禺等海洋领域的安全应用,为海洋环境双金属复合管材防腐应用打开了新局面。

(5)创新了海底管道运行风险检测及快速维修技术:开发的海管防腐层检测装置填补了国内空白,首次实现对外防腐层非接触式连续检测评价,编制的维抢修施工作业规范,进一步提升了管道维护水平。研究成果保证了冀东油田南堡作业区两条海底输油管道近 10km 外防腐层状况检测,解决了管道防护状况无法全面掌握的难题。

本项目申请国家专利 17 件,其中发明专利 11 件(授权 7 件)、实用新型专利 6 件,制定行业及中油集团企业标准 4 项,在国内外核心期刊共发表文章 20 余篇(SCI/EI 收录 4 篇),出版专著 2 部。

项目研究成果先后在西气东输二线香港支线、冀东油田南堡作业区及中海油平湖、崖城和番禺得到应用,建立了从管道防腐设计、性能评价、施工工艺到后期运行保障技术体系,开发了系列管道防腐产品及检测评价装置,丰富和完善了海底管道安全运行管理机制,提升了海底管道本质安全保障水平,保障了我国油气战略开发进程,产生了显著的社会效益和经济效益,项目实施以来产生直接经济效益超过 3.2 亿元。

相关成果获 2016 年中国石油天然气集团公司科技进步奖二等奖。

特殊用途管线钢管应用技术研究

本项目为中国石油天然气集团公司重大科技专项研究课题"西气东输二线管道断裂与变形控制关键技术研究"的关键研究内容，是材料科学与工程和石油管工程领域的交叉学科，属于材料科学技术领域，适用于高强度大口径天然气特殊用途钢管（抗大变形钢管、低温环境用钢管和管件、高屈强比 X80 钢管、冷弯管）的生产开发、质量改进和质量控制，确保管道运行安全。

本项目针对基于应变设计地区用抗大变形钢管、低温环境用钢管和管件、高屈强比 X80 钢管、冷弯管等特殊用途管线钢管开展应用技术研究，通过系统性的理论探索、技术开发和工程应用，结合我国天然气管道服役环境及运行条件的特点，形成了特殊用途管线钢管的性能控制、生产制造方面的系列应用技术。解决了抗大变形钢管，低温用钢管、弯管及管件，冷弯管的多项关键技术问题，包括 X70 抗大变形钢管全尺寸弯曲试验技术、管道应变控制时屈曲应变预测方法、高寒地区站场裸露钢管及管件低温脆性断裂控制技术、X80 钢管屈强比安全性影响、X80 冷弯管应变时效规律、X80 冷弯管用母管关键技术指标、X80 冷弯管质量控制等核心技术和方法。

本项目发表论文 17 篇、发布标准 3 项、获得授权专利 10 件。利用课题研究成果，为管道建设公司提供了及时有效的技术支持，同时为厂家提供了全面的技术服务，促进了管线建设工程的进展。通过课题研究，取得了一系列创新性的研究成果，主要创新成果如下：(1) 揭示了承压管道应变容量与钢管规格、工作压力和管材基本性能的相关性，创立了管道应变控制时屈曲应变预测公式，并确定了工程应用的控制指标，建立的钢管屈曲应变容量预测公式包含了应力比和屈服强度，经全尺寸内压弯曲试验证实，其准确度优于国外现行预测公式。(2) 研究建立了高寒地区站场裸露钢管和管件的最低壁温计算方法，制定了低温脆断控制方案。(3) 攻克了屈强比影响管道安全的技术难题，得出了屈强比和管道壁厚的简化公式，利用该模型掌握了屈强比对管道承载能力和管道安全性的影响规律，解决了屈强比影响管道安全的问题。(4) 研究获得冷弯工艺和应变时效对 X80 钢管性能的影响规律，提出了 X80 钢级冷弯管母管技术要求和 X80 冷弯管质量控制指标。(5) 通过课题研究，撰写完成了多个抗大变形钢管相关的工程技术标准，应用于西气东输二线等工程中。

本项目取得的研究成果已被纳入西气东输二线、三线，中亚 C 线、D 线，中缅管线等系列技术标准中，并获得了工程应用。本项目成果解决了长期困扰高压天然气管道工程建设及运行安全的技术难题，避免了西二线、三线及中缅管线工程中管道敷设长距离绕行的问题，节约了大量的管道建设成本。本成果可在油气管道建设和运行过程中全面应用和推广，不仅可用于新建油气管道的可行性研究、设计、管材采购、质量控制等多个环节，也可用于在役管线的安全评价和完整性管理。我国正处于油气管线建设的高峰期，因此本成果具有广阔的应用前景。

相关成果获 2016 年陕西省科技进步奖二等奖。

输气管道提高强度设计系数工程应用研究

该成果为2012年中国石油天然气集团公司"第三代大输量天然气管道工程关键技术研究"重大专项课题八"2012E-2801-08"的主要研究成果,属于油气储运工程学科和能源安全技术领域。

课题以西气东输三线为依托工程,通过国内外设计标准及工程应用现状调研、现场检测、统计分析、试验研究、理论计算等研究手段,对输气管道提高设计系数的断裂控制、设计及施工技术、安全可靠性、风险水平、完整性管理技术等开展配套技术研究,编制0.8设计系数管材关键技术指标体系、现场焊接及施工技术规范、完整性管理措施。

欧美国家也已开展了相关研究及探索,美国和加拿大已建设0.72以上设计系数管道工程,设计系数主要为0.72~0.78,但0.8设计系数的应用却很少,仅Alliance在0.72设计系数通过后期提压至0.8设计系数。自1994年我国首次颁布实施GB 50251—1994《输气管道工程设计规范》以来,一直沿用了一级地区0.72设计系数,在较高设计系数技术方面则未开展相关研究,更没有相关工程经验。

课题调研了0.8设计系数在国外的标准、研究和应用情况,对比了国内外X80钢级钢管实物质量,评估了提高我国输气管道强度设计系数的安全可行性,开展了0.8设计系数X80管道断裂控制技术、管材质量评价技术、现场焊接技术、100%最小屈服强度试压技术、施工质量控制、完整性管理措施等关键技术攻关,指导开发出0.8设计系数管道用X80螺旋缝埋弧焊管,并为西三线0.8设计系数示范工程提供$12.5×10^4$t钢管,制定了《西气东输三线0.8设计系数管道用X80螺旋缝埋弧焊管技术条件》《西气东输三线西段一级地区荒漠无人区提高强度设计系数技术规范》等企业标准及0.8设计系数管道现场焊接工艺规程,并在此研究基础上修订了GB 50251《输气管道工程设计规范》。共申请专利5件,制/修订标准3项,焊接工艺规程8项,发表论文16篇。

课题采用边研究、边应用的模式,为X80钢级0.8设计系数示范工程建设提供技术支撑和保障。X80钢级260km示范工程强度设计系数从0.72提高到0.8,钢管壁厚由18.4mm减小到16.5mm,壁厚减薄10.3%,直接节约管材$1.25×10^4$t,节约投资约1亿元,另外间接降低了运输、焊接等费用。0.8设计系数技术在西四线、西五线、中俄输气管线一级地区都要进行全面推广,直接经济效益可达数十亿,应用前景广阔。

本项目获2016年陕西省科学技术奖二等奖。

100亿立方米调峰能力储气库重大关键技术及应用

进入新世纪，我国天然气工业发展迅速，2016年消费量突破$2000×10^8 km^3$，天然气长输管道里程达$7.6×10^4 km$，用户遍及全国，安全平稳供气成为国计民生重中之重。而我国主要气源远离市场、对外依存度高、冬夏峰谷差大，调峰保供面临严峻挑战。地下储气库是解决天然气供需矛盾的有效手段，已成为国家能源安全的战略性基础设施。

我国地质条件极其复杂，多期构造运动与陆相沉积环境导致构造破碎、埋藏深、非均质性强，严重制约储气库选址与建设。在国外经验无法有效指导前提下，中国石油经过十多年持续攻关，取得了四项重大关键技术创新，创建了具有自主知识产权的成套技术和标准体系，形成了我国储气库新型产业。专家组鉴定认为，在复杂地质条件下大型储气库选址与建库技术国际领先。主要创新成果有以下几个方面。

（1）创建了复杂地质条件储气库圈闭动态密封理论和库容分区动用新方法。建立了以盖层"动态突破、交变疲劳"和断层"柔性连接、剪切滑移"为核心的动态密封理论，实现了选址评价由静态定性到动态定量评价的根本转变；揭示了储气库"分区差异动用、高速有限供流"注采机理，建立了库容和井网预测模型，形成了建库地质方案设计方法和标准体系，关键指标符合率90%以上。

（2）创新了适应超低压地层、储气库交变载荷工况的钻完井技术。研发的吸水树脂复合凝胶承压堵漏材料及无固相储层保护钻井液体系，实现了低压地层、30MPa压差下安全钻井和储层保护；发明的抗200℃晶须纳米材料高强低弹韧性水泥浆体系，攻克了世界最深、温度最高的储气库固井难题，固井质量合格率100%，保障了高速交替注采条件下井筒密封。

（3）创新了地面高压大流量注采核心装备与技术。研制了大功率高压高转速往复式注气压缩机组和国内单套规模最大的高压采气处理装置；首创了水下爆燃制管和振动模态无损检测技术，制造了国内最大口径最高钢级双金属复合管；形成了"变流量精准注采、超高压密相输送"为核心的地面工艺技术及标准体系。

（4）创新了储气库风险评价与控制关键技术。建立了储气库往复注采过程岩石形变微地震信号识别与定位技术；首创了拉伸/压缩交替变化下管柱密封评价准则与设计方法；形成了多因素耦合的注采设施泄漏风险定量评价技术；构建了储气库地层—井筒—地面"三位一体"风险管控技术体系，保障了储气库安全运行。

获授权发明专利29件、实用新型专利35件、技术秘密7项、软件著作权7项，制定标准15项（行标2项、企标13项），出版专著11部，论文55篇。

该重大成果推动了储气库规模化建设，使我国成为复杂地质条件储气库建设的引领者。建成22座储气库年$100×10^8 m^3$调峰保供能力，惠及京津冀等10余省市2亿人口，缓解天然气冬季供需矛盾；在应对中亚管道进口气减供、西气东输管道中断等重大突发事件中发挥了关键作用。近三年创造直接经济效益31.56亿元，替代标煤$2291×10^4 t$，综合减排$4089×10^4 t$，社会和环保意义特别重大。将对我国2030年实现$500×10^8 m^3$调峰保供能力规划目标起重大支撑作用。

相关成果获2017年中国石油天然气集团公司科技进步奖特等奖，石油管工程技术研究院和国家重点实验室主要参与了储气库风险评价与控制关键技术创新相关研究内容。

西部油田非金属管规范化选材与应用研究

随着油气输送介质环境日益苛刻，传统碳钢管腐蚀加剧，泄漏事故频发，造成环境污染和不良的社会影响，中国石油天然气集团公司安全环保形势愈加严峻。非金属及复合材料管材（以下简称非金属管）成为解决地面集输管网腐蚀问题的重要方案之一，当前非金属管在中国石油天然气集团公司地面管网用量已超过三万千米，且数量逐年递增。由于非金属管属于新型管材，因设计选材及应用过程不规范导致非金属管失效次数明显增加，该两类失效分别占非金属管道总失效次数的38%和47%。本项目针对腐蚀环境最为严苛的西部油田工况条件（酸性介质、高温、高压）及地形地貌（沙漠戈壁、丘陵沟壑等），系统研究非金属管的设计选材及应用，建立完善的非金属管产品、测试方法、质量控制、施工验收等领域的标准体系，从而降低非金属管失效风险，保障西部油田"安、满、稳"运行，并为非金属管在中国石油天然气集团公司的推广应用奠定基础。

本项目在系统研究国内外非金属管标准规范及其应用情况的基础上，通过现场失效分析和模拟实验相结合，创建了油气集输用非金属管的选材评价技术体系，为油田非金属管设计选材提供了指南。首次开发设计了模拟油田现场应用环境的6项检测评价技术，并研发了配套的实物管材评价设备，填补了多项非金属管性能检测评价技术的空白。首次明确提出了油气集输用非金属管对应的关键技术指标，建立了石油天然气行业用非金属管性能安全可靠性评价体系。提出了油田现场非金属管设计、施工、验收等关键环节控制点，制定非金属管系列国家、行业和油田企业标准14项。

通过以上技术内容的研究，建立了"源头防、重点治、过程控"为核心的非金属管应用技术体系。该体系已在我国石油工业用管材的标准体系建设中得到应用，表明本项目建立的标准体系已经处于国内领先。经检索及专家评审表明，该项目主要科技创新成果与当前国际先进水平相当，部分技术处于国际领先，核心技术填补了该领域多项国内空白。制定系列标准14项；发表学术论文10篇；授权专利8件。项目成果2016年获中国石油集团石油管工程技术研究院科技进步奖一等奖。

项目研究成果在油田用户单位（塔里木油田、长庆油田、新疆油田等）、油田设计单位、石油管材专业标准化技术委员会、非金属管生产厂家等进行了广泛宣贯交流及推广应用，获得了极好的社会效益，显著促进了非金属管在西部油田及国内其他油田的规模化应用，推动了我国石油工业用非金属管的生产与应用技术进步。同时，也取得良好的经济效益，产生的总经济效益约为5.2亿元。该项目研究成果不仅可为中国石油天然气集团公司油气地面集输用非金属管规模化、规范化应用奠定了基础，也为非金属管在井下、海洋等领域的应用提供了技术储备，对实现我国油气田开发过程中的降本增效、腐蚀防护和安全运营具有重大意义。

相关成果获2017年中国石油天然气集团公司科技进步奖一等奖。

管道全尺寸爆破试验场建设及试验关键技术

随着西气东输一线、二线、三线的建设运行，我国天然气管道逐步迈入"高钢级、大口径、高压力"时代，φ1422mm X80 钢管的研制应用使得管道年输量由二线、三线的 $300×10^8m^3$ 提高到中俄东线的 $450×10^8m^3$，为我国天然气资源的引进和输送提供了有力的保障。管道建设和服役中产生的损伤和缺陷，大大增加了裂纹萌生的可能性，高压大口径输气管道一旦发生裂纹起裂，就可能引发灾难性后果，因此确定准确的管线钢止裂韧性指标成了管道延性断裂控制研究领域的热点和难点。对于 X80 及以上级别高钢级管线钢，全尺寸实物爆破试验是世界上公认的确定止裂韧性最为有效、最为准确的方法。国外仅三个国家建有全尺寸爆破试验场，西二线全尺寸爆破试验由意大利 CSM 公司完成。

为了推动我国天然气管道管材技术研究的进步，掌握高钢级、大口径管材关键性能技术指标，为管道建设提供可靠的技术支持，集团公司决定建设管道断裂控制试验场，开展高钢级、大口径、高压力管材全尺寸气体爆破试验，逐步建立领先世界先进水平的管道材料数据库。与此同时，基于天然气管道爆破试验，开展管道爆破和天然气爆炸对周围环境和同沟并行管道的危险影响评估研究，也将为管道建设和运行提供安全评价参考指标。该项目历时五年，解决了试验场建设、全尺寸爆破试验、管道爆破（爆炸）危害效应测试等一系列技术难题，成功实施了 φ1422mm X80 直缝、螺旋缝焊管天然气全尺寸爆破，准确掌握了管道止裂韧性指标，取得了一系列创新成果。经中国石油天然气集团公司组织的专家鉴定：该技术成果填补了我国在该领域的技术空白，整体达到国际先进技术水平。

（1）主要创新成果。

① 创新形成了天然气管道全尺寸爆破试验场设计建设成套技术。包括全尺寸爆破试验建设技术、总平面布置、储气管锚固技术、安全间距模拟计算方法、多通道并行同步高速连续数据采集系统、适合上下游大压差及压力大幅度波动条件下的供气及调压装置。

② 研发了天然气管道全尺寸爆破试验技术和方法，包括线性聚能切割器、管道爆破初始裂纹引入技术、管道断裂速度和减压波等试验测试方法，实现了不同类型信号数据的高速海量采集和存储，建立全尺寸爆破试验流程、规范和标准。

③ 开发了管道爆破试验场天然气管道爆破（爆炸）冲击波、地震波、热辐射危害效应及对同沟敷设管道影响的测试技术和方法。包括爆破冲击波与地震波、爆燃冲击波与热辐射测试方法与技术、空气与岩土介质能量转换模型，气云燃爆冲击波超压和热辐射伤害的动态理论计算模型，冲击波超压和热辐射三维场及图像解析构建方法与程序，高压力气体瞬间膨胀冲击波传播过程和超压幅值变化规律。

（2）该成果申报发明专利 7 件、授权实用新型专利 2 件，形成中国石油天然气集团公司技术秘密 4 项，编写企业标准 1 项、规程 2 项，发表论文 7 篇。

（3）设计建成了我国首座设计压力为 20MPa 的高压大口径天然气管道全尺寸爆破试验场。在国际上首次进行了在 12MPa 条件下的 φ1422×21.4mm、X80 直缝埋弧焊管，以及在 13.3MPa 条件下的 φ1422×21.4mm、X80 螺旋缝埋弧焊管天然气全尺寸爆破试验，验证了中俄东线管道的止裂韧性指标。同时，进行了管道爆破试验场天然气管道爆破（爆炸）冲击波、地震波、热辐射及同沟敷设管道的危害效应试验。

相关成果获 2017 年中国石油天然气集团公司科技进步奖一等奖。

新型气密封特殊螺纹套管研发及应用

依托中国石油天然气集团公司课题"油井管柱完整性技术研究(2011A-4208)"。中国石油长庆油田等油气田单井产量低,都必须进行大型的压裂、酸化改造才能出气。"三低"油藏,主要靠多打井来实现上产,对套管的需求数量都在$50×10^4$t以上。生产中存在的主要问题有两个。(1)标准化及降低管理成本需要:非API套管扣型种类太多,立项前油田共使用特殊螺纹金属密封扣套管种类有7种,给现场管理带来困难;扣型的特殊性,套管附件不能互换,浪费很大;检验工具无法满足现场及时检验需要;附件螺纹加工也很麻烦。(2)经济可靠性需要:高端气密封螺纹接头套管,性能过剩,价格高,油田成本居高不下。API长圆螺纹套管加特殊螺纹脂达到所需扭矩值后,就不能再实施卸扣作业,上扣速度较慢,容易粘扣,且不能满足多段循环压裂和大弯曲狗腿度需要。因此,需要有一种螺纹套管能够实现快速上扣、承受大的弯曲拉伸压缩载荷、过扭矩,不使用特殊密封脂就能够实现气密封,且价格又便宜,满足低压气田低成本高效率开发需要。

1 研究开发内容

API圆螺纹气密封机理分析:以气体变密度雷诺方程为基础,以气密封泄漏速率$0.9cm^3/15min$为准则,分析研究在拉伸载荷下螺纹牙顶与牙底通道间隙面积与有效啮合长度下低压(40MPa)气密封性能。金属对金属特殊螺纹接头气密封机理分析:探讨密封面接触压力和长度及表面粗糙度和涂层的影响。抗粘扣机理分析:采用过应力模型分析上扣转速与结构塑性应变的关系。螺纹抗拉伸机理分析:比对螺纹不同承载面牙型角和螺纹过盈量对拉伸断裂和密封失效的影响。螺纹抗压缩机理分析:比对螺纹不同导向面牙型角和分离或接触方式对压缩下台肩塑性应变和密封失效的影响。密封结构形式分析:比对主密封位置及结构形式(锥面/锥面和锥面/球面以及负角度和直角台肩)密封接触压力和长度变化;循环包络线载荷谱和高温(150℃)热循环下材料力学行为对主密封接触压力及长度变化的影响。金属/金属气密封特殊螺纹连接套管满足水平井大曲率多段体积压裂气密封螺纹连接套管,满足弯曲狗腿度(20°/30m),压裂内压(90MPa),高温150℃工况。低压用经济型气密封螺纹连接套管满足低压气田经济开发需求的非API长圆螺纹套管,满足弯曲狗腿度(20°/30m),气密封内压(40MPa),高温135℃工况。

2 技术创新点

(1)引入表面粗糙度及涂镀层影响,提出新的气密封螺纹设计判据。(2)建立了特殊螺纹上扣转速抗粘扣材料过应力分析模型,形成特殊螺纹密封面优化设计方法。(3)引入材料循环软化对螺纹气密封性影响,建立了热采套管气密封螺纹配套材料选用方法。(4)设计了无台肩无密封面经济型气密封螺纹连接结构,实现拉伸下低压气密封,获得自主创新产品1项。(5)设计了密封面与台肩远距离自过盈高压气密封螺纹连接结构,实现95%等效应力大弯曲螺纹密封,获得自主创新产品1项。(6)针对页岩气井套管柱非均匀外挤特征,设计开发了专用特殊螺纹结构。(7)新技术新产品在宝鸡钢管、渤海装备、达利普、山东墨龙、延安嘉盛等厂家获得转化,实现规模化生产。(8)新技术新产品在长庆油田、新疆油田、延长油田等公司获得批量推广应用，正在开展大庆油田推广应用。

3 应用情况及经济效益

高性能气密封螺纹接头供货新疆油田、长庆油田、延长油田总计8000t；宝鸡石油钢管公司向长庆油田供货1000t，两项总计实现产值1亿元。经济型气密封螺纹山东墨龙石油钢管公司向长庆油田供货4000t，大庆油田供货2000t，实现产值3600万元。项目截至2016年底实现产值1.5亿。石油管工程技术研究院与企业开展横向科研技术服务实现直接经费1000万元。

相关成果获2017年中国石油天然气集团公司技术发明奖二等奖。

高钢级大口径焊管残余应力及控制技术研究

在油气输送焊管的成型及焊接过程中，焊管内部均会产生残余应力。随着钢级的提高、管径及壁厚的增大，焊管的制造难度加大，螺旋缝埋弧焊管的残余应力问题更加突出，其控制难度亦加大。

螺旋缝埋弧焊管为我国油气输送钢管的主要生产方式，螺旋缝埋弧焊管的生产成本较直缝焊管低，国内拥有众多的螺旋缝焊管生产线。但是，螺旋缝埋弧焊管的主要缺点是残余应力较大，不同钢级、不同管型、不同厂家的螺旋缝埋弧焊管的残余应力的分布状态及水平差别较大，严重制约了螺旋缝焊管的使用。在我国油气管道建设，尤其是重大管道工程建设中能否大量使用螺旋缝埋弧焊管，降低工程成本，避免或减少进口，节约外汇，推动国内制管行业的发展，降低和控制高钢级大口径螺旋缝焊管的残余应力就成为关键环节之一。

围绕我国油气管道工程建设需求，中国石油在不同阶段的科学研究与技术开发项目中对油气输送焊管的残余应力进行研究。申报单位开展了科研攻关，实现了科技创新。主要研究内容及创新点包括以下几个方面。

（1）采用不同残余应力测试方法对油气输送焊管的残余应力进行了测试，优选出了适用于油气输送焊管的残余应力测试方法，并开发了厚壁大直径焊接钢管残余应力值及其分布的盲孔测试方法。

（2）首次建立了适用于螺旋焊管的高钢级大口径焊管残余应力预测模型，并提出了不同钢级、不同规格高钢级大口径螺旋缝焊管的残余应力控制指标，纳入了西气东输二线等重大管道工程用螺旋缝埋弧焊管技术条件。

（3）对不同钢级、不同管型的油气输送焊管的残余应力进行了测试分析，系统掌握了不同厂家、不同钢级高钢级大口径焊管螺旋缝埋弧焊管及直缝埋弧焊管残余应力的水平及分布规律。

（4）系统分析了影响高钢级大口径螺旋缝埋弧焊管残余应力的关键成型参数及主要因素，提出了残余应力控制措施，实现了对高钢级大口径螺旋焊管残余应力的有效控制，其残余应力水平达到或接近直缝焊管的水平。

本项目的研究成果已应用于西气东输二线、西气东输三线、中缅天然气管道工程、中亚C线、陕京四线等重大管道工程，并已纳入相关管材技术条件等，有效地控制了螺旋焊管的残余应力，用于高钢级大口径螺旋缝埋弧焊管的制造和质量控制过程，保证了重大管道工程用焊管的质量，扩大了螺旋焊管的使用范围，降低了工程成本，减少或替代了进口焊管，提升了焊管生产厂的生产水平，推动了行业进步。本项目解决了高钢级大口径螺旋缝埋弧焊管应用关键技术难题，对保证重大管道工程的安全可靠性起到了重要的技术支持作用，产生了显著的经济效益和社会效益。

本项目共授权专利7项，发表专题论文19篇，涉及标准7项。高钢级大口径焊管残余应力控制技术已在宝鸡石油钢管有限责任公司、渤海石油装备制造有限公司华油钢管有限公司等多条螺旋缝焊管生产线上推广应用。近三年，完成单位新增产值1.258亿元，新增利润1132.4万元。低应力高钢级大口径螺旋缝焊管在西气东输二线、西气东输三线、中缅天然气管道工程、中亚C线、陕京四线等重大管道工程中成功应用，满足了重大管道工程对高品质管材的急需，提升了关键材料的自主保障能力。

相关成果获2017年陕西省科技进步奖二等奖。

油气管道系统完整性关键技术与工业化应用

我国油气管道总里程已达到 12×10^4 km，随着我国建设的管道进入老龄期及新建管道规模加大，因系统完整性引发的事故占比 80% 以上，系统完整性问题成为国际公认的技术难题。本成果自 2005 年以来，历经 10 年理论、试验和应用研究，建立了油气管道完整性新理论，研发了不停输修复等关键技术与装备，发明点包括以下几个方面。

（1）建立了应力和应变双重判据的管道失效评估模型，发明腐蚀缺陷管道补偿器弯管剩余强度评价和剩余寿命预测方法。

（2）研制管道内涂层热应力模拟装置，建立全尺寸管道内涂层粉尘磨损试验方法和生产检验、试验、在线运行关键指标体系，提出延长管道内涂层寿命的风险控制措施。

（3）发明管道碳纤维复合材料技术与应用方法，包括浸润式碳纤维工艺，胶黏剂、抗剥离绝缘底胶及填料体系；缺陷填充材料工作温度范围为 $-60\sim150$℃，压缩强度为 $85\sim100$ MPa，抗阴极剥离绝缘底胶的甲、乙双组成固化时间小于 12h，剥离强度达 170N/mm，层间胶黏剂为浸润式复合材料基体。

上述研究成果在西气东输二线、三线，中俄东线，新疆油田克炼—风城首站、红浅—克炼输油管道、彩火线等 30 余条管道上得到应用，发现管道重大安全隐患 200 余次，显著降低了管道失效频率，实现了油气能源战略通道本质安全，直接经济效益近亿元。

相关成果获 2018 年国家技术发明奖二等奖。

石油管材及装备防腐涂镀层开发与应用关键技术

随着我国油气资源开发战略的深入推进,石油管材及装备面临着前所未有的高温、高压、高含 H_2S/CO_2、交变载荷、高流速、复杂作业工艺等苛刻服役环境带来的腐蚀难题,腐蚀安全风险不断增加。涂镀层防腐技术因其具有性价比高、可设计性强、应用工况范围广等特点而成为解决石油管材及装备腐蚀问题的研究热点和攻关方向。本项目针对涂镀层防腐技术,通过理论研究、产品开发、技术攻关和现场应用,取得了多项创新性的研究成果。

(1)提出了涂镀层失效退化模型,全面、系统分析了涂镀层在多因素交互作用下的防腐和失效机制,相关研究成果在 Engineering Failure Analysis、ASTM International 等国际知名期刊及会议上发表。

(2)开发了系列高性能涂镀层产品,自主研发的抗 CO_2/H_2S 环氧酚醛涂层、高耐蚀 Ni-W 合金镀层、环保型无溶剂防腐涂层、平台长效聚氨酯涂层等产品在国内首次应用于多元热流体、CO_2 驱、高含 CO_2/H_2S 油气井及高温、高湿、高盐雾海洋环境等苛刻工况。

(3)建立了涂镀层检测评价技术体系,形成了涂镀层全尺寸实物试验方法,自主设计的国内首套实物拉伸腐蚀系统及冲刷腐蚀环路试验系统试验能力处于国际领先水平。

(4)建立了涂镀层应用技术体系,确立了涂镀层关键性能指标,规范了涂镀层设计、选材、施工、验收的应用过程管理,创建了涂镀层防腐技术多层次、多元化、全方位的标准规范体系,填补了国内空白。

项目研究成果在塔里木油田、新疆油田、长庆油田等重点油气田企业,以及大港石化、海洋工程公司、冀东油田、大港油田等临海石化企业和油田企业成功进行了应用,突破了制约涂镀层开发及应用的系列瓶颈技术,提高了防腐效果,延长了石油管材及装备的使用寿命,创造了良好的经济效益和社会效益。形成的防腐涂镀层系列产品、检测评价技术、标准规范及应用关键技术,为国家和中国石油天然气集团有限公司石油管材及装备防腐涂镀层的开发和规范化应用提供了重大技术支持。

本成果获 2018 年中国石油天然气集团有限公司科技进步奖一等奖。

高强度铝合金钻杆关键技术研究及应用

本项目基于中国石油天然气集团有限公司重大项目"油气管道工程建设新技术、新产品研究"核心内容之一的"石油管应用基础研究"课题及中国石油塔里木油田公司"φ147×13mm 钢接头铝合金钻杆的国产化研究"课题。依托塔里木盆地大型碳酸盐岩油气田勘探开发示范工程,针对我国西部油气田深井、超深井等复杂工况油气井数量不断增多,钻具失效频繁、完钻周期、长效率低等问题,开展了高强度铝合金钻杆关键技术研究,建立了高强度铝合金钻杆性能评价方法,发展了高强度铝合金钻杆设计、选材及制备技术,开发了 1 种规格的高强度铝合金钻杆实物产品及配套处理技术。主要创新点包括以下几个方面。

(1) 在国内首次完成了基于铝合金服役工况的整体结构设计与关键性能指标制定,系统研究提出了高强度铝合金钻杆管体材料的预拉伸变形欠时效处理生产制造技术;重点研究合金化的管体材料配方、强化固溶热处理工艺等,揭示了高强度铝合金钻杆材料组织—成分—工艺—性能之间的关系,形成了高强度铝合金钻杆用管体材料理论体系。

(2) 在国内首次系统研究形成了高强度铝合金钻杆用接头的高密封耐疲劳 TT 型特殊螺纹接头加工及热装配连接技术;重点围绕钢接头铝合金钻杆特殊螺纹接头的设计、加工及低扭矩热装配连接,形成了高强度铝合金钻杆的接头设计及连接技术。

(3) 在国内首次系统研究形成了高强度铝合金钻杆性能评价技术。针对深井、超深井等复杂工况条件及钢接头铝合金钻杆结构特点,采用小试样室温力学试验、高温力学试验、疲劳及腐蚀试验模拟结合带螺纹连接的全尺寸抗内压、抗外挤、拉伸至失效、实物旋转弯曲疲劳试验等实物评价方法,系统研究获得了高强度铝合金钻杆的理化性能及实物性能数据,揭示了各种工况条件下钻杆的结构—性能—寿命关系规律,形成了高强度铝合金钻杆性能评价方法。

上述相关成果打破了铝合金钻杆技术领域国外的长期技术垄断,对于铝合金钻杆技术在我国油气田的全面推广应用,保障塔里木油田等我国重点油气田油气资源勘探开发及西气东输等重大工程建设,促进我国装备制造业水平提升,推动行业技术进步具有重要的战略意义。

相关成果获 2018 年陕西省科技进步奖二等奖。

枯竭气藏型储气库管柱设计、选材与评价技术

我国储气库建设起步较晚，主要以枯竭气藏型储气库为主，并有部分盐穴型储气库。按照中国石油储气库发展初步规划，到 2020 年，中国石油储气库工作气量要达到 $250×10^8 m^3$，而目前工作气量仅约 $42×10^8 m^3$，差距很大，储气库建设力度仍将加大。但在储气库建设和运行过程中，因管柱选材未凸显工况的差异性，引起系列失效和隐患风险，进一步影响了储气库安全生产运行。

本项目针对枯竭气藏型储气库井环空带压、泄漏、腐蚀等现象，重点对储气库井管柱接头和材质的选用以及在役井的检测评价和安全分析问题，利用全尺寸实物复合加载试验机、高温高压釜和扫描电镜等试验设备，进行了大量的实物试验、腐蚀试验和微观试验分析，同时结合弹塑性力学理论和数值模拟手段，最终获得的创新理论方法和技术有：

（1）首创地下储气库管柱密封设计方法与技术，实现了储气库管柱结构完整性和密封完整性，提高了储气库管柱安全运行周期。

（2）自主研发了低含水率下枯竭气藏型储气库管柱腐蚀选材技术，确保储气库管柱安全且经济选材，实现了储气库管柱选材成本降低 10% 以上。

（3）自主研发了储气库套管柱安全评价技术方法，提出了基于全井段测井数据的储气库套管柱剩余强度评价方法和寿命预测方法，打破国外技术垄断，实现国内独立自主检测评价。

本项目成果已在长庆油田、辽河油田、新疆油田、大港油田及北京管道等公司九座储气库建设和运行中，获得较好的应用，既提高了储气库用油套管的质量，保障储气库管柱的密封性、耐蚀性，又做到了管材成本合理控制。

相关成果获得 2018 年陕西省科技进步奖二等奖。

X70 大应变管线钢管及应用关键技术

本项目依托中国石油"特殊地区管道建设关键技术研究"等重大科技项目，历经多年的研究攻关，攻克了 X70 大应变管线钢管关键技术指标设计与评价、X70 大应变管线钢设计与制造、X70 大应变管线钢管制造、实物模拟变形装置及试验评价等技术难题，形成了 X70 大应变管线钢管及应用关键技术，首次大批量应用于中缅油气管道工程。主要技术发明和创新点有以下几个方面。

（1）系统研究建立了钢管材料应力比等关键技术指标与钢管临界屈曲应变的关系，发明了钢管临界屈曲应变能力的预测方法；创新提出用多个应力比、屈强比、均匀塑性变形伸长率等多参量联合表征评价和控制钢管变形行为的方法；首次提出 X70 大应变管线钢和钢管新产品技术指标体系和标准，并纳入美国石油学会（API）管线钢管标准 API RP 5L 附录 N Pipe ordered for applications requiring longitudinal plastic strain capacity。

（2）在国际上首次自主研发攻克 X70 大应变管线钢板制造技术，通过成分优化设计，洁净化冶炼、控轧控冷工艺参数精确控制，实现了块状铁素体+粒状贝氏体双相组织调控，发明了 X70 大应变管线钢板生产工艺和质量性能控制技术。

（3）在国际上首次自主研发形成大应变 JCOE 直缝埋弧焊管均匀协调变形成型、低热输入多丝埋弧焊接、低温 3PE 涂覆等关键技术，发明了 X70 大应变管线钢管生产工艺和质量性能控制技术。

（4）发明并设计制造了钢管内压+弯曲大变形实物试验装置，该装置具有可更换式法兰盘、可拆卸式承载框架、轴承式力臂支撑等特征；研发形成钢管实物模拟变形试验技术，发明了钢管特定截面弯曲角以及应力应变实时测量的装置和方法。

该项成果突破了我国 X70 大应变管线钢管研发及应用中的重大技术瓶颈，实现了从无到有到大批量生产应用的跨越式发展。同时取得重大技术创新和显著经济效益，总体达到国际先进水平，在大应变管线钢管关键技术指标建立和实物试验装置及技术两方面达到国际领先水平。

相关成果获得 2018 年中国石油天然气集团有限公司技术发明奖二等奖。

复杂多场耦合工况环境中油井管柱腐蚀控制理论研究

高温高压气井的开发对油井管柱的长期安全服役提出了苛刻的技术要求，主要存在以下技术理论问题：力学场、温度场、化学场、时域场等多场耦合导致油井管柱频繁发生腐蚀失效，但多场耦合的作用机制和主控因素缺乏清晰的认识；常规油井管柱小尺寸试样研究结果的局限性越来越突出，因尺寸效应导致的小试样与全尺寸实验研究结果的差异性有待深入研究；传统的酸化缓蚀剂无法满足高温高压环境超级13Cr油井管柱酸化压裂的防腐技术要求，亟需研究新型的缓蚀理论模型。

本项目以高温高压油气井管柱材料（N80、P110、13Cr）的局部腐蚀和应力腐蚀开裂两大类失效为切入点，针对油井管柱在酸化过程中的严重腐蚀难题，提出"空间多分子层多吸附中心"的理论模型，通过结构优化开发13Cr油管专用高温酸化缓蚀剂。主要创新点包括以下几个方面。

（1）揭示了油井管柱材料在复杂苛刻环境中的腐蚀失效机制，明确了碳钢和马氏体不锈钢在酸化压裂过程中的腐蚀机理，首次揭示了马氏体不锈钢油管在环空保护液中的应力腐蚀开裂机制，相关成果在 Corrosion Science 和 Electrochimica Acta（均为中国科学院1区top期刊）上发表，得到俄亥俄州立大学 Nesic、北京科技大学李晓刚等国内外著名专家引用。

（2）建立了基于高温高压气井全生命周期的油井管柱腐蚀研究方法及理论，首次揭示了油井管柱在气井全生命周期服役过程中时域场与化学场的交互作用机制和耦合腐蚀效应。相关成果在2016年欧洲腐蚀大会上发表，受到广泛关注。

（3）依托国内首台油井管柱全尺寸实物腐蚀实验系统，开展"化学场、力学场、温度场、时域场"多场耦合环境中全尺寸油井管柱的服役行为研究，揭示了典型油井管柱材料在酸化压裂过程中点蚀发展过程及应力腐蚀开裂机制。相关研究成果在 Engineering Failure Analysis 上发表。

（4）提出了"空间多分子层多吸附中心"的理论模型，设计合成了喹啉季铵盐和曼尼希碱化合物体系。相关研究成果在 ACS 旗下权威期刊上发表，得到美联社、路透社和中国中央电视台科技之光栏目等媒体报道，引起同行的广泛关注。

本项目形成的油井管柱选材标准为油井管柱设计、选材、评价提供依据；建立的全尺寸油井管柱腐蚀实验系统为长庆、大庆、塔里木等油田开展实物油管的适应性评价及选材研究；形成的缓蚀剂产品已在塔里木油田90多口井得到推广应用，大幅降低油井管柱的失效频次。

相关成果获2018年中国石油天然气集团有限公司基础研究奖二等奖。

石油钻采机械用材料技术体系研究

石油钻采装备构件种类繁多，装备制造企业鳞次栉比，技术水平参差不齐，而钢铁冶金企业与装备制造企业之间科技信息和研究成果共享不及时，缺乏相应信息沟通渠道，石油装备制造企业主要依据自身生产经验选择材料和加工工艺，各企业的材料选择和加工工艺都大不相同，产品性能良莠不齐，钻采装备失效事故频繁发生，给石油钻采装备的研发、生产制造和使用管理都造成了极大困扰，也造成了极大的资源浪费。

本项目通过对国内外主要石油钻采装备制造企业、钢铁冶炼企业、油田公司、钻探公司及科研单位的深入调研和国内外相关技术和文献资料的系统检索分析，形成了国内外石油钻采机械用材料调研分析报告。并以石油钻采装备为主线，编写出版了我国石油钻采装备金属材料领域第一部实用性手册——《石油钻采装备金属材料手册》，制定国家及石油行业标准4项。

（1）完成了国内外石油钻采机械用材料调研工作，形成了石油钻采机械用材料调研分析报告，为我国石油钻采装备材料的选择和规范化应用提供了决策支持。

（2）完成了石油钻采机械构件分类研究，首次建立了符合我国石油机械用材料特点的材料技术体系，主要包括铸铁、碳钢、合金钢、铝合金、镍基合金、钛合金和硬质合金等类别，完善了石油装备材料试验评价技术。

（3）以石油钻采装备为主线，针对钻机提升系统、井控装备等十大类别钻采装备的关键构件使用的119种金属材料，建立了关键性能指标。研究成果形成的《石油钻采装备金属材料手册》受到业内专家一致的好评，在油气钻采装备制造企业、油田、钻探公司、科研机构及高等院校广泛推广应用，取得了显著的经济效益和社会效益。项目研究成果有力支撑了我国油气装备材料质量提升和规范化应用。

相关成果获得2018年中国石油天然气集团有限公司科技进步奖二等奖。

X80 高强度管道服役安全关键技术研究及应用

石油天然气管道已成为我国能源供应的大动脉，其战略地位举足轻重，其安全运行直接关系到国民经济发展和社会稳定。油气管道向着高强度高压方向发展的同时，管道运行安全问题也备受关注。针对 X70 及以下钢级的油气管道服役安全相关技术已开展了大量研究工作，技术积累较完善。但是，随着 X80 高强度管线钢管在西气东输二线、三线、陕京四线以及中俄东线上的大批量应用，对管道服役安全提出了新的技术挑战。

本项目在中国石油天然气集团有限公司十二五重点应用基础研究项目以及中国石油西部管道公司、西气东输管道公司三个科研项目的资助下，经过研究攻关，突破了 X80 高强度管道服役安全多项关键技术，取得了以下五个方面的创新成果：(1)通过 X80 管线钢的断裂韧性厚度效应研究，基于裂尖塑性区理论建立了高强度管线钢的平面应变临界壁厚确定方法，给出了 X80 管线钢管的工程安全临界壁厚。(2)首次建立了高强度管道提高试压系数的计算模型，以及试压过程中的风险控制措施，充分释放了管道强度设计裕量，挖掘了管道输送潜力。(3)通过在 NS_4 溶液中模拟近中性土壤应力腐蚀开裂试验，研究了 X80 管线钢的近中性土壤应力腐蚀开裂规律及其影响因素，研究发现管线钢强度、管道阴极保护电位、压力波动对高强度管道土壤应力开裂有显著影响，建立了管道近中性土壤应力腐蚀开裂的完整性评价和风险评估方法（风险指数法）。(4)采用力学叠加原理，建立了焊缝复合缺陷断裂分析的理论模型，推导出了焊缝复合缺陷应力强度因子的理论计算公式，建立了焊缝复合陷疲劳裂纹扩展与寿命预测模型。(5)针对复合材料和环氧钢套筒修复技术的质量控制难题，改进了复合材料修复技术，建立了环氧钢套筒修复评价指标体系，满足了 X80 管道不动火修复需求。

本项目授权国家专利 18 项，其中发明专利 16 项，实用新型专利 2 项，制定标准规范 4 项，取得软件著作权 1 项，发表论文 20 余篇。项目采用边研究、边应用的模式，研究成果已经在西气东输二线、三线、陕京四线和伊霍线等 X80 高强度高压天然气管线上应用。取得直接经济效益 3 亿多元，经济效益显著。同时，研究成果的推广应用有效降低了高强度管道失效事故率，有力保障了高强度管道服役安全和平稳运行，社会效益更加显著。研究成果将进一步丰富和完善高强度管道失效控制和完整性评价方法和技术，推动中国石油管道工程技术进步。

相关成果获 2019 年陕西省科技进步奖一等奖。

西部油气田地面管网腐蚀评价及治理技术

随着我国油气资源勘探开发不断向西部深入，复杂苛刻的油气作业和生产环境导致油气管网系统腐蚀泄漏频发，油气管网作为输送油气资源的重要通道，其频繁失效严重影响油气正常生产，同时，也带来了巨大的安全风险和环保隐患。目前，地面管网系统的腐蚀防护主要存在以下技术难题：（1）影响地面管网的腐蚀因素众多，多因素耦合作用下的腐蚀失效机制和主控因素尚缺乏清晰的认识。（2）常规小尺寸试样研究结果的局限性突出，亟需建立针对地面系统的全尺寸实物腐蚀试验系统及评价方法。（3）小口径地面管道（$\phi<200mm$）内腐蚀检测受尺寸限制无法开展，使得小口径管网成为内腐蚀检测盲区，存在较大安全隐患。（4）发生严重腐蚀的在役小口径地面管道的修复治理，存在开挖更换难和成本高的难题。

本项目针对以上技术问题开展攻关研究，主要内容及技术成果如下。

（1）揭示了地面管网材料在"材料—化学—结构—力学—电学"多因素耦合作用下的腐蚀失效机理。以双金属复合管为例，建立了腐蚀预测模型和剩余寿命评价方法。

（2）自主研发了全尺寸循环冲刷腐蚀试验系统，建立了油气管道"小试样—全尺寸—现场实验段"的腐蚀评价技术。

（3）攻克了小口径管道内腐蚀检测技术，研发了小口径管道内窥检测、内防腐层测厚、电磁涡流检测机器人，实现了对$\phi114mm$以下小口径管道的内防腐层测厚和内腐蚀检测。

（4）建立了"在线风送挤涂"和"聚乙烯内衬"修复工艺，成功实现了小口径在役油气管道的防腐修复与治理。

应用推广情况：本项目研究成果在长庆油田38个主力井区得到推广应用，共计开展集输管道腐蚀综合治理12400km，管道泄漏次数大幅减少（分别降低65%~80%）。同时，该成果还在西安图博可特石油管道涂层有限公司、陕西天普石油技术有限公司和克拉玛依市科能防腐有限公司等防腐企业和技术服务公司得到推广。自2016年以来，在管道维护、安全环保等方面共创造经济效益3.6亿元，为防腐企业新增利润6500余万元，共计产生经济效益超4亿元，应用效果显著。

相关成果获2019年中国腐蚀与防护学会科技进步奖一等奖。

稠油蒸汽热采井套管柱应变设计及工程应用配套技术

稠油是我国油气资源主要类别之一，分布在新疆、辽河、吐哈、胜利及环渤海湾等地区，年产量逾 $3000×10^4$ t。国内外一直缺乏热采井套管技术标准，各大稠油油田套损率平均 15%~30%，局部更高，如新疆油田百重七区在 2002 年曾达到 70%。

项目在揭示套损模式与机理基础上，提出并建立了新型套管应变设计方法，开发专用产品，形成管材选用技术体系，通过现场试验验证，并结合现场作业需求，开发应用配套技术，形成工业标准体系，实现了工业推广应用，可有效预防套损。

项目协助制造厂家开发了 80SH、90SH 及 110SH 三个钢级套管新产品。同时，对现有产品进行优化设计，覆盖最高 125ksi 钢级，实现了产品制造工艺的多样性，覆盖国内稠油油井需求。针对热采井套管柱气密封要求，开发了台肩型和气密封特殊螺纹型新接头产品及多种螺纹新结构，实现了螺纹连接的多样性。

研究成果在天津钢管公司、华北一机油井管厂、南通永大管业等公司获得规模生产 $5.83×10^4$ t，在新疆油田规模应用 680 井，在辽河油田、中国海油等公司获得应用，新增产值 4.98 亿元，油田现场综合修井节省成本 7850 万元，总体经济效益达 5.8 亿元，显著支撑了油田高效、安全、经济开发。

项目累计获得国家授权发明专利 14 项，实用新型专利 9 项，开发套管新产品 3 项、新型螺纹接头新产品 2 种及多种新结构、形成管柱强度错配新工艺 2 项，制定国家标准 1 项、石油行业标准 4 项，开发工程应用软件 2 套（注册登记），发表学术论文 25 篇，其中 SCI/EI 收录论文 15 篇。相关成果在《中国石油报》进行了专题报道，项目组成员受邀在 2013 年世界石油大会和 2016 年 SPE 加拿大稠油国际会议上进行学术报告，获得国内外专家高度评价。2018 年，以高德利、陈学东院士等组成的科技成果鉴定专家评审委员会认定相关成果达到国际领先水平。

相关成果获 2019 年陕西省科技进步奖二等奖。

厚壁高钢级管线钢与钢管应用基础理论研究

项目以重大天然气管道工程对厚壁高钢级管线钢管应用关键技术的需求为导向,针对厚壁 X70、X80 高强度管线钢材料应用方面的难题,依托与我国天然气管道建设重点工程紧密相关的课题,研究突破了厚壁 X70/X80 大口径高压输气管线钢及钢管(弯管、管件)应用基础理论和关键技术,揭示了厚壁 X70/X80 管线钢及钢管成分、组织、性能、工艺的相关性,阐明了拉伸性能影响因素、建立了控制指标,形成厚壁弯管与管件成分设计理论,揭示了焊接热影响区的脆化机理,提出管材强韧性合理匹配等应用基础理论。

研究形成管线钢针状铁素体组织分析鉴别与评判,管线钢管强度、塑性、韧性和屈强比测试与控制,弯管和管件成分优化设计和热加工工艺,厚壁钢管 DWTT 异常断口分析评判,高强度管线钢焊接热影响局部脆化机理、断裂规律与控制,高钢级管材强韧性合理匹配及关键性能指标体系建立等六项关键技术。

研究成果全面应用于西气东输二线和三线、中亚管线等重大天然气管道工程钢管、弯管、管件的质量控制和安全可靠性保障,推动了国内高压输气管道建设的技术进步,促使中国石油成为国际上 X80 管道研究与应用的领跑者,推动了厚壁 X70、X80 高钢级管材的全面国产化和质量性能提升,带动了高钢级管线钢、钢管冶金和制造业的发展。

相关成果获 2019 年中国石油天然气集团有限公司基础研究奖二等奖。

高钢级管道断裂控制技术

高钢级管道延性裂纹的长程扩展,会造成灾难性后果。然而 API 5L 和 ISO 3183 中规定的 4 种钢管延性断裂止裂韧性预测方法均无法准确预测高韧性(100J 以上)、高钢级(X80 以上)、高压力(12MPa 以上)、大口径(ϕ1219mm 以上)管道的止裂韧性。本项目针对 X90/X100 管道动态断裂控制关键技术问题,开展了大量的试验研究、理论建模、数值分析和技术开发工作,形成的研究成果包括以下几种。

(1)在国际首次开展了一次单管空气、两次三管空气及一次全尺寸天然气 X90 爆破试验,证明 X90 管道可以依靠自身韧性进行止裂,填补了国际全尺寸爆破试验数据库无 X90 试验数据的空白。

(2)在国内首次建立了 X100 全尺寸气体爆破试验数据库,改进了 BTC 断裂阻力曲线。基于单管止裂韧性计算、止裂概率计算和止裂器设计研究成果,分别提出了 ϕ1219mm X90 和 ϕ1219mm X100 断裂控制方案。

(3)采用单位面积损伤应变能(拉伸曲线中最大载荷点后的积分面积)作为 X90 钢断裂准则,采用有限元方法计算得到了含初始裂纹 X90 管道的裂纹扩展及止裂过程,计算结果与全尺寸爆破试验结果吻合。

(4)国内首次攻克了高钢级管道整体和外部止裂器设计、制造、施工关键技术。开发了玻璃纤维止裂器、碳纤维止裂器、钢套筒止裂器系列产品,制定了高钢级管道止裂器设计导则。

基于本项目获得授权发明专利 8 项,实用新型 5 项,发布标准 2 项(其中行业标准 1 项,企业标准 1 项),发表文章 6 篇(其中 SCI 1 篇,EI 2 篇),研发软件 1 套。确定的 X90/X100 钢管止裂韧性指标纳入管材技术标准,并应用于 X90/X100 管材的试制、检验和质量评价。

相关成果获 2019 年中国石油天然气集团有限公司技术发明奖二等奖。

OD1422mm X80 管线钢管研制及应用技术

随着国民经济发展和能源战略的实施，以及受土地、环保、建设与运营等因素制约，发展年输气量近 $400×10^8m^3$ 大输量天然气管道工程迫在眉睫。管材钢级的提高、输送压力和钢管口径增大等技术手段，是降低天然气长距离输送成本的有效途径。其中 OD1422mm、12MPa、X80 天然气管道方案的经济输量范围为 $(320\sim440)×10^8m^3/a$，最大输气量可达 $500×10^8m^3/a$。对于直径 1422mm，厚壁 X80 管材，我国尚没有开发和应用的经验。2012 年中国石油天然气集团公司设立"第三代大输量天然气管道工程关键技术"重大专项，系统开展高钢级、大口径管线钢的研制，经过科研攻关，取得了三项重大创新成果。

（1）提出了 OD1422mm X80 管材化学成分要求，研究确定了管材的关键技术指标和试验方法，形成了管材系列技术标准。

① 研究设计了 OD1422mm X80 管线钢的化学成分。通过降低 C 含量、控制 Nb 含量 0.05%~0.08%（技术指标对比见表1），严格控制微合金元素波动范围，使管线钢具有较低的碳当量，经反复工业化试验验证，可有效兼顾管线钢管的强韧性和焊接性。

表 1 技术指标对比表

标准		C 含量（%）	Si 含量（%）	Mn 含量（%）	Mo 含量（%）	Ni 含量（%）	Cu 含量（%）	Cr 含量（%）	Nb+V+Ti 含量（%）
中俄东线标准	要求值	≤0.09	≤0.42	≤1.85	≤0.35	≤0.50	≤0.30	≤0.45	≤0.15
	直缝推荐值	≤0.07	≤0.30	≤1.80	0.08~0.30	0.10~0.30	≤0.30	≤0.30	0.095~0.135
	螺旋推荐值	≤0.07	≤0.30	≤1.80	0.12~0.27	0.15~0.25	≤0.30	0.15~0.30	0.105~0.135
西二线/三线标准	要求值	≤0.09	≤0.42	≤1.85	≤0.35	≤0.50	≤0.30	≤0.30	≤0.15
中国石油通用标准（CDP）	要求值	≤0.12	≤0.45	≤1.85	≤0.5	≤1.0	≤0.50	≤0.50	≤0.15
API 5L X80	要求值	≤0.12	≤0.45	≤1.85	≤0.5	≤1.0	≤0.50	≤0.50	≤0.15

② 通过对 OD1422mm 钢管的包申格效应、韧脆转变行为、各向异性研究，确定了 OD1422mm X80 钢管的强韧性指标，提出了拉伸、DWTT 试验取样位置和方法。

③ 国内首次针对 OD1422mm X80 螺旋缝焊管和直缝焊管建立了管材/板材化学成分、启裂和止裂韧性等关键技术指标，形成了系列技术标准，用于指导国内钢厂、管厂进行管材开发和试制。

（2）攻克了 OD1422mm X80 管材、弯管、管件的制造工艺，研制出 OD1422mm X80 焊管、感应加热弯管等系列产品，形成了性能测试与表征技术。

① 开发了螺旋缝焊管成型包角角度显示仪和周长自动测量专用工装、大径厚比直缝钢管的成型对中定位装置和 JCOE 成型专用模具，解决了大径厚比钢管尺寸控制精度难题。

② 开发了适合 X80 钢级 OD1422mm×214mm 规格螺旋缝埋弧焊管低应力成型工艺技术和 OD1422mm 厚壁直缝钢管的残余应力控制技术，显著降低了钢管的残余应力，保证了钢管服役的安全性。

③ 通过开发新型焊丝、焊剂，优化坡口形式、焊接参数，形成了 OD1422mm X80 螺旋缝/直缝埋弧焊管高速、高强韧性埋弧焊接技术。

④ 开发了 OD1422mm X80 钢级感应加热弯管专用工装，包括专用前夹具、加热线圈等。通过优化煨制工艺，增加扶正辊轮等，开发出了 OD1422mmX80 钢级感应加热弯管椭圆度控制技术。

⑤ 开发了 OD1422mm、X80 板材、焊管、弯管及管件产品。开展了板材、焊管、弯管、管件的单炉、小批量试制及质量评价和研究，制定了 OD1422mm X80 板材、焊管及配套弯管、管件的综合评价方案，形成了管材性能测试和表征技术。

（3）修正了高压输气管道的止裂韧性预测模型及方法，发展了高压输气管线止裂预测技术。

① 分析了 BTC、HLP、Sumitomo 模型对 OD1422mm X80 管道止裂韧性计算的适用性，首次引入土壤因子，修正了高压输气管道止裂韧性 BTC 中管材流变应力、能量释放率与断裂韧性之间的关系。

② 完善了高压输气管道的止裂韧性预测模型及方法，提高了 BTC 方法的计算精度，发展了高压输气管线止裂预测技术。

③ 结合现有 X80 气体爆破试验数据库，提出了 OD1422mm X80 管道的止裂韧性指标。

④ 针对管道具体的服役条件，通过断裂力学、气体减压波特性，制定了 OD1422mm X80 高压输气管道断裂控制方案，解决了高钢级、大口径管道安全应用的核心问题。

本项目获发明专利 15 件，实用新型专利 9 件，发布产品和试验技术标准 14 部，发表论文 21 篇，累计供应研制规格的钢管近 80×10^4 t，租赁及销售研制施工装备 97 台（套），近 2 年获得直接经济效益 25.5939 亿元。该研究成果填补了国内 OD1422mm X80 管材开发应用技术空白，成果已应用于中俄东线等国家重大管道建设，应用效果证明研究成果完全符合大输量管道建设工程需求，未来将具有广阔的推广应用价值。该成果刷新了我国高压大口径天然气管道建设记录，提升了高强度大口径油气管道建设技术水平，推动了我国钢铁冶金、材料加工、机电学科和油气输送管道领域的技术进步。

相关成果获 2020 年中国石油天然气集团有限公司科技进步奖特等奖。

复杂工况油气井管柱腐蚀控制技术及工程应用

油气井管柱作为油气生产的唯一通道,是油气勘探开发和安全生产的基础。我国每年消耗油气井管柱$(300\sim350)\times10^4$ t,耗资$(250\sim300)$亿元。随着我国油气勘探开发向"深、海、低、非"发展,油气井管柱服役工况环境日益复杂苛刻,失效事故频发,据统计,我国60%以上的油气井管柱失效是因腐蚀造成的。我国油气井主要有两大类复杂苛刻工况,一类是一次开发井工况,主要表现为高温(井底温度最高达200℃)、高压(地层压力最高达138MPa)、高矿化度地层水(Cl^-含量高达160000mg/L)、高含CO_2/H_2S(含量超过15%);另一类是二次增产井工况,随着我国大部分油气田开发进入中后期,大规模应用酸化压裂、CO_2驱、高温蒸汽驱、空气泡沫驱等增产工艺引入新的高腐蚀工况。复杂苛刻工况造成油气井管柱腐蚀失效,严重影响井筒的完整性和油气安全高效生产,国外复杂苛刻工况油气井也面临类似问题,如美国墨西哥湾高温高压气井、加拿大稠油井、英国和挪威北海油井、东南亚海上油气井等,属于世界性难题。本项目从复杂工况油气井管柱腐蚀失效难题入手,联合重点油气田和优势高校开展多学科联合攻关,阐明了油气井全生命周期内的多场耦合作用下的管柱腐蚀失效机制,建立了跨尺度腐蚀评价及选材技术,形成了一体化腐蚀控制技术体系,支撑我国重点油气田高效勘探开发和安全生产。

(1)揭示了油气井全生命周期内"材料(组织)—力学(载荷)—化学(环境)—工艺(增产作业)"多场耦合作用下的管柱腐蚀失效机制。

多环境场连续作用下点蚀失效机制:创新提出了基于油气井全生命周期服役环境("鲜酸→残酸→凝析水→地层水")的点蚀诱发及生长机制,主要表现为高酸度促使钝化膜减薄(鲜酸酸化阶段)、高浓度Cl^-攻击钝化膜薄弱部位并诱发点蚀萌生(残酸返排阶段)、高分压CO_2阳极溶解形成闭塞环境引发点蚀扩展(凝析水阶段)、高分压CO_2和高浓度Cl^-双重作用加速点蚀生长(地层水阶段)。揭示了13Cr不锈钢管柱点蚀发展过程的遗传效应,主要表现为点蚀坑内部微环境的改变滞后于外部环境的变化,使得前序腐蚀环境得以遗传并诱导点蚀自催化生长,加速点蚀进程。探索了微观组织对13Cr不锈钢亚稳态/稳态点蚀行为的影响,揭示了逆变奥氏体抑制界面贫铬从而提高耐点蚀性能的微观机制。多尺度结构界面失稳诱发应力腐蚀开裂失效机制:基于马氏体组织复杂多尺度结构界面(原奥氏体晶界、域界、块界、板条界),揭示了界面失稳对13Cr不锈钢管柱内(酸性环境)/外(碱性环境)表面应力腐蚀开裂诱发机制。阐明了13Cr不锈钢管柱在大排量酸液环境中出现的"阳极溶解"向"氢致开裂"的转变机制,即管柱表面缓蚀剂吸附膜局部破损导致"大阴极—小阳极"引发点蚀萌生,蚀坑内"低pH-高应力"微环境进一步诱发裂纹并沿原奥氏体晶界扩展。首次发现了13Cr不锈钢管柱在碱性环境中的应力腐蚀开裂行为,揭示了缓慢"阳极溶解"开裂机制,即马氏体多尺度结构界面发生铬元素的贫化,导致界面耐蚀性显著降低,在高应力下易发生裂纹萌生及扩展。基于上述理论,形成了对马氏体不锈钢在酸性及碱性溶液体系中应力腐蚀开裂敏感性的新认识,明确了13Cr不锈钢管柱在复杂苛刻工况环境中的应用边界条件。

电偶/应力促进作用下的缝隙腐蚀失效机制:针对油气井管柱连接结构(如螺纹连接、管柱与井下工具接触等)存在的缝隙—电偶—应力等多因素耦合作用下的腐蚀失效,发现了异种金属、同种金属、金属—非金属在油田产出水中的缝隙腐蚀敏感性规律。揭示了非金属材料在扭矩作用下的轻微变形诱发闭塞电池形成从而加速缝隙腐蚀的失效机制（13Cr-PTFE的加速幅度约为140%，TP140-PTFE的加速幅度约为21%）；揭示了异种金属连接中电偶加速缝隙腐蚀的失效机制（13Cr-TP140的加速幅度约为70%，13Cr-G3的加速幅度约为210%）；阐明了腐蚀产物在缝隙口沉积导致传质阻滞进而形成金属离子浓差电池的缝隙腐蚀机制；揭示了应力对缝隙腐蚀的倍增式加速效应（缝隙对腐蚀的加速幅度约为800%，应力对腐蚀的加速幅度约为170%，而缝隙与应力的综合加速幅度约为530%）。

（2）研发了国际上首套复杂工况油气井管柱全尺寸实物腐蚀试验系统，建立了跨尺度的腐蚀评价及选材技术。

基于油气井全生命周期的管柱腐蚀评价技术：针对油气井全生命周期的服役工况，建立了"动态多环境"腐蚀试验评价技术，用于表征"油气增产→生产工况"对管柱腐蚀过程的影响，形成了基于油气井全生命周期的管柱腐蚀选材技术标准，支撑了复杂工况油气井管柱的腐蚀选材及评价。全尺寸实物管柱应力腐蚀试验评价技术：针对传统小试样腐蚀实验方法因尺寸小、无管柱结构特征、无内压和应力等不足，自主设计研发了集环境、载荷、结构于一体的全尺寸油气井管柱应力腐蚀试验系统，实现了100MPa内压、1000t拉伸载荷、200℃高温、油气水多相腐蚀介质等复杂工况的多参量实验模拟。试验可获得管柱的极限承载能力、连接结构失效模式、表面处理工艺可靠性等关键信息。突破了油套管在应力作用下腐蚀和密封耦合作用试验瓶颈，解决了应力腐蚀试验中的结构和尺寸效应难题。为了进一步让室内实验结果接近现场，建立了油气井管柱"小尺寸试验筛选（Screening Test）→全尺寸试验适用性评价（Fitness for Service）→现场试验段验证（Validation）"相结合的跨尺度试验评价技术体系，实现了复杂工况油气井管柱全生命周期内"多环境""多尺度""多因素"的综合评价。该技术先后在长庆、塔里木、大庆等油气田的 CCUS 驱采井、高温高压气井、高含硫气井的管材适用性、连接结构选型及优化、涂层/镀层工艺可靠性等评价研究中得到应用。

（3）建立了以"高温酸化缓蚀剂技术""合金镀层技术""甲酸盐完井液防腐技术"为核心的复杂工况油气井管柱"一体化"腐蚀控制技术体系。高温酸化缓蚀剂技术（适用于增产改造阶段的内防腐）：针对高温酸化压裂增产工艺中的酸液腐蚀问题，建立了"空间多分子层多吸附中心"酸化缓蚀剂设计模型，以喹啉季铵盐化合物和曼尼希碱化合物为空间多分子层架构，植入多种金属离子（Sb^{2+}、Cu^+、Ca^{2+} 及 Al^{3+}）进行复配，形成系列高温酸化缓蚀剂配方和产品（TG201、TG201-II、TG202）。其中 TG201-II 缓蚀剂产品被认定为中国石油天然气集团公司自主创新重要产品。第三方检测机构的检测结果表明，在180℃的酸化液中，TG202缓蚀剂的加入将13Cr不锈钢的腐蚀速率控制在 $18g/(cm^2·h)$，远优于 SY/T 5405—1996 标准的一级指标 $[70g/(cm^2·h)]$。TG 系列酸化缓蚀剂已经量产，并在高温酸化压裂增产井中广泛使用。

合金镀层技术（适用于长期生产阶段的内防腐）：针对油气井生产阶段的严苛腐蚀环境，发明了机械研磨技术制备纳米镍基合金镀层的装置和方法，开发了 Ni-W 镀层和 Ni-Sn-P 镀层。其中，Ni-W 镀层在返排酸化液和地层水中具有优异的耐蚀性能，其耐蚀性能优于马氏体不锈钢，接近于 G3 合金；Ni-Sn-P 镀层在240℃的 $CO_2-O_2-H_2O$ 多元热流体中具有优异的耐蚀性能，其耐蚀性能与 310S 不锈钢相当。全尺寸实物 Ni-W 镀层油管具有优异的抗黏扣和密封性能。甲酸盐完井液防腐技术（适用于长期生产阶段的外防腐）：针对油套环空管柱应力腐蚀难题，采用高温高压慢应变速率拉伸法（SSRT）筛选和 C 型环法长周期验证，探究了"K_2CO_3-$KHCO_3$-甲酸盐"缓冲配方体系的服役可靠性，首次提出了以该体系替代磷酸盐作为高温高压气井"一体化"作业用完井液，建立了甲酸盐完井液质量控制和现场应用技术规范，出版了国内首部甲酸盐完井液技术专著，建立了完井液"优选—质控—应用"环空防腐技术体系。甲酸盐完井液于2017年起在塔里木高温高压区块推广应用85口井，新井A环空压力异常比例由50%降到0，至今未出现管柱腐蚀失效，全油田高温高压气井完整性提高至79%，高于墨西哥湾井完整性55%的国际先进水平。

相关成果获2020年陕西省科技进步奖一等奖。

西部油气田集输管线内腐蚀控制技术及工程应用

随着塔里木（"三高"油气田）和长庆（"三低"油气田）等西部重点油气资源的深入开发，复杂苛刻采出介质导致集输管道腐蚀泄漏频发，严重影响油气正常生产和有油田经济效益。特别是自2015年新的"两法"实施以来，油气管道腐蚀泄漏不仅是经济问题，更是安全和环境问题。为了适应安全环保新形势和新要求，本项目针对油气田复杂苛刻环境中油气管道腐蚀泄漏难题，突破关键技术瓶颈，最终形成油气田管道"腐蚀失效识别→实验研究方法→防腐技术产品→腐蚀检测技术→现场防腐治理技术"体系，大大降低油田地面集输管道腐蚀穿孔频次，对保障我国"西气东输"主要气源安全输送具有重要战略意义。本项目取得了五项主要创新成果。

（1）揭示了油气管道在复杂工况环境中"材料—化学—工艺—力学—电学"多因素耦合腐蚀失效机理。基于油气生产工艺，立足管道材料本构特性，结合油气管道的化学环境和受力状态以及电场干扰，系统研究了油气田地面集输系统用低碳钢、双金属复合管、2205双相不锈钢、内涂层管等管道的失效退化机理，建立了含不同缺陷管道的腐蚀预测模型和剩余寿命评价方法。

（2）自主研发了全尺寸油气管道腐蚀实验系统，建立了油气管道"小试样筛选→全尺寸适用性评价→现场试验段验证"腐蚀评价技术。创新研发了集环境、应力、流态、结构于一体的全尺寸油气管道腐蚀实验系统，可实现极端工况多参量腐蚀实验模拟。为了进一步接近现场，建立了油气管道现场试验段橇装装置。最终形成了油气管道"小试样—全尺寸—现场试验段"的腐蚀评价技术体系。该技术先后在长庆、塔里木、大庆等油田管道评价研究中得到应用。

（3）攻克了"三高一低"复杂苛刻环境油气管道的缓蚀难题，自主研发了桐油松香咪唑啉曼尼西碱系列缓蚀剂产品及配套技术。首创提出了"多分子层多吸附中心缓蚀模型"，采用量子化学计算优化了缓蚀分子自组装结构，借助在线红外和量热合成技术精准合成了以桐油松香咪唑啉曼尼西碱为主剂的预膜、高H_2S、高CO_2、含氧等系列集输缓蚀剂产品，配套开发了缓蚀剂智能加注技术，建立了缓蚀技术"研究—生产—应用—服务"完整性体系，在塔里木油田碳钢管线推广应用5000多千米。

（4）攻克了小口径油气管道内腐蚀检测技术。为了解决油气田小口径管道（<DN200mm）内腐蚀检测的世界性技术难题，首创研发了小口径管道内窥检测、内防腐层测厚、电磁涡流检测为核心的检测技术，实现了机器人化、计算机远程控制及无线传输功能，填补了国内外技术空白。实验室和现场结果表明检测准确率达90%以上，在长庆油田示范应用260km。

（5）创新研发了油气田在役管道"在线风送挤涂"和"聚乙烯内衬"防腐修复工艺及标准规范。针对油气田在役管道开挖更换难且成本高的难题，建立了油气田地面系统在役管道"在线风送挤涂"和"聚乙烯内衬"防腐修复工艺及标准规范。以上两种技术在长庆油田累计应用超过11000千米，成功实现了油气田小口径在役油气管道的防腐修复与治理。长庆油田管道泄漏频次由2016年的0.18次/（km·a）下降至2019年的0.065次/（km·a），年腐蚀泄漏频次减少70%。

本项目获授权专利 18 项(发明专利 8 项)、制修定标准 9 项(国/行标 4 项)、发表论文 24 篇(SCI/EI12 篇)、专著 3 部。自 2016 年以来,累计产生经济效益 10.3872 亿元。项目总体成果达到国际先进水平,其中抗 CO_2/H_2S 缓蚀剂产品、全尺寸油气管道腐蚀实验系统及小口径管道内腐蚀检测技术领先于国内外同类技术和产品。该项目成果已在塔里木油田和长庆油田得到成功推广应用,为我国油气集输管道腐蚀研究和防腐技术应用提供了重要技术支持,推动了行业技术进步和产业发展。

相关成果获 2020 年中国石油天然气集团有限公司科技进步奖一等奖。

石油工业用高性能膨胀管及其性能评价技术

该项目率先提出基于应变诱发塑性机制的膨胀管管材设计理念，即利用相变诱发塑性（Transformation induced plasticity，TRIP）机制和孪晶诱发塑性（Twinning induced plasticity，TWIP）机制，显著提高管材的加工硬化能力和均匀延伸率，以金属塑性变形导致的应变（加工）硬化为突破口，提高低屈服管材的强度，有效解决易胀与高强度之间的突出矛盾。综合利用微合金化技术、炉外精炼技术、径向锻造技术、先进管材成形与热处理技术实现对管材几何尺寸精度、微观组织与性能的精准控制，开发出系列化高性能膨胀管管材。

在获得管材高尺寸精度的同时，微观组织内引入亚稳奥氏体，其在塑性变形过程中将持续转变为马氏体或形成机械孪晶，通过 TRIP 和 TWIP 机制获得持续稳定的高加工硬化率提高管材的均匀塑性变形能力和胀后强度，满足高性能膨胀管对管材高均匀延伸率与加工硬化能力的要求，并且管材低的原始屈服强度满足了其易膨胀的要求。开发的新型系列膨胀管管材均匀延伸率大于 20%，最高均匀延伸率可达 50% 以上，经测试其内径膨胀率达 25.75%，可满足"单一井径"技术对膨胀管超大变形率的要求。膨胀前管材屈服强度约 350MPa，经过约 15% 膨胀，管材的屈服强度可迅速提高至 80ksi，管材综合性能超过美国亿万奇公司的水平，达到国际领先水平。

项目设计开发了两种膨胀管专用特殊螺纹接头，并形成相应的配套加工技术，其中单钩形端面金属自密封膨胀管螺纹接头外螺纹端和内螺纹端具有负角度的钩形内外螺纹，拧接后紧密咬合，接触应力分布均匀，具有较强的抗拉强度。单钩形螺纹的圆柱形金属密封面和橡胶密封圈共同密封，提高了膨胀管螺纹接头的密封可靠性，该结构具有膨胀后密封性良好，加工成本低等优点。双钩形膨胀管螺纹采用变导向面偏角和螺纹锥度设计使得膨胀管接头具有双钩内外螺纹紧密啮合，消除了应力集中，具有更高的强度，且采用多重金属—金属、金属—橡胶—金属密封设计，具有膨胀后接头强度分布均匀与密封强度高的特点。

基于螺纹结构设计形成一套完整的膨胀管特殊螺纹加工与检验技术，确保了螺纹质量。该项目设计的膨胀管螺纹性能达到国外同类产品技术水平。

膨胀工艺性能是膨胀管首要的服役性能，膨胀力的大小是衡量管材膨胀工艺性能的重要参数，很多失败案例都是由于膨胀压力过高而导致的。在现场进行膨胀作业时，首先通过水压实现膨胀锥的运动，将钢管的内径扩开。当膨胀锥通过预设在膨胀管底端的锚定橡胶后会将膨胀管固定在井眼内。此后，作业管柱向上提拉膨胀锥，膨胀将由水压与机械拉力共同完成。特殊情况下，膨胀作业则是单纯由机械拉力完成的。

该项目自主设计开发了我国首台实体膨胀管实物膨胀评价装备。膨胀管实物膨胀评价系统主要参数：液压油缸行程 5m；膨胀试验管材规格为外径 89～340mm；机械拉伸载荷为 0～1000kN，速度为 0～9.15m/min，液力泵最大液压输出为 70MPa，排量为 0～17.2m³/h，液压膨胀试验条件下，试验样管的长度不受液压缸行程约束。可进行机械膨胀、水压膨胀和机械/水压混合膨胀试验，可模拟试验各类膨胀管产品在套管井与裸眼井内的膨胀，可对膨胀管施工作业过程中遇到的各种复杂工况进行模拟试验，并完成相关试验数据的自动采集。设备关键技术指标，如试验范围、载荷能力、试验方式等方面均优于国外同类设备达到国际先进水平，填补了我国膨胀管实物膨胀性能检测评价装备领域的空白。

相关成果获 2020 年陕西省科技进步奖二等奖。

第二篇 论文篇

输送管产品开发与工程应用技术支撑体系

李鹤林 霍春勇 池 强 杜 伟

(中国石油集团石油管工程技术研究院·石油管材及装备材料服役行为与结构安全国家重点实验室)

摘　要：随着天然气需求的持续增长，中国高钢级油气输送管道建设快速发展，已进入国际领跑者行列。在高钢级油气输送管材应用过程中，产品开发与工程应用技术支撑体系发挥了重要作用。该技术支撑体系由质量监督与评价、标准化、科学研究、失效分析4个部分组成，分别阐述了各组成部分的内涵、发展历程及现状。该技术支撑体系在中国油气输送管产业发展与工程应用及中俄东线天然气管道工程 X80 输送管产品开发与工程建设中发挥了重要作用，并将继续为本质安全的管道工程建设提供技术支撑。

关键词：输送管；技术支撑体系；质量监督与评价；标准化；科学研究；失效分析

自1959年建成新疆克拉玛依—独山子输油管道以来，中国油气管道建设已经过60余年的发展历程。2012年建成的西气东输二线，主干线管径1219mm、钢级X80、输气压力12MPa/10MPa，干线全长4895km，标志着中国油气管道建设技术水平已跻身世界先进行列。2019年12月，中俄东线天然气管道工程投产通气，其主干线管径1422mm、钢级X80，输气压力12MPa，代表了目前国际上X80天然气管道建设的最高水平[1-12]。截至2018年，中国已建成油气管道总里程 13.6×10^4 km，其中天然气管道 7.9×10^4 km，形成了"北油南运""西油东进""西气东输""海气登陆"的油气输送格局，初步实现了横跨东西、纵贯南北、联通海外、覆盖全国的油气管道运输体系。

20世纪50—70年代，中国管线钢主要采用鞍钢等厂家生产的A3、16Mn；20世纪70年代后期及80年代，采用从日本进口的TS52K(相当于X52)[13]。"六五"到"七五"10年间，武钢牵头开展了X系列管线钢科技攻关。"八五"期间，进一步开发了高性能X52—X70API系列管线钢。进入21世纪，在西气东输及西气东输二线等重大管道工程的推动下，石油管工程技术研究院(以下简称管研院)联合国内多家钢厂、管厂、科研院所开展多项科研攻关，制定了X70/X80管线钢及钢管标准，解决了X70/X80钢管应用中的诸多问题，推动了X70/X80钢管产品的国产化与规模化应用，使中国管线钢管从跟仿国外技术发展至加入全球领跑

基金项目：中国石油科学研究与技术开发项目"公司发展战略与科技基础工作决策支持研究"专题"石油管工程学科体系建设研究"(2017D-5001-10)。

作者简介：李鹤林，男，1937年生，教授级高工，中国工程院院士，1961年毕业于西安交通大学金属材料及热处理专业，现主要从事石油管及装备材料工程技术方向的研究工作。地址：陕西省西安市锦业二路89号(710077)，电话：029-81887866，E-mail: lihelin@cnpc.com.cn。

者行列[14-29]。目前,中国已启动开展X90、X100管线钢管的开发与应用技术研究[30-33]。中国在高钢级管线钢及钢管研发应用方面起步较晚,但研发与应用速度较快,近20年走完了发达国家管线钢管50多年的研发进程。究其原因,油气输送管产品开发与工程应用技术支撑体系的建立与完善起到了举足轻重的作用。

1 技术支撑体系的形成与发展

管研院及其前身石油管材研究所、石油管材研究中心是推动油气输送管开发与工程应用的重要机构。建所伊始,便提出了适合石油管研制与工程应用的失效分析、科研攻关、技术监督"三位一体"的工作模式,也是最初石油管材开发与工程应用的技术支撑体系。在该体系中,通过失效分析发现问题,通过科研攻关找出答案,通过技术监督解决问题。充分利用该体系,不仅提高了技术监督与失效分析水平,而且从根本上实现了科研成果向生产力的转化。几十年的实践证明,该体系为油气输送管产业发展及管道安全运行提供了保障。标准化工作最初被划分在技术监督体系内,但随着技术监督工作的不断扩展与升级,标准化工作范围愈加广泛,发挥的作用也更大,有必要将其独立出来。基于此,将油气输送管产品开发与工程应用技术支撑体系的内涵调整为质量监督与评价、标准化、科学研究、失效分析四个方面(图1)。

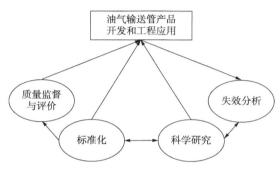

图1 油气输送管产品开发与工程应用技术支撑体系示意图

2 技术支撑体系的内涵

2.1 质量监督与评价

油气输送管的质量监督与评价包括三个方面:(1)生产许可证相关工作,即压力管道元件制造行政许可评审,由国家质检总局授权的机构负责;(2)产品的第三方检测与评价,由国家或行业的石油管材质量监督检验中心负责;(3)驻厂监造,对产品生产及厂内检验过程进行全程跟踪监督,由业主委托第三方监理机构执行。经过多年发展,油气输送管材质量监督机构已较为完善,质量监督与评价在管材开发与工程应用中的作用凸显。

2.1.1 生产许可证相关工作

为了提高压力管道的整体质量水平,从根本上降低安全风险,国家质量监督检验检疫总局特种设备局对压力管道实行制造许可制度。压力管道元件制造许可包括申请、受理、产品试制、型式试验、鉴定评审、审批、发证共七个环节,其中型式试验与鉴定评审是企业获得制造许可的两个关键环节。型式试验是对压力管道元件的设计、制造工艺、产品功能进行验证,目的是审查被设计、制造的产品是否存在不满足安全性能的缺陷,验证制造企业生产符合安全性能产品的能力,属于产品的可靠性试验范畴。鉴定评审是对申请单位的法定资格、资源条件、体系建立及实施情况、型式试验结果、产品质量等进行审查,判定其是否具备行政许可条件。国家质检总局核准授权行业的权威技术机构来承担型式试验及鉴定评审任务。管研院以及依托于管研院的国家石油管材质量监督检验中心分别是国家质量监督检验检疫总局授权的压力管道元件制造行政许可鉴定评审机构及国家压力管道元件制造行政许可型式试验机构。

20世纪90年代初，管研院承担焊接钢管生产许可评审工作，2006年之后，又作为首批特种设备行政许可鉴定评审机构之一，开展压力管道元件制造许可评审工作。经过多年发展，管研院建立并完善了型式试验、鉴定评审实施细则与作业程序，培养了一支业务精通、技术过硬、原则性强的专家队伍，积累了丰富的经验。此项工作对于促进企业加强质量管理、提高生产技术水平发挥了重要作用，整体上提升了产品质量层次，保障了油气管道安全运行。

2.1.2 产品的第三方检测与评价

在油气输送管产业迅速发展的背景下，成立了与管材相关的国家及行业的第三方检测与评价机构，最主要的机构是国家石油管材质量监督检验中心（以下简称国家质检中心）。国家质检中心的依托单位是管研院，其拥有先进仪器设备300余台套，已形成完备的石油管材检验与检测能力。在金属材料方面，具备化学成分、金相分析、无损检测、力学性能、防腐检测、实物性能六类检测能力，满足油气输送管产业发展及管道工程建设的需要。

2.1.3 驻厂监造

中国油气输送管材生产的大规模驻厂监造始于20世纪90年代，管研院借鉴国外经验，最早建立了输送管监理体系与实施细则，经过20余年的发展，已形成完善的油气输送管材第三方监理体系。从事此项工作较早、具有代表性的单位是由管研院出资成立的北京隆盛泰科石油管科技有限公司（以下简称隆盛公司）。驻厂监造是甲方委托具有资质的第三方监理机构，按照国家有关法规、规章、技术标准及合同规定，对乙方承包商产品的制造过程实施质量监督。输送管监造工作范围包括检查制造单位质量体系运行的有效性；抽查确认钢管产品质量的符合性；从原材料入厂，到生产、检验完成，直至钢管发运的全程监督。常见监造方式包括文件审查、现场监控、现场见证、停止点检查、随机抽查检验等。驻厂监造工作在保证发运钢管质量的同时，也促进了承包商质量控制水平的提升，确保油气管道项目建设进度与运行安全。

2.1.4 中俄东线管材产品质量监督

中俄东线天然气管道工程采用外径1422mm、X80钢管，此规格产品为国内首次研发生产，钢管制造厂家需扩大制造许可范围，重新进行型式试验与鉴定评审。管研院承担了型式试验与鉴定评审工作，严格执行程序，客观评价厂家制造能力，为中俄东线天然气管道工程建设保驾护航。

在钢管生产过程中，管研院制定了完备的产品第三方检测评价方案，具体试验由国家质检中心承担。中俄东线天然气管道工程涉及的管材产品主要包括直缝钢管、螺缝钢管、弯管、管件等，通过严格把关，提升了产品开发水平，保障了产品质量的稳定性。隆盛公司承担了驻厂监造工作，截至2020年，监造钢管超过$80×10^4$t，弯管管件与绝缘接头4000余个，涉及宝鸡、宝世顺、巨龙、华油、宝钢等九家焊管厂商，中油管道机械、恒通、隆泰迪、西安泵阀等十家弯管管件及绝厂商，天钢、衡钢等三家站场无缝管厂商。

2.2 标准化

2.2.1 中国石油管材标准化组织及其工作

中国石油管材标准体系由国家标准（GB）、石油天然气行业标准（SY）、中国石油天然气集团有限公司企业标准（Q/SY）共同组成。中国的石油管材标准主要等同或等效采用API/ISO标准。标准制修订等工作的组织机构由全国石油天然气标准化技术委员会石油专用管材分技术委员会、石油工业标准化技术委员会石油管材专业标准化技术委员会、中国石油天然气集团公司标准化委员会石油石化设备与材料专业标准化技术委员会石油管材分技术委员会组成。这些机构的秘书处均设立在管研院。

近几年，针对重大管道工程，结合科研成果，制定了油气输送管道用管材通用系列标准，包括钢管（含板、卷）及弯管、管件等系列标准，对规范管道建设用管材订货技术要求，保证管材质量与经济性发挥了重要作用。针对中俄东线天然气管道工程的具体情况，总结近几年 X80 高钢级管材关键技术指标最新研究成果，编制了中俄东线天然气管道工程用板材（外径为 1422mm、X80）、螺旋缝埋弧焊管、直缝埋弧焊管、弯管、管件等 11 项管材产品标准，极大地支持了工程建设。

2.2.2 国际标准化组织及其工作

管研院是国际标准化组织 ISO/TC67/SC2 石油、石化、天然气工业用材料设备与海上结构——管道输送系统分委会的国内技术归口单位，同时也是 ISO/TC67/SC2 并行秘书处单位。管研院始终与 ISO/TC67/SC2 保持密切关系，积极参加国际标准化项目，及时投票，并每年派人参加国际标准化会议，承担 SC2 专项标准化工作组的工作，成立专家专业组并细分专业，面向石油管道规划、设计、建设、运营管理各领域，广泛、有效传播 ISO/TC67/SC2 标准化动态及标准技术。近几年，针对油气输送管方向，提出 6 项 API 提案（表1），组织提交 ISO 标准提案并成功立项 6 项（表2）。

表1 油气输送管 API 国际标准提案项目统计表

从属领域	项目名称	进展情况	提交时间
API Spec 5L	异常断口评定	API 成立了新项目组，负责 API 5L3 修订	2010.06
API Spec 5L	SAW 钢管焊缝焊偏测量	已纳入 ISO/API 标准	2010.06
API Spec 5L	焊偏时埋弧焊管焊接热影响区冲击试样缺口位置	焊偏时 HAZ 冲击取样：根据要求提交了新的取样图，采纳了理想状态下取样图	2010.06
API Spec 5L	钢管横向屈服强度拉伸试样的选择	横向拉伸试样：超出标准范围，未采用	2010.06
API Spec 5L	管线钢管拉伸试验伸长率	API 成功立项 WI 4238，经过 2 年工作，由于绝大多数成员认为相关指标的修订影响不大，决定暂不做更改	2015.01
API 5L3	试样断口"三角区"评定方法	API 5L3 修订考虑	2012.01

表2 油气输送管 ISOTC67SC2 国际标准提案项目统计表

从属领域	项目名称	进展情况	提交时间
ISO 3183	基于应变设计地区使用的 PSL2 级钢管	2013 年 SC2 年会上进行了汇报与讨论，拟作为 ISO 3183 一个新附录。后 API 成立相关工作组，相关成果纳入 API 新增附录。ISO 3183：2019 规范性引用 API 全文	2012.11
新工作项目	管道完整性管理规范	2013 年 SC2 年会上做了报告，通过会议决议纳入工作组讨论，2014 年通过立项投票。经过 5 年工作，2019 年正式发布 ISO 19345-1：2019、ISO 19345-2：2019	2012.11
新工作项目	油气管道地质灾害风险管理技术	2014 年经 SC2 年会讨论确认可以立项，2014 年通过立项投票。经过 5 年工作，2019 年正式发布 ISO 20074：2019	2013.11

续表

从属领域	项目名称	进展情况	提交时间
新工作项目	油气管道直流杂散电流防护技术	2016年通过立项投票，2017年SC2年会决定由英国代表出任召集人，中国专家参与制定，已完成DIS稿（ISO/DIS 21857）	2016.01
新工作项目	管道完整性评价	2017年经SC2年会讨论确认可以立项，2017年通过立项投票，正在编制工作组草案（ISO/AWI 22974）	2017.05
新工作项目	管道输送系统用耐蚀合金内覆复合弯管及管件	2019年通过立项投票，正在编制工作组草案（ISO/AWI 24139-1、ISO/AWI 24139-2）	2018.12

2.3 失效分析

2.3.1 油气输送管失效分析案例

20世纪50年代以来，随着油气管道的大量敷设，管道事故屡有发生，并造成灾难性后果。迄今为止，破裂裂缝最长的管道失效事故是在1960年美国Trans-Western公司的输气管道脆性破裂事故，该管道直径为30in（1in=25.4mm），钢级为X56，裂缝长度达13km[34]。损失最惨重的是1989年苏联乌拉尔山隧道附近的输气管道爆炸事故，烧毁两列列车，伤亡1024人，其中约800人死亡[35]。

据美国管道与危险物资安全管理局数据，1999—2010年，美国共发生2840起重大天然气管道失效事故，包括992起致死、致伤事故，323人死亡，1327人受伤。近20年，加拿大油气管道干线平均每年发生30~40起失效事故。1971—2000年，欧洲油气管道干线平均每年发生13.8起失效事故。

中国油气管道建设起步较晚，但失效事故也屡见不鲜。1966年，威远气田内部集输管道通气试压时，4天时间内连续爆裂3次。经失效分析及再现性试验，确认爆裂是天然气所含H_2S在含水条件下引起的应力腐蚀开裂所致[36]，这是中国油气管道的第一起重大失效事故。1971—1976年，东北曾发生3次输油管道破裂事故，其中一次发生在1974年冬季大庆—铁岭输油管道复线的气压试验过程中。当时气温为-30~-25℃，裂缝长度为2km，断口几乎全部为脆性断口。四川气田在1970—1990年共发生108次输气管道爆裂事故[37]。1992年，轮库输油管道试压时发生14次爆裂事故。1999年，采石输油管道发生12次试压爆裂事故。

管研院承担了大部分国内油气输送管道失效分析项目，近10年来，较为典型的事故有以下几起：(1)2010年7月20日—9月22日，西气东输二线东段18标段桩间管道试压后排水作业过程中，发生管道破裂事件，其原因是管道高度起伏变化导致试压过程中产生弥合水击作用。(2)2012年2月，中俄原油管道漠河—大庆段工程发生环焊缝开裂，其原因是管道承受轴向弯曲应力，在该事故调查中发现的管道质量情况也引起中国石油天然气集团有限公司的重视。(3)2013年11月22日，青岛东黄输油复线发生管道破裂，其原因是管道腐蚀减薄，泄漏原油进入市政涵道，导致市政涵道内发生爆炸，造成严重后果。(4)2017年7月2日，中缅管道晴隆段发生环焊缝断裂。国家应急管理部特别重视这次失效分析工作，多次组织专家进行研讨。失效分析有助于管理人员与技术人员总结经验，提升管理与技术水平，并对新建管道工程提供重要借鉴。

2.3.2 失效分析机构

中国的石油管材失效分析开始于1966年对四川油田天然气输送管道严重爆破事故的分

析，1979年，华北油田某井接连发生两起G105钻杆断裂事故，经失效分析进一步认识到石油管材研究及失效分析的重要性。1981年，石油部成立石油管材试验研究中心，石油管材的失效分析工作普遍开展起来。石油部要求"凡失效，皆分析"，对失效分析工作起到了极大的促进作用。1986年，石油管材试验研究中心成立失效分析与预防研究室，专门从事石油管失效分析工作。

1996年初，石油管材研究所建立的失效分析质量管理体系通过ISO 9002认证，这是中国技术服务领域首次通过该认证，标志着石油管材失效分析质量管理与国际接轨，走向科学化与规范化。2006年，中国科学技术协会工程学会联合会失效分析与预防中心同意在石油管材研究所成立中国科协工程联失效分析和预防中心石油管材与装备分中心。2010年，根据中国石油学会石油管材专业委员会油管专字〔2010〕02号通知，决定成立中国石油学会石油管材与装备失效分析及预防中心，秘书处设在管研院。

2.3.3 失效分析与失效控制

机械产品的零件或部件处于下列3种状态之一时，定义为失效：(1)完全不能工作；(2)仍可工作，但不能满意地实现预期功能；(3)受到严重损伤不能安全可靠地工作，必须进行修理与更换。对于失效事件，按一定的思路与方法判断失效性质、分析失效原因、研究失效事故处理方法与预防措施的技术活动及管理活动，统称失效分析。失效分析预测预防是产品或装备安全可靠运行的保证，是提高产品质量的重要途径，目的在于不断降低装备的失效率，提高可靠性，防止发生重大失效事故，促进经济持续稳定发展。

基于失效分析，提出防止失效的措施，对失效分析结果进行反馈，即失效控制[38]。油气管道失效控制(图2)的主要内容包括：(1)大量搜集国内外失效案例，建立油气管道失效信息案例库；(2)对失效案例进行综合统计分析，确定油气管道的主要失效模式；(3)研究各种失效模式的发生原因、机理及影响因素；(4)研究并提出各种失效模式的控制措施与方法。失效信息案例库是失效控制的基础，应具有较强的数据处理与统计分析功能，并拥有尽可能多的案例，一方面广泛搜集国内外油气管道已发生的重大失效案例，另一方面加强对新发生的油气管道失效事故的搜集分析。在大量失效分析的基础上，总结一些重大共性科学问题进行较深入的研究。随着失效案例的不断增多，失效信息案例库不断充实，失效模式及其原因、机理、影响因素随之动态变化与调整，失效控制措施和方法也不断完善。

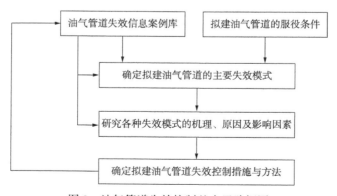

图2 油气管道失效控制基本思路框图

2.4 科学研究

油气输送管的科学研究包括两个方面：(1)油气输送管工程应用与应用基础研究；

(2)新产品开发与制管工艺研究。工程应用与应用基础研究由用户的研究机构及一些独立的研究机构承担,油气输送管新产品开发与制管工艺研究的主体是生产企业,同时联合研究机构、高校等。

2.4.1 油气输送管工程应用与应用基础研究

20世纪90年代,管研院开始了石油管工程应用与应用基础研究,承担了国家及中国石油的许多重大科研课题,在取得一系列重要研究成果的同时,总结、梳理形成了石油管工程学(图3)。

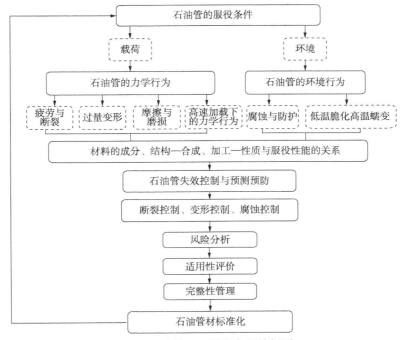

图3 石油管工程学研究领域框图

石油管工程学致力于研究不同服役条件下石油管的失效规律、机理及克服失效的途径。石油管的服役条件主要体现在载荷与环境两方面,石油管的服役行为包括力学行为、环境行为及两者的复合。通过研究得出材料的成分/结构、合成/加工、性质与服役性能的关系,解决失效控制与失效的预测预防问题。所有研究成果均转化为技术标准与规范。

2015年,国家科技部批准,挂靠管研院建设国家石油管材与装备服役行为与结构安全重点实验室,进一步推动了中国油气输送管的工程应用与应用基础研究,为油气管道安全运行保驾护航。基于石油管工程学技术脉络,管研院从20世纪90年代开始,针对油气输送管,开展了大量应用基础研究工作,承担或参加的课题包括:油气输送管止裂韧性试验方法的研究、石油天然气用焊管、石油天然气输送用X80级管线钢管研制、长距离输气管道材质和管型选用技术研究、油气管线钢的止裂韧性以及与动态断裂韧性的关系、油气输送用焊管包申格效应的研究、高性能管线钢的重大基础研究、西气东输管道使用国产螺旋缝埋弧焊管的可行性研究、X80管线钢管的开发、西气东输二线工程关键技术研究等,在管材组织成分、关键技术指标、管材失效控制与预测预防等方面,取得了系列研究成果,支撑了陕京管道、西气东输管道、中亚管道等重大管道工程的建设。2012年,中国石油天然气集团有限公司设立重大专项"第三代大输量天然气管道工程关键技术",管研院系统开展了外径1422mm、X80钢管应用技术研究,综合考虑管材加工性能及焊接性,提出了产品化学成分范围;通过理论计算与全尺寸试验,提出并验证了管道延性止裂韧性;基于试验研究与分析,提出管材产品技术指标,制定产品检测方法。以上科研成果均应用于中俄东线天然气管道工程管材技术标准编制与产品生产。

2.4.2 新产品开发与制管工艺研究

油气输送钢管(特别是长距离输送钢管)的主体是焊管。中国油气输送焊管的主要开发模式是:由业主组织,研究机构技术牵头并对产品进行评价,钢铁企业研究开发管线钢(钢板、板卷),制管企业从钢铁企业购置钢板(板卷),加工生产焊管产品。目前,仅上海宝钢既开发管线钢钢板(板卷),又生产油气输送焊管。油气输送焊管新产品开发包括管线钢的研究开发及制管工艺研究,前者主要由宝钢、武钢、鞍钢、首钢、沙钢、太钢等大型钢铁企业的科研机构承担,后者主要由宝钢、渤海装备等焊管企业承担。

中俄东线天然气管道工程用管材产品研发过程中,基于工程需求与产品技术标准,由管道公司、管研院、钢铁与制管企业联合协作,按照管研院提出的油气输送管材试制过程质量控制体系(图4),开展产品试制工作。第一阶段主要是考察钢铁与制管企业的工艺可行性,为单炉试制;第二阶段是验证产品生产工艺的可靠性,通常为500~1000t的批量试制,在该阶段,严格按正式产品生产流程进行试制,实施全程驻厂监造。基于研究成果及以上工作,成功开发了中俄东线天然气管道工程用系列管材产品,包括外径1422mm,壁厚21.4mm、25.7mm、30.8mm、32.1mm焊管,以及外径1422mm,壁厚25.7mm、30.8mm、33.8mm、35.2mm弯管等产品,保障了中俄东线天然气管道工程的顺利实施。

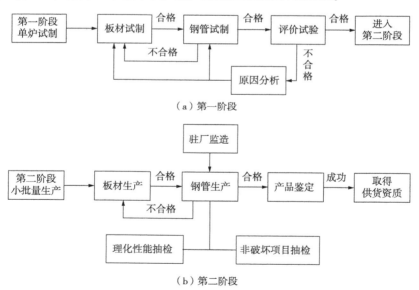

图4 油气输送管材试制过程质量控制体系框图

3 结束语

油气输送管产品开发与工程应用技术支撑体系涵盖质量监督与评价、标准化、科学研究、失效分析四部分内容,四位一体,协调配合,在油气输送管材国产化与工程应用全周期中发挥了重要作用,其成功经验可供其他行业或领域参考。随着长输油气管道工程建设的高速发展,油气输送管产品开发与工程应用技术支撑体系也将不断完善,持续促进油气输送管产品研发与应用技术的进步,进而为油气管道工程建设提供技术支撑。

参 考 文 献

[1] 姜昌亮.中俄东线天然气管道工程管理与技术创新[J].油气储运,2020,39(2):121-129.
[2] 程玉峰.保障中俄东线天然气管道长期安全运行的若干技术思考[J].油气储运,2020,39(1):1-8.
[3] 蒲明,李育天,孙骥姝.中俄东线天然气管道工程前期工作关键点及创新成果[J].油气储运,2020,39(4):371-378.
[4] 张振永.中俄东线X80钢级φ1422mm管道工程设计关键技术应用[J].焊管,2019,42(7):64-71.
[5] 陈小伟,嵇峰,白学伟.中俄东线X80钢级φ1422mm×30.8mm钢管理化性能研究[J].焊管,2019,42(5):10-17.
[6] 赵新伟,池强,张伟卫,杨峰平,许春江.管径1422mm的X80焊管断裂韧性指标[J].油气储运,2017,36(1):37-43.
[7] 张振永,周亚薇,张金源.现行设计系数对中俄东线OD1422mm管道的适用性[J].油气储运,2017,36(3):319-324.
[8] 张振永,张文伟,周亚薇,等.中俄东线OD1422mm埋地管道的断裂控制设计[J].油气储运,2017,36(9):1059-1064.
[9] 刘迎来,许彦,王高峰,等.中俄东线-45℃低温环境油气管道工程用X80钢级φ1422mm×33.8mm感应加热弯管研发[J].焊管,2019,42(7):48-54.
[10] 尤泽广,王成,傅伟庆,等.中俄东线站场用直径1422mm×1219mm三通设计[J].油气储运,2020,39(3):347-353.
[11] 蒋庆梅,张小强,钟桂香,等.中俄东线黑龙江穿越段管材关键性能指标对比与确定[J].油气储运,2020,39(1):92-98.
[12] 张小强,侯宇,蒋庆梅,等.中俄东线直径1422mm X80钢级冷弯管设计参数的确定[J].油气储运,2020,39(2):222-225.
[13] 李鹤林,冯耀荣,霍春勇,等.关于西气东输管线和钢管的若干问题[J].中国冶金,2003(4):36-40.
[14] 张圣柱,程玉峰,冯晓东,等.X80管线钢性能特征及技术挑战[J].油气储运,2019,38(5):481-491.
[15] 任俊杰,马卫锋,惠文颖,等.高钢级管道环焊缝断裂行为研究现状及探讨[J].石油工程建设,2019,45(1):1-5.
[16] 庄传晶,李云龙,冯耀荣.高强度管线钢环焊缝强度匹配对管道性能的影响[J].理化检验(物理分册),2004,40(8):383-386.
[17] 张宏,吴锴,刘啸奔.直径1422mm X80管道环焊接头应变能力数值模拟方法[J].油气储运,2020,39(2):162-168.
[18] 毕宗岳,黄晓辉,牛辉.X80级φ1422mm×21.4mm大直径厚壁焊管的研发及性能研究[J].焊管,2017,40(4):1-7.
[19] 冯耀荣,霍春勇,吉玲康,等.我国高钢级管线钢和钢管应用基础研究进展及展望[J].石油科学通报,2016,1(1):143-153.
[20] 熊庆人,杨扬,许晓锋,等.切环法和盲孔法测试大口径厚壁X80钢级埋弧焊管的残余应力[J].机械工程材料,2018,42(12):27-30,67.
[21] 杨坤,池强,李鹤,等.高钢级天然气输送管道止裂预测模型研究进展[J].石油管材与仪器,2019,

5(4):9-14.

[22] 霍春勇,李鹤,张伟卫,等.X80钢级1422mm大口径管道断裂控制技术[J].天然气工业,2016,36(6):78-83.

[23] 崔天燮,张彦睿.太钢X80级管线钢热轧卷板技术开发与应用[J].焊管,2009,32(12):32-39.

[24] 杨忠文,毕宗岳,牛辉.高钢级管线钢焊管研制[J].焊管,2011,34(4):5-11.

[25] 李少坡,姜中行,李永东,等.超低碳贝氏体厚壁X80宽厚板的研发[J].钢铁,2012,47(4):55-59.

[26] 陈小伟,王旭,李国鹏,等.我国高钢级、大直径油气输送直缝埋弧焊管研究进展[J].钢管,2016,45(5):1-8.

[27] 谢仕强,桂光正,郑磊,等.X80钢级大直径UOE直缝埋弧焊管的开发及应用[J].钢管,2011,40(4):29-36.

[28] 张伟卫,吉玲康,陈宏远,等.一种X70或X80抗大变形钢管生产方法:201010550557.4[P].2013-07-31.

[29] 刘文月,任毅,张禄林,等.一种X80抗大变形管线钢及制造方法:201510336908.4[P].2018-08-31.

[30] 史立强,牛辉,杨军,等.大口径JCOE工艺生产X90管线钢组织与性能的研究[J].热加工工艺,2015,44(3):226-229.

[31] 刘刚伟,毕宗岳,牛辉,等.X90高强度螺旋埋弧焊管组织性能研究[J].焊管,2015,38(10):9-13.

[32] 王红伟,吉玲康,张晓勇,等.批量试制X90管线钢管及板材强度特性研究[J].石油管材与仪器,2015,1(6):44-51.

[33] 李鹤,封辉,杨坤,等.断口分离对X90焊管断裂阻力影响试验[J].油气储运,2019,38(10):1104-1108.

[34] 潘家华.油气管道断裂力学分析[M].北京:石油工业出版社,1989:2-10.

[35] Starostin V. Pipeline disaster in the USSR: It had to happen, yet it could have been averted[J]. Pipes Pipelines Int, 1990, 35(2):7-8.

[36] 李鹤林.某管线试压爆破原因分析[M]//李鹤林文集——石油管工程专辑.北京:石油工业出版社,2017:601-607.

[37] MAO H G. Failure analysis of the natural gas pipelines in Sichuan[C]. Beijing: International Symposium on Structural Technique of Pipeline Engineering, 1992:20-30.

[38] 李鹤林.油气管道失效控制技术[J].油气储运,2011,30(6):401-411.

本论文原发表于《油气储运》2020年第39卷第7期。

石油科研院所推进高质量发展探索与实践

刘亚旭　瞿云萱　谢文江　陈娟利　崔　巍

(中国石油集团石油管工程技术研究院)

摘　要：企业是实施创新驱动发展战略的主体，大企业集团的科研院所是国家和行业层面相关领域科技创新的主力军。中国石油集团石油管工程技术研究院作为中国石油集团改革试点单位，开展了一系列以改革开放促进高质量发展的探索与实践：明确目标，全力打造科技创新、质量标准、成果创效3个平台，建立高质量发展的创新体系；深化改革，加大人才培养及创新激励力度，建立双序列职级体系，营造创新创效良好氛围；开放办院，搭建国内外专业技术合作研究平台，提升国际专业研究和标准领域话语权，形成合作共享的创新机制；服务主业，围绕重大工程和生产需求推进研发创新，高强度管线钢应用关键技术取得新突破，石油管新产品新技术开发取得新进展，石油石化材料领域重大失效事故分析取得新认识，原始创新和超前储备技术研究取得新成果。这些举措的有效实施，为集团公司主营业务发展提供了有力技术支撑，并探索出一条独具特色的石油科研院所高质量发展之路。

关键词：科研院所；科技创新；深化改革；高质量发展

习近平总书记指出："党的十八大提出实施创新驱动发展战略，强调科技创新是提高社会生产力和综合国力的战略支撑，必须摆在国家发展全局的核心位置"[1]。"十三五"以来，中国石油天然气集团有限公司(以下简称中国石油)全面贯彻落实党中央的要求，大力实施创新驱动发展战略[2]。中国石油集团石油管工程技术研究院(以下简称管研院)以习近平新时代中国特色社会主义思想为指引，认真贯彻落实中央指示精神及集团公司党组的部署要求，紧紧围绕中国石油"建设世界一流综合性国际能源公司"和"创建世界一流示范企业"战略目标，坚持以改革促进高质量发展，在科技创新体系优化、关键核心技术突破、科技成果转化创效、高水平人才队伍建设等方面取得了显著成效，破解了一批石油管工程技术领域的重大技术瓶颈问题，为推动中国石油高质量发展作出了应有贡献。

1　明确目标，建立推进高质量发展的创新体系

管研院主要从事石油管及装备材料工程技术工作，是我国唯一专业从事石油管工程技术的研究院，也是石油石化行业的材料工程研究机构[3]，具有先进的科研设备、丰富的人才

基金项目：中国石油天然气集团有限公司管理创新研究与实践项目"以市场化激励方式提升科研能力的管理创新实践"(编号：中油企201904)；中国石油天然气集团有限公司科学研究与技术开发项目"科技成果价值化及转化应用研究"(编号：2019D-5005-29)。

作者简介：刘亚旭，1968年生，2007年毕业于西安交通大学管理科学与工程专业，博士，正高级工程师，现任中国石油集团石油管工程技术研究院党委书记、院长，主要从事石油管工程技术和科技管理等工作。

储备、雄厚的科研实力。但由于体制机制的制约使许多优势资源未能实现"物尽其用",业务板块之间不能有效协同,科技成果创新性不足,成果转化及产业化能力较低,缺少做强做大的核心人才和机制环境,国内外影响力也有待进一步提升。结合自身发展历程、未来发展需要和石油工业发展需求,管研院明确提出紧紧围绕全面建成国际一流研究院的战略目标,着力打造科技创新、质量标准、成果创效"三个平台",努力构建成果、技术、创效、人才"四个高地",大力建设精干高效、独具特色的国际一流科技创新体系,确立了引领管研院高质量发展的战略目标和战略重点。

1.1 打造原始创新能力强、成果转化效率高的科技创新平台

以国家重点实验室为依托,面向社会和石油天然气行业未来发展需求,开展石油管材及装备材料服役行为与结构安全领域的应用基础研究和竞争前共性技术研究,以提升国家石油管材及装备科技支撑能力为目标,发展石油管材及装备的应用关键技术,按照中国石油科技创新"三大工程"的整体部署,在油井管与管柱、输送管与管线优势领域持续保持领先,在腐蚀与防护、炼化装备材料、安全评价与完整性、石油装备材料等领域实现跨越式提升,抢占非金属与复合材料领域制高点,学科体系和核心技术体系不断完善,支撑和引领相关业务发展[4-5]。

1.2 打造权威公正、技术精湛、具有国际影响力的质量标准平台

以国家质检中心为依托,建立具有国际先进水平、系统配套的重大试验装备体系,成为国内唯一、世界一流的石油管材及装备监督检验中心,有效保障了石油管与装备材料的质量和安全。以归口管理的标准化机构为依托,增强主导和影响制定国际标准与技术规则的能力,不断提升在国际标准化组织的话语权,推动我国更多标准成为国际标准;积极构建我国石油标准体系,以"一带一路"建设为契机,把我国的标准体系和装备产品推向世界[6]。

1.3 打造体制机制灵活、效益显著提升的成果创效平台

以院所属的两个成果转化公司为主体,将国家重点实验室研发的创新成果孵化、转化,实现规模效益,回馈研发投入,奖励研发骨干,形成良性循环。着力打造国内一流、国际知名的第三方技术服务品牌和高附加值创新产品转化基地。

"三个平台"相互支撑、三位一体、协调发展,逐步建成"四个高地",即尖端实用创新成果不断涌现的成果高地、凸显第三方权威公正地位和标准化主导地位的技术高地、科技成果转化和推广应用的创效高地、培养和集聚高层次专家的人才高地。

2 深化改革,营造创新创效的良好氛围

多年来,科研院所员工关心的热点问题主要集中在成果转化激励机制、薪酬分配制度和绩效考评制度等方面的改革上。这些改革,涉及员工切身利益,更关系到管研院的长远发展,对于激发员工工作积极性和创新热情至关重要,是深化综合改革、解决制约发展瓶颈的着力点和突破口。2016年,管研院作为中国石油综合改革试点单位,按照集团公司改革工作的总体部署,实施了一批改革举措,取得了初步成效。但改革推进还不够深入,有的举措仍然处在磨合阶段,未取得实质性突破。部分干部职工实施改革的信心不够坚定,对改革过程中出现的矛盾、困难和问题存有顾虑和畏难情绪,全面推进改革任务落实的工作机制也尚不健全,推进深化改革的任务依然艰巨繁重。为解决这些问题,管研院以改革开放40周年为契机,推动思想再解放、改革再深入,统一认识、坚定信心,积极落实国家相关政策及集团公司系列制度和举措,认真研究、及时调整配套实施方案,把握好改革的整体推进节奏和

力度，取得了持续深化改革的新突破新进展。

2.1　加强党的建设，弘扬石油精神激励创新创效

院党委提高政治站位，加强党的建设，切实发挥把方向、管大局、保落实的重要作用。特别是高度重视人才建设，努力为人才建台子、搭梯子、抬轿子，通过经常性思想政治工作和"四合格、四诠释"岗位实践活动，发挥党员干部的先锋模范作用，用事业、感情留人，增强队伍凝聚力；大力弘扬石油精神和大庆精神铁人精神，凝聚创新创效的精神力量，将爱国、爱党、爱油、爱院热情转化为创新创效激情，形成创新创效的强大动力。

2.2　建立双序列职级体系，畅通人才成长通道

管研院在集团公司内部率先搭建并不断完善技术和管理双序列职级体系，建立八个专业技术层级和九个管理层级的岗位序列，形成大 H 型人才发展双通道，激励两支队伍共同围绕创新创效努力工作，形成了层次合理、专业配套、素质优良的科技创新团队。

"十三五"以来，管研院新增省级创新团队 4 个、获批陕西省首批院士工作站；新增国家有突出贡献中青年专家 1 人，国务院特殊津贴专家 2 人，石油科学家培育对象 2 人，青年科技英才培育工程 16 人；陕西省杰出青年 1 人、陕西省中青年领军人才 5 人、陕西省科技新星 11 人；"孙越崎"青年奖 4 人。博士占员工总数的 17%，硕士以上占比达 50%；副高级以上职称占比达 45%。全院人员素质进一步提升，专业结构更趋合理，为创新创效提供坚实的人才保障。

2.3　突出业绩导向，激发全员创新创效活力

院领导班子成员、中层干部和普通管理岗位实行任期连续末位淘汰制，技术序列实行任期考核晋升制和末位淘汰制，使"先进有动力，后进有压力，中游有标杆"。探索建立了以岗位管理为基础的"易岗易薪"薪酬制度，突出业绩在薪酬分配中的主导作用，拉大收入差距。薪酬分配向一线核心骨干倾斜，保持技术序列收入高于管理序列，部分技术骨干收入高于领导班子成员，增强了科技人员的获得感、归属感和责任感。

员工实行动态管理，抬高进人"高门槛"，建立不合格员工的淘汰机制。"十三五"以来，先后有 20 名表现较差、考核不合格的员工离职或被解除劳动合同，占员工总数的 5%。能上能下、能高能低、能进能出的常态化初步形成，提高了人力资源配置的效率和价值，充分激发了骨干人才创新创效的积极性和创造性。目前，科技队伍稳定，并出现人才回流现象。

2.4　加大激励力度，让科研骨干"名利双收"

设立中青年拔尖人才计划，两批选拔 34 名优秀骨干，按月发放技术津贴，促进人才脱颖而出、快速成长；设立青年创新奖，奖励贡献突出青年英才，激发创新创效热情。师徒结对培养，发挥高层级专家的传帮带作用。多渠道、多方位加大创新激励力度，完善《科技创新激励办法》，提高知识产权奖励标准和范围，近两年奖励额度达 500 万元以上。落实国家和集团公司促进科技成果转化激励政策，设立成果转化项目 5 项，按转化收益比例奖励，增强科技创新和成果转化的动力活力。在集团公司率先试点科技型企业岗位分红激励改革，积极推动科技骨干收入与市场接轨。

通过持续深化改革，革除了多年积弊，破解一些深层次机制性障碍，全院员工精神面貌焕然一新，科技人员创新创效积极性高涨，科研能力得到有效提升，解决现场技术问题能力显著增强，尊重人才、鼓励创新创效的机制初步建立，"人人愿意创新、处处可见创新、时时体现创新"的良好局面和氛围日益形成。

3 开放办院，形成合作共享的创新机制

开放也是改革，开放合作是后来者追赶先行者的必由之路。按照全面建成国际一流研究院的战略目标，管研院尽管具有基础研究的长期储备和雄厚积累，有国内行业最权威的资质和创新平台，部分技术与装备达到国际领先和先进水平，综合一体化优势、政治文化优势独具特色，但通过与国际上同类研究机构对标比较，其创新创效能力还不够强，跨越发展的新格局还没有形成；设备自主开发能力还不具备，试验装备整体实力与国际先进水平存在一定差距；针对国家能源发展需求的前沿技术攻关和创新不足，市场优势未能完全转化为话语权优势；科研和技术服务的国际化程度偏低。目前管研院正以时不待我的紧迫感，坚持开放办院，扩大开放合作，紧盯差距，抓重点、补短板、强弱项，不断向既定目标迈进。

3.1 搭建国内外专业技术合作研究平台

管研院联合国内外行业技术力量，牵头成立中国管线研究组织（CPRO），成为北美PRCI、欧洲EPRG、澳洲APIA之后第四家管道领域的知名研究组织；成立油气井管材及管柱技术创新联盟（COSTA），为油井管设计、生产、使用搭建交流合作平台；成立国际腐蚀工程师协会NACE新疆分会，扩大了腐蚀领域的影响力；与巴西矿冶公司（CBMM）联合成立国际焊接研究中心，获得海外科技投入3000余万元；成立中国材料与试验团体标准（CSTM）石油石化材料领域委员会，材料领域影响进一步扩大。鉴于国际合作取得的成效，2017年，管研院获批陕西省国际合作基地。

3.2 与国际知名科研机构开展战略合作

管研院与美国西南研究院（SWRI）、俄罗斯天然气研究院（GAZPROM）、英国焊接研究所（TWI）、韩国天然气研究院（KOGAS）、加拿大C-FER、意大利CSM、挪威SINTEF、日本JFE等十几个国家二十多个国际著名科研机构和企业签订了战略合作协议。以这些机构为对标单位，通过加强交流合作，缩短了差距，提高了水平。

3.3 提升国际标准领域的话语权

超前策划ISO/TC67绿色制造分委会，筹备工作顺利推进。提出国际标准培育计划5项，首次主导制定的3项ISO国际产品标准正式立项，提交14项国际标准制修订提案被采纳。多名专家担任国际标准化组织相关委员会职务，国际标准领域的话语权和影响力大幅提升。

3.4 多途径培育国际化专业技术人才

为使广大科研人员及时了解国际前沿技术，全方位、多途径培养国际化专业技术人才，院里推荐优秀博士到国外知名机构做博士后，推荐优秀人才在国外高校攻读博士学位，鼓励科研骨干到国外访问交流、联合攻关、参加国际会议，定期举办石油行业具有影响力的国际品牌会议"石油管及装备材料国际会议（TEC）"。这些都已成为广大科研人员重要的学习交流平台。

通过开放办院，扩大交流合作，"引进来"和"走出去"并举，提高了资源配置能力，提升了创新创效水平，加快了石油管工程科技进步，增强了行业话语权和影响力，推进了管研院国际化程度，合作共享的机制不断完善。

4 服务主业，围绕公司重大需求推进研发创新

目前，国际油气市场供需格局发生重大调整，特别是受新冠肺炎全球蔓延的冲击，世界

经济大幅走低，油气需求明显萎缩，国际油价低位震荡运行并可能持续较长时间，同时风能、太阳能、核能及燃料电池等新能源技术不断取得突破，对传统油气产业带来新的挑战，油气行业的科技创新比以往任何时候都更加迫切。管研院作为中国石油集团直属科研院所，肩负着公司石油管工程领域技术创新的重任，近些年来努力适应新形势新要求，坚持以服务国家和中国石油集团油气田开发、管道储运、炼油化工等主营业务为使命，以解决石油管工程技术领域重大、共性、关键、瓶颈技术难题为己任，以"增效益、降成本、保安全"为重点，在重大管道工程建设、重点油气田开发、炼化安全运行等方面，取得了一系列技术创新成果。

4.1 高强度管线钢应用关键技术取得新突破

通过第三代大输量天然气管道关键技术重大专项攻关，突破了 OD1422mm X80 管线钢应用关键技术，支撑了中俄东线管道建设；超前研究 X90/X100 超高强度管线钢，为超大输量管道建设提供技术储备；联合建设国内首个、世界第三个全尺寸气体爆破试验场，自主掌握全套试验技术，为油气管道断裂控制研究创造条件，保持了关键技术领跑地位[7]；通过抗大变形钢管应用技术攻关，提出指标体系并被纳入美国石油学会（API）标准附录，联合开发了 X70 和 X80 大应变钢管，在中缅、西气东输、陕京等管线工程应用；研究构建储气库完整性技术与标准体系，在大港、相国寺、呼图壁等储气库成功应用。全尺寸气体爆破试验场、第三代大输量技术、大变形钢管 3 项成果连续 3 年被评为中国石油十大科技进展。

2019 年以来，按照中国石油隐患治理和升级管理要求，积极开展含缺陷环焊缝的套筒修复、换管焊接等工艺研究及环焊缝完整性评价，全力配合长输管道管理提升，为保障管道运行安全作出了积极贡献。

4.2 油田所需石油管新产品新技术开发取得新进展

针对塔里木"三超"气田开发，管研院建立了油套管安全可靠性设计评价方法，开发了酸化压裂和地面集输专用系列缓蚀剂，突破了 13Cr 油管不能适应高温酸化作业的禁区；2018 年，集输管网腐蚀失效事故十年来最低，是此类事故次数最高年度时的 1/4，实现了高温高压气井井筒—地面防腐一体化，为"三超"气田开发作出新贡献；针对西南页岩气、新疆页岩油开发套管变形难题，研发了试验评价装备，建立选用及评价方法，为解决套管变形提供了工艺方案；针对长庆"三低"气田，研发了经济型特殊螺纹套管，解决了长庆油田兼顾经济性和安全性的技术难题，助力中国石油宝鸡石油钢管有限责任公司（以下简称宝鸡钢管）扭亏为盈；推广非金属管设计及制造技术，将下游炼化企业的化工原料制管后用于上游油田集输管网，并支持宝鸡钢管建成非金属管生产线；针对天然气水合物、深层油气开发，超前研究钛合金、铝合金等轻量化材料，为新能源和万米深层油气开采奠定技术基础。

4.3 石油石化材料领域重大失效事故分析取得新认识

高质量完成中缅管道"7·2""6·10"天然气燃爆[8]、西二线"7·28"同心段管道环焊缝渗漏、泰青威管道"3·20"燃爆、中国海油尼克森输油砂管道断裂、昭通压裂车曲轴断裂起火等重大事故失效分析，分析结论和新的认识，为事故调查、责任认定提供科学依据，为事故预防提供技术支持，有效发挥了国家级机构的公正和权威作用，得到国家主管部门和同行的认可。

4.4 原始创新和超前储备技术研究取得新成果

积极开展增材制造、石墨烯应用等超前储备技术攻关。首次将增材制造技术应用于石油管材制造领域，试制出高强度、高韧性三通，为低温环境用高强度厚壁管件制造开辟了革命性新途径[9]。形成石墨烯防腐新技术,为油井管高性能涂层改进提供了技术储备。

服务国家和中国石油主营业务发展,科技创新取得丰硕成果。先后获国家、中国石油和省部级以上科技成果奖励36项,其中国家技术发明二等奖1项,集团公司特等奖2项、一等奖3项。制修订国家标准9项、行业和中国石油企业标准54项,授权国家发明专利122件,发表论文746篇,出版专著12部,软件著作权18项。

5　结束语

当今世界正处于百年未有之大变局,全球能源行业进入大变革大调整时期。中国石油发展正处于重要战略机遇期,也面临着巨大的风险与挑战。面对复杂多变的宏观环境,科技创新将成为影响和改变未来发展格局的关键力量,科研院所发展建设的重要性更加凸显。石油科研院所作为中国石油科技创新的主战场,必须适应新形势、迎接新挑战。一是立足特色专业技术,优化内部资源配置,创新发展新技术,加强内部整合与外部合作;二是建立高效科研运营机制,在全面深化科研院所体制机制改革中,健全激励创新的政策制度,推动科技创新创效;三是与俱进,砥砺前行,努力实现高质量跨越式发展。今天的创新就是明天的过去,科研院所的发展创新永远在路上,必须坚持业务主导、自主创新、强化激励、开放共享,培育新优势,创造新业绩,在中国石油建成世界一流综合性国际能源公司的进程中当好主力军。

参 考 文 献

[1] 中共中央文献研究室. 习近平关于科技创新论述摘编[M]. 北京:中央文献出版社,2016.
[2] 中国石油天然气集团有限公司党组. 奋力建设具有全球竞争力的世界一流企业[J]. 求是,2018(12):52-54.
[3] 冯耀荣,张冠军,李鹤林. 石油管工程技术进展及展望[J]. 石油管材与仪器,2017,3(1):1-8.
[4] 张冠军,齐国权,戚东涛. 非金属及复合材料在石油管领域应用现状及前景[J]. 石油科技论坛,2017,36(2):26-31.
[5] 杨放,雷琳,高蓉,等. 科技项目管理体系创新与应用[J]. 石油科技论坛,2019,38(3):17-22.
[6] 高建忠,秦长毅,吕华,等. API标准的发展对石油管材产品演化和标准化研究的启示[J]. 中国标准化,2019(17):121-128.
[7] 杨坤,池强,李鹤,等. 高钢级天然气输送管道止裂预测模型研究进展[J]. 石油管材与仪器,2019,5(4):9-14.
[8] 杨锋平,张硕,王明波,等. 油气管道环焊缝缺陷检测评价方法研究[J]. 石油管材与仪器,2019,5(6):11-15.
[9] 胡美娟,吉玲康,马秋荣,等. 激光增材技术及现状研究[J]. 石油管材与仪器,2019,5(5):1-6.

本论文原发表于《石油科技论坛》2020年第39卷第4期。

石油管及装备材料科技工作的进展与展望

刘亚旭　李鹤林　杜　伟　陈娟利　沈　沉

（中国石油集团石油管工程技术研究院）

摘　要：石油管及装备材料科技创新对于制造强国建设和石油工业高质量发展意义重大。总结了我国石油管及装备材料科技工作取得的进展，分析了石油管及装备材料发展中存在的问题和不足，围绕石油工业提质增效和长远发展提出了科技攻关的重点方向。针对我国石油工业标准和认证体系尚不健全的问题，建议借鉴API成功经验，加快形成行业广泛认可的团体标准和认证体系。围绕提质增效，建议构建创新联合体开展针对性攻关，重点解决油气田套管变形和损坏、集输管网泄漏、炼化设施腐蚀与磨损、长输管道环焊质量提升、海洋石油装备材料国产化等突出问题。着眼长远发展，应加快推进石油管及装备领域的智能化应用技术攻关、超前布局新能源发展中的管材应用关键技术、开展新型石油管及装备材料的研发与应用，助力推动我国能源技术变革和产业升级。

关键词：石油管；石油装备材料；提质增效；人工智能；新能源

材料是制造业的基础和关键，材料创新发展是实现制造业强国的必要条件。工业强基工程（以下简称"四基"），包括核心基础零部件（元器件）、先进基础工艺、关键基础材料和产业技术基础，是我国制造业的薄弱环节。国家制造强国建设战略咨询委员会明确指出，我国从实现工业大国向工业强国的转变，亟需加快推进工业强基，进而提升我国工业整体水平，建设制造强国[1]。其中，关键基础材料作为各个产业链的最上游环节，被誉为制造业"底盘"，是支撑现代产业体系不可或缺的物质基础[2]；产业技术基础以标准、计量、检验检测、认证认可等为核心要素，贯穿于核心基础零部件（元器件）、关键基础材料和先进基础工艺发展的全过程，是关键技术基础和支撑。因此，加快推进材料标准、认证认可、新产品开发、检测评价与产业化发展意义重大。

石油装备，包括石油管，作为油气勘探、开发、储运和炼化不可或缺的工具利器，是石油工业重要的物质保障[3-4]，发展石油管及装备材料是制造强国战略的重要内容。十九届五中全会强调，坚持创新在我国现代化建设全局中的核心地位，把科技自立自强作为国家发展的战略支撑。为贯彻落实十九届五中全会精神、加快推进石油管及装备材料科技创新，在总结石油管及装备材料科技工作进展基础上，重点围绕石油管及装备解决薄弱环节、支撑驱动

基金项目：中国石油天然气集团有限公司管理创新研究与实践项目"以市场化激励方式提升科研能力的管理创新实践"（编号：中油企201904）；中国石油天然气集团有限公司科学研究与技术开发项目"科技成果价值化及转化应用研究"（编号：2019D-5005-29）。

作者简介：刘亚旭，1968年生，2007年毕业于西安交通大学管理科学与工程专业，博士，正高级工程师，现任中国石油集团石油管工程技术研究院党委书记、院长，主要从事石油管工程技术和科技管理等工作。

石油工业提质增效和长远发展，提出相关标准化以及新产品和新技术开发、评价与工程应用等方面的重点方向。

1 石油管及装备材料科技工作进展

经过60多年发展，我国石油管及石油装备材料从无到有、从低端到高端，基本形成门类齐全、产业完整、质量可靠的产品体系，其中油井管实现了大量出口，高钢级输送管的质量和用量已经走在了世界前列，有力支撑和保障了我国石油工业的发展壮大[5-6]。中国石油集团石油管工程技术研究院（以下简称"管研院"）作为从事石油管及装备材料应用基础和工程应用研究的专业机构，对于推动我国石油管及装备材料技术进步发挥了至关重要的作用。

1.1 研发应用新型钢铁材料，带动石油装备升级换代

20世纪六、七十年代，针对我国石油装备傻大笨粗的突出问题，李鹤林院士带领科研团队从石油机械的服役条件出发，分析、研究了一批关键的有代表性产品的服役条件和失效判据，有针对性地研究开发了20SiMn2MoVA等十余种新型钢铁材料，并充分发挥现有材料的性能潜力，把节约铬、镍与提高石油机械产品的质量和寿命结合起来，取得了很好的效果，使一大批石油机械产品减轻了重量、延长了寿命、提高了服役性能，降低了综合成本，推动了石油装备升级换代[7]。

1.2 与冶金和制管企业协同攻关，实现油井管大规模国产化

20世纪90年代前，我国石油工业所用的油井管90%以上依赖进口，且失效事故频发。管研院团队从服役工况出发，通过大量的失效分析，研究构建了油井管标准体系，建立了油井管选材评价和应用关键技术体系；建立了能够模拟油井管复杂力学与腐蚀环境条件的全尺寸模拟试验平台及方法；协助冶金和制管企业，生产出了全系列更适合我国油气开发工况的国产化产品，全面替代了进口，并在部分产品和指标方面超越了进口产品。推动实现油井管国产化率由1990年之前的10%提高至2012年的近100%，支撑了长庆"三低"、塔里木"三超"、新疆"稠油"和西南"高含硫"等重点油气田开发[8]。

1.3 持续推动天然气管道高钢级、高压、大输量输送

围绕我国重大天然气管道建设项目，积极开展超前研究，持续推动天然气长输管道提高输送压力和管线钢强度级别。根据管研院的超前研究成果并面向国际管道建设技术前沿，以李鹤林院士和黄志潜教授为代表的科研团队2000年提出了西气东输管道采用X70钢级、10MPa输送压力的技术方案及其科学依据，被中国石油决策层采纳，实现了西气东输设计年输量$120 \times 10^8 m^3$的目标（实际年输量曾达到$170 \times 10^8 m^3$），管道技术实现跨越发展，大大缩短了与国际先进水平的差距。2006年提出西二线采用X80钢级、12MPa输送压力，较原X70钢级双线方案节省投资130亿元，实现年输量$300 \times 10^8 m^3$的同时，标志着我国管道建设关键技术进入领跑者行列。2015年，中俄东线开工建设，采用了X80钢级、1422毫米管径、12MPa输送压力，设计年输气量达到$380 \times 10^8 m^3$。管研院研究建立的高钢级管道失效控制技术和适合实际工况的标准体系支持了西一线、西二线、西三线以及中俄东线等重大项目的建设与运行安全。

1.4 建立和完善应用技术支撑体系，推动实现管线钢及钢管全面国产化

20世纪末，伴随国民经济发展和对天然气等清洁能力的需求增长，我国油气管道建设迎来了高速发展期。针对当时管线钢全部依赖进口的困境，建立并完善了"失效分析—标准化—科学研究—检测评价"的应用技术支撑体系[9]。根据工况环境研究制定了X70、X80管

线钢及钢管系列标准，提出了经过严格质量控制的螺旋埋弧焊管可以用于高压大口径天然气管线，确立了国产螺旋埋弧焊管在重要大口径高压输气管道建设中的重要地位。建成世界上第三家可以完成天然气爆破试验的全尺寸气体爆破试验场，研究开发了输送管全尺寸实物性能测试平台及检测评价技术，联合冶金、制管企业协同攻关，推动实现了 X70、X80 管线钢及钢管的全面国产化，在西一线 X70 钢管 50% 国产化的基础上，西二线全面实现国产化，节约采购资金 90 多亿元，带动和引领了我国冶金和制管技术的快速发展。

1.5 开展失效分析和预测预防技术研究，为石油管及装备运行安全提供技术支撑

开展失效分析和预测预防技术研究，判明失效模式、机理和影响因素，反馈到设计、材料、工艺、使用等过程，并采取有效措施预防事故重复发生，对于提高石油管及装备的安全可靠性意义重大。管研院先后开展了石油管及装备失效分析 1600 余项，包括中缅管道贵州晴隆段天然气两次泄漏燃爆、西气东输二线同心段管道环焊缝渗漏、"11·22"黄岛输油管道泄漏爆炸、加拿大尼克森输油砂管道失效、克深 2-1-3 井套管断裂、西二线东段 76# 阀室爆管失效、土库曼斯坦直缝埋弧焊管泄漏失效等重大失效项目，奠定了行业的失效分析权威地位，为石油工业的安全和高效提供了技术支撑。

1.6 持续开展安全评价和完整性管理，保障油气管道和储气库全生命周期风险受控

20 世纪 90 年代，国内率先开展油气管道完整性技术研究和应用，通过在剩余强度评价、剩余寿命预测、风险评估、完整性评价、复合材料和套筒修复补强等方面持续开展科研攻关，建立了油气管道完整性技术和管理体系。研究成果在西气东输一线、二线、三线、陕京管道等所有重大管道工程和油气田地面管道推广应用，显著降低了油气管道失效率，保障了国家能源安全。针对我国储气库地质条件复杂，风险点多面广难题，"十一五"期间，率先攻关储气库风险评估技术，构建了储气库"地下—井筒—地面"三位一体的全生命风险管控体系，有效保障了储气库运行风险受控，为天然气保供调峰和储气库大规模建设提供了强有力的技术支撑。

1.7 构建石油管材标准体系，引领行业快速发展

构建和完善石油管材标准体系，形成涵盖团体标准、行业标准、国家标准和国际标准等各个层级的标准化工作机构，推动和引领了石油管及装备材料的技术进步。在国际标准化方面，自从 1987 年在 API 年会上一篇有关钻杆的论文《钻杆失效分析及内加厚过渡区结构对钻杆使用寿命的影响》"一举敲开 API 大门"后，国际标准化工作取得长足进步。组织制修订国际标准 6 项，新立项并承担国际标准项目 4 项（其中产品标准 3 项），承担了 ISO/TC67/SC2 并行秘书处并担任副主席，同时担任 ISO/TC 67 AHG 绿色制造特别工作组召集人，引领和推动了管材新产品的研发及应用。

综上，石油管及装备材料科技创新取得了重要进展，陆上石油管材通过创新链与产业链的紧密融合，解决了一系列卡脖子问题，实现了自立自强，支撑和保障了我国石油工业的快速发展。随着石油工业的持续深入，石油管及装备的服役工况日益复杂严苛，对石油管及装备材料的质量和性能水平提出了更高的要求，而在海洋油气装备材料方面国产化程度较低，耐高温的油田用非金属材料也与国外有较大的差距。与此同时，在第四次工业革命背景下，世界能源技术创新进入活跃期，人工智能、新能源、新材料等技术蓄势待发，有望深刻影响石油工业发展格局。面对新形势，亟需进一步梳理石油管及装备材料科技创新的重点方向和发展策略，从而形成行业共识、协同攻关、重点突破。

2 借鉴 API 成功经验，建立我国石油工业标准和认证体系

美国石油学会（API）成立于 1919 年，是一家代表美国石油和天然气行业的非营利性贸易协会，主要开展标准、认证与培训服务。目前 API 已建立形成完善的石油工业标准和认证体系，制订的技术标准超过 700 项，获得 100 多个国家认可和使用。

20 世纪 80 年代，我国开始广泛使用 API 标准。石油工业现行有效的 342 项国家和 1629 项行业标准中，约 50% 依赖 API 标准。API 以完善的标准体系为基础，在全世界开展认证，已成为石油产品和服务在国际市场的通行证，我国共有 1540 多家石油装备企业获得了 API 认证，涉及年出口额达 700 多亿元。

目前，国内石油管及装备材料领域的权威性团体标准和认证缺失，在取得国外广泛认可方面存在难度。2017 年修订的《标准化法》明确了团体标准的法律地位，但是目前团体标准体系缺乏顶层设计、统一谋划，已出台的标准零散归口于 20 多个学会和协会，存在交叉重复和无序竞争，难以形成像美国 API、ASTM、ASME、NACE 等被用户广泛认可和执行的团体标准体系，也难以开展像 API 一样权威的认证。

应当借鉴 API 成功经验，从以下几方面开展工作。

2.1 积极构建团体标准体系并形成权威认证体系

由行业领军企业（中国石油）牵头，以石油装备（管材）制造和油气工程技术服务两个领域为重点，依托最具实力的科研院所或科技型企业，联合行业的其他石油企业、生产厂家，可以以中国石油学会的名义，借鉴已有的相关成熟的行业或企业标准，并补充制定相关标准，尽快建立我国石油工业自主的、符合市场经济规律、系统的团体标准体系。在此基础上，在石油装备企业推行自主团体标准体系以及产品和服务的认证，不断提升我国自主石油团体标准的影响力和权威性。

2.2 加快推进标准和认证体系的国际化进程

目前我国约有 500 多支队伍在 40 多个国家从事油气勘探开发，钻采装备和管材产品与工程技术服务的海外市场潜力很大。应当依托"一带一路"建设，在沿线国家广泛开展 CPS 标准及认证的合作与互认，并逐步辐射至全世界认同和认可采用；另一方面，积极争取与 IOGP（国际油气生产商协会）、ISO 等组织的合作，将我国石油行业标准和认证带出国门。通过不断的努力，使我国自主的石油工业团体标准成为国际有影响的、权威的国际先进标准。

2.3 加大标准化投入，抢占石油标准技术制高点

提前谋划和布局，将海洋油气、非常规油气、智慧管道、绿色制造、新能源及储能材料等领域的自主创新和标准研制纳入"十四五"发展规划，投入专项资金，支持形成一批前沿技术领域具有权威性的国际标准，积极抢占标准技术制高点。

3 围绕提质增效，着力石油管及装备材料的重大技术难题

油气田开发、炼油化工和管道储运等领域涉及的石油管材及装备数量规模庞大、投资占比高，一旦发生失效，往往会造成巨大经济损失、严重环境污染甚至灾难性伤亡事故。

3.1 开展科技攻关，有效推动油气田套管变形和损坏预防及治理

油气田套管损坏严重影响油气田开发和效益，套管变形又是页岩气、页岩油开发中遇到的重大技术瓶颈，每年因套损和套变造成的经济损失巨大。近年来随着开采强度的提高和开

采难度的加大，套损率上升趋势明显；2018年四川某些地区的页岩气井在压裂过程中的套变比例更是高达49%；此外，西南和塔里木油田的高压气井的特殊螺纹连接的气密封性难以保证，造成环空带压，成为卡脖子问题。管研院曾在20世纪八、九十年代针对大庆油田的套损进行了调查和治理，一度使套损率明显下降。2018年起，针对长宁、威远区块页岩气开发中的严重套变问题进行了理论研究、数值模拟和实物实验，现已开展了预防套变技术的现场试验，取得预期成果，有望大幅度降低页岩油气开发中的套变比例。同时，管研院围绕油气田需求，开展套管损坏的预防和治理工作。"十四五"期间，开展国产高气密封特殊螺纹连接的套管开发也势在必行。

3.2 加强新产品、新技术研究，防治油气集输管网泄漏

油田集输管网是地表下最广泛的线性设施，更是油气输送的必经通道。集输管网由于里程长、数量大、分布广，往往因腐蚀、第三方破坏等原因，造成跑冒滴漏现象时有发生。针对该问题，管研院近几年开展了冶金复合管、环氧套筒、非金属管穿插、缓蚀剂、管网智能检测等新产品、新技术研究，并取得了初步成果。"十四五"将在上级和长庆、大庆、塔里木等油气田公司的支持下，进一步加大新技术、新产品推广应用力度，解决集输管网腐蚀、运行监测和快速修复难题，积极参加无泄漏示范区建设，提高油气田效益，避免环境事故。

3.3 应用基于风险的检测和故障诊断技术，解决炼化设施腐蚀与磨损

炼化设施和压力容器容易发生腐蚀、冲蚀和磨损，许多关键材料仍依赖进口。需要组织力量对炼化设施材料深入研究，延缓腐蚀和磨损，并实现国产化替代。在此基础上，推广应用基于风险的检测和故障诊断等技术，在保障安全的前提下，尽量延长设施的大修周期，提升炼化业务的效益。

3.4 以非金属管材为突破口，实现上下游一体化运作

2015年，管研院提出了加强上下游一体化运作、推动化工原料在油田集输管网中应用的建议。2017年，中国石油设立了重点工业试验项目"油田用耐温聚乙烯管材新产品在长庆油田工业化试验"，由长庆油田、石化院、管研院、大庆石化、吉林石化等多家单位承担，目前已经开发了成功了聚乙烯管材新产品，并在长庆油田铺设了12条管线（20.8km），管材性能稳定，运行状况良好。此工作推动了油气工业从聚乙烯开发生产、复合管制备到油田应用的全产业链、一体化运作，实现了效益最大化。"十四五"将深入开展橡塑原料和非金属管材的研究、开发和评价等工作，制定科学合理的标准体系，扩大橡塑管材的应用范围，实现在油气混输、天然气输送以及井下等多个领域的应用。

3.5 研究高钢级天然气管道焊接技术和检测技术，保证长输油气管道运行安全

截至2019年底，国内油很气长输管道总里程达到$13.9\times10^4 km$[10]。根据中缅管道晴隆段两次事故以及其他的管道事故结果分析来看，环焊缝焊接质量和技术问题已成为影响高钢级天然气管道安全服役的主要问题。因此应继续针对高钢级管道环焊缝，系统开展强度匹配设计、钢管母材、焊接材料、焊接工艺、断裂控制、缺陷修复等方面的科研工作，攻克高性能焊机和焊接材料依赖进口、工艺质量不稳定、管道应力状态监测、焊接缺陷检测和容限评估技术难题，在高钢级管道建设和安全运行领域持续形成自主创新的领先科技成果，保障高钢级管道运行安全，保持领跑者之一的地位。

3.6 组织关键技术攻关，推动海洋石油装备材料国产化

我国海洋石油装备存在自主创新能力不足，关键核心材料研发滞后等问题[11]。水下井口、水下采油树、水下阀门等海洋钻采关键装备用马氏体不锈钢、超级奥氏体不锈钢、超级双相不锈钢、镍基和铁镍基合金等的加工制造技术不成熟，主要依靠进口。海洋柔性管主要由法国和美国的三家企业垄断，国内企业尚无生产能力。"十四五"将借鉴高钢级输送管国产化建立的"失效分析—标准化—科学研究—检测评价"应用支撑模式和"产学研用检"一体化开发机制，组织对海洋石油装备材料的国产化攻关，提升我国海洋石油装备材料的自主保障能力和整体水平。

4 着眼长远，超前布局人工智能、新能源及其新材料技术发展

以人工智能、大数据、云计算等为核心的信息技术以及新能源、新材料等技术不断渗透到油气领域，各大油公司纷纷抢占能源技术进步先机，谋求新一轮科技革命和产业革命竞争制高点。

4.1 构建智能决策平台，实现石油管材及装备的健康管理和预测性维护

管研院在多年工作中，积累了海量的石油管及装备材料、服役环境和失效特征等方面的数据和资料。20世纪90年代，"钻柱失效分析案例库和计算机辅助失效分析"成果获陕西省科技进步二等奖，是一次利用计算机技术提高失效预测预防水平的尝试，为人工智能技术在石油管及装备质量管理中的应用奠定了基础。人工智能应用于石油管材及装备全生命周期健康管理，建立实时诊断、智能决策的数字孪生体和失效预测预警智慧化平台，可有效降低失效事故，保障油井管柱、钻采装备、地面管道及炼化设备的安全。

4.2 布局突破新能源发展中的材料应用关键技术，为新能源发展创造条件

氢气作为未来发展的清洁能源已成为共识，而氢的集输是关键瓶颈之一。国际能源署预测，在距离1500km以内，管道输送氢气是经济性选择。因此，开展高压氢气储运材料开发及应用技术、氢气管道建设和安全运行关键技术以及在役天然气管道混氢输送适用性研究尤为迫切。

为实现2030年达峰，2060年碳中和的目标，中国石油正在开展碳捕获、利用与封存技术（CCUS）攻关和现场试验，实现化石能源所产生CO_2的埋存和有效利用，届时CO_2高效集输将形成规模产业。而CO_2集输存在腐蚀、泄漏和爆炸等安全风险，因此着眼CO_2的规模应用，超前储备CO_2安全集输关键技术十分必要，这也是石油管和装备材料领域的研究重点。

4.3 开展新型石油管及装备材料的研发及应用，抢占技术制高点

围绕"深、低、海、非"油气开发，钛合金、铝合金、镁合金以及碳纤维管材具有密度低、比强度高、耐腐蚀和抗疲劳等优点，有望成为超深井、水平井、大位移井等开发的利器；冶金结合双金属复合管、中低Cr耐蚀合金管、新型涂镀层管材、耐高温（耐温80度以上）非金属及复合材料管材兼具良好的耐蚀性和经济性，这些产品的开发可推广应用于腐蚀油气田；3D打印（增材制造）、绿色再制造、高性能新型结构钢、先进复合材料等技术的快速发展，将为石油装备的轻量化、绿色化和智能化奠定基础。

针对天然气水合物开发，开发应用钛合金管材、新型表面涂层材料，解决大狗腿度通过以及水合物堵塞等瓶颈难题；针对煤炭地下气化，开发应用耐350℃高温腐蚀材料和定温可燃套管，解决套管变形、腐蚀穿孔、环空带压等气化炉完整性问题；地热开发方面，需要针对深层（大于5000m）、高温（大于450℃）环境，研发建立耐高温钻测工具及材料体系。这都是着眼于未来发展的研究方向和重点工作。

5 结束语

多年来，管研院一直致力于石油管及装备材料的科技创新，聚焦解决了油气田开发、炼油化工、管道储运、装备制造、新能源发展等领域的诸多共性、关键和瓶颈技术难题。当前国家大力实施创新驱动发展战略，面对我国石油管及装备材料发展中的新问题、新挑战，油气行业的石油管和装备材料领域科研机构应加强与国内外冶金企业、高等院校以及科研院所的合作，积极构建创新联合体，共同推动新材料、新技术在油气田开发、管道建设及运行、新能源利用等领域的创新发展，助力我国能源技术变革和升级，为制造强国建设和国家能源安全贡献科技力量，实现石油管和装备材料领域的科技自立自强和产业链供应链的自主可控。

参 考 文 献

[1] 国家制造强国建设战略咨询委员会，中国工程院战略咨询中心．工业强基[M]．北京：电子工业出版社，2016.
[2] 刘志强，丁怡婷．做强材料"口粮"端稳工业"饭碗"[N]．人民日报，2021-02-03(18)．
[3] 李鹤林，张冠军，杜伟．"石油管工程"的内涵及主要研究领域[J]．石油仪器，2015，1(1)：1-4．
[4] 冯耀荣，马秋荣，张冠军．石油管材及装备材料服役行为与结构安全研究进展及展望[J]．石油管材与仪器，2016，2(1)：1-5．
[5] 冯耀荣，张冠军，李鹤林．石油管工程技术进展及展望[J]．石油管材与仪器，2017，3(1)：1-8．
[6] 张国信，苏月．石油化工装备用金属材料未来发展趋势展望[J]．石油化工腐蚀与防护，2018，35(3)：1-5．
[7] 李鹤林．开展金属材料强度的研究——提高石油钻采机械质量和寿命[J]．石油钻采机械通讯，1979，18(3)：1-17．
[8]《中国钢管70年》编写组．中国钢管70年[M]．北京：冶金工业出版社，2019．
[9] 李鹤林，霍春勇，池强，等．输送管产品开发与工程应用技术支撑体系[J]．油气储运，2020，39(7)：721-729．
[10] 高鹏，高振宇，刘广仁．2019年中国油气管道建设新进展[J]．国际石油经济，2020，28(3)：52-58．
[11] 杜伟，李鹤林．海洋石油装备材料的应用现状及发展建议(上)[J]．石油管材与仪器，2015，1(5)：1-7．

本论文原发表于《石油管材与仪器》2021年第7卷第1期。

复杂工况油套管柱失效控制与完整性技术研究进展及展望

冯耀荣[1] 付安庆[1] 王建东[1] 王 鹏[1] 李东风[1]
尹成先[1] 刘洪涛[2]

(1. 中国石油集团石油管工程技术研究院·石油管材及装备材料服役行为与结构安全国家重点实验室；2. 中国石油塔里木油田公司)

摘 要：随着对深层碳酸盐岩、新区、页岩等油气勘探开发力度的加大，油气田的地层条件和介质环境变得更为苛刻复杂，油套管柱变形、泄漏、腐蚀、挤毁、破裂等失效事故时有发生；加上特殊结构井和特殊工艺井、特殊增产改造措施等对油套管柱提出的新要求，油套管柱面临着一系列新的挑战和难题需要破解。为此，围绕我国石油天然气工业增储上产的迫切需求和深井超深井、特殊结构和特殊工艺井、强酸/大排量高压力反复酸化压裂增产改造等复杂工况油套管柱失效频发的技术难题。历经十余年攻关。取得了一系列试验研究进展与重要技术成果，主要包括：（1）形成了基于气井全生命周期的油套管腐蚀选材评价、腐蚀控制和油套管柱完整性技术；（2）形成了油套管柱螺纹连接结构和密封可靠性设计评价及配套技术；（3）研发并形成了低渗透致密气井经济高效开发"API长圆螺纹套管+CATTS101高级螺纹密封脂"套管柱技术；（4）建立了高温高压等复杂工况油套管柱结构和密封完整性试验平台和评价技术。进而根据当前国家大边提升油气勘探开发力度的新要求和面临的新问题，提出了继续深入开展深层、酸性环境、面岩气等复杂工况油套管柱失效控制与完整性研究攻关的若干建议。结构认为，上述技术成果有力地支撑了我国重点油气田的经济有效开发。

关键词：复杂工况；套管；油管；失效控制；腐蚀控制；特殊螺纹；完整性评价；实物试验；井完整性

我国每年油套管消耗量介于$(300\sim350)\times10^4$ t，耗资$(250\sim300)$亿元[1]。在油气勘探开发和生产过程中，油套管柱承受拉伸/压缩、内压/外压、弯曲等复杂载荷作用，同时会遭受油/气/

基金项目：中国石油天然气集团有限公司应用基础研究项目"苛刻服役条件油井管工程应用基础研究"（编号：2016A-3905）、中国石油天然气股份有限公司重大科技专项"超深高温高压气井井完整性及储层改造技术研究与应用"（编号：018E-1809）、中国石油天然气集团有限公司科技基础条件平台建设项目"石油管材及装备材料服役行为与结构安全国家重点实验室建设"（编号：016E-5104）。

作者简介：冯耀荣，1960年生，正高级工程师，博士，一直从事石油工程材料应用基础研究及重大工程技术支持工作，现任石油管材及装备材料服役行为与结构安全国家重点实验室主任，地址：陕西省西安市雁塔区锦业二路89号（710077），ORCID：0000-0002-3325-6067，E-mail：fengyr@cnpc.com.cn。

通信作者：付安庆，1981年生，高级工程师博士，地址：陕西省西安市雁塔区锦业二路89号（710077），E-mail：fuanqing@cnpc.com.cn。

水、$H_2S/CO_2/Cl^-$等井下介质和温度作用。随着深井超深井、特殊结构和特殊工艺井、强酸/大排量高压力反复酸化压裂增产改造等工况条件日益复杂，油套管柱失效频发，严重制约油气田正常生产。近年来，我国油套管柱失效概率介于10%～20%，高温高压气井油套管柱泄漏一度超过40%。例如我国西部某油田2008—2012年油管柱发生腐蚀断裂失效123井(次)[2,3]，在完井过程中因油管柱失效造成的经济损失达7.24亿元，高产天然气井每口井修井费用高达(3000～5000)万元。每年因油套管柱失效造成的经济损失高达数十亿元[1]。然而现有的油套管柱失效控制技术只能解决常规油气井勘探开发过程中的失效问题，但是不能有效控制高温高压气井、非常规、特殊工艺和特殊结构井等复杂工况油套管柱的失效问题。复杂工况油套管柱的完整性与失效控制是国际上研究的热点和重大难题，也是我国石油天然气工业增储上产的瓶颈问题。这是一项十分复杂的系统工程问题，迫切需要通过系统研究加以解决。

针对我国复杂油气田开发中的油套管柱完整性技术需求，中国石油集团石油管工程技术研究院联合相关单位，依托"油井管柱完整性技术研究"等[4-6]多项重大科技项目，从油套管柱服役工况和失效分析入手，重点围绕西部高温高压气井油套管柱严重腐蚀泄漏和断裂、低渗透致密油气井油套管柱泄漏和经济高效开发等技术难题开展系统研究攻关，建立了复杂工况油套管柱结构和密封完整性试验平台，攻克了复杂工况油套管柱完整性技术与失效控制难题，为我国重点油气田高效勘探开发和安全生产提供了重要技术支撑。

但是随着深层碳酸盐岩、新区、页岩油气等勘探开发力度的加大，油气田的地层条件和介质环境变得更为苛刻复杂，油套管柱变形、泄漏、腐蚀、挤毁、破裂等失效事故时有发生，有的区块还十分严重；特殊结构井和特殊工艺井、特殊增产改造措施等对油套管柱提出了新的要求，油套管柱面临着一系列新的挑战和难题仍需要破解。

1 高温高压气井油套管柱腐蚀与防护技术

针对高温高压气井油套管柱均匀腐蚀、点蚀、缝隙腐蚀、应力腐蚀及化学—力学协同作用技术难题，形成了基于高温高压气井全生命周期的油套管腐蚀选材评价技术，研发了超级13Cr酸化缓蚀剂，自主研制了油套管实物应力腐蚀试验系统，形成高温高压气井油套管腐蚀与应力腐蚀控制技术、油套管柱完整性技术和管理规范[4-7]。

1.1 高温高压气井全生命周期的油套管腐蚀选材评价及控制技术

系统考虑油气井在作业生产过程中鲜酸酸化—残酸返排—凝析水—地层水四种典型的服役环境和具体的工况参数，以均匀腐蚀速率为参考、局部腐蚀速率为依据进行综合试验评价，形成油气井管柱全生命周期腐蚀评价新方法。首次揭示了13Cr油管、15Cr油管、V140低合金高强度套管等八种材料在鲜酸酸化—残酸返排—凝析水—地层水各个作业过程中的单环境和连续多环境中的腐蚀规律和协同作用机制。鲜酸—残酸—凝析水—地层水全过程的平均腐蚀速率及点蚀速率均大于各个独立过程之和，试验时间超过60天后实验数据趋于稳定[图1(a)]。揭示了温度、CO_2分压、Cl^-含量和流速等主要因素对油气井管材腐蚀的影响规律，构建了高温高压气井油管选材图[图1(b)]。基于多年的失效分析和试验研究，建立了高温高压气井油套管腐蚀数据库，包括井筒和工况数据采集、失效数据统计、失效数据分析，以及失效预警等四大功能。发明了油管腐蚀程度预测方法[8]，建立了含非均匀腐蚀缺陷油套管强度评价和寿命预测模型及软件，用于腐蚀油套管的完整性评价及安全预警。

1.2 超级13Cr油管系列酸化缓蚀剂

建立了适用于高温高压气井酸化增产工况的"空间多分子层多吸附中心"高温酸化缓蚀

模型(图2),该模型有效耦合多层大分子的屏蔽效应和小分子的填充效应,从而有效抑制酸化增产工艺中的酸腐蚀。通过百余次的喹啉季铵盐、曼尼希碱与多种金属离子(如Cu^+、Ca^{2+}、Al^{3+}等)复配试验,揭示了多分子多离子复配的酸化缓蚀剂的混合型缓蚀机理:季铵盐能有效抑制酸液中氢的作用,曼尼希碱能有效抑制酸液中氯的作用,金属离子则能提高缓蚀剂的耐温性和成膜性,因此复配后的酸化缓蚀剂呈现出同时抑制阴极和阳极的混合型缓蚀效果,如图3所示。应用上述模型和理论,研发了TG201、TG201-Ⅱ、TG202等超级13Cr油管系列酸化缓蚀剂产品[9],列入中国石油自主创新产品,对于超级13Cr油管酸化环境的缓蚀效率良好,从2007年起在塔里木库车山前高温高压气井应用120余井次,有效解决了塔里木油田高温高压气井酸化压裂过程中超级13Cr油管柱严重腐蚀问题。

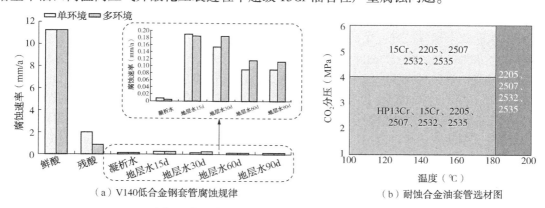

图1 高温高压气井油套管腐蚀规律及选材图

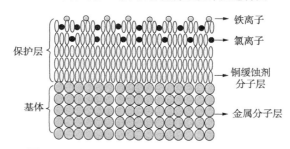

图2 "空间多分子层多吸附中心"模型示意图

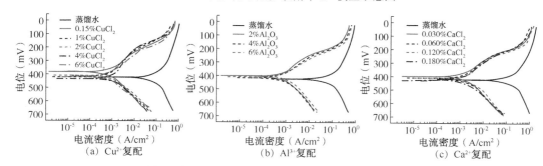

图3 多分子多离子复配酸化缓蚀剂的极化曲线图

1.3 油套管全尺寸应力腐蚀试验系统

针对我国西部油气田超深、超高压、高含CO_2和Cl^-、强酸大排量增产改造等严酷工况条件,自主设计研发了集环境介质、载荷、结构、材料于一体的全尺寸油套管应力腐蚀试验

系统[10]，如图4所示，建立了油套管实物腐蚀试验流程和方法，实现了100MPa内压、1000kN拉伸载荷、200℃高温、油/气/水多相腐蚀介质等极端工况的多参量实验模拟，解决了油套管螺纹接头在应力作用下腐蚀和密封耦合作用试验评价难题及尺寸效应问题。系统开展了N80和V140低合金钢油套管、超级13Cr油管、W-Ni-P内涂层油管的实物应力腐蚀实验研究，首次发现了超级13Cr油管在模拟酸化压裂复杂工况下的点蚀—应力腐蚀失效过程和机制，如图5所示。研究揭示了V140低合金套管和超级13Cr油管的缝隙腐蚀特性，综合考虑腐蚀介质对密封面长度、表面洁净度、密封面接触压力的影响，提出采用折减系数来表征腐蚀介质对螺纹接头密封影响的方法，建立了腐蚀环境下特殊螺纹接头的密封评价准则。

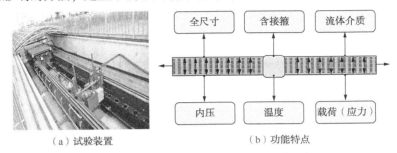

（a）试验装置　　　　　（b）功能特点

图4　油套管全尺寸应力腐蚀试验系统[11,12]图

1.4　超级13Cr油管应力腐蚀断裂控制技术

超级13Cr油管在甲酸盐环境中曾多次发生腐蚀穿孔或断裂，使用时间最短的只有11天。通过系统失效分析和实验研究，揭示了超级13Cr油管在磷酸盐体系中的失效规律、影响因素和断裂机理。其断裂机制表现为阳极溶解膜致损伤机理和裂纹沿马氏体多尺度结构界面扩展及有害第二相促进裂纹扩展[图5(b)]。研发了超级13Cr油管应力腐蚀断裂控制技术，以低开裂敏感的甲酸盐完井液体系替代磷酸盐完井液体系，2015年以来在塔里木油田应用36口井，至今未出现环空带压或油管柱腐蚀断裂失效。

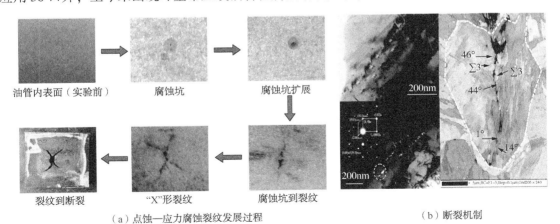

（a）点蚀—应力腐蚀裂纹发展过程　　　　　（b）断裂机制

图5　超级13Cr油管点蚀—应力腐蚀断裂过程[11,12]及机制图

1.5　油气井管柱完整性技术和管理规范

在系统总结高温高压及高含硫气井油套管柱研究成果和实践经验的基础上，制定了《油气井管柱完整性管理》行业标准[13]，编制了《高温高压及高含硫井完整性指南》[14]《高温高压及高含硫井完整性设计准则》[15]《高温高压及高含硫井完整性管理规范》[16]并推广应用，

塔里木油田井完整性从 70% 提高到 79%。

2 油套管柱优化设计与可靠性技术

建立了套管柱结构和密封可靠性设计与评价方法，发明了水平井用新型特殊螺纹套管及制备技术，研发了低渗透致密气直井"API 长圆螺纹套管+CATTS101 高级螺纹密封脂"套管柱技术，在重点油气田得到应用[4-6]。

2.1 高温高压气井套管柱结构可靠性设计与评价方法

研究确定了高温高压气井套管柱的主要失效模式，构建了用故障树法计算套管柱失效概率的方法，建立了用分项系数法计算套管柱可靠性的方法[17]和软件，如图 6 所示，制定了《油气井套管柱结构与强度可靠性评价方法》[18]行业标准。

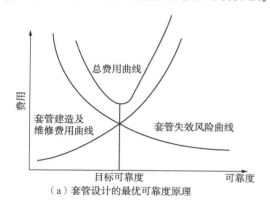
（a）套管设计的最优可靠度原理

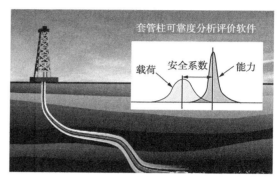
（b）套管柱可靠性分析评价方法和软件

图 6 套管柱可靠性设计分析与评价图

2.2 高温高压气井油套管柱密封可靠性设计与评价方法[4]

建立了特殊螺纹接头密封准则，即密封抗力（W_a）：

$$W_a = \int_0^l p_c^n(l)\,dl \geqslant B \left(\frac{p_{gas}}{p_{atm}}\right)^m$$

式中：p_c 为密封面接触压力，MPa；l 为接触长度，m；p_{gas} 为气密封内压力，MPa；p_{atm} 为大气压力，0.1MPa；$n=1.4$；$B=0.01$；$m=0.838$。

采用径向基函数和蒙特卡洛模拟方法，建立了高温高压气井油套管螺纹连接密封可靠性设计的极限状态方程、计算程序及判据，系统研究揭示了套管螺纹结构尺寸、材料性能、工作应力等对螺纹密封抗力的影响规律（图 7）。

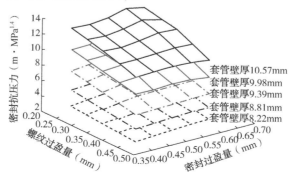

图 7 套管壁厚、螺纹和密封过盈量对密封抗力的影响图

2.3 新型特殊螺纹套管设计研发及应用[4,19-22]

在深入研究油套管特殊螺纹密封机理、进行理论分析和试验研究的基础上，建立了考虑表面涂层和粗糙度影响的油套管特殊螺纹临界气密封压力计算方法（图8），用于特殊螺纹设计。

综合研究了拉伸/压缩、内/外压、弯曲等复合载荷作用下螺纹结构参数和公差、材料性能、密封面和螺纹过盈量等因素对其承载能力和密封可靠性的影响，设计研发了既安全又经济的水平致密气井用新型特殊螺纹套管及制备技术（图9），满足了4200m深、弯曲狗腿度20°/30m、液体压裂内压90MPa、气体生产压力50MPa、150℃水平井压裂改造和生产井工况下螺纹连接的强度和密封可靠性。

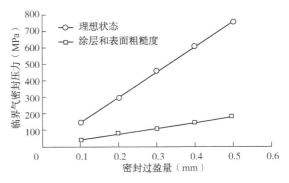

图8 临界气密封压力与密封过盈量的关系图

该套管螺纹密封面不易碰伤、螺纹易于加工和清洗，在宝鸡石油钢管有限责任公司、延安嘉盛石油机械有限责任公司等多个制造厂批量生产超过了$5×10^4$t，在长庆、延长、新疆等油气田推广应用，将长庆油田水平气井开发套管特殊螺纹由原来的7种统一为该螺纹，有效降低了管柱管理和使用成本。

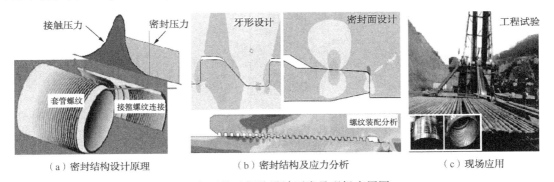

图9 新型特殊螺纹设计开发及现场应用图

2.4 "API长圆螺纹套管+CATTS101高级螺纹密封脂"套管柱技术[23,24]

系统开展了API长圆螺纹、偏梯形螺纹、特殊螺纹套管与API标准螺纹脂、CATTS101高级螺纹密封脂适用性试验研究，开发了"API长圆螺纹套管+CATTS101高级螺纹密封脂"套管柱技术，在中国石油长庆油田公司苏里格气田直井应用超过10000口井，在保证套管柱使用安全的前提下，套管成本降低了20%~30%。

3 油套管柱结构和密封完整性试验平台建设

建立了2500t全尺寸油套管复合载荷和环境试验系统，轴向+外压复合载荷挤毁试验系统，立式挤毁试验系统，热气循环试验系统，连接螺纹上卸扣试验系统等油套管柱结构与密封完整性试验平台，形成了高温高压气井、页岩气井、致密气井等油套管柱结构与密封完整性试验评价技术[7]。

该试验平台由9台试验系统构成（图10），是目前国际上最先进的平台之一，也是国内功能最为强大的试验平台,最大拉伸和压缩载荷为25000kN,最大内压为276MPa,最大弯矩为700kN·m,最大外压为210MPa,最高试验温度为500℃。可对外径介于88.9~406.4mm的所有钢级油套管在拉伸/压缩、内压/外压、弯曲复合载荷、温度循环条件下进行结构和密封完整性评价。其中,25000kN复合加载试验系统可进行高温外压试验,采用大口径单液缸加载方式,具有加载均匀、摩擦阻力小、寿命长等特点,将载荷轴直接作为力传感器,数据测量更为准确。发明了壁厚200mm抗外压至失效试验用挤毁缸的制造方法[25],实现了挤毁试验机核心部件的国产化;发明了系列非API标准油套管实物评价试样制备方法[26-28]。该试验平台为塔里木、西南、长庆等十余个油气田和中国宝武钢铁集团有限公司、天津钢铁集团有限公司、湖南衡阳钢管(集团)有限公司等十余个制造厂提供试验评价技术服务超过15年,为油套管国产化、高性能油套管新产品研发及规模应用提供了技术支撑。

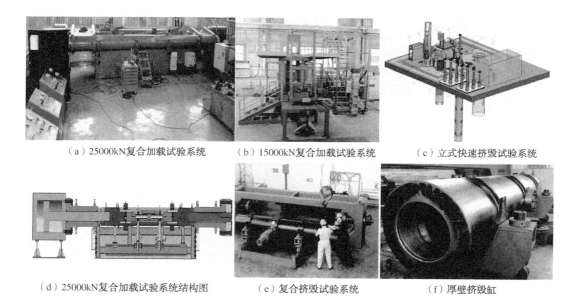

(a)25000kN复合加载试验系统　　(b)15000kN复合加载试验系统　　(c)立式快速挤毁试验系统

(d)25000kN复合加载试验系统结构图　　(e)复合挤毁试验系统　　(f)厚壁挤毁缸

图10　油井管柱结构和密封完整性试验平台及实验技术图

4　油套管柱面临的新挑战及建议

2018年以来,面对我国国民经济发展和人民生活对油气资源的巨大需求,党和国家领导人要求大力提升油气勘探开发力度、保障国家能源安全。国内4大石油天然气公司立即安排部署,实施了加大油气业务发展的计划,在塔里木盆地、准噶尔盆地、四川盆地、渤海等多个油气田取得了重大油气发现。随着古老碳酸盐岩、深层、新区、页岩油气等勘探开发力度的加大,油气田的地层条件和介质环境变得更为苛刻复杂,油套管柱变形、泄漏、腐蚀、挤毁、破裂等失效事故时有发生,在有的区块还十分严重,特殊结构井和特殊工艺井、特殊增产改造措施等对油套管柱提出了新的要求,油套管柱完整性面临着一系列新的挑战和难题仍需要破解。建议如下。

(1)持续深化复杂工况油套管柱失效控制与完整性理论和技术研究。持续发展油套管柱可靠性设计、应变设计方法及评价技术,油套管柱失效分析诊断预测预防技术,智能管柱技术(含专家系统),高强、高韧、高抗挤、耐腐蚀、特殊螺纹、铝(镁)合金、钛合金、复合材料等高性能油套管及应用技术，特种缓蚀剂和表面防护技术等研究攻关。

（2）持续开展复杂深层、高温高压、酸性气田等油套管柱失效控制与完整性技术研究。我国西部油气田是油气增储上产的主战场，7000m以上的超深井逐渐增多甚至超过8000m，井底温度大于180℃甚至超过200℃，压力大于100MPa甚至超过130MPa，且处于高含CO_2/Cl^-或$H_2S/CO_2/Cl^-$环境，面临复杂地质条件，强酸酸化、水平井大排量高压力反复压裂等复杂工况条件，对油套管柱的完整性和安全可靠性提出了更高要求，亟待深入开展高温高压、酸性气田等复杂工况油套管柱结构和密封完整性及适用性评价技术，复杂工况油套管柱腐蚀规律与机理、选材评价、腐蚀控制及预防技术等研究攻关。

（3）持续开展页岩气井套变机理与预防技术研究。页岩气开发是我国油气工业新的增长点，但深层页岩气开发井套管柱变形问题十分突出，套变率一度达到50%，导致桥塞等工具无法正常下入，局部区块丢弃长度达到1/3，严重影响了页岩气开发及产能建设。初步分析研究表明[6]，套管柱变形是套管柱使用性能、高压力大排量反复体积压裂工艺、地层裂缝等因素综合作用的结果。虽然近年来一直在持续研究，也采取了一些措施，但套损问题仍未得到很好的解决。从根本上来说，页岩气井套变取决于套管柱的承载能力与载荷的相对大小及分布。拟以套管柱为核心，优化套管柱设计，建立基于应变的套管柱设计新方法；提高套管柱的服役性能，选用抗剪切和抗外挤等使用性能优良的高钢级厚壁套管，发展模拟套管实际使用工况的使用性能评价技术；优化页岩气井井位选择、井距排布及井眼轨迹控制，减小天然裂缝、断层等对套变的影响；优化压裂工艺，减小断层滑移、天然裂缝扩展、水力裂缝扩展及相互干扰等对套变的影响。通过系统研究，建立相应的设计、选材、评价、工程应用和现场作业技术和标准体系。

（4）深井超深井复杂工况套管柱疲劳断裂失效及预防措施研究。2018年以来，新疆、长庆等油气田多次发生套管疲劳断裂事故，严重影响油气田开发。初步分析研究表明：套管断裂是由于水平井大排量高压力反复水力压裂及套管反复上提—下放引起的疲劳失效。需要进一步研究套管断裂的机理和原因、套管结构强度和寿命、作业工艺对断裂的影响、套管承受的动态载荷及控制措施、套管扣型选用评价方法、套管选用技术规范、现场作业规范等，系统解决套管柱的疲劳断裂失效问题。

（5）老油气田套损机理与防治技术研究。套损一直是制约油气田产量和效益的顽疾，多数油气田套损率超过15%。主要失效模式有变形、腐蚀、挤毁、泄漏、错断、破裂等，多是力学因素和/或化学因素对套管柱耦合作用的结果。老油气田复杂的地层和环境条件及其演变对套损有重要影响，稠油热采、火驱、CO_2驱、空气驱/空气泡沫驱等套管柱的变形和腐蚀问题仍然比较严重，这些问题需要针对性地持续研究加以解决。

（6）开展大数据和人工智能在油套管柱失效控制及预测预防中的应用研究。近年来，数据科学、人工智能、机器学习、材料信息学等发展迅速[29,30]，将这些学科的最新技术应用到石油管工程和材料服役安全领域，将会产生事半功倍的效果。重点是在本领域知识库的基础上，通过油套管柱材料性能和服役性能、力学状态分析、运行状态监检测等数据采集和失效数据库及案例库构建，数据解释，机器学习建模，模型评估，实验/计算设计，实验/计算，再反馈到前端的反复循环，辨识规律、机理及影响因素，提出控制和预防措施，确保油套管柱的运行安全。

应当强调指出，油套管柱的完整性保障与失效控制，就是要保障油套管柱全生命周期的安全可靠性和经济性，而70%以上的油套管柱失效是由于前期的设计/选材/评价和不恰当的工程作业及增产改造措施造成的。所以，加强油套管柱优化设计、选材、试验评价，综合考虑油套管柱设计寿命、失效概率、失效后果等因素，发展基于风险的油套管柱可靠性设计新方法及配套技术，是控制油套管柱失效、保障管柱完整性的关键。

致　谢

中国石油集团石油管工程技术研究韩礼红、白真权、韩新利、刘文红、张娟涛、韩燕、龙岩等，中国石油大学（华东）闫相祯，加拿大 C-FER 公司谢觉人，中国石油塔里木油田公司谢俊峰等参加了相关试验研究工作，特此致谢！

参 考 文 献

[1] 冯耀荣，马秋荣，张冠军．石油管材及装备材料服役行为与结构安全研究进展及展望[J]．石油管材与仪器，2016(1)：1-5．

[2] 吕拴录．塔里木油田油套管失效分析及预防[C]．塔里木油田井筒完整性会议，2013．

[3] 冯耀荣，韩礼红，张福祥，等．油气井管柱完整性技术研究进展与展望[J]．天然气工业，2014，34(11)：73-81．

[4] 中国石油天然气集团公司石油管工程重点实验室．油井管柱完整性技术研究[R]．西安：中国石油集团石油管工程技术研究院，2014．

[5] 中国石油天然气集团公司石油管工程重点实验室．复杂工况气井油套管柱失效控制与完整性技术研究[R]．西安：中国石油集团石油管工程技术研究院，2016．

[6] 中国石油天然气集团公司石油管工程重点实验室．苛刻服役条件油井管工程应用基础研究[R]．西安：中国石油集团石油管工程技术研究院，2019．

[7] 冯耀荣，张冠军，李鹤林．石油管工程技术进展及展望[J]．石油管材与仪器，2017(1)：1-8．

[8] 王鹏，陈光达，宋生印，等．一种油管腐蚀程度预测方法及装置：中国，201210548324.X[P]．2017-10-17．

[9] 尹成先，冯耀荣，白真权，等．一种用于含 Cr 油管的高温酸化缓蚀剂：中国，200710178677.4[P]．2010-09-29．

[10] 白真权，韩燕，张娟涛，等．一种管材实物应力腐蚀试验机：中国，201210165009.9[P]．2015-10-14．

[11] Lei XW, Feng YR, Fu AQ, et al. Investigation of stress corrosion cracking behavior of super 13Cr tubing by full-scale tubular goods corrosion test system[J]. Engineering failure analysis. 2015, 50: 62-70.

[12] 付安庆，史鸿鹏，胡垚，等．全尺寸石油管柱高温高压应力腐蚀/开裂研究及未来发展方向[J]．石油管材与仪器，2017，3(1)：40-46．

[13] SY/T 7026—2014 油气井管柱完整性管理[S]．

[14] 吴奇，郑新权，张绍礼，等．高温高压及高含硫井完整性指南[M]．北京：石油工业出版社，2017．

[15] 吴奇，郑新权，张绍礼，等．高温高压及高含硫井完整性设计准则[M]．北京：石油工业出版社，2017．

[16] 吴奇，郑新权，邱金平，等．高温高压及高含硫井完整性管理规范[M]．北京：石油工业出版社，2017．

[17] 樊恒，闫相祯，冯耀荣，等．基于分项系数法的套管实用可靠度设计方法[J]．石油学报，2016，37(6)：807-814．

[18] SY/T 7456—2019 油气井套管柱结构与强度可靠性评价方法[S]．

[19] Wang Jiandong, Feng Yaorong. Economy and reliability selection of production casing thread connections for low pressure and low permeability gas field[J]. Procedia Environmental Sciences, 2011, 11: 989-995.

[20] 王建东，冯耀荣，林凯，等．特殊螺纹接头密封结构比对分析[J]．中国石油大学学报（自然科学版），

2010, 34(5): 126-130.
[21] 王建东, 冯耀荣, 林凯, 等. 高气密封油套管特殊螺纹接头: 中国, 200910092027.7[P]. 2012-11-14.
[22] 王建东, 杨力能, 冯耀荣, 等. 低压用气密封特殊螺纹接头: 中国, 201110229616.2[P]. 2015-12-02.
[23] 王建东, 林凯, 赵克枫, 等. 低压低渗苏里格气田套管柱经济可靠性优化[J]. 天然气工业, 2007, 27(12): 74-76.
[24] 王建东, 林凯, 赵克枫, 等. 低效气田套管经济可靠性选择[J]. 石油钻采工艺, 2007, 29(5): 98-101.
[25] 王蕊, 李东风, 张森, 等. 一种抗外压至失效试验用挤毁缸的制造方法: 中国, 20111 0251302.2[P]. 2014-09-03.
[26] 王蕊, 李东风, 韩军, 等. V150钢级油井管全尺寸实物试验试样制备方法: 中国, 20131 0109959.4[P]. 2016-07-13.
[27] 娄琦, 张森, 李东风, 等. 一种V140钢级油井管试验实物的制备方法: 中国, 2011100437217[P]. 2014-08-06.
[28] 李东风, 张森, 韩新利, 等. 一种G3合金油井管试验实物焊接制备方法: 中国, 200910092835.3[P]. 2011-4-20.
[29] 王鹏, 孙升, 张庆, 等. 力学信息学简介. 自然杂志, 2018, 40(5): 313-322.
[30] 李晓刚. 材料腐蚀信息学[M]. 北京: 化学工业出版社, 2014.

本论文原发表于《天然气工业》2020年第40卷第2期。

石油管工程技术进展及展望

冯耀荣　张冠军　李鹤林

（中国石油集团石油管工程技术研究院·
石油管材及装备材料服役行为与结构安全国家重点实验室）

摘　要：伴随着中国石油集团石油管工程技术研究院（及其前身）的建立和发展，"石油管工程（学）"学科应运而生。30余年来，"石油管工程（学）"学科得到了快速发展。研究院建立了石油管材及装备材料服役行为与结构安全国家重点实验室、中国石油学会管材专业委员会等科研和学术技术交流平台，构建了从微观组织分析到全尺寸模拟试验完整的石油管试验研究装备体系，形成了以院士、国家和省部级专家为骨干、专业和年龄结构合理的创新团队，基本形成石油管工程核心技术体系，先后获得国际、国家、石油天然气行业授予的安全、质量、计量、标准、失效分析等方面的权威资质和授权25项，开创了科学研究、质量监督、技术服务三位一体、协调发展的模式，取得重要科技成果100余项，有力支撑了国家西气东输等重大管道工程、重点油气田勘探开发，促进了石油管的全面国产化和质量性能水平提升。在当前我国经济发展"新常态"和低油价形势下，石油管工程技术面临诸多挑战，石油管工程科技创新比以往任何时候都更加迫切。亟待进一步传承和发展"石油管工程（学）"学科，发展石油管失效控制和服役安全理论，进一步加强石油管工程超前储备和应用基础研究，突破石油管工程应用关键技术，建立或完善石油管材料服役行为与结构安全核心技术体系，有力支撑石油天然气工业发展。

关键词：石油管工程（学）；油井管；油气输送管；材料服役行为；结构安全；失效控制

1　概述

经过35年的发展，中国石油集团石油管工程技术研究院从无到有、从小到大、由弱变强，创立了石油管工程（学）学科[1-2]；建立了石油管材及装备材料服役行为与结构安全国家重点实验室、陕西省重点实验室、中国石油集团石油管工程重点实验室、国家安全生产石油管及装备安全技术研究中心、中国石油学会管材专业委员会等科研和学术技术交流平台；构建了从微观组织分析到全尺寸模拟试验完整的石油管试验研究装备体系，试验仪器和装备

基金项目：中国石油天然气集团公司应用基础研究项目（苛刻服役条件油井管工程应用基础研究编号：2016A-3905）。

作者简介：冯耀荣，1960年生，教授级高工，博士；孙越崎能源大奖获得者；一直从事石油管材及装备材料服役行为与结构安全研究及重大工程技术支持工作，现任中国石油集团石油管工程技术研究院总工程师、石油管材及装备材料服役行为与结构安全国家重点实验室主任。电话：（029）81887699；E-mail：fengyr@cnpc.com.cn。

488台（套），其中全尺寸模拟试验研究的标志性大型设备有10台（套）；形成了以院士、国家和省部级专家为骨干、专业和年龄结构合理的创新团队；基本形成石油管工程核心技术体系，涉及5个主要技术领域，包含一级技术12项，二级技术36项；先后获得国际、国家、石油工业和中国石油集团授予的安全、质量、计量、标准、失效分析等方面的权威资质和授权25项；开创了科学研究、质量监督、技术服务三位一体、协调发展的模式，成效显著。

（1）围绕国家重大油气管道工程需求，研究形成了X70/X80管线钢和钢管关键技术指标体系、管道断裂和变形失效控制技术、X70/X80大口径输气管道风险评估技术；研究制订了X70/X80管线钢和钢管系列标准70余项，为西气东输等重大管道提供决策支持和技术方案；推动了X70、X80高强度管材的规模化应用；支持和促进了西气东输$800×10^4$t管材国产化，其中西气东输管线国产化率达到50%，西气东输二线国产化率达到90%，实现了钢级从X52到X80、输送压力从6.4MPa到12MPa的重大跨越。我国用20年时间赶上和超过了发达国家，实现了国际领跑，有力支撑了国家重大管道建设及安全运行，为我国石油天然气工业做出了重要贡献[3-5]。

（2）建立了深井超深井、高温高压气井、致密油气井用高强度、高韧性、高抗挤、耐腐蚀、特殊密封结构等非API油井管标准体系和技术体系，形成基于气井全寿命周期的管材选用和适用性评价方法，研发了系列缓蚀剂，为塔里木、长庆等西部大庆、新疆大庆建设提供了有力的技术支撑，促进了油井管国产化率和质量技术水平持续提高。油井管国产化率由20世纪80年代初的不足10%上升到目前的98%以上，保障了重点油气田勘探开发安全[3,5,6]。

（3）系统研究了石油管材的失效模式、机理、规律及原因，建立了石油管材失效数据库和案例库，积累石油管材及装备失效案例1600例，其中包括"11·22"黄岛输油管道泄漏爆炸事故等在内的重大失效事故100余起，为事故处理和预防提供了决策和技术支持。

（4）完成了数以万计的石油管材质量监督检验、新产品质量鉴定评价、制造许可和型式试验；完成了数千万吨的石油管材驻厂质量监督和现场检验，保障了管材质量和使用安全；建立了比较完善的石油管材标准体系，围绕重大管道工程、重点油气田勘探开发制修订近200项国家、行业、重大工程标准；成为ISO TC67/SC2副主席、并行秘书处，SC2、SC5投票委员单位和中方技术对口单位，为石油管材的质量和安全保障做出了贡献。

2 石油管工程技术主要进展

35年来，石油管工程技术研究院和石油管及装备材料服役安全国家重点实验室围绕西气东输等国家重大管道建设工程、塔里木和长庆等重点油气田勘探开发工程潜心研究，攻坚克难，取得了100余项重要科研成果。发表论文2100余篇，出版专著及研究文集20余部，获得授权专利400余件（其中发明专利152件），研发软件54套，制修订国家、行业、企业和重大工程标准270余项，向ISO/API提出标准制修订提案23件，100余项成果获得省部级科技成果奖励，其中西气东输工程技术及应用、我国油气战略通道建设与运行关键技术等13项成果获国家科技进步一、二等奖，油气输送管道失效控制及工程应用、石油管工程试验平台建设及关键技术创新等10项成果获省部级特等奖和一等奖。

2.1 X70/X80管线钢和钢管应用关键技术

针对国家重大管道工程和高钢级管线钢及钢管工程应用和应用基础问题，开展了系统研究，取得了一系列重要成果，为重大管道工程提供了有力的技术支撑，推动了X70/X80管线钢和钢管在重大管道工程中的规模化应用[3-5,7-8]。

（1）研究提出了西气东输系列管线 X70/X80 管材关键技术指标，形成高钢级管线钢和钢管材质选用及针状铁素体组织分析鉴别与评定技术、管型选用及螺旋埋弧焊管残余应力控制技术、高钢级厚壁管线钢及钢管综合性能评价和质量控制技术；制订了兼顾安全性与经济性的系列标准 70 余项；研究提出了 X80 弯管和管件成分设计，研发了感应加热弯管及管件的制造工艺。

（2）联合研发了国际先进的 50000J 大摆锤试验系统，建成了高压输气管道全尺寸气体爆破试验场，建立了相应的试验方法。自主研发了适用于富气组分的天然气减压波分析和高压输气管道止裂预测软件。研究提出了西气东输系列管线安全运行参数控制要求和管材止裂韧性要求。首次开展了 X80 螺旋埋弧焊管管道实物气体爆破试验，验证了西气东输二线管材的止裂能力（图 1）。

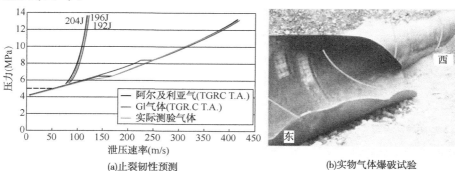

(a) 止裂韧性预测　　　　(b) 实物气体爆破试验

图 1　X80 螺旋焊管止裂韧性预测及实物气体爆破试验

（3）联合研发了国际先进水平的大口径钢管在模拟工况条件下整管内压+弯曲复合载荷变形试验系统及试验方法。研究提出了基于屈曲应变数值模拟和量纲分析的管线压缩应变容量预测方法。研究揭示了多种规格 X70 大变形钢管的压缩应变容量与钢管几何尺寸、材料组织性能之间的关系（图 2），提出了确定钢管屈曲应变能力的方法。发明了大应变钢管的制造方法。研究制订了西气东输二线、中亚管线、中缅管线等基于应变设计地区使用直缝埋弧焊管技术条件。

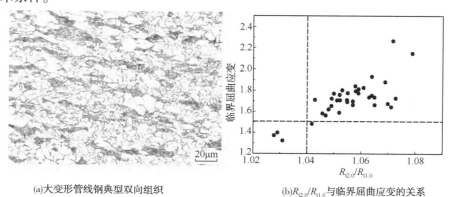

(a) 大变形管线钢典型双向组织　　　　(b) $R_{t2.0}/R_{t1.0}$ 与临界屈曲应变的关系

图 2　X70 抗大变形管线钢管的组织和性能

2.2　油气管道和储气库安全评价与完整性技术[9-12]

围绕油气管道安全评价、风险评估、修复补强、储气库管柱选材评价、储气库安全风险评价等几个方面开展技术攻关，取得了系列创新成果。

（1）研究形成油气管道安全风险评价技术及软件，建立了油气管道失效概率计算模型；提出了油气管道目标可靠度确定方法，制订了我国油气管道风险可接受推荐准则，形成了系统的油气管道安全评价和定量风险评估技术体系。

（2）形成基于风险的油气管道检测技术，发明了管道修复用复合材料体系和施工方法，研发了带锈转化、预浸料修复和均匀加压固化等工艺技术，建立了管道复合材料修复补强技术评价指标体系。

（3）研究建立了基于故障树理论的注采井风险评估方法；建立了基于全井段测井数据的地下储气库套管柱剩余强度评价与剩余寿命预测方法；形成储气库管柱优化设计、选材、适用性和完整性评价技术，建立了基于层次分析法和数值模拟的在役老腔稳定性评价方法；形成了储气库注采气站及管道检测、风险评估及适用性评价技术。

研究成果应用于西气东输系列管线（图3）、陕京管道等22条重要油气管道和塔里木油田、陕西天然气等多个油气田集输管网和地方天然气管网，金坛、大港、长庆、辽河、新疆等储气库也得到应用。研究成果为油气管道和储运设施的安全运营提供了有力的技术支撑。

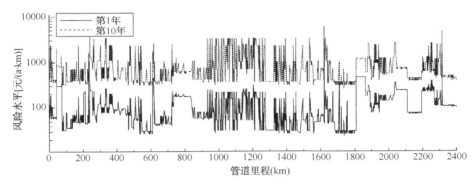

图3 西气东输二线西段总风险水平全线分布

2.3 第三代大输量天然气输送管应用技术

为了满足建设 $380×10^8 m^3$/年以上超大输气量管道需求，提出了采用超高钢级X90及以上钢管、提高管道设计系数、增大输送管口径等三种技术方案，称为第三代大输量天然气管道应用技术[13-19]。

（1）系统分析研究了提高强度设计系数对管道安全性及风险的影响（图4），研究确定了0.8设计系数管材关键性能指标和质量控制要求，制订了西气东输三线0.8设计系数管道用X80螺旋缝埋弧焊管技术条件，在西气东输三线示范工程中成功敷设261km管道，节约管材12.6万吨，节约采购成本约1亿元。分析研究了X80厚壁三通的断裂抗力和极限承载能力，在确保三通极限承载能力不小于3.5倍管道设计压力的情况下，管件壁厚减薄20%~30%，减小了厚壁管件设计的过度保守性，降低了制造难度，在西气东输三线管道工程中得到应用。

（2）研究形成φ1422mm X80 12MPa管道断裂控制技术，在国际上首次开展φ1422mm X80 12MPa使用天然气介质管道全尺寸爆破试验；研究制订了φ1422mm X80管材系列技术标准，完成了螺旋埋弧焊管、直缝埋弧焊管、弯管和管件的试制评价，为中俄东线建设奠定了基础。

（3）研究提出了X90管线钢管断裂控制技术及止裂韧性指标。系统研究了X90管材成分、组织、性能、工艺之间的相关性，提出了X90管材的关键技术指标和检测评价方法及配套系列标准。基本完成了X90焊管和管件的研发及试验评价，为X90管线钢和钢管的工程应用奠定了基础。

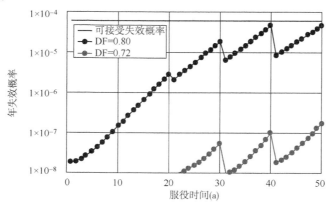

图4 两种设计系数X80管道年失效概率与服役时间的关系

2.4 先进钻柱构件设计开发与工程应用关键技术

针对钻柱构件频繁失效现状，开展失效机理、规律及优化设计研究，形成系列高性能钻柱构件优化设计及工程应用关键技术[20-25]。

（1）系统研究揭示了钻杆内加厚过渡区的失效机理和影响因素，分析研究了钻杆内加厚过渡区结构对使用寿命的影响规律（图5），提出内加厚过渡区锥面长度 M_{iu} 的控制指标，被API采纳修改 API SPEC 5D 标准。

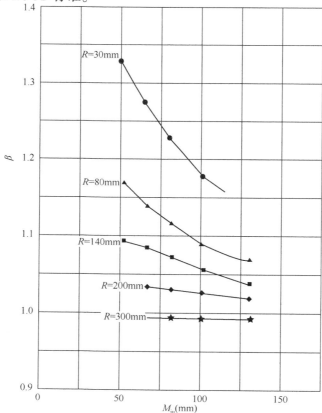

图5 弯曲载荷下钻杆内加厚过渡区锥面长度 M_{iu} 和过度圆角半径 R 对象对应力集中系数 β 的影响

（2）联合研发了钻杆实物疲劳试验系统及试验方法。在研究揭示钻柱构件失效机理及原因的基础上，提出了钻铤、钻杆接头和转换接头的选材及热处理工艺。在钻柱构件安全韧性判据研究的基础上，提出了钻柱构件的韧性控制指标和标准。

（3）研究形成了钻杆安全可靠性评价技术、软件及标准，包括：损伤钻杆的安全性评价（FAD评价与极限缺陷尺寸曲线）、钻杆疲劳寿命预测、钻杆安全可靠性及风险评价、钻杆的操作极限（包括API RP 7G的内容）、钻杆断裂事故原因的定量分析等。

（4）针对高含硫气田勘探开发，完成抗硫钻杆材料成分、组织、工艺及性能优化设计，联合开发了SS105钢级抗硫钻杆产品，形成产品检测评价技术规范，相关成果被ISO国际标准委员会采纳。对现有ISO 11961标准进行修订。技术成果在中石化普光等酸性气田获得成功应用，为高含硫化氢气田钻井安全提供了技术支撑。

（5）针对钻机负荷轻量化、大位移水平井摩阻控制及耐蚀性需求，联合研发了460MPa级铝合金钻杆，在塔里木油田获得成功应用，为钻机减负及大位移水平井钻井提供了技术支撑。

（6）围绕西气东输二线等长输管道穿越工程需求，开展钻杆结构及材料优化设计，联合研发了6⅝in-V150定向穿越专用钻杆，在长江、黄河、渭河等大型穿越工程中获得应用，为石油行业穿越工程提供了技术支撑。

2.5 复杂工况油套管柱失效控制与完整性评价技术

针对高温高压气井、稠油热采、低渗透、页岩气开发等复杂油气田工况套管的失效预防和安全保障，开展了系统研究，形成一系列创新成果[26-29]。

（1）合作研发了2500t油套管复合载荷试验系统、轴向+外压复合载荷挤毁试验系统，形成了相应的实验方法。

（2）系统研究揭示了注水开发油田油层套管射孔开裂的影响因素，提出了套管射孔开裂判据及预防措施（图6）。

（3）研究确定了高温高压气井套管柱失效模式与失效概率计算方法，形成了套管柱系统可靠性设计基本流程，建立了基于可靠性的油气井管柱强度和密封性计算模型、方法和判据，形成了高温高压气井套管柱结构强度和密封性设计评价方法，建立了高温高压气井套管密封可靠性设计的极限状态方程和计算程序。研究制定了高抗挤套管的技术标准，并被API采纳修改套管标准。

（4）针对稠油蒸汽热采井套管循环高温服役环境，形成稠油蒸汽热采井套管柱应变设计、选材与适用性评价方法及配套标准。采用强度错配理念与工艺，成功设计开发了80SH、90SH及110SH热采套管，在新疆油田风城及红浅等区域获得成功应用，实现了5年15轮次注汽生产零套损的优异效果。

（5）针对长庆等低压低渗低产气田，研发了经济型特殊螺纹连接套管及配套技术（图7），在长庆、大庆等油气田获得应用。

（6）针对西南长宁、威远等页岩气开发套管变形、泄漏、脱扣等问题，研究了水平井复杂压裂条件下套管柱的力学行为、管材特性及关键技术指标、管柱和螺纹优化设计选用与适用性评价方法。

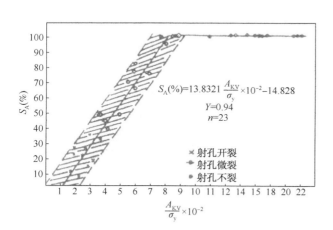
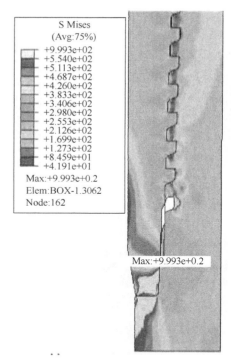

图6 射孔套管 A_{KV}/σ_Y 与开裂倾向的关系

图7 经济型特殊螺纹套管优化设计

2.6 高温高压气井油套管柱腐蚀完整性技术

针对高温高压、严酷腐蚀气井、复杂酸化压裂工况,研究院在油套管柱选材、腐蚀评价方法体系、井筒腐蚀完整性、耐高温酸化缓蚀剂等方面取得了重要研究成果[3,5-6]。

(1)联合研发了功能强大的实物拉伸+腐蚀试验系统(图8),冲刷腐蚀试验系统。建立了石油管实物应力腐蚀和冲刷腐蚀模拟评价试验方法。

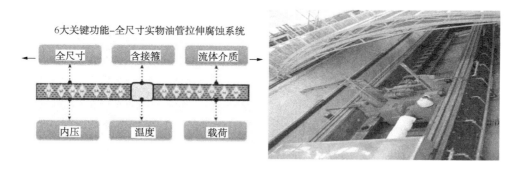

图8 油管全尺寸实物拉伸+腐蚀试验系统及功能

(2)系统研究获得了高温高压气井油管腐蚀失效特征及影响因素,揭示了超级13Cr油管的耐蚀性随温度、CO_2分压、Cl^-浓度和流速、表面状态以及酸化环境、完井液、加载应力的变化规律、腐蚀行为和特征。

(3)研究建立了基于气井全寿命周期的油套管选材与评价方法,包含酸化压裂、完井生产2个作业过程,鲜酸酸化、残酸返排、凝析水、地层水、完井液5个工况环境与恒定载荷、交变载荷、管柱震颤3个力学因素，断裂、腐蚀、泄漏3种主要失效模式。形成了一套基于井筒全寿命周期的腐蚀完整性选材评价技术，为塔里木高温高压气田油套管选材提供决策依据。

（4）发明了超级13Cr酸化缓蚀剂，有效解决了高温高压气井超级13Cr腐蚀难题。

上述成果已全面应用于塔里木油田高温高压气井油套管的选材评价和失效控制，超级13Cr酸化缓蚀剂系列产品，在塔里木油田应用100多井次，累计使用800多吨，创造了良好的经济效益和社会效益。

2.7 实体膨胀管应用关键技术

经过系统研究，研究院形成了实体膨胀管全尺寸实物评价系统、实体膨胀管膨胀工艺及关键参数模拟仿真预测、实体膨胀管实物评价方法及标准、高性能膨胀管材料及螺纹优化设计、实体膨胀管实验及施工配套工具以及膨胀套管力学性能预测分析软件，形成了从理论研究、实(试)验装备及评价方法、配套工具、高性能材料和螺纹、现场施工、预测软件等实体膨胀管应用技术体系[3,30]。

2.8 油气田与管道用非金属及复合材料管及应用关键技术

针对油气田地面管道、长输管道、海洋非金属复合管新材料及应用[3,31-32]方面的技术难题，研究院开展了系统研究，取得了多项重要创新成果。

（1）研究提出了油气田用非金属管的关键技术指标和质量性能评价方法，制订了系列标准，形成油气田用非金属管选材与质量控制评价技术体系。

（2）攻克了非金属材料复合管的选材、结构设计、制备工艺、关键性能测试评价等技术难题，制备了适用于水深500m的海洋非金属材料复合管，在黄海海域通过海试(图9)。

图9 水深500m的海洋非金属材料复合管

（3）研发了包括内层钢管、过渡层、复合材料增强层及外保护层四层结构的复合材料增强管线钢管，制备了玻璃纤维复合材料增强X80 ϕ1219mm管线钢管，爆破压力由23MPa提高至37MPa。

（4）研究提出了抗硫非金属管关键技术指标，建立了抗硫非金属管选材评价方法、非金属材料气体渗透性检测及控制方法，形成含硫油气环境中的介质适应性评价技术，研发出抗硫非金属复合管新产品，在塔中Ⅰ号气田TZ 262-H1得到应用。

上述成果已广泛应用于国家和石油行业标准制定、制造企业的产品优化和国内油气田地面管线的选材设计、质量控制、施工验收等，取得了良好的经济效益和社会效益，具有良好的推广应用前景。

3 石油管工程技术发展展望

经过35年的发展，石油管工程技术取得了长足进步。在当前我国经济发展"新常态"和低油价形势下，石油管工程技术面临诸多挑战。油气勘探开发向低渗透、深层、海洋、非常规扩展，炼化系统原油劣质化、多元化趋势明显，油气长输管道向大口径、高压力、高钢级方向发展。复杂的力学环境、高温高压、严酷腐蚀环境等对石油管的质量性能和使用的安全可靠性提出了更高要求，确保安全可靠性和提高经济型的矛盾更加突出，石油管工程科技创新比以往任何时候都更加迫切。亟待进一步传承和发展"石油管工程（学）"学科，发展石油管失效控制和服役安全理论，进一步加强石油管工程超前储备和应用基础研究，突破石油管工程应用关键技术，建立或完善石油管材料服役行为与结构安全核心技术体系，有力支撑油气工业发展。

3.1 传承和发展"石油管工程（学）"

经过35年的发展，初步形成"石油管工程（学）"学科体系。石油管工程（学）涉及石油天然气工程、材料科学与工程（材料学、材料物理、材料化学、材料加工工程）、冶金工程、机械工程、力学（弹塑性力学、断裂力学、管柱力学、流体力学、岩土力学）、安全工程、计算机科学与技术、化学、物理等学科，是多学科交叉在工程领域的体现。

"石油管工程（学）"是从石油管的服役条件出发，重点研究石油管的力学行为、环境行为、先进材料及其成分/结构—合成/加工—性质与服役性能的关系、石油管的失效控制与预测预防、安全评价与完整性，揭示石油管的失效机理和规律，提出失效控制和预防方法，发展全寿命周期的安全可靠性和完整性技术，确保石油管的长期服役安全。

在已有工作的基础上，逐步发展完善"石油管工程（学）"，建立了石油管材服役行为与结构安全理论和技术体系。针对油井管与管柱失效预防、输送管与管线安全评价、腐蚀与防护、先进材料及应用技术四大方向，发展高钢级管道服役安全与失效控制理论，建立适合我国国情的高钢级大口径油气输送管材技术和标准体系；形成复杂工况油气井管柱失效控制及完整性技术，为重点油气田勘探开发提供技术支撑；形成高温高压及严酷腐蚀环境管材腐蚀机理及综合防治技术、炼化管道腐蚀机理与评价技术；解决先进石油管材料应用技术难题，建立新型管材测试与评价核心技术体系。

3.2 发展石油管失效控制理论

（1）从石油管的失效模式入手控制失效[33]。石油管的失效包括结构失效、功能失效、工艺失效。主要失效模式有变形（过量弹性和塑性变形）、断裂（过载断裂、脆性断裂、疲劳和腐蚀疲劳、氢脆和应力腐蚀）、失稳（屈曲、挤毁）、损伤[磨损（黏着、磨粒、腐蚀磨损）、腐蚀（均匀、点蚀、缝隙腐蚀）、表面损伤（含机械损伤）]、泄漏、滑脱等。通过对典型石油管的服役条件和失效分析入手，归纳总结其失效模式、规律和原因，提出失效抗力指标、失效判据及具体的预防措施。

（2）从石油管失效的宏观规律入手控制失效。在复杂工况下石油管的失效不可能完全避免，或完全避免会付出很高代价，但可以控制。失效过程可以用"浴盆曲线"[34]来描述（图10），由早期失效、偶发失效、耗损失效三条曲线合成，可分为失效事故率由高到低、近乎常数、再由低到高三个阶段。失效控制可以从三个方面入手：一是控制失效概率，提高安全可靠性，主要是通过安全可靠性设计与评价，减少偶发失效；二是控制失效后果，减少失效影响，主要是通过风险分析评价与控制来实现；三是控制失效发生的时间，延长使用寿命，主要是通过优化设计、合理选材、制造质量控制、严格检验验收，减少早期失效，通过过程检验、维修维护、使用科学管理，减少耗损失效，发展寿命设计评估和预测技术。

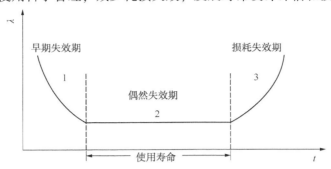

图 10　用于描述失效事故率 λ 与使用时间 t 的"浴盆曲线"

3.3　发展石油管服役安全理论

石油管的服役安全取决于其在不同服役条件下的性能。石油管的性能包括力学性能、物理性能、化学性能、工艺性能、使用性能（服役性能），力学、物理、化学、工艺性能间接影响结构安全，而使用性能（服役性能）直接影响结构安全。安全评价、风险评价、可靠性评估、寿命预测等是石油管服役安全的主要手段，也是石油管完整性的主要内容。结构完整性（评价）是基于断裂力学和极限载荷理论的含缺陷结构能否使用的定量科学评价，完整性评价和完整性管理早期主要是针对在役结构，随后向前延伸形成全过程的完整性评价和完整性管理。完整性的本质和核心是全寿命周期的安全可靠性，包括完整性设计即安全可靠性及寿命设计（全寿命周期的安全可靠性设计）、完整性评价、完整性管理。一方面要进一步加强不同服役条件下石油管服役行为研究，另一方面要发展石油管完整性即全寿命周期的安全可靠性研究，确保石油管的服役安全。

3.4　突破石油管应用基础和工程应用关键技术

（1）针对我国大口径、长距离、高压力、高钢级管道和储运设施建设和长期安全运行需求，重点研究高钢级管材关键技术指标及表征评价方法、高压大口径天然气管道断裂控制技术、高强度管道变形控制技术、油气管道失效机理与失效控制理论、油气管道安全风险评价技术、油气储运设施完整性技术，建立油气输送管道失效控制理论，形成完整性技术体系。

（2）围绕特殊结构和特殊工艺井，超深、超高温、超高压、超长水平井及高压大排量分段压裂改造对油气井管柱服役性能、安全可靠性、质量和寿命提出的新的更高的要求，重点研究油井管的失效机理和规律、失效预测预防技术、管柱优化设计及适用性评价技术、油气井管柱结构完整性和密封完整性、高性能油井管关键技术指标及表征评价方法，建立油气井管柱失效控制理论，形成完整性技术和标准体系。

（3）针对高温、高压、严酷腐蚀介质油气田石油管材存在的严重腐蚀问题，重点研究高温高压及严酷腐蚀环境油套管腐蚀机理及防治技术、高酸性气田管材腐蚀机理及采集系统腐蚀综合防治技术、炼化管道腐蚀行为与评价方法、石油管材腐蚀检测、监测、预测和预防技术，揭示失效机理和规律，提出有效预防措施，确保管材使用安全。

（4）针对复杂工况和特殊服役环境用石油管材，重点研究高性能管线钢管成分/组织/性能/工艺相关性，高强高韧、高抗挤、高抗扭、耐腐蚀、长寿命油井管成分/组织/性能/工艺相关性，高性能非金属管材、复合管材成分/组织/性能/工艺相关性，发展先进和特殊专用材料应用关键技术，保障其服役安全。

总之，要在国家创新驱动战略引领下，通过大力推进原始创新，进一步发展石油管工程技术，形成比较完善的油气输送管道失效控制技术、高性能管线钢和钢管应用关键技术、油气管道完整性技术、新型钻柱构件材料及安全可靠性技术、非API油套管应用及管柱完整性技术、高温高压气井油套管柱腐蚀防护技术、油气田地面管道失效控制及预防技术，形成一批国家标准、发明专利、系列高新技术和产品、高水平学术论文和专著等载体化有形化成果，并加大成果转化推广力度，创造良好经济社会效益，为我国油气工业作出新贡献。

参 考 文 献

[1] 李鹤林."石油管工程"的研究领域、初步成果与展望[M].北京：石油工业出版社，1999：1-10.

[2] 李鹤林，张冠军，杜伟."石油管工程"的内涵及主要研究领域[J].石油管材与仪器，2015，1(1)：1-4.

[3] 张冠军.中国石油集团石油管工程技术研究院"十二五"科技成果汇编[M].北京：石油工业出版社，2016：1-41.

[4] 冯耀荣，霍春勇，吉玲康，等.我国高钢级管线钢和钢管应用基础研究进展及展望[J].石油科学通报，2016，1(1)：143-153.

[5] 冯耀荣，马秋荣，张冠军.石油管材及装备材料服役行为与结构安全研究进展及展望[J].石油管材与仪器，2016，2(1)：1-5.

[6] 冯耀荣，韩礼红，张福祥，等.油气井管柱完整性技术研究进展与展望[J].天然气工业，2014，34(11)：71-81.

[7] "西气东输二线X80管材技术条件及关键技术指标研究"技术报告[R].西安：中国石油集团石油管工程技术研究院，2009.

[8] "西气东输二线管道断裂与变形控制关键技术研究"技术报告[R].西安：中国石油集团石油管工程技术研究院，2013.

[9] 冯耀荣，陈浩，张劲军，等.中国石油油气管道技术发展展望[J].油气储运，2008，27(3)：1-8.

[10] 赵新伟，罗金恒.油气管道完整性评价技术[M].西安：陕西科技出版社，2010.

[11] 赵新伟，张华，罗金恒.油气管道可接受风险准则研究[J].油气储运，2016，35(1)：1-6.

[12] 赵新伟，李丽锋，罗金恒，等.盐穴储气库储气与注采系统完整性技术进展[J].油气储运，2014，33(4)：347-353.

[13] 吴宏，张对红，罗金恒，等.输气管道一级地区采用0.8设计系数的可行性[J].油气储运，2013，32(8)：799-804.

[14] 赵新伟，罗金恒，张广利，等.0.8设计系数下天然气管道用焊管关键性能指标[J].油气储运，2013，32(4)：355-359.

[15] 吴宏，刘迎来，郭志梅.基于验证试验法的X80钢级大口径三通设计[J].油气储运，2013，32(5)：513-516.

[16] 刘迎来，吴宏，井懿平，等.高强度油气输送管道三通试验研究[J].焊管，2014，37(3)：28-33.

[17] 霍春勇，李鹤，张伟卫，等.X80钢级1422mm大口径管道断裂控制技术[J].天然气工业，2016，36(6)：78-83.

[18] 李丽锋，罗金恒，赵新伟，等.φ1422mmX80管道的风险水平[J].油气储运，2016，35(4)：25-29.

[19] 王红伟，吉玲康，张晓勇，等.批量试制X90管线钢管及板材强度特性研究[J].石油管材与仪器，2015，1(6)：44-51.

[20] 冯耀荣,马宝钿,金志浩,等.钻柱构件失效模式与安全韧性判据的研究[J].西安交通大学学报,1998,32(4):54-58.
[21] 李鹤林,冯耀荣,李京川,等.钻杆接头和转换接头材料及热处理工艺的研究[J].石油机械,1992,20(3):1-6,42.
[22] 王航,韩礼红,胡锋,等.回火温度对抗硫钻杆钢析出相形貌及力学性能的影响[J].材料热处理学报,2012,33(3):88-93.
[23] 冯春,张冠军,韩礼红,等.热处理对超高强度铝合金钻杆用Al-Zn-Mg-Cu系合金力学性能和组织的影响[C]//第十一次全国热处理大会.太原:中国机械工程学会热处理分会,2015.
[24] 韩礼红,王航,李方坡,等.酸性油气田开发用钻杆关键技术研究[C]//油气井管柱与管材国际会议.西安:中国石油学会石油管材专业委员会,2014.
[25] 张平生,韩晓毅,罗卫国,等.钻杆适用性评价及其软件[M]//石油管工程应用基础研究论文集.北京:石油工业出版社,2001.
[26] 王建东,冯耀荣,林凯,等.特殊螺纹接头密封结构比对分析[J].中国石油大学学报(自然科学版),2010,34(5):126-130.
[27] 张毅,吉玲康,宋治,等.油层套管射孔开裂的安全韧性判据[J].西安石油大学学报(自然科学版),1998(6):46-49.
[28] 韩礼红,谢斌,王航,等.稠油蒸汽吞吐热采井套管柱应变设计方法[J].钢管,2016,45(3):11-18.
[29] 中国石油天然气集团公司石油管工程重点实验室.油井管柱完整性技术研究[R].西安:中国石油集团石油管工程技术研究院,2014.
[30] 刘强,宋生印,白强,等.实体膨胀管性能评价方法研究及应用[J].石油机械,2014,42(12):39-43.
[31] 张冬娜,戚东涛,魏斌,等.纤维缠绕钢管道的研究现状及展望[J].油气田地面工程,2016,35(5):5-8.
[32] 丁楠,戚东涛,魏斌,等.海洋柔性管气体渗透机理及其防护措施的研究进展[J].天然气工业,2015,35(10):112-116.
[33] 李鹤林.油气管道失效控制技术[J].油气储运,2011,30(6):401-410.
[34] 李鹤林,李平全,冯耀荣.石油钻柱失效分析及预防[M].北京:石油工业出版社,1999:2.

本论文原发表于《石油管材与仪器》2017年第1期。

X80 钢级 ϕ1422mm 大口径管道断裂控制技术

霍春勇[1,2] 李 鹤[1,2] 张伟卫[1,2] 杨 坤[1,2] 池 强[1,2] 马秋荣[1,2]

(1. 中国石油集团石油管工程技术研究院; 2. 石油管材及装备材料服役行为与结构安全国家重点实验室)

摘 要：在中俄东线天然气管道工程中采用高钢级(X80 钢级)、大口径(外径为 1422mm)的管道进行高压输送(压力为 12MPa)可以有效增加天然气输送量，满足我国能源战略的需要。然而随着钢级、输送压力、管径及设计系数的不断提高，管道的延性断裂成为断裂的主要方式，止裂控制便成为研究的重点。为此，对止裂韧性计算的主要方法 Battelle 双曲线(BTC)方法以及断裂阻力曲线和减压波曲线进行了深入研究，明确了 BTC 方法的原理及其适用范围，同时分析了 BTC 方法应用于高强度、高韧性管线钢时所存在的问题，给出了目前国际上针对 BTC 计算结果常用的修正方法。回顾了俄罗斯 Bovanenkovo-Ukhta 外径为 1422mm X80 管道断裂的控制方案，针对中俄东线管道的设计参数，对 BTC 计算结果进行了修正，进而制订出符合中俄东线管道安全要求的止裂韧性值(245J)。

关键词：中俄东线天然气管道工程外径 1422mmX80 钢级断裂控制；止裂韧性；BTC 方法；减压波

随着天然气需求量的与日俱增和管线钢技术的不断进步，采用 X80 及以上级别高强度钢管进行高压、大输量、长距离输送天然气已经成为世界天然气管道输送技术发展的主流趋势。然而随着钢级、输送压力、管径及设计系数的不断提高，管道的延性断裂成为了断裂的主要方式，已经严重威胁管道安全并成为制约高钢级焊管广泛应用的瓶颈问题[1]，止裂控制也成了研究的重点。

通常采用 Battelle 双曲线方法(BTC)对输气管道的止裂韧性进行预测。然而当管道钢级达到 X80 以上，止裂韧性达到 100J 以上时，BTC 方法预测的准确性便会急剧下降。因此，对于高压(10MPa 及以上)、大口径(外径为 1219mm 及以上)、富气 X80 钢管道，就需要对 BTC 结果进行修正来确定止裂韧性。为此笔者就 BTC 的方法原理、存在的问题及中俄东线 ϕ1422mm 的 X80 钢管道止裂韧性确定方法进行深入的分析和阐述。

1 止裂韧性预测技术

1.1 BTC 方法

API 5L 和 ISO 3183 规定了计算钢管延性断裂止裂韧性的 4 种方法，如表 1 所示。可见

作者简介：霍春勇，男，1966 年生，教授级高级工程师，博士；主要从事长输管道断裂控制方面的研究工作。地址：陕西省西安市锦业二路 89 号。电话：(029)81887999，ORCID：0000-0002-0028-3958，E-mail：huochunyong@cnpc.com.cn。

当压力达到12 MPa，钢级达到X80时只有BTC方法适用，但是当BTC的计算值超过100J时，则需要对止裂韧性计算结果进行修正。

表1 钢管延性断裂止裂韧性计算方法

方法	适用范围
EPRG	$p \leq 8$ MPa，$D \leq 1430$ mm，$t \leq 25.4$ mm，钢级$\leq$X80，贫气
Battelle简化公式	$p \leq 7$ MPa，$40 < D/t < 115$，钢级$\leq$X80，贫气 预测$C_v > 100$J需修正
BTC	$p \leq 12$MPa，$40 < D/t < 115$，钢级$\leq$X80，贫气/富气 预测$C_v > 100$J需修正
AISI	$D \leq 1219$ mm，$t \leq 18.3$ mm，钢级$\leq$X70，贫气 预测$C_v > 100$J需修正

注：表中p表示压力；D表示钢管外径；t表示钢管壁厚；C_v表示夏比冲击功

BTC方法的原理是通过比较材料阻力曲线（J曲线）和气体减压波曲线来确定止裂韧性。当这两条曲线相切，代表在某一压力下裂纹扩展速率与气体减压波速率相同，达到止裂的临界条件，与此条件相对应的韧性（C_v，夏比冲击功）即为BTC方法确定的止裂韧性。

1.1.1 材料阻力曲线

BTC方法中用来计算材料阻力曲线的基本模型如下所示。

$$v_c = c \times \frac{\sigma_{\text{flow}}}{\sqrt{R}} \times \left(\frac{p_d}{p_a} - 1\right)^m \tag{1}$$

$$p_a = \frac{t}{r}\sigma_{\text{arrest}} = \frac{t}{r}\frac{2\sigma_{\text{flow}}}{\pi M_T}\arccos\left[\exp\left(\frac{-10^3 \times \pi \times E \times R}{8\sigma_{\text{flow}}^2 \times C_{\text{eff}}}\right)\right] \tag{2}$$

式中：v_c为裂纹扩展速度，m/s；R为断裂阻力，J/mm²；σ_{flow}为流变应力，MPa；p_d为裂纹尖端动态压力，MPa；p_a为止裂压力，MPa；c、m为回填常数；t为钢管壁厚，mm；r为钢管半径，mm；σ_{arrest}为止裂应力，MPa；M_T为膨胀因子；E为钢管弹性模量，MPa；C_{eff}为有效裂纹长度，mm。

其中R、σ_{flow}、c、m按表2进行计算和取值。

表2 BTC方法中对于不同参数的定义表

参数	R	σ_{flow}	C_{eff}	M_T	m	c
定义/取值	C_v/A_c	$\sigma_{\text{YS}}+68.95$MPa	$3\sqrt{Dt/2}$	3.33	1/6	有土壤回填情况下取0.2750； 无回填情况下取0.3795； 海底管道取0.2350

注：表中A_c表示冲击试样韧带面积；σ_{YS}表示钢管的屈服强度

BTC方法在20世纪70年代由美国著名的研究机构巴特尔纪念研究所（BattelleMenorial Institute，以下简称为Battelle）建立，成功地预测了在不同工况条件下X70及以下级别管线钢延性断裂止裂所需 C_v。然而对于现代高强度(X80 及以上)、高韧性(100J 以上)的管线钢,BTC 方法预测的准确性则随着钢级和止裂韧性的升高而下降,对此国内外学者进行了大量的研究,主要观点如下:

(1)BTC 方法发展之初,主要用于 X70 及以下级别管线钢的止裂韧性预测,其流变应力定义为屈服强度加上 68.95MPa。随着管道钢级的提高,材料的屈强比也随之上升,采用传统方法计算的流变应力将接近或者超过材料的抗拉强度。因此对于 X80 及以上级别管线钢,采用屈服强度和抗拉强度的平均值或引入加工硬化指数来重新定义流变应力将更为合理。同时,大量的研究成果也表明不同的流变应力定义会优化计算结果,但不是 BTC 方法不准确的根本原因。

(2)BTC 方法中的土壤回填常数 m 和 c 是 Battelle 根据早期全尺寸气爆试验结果回归计算得到,如图 1 所示。回归计算结果如表 2 所示,在任何情况下 m 都为 1/6,而当回填时 c 为 0.2750,无回填时 c 为 0.3795。早期的气爆实验主要采用 X52 和 X65 管线钢管,其韧性都低于 100J,回填时采用黏土和沙土,回填高度为 0.762m。而高钢级(X70、X80、X100 及 X120)、高韧性(大于 100J)管线钢问世之后进行的全尺寸爆破试验中,一方面钢级和韧性不同于 BTC 方法建立时的研究对象(低强度、低韧性管线钢管);另一方面回填深度和回填土类型也有所不同,例如回填深度增加到 1.2m。试验钢级和回填参数的不同会导致不同的回填常数,例如日本高强度管线钢管委员会通过进行 7 次直径为 1219mm,壁厚为 18.3mm 的 X70 管线钢管全尺寸气爆试验而确定的 m 和 c 分别为 0.67 和 0.393(回填情况)[2]。回填常数既反映了回填土对于裂纹扩展的约束作用,也反映了不同性能钢管对于约束的不同响应。在无约束(例如不回填)和低约束(例如低的回填深度)的情况下,裂纹将扩展得更快。因此 Battelle 早期计算得到的 m 和 c 已不再适用于高钢级管线钢管以及新的回填工艺。

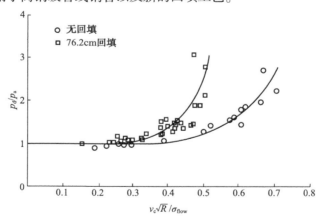

图 1 BTC 方法回填参数回归计算图

(3)BTC 方法中材料断裂阻力 R 是一个非常重要的参数,它反映了材料对于裂纹扩展的阻力。C_v 本来不是断裂力学中的断裂韧性参数,但是 Battelle 在早期的研究中,发现低钢级、低韧性管线钢(100J 以下)单位面积 C_v 与平面应力下的应变能释放率 G_c 呈 1:1 的线性关系[3]。因此,用 C_v 替代 G_c 将断裂阻力 R 表征为 C_v/A_c。然而众多的研究结果表明,随着现代管线钢韧性的增加,当超过 100J 后单位面积 C_v 与平面应力下的应变能释放率 G_c 不再表现为 1:1 的线性关系[3-4],如图 2 所示。因此如何正确地表征 BTC 公式中的断裂阻力 R 成为了当前研究的热点和难点问题。

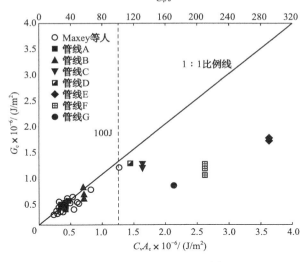

图 2　G_c 与 C_v/A_c 关系图[3]

1.1.2　减压波曲线

减压波的预测模型基于一维等熵流动，因此与管径无关，只与气质组分、温度和压力有关。减压波速度定义为：

$$W = C - U \tag{3}$$

式中：W 为减压波波速（m/s）；C 为局部声速（m/s）；U 为介质流出的平均速度（m/s）。

声速和局部平均速度可用压力和密度的增量来计算。在给定压力下的平均速度 U 是各个变化量的总和：

$$u = \sum_{\rho_1}^{\rho_2} (\Delta U)_s \quad \Delta u = c \frac{(\Delta \rho)_s}{\rho} \tag{4}$$

式中：Δu 为介质的速度差（m/s）；ρ 为介质密度（kg/m³）；$\Delta \rho$ 为密度差（kg/m³）。

密度和声速都需要由状态方程来确定，目前已发表的状态方程包括 AGA-8、BWRS、SRK、PR 和 GERG 等[5-10]，其中 BWRS 和 GERG 状态方程计算精度最高，常用于天然气减压波计算。在计算减压波时，主要有两点需要注意：①对于大口径钢管（如外径为 1422mm 的钢管），外流气体与管壁摩擦的影响可以忽略不计，因此对于大口径钢管无需考虑钢管内表面的粗糙度；②含有重烷烃的富气在减压过程中会出现由气相到液相的相变，造成气液两相共存的现象，从而产生减压波平台。减压波平台的出现使裂纹尖端的压力长时间保持在高位而无法降低，增加了止裂的难度。

1.2　基于 BTC 模型的主要修正方法

当钢管止裂韧性超过 100J 时，需要对 BTC 方法进行修正，主要的修正方法如下。

（1）Leis 修正，如式（5）和式（6）所示，其中式（5）适用于 X70 管线钢，式（6）适用于 X80 管线钢。

$$\text{Leis1} \, \text{CVN}_{\text{arrest}} = \text{CVN}_{\text{BTCM}} + 0.002 \text{CVN}_{\text{BTCM}}^{2.04} - 21.18 \tag{5}$$

$$\text{Leis-2} \, \text{CVN}_{\text{arrest}} = 0.985 \text{CVN}_{\text{BTCM}} + 0.003 \text{CVN}_{\text{BTCM}}^2 \tag{6}$$

式中：CVN_{BTCM} 为 BTC 方法计算的 CVN 能量；$\text{CVN}_{\text{arrest}}$ 为经修正后的止裂韧性值。

（2）Eiber 修正，如式（7）所示，适用于 X80 管线钢。

$$CVN_{arrest} = CVN_{BTCM} + 0.003CVN_{BTCM}^{2.04} - 21.18 \qquad (7)$$

(3) Wilkowski 修正，如式(8)所示，适用于 X80 管线钢。

$$CVN_{arrest} = 0.056 \times (0.1018 \times CVN_{BTCM} + 10.29)2.597 - 16.8 \qquad (8)$$

(4) 线性修正，如式(9)所示，K 为常数：

$$CVN_{arrest} = KCVN_{BTCM} \qquad (9)$$

2 ϕ1422 mm 的 X80 钢管道断裂控制方案

2.1 俄罗斯 ϕ1422mm 的 X80 钢管道断裂控制方案

为了建设总长度达 1106 km 的 Bovanenkovo-Ukhta 长输管线，从 2008 年 3 月至 2009 年 1 月，俄罗斯共进行了 10 次 X80 钢管道全尺寸气爆实验[11,12]。实验钢管包括壁厚为 23.0mm、27.7mm 和 33.4 mm 的 ϕ1420mm 的 X80 钢直缝埋弧焊管，实验中在起裂管两侧各排列 3 根等 CVN 能量测试钢管，根据 EPRG 标准要求在 3 根钢管内止裂。在第 4 次实验中，裂纹穿过所有 3 根测试钢管而无法止裂。这些钢管的断口特征如图 3 所示，表现为韧脆混合断口而不具备典型的 45°剪切断裂特征。这些钢管的 DWTT 形貌如图 4 所示，可见具有明显的断口分离特征。

图 3 俄罗斯扩展管的爆破试验断口特征　　图 4 俄罗斯扩展管的 DWTT 断口形貌

Bovanenkovo-Ukhta 管道的设计温度为 -20℃，BTC 修正计算后的止裂韧性为 200J，在此温度下进行的 CVN 试验表明裂纹扩展管和止裂管的 CVN 能量区别不大，而在 -40℃ 进行试验则会使裂纹扩展管 CVN 试样的断口分离现象明显增强，并表现为 CVN 能量的急剧下降。而止裂管即便在 -40℃ 进行试验，其 CVN 能量下降也很小。因此，为了能有效鉴别具有断口分离特征的扩展管，在 Bovanenkovo-Ukhta 管道的技术指标中，将 CVN 试验温度规定为 -40℃。

2.2 中俄东线外径为 1422mm 的 X80 钢管道断裂控制方案

中俄东线天然气管道输送的天然气组成为：C_1 的摩尔分数为 91.41%，C_2 的摩尔分数为 4.93%，C_3 的摩尔分数为 0.96%，C_4 的摩尔分数为 0.41%，C_5 的摩尔分数为 0.24%，N_2 的摩尔分数为 1.63%，CO_2 的摩尔分数为 0.06%，He 的摩尔分数为 0.29%，H_2 的摩尔分数为 0.07%。一类地区设计参数为：管道钢级为 X80，管径为 1422 mm，压力为 12MPa，管道壁厚为 21.4mm，设计系数为 0.72。

中俄东线的设计压力为 12MPa，但是只有在压气站的出气口处压力才会达到 12MPa，而在下一个压气站的进气口处压力会显著降低。中俄东线最低冻土层温度为 $-1.5℃$，在正常输送的情况下，压气站出气口的温度会显著高于 $-1.5℃$，而下一个压气站进气口处的温度也会在 $-1.5℃$ 以上。只有在最恶劣的情况下(管道埋于冻土层内及长时间停输)，管道内的气体温度才会降至地温，但此时管道内压力也会下降(约 1MPa)。综合考虑，在进行止裂韧性计算时选取 12MPa 和 0℃ 作为计算参数。

图 5 为中俄东线 BTC 计算结果，计算中采用 BWRS 状态方程进行减压波计算。可见中俄东线气质组分存在明显的减压波平台，止裂韧性 CVN 能量计算值为 167.97J，由于 BTC 计算值超过 100J。因此必须进行修正。

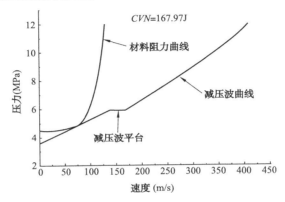

图 5　中俄东线 BTC 计算结果图

经不同方法修正后得到的止裂韧性结果为：BTC 预测值为 167.97 J；1.46 倍修正值为 245 J；Leis-2 修正值为 250 J；Eiber 修正值为 251 J；Wilkowski 修正值为 286 J。其中 Leis-2、Eiber 和 1.46 倍修正的结果基本一致。考虑到 1.46 倍修正可以较好地将全尺寸爆破试验数据库中的裂纹扩展点和止裂点分开，如图 6 所示[13]。进而最终将止裂韧性指标确定为 245 J，此指标为单根止裂韧性指标。

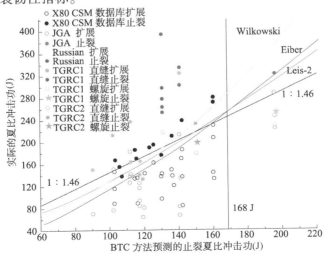

图 6　X80 钢管道全尺寸气体爆破试验数据库图

图7为单炉试制钢管的断口形貌,可见试样不存在严重的断口分离。最近在国内全尺寸爆破试验场开展的外径为1422mm的X80钢管道全尺寸爆破试验同样表明,国内生产的外径为1422mm的X80钢管爆破断口形貌为45°剪切断口,可以依靠自身韧性进行止裂。

图7 中俄东线单炉试制钢管DWTT断口形貌图

3 结论及建议

(1)Battelle建立的BTC方法不能直接应用于现代高强度高韧性管线钢的止裂韧性计算,其模型的改进有待进一步深入研究解决。

(2)采用BTC结合1.46倍修正的方法确定中俄东线外径为1422mm、X80钢管道的止裂韧性指标为245J。

参 考 文 献

[1] Mannucci G, Demofonti G. Control of ductile facture propagation in X80 gas linepipe[C]//Pipeline Technology Conference, 12-14October, 2009, Ostend, Belgium.

[2] Makino H, Takeuchi I, Higuchi R. Fracture propagation and arrest in high-pressure gas transmission pipeline by ultra high strength line pipes [C]//2008 7th International Pipeline Conference, Calgary, 2008. DOI: http://dx.doi.org/10.1115/IP C2008-64078.

[3] Kawaguchi S, Ohata M, Toyoda M, et al. Modified equation to predict leak/rupture criteria for axially through-wall notched X80 and X100lipepipes having a higher charpy energy[J]. Journal ofPressure Vessel Technology, ASME 2006, 128(4): 572-580. DOI: http://dx.doi.org/10.1115/1.2349570

[4] Maxoy WA. Fracture initiation, propogation and arrest[C]//Proceedings of 5th Syrrposium on Line Pipe Research. 1974, AGA Catalogue no L30174, J1-J31.

[5] Groves T K, Bishnoi P R, WallvridGe J M E. Decompression wave velocities in natural gas in pipelines[J]. The Canadian Journal of Chemical Engineering, 1978, 56 (6): 664-668. DOI: http://dx.doi.org/10.1002/cjce.5450560602/full

[6] Picard D J, Bishnoi PR. The importance of real-fluid behavior and non-isentropic effects in modeling decompression characteristics of pipeline fluids for application in ductile fracture propagation analysis[J]. The Canadian Journal of Chemical Engineering, 1988, 66 (1): 3-12. DOI: http://dx.doi.org/10.1002/cjce.5450660101.

[7] Botros K K, Geerligs J, Glover A, et al. Expansion tube for determination of the decompression wave speed for dense/rish gases at initial pressures of up to 22MPa[C]//2001 International Gas Research Conference, Amsterdam, Netherlands, 2001.

[8] Makino H, Kubo T, Shiwaku T, et al. Prediction for crack propagation and arrest of shear fracture in ultra high pressure natural gas pipelines[J]. ISIJ International, 2001, 41(4): 381-388.

[9] Eiber R J, Bubenik A T, Maxey W A. Gasdecom computer code for the calculation of gas decompression speed that is included in fracture control technology for natural gas pipelines[R]. Houston: American Gas Association Catalog, 1993.

[10] Kunz O, Wagner W. The gerg-2008 wide-range equation of state for natural gases and other mixtures: An expansion of gerg-2004[J]. Journal of Chemical &Engineering Data, 2012, 57(11): 3032-3091. DOI: http://dx.doi.org/10.1021/je300655b

[11] Pyshmintsev I Y, Lobanova T P, Arabey A B, et al. Crack arrestability and mechanical properties of 1420mm X80 grade pipes designed for 11.8MPa operation pressure[C]//Pipeline Technology Conference, 12-14October, 2009, Ostend, Belgium.

[12] Pyshmintsev I Y, Arabey A B, Gervasyev A M, et al. Effects of microstructure and texture on shear fracture in X80 linepipes designed for 11.8MPa gas pressure[C]//Pipeline Technology Conference, 12-14October, 2009, Ostend, Belgium.

[13] 李鹤, 王海涛, 黄呈帅, 等. 高钢级管线焊管全尺寸气体爆破试验研究[J]. 压力容器, 2013, 30(8): 21-26.

本论文原发表于《天然气工业》2016年第36卷第6期。

316L 衬里复合管道主要失效形式及其完整性检测技术研究

赵新伟　魏　斌　杨专钊　聂向晖　李发根　李为卫

(中国石油集团石油管工程技术研究院·石油管材及装备材料服役行为与结构安全国家重点实验室)

摘　要：316L 衬里复合钢管在国内油气集输管线上已大批量推广应用，在高含 CO_2/H_2S 和 Cl^- 油气集输管道腐蚀控制方面取得了良好的效果，但也发生过多起泄漏、开裂失效的事故。简要介绍国内外 316L 衬里复合钢管应用现状，分析 316L 衬里复合管道在服役过程中发生的主要失效形式。讨论了 316L 衬里复合管道在完整性检测方面面临的技术挑战，试验验证并分析了漏磁检测技术和爬行机器人视频检测技术的适用性，研究提出了基于风险的 316L 衬里复合管道检测方法。

关键词：316L 衬里复合管道；失效形式；内检测；基于风险的管道检测

当油气介质中 CO_2 和/或 H_2S 含量较高时，普通碳钢或低合金钢管难以满足内防腐要求。因此，选择技术可行、经济合理、安全可靠的油气集输管道材料是油气田地面工程领域的一项关键技术。双金属复合钢管兼顾了碳钢或低合金钢的高强度与不锈钢或耐蚀合金良好的耐蚀性，同时与纯不锈钢或耐蚀合金钢管相比更为经济，在高含 CO_2 和/或 H_2S 油气田的油气集输管道上已大量应用[1-5]。

双金属复合钢管的外层管(称为基管)，由碳钢或低合金钢焊管或无缝管构成，起到承压和支撑内层管的作用，用以保证管道的各项力学性能。内层材料为不锈钢、铁-镍基合金、镍基合金或其他耐蚀合金材料，若与基管之间为冶金结合，称为内覆层(Clad layer)；若与基管之间为机械结合，称为衬里层或内衬层(Lined layer)。内层材料的主要功能是提高管道耐腐蚀与抗冲刷性能，延长管道使用寿命。双金属复合钢管按基管与内层材料复合方式不同，分为机械复合和冶金复合，分别称为衬里耐蚀合金复合钢管和内覆耐蚀合金复合钢管[6]。20 世纪 90 年代初，国际上开始应用双金属复合管，我国从 21 世纪初开始双金属复合管的开发及应用，目前已是双金属复合管的制造大国，在油气集输管线上的应用已超过 1000km。在双金属复合管中，316L 不锈钢衬里复合钢管的用量最大，占到 80%以上，在高含 CO_2 和/或 H_2S 油气集输管道上取得良好的防腐效果。但是，316L 不锈钢衬里复合管道也发生过多起泄漏、开裂失效事故，未能完全做到免维护，管道检测、修复等管道完整性技术面临一些技术难题和挑战。

本文综述 316L 不锈钢衬里复合管道的主要失效形式，分析 316L 不锈钢衬里复合管道在

作者简介：赵新伟(1969)，男，教授级高级工程师，主要从事油气输送管和管道完整性技术研究工作。地址：陕西省西安市锦业二路 89 号中国石油集团石油管工程技术研究院(710077)，E-mail: zhaoxinwei001@cnpc.com.cn。

管道检测方面面临的技术挑战,并验证漏磁内检测技术、爬行机器人视频检测技术以及射线、超声常规无损检测技术在316L不锈钢衬里复合管道的适用性,研究提出基于风险的管道检测方法。

1 316L衬里复合管道主要失效形式

从20世纪90年代开始,国际上在油气输送领域应用双金属复合钢管,基管材料主要为X52和X65,耐腐蚀合金层主要为316L和镍基合金825,规格为$\phi114mm\sim\phi998mm$,主要用于含H_2S/CO_2和Cl^-苛刻腐蚀环境。在国内,2005年塔里木油田开始采用国产中小口径($\phi60mm\sim\phi168mm$)衬里复合钢管(20G/316L)用于高含CO_2和Cl^-油气集输,并良好的防腐效果,这是复合钢管在国内油气田的首次使用。2011年,国产316L内衬复合钢管开始应用于海洋油气集输管道,先后在多个海上油气田应用超过100km,材料为X65/316L,规格$\phi168mm\sim\phi219mm$。2013年开始,大口径316L衬里复合钢管在塔里木油田集气干线得到试用与推广,规格主要为$\phi508mm$和$\phi355.6mm$,材料X65/316L。截至目前,双金属复合钢管在国内油气田的使用里程已近2000km[7],在中国石油、中国石化和中国海油等油气公司都有应用,而且以316L衬里复合钢管应用最多,占到双金属复合管应用量的80%。

尽管316L衬里复合钢管应用于油气集输管线有良好的防腐效果,但316L衬里复合管道也发生了多起失效事故。从316L衬里复合管道失效案例统计分析结果来看,主要的失效形式包括环焊缝腐蚀刺漏、环焊缝开裂、衬里层塌陷和衬里层腐蚀,尤其是前三种失效形式占比较高。

1.1 环焊缝腐蚀刺漏

环焊缝腐蚀刺漏主要发生于采用"封焊+多层焊"焊接工艺的管道,腐蚀主要发生于焊接热影响区以及熔合线部位,少量发生于焊缝中心部位,腐蚀形貌如图1所示。端部若采用封焊工艺设计(图2),封焊过程中不锈钢焊材不断往316L衬里层上熔焊,衬里层会反复受热,一旦封焊电流使用过大,则会造成管端衬里层热影响区晶粒粗大[8];现场环焊采用药芯焊丝,背面未充氩气保护,保护效果不佳,会导致衬里层局部严重贫Cr(例如失效样品焊缝及热影响区Cr含量分别仅为10.96%和5.77%)。上述因素使复合钢管焊缝及热影响区的耐腐蚀能力严重降低,现场检验手段又难以发现,最终导致焊缝刺漏。

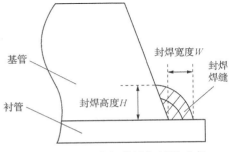

(a)焊缝中心刺漏　　(b)热影响区刺漏

图1　环焊缝腐蚀刺漏宏观形貌　　　图2　衬里复合钢管端部封焊示意

1.2 环焊缝开裂

环焊缝开裂失效的宏观形貌如图3所示。失效分析发现采用"封焊+多层焊"焊接工艺的焊接接头,在基管、衬里层和封焊交界位置易出现孔洞和裂纹缺陷(图4),构成了启裂源区，另外，焊缝存在高硬度(超过400HV10)的马氏体组织，韧性差，一旦有裂纹萌生就会很快扩展，最终导致环焊缝开裂失效。

图3　环焊缝开裂宏观形貌

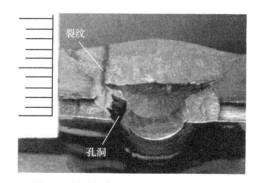
图4　环焊缝孔洞缺陷及裂纹扩展路径

1.3　衬里层塌陷

国内油气田316L衬里复合钢管发生了较多的衬里层塌陷失稳现象，有两种情况：一种情况是发生在复合钢管3PE外防腐过程中，如图5(a)所示；另一种情况是塌陷发生在管道服役过程，如图5(b)所示。造成防腐施工中衬层塌陷的原因是，复合工艺管控不到位，基管和衬管间隙中存在气体、水分，外防腐过程中钢管温度升高造成间隙高压，从而导致衬里层失稳。

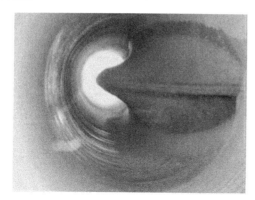
(a)外防腐过程衬层塌陷

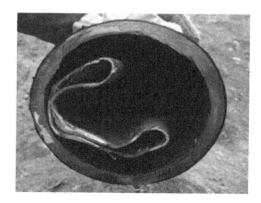
(b)运行中程衬层塌陷

图5　316L内衬层塌陷照片

运行中衬里层塌陷的机理和原因尚不明确，可能与以下因素有关：(1)衬里层径厚比设计与管材截面圆度等控制不到位，影响了衬里层结构稳定性和抗塌陷能力；(2)运行过程中，环焊缝刺漏导致高压输送介质进入基管与衬里层间隙，管线因停输检修快速泄压时，由于间隙中高压介质来不及释放，存在较高压差，从而导致衬里层塌陷失稳。

1.4　衬里层腐蚀

国内某油田使用的316L衬里复合钢管发生了衬里层腐蚀，典型的腐蚀形貌如图6所示。工况参数为：天然气中CO_2含量为0.59%~1.04%，CO_2分压最高为0.19MPa，Cl^-含量为125000mg/L，不含H_2S，运行过程中温度低于60℃。运行工况参数在国际上推荐的316L适用服役环境范围内[9]，理论上，316L衬里层不应该出现点腐蚀。经排查管线服役历史，投产初期运行温度超过设计温度,最高达到93℃,而且在酸化作业后有残酸进入管线,从而导致衬里层管体发生腐蚀。在实验室模拟投产初期工况进行30天的腐蚀试验,观察到316L材料表面有腐蚀坑存在,证实了上述结论。

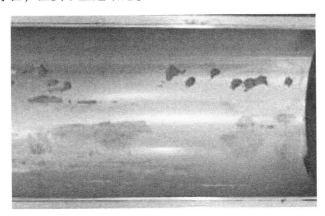

图6 316L衬里层腐蚀形貌

2 316L衬里复合管道检测技术研究

因为316L衬里复合管道发生多起失效事件,必须重视其完整性管理。损伤缺陷检测技术是316L衬里复合管道完整性管理急需解决的关键技术。对于衬里层塌陷,可以通过清管通球或者通径检测发现。对于316L衬里复合管道环焊缝腐蚀、开裂及衬里层腐蚀的检测,面临技术挑战和困难,在现有内检测技术中,有些技术不适用于316L衬里复合管道,有些技术应用于316L衬里复合管道有局限性。如目前应用较广泛也较成熟的管道漏磁检测器(MFL),因为316L奥氏体不锈钢为非铁磁性材料,使该技术不能检测内衬管缺陷;超声内检测器因为耦合问题,不能应用于输气管线;电磁超声(EMAT)内检测器在国外已经开发成功且商业化[10,11],但该技术在管道环向缺陷以及非轴向缺陷定量化检测方面存在局限性[11],在国内,目前该技术尚处在开发阶段;爬行机器人视频内检测技术,只能检测内表面缺陷,而且必须停输。另外,由于316L衬里复合管道塌陷情况较普遍,存在内检测器无法通过的可能。以下试验验证漏磁检测技术和爬行机器人视频内检测技术在316L复合管道上应用的可行性,结合现有技术条件,研究提出基于风险的开挖检测方法,以满足316L衬里复合管道现场检测需要。

2.1 内检测技术的适用性试验验证

2.1.1 漏磁检测技术(MFL)

采用φ508mm三轴高清漏磁检测器在长度约100m的试验管段上进行牵拉试验。试验管段由普通碳钢管和2段316L衬里复合管道组成,中间用法兰连接。316L衬里复合管道取自油田现场,规格为φ508mm,基管材料为L245,公称壁厚为15mm,衬里层为316L不锈钢,公称壁厚为2.5mm。2个316L衬里复合管段均包含环焊缝,1#管段的环焊缝有一处腐蚀刺漏,在法兰连接处有衬里层塌陷,2#管段上无原始缺陷。在基管内外表面预制8个人工缺陷,缺陷直径为30mm和50mm,深度为1.7~7.0mm。在316L衬里层上预制3个人工缺陷,为环向刻槽,轴向宽度1mm,环向长度40mm,深度为0.75~2.5mm。分别进行了0.5m/s,1m/s,2m/s,3m/s等4种速度下的牵拉试验,如图7所示速度为0.5m/s下牵拉试验给出的特征信号。图7中，1#~8#为基管上预制缺陷的特征信号，衬里层上预制的3个缺陷(编号为9#~11#)无特征信号显示。可见，MFL可检测到环焊缝位置、环焊缝腐蚀刺漏、衬里层塌陷、基管内外表面缺陷，但无法检测到316L不锈钢衬里层上的缺陷。实际工程中，由于油田集输管线服役环境苛刻，一旦衬里层腐蚀并发展到基管，很快就会腐蚀穿孔，因此，对于不锈钢衬里复合管道，及早检测发现衬里层缺陷比检测发现基管缺陷更有意义。另外，衬里层严重塌陷可能会导致MFL检测器无法通过。

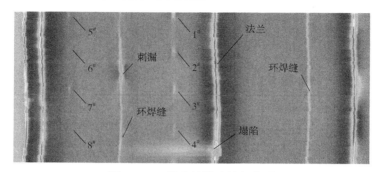

图7 MFL给出的缺陷特征信号

2.1.2 爬行机器人视频检测技术

采用爬行机器人视频检测技术对某油田集气干线的 $\phi 508mm \times (14.2+2.5)mm$ 316L衬里复合管道进行内检测，检测在停输条件下的2个管线开口处进行。检测发现，该集气干线复合管道的316L内衬层塌陷比较严重，而且多处环焊缝及其附近发生腐蚀。表1和表2列出爬行机器人视频检测的部分结果。图8为检测到的管道典型缺陷形貌。检测试验结果表明，爬行机器人视频检测技术可以有效地检测316L复合管道内部的塌陷及表面损伤情况，但管道内壁塌陷变形程度、内壁积液、机器人续航等因素对检测有一定的影响。由于该技术必须在管道停输条件下使用，限制了其应用。另外，衬里层塌陷严重时，检测器将无法通过。在管道建设期，可以采用该方法检查环缝表面质量和缺陷。最近，塔里木油田已将该技术应用于2205不锈钢管线施工质量检查中。

表1 爬行机器人视频检测发现的衬里层塌陷

缺陷类型	离开口处距离(m)	周向方位(钟点位置)	长度(m)
纵向塌陷	22.0	1:00	4.6
纵向塌陷	33.6	11:00	9.0
纵向塌陷	58.8	12:00	10.0

表2 爬行机器人视频检测发现的内表面腐蚀损伤

离开口处距离(m)	缺陷特征	离开口处距离(m)	缺陷特征
9.9	环焊缝表面腐蚀	60.8	衬里层腐蚀
22.0	环焊缝附近腐蚀	68.4	环焊缝附近疑似腐蚀穿孔
33.6	环焊缝附近腐蚀	81.0	环焊缝附近腐蚀
45.6	环焊缝附近衬里层损伤		

（a）衬里层塌陷　　　　　　（b）衬里层腐蚀缺陷　　　　　　（c）环焊缝腐蚀缺陷

图8　采用爬行机器人视频检测技术检测到的管道缺陷

2.2　基于风险的环焊缝开挖检测技术

316L衬里复合管道环焊缝腐蚀和环焊缝开裂，在不停输条件下，缺乏有效的内检测手段，根据管道检测技术发展现状，解决316L内衬复合管道检测在工程上可行的方法是基于风险的开挖检测，即在风险评估的基础上，对高风险区段的环焊缝进行开挖，然后采用射线（RT）、超声（UT）等常规无损检测手段进行环焊缝缺陷检测。

基于风险的管道环焊缝开挖检测的基础是管道风险评估。由于双金属复合管线应用时间还比较短，工程经验少，积累数据非常有限，不具备建立定量风险评估方法的基础，建立半定量的316L衬里复合管道风险评估方法（即评分法），在工程上更可行。在316L衬里复合管道失效调查的基础上，充分征询设计、制造、施工、运行管理等方面专家意见和经验，针对腐蚀刺漏和环焊缝开裂失效风险，从设计、制造、施工、运行和环境5个方面，分别识别出28个和17个风险因素，并给出了相应的打分权重和赋值方法，建立了半定量的风险评分法。该方法能够较全面地考虑管道设计、制造、施工和运行等全过程风险因素，打分权重和分值充分结合了管道设计、制造、施工及运行管理等方面的专家经验，工程适用性强，为开展基于风险的开挖检测奠定了基础。

以某油田KS2集气干线ϕ508mm的316L衬里复合管道的风险评估为例。该管线基管材料为L245，壁厚为14.2，20mm，316L衬里层厚度为2.5mm，管线长度为29.16km。在管道属性和服役环境调查基础上，划分了4个评价单元，在风险因素识别基础上，采用半定量风险打分方法对4个评价单元的环焊缝腐蚀刺漏和环焊缝开裂风险分别进行评估。作为算例，表3和表4列出评价单元1环焊缝腐蚀刺漏和环焊缝开裂风险因素识别及风险评分结果。表5列出KS2集气干线两种风险的综合评估结果。对环焊缝腐蚀刺漏和环焊缝开裂每一类风险，最优为10分，分值越低、风险越高。根据风险评估结果，管道风险由高到低的排序为单元1（11.61分）、单元4（12.35分）、单元2（12.95分）和单元3（13.55分），对应的历史环焊缝失效次数分别为8次，3次，1次和0次。风险评估结果和历史失效次数统计结果是一致的。

表3　KS2集气干线评价单元1环焊缝刺漏风险因素识别及评分结果

类别及权重	风险因素	二级风险因素	权重	调查结果	打分[①]	加权计分
设计因素（15%）	管端处理方式	—	60%	封焊	2	0.18
	衬层壁厚		40%	不符合设计标准要求	0	0

续表

类别及权重	风险因素	二级风险因素	权重	调查结果	打分[①]	加权计分
制造因素（25%）	封/堆焊工艺评定	—	35%	有工艺评定	10	0.88
	端部封/堆焊厚度	—	30%	有控制	10	0.75
	封/堆焊后PT检验	—	15%	未检验	0	0
	封/堆焊后RT检验	—	10%	检验合格	10	0.25
	是否内检测	—	10%	是	10	0.25
施工因素（35%）	组对情况	是否死口	7%	非死口	10	0.25
		工装情况	5%	有组对工装	10	0.25
	是否有管端修补	—	10%	无	10	0.25
	焊接工艺	—	40%	老工艺：R316LT1-5打底，ER309LMo过渡，E5015填充盖面	3	0.08
	是否有焊接工艺评定	—	9%	有焊评，结果合格	10	0.25
	施工环境条件	—	14%	冬季施工，有保护措施	5	0.13
	气保护措施	外表面气体保护	3%	有	10	0.25
		内表面气体保护	3%	有	10	0.25
	焊后检验情况	—	2%	有检验记录	10	0.25
	焊接返修情况	—	7%	焊接返修率在30%~50%	3	0.08
运行因素（20%）	输送介质（腐蚀性介质）	含水量	4%	无凝析水	10	0.08
		pH值	8%	在4~7之间	7	0.11
		CO_2含量	10%	CO_2分压<0.021MPa	10	0.20
		Cl^-含量	10%	小于10000mg/L	10	0.20
		H_2S含量	4%	H_2S含量≤20mg/m³	10	0.08
		介质流速	19%	介质流速≥3m/s	10	0.38
	压力波动和超压情况	—	5%	输送压力无显著波动、无超压现象	10	0.10
	历史类似案例30%	—	30%	曾经具有相似案例3起以上	0	0
	输送温度	—	5%	温度无波动或波动小于30℃	10	0.10
	线路类型	—	5%	干线	10	0.10
环境因素（5%）	管道走势（是否存在低注）	—	100%	存在低注1处	5	0.25
综合得分						5.95

[①] 最优为10分，最差0分。

表4 KS2集气干线评价单元1环焊缝刺漏风险因素识别及评分结果

类别及权重	风险因素	二级风险因素	权重	调查结果	打分[①]	加权计分
设计因素（15%）	管端处理方式	—	70%	封焊	2	0.21
	衬层壁厚	—	30%	不符合设计标准要求	0	0
制造因素（20%）	封/堆焊工艺评定	—	35%	有工艺评定	10	0.70
	端部封/堆焊厚度	—	30%	有控制	10	0.60
	封/堆焊后PT检验	—	15%	未检验	0	0
	封/堆焊后RT检验	—	20%	检验合格	10	0.40

续表

类别及权重	风险因素	二级风险因素	权重	调查结果	打分[①]	加权计分
施工因素（50%）	组对情况	是否死口	6%	非死口	10	0.30
		工装情况	4%	有组对工装	10	0.20
	焊接工艺	—	30%	老工艺：R316LT1-5打底，ER309LMo过渡，E5015填充、盖面	3	0.45
	施工环境条件	—	20%	冬季施工，有保护措施	5	0.50
	气保护措施	外表面气体保护	10%	有	10	0.50
		内表面气体保护	10%	有	10	0.50
	焊接检验情况	—	10%	有检验记录	10	0.50
	焊接返修情况	—	10%	焊接返修率在30%~50%	3	0.15
运行因素（10%）	压力波动和超压情况	—	40%	输送压力无显著波动、无超压现象	10	0.40
	历史类似案例30%	—	60%	曾经发生类似案例3起以上	0	0
环境因素（5%）	管道走势（是否存在低洼）	—	100%	存在低洼1处	5	0.25
综合得分						5.66

[①] 最优为10分，最差0分。

表5　φ508mm316L衬里复合管道风险评分结果

风险类型	评分			
	评价单元1(8.36km)	评价单元2(6.64km)	评价单元3(7km)	评价单元4(7.16km)
环焊缝腐蚀刺漏	5.95	6.40	6.70	6.10
环焊缝开裂	5.66	6.55	6.85	6.25
总分	11.61	12.95	13.55	12.35
历史失效次数(次)	8	1	0	3

基于风险评估结果，对该KS2集气干线的φ508mm 316L衬里复合管道中风险最高区段（评价单元1）的5道环焊缝进行开挖检测，验证了X射线检测和超声检测两种方法对316L衬里复合管道环焊缝检测的适用性。如图9所示φ508mm 316L衬里复合管道环焊缝部分RT检测结果，图中的环焊缝腐蚀缺陷和开裂缺陷均为射线Ⅳ级片。另外，在同一位置，采用UT检测，发现缺陷回波信号位于Ⅲ区，超过判废线。现场RT和UT检测试验结果表明，采用RT和UT方法可以检测316L衬里复合管道环焊缝腐蚀和裂纹缺陷，在开挖条件下，RT和UT检测是316L衬里复合管道环焊缝检测的有效手段。

（a）环焊缝腐蚀　　　　　　　　　（b）环焊缝开裂

图9　采用RT方法检测到的316L衬里复合管道环焊缝缺陷

3 结语

（1）316L衬里复合钢管在国内高含H_2S/CO_2和Cl^-油气集输管线上的应用取得了良好的防腐效果，但也发生了多次失效事件。主要失效形式包括环焊缝腐蚀刺漏、环焊缝开裂、衬里层塌陷和衬里层腐蚀等。

（2）损伤缺陷检测技术是316L衬里复合管道完整性管理亟待解决的关键技术，目前缺乏有效的内检测手段。在管道上应用较成熟的漏磁检测（MFL）技术可以检测316L衬里复合管道环焊缝刺漏、内衬塌陷、基管损伤等缺陷，但无法检测316L不锈钢衬里层表面缺陷。管道爬行机器人视频检测技术可检测衬里层塌陷、表面损伤以及焊缝腐蚀缺陷，但必须在停输条件下进行。

（3）基于现有技术条件，研究建立了基于风险的环焊缝开挖检测方法，可以满足316L衬里复合管道检测的需要。

参 考 文 献

[1] 魏斌，李鹤林，李发根．海底油气输送用双金属复合管研发现状与展望[J]．油气储运，2016，35（4）：343-355.

[2] Wei B，Bai Z Q，Yin C X，et al. Research and application of clad pipe for gathering pipelines in yaha gas condensate field：the proceedings of NACE 2010，2010[C]．The Proceedings of NACE 2010，2010.

[3] 罗世勇，贾旭，徐阳，等．机械复合钢管在海底管道中的应用[J]．管道技术与设备，2012（1）：32-34.

[4] Spence M A，Roscoe C V. Bimetal CRA lined pipe employed for north sea field development[J]．Oil & Gas Journal，1999，97（18）：80-88.

[5] 张捷，聂新宇，徐平，等．模拟海洋复杂载荷条件复合管道性能测试装备[J]．压力容器，2017，34（1）：1-6.

[6] 国家市场监督管理总局，中国国家标准化管理委员会．石油天然气工业用内覆或衬里耐腐蚀合金复合钢管：GB/T 37701—2019[S]．北京：中国标准出版社，2019.

[7] 李发根．高腐蚀性油气集输环境用双金属复合管[C]//压力容器先进技术——第九届全国压力容器学术会议论文集．合肥：合肥工业大学出版社，2017：131-135.

[8] 李发根，孟繁印，郭霖，等．双金属复合管焊接技术分析[J]．焊管，2014，37（6）：40-43.

[9] Craig B D，Smith L. Corrosion resistant alloys（CRAs）in the oil and gas industry-selection guidelines update，Nickel Institute Technical Series No. 10073（3rd Edition）[R]．Belgium：Nickel Institute，2011.

[10] Roy van Elteren，Ian Diggory，Jochen Spalink，et al. Pipeline integrity framework："Mind the gap!"[C]．15th Pipeline Technology Conference，Hannover Germany：Euro Institute for Information and Technology Transfer in Environmental Protection Gmbh，2020.

[11] Matt Romney，Dane Burden. Detection of non-axial stress corrosion cracking（SCC）using MFL technology[C]．The Proceedings of 15th Pipeline Technology Conference，Hannover Germany：Euro Institute for Information and Technology Transfer in Environmental Protection Gmbh，2020.

本论文原发表于《压力容器》2020年第37卷第11期。

高性能钻杆研发及应用进展

秦长毅[1]　蒋家华[2]　蒋存民[2]　张震宁[2]

(1. 中国石油集团石油管工程技术研究院；2. 江苏曙光集团股份有限公司)

摘　要：本文从钻杆材料的韧性、抗腐蚀性能、疲劳寿命的研究，钻杆特殊螺纹接头研发，钻杆几何尺寸改进等5个方面，分析了钻井生产需求和钻杆失效预防对高性能钻杆使用性能的需求，以及对其研发应用的推动，阐述了高性能钻杆的研究进展和应用现状，并提出了持续开展高性能钻杆研究的建议。

关键词：高性能钻杆；韧性；疲劳；特殊螺纹接头

钻杆是钻柱的主要组成部分，以达到在旋转钻井中传递扭矩，在钻井过程中作为钻井液的循环通道，并通过不断连接来加长钻柱进而达到不断加深井眼的目的。钻杆在钻井设备中占有十分重要的地位，其使用寿命对钻井速度和质量具有至关重要的影响。钻井过程是石油天然气勘探开发的关键环节，其技术和水平直接影响油气勘探开发的质量和效益，其费用占勘探开发总投资的50%以上，其中钻柱的质量和安全使用寿命起着至关重要的作用[1,2]。钻杆在钻井过程中服役条件苛刻，不仅承受拉压、弯曲、扭转、振动等多种载荷的复合作用，而且受到钻井液、地层水以及油气中的腐蚀性气体介质的腐蚀，在钻进过程中常常发生失效，从而严重影响钻井进度并增加钻井成本。随着深层油气资源的勘探开发，水平井、定向井、大位移井、深井超深井以及欠平衡和气体钻井技术不断推广应用。这些钻井技术的应用使钻井条件和井下情况出现了新的变化，对钻杆的质量和使用性能提出了更高的要求，也使钻杆面临着许多新的具有挑战性的问题。例如，复杂的地质条件和钻井深度的增加，使钻柱承受更大的拉伸、扭转和冲击等交变载荷，整体受力更为复杂，对钻杆的疲劳和腐蚀疲劳抗力提出了新的更高的要求；高含硫气田开发中地层流体富含腐蚀性很强的硫化氢和卤性盐水，对钻杆的抗腐蚀性能提出了更高的要求[3,4]。为了满足最新钻井技术要求和日益苛刻的钻井作业要求，钻杆生产企业不断地进行钻杆的理论研究、不断地提高钻杆的质量、不断地研究开发新产品[5]，越来越多具有高强度高韧性及特殊螺纹接头的高性能钻杆应用到钻井过程中。高性能钻杆在使用过程中也逐渐暴露出一些问题甚至发生了失效事故。国内外相关研究人员为此展开了一系列研究，旨在解决问题、预防失效，切实提高钻杆质量、提升使用性能、延长使用寿命，从而保证整个钻柱系统的结构完整性和密封完整性，确保钻井生产的正常运行，对于保证石油天然气钻井安全具有重要意义。

1　高性能钻杆材料研究

1.1　高强度钻杆的韧性

目前，API规范中的G105和S135高强度钻杆已广泛应用于石油天然气钻井过程中。经

作者简介：秦长毅，男，1962年生，教授级高级工程师，主要从事质量计量标准化管理工作。E-mail: qincy@cnpc.com.cn。

过近几十年来的不断研究和改进，钻井技术和钻杆性能相互促进，钻杆的产品质量已有了大幅提升。在严苛的使用条件下，钻杆依然会出现失效事故。统计分析发现，失效形式主要为刺孔和断裂[6-13]，其主要的失效机理为疲劳和腐蚀疲劳，属于低应力断裂[14]。钻杆刺穿或断裂的失效过程：钻井液腐蚀坑形成→疲劳裂纹萌生→裂纹扩展穿透壁厚→高压钻井液刺出→形成刺孔→断裂。从本质上讲，刺穿是早期疲劳或腐蚀疲劳失效，裂纹穿透钻杆壁厚以后，最后的表现形式是发生刺孔失效还是断裂失效取决于钻杆所承受的外力载荷和钻杆材料的性能[11,15]。在一定的工作应力下，裂纹能否扩展穿透壁厚和刺穿后裂纹是否稳定扩展以及稳定扩展裂纹尺寸的大小取决于材料的韧性[16]。高韧性材料有高的抵抗裂纹扩展的能力，在钻杆断裂前允许有更长的裂纹存在，因此提高材料的韧性不但可以防止钻具的脆断，更重要的是延长钻具的使用寿命[17]。

随着深井、超深井和盐层复杂地层井的开发，国内外钻杆生产厂家均致力于S135高韧性钻杆和超高强度钻杆的开发。新产品已应用到油气钻井中，并取得了增加钻井深度，缩短完钻周期、节约钻探成本的突破性进展。例如，NOV Grant Prideco公司的S135-T，Z-140，V-150和UD-165钻杆，提高的屈服强度提供了优越的抗扭和抗拉强度，使钻柱重量降低，钻柱设计简单。在国内，天管和海隆合作研发出165钢级的钻杆，宝钢和渤海装备也均生产了150钢级的钻杆。通过某V150钢级钻杆的断裂失效，分析其失效原因是韧性降低导致的脆性断裂[18]。可见超高强度钻杆能否大规模投入应用，关键还是在保证屈服强度的同时，能否具备符合复合生产需求的韧性。

钻杆的最低韧性要求不但要与其强度相配合，而且与其服役环境有关。为避免脆性断裂事故的发生，首先应使材料的韧脆转变温度低于钻井现场的地面气温。对于寒冷地区以及腐蚀环境，则需要更高的材料韧性。

1.1.1 安全韧性指标的确定

当钻杆内形成裂纹后，如果钻杆所承受的外载荷以及材料自身性能使疲劳裂纹扩展速度较慢，临界裂纹尺寸较大，高压钻井液的冲刷作用会将裂纹尖端钝化，降低裂纹尖端应力强度因子，从而控制裂纹的扩展速度。裂纹沿周向和径向扩展，穿透壁厚后形成刺孔，钻井液通过刺孔流出。随着刺孔的形成和扩大，钻井工作者可以通过地面泵压的明显下降判断钻杆已发生刺穿失效，及时采取措施，防止钻杆进一步发生失稳断裂。反之，当钻杆内形成裂纹后，如果钻杆所承受的外载荷以及材料自身性能使疲劳裂纹扩展速度较快，临界裂纹尺寸较小，裂纹直接演变为断裂失效[15]。因此，为保障疲劳和腐蚀疲劳断裂的安全性，应采用"先漏后破"准则。

当材料的韧性达到一定值时，钻杆的失效模式可由"断裂"转变为"刺穿"[16]。统计分析刺孔的尺寸在20~40mm之间集中分布。根据断裂力学原理，利用断裂韧性与夏比冲击韧性之间的关系[19]，可以计算出钻杆夏比冲击韧性的最低要求值。应力强度比系数的提出，可进一步推导出钻杆发生"先漏后破"失效模式的材料韧性指标的计算公式。统计分析失效钻杆的断口形貌得出疲劳裂纹稳定扩展长度为60~80mm，可以计算高强度钻杆的夏比冲击要求值[15]。

1.1.2 提高韧性的途径

目前，实现高强度钻杆具有高韧性的主要途径为控制化学成分和淬火—回火热处理工艺。有学者总结了我国钻杆用钢的成分及热处理工艺的研究情况，统计分析了近年高性能钻杆用钢的主流钢种及其力学性能，发现当前我国高强度钻杆用钢性能可以达到使用要求并具有较好的强度和韧性匹配[20]。高强度和超高强度钻杆用钢的冶炼中采用纯净钢生产技术，降低钢种的 P 和 S 含量，不仅可以明显提升材料的冲击韧性，还可以明显降低对硫化物应力腐蚀开裂的敏感性[5]。

关于钻杆用钢的热处理工艺和材料力学性能，研究 26CrMo4s/2 钻杆用钢在不同回火温度处理后的拉伸强度和冲击韧性，分别确定了得到较高冲击韧性的回火温度范围和拉伸强度不发生明显降低的温度范围，从而得到同时保证材料强度和韧性文献的精确回火温度控制范围[21]。对 S135 钢级钻杆进行了分级淬火及回火热处理，获得下贝氏体/马氏体型复相组织。研究表明，在保证强度的前提下，下贝氏体相的引入可以提高钻杆钢的韧性，并可以有效提高微裂纹萌生及扩展的阻力[22]。对 V150 钢级钻杆进行系统的回火温度优化试验，找出回火温度的拐点，在拐点温度以下，可以在保证强度的基础上实现冲击韧性的显著提高。统计分析第二相粒子析出形态和数量，优化回火温度，增强第二相强化并提高材料韧性[23]。研究析出相的形貌及结构尺寸发现其具有多重特征，表现出对回火温度的依赖性。通过对不同回火温度处理后材料微观组织的观察，分析析出相的组织和形貌，得出析出相形貌演变，进而得到使材料具有良好的强度/韧性匹配的回火温度[24]。

钻杆管体与接头通过摩擦焊接的方式连接成整体。为了保证钻杆的整体性能，焊缝的质量不容忽视。通过精准的回火温度测控和焊缝热处理宽度的控制，可以制造出焊缝区域性能参数均匀稳定的高韧性的 S135 钢级钻杆[25]。

1.2 钻杆的抗腐蚀性能

在钻井作业中，为适应各种钻井工艺的需要，使用了盐水、钾基聚合物等钻井液体系，并含有多种添加剂，在井下高温高压作用下具有强烈的腐蚀性[26]。在酸性气田开发中，高强度钻杆极易被 H_2S 腐蚀，发生钻杆的过早失效，从而增加了钻井的风险和不安全因素。高强度钻杆因其材料强度的提高，必然会导致材料硬度的提高，而材料硬度的提高，对腐蚀环境的敏感性增强，发生腐蚀和应力腐蚀的风险性越高[27]。

采取优化钢的化学成、严格控制淬火和回火的热处理技术，可以提高钻杆的抗硫腐蚀性能。抗硫钻杆钢的强/韧性匹配依赖于析出相—位错交互作用和析出相—基体变形协调两种机制的相互竞争[24]。有学者对两种材质的钻杆进行腐蚀性能对比试验，包括氧腐蚀试验、CO_2 腐蚀试验、硫化物应力腐蚀试验，系统分析了两种材质对抗不同腐蚀的性能[28]。该研究不仅为进一步提高钻杆材质的使用性能提供依据，还为高性能钻杆在不同用途的选材提供依据。

1.3 疲劳寿命研究

通过统计分析我国主要油田的钻杆失效事故，发现失效事故中的 80% 都与疲劳有关[19]。对钻杆疲劳寿命进行科学预测，避免钻杆超寿命服役，对于预防钻杆失效具有重要意义。

关于采用小试样试验研究钻杆材料的疲劳寿命研究，集中在裂纹扩展速率和机理的研究。通过对不同化学成分、组织状态及钢级的钻杆材料在不同应力集中程度下和典型环境介质中腐蚀疲劳行为进行研究表明[29]：随着强度的升高，钻杆材料的腐蚀疲劳敏感性增大，应力集中对腐蚀疲劳寿命的降低更严重；载荷频率降低，腐蚀疲劳裂纹扩展速度加快，应力集中对腐蚀疲劳寿命降低更明显；介质的温度升高和饱和 O_2 或 CO_2 都促进腐蚀疲劳过程；钻杆上先期形成的点蚀坑越深，疲劳寿命的降低越显著。采用单点弯曲单边缺口试样对 165 钢级钻杆用钢在 H_2S 酸性环境中的腐蚀疲劳裂纹扩展速率进行研究并指出[30]：当硫化氢分压较大时，疲劳裂纹的扩展速率并不随应力频率而变化。对拉扭载荷作用下的 S135 钻杆疲劳寿命曲线进行了试验检测,分析疲劳断口特征发现[31]:扭转疲劳裂纹形成于光滑试样表面,且疲劳源的大小随切应力幅值的增加而增加;在疲劳源与疲劳裂纹扩展处的微观形貌特征为典型纯滑移型断裂,在裂纹扩展处涟波状花样区域的大小随切应力幅值的增加而增大。对4种化学成分及材料性能均符合标准要求且材料的拉伸强度基本相同的钻杆进行腐蚀疲劳试验,发现钻杆材料的腐蚀疲劳寿命相差很大。研究表明[32]:腐蚀疲劳裂纹扩展机理是阳极溶解与氢致开裂共存,氢致开裂加快了疲劳裂纹扩展;成分偏析及夹杂物导致材料的阳极溶解、特别是氢致开裂速度加快,所以是腐蚀疲劳寿命减少的主要原因;钻杆材料的低倍组织酸洗检验可作为间接评定钻杆材料腐蚀疲劳性能的方法。

关于钻杆疲劳寿命预测的理论计算研究,集中在疲劳裂纹扩展公式或S-N曲线对疲劳寿命的分析计算,以及通过模型的改进使计算结果更好地指导钻杆在实际钻井工程中的应用。对钻柱进行应力状态分析和等效应力合成,在此基础上运用可靠性理论确定影响钻杆疲劳寿命的随机性参数。结合裂纹扩展速率Forman模型建立钻柱I、III复合型疲劳裂纹扩展速率的计算模型,可以计算相应可靠度下钻柱疲劳裂纹的循环寿命[33]。采用疲劳寿命对数正态分布模型和Basquin公式,可以得出中短寿命区完整的概率—应力—疲劳寿命(P-S-N)曲线,并利用$2m-1$外推法给出长寿命区疲劳寿命表达式及疲劳极限[34]。基于可靠性理论对钻杆疲劳裂纹萌生寿命预测进行研究,发现钻杆管体部位承受的弯曲及拉伸应力载荷均明显大于接头螺纹连接部位。通过不同应力幅条件下的疲劳寿命试验,发现钻杆疲劳裂纹萌生寿命分布符合正态分布,且随着应力水平的降低,疲劳裂纹萌生寿命分布概率密度函数的峰值逐渐降低,疲劳裂纹萌生寿命的离散性逐渐增强。可以计算不同可靠度下的疲劳裂纹萌生寿命对数值,得到寿命预测方程[35]。

2 高性能钻杆结构研究

高性能的钻杆不仅需要使用具有高强度高韧性耐腐蚀的材料,还需要具有合理的适于使用的结构。日益苛刻的钻井工况环境一方面对钻杆的使用性能提出了更高的要求,增大上扣扭矩、提高上扣速度、提高机械性能和水力性能、具有良好的气密封性能等。另一方面,对钻杆的使用寿命和使用安全也提出了更高的要求。钻井生产对钻杆性能和寿命的需求促进了钻杆结构设计的优化。

2.1 特殊螺纹接头的研发

API钻杆接头的抗扭强度普遍低于管体,在使用时,靠外螺纹根部的直角台肩面与内螺纹端面的直角台肩互相挤压,形成密封。当钻杆受到拉伸载荷作用时,密封面上的接触压力就会显著下降,从而失去密封作用[36]。为了满足高强度钻杆钻探深井、水平井或大斜度定向井的需要以及解决钻杆接头在使用中出现的问题,特殊螺纹钻杆接头应运而生。其主要特点是具有更小的外径,更大的内径,更高的抗扭强度和更好的密封性能,可以承受更为苛刻的复合载荷。特殊螺纹接头基本采用双密封台肩的设计形式,高抗扭接头的扭转强度相比API NC螺纹接头增加40%左右,超高抗扭接头可增加70%左右;较小的外径和较大的内径接头可用于小井眼钻井并提高液压;更高的扭转强度,可施加更大的扭矩操作;磨损留量增加。

考虑接头的使用性能,特殊螺纹接头在研发设计中还考虑了接头外径、内径、镗孔段长度、应力分散槽、螺纹锥度等的改进。例如,钻杆接头的标准锥度为1:6。NOV Grant Predeco公司开发了和标准油套管接头相同的1:16锥度,明显增加了辅助内台肩的接触面积,大幅提高了接头的抗扭能力。有研发人员[37]改变了外螺纹的锥度,使得螺纹应力分布更加均匀,并降低了主台肩过渡圆角处的应力集中。

评价螺纹接头的抗扭性能、密封性能和承受复合载荷的能力,主要有两种方法：实物试验和有限元分析。通过实物上/卸扣试验,可以检验接头的抗粘扣性能和抗扭能力。通过实物复合载荷试验,可以检验接头在轴向载荷、内/外压力、弯曲等多种载荷复合应力作用下的密封性能,以及温度环境对接头密封性和结构完整性的影响。通过实物弯曲疲劳试验,可以检验钻杆管体和接头的抗疲劳性能。实物试验可以检验产品的使用性能是否满足设计要求,及时发现设计和加工中存在的问题。钻杆接头在使用中的应力分布状态直接决定了钻杆的连接强度和密封性能等[38]。因此,对钻杆接头的应力分布规律进行分析,是提高钻杆使用性能的关键,有限元分析方法被广泛运用。采用有限元分析方法,对比某双台肩高抗扭接头与API接头,分析发现双台肩高抗扭接头的抗扭强度为API接头的1.5倍,前者在轴向拉力和扭矩的共同作用下,等效应力明显好于后者[39]。建立力学模型分析接头在不同紧扣圈数及拉伸载荷作用下的受力情况表明,合理的紧扣圈数可以显著提高接头的密封性能,能有效降低拉伸载荷作用下螺纹牙轴向载荷增长幅值,延长接头的使用寿命[40]。采用三维有限元模型分析双台肩钻杆接头在上扣扭矩、轴向拉力和弯矩作用下的受力特性,可以得出螺纹接头在不同作用下承受最大载荷的部位[41]。有学者[42]针对螺纹粘扣时接触压力过大引发的塑性变形的问题,分析了钻杆接头接触压力分布规律和Mises应力分布规律,发现外螺纹大端第一个螺纹牙是最易发生粘扣的部位,应在螺纹结构优化中重点考虑。有学者[43]通过三维有限元模型,研究大轴向载荷作用下钻杆接头的密封机理发现：合理的接触压力才能保证接头的密封性能；当轴向载荷超过一定值后,主、副台肩失去密封性能,会增加接头发生应力腐蚀的风险。有限元分析方法基侧重应力分布和理论机理研究,可与实物实验的相互验证。两种方法共同使用,对钻杆接头的优化设计起到了积极的促进作用。

2.2 钻杆规格和几何尺寸的改进

钻杆的腐蚀疲劳和硫化物应力腐蚀失效,除材料的抗腐蚀性能影响外,集中出现在应力集中处。钻杆的应力集中主要位于螺纹接头的台阶圆角和螺纹牙底,以及管体的内加厚过渡处。中国石油管材研究院对大量失效钻杆分析研究,发现随着内加厚过渡区长度的增加和过渡圆角半径的增大,内加厚过渡区消失点的应力逐渐减少,应力集中逐渐降低。通过有限元分析和实物试验验证,提出了改进内加厚过渡区结构和尺寸,即内加厚过渡区最小长度为100mm,过渡圆角半径最小为300mm,该研究成果被API采纳。现场数据证明,钻杆的内加厚过渡区增长,可有效提升钻杆寿命[44]。近年来,又发现内加厚过渡区长度大于100mm的钻杆在深井和水平井钻探中失效事故频发,将过渡区长度增加到140mm以上后,钻杆的疲劳寿命大幅提高[5]。

通过分析我国西部油田多年钻杆大量疲劳刺漏的原因,研究人员[45]发现除了增加内加厚过渡长度之外,增加壁厚和延长外加厚长度均可提高钻杆的疲劳强度。研究还发现,127mm钻杆不适合深井、超深井的大拉力、高扭矩、高转速、大狗腿、强腐蚀等复杂工况,并提出了用于深井的139mm和非API的149mm大直径钻柱设计思想。

3 结语

先进钻井技术的推广应用和安全高效钻井作业的生产需求,推动了钻杆制造技术的进步和高性能钻杆的研发。国内外研究人员在高强度高韧性钻杆材料的研制、特殊螺纹接头开发、钻杆疲劳断裂机理以及寿命预测技术等方面均取得了重要的研究成果，并取得了显著的应用效果。"三超"、非常规以及新型油气资源的勘探开发，对钻杆服役性能提出了更高要求，高性能钻杆研发还需要持续开展。为了延长钻杆的使用寿命，有效控制和减少钻杆失效事故的发生，需要进一步完善超高钢级钻杆材料的韧性、抗腐蚀性能和疲劳寿命预测研究；开发性能更优越的特殊螺纹接头；提升钻杆制造技术，避免因制造过程的问题引起钻杆的早期失效；提升高性能钻杆的检验和评价技术，充分发挥实物试验和三维有限元分析的作用；加强高性能钻杆的选用和管理。

参 考 文 献

[1] 沈忠厚,黄洪春,高德利.世界钻井技术新进展及发展趋势分析[J].中国石油大学学报(自然科学版),2009,33(4):64-70.

[2] 张毅,赵仁存,张汝忻.国内外高强度钻杆的技术质量评述[J].钢管,2000,29(5):1-8.

[3] 李鹤林,韩礼红,张文利.高性能油井管的需求与发展[J].钢管,2009,38(1):1-9.

[4] 刘永刚,陈绍安,李齐富,等.复杂深井钻具失效研究[J].石油矿场机械,2010,39(9):13-16.

[5] 赵鹏,黄子阳,朱世忠.宝钢钻杆技术的最新发展[J].钢管,2009,38(6):9-14.

[6] 朱丽娟,刘永刚,李方坡,等.G105钢制钻杆腐蚀失效的原因[J].腐蚀与防护,2016,37(9):775-780.

[7] 李方坡,刘永刚,林凯,等.G105油井失效分析[J].金属热处理,2009,34(10):94-97.

[8] 韩勇,冯耀荣,李鹤林.G105钻杆管体刺穿原因分析[J].石油机械,1990,18(1):37-43.

[9] 陈长青,刘聪,钱强.S135钢级ϕ139.7mm钻杆管体断裂失效分析[J].焊管,2016,39(10):42-47.

[10] 寇菊荣,杜志杰,张国正,等.G105钻杆刺穿原因分析[J].热加工工艺,2016,45(8):256-258,261.

[11] 张春婉,张国正,董会,等.S135钻杆本体刺穿失效分析[J].石油矿场机械,2009,38(12):65-75.

[12] 龚丹梅,余世杰,袁鹏斌,等.水平井中S135钻杆断裂原因分析[J].理化检验—物理分册,2015,51(1):69-72.

[13] 刘永刚,苏建文,林凯,等.一例S135钻杆本体断裂原因分析[J].石油矿场机械,2007,36(5):58-61.

[14] 王新虎,薛继军,谢巨良,等.钻杆接头抗扭强度及材料韧性指标研究[J].石油矿场机械,2006,35(增刊):1-4.

[15] 李方坡,韩礼红,刘永刚,等.高钢级钻杆韧性指标的研究[J].中国石油大学学报(自然科学版),2011,35(5):130-133.

[16] 冯耀荣,马宝钿,金志浩,等.钻柱构件失效模式与安全韧性判据的研究[J].西安交通大学学报,1998,32(4):54-58.

[17] 孔寒冰,王新虎.提高深井和超深井钻柱使用寿命的途径[J].石油机械,1998,26(11):37-39.

[18] 龚丹梅,余世杰,袁鹏斌,等.V150高强度钻杆断裂失效分析[J].金属热处理,2015,40(10):205-210.

[19] 李鹤林,李平全,冯耀荣.石油钻柱失效分析及预防[M].北京:石油工业出版社,1999:63,68.

[20] 王新虎,张冠军,李方坡,等.钻杆钢的成分、热处理工艺及其力学性能[J].石油管材与仪器,2015,1(2):33-36,40.

[21] 钱强,曹贵贞,刘聪,等.回火温度对26CrMo4s/2钢钻杆用管性能的影响[J].钢管,2012,41(5):36-39.

[22] 韩礼红,姜新越,冯耀荣,等.分级热处理对钻杆材料力学行为的影响[J].材料热处理学报,2011,

[23] 姜新越,胡峰,庄大明,等.回火温度对V150钻杆钢的强韧性匹配的影响[J].钢管,2012,41(5):22-27.

[24] 王航,韩礼红,胡锋,等.回火温度对抗硫钻杆钢析出相形貌及力学性能的影响[J].材料热处理学报,2012,33(3):88-93.

[25] 朱世忠.石油钻杆的摩擦焊接和焊缝热处理工艺研究[J].宝钢技术,2006,(1):52-55.

[26] 侯彬,周永璋,魏无际.钻具的腐蚀与防护[J].钻井液与完井液,2003,20(2):48-50,53.

[27] 赵金,陈绍安,刘永刚,等.高性能钻杆研究进展[J].石油矿场机械,2011,40(5):96-99.

[28] 赵鹏,于杰,郭金宝.钻杆腐蚀性能对比分析[J].钢管,2015,44(3):16-19.

[29] 李鹤林,宋余九,赵文轸.钻杆用钢腐蚀疲劳试验研究[J].石油管材与仪器,2016,2(2):22-27,34.

[30] Ramgopal T, Gui F, HaFK J. Corrosion fatigue performance of high strength drill pipe in sure environments [C]. 2011 NACE. Dallas, 2011: 11108.

[31] 刘古峰,王荣,雒设计.S135钻杆钢扭转疲劳寿命及断裂特征[J].石油机械,2011,39(4):4-6.

[32] 王新虎,邝献任,吕拴录,等.材料性能对钻杆腐蚀疲劳寿命影响的试验研究[J].石油学报,2009,30(2):312-316.

[33] 李文飞,管志川,赵洪山.基于可靠性理论的钻柱疲劳寿命预测[J].石油钻采工艺,2008,30(1):12-14,18.

[34] 林元华,李光辉,胡强,等.钻杆应力—疲劳寿命曲线试验研究[J].石油钻探技术,2015,43(4):124-128.

[35] 李方坡,王勇.钻杆疲劳寿命预测技术的研究现状与展望[J].材料导报A,2015,29(6):88-91.

[36] 任辉,高连新,鲁喜宁.适用于弯曲井段的高效密封钻杆接头研究[J].石油机械,2010,38(11):20-22,27.

[37] 焦文鸿,高连新,鲁喜宁.一种新型钻杆螺纹设计及应力分析[J].石油机械,2012,40(6):17-21.

[38] 袁鹏斌,陈锋,王秀梅.应力释放槽对钻杆接头力学性能的影响[J].石油钻探技术,2010,38(3):32-35.

[39] 张毅,王治国,刘甫清.钻杆接头双台肩抗扭应力分析[J].钢管,2003,32(10):7-10.

[40] 庄泳,鲁喜宁,高连新,等.气密钻杆接头受力分析与紧扣圈数优选[J].石油机械,2015,43(7):30-35.

[41] 狄勤丰,陈锋,王文昌,等.双台肩钻杆接头三维力学分析[J].石油学报,2012,33(5):871-877.

[42] 陈锋,狄勤丰,王文昌,等.双台肩钻杆丝扣粘扣失效的力学机制探究[J].应用力学学报,2012,29(6):730-734.

[43] 陈锋,狄勤丰,王文昌,等.基于密封性能要求的超深井钻柱极限提升力的确定[J].石油学报,2013,34(6):1176-1180.

[44] 刘永刚,崔顺贤,路彩红,等.失效分析促进了钻杆结构设计优化[J].石油矿场机械,2010,39(11):87-90.

[45] 王新虎,卢强,苏建文,等.深井钻杆疲劳刺漏原因与钻柱设计[J].石油管材与仪器,2016,2(2):43-46,52.

本论文原发表于《石油管材与仪器》2017年第3卷第1期。

复合材料增强管线钢管研究进展

马秋荣

(中国石油集团石油管工程技术研究院·
石油管材及装备材料服役行为与结构安全国家重点实验室)

摘　要：复合材料增强管线钢管融合了金属管线钢管良好的强韧性和非金属增强材料的特性，可以有效地改善管线承压能力水平和抗轴向裂纹扩展的能力，有效降低油气管输成本，又可提高管线的安全系数。国内通过复合材料增强管线钢管关键技术指标的确定，复合材料增强管线钢管的设计技术和合材料增强管线钢管的制造与评价技术的研究，试制出 X65 和 X80 复合材料增强管线钢管，承压能力分别可以达到 X80 和 X120 的钢管的承压水平。

关键词：复合材料增强；管线钢管；承压能力

随着国民经济和能源战略的发展，天然气的需求与日俱增。截至 2018 年底，我国长输管道为 $16.64×10^4$ km，居世界第三位。其中，境外油气管道 $2.13×10^4$ km，包括中俄、中亚、中哈、中缅等十条管线，输油能力 $7200×10^4$ t/a，输气能力 $1350×10^8 m^3/a$。为确保天然气供应量满足未来国民经济发展的要求，高压大流量长距离输送成为我国天然气输送管道技术发展的必然趋势。

西气东输二线以及三线采用的 $\phi1219$ mm、12MPa、X80 方案，最大设计输气量达到 $300×10^8 m^3/a$。中俄东线 $\phi1422$ mm、12MPa、X80 方案，设计最大输量为 $380×10^8 m^3/a$。为了满足未来对接西伯利亚力量 2 项目对应的 $500×10^8 m^3/a$ 的大输量天然气管线建设需求，还需要进一步研究输量在 $500×10^8 m^3/a$ 及以上超大输量天然气管道技术方案，对于超大输量目前国际上的主要方案包括：采用超高钢级管线钢（如 X100、X120 等），采用 0.8 等较高设计系数，或采用更大口径钢管等[1]。这些方案的主要难度在于管线钢新产品的材料开发应用和如何确保管道服役安全。

近年来，复合材料的研究应用取得了飞速的发展，复合材料增强管线钢管技术逐渐成为一种超大输量天然气管道工程的解决方案[2]。该技术基本原理是在管线钢管（如 X80 管线钢管）基体外表面缠绕复合材料增强层。其中管线钢管提供全部的纵向强度和环刚度，复合材料增强层除了提供部分环向强度（理想状态下承担 50%）之外，可同时提供外防腐和止裂功能。复合材料增强管线钢管融合了金属管线钢管良好的强韧性和非金属材料优异的耐蚀性，且复合材料增强层和管线钢管的壁厚比例可根据不同使用工况条件进行自由设计（图 1）。

1　国外复合材料增强管线钢管技术现状

20 世纪 80 年代末，复合材料增强技术开始应用于金属管线钢管领域，最初是为了修复钢管。经过三十多年的发展，这种修复技术也成为管线缺陷修复的常用方法，而 ASME-PCC-2 以及 ISO24817 等标准的陆续发布，也为这种方法的广泛推广提供了指导依据。随着

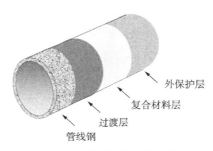

图1 复合材料增强管线钢管结构示意图

复合材料在管道缺陷修复过程中应用的日益普遍，TransCanada、NCF等公司开始对玻璃纤维等复合材料的结构设计、极限承压能力、止裂能力、抗机械损伤能力、制造工艺和长期服役性能等进行全面研究，并在这些研究基础上开发出了复合材料增强管线钢管（Composite Reinforced LinePipe，CRLP）[3-5]。这种钢管在油气管线的应用最早始于20世纪90年代初期，1991年，美国安然公司第一次在天然气管道建设中试验性铺设了复合材料增强钢管。随后TransCanada分别于1998年，2001年以及2002年在管线建设中铺设了复合材料增强钢管试验段，并已成功开发出承压能力从8.275MPa至24.800MPa不等的复合材料增强钢管产品。

国外三十多年来的研究以及应用情况表明，复合材料增强管线钢管可以安全有效提高管线的油气输送压力，同时由于复合材料隔离了外部环境对钢管材料腐蚀，改善了管线运行时抵抗轴向裂纹扩展的能力，因此既可有效降低油气管输成本，又可提高管线的安全系数。加拿大在2011年对CSA Z662《油气管道系统》标准的修订中，专门添加了第17章内容，对复合材料增强管道的设计、材料和制造、安装、连接、压力测试和操作维护进行了规定。标志着复合材料增强管线钢管在加拿大正式纳入管道输送系统标准体系。

2 国内复合材料增强管线钢管研究需要解决的关键技术

复合材料增强管线钢管具有承压能力高、止裂性好、耐腐蚀、价格低等优点，但在国内的研究才刚刚起步，需要解决管道设计、制造及评价过程中一系列问题，最终成功制得复合材料增强管线钢管样管，填补国内研究应用的空白，需要解决的关键技术主要如下。

（1）复合材料增强管线钢管关键技术指标的确定：确定复合材料增强管线钢管关键技术指标将为研发该管材提供明确方向。钢管等金属材质和复合材料增强层等非金属材质的强度、韧性、塑性、导热系数等关键技术参数差异较大。只有明确适用于特定管线钢管的复合材料增强层技术指标，才能最大可能地发挥各自优势，做到性能互补。在此基础上，明确复合之后的增强管线钢管各结构层（尤其是基体钢管和复合材料层）承载变形时的均衡协调能力、止裂能力等关键技术指标，可最终确保复合材料增强管线钢管各结构层的性能优势充分体现。

（2）复合材料增强管线钢管的设计技术：基于复合材料增强管线钢管关键技术指标，对于特定的管线钢管（X80/X90）应首先进行复合材料增强层的结构设计，包括增强纤维类型、增强纤维铺设结构、复合材料基体类型、复合材料制备工艺等，确保复合材料增强层性能指标与管线钢管匹配。同时开展过渡界面层和外防护层材料选型及黏接性能设计，确保管道应力均匀传递到复合材料增强层[6]。对于不同材质的管线钢管，应针对不同研究目标，综合考虑管道设计因素确定复合材料的材质、尺寸和结构，使钢管与复合材料层间的承载能力和载荷配比、变形响应、强度增加幅度等满足设计指标要求[7]。另外,应重点研究管线钢管连接处(焊缝)的修复补强工艺,在确保复合材料增强层连续性和完整性的基础上,保证焊缝连接部位的性能指标满足设计要求。

(3) 复合材料增强管线钢管的制造与评价技术:管线钢管在制备过渡界面层之前需进行表面处理,以提高界面层与管线钢管外表面的结合性能。过渡界面层选材完成之后如何连续地制作在管线钢管外表面也成为提高界面结合性能和传载能力的关键因素。其次,需研发适用于大口径管线钢管复合材料增强层现场缠绕设备,并系统研究复合材料结构层的缠绕方式、缠绕预紧力、缠绕角度、固化或加热工艺等,为管线钢管外表面复合材料结构层的制备提供技术支撑。制备的外保护层是防止管线外腐蚀的保障,同时应能避免复合材料增强层受到外力冲击和磨损,还应防止吸潮,保证其具有良好的长期服役性能。在复合材料增强层制备完成并充分固化后,通常采用施加内压的方式使管线钢管产生塑性变形,进而与复合材料增强层紧密贴合,其目的是使管线钢管产生预紧力[8],避免出现两张皮现象。复合材料增强管线钢管的强度、塑性、韧性等力学性能评价、焊缝连接处性能保障、变形响应效果分析、止裂能力评价、失效模式预测分析等成为复合材料增强管线钢管结构优化和性能改进的前提。复合材料增强管线钢管产品如何开展系统的测试与评价也是工程应用的关键技术问题之一。

3 国内复合材料增强管线钢管研究进展

国内从2015年开始,进行了高钢级管线钢管的复合增强实验研究,首先确定了复合材料增强管线钢管结构形式,优选了结构层材料。通过树脂、纤维和过渡层材料的力学性能、理化性能和工艺性试验和对比,优选了环氧树脂E51、聚氨酯/KH550和玻璃纤维158B作为CRLP各结构层主要材料;完成了复合材料增强管线钢管结构设计。通过有限元模型和宏观力学计算相结合的方法,开展了CRLP的主要承载模式和力学分析研究,确定了CRLP的主要结构形式和设计方案,明确了表征CRLP承压能力的关键技术指标。形成了复合材料增强管线钢管评价技术。测试了CRLP强度和复合材料性能,确定了CRLP样品管基本力学性能;分析了预应力处理和CRLP性能的关系,并结合试验结果和模型计算,对CSA Z662中CRLP设计压力计算方法进行了优化。在此研究基础上,试制出X65和X80复合材料增强管线钢管,承压能力分别可以达到X80和X120的钢管的承压水平。

4 复合材料增强管线钢管(CRLP)样管承压能力试验

钢和复合材料的应力/应变曲线如图2所示,管线钢管随应力增加会先出现屈服,屈服后钢管发生变形,随后应力继续增加直至最后断裂;复合材料的断裂伸长率在1.5%~2.0%之间,在整个拉伸阶段应力与应变为线性关系,无屈服现象。

对制成的两种规格的CRLP样管进行了水压爆破试验。通过有限元计算及试验分析得出预应力不影响CRLP的爆破压力,失效时的应变为复合材料的断裂伸长率,此时钢层和复合材料层的应力都达到了最大值,因此CRLP的爆破压力计算公式为

$$p_p = \frac{2}{D}(\sigma_b t + T_h W) \tag{1}$$

式中:p_p为复合管道的爆破压力,MPa;D为钢管外径,mm;σ_b为钢管的拉伸强度,MPa;t为钢管设计壁厚,mm;T_h为复合材料环向拉伸强度,MPa;W为纤维增强复合材料

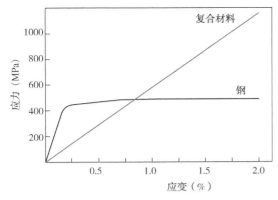

图2 复合材料和钢的应力/应变曲线

层的设计壁厚，mm。

试验结果表明，外径508mm、壁厚9.5mm X65钢级钢管用厚度2.6mm复合材料增强后爆破压力已超过同规格X80钢级钢管（表1、图3）。与同规格管线钢管相比，外径508mm的钢管屈服压力提高了25%以上，爆破压力提高了40%以上。复合增强钢管实际爆破压力与计算爆破压力符合性很好。

由表2实验结果可知，外径1219mm、壁厚18.4mm X80钢级钢管用厚度为8.9mm增强层增强后爆破压力达37MPa，超过了同规格X120钢级钢管的承压能力（图4）。

表1 外径508mm、壁厚9.5mm X65钢级钢管增强后爆破压力

样管编号	增强层厚度（mm）	裸管屈服压力（MPa）	预应力（MPa）	屈服压力（MPa）	爆破压力（MPa）	计算爆破压力（MPa）
P-508-1	2.6	19.2（同规格X80的屈服压力为24.2）	—	—	30.4	30.2
P-508-2	3.58		—	—	34.5	34.1
P-508-3	4.15		25	25	37.1	36.3
P-508-4	7.06		33	33	43.8	47.8
P-508-5	7.3		37	37	44.7	48.8

表2 外径1219mm、壁厚18.4mm X80钢级钢管增强后爆破压力（增强层厚度8.9mm）

管道类型	屈服强度（MPa）	屈服压力（MPa）	拉伸强度（MPa）	爆破压力（MPa）
X80管线钢管	555~705	16.8~21.3	625~825	18.9~24.9
X100管线钢管	690~840	20.8~25.3	760~990	22.9~29.9
X120管线钢管	840~990	25.3~29.8	910~1110	27.4~33.5
P-1219CRLP	—	25	—	37

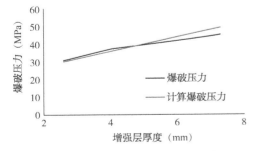

图3 X65钢级复合增强钢管计算爆破压力与实际爆破压力对比

图4 外径1219mm，X80CRLP复合增强钢管爆破样品

5 结论

研究确定了CRLP的结构形式，完成了复合材料增强管线钢管结构设计，优选了结构层

材料，形成了复合材料增强管线钢管制备工艺，分别制得了外径508mm和1219mm的CRLP样管。制得508mm和1219mm两种规格CRLP样管，外径1219mm、壁厚18.4mmX80钢级钢管用厚度为8.9mm增强层增强后爆破压力达37MPa，超过了同规格X120钢级钢管的承压能力。为未来超大输量管线建设提供了一个新的管材选用方案。

参 考 文 献

[1] 李鹤林，吉玲康，田伟. 高钢级钢管和高压输送：我国油气输送管道的重大技术进步[J]. 中国工程科学，2010，12(5)：84-90.

[2] Laney P. Use of composite pipe materials in the transportation of natural gas[J]. Idaho International Engineering and Environmental Laboratory, Bechtel BWXT Idaho, 2002：1-22.

[3] Zimmerman T., Stephen G., Glover A. Composite Reinforced Line Pipe (CRLP) for Onshore Gas Pipelines[C]. 2002 4th International Pipeline Conference, 2002, American Society of Mechanical Engineers：467-473.

[4] Wolodko J. Simplified methods for predicting the stress-strain response of hoop wound composite reinforced steel pipe[C]. 2006 International Pipeline Conference, 2006, American Society of Mechanical Engineers：519-527.

[5] Salama M. M. Qualification of Fiber Wrapped Steel Pipe for High Pressure Arctic Pipeline[C]. OTC Arctic Technology Conference, 2011, Offshore Technology Conference：1-11.

[6] Altunisik A. Dynamic response of masonry minarets strengthened with Fiber Reinforced Polymer (FRP) composites[J]. Natural Hazards and Earth System Science, 2011, 11(7)：2011-2019.

[7] 黄再满，蒋鞠慧，薛忠民，等. 复合材料天然气气瓶预紧压力的研究[J]. 玻璃钢/复合材料，2001，5：29-32.

[8] 赵立晨. 金属内衬纤维增强复合材料筒体设计[J]. 宇航材料工艺，2007，37(2)：45-47.

X90 管线钢母材和焊缝在近中性模拟溶液中不同加载电位下的应力腐蚀行为

罗金恒[1,2]　雒设计[3]　李丽锋[1,2]　张　良[1,2]　武　刚[1,2]　朱丽霞[1,2]

(1. 中国石油集团石油管工程技术研究院；
2. 石油管材及装备材料服役行为与结构安全国家重点实验室；3. 西安石油大学)

摘　要：管道运输是石油天然气的主要输送方式，土壤应力腐蚀开裂是影响油气管道安全的主要因素之一。为了保证输送管道的安全，通常采用涂层+阴极保护的防护措施以预防埋地管道的腐蚀，为了深入了解外加阴极保护电位对X90管线钢土壤应力腐蚀行为的影响，本文采用电化学测试技术和慢应变速率拉伸试验(SSRT)研究了X90管线钢直缝焊管母材和焊缝在近中性模拟溶液(NS4)中不同外加电位下的应力腐蚀行为，并采用扫描电子显微镜(SEM)对不同外加电位下的断口微观形貌进行了分析。结果表明：X90管线钢直缝焊管母材和焊缝在NS4溶液中的极化曲线具有典型的阳极溶解特征，无活化-钝化现象。X90管线钢直缝焊管母材和焊缝在NS4溶液中存在一定的应力腐蚀敏感性，随着外加电位的负移，用延伸率的损失率$I_δ$和断面收缩率的损失率$I_ψ$表示的应力腐蚀敏感性指标呈现出先减小后增大的趋势，且焊缝的应力腐蚀敏感性大于母材。在E_{ocp}下，X90管线钢的SCC机制为阳极溶解机制；在-850mV(vs SCE)下，X90管线钢的SCC机制为阳极溶解+氢脆机制；在-1000mV(vs SCE)和-1200mV(vs SCE)下，X90管线钢的SCC机制为氢脆机制。

关键词：管道运输；X90管线钢；焊缝；应力腐蚀；外加电位；阳极溶解；氢脆

高压天然气输送管线是一种经济、快捷的天然气长距离输送方式[1]。目前，绝大多数管道采用埋地铺设，并采用涂层+阴极保护的联合防护措施，以有效的减缓或防止管线钢在土壤环境中的腐蚀。埋地管道在运行过程中，因涂层自然老化和其他原因，管道涂层破损不可避免，涂层破损处管线在土壤环境和应力作用下发生应力腐蚀开裂(Stress Corrosion Cracking，简称SCC)在所难免[2,3]。管线钢在土壤环境中的SCC破坏主要可分为高pH-SCC和近中性pH-SCC[4,5]，自从1965年美国发生第一例管线钢高pH-SCC和1985年加拿大发生第一例近中性pH-SCC以来，国内外学者对管线钢的应力腐蚀开裂行为进行了广泛的研究，已在X80以下钢级的研究上取得了一定的成果[6-9]。研究者普遍认为，管线钢高pH-SCC为膜破裂+阳极溶解机制，然而管线钢近中性pH-SCC的机理至今未形成统一的认识。

基金项目：国家"十三五"重点研发计划课题"危险化学品储存设施燃爆毁伤效应及事故调查技术"(编号：2016YFC0801204)。

作者简介：罗金恒，1972年生，教授级高级工程师，博士；主要从事油气管道及储运设施完整性技术方面的研究与工程服务工作。地址：陕西省西安市雁塔区锦业二路89号(710077)，电话：(029)81887989，E-mail：luojh@cnpc.com.cn。

随着我国"西气东输"战略的实施,大口径、长距离、高压输送已经成为我国天然气工业管道输送的发展方向。采用高强度管道钢有利于提高天然气长输管道的输送能力,但对管道钢的强度提出了更高的要求[10,11]。X70 和 X80 管线钢已成功应用于长输管道建设,X100、X120 等超高强管线钢虽然已有试验段的应用,但受其安全性评估的影响未能实现大批量工程应用[12,13]。X90 是继 X80 和 X100 管线钢后开发的新一代管线钢,已成为国内外研究的新热点[14,15],研究主要集中在 X90 管线钢的开发试制[16-18]和组织性能测试[19-21],关于 X90 管线钢在服役环境中的安全性研究鲜见报道。因此,本文采用慢应变速率拉伸试验(SSRT)、电化学测试技术和断口分析相结合的方法,研究不同外加电位下 X90 管线钢直缝焊管母材和焊缝在近中性模拟溶液(NS4)的应力腐蚀行为,探讨其应力腐蚀机理,以期为 X90 管线钢的安全使用提供理论依据和技术参考。

1 试验方法

实验材料为国内某钢管厂生产的 X90 直缝焊管,其规格为 $\phi1219mm \times 19.6mm$,焊管母材的化学成分(质量分数,%):0.056 C,0.21 Si,1.92 Mn,0.01 P,0.0018 S,0.33 Cr,0.21Mo,0.081Nb,0.012Ti,0.22 Cu,0.029Al 余量 Fe;焊缝的化学成分(质量分数,%):0.056 C,0.29 Si,1.91 Mn,0.013 P,0.0031 S,0.23 Cr,0.28Mo,0.050Nb,0.017Ti,0.17 Cu,0.015Al 余量 Fe;图 1 为 X90 直缝焊管母材和焊缝的显微组织。可见,母材的显微组织为 B 粒+少量 PF+P,焊缝的显微组织为 PF+MA+P。

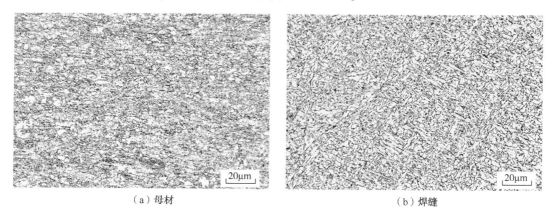

(a) 母材　　　　　　　　　　　　　　(b) 焊缝

图 1　X90 直缝焊管母材和焊缝的显微组织

实验溶液为近中性模拟土壤溶液(NS4 溶液),采用蒸馏水和分析纯化学试剂配制而成,溶液的组成为(mg/L):122 KCl,483 $NaHCO_3$,181 $CaCl_2 \cdot 2H_2O$,131 $MgSO_4 \cdot 7H_2O$,pH 值约为 7。

将 X90 管线钢直缝焊管母材和焊缝加工成面积为 $1cm^2$ 的电化学试样,试样背面焊接 Cu 导线,非工作表面用环氧树脂密封使其和腐蚀介质隔绝。实验前将工作电极用 100~1000#砂纸逐级进行打磨,然后用去离子水和酒精清洗试样表面的油污。动电位极化测试采用标准的三电极体系,工作电极为 X90 管线钢,参比电极为饱和甘汞电极(SCE),辅助电极为 Pt 电极。电化学测试采用 PARSTAT2273 电化学测试仪,扫描速度为分别为 1mV/s 和 10mV/s,扫描电位范围为 -1.3~0.4V(vs SCE),电化学测试前先向溶液中通入纯度为 99.5% 的 N_2 除氧 2 h 使溶液的 pH 值稳定在 6.6~6.8,试验过程中持续通入 95%N_2+5%CO_2 混合气体以维持近中性 pH 环境,所有实验在室温下进行。

拉伸试样取材方向应沿实际管线的环向（即试样轴向为实际管线环向），以保证拉伸时试样的主受力方向与实际受力方向一致。按 GB228—2002 加工成 6mm×4mm 的板状慢拉伸试样，其尺寸如图 2 所示。慢应变速率拉伸实验按照 GB/T15970.7—2000，在 PLT-5 型微机控制慢应变速率应力腐蚀试验机上进行，试验机的控制精度为±2N。腐蚀溶液为 NS4 近中性土壤模拟溶液。实验前通入纯度为 99.5% 的 N_2 除氧 2h，使溶液的 pH 值稳定在 6.6~6.8，实验过程中持续通入 95%N_2+5%CO_2 混合气体，以维持近中性 pH 环境。慢应变速率拉伸实验采用三电极体系，工作电极为 X90 管道钢，参比电极为饱和甘汞电极(SCE)，辅助电极为 Pt 电极。应变速率为 10^{-6}/s，利用 PS168 型恒电位仪对试样施加外加电位，其电位分别为：开路电位（E_{OCP}）、-850mV、-1000mV 和 -1200mV。断口分析在 JSM-6390 型扫描电子显微镜上进行。所有实验在室温下进行，所有电位值均相对于 SCE。

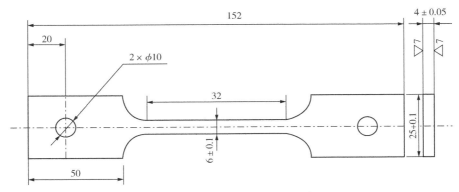

图 2　慢应变速率拉伸试样尺寸

通常采用对暴露到实验环境中和惰性环境中的相同试样进行比较的方法来评定应力腐蚀的敏感性[22]，为了量化讨论 X90 管线钢在 NS4 溶液中的应力腐蚀敏感性，以延伸率损失（I_δ）和断面收缩率损失（I_ψ）作为管线钢应力腐蚀敏感性的评价指标：

$$I_\delta = \frac{\delta_a - \delta_{SSRT}}{\delta_a} \times 100\% \quad (1)$$

$$I_\psi = \frac{\psi_a - \psi_{SSRT}}{\psi_a} \times 100\% \quad (2)$$

式中：δ_a 和 ψ_a 分别为大气中拉伸断裂试样的延伸率和断面收缩率，δ_{SSRT} 和 ψ_{SSRT} 分别为腐蚀溶液中慢应变速率拉伸断裂试样的延伸率和断面收缩率。

2　试验结果与分析

图 3 为了 X90 管线钢直缝焊管母材和焊缝试样在 NS4 溶液中开路电位（E_{OCP}）随实验时间的变化关系。可见，在实验初期的 0~0.3h 内，母材和焊缝试样的 E_{OCP} 呈快速下降趋势；从 0.3h 开始，母材和焊缝试样的 E_{OCP} 随时间延长变化程度有所减缓。当实验时间为 4h 时，母

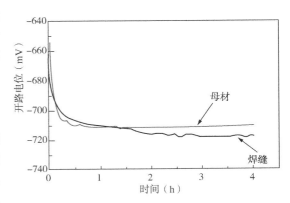

图 3　X90 管线钢直缝焊管母材和焊缝在 NS4 溶液中的开路电位

材的开路电位 E_{OCP} 稳定在 -708mV；焊缝的开路电位 E_{OCP} 稳定在 -718mV。这表明在 NS4 溶液中，母材的电化学活性较小，腐蚀热力学趋势减小，即母材的电化学稳定性优于焊缝。

2.1 慢应变速率拉伸实验

图 4 和表 1 为 X90 直缝焊管母材和焊缝在不同电位下的 SSRT 曲线和实验结果。可见，与空气中的慢拉伸实验结果相比，不同电位下 X90 管线钢母材和焊缝的断裂时间、断后延伸率 δ、断面收缩率 ψ 均有不同程度的减小，说明 X90 管线钢母材和焊缝在不同电位下均表现出一定的 SCC 敏感性。随着外加电位的负移，X90 管线钢母材和焊缝的应力腐蚀敏感性指标 I_δ 和 I_ψ 呈现出先减小后增大的趋势，用断面收缩率表示的应力腐蚀敏感性指标 I_ψ 大于用延伸率表示的应力腐蚀敏感性指标 I_δ，且焊缝的应力腐蚀敏感性指标均大于母材的应力腐蚀敏感性指标，这说明焊缝相比于母材更容易发生 SCC。分析其原因是在焊接过程中，焊缝的形成是一种冶金过程，焊缝附近的区域金属相当于受到一次不同工艺的热处理。由于焊接热的瞬时性和局限性，焊接过程中温度场分布的不均匀性，导致焊缝和热影响区的组织结构发生改变会显著改变钢的固有性能。此外，焊接过程中由于焊缝具有较高的冷却速度，焊缝和热影响区含有较高的晶格缺陷，这会导致焊缝和热影响区具有较高的活性。热影响区由于受到焊接热循环作用致使组织和性能发生变化，发生局部的硬化、脆化和韧性降低，在焊接残余拉应力的作用下，使拉伸试样的热影响区成为 SCC 的敏感区域。X90 管线钢母材和焊缝的 SCC 敏感性顺序为：-1200mV > -1000mV > E_{OCP} > -850mV > NS4 溶液 > 空气。

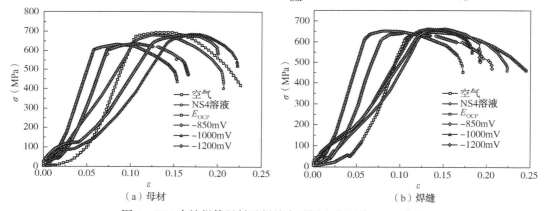

图 4 X90 直缝焊管母材及焊缝在不同电位下的 SSRT 曲线

表 1 X90 直缝焊管母材及焊缝在不同电位下 SSRT 试验结果

试样	试验条件	断裂时间(h)	σ_b(MPa)	δ(%)	ψ(%)	I_δ(%)	I_ψ(%)
母材	Air	76.4	683	13.56	74.99	0	0
	NS4 溶液	64.4	686	12.40	45.98	8.5	38.6
	E_{OCP}	43.5	632	9.53	38.49	29.7	48.6
	-850mV	44.5	635	10.12	40.79	25.3	45.6
	-1000mV	69.2	686	8.98	37.80	33.8	49.6
	-1200mV	62.5	684	8.00	25.70	41.0	65.7
焊缝	Air	76.2	653	14.23	67.44	0	0
	NS4 溶液	68.7	651	12.44	36.50	12.5	45.8
	E_{OCP}	48.4	652	9.82	34.0	30.9	49.6
	-850mV	54.8	642	10.21	35.21	28.3	47.8
	-1000mV	56.3	661	9.20	28.70	35.3	57.4
	-1200 mV	54.6	668	8.04	26.52	43.4	65.7

2.2 断口分析

图5为X90直缝焊管母材在空气、NS4溶液中未加电位和不同外加电位下的慢拉伸断口微观形貌。可见,X90管线钢母材在空气中慢拉伸属于典型的韧性断裂,断口形貌为等轴韧窝,韧窝的大小和深度分布较均匀,韧窝底部可观察到微孔。在NS4溶液中未加电位时,其慢拉伸断裂属于韧性断裂,断口形貌以韧窝为主,韧窝上部存在着蛇形滑移的特征。当外加电位为E_{OCP}和−850mV时,其断口形貌以浅韧窝为主,在断面上存在少量的具有脆性特征的小平面。当外加电位为−1000mV时,其断口形貌以解理断裂为主,在解理面上存在撕裂棱,为典型的韧−脆混合断裂,具有准解理断裂的特征。当外加电位为−1200mV时,其断口形貌为典型的解理断裂,在解理面上可明显的观察到二次裂纹,二次裂纹与拉伸方向呈一定的角度。

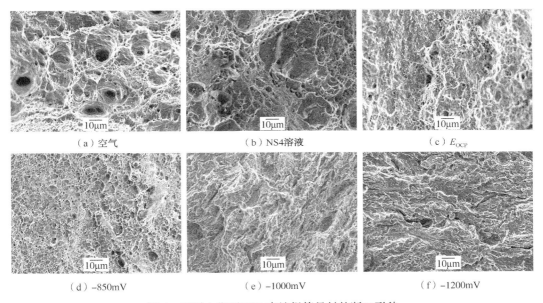

(a)空气　　(b)NS4溶液　　(c)E_{OCP}
(d)−850mV　　(e)−1000mV　　(f)−1200mV

图5　不同电位下X90直缝焊管母材的断口形貌

图6为X90直缝焊管焊缝在空气、NS4溶液中未加电位和不同外加电位下的慢拉伸断口微观形貌。可见,X90管线钢焊缝和母材在空气中慢拉伸的断口相似,也属于典型的韧性断裂,断口形貌为等轴韧窝,与母材相比,焊缝的韧窝数量多,尺寸小而浅。NS4溶液中未加电位,外加电位为E_{OCP}和−850mV时,其断口形貌以浅小韧窝为主,韧窝底部存在少量的解理小平面。当外加电位为−1000mV时,其断口形貌以解理断裂为主,在解理面上存在撕裂棱,为典型的韧—脆混合断裂,具有准解理断裂的特征。当外加电位为−1200mV时,其断口形貌为解理台阶,在解理面上可明显的观察到二次裂纹。

2.3 应力腐蚀过程及机理分析

Parkins理论[23]指出,可以通过快扫和慢极化曲线测试来表征裂纹尖端(金属表面无腐蚀产物膜)和非裂纹区域(金属表面有腐蚀产物膜)的电化学行为。图7为X90直缝焊管母材和焊缝在NS4溶液中快、慢扫的极化曲线。可见,X90管线钢母材和焊缝在快扫(10mV/s)和慢扫(1mV/s)条件下均表现为典型的阳极溶解特征,无活化—钝化现象。对于X90管线钢母材来说,当外加电位高于−750mV、裂纹尖端和非裂纹区域均为阳极极化区,阳极反应速率大于阴极反应速率,金属处于活化溶解状态,溶解电流较大,金属腐蚀严重;当外加电

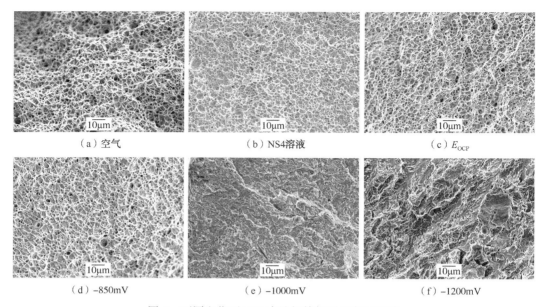

图6 不同电位下X90直缝焊管焊缝的断口形貌

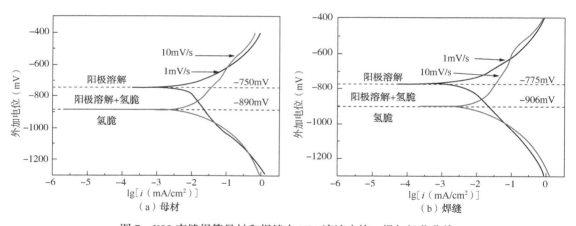

图7 X90直缝焊管母材和焊缝在NS4溶液中快、慢扫极化曲线

位在-890~-750mV时，裂纹尖端区域发生阳极溶解反应，非裂纹尖端区域发生阴极析氢反应，反应过程生成的H扩散进入金属内部，改变了晶格的畸变能，增加了裂纹扩展的动力，加速了SCC的发生，在该外加电位区间内SCC的机制为阳极溶解+氢脆的混合机制；当外加电位低于-890mV时，裂纹尖端和非裂纹尖端区域均发生阴极析氢反应，此时裂纹尖端的氢脆作用决定了应力腐蚀的机制，在该电位以下SCC的机制为氢脆机制；随着外加电位的进一步降低，当外加电位降至-1000mV甚至是-1200mV，阴极反应的速率加快，氢的析出量增加，析出的氢不断在裂纹尖端富集并达到临界氢浓度，造成材料内聚力下降，材料的SCC敏感性增加。对于X90管线钢焊缝来说，当外加电位高于-775mV时，裂纹尖端和非裂纹区域均为阳极极化区，阳极反应速率大于阴极反应速率，金属处于活化溶解状态，溶解电流较大，金属腐蚀严重。当外加电位在-906~-775mV时，裂纹尖端区域发生阳极溶解反应，非裂纹尖端区域发生阴极析氢反应，反应过程生成的H扩散进入金属内部，改变了晶格的畸变能，增加了裂纹扩展的动力，加速了SCC的发生，在该外加电位区间内SCC的机制为阳

极溶解+氢脆的混合机制；当外加电位低于-906mV时，裂纹尖端和非裂纹尖端区域均发生阴极析氢反应，此时裂纹尖端的氢脆作用决定了应力腐蚀的机制，在该电位以下SCC的机制为氢脆机制；随着外加电位的进一步降低，当外加电位降至-1000mV甚至是-1200mV，阴极反应的速率加快，氢的析出量增加，析出的氢不断在裂纹尖端富集并达到临界氢浓度，造成材料内聚力下降，材料的SCC敏感性增加。

实验前通入N_2除氧，可忽略氧的去极化过程，实验过程中持续通入95%N_2+5%CO_2混合气体，可增加CO_2的含量，因此X90管线钢在NS4溶液中存在下列反应：

$$H_2CO_3 = H^+ + HCO_3^-, \quad HCO_3^- = H^+ + CO_3^{2-}$$

阳极反应：$Fe \rightarrow Fe^{2+} + 2e$，$Fe^{2+} + CO_3^{2-} \rightarrow FeCO_3$

阴极反应：$H^+ + e \rightarrow H$

通过以上分析可知，X90管线钢母材的E_{corr}高于-750mV、焊缝的E_{OCP}高于-775mV，当外加电位为E_{OCP}时，X90管线钢的SCC为阳极溶解机制；当外加电位为-850mV时，X90管线钢的SCC为阳极溶解+氢脆机制；当外加电位为-1000mV和-1200mV时，X90管线钢的SCC为氢脆机制。

3 结论

（1）X90直缝焊管母材和焊缝在NS4溶液中的开路电位E_{OCP}分别为-708mV和-718mV，其极化曲线具有典型的阳极溶解特征，无活化—钝化现象。

（2）X90直缝焊管母材和焊缝在NS4溶液中存在一定的应力腐蚀敏感性，随着外加电位的负移，其应力腐蚀敏感性指标I_δ和I_ψ呈现出先减小后增大的趋势，且焊缝的应力腐蚀敏感性大于母材。

（3）X90直缝焊管母材和焊缝在NS4溶液中的SCC行为存在三种机制：在E_{OCP}下，X90管线钢的SCC机制为阳极溶解机制；在-850mV下，X90管线钢的SCC机制为阳极溶解+氢脆机制；在-1000mV和-1200mV下，X90管线钢的SCC机制为氢脆机制。

参 考 文 献

[1] 高惠临. 管道钢与管道钢管[M]. 北京：中国石化出版社，2012.

[2] 王志英，王俭秋，韩恩厚，等. 涂层剥离条件下X70管线钢的应力腐蚀裂纹萌生行为[J]. 金属学报，2012，48(10)：1267-1272.

[3] YAN Mao-cheng, SUN Cheng, XU Jin, et. Stress corrosion of pipeline steel under occluded coating disbondment in a red soil environment[J]. Corrosion Science, 2015, 93(1): 27-38.

[4] JAVIDI M, BAHALAOU HOREH S. Investigating the mechanism of stress corrosion cracking in near-neutral and high pH environments for API 5L X52 steel[J]. Corrosion Science, 2014, 80(3): 213-220.

[5] SALEE B, AHMED F, RAFIQ M A, et al. Stress corrosion failure of an X52 grade gas pipeline[J]. Engineering Failure Analysis, 2014, 46(11): 157-165.

[6] 朱敏，刘智勇，杜翠薇，等. X65和X80管线钢在高pH值溶液中的应力腐蚀开裂行为及机理[J]. 金属学报，2013，49(12)：1590-1596.

[7] 范林，刘智勇，杜翠薇，等. X80管线钢高pH应力腐蚀开裂机制与电位的关系[J]. 金属学报，2013，49(6)：689-698.

[8] CONTRERAS A, HERNANDEA S L, OROZCO CRUZ R, et al. Mechanical and environmental effects on stress corrosion cracking of low carbon pipeline steel in a soil solution[J]. Materials and Design, 2012, 35

（3）：281-289.

［9］ LIU Z Y, Li X G, CHENG Y F. Mechanistic aspect of near-neutral pH stress corrosion cracking of pipelines under cathodic polarization［J］. Corrosion Science, 2012, (55): 54-60.

［10］ HARA T, SHINOHARA Y, TERADA Y, et al. Metallurgical design and development of high-deformable high strength line pipe suitable for strain-based design［C］//paper ISOPE-I-09-443 presented at the SPE Proceedings of the Nineteenth (2009) International Offshore and Polar Engineering Conference, 21-26 June 2009, Osaka, Japan, ISOPE, 2009.

［11］ 李延丰, 王庆强, 王庆国, 等. X90钢级螺旋缝埋弧焊管的研制结果及分析［J］. 钢管, 2011, 40（2）：25-28.

［12］ TANGUY B, LUU T T, PERRIN G, et al. Plastic and damage behaviour of a high strength X100 pipeline steel: Experiments and Modelling［J］. International Journal of Pressure Vessels and Piping, 2008, 85(05): 322-335.

［13］ YAKUBTSOV I A, DORUKS P, BOYDJ D. Microstructure and mechanical properties of bainitic low carbon high strength plate steels［J］. Materials Science and Engineering A, 2008, 480(2): 109-116.

［14］ 黄晓辉, 张冬冬, 符利兵, 等. X90螺旋埋弧焊管抗硫化氢性能的研究［J］. 北京联合大学学报, 2014, 28(4): 81-85.

［15］ NAGAYAMA H, NAKAMURA J, HAMADA M, et al. Development of double joint welding procedure for X90 grade seamless pipe in rise application［C］. Offshore Technology Conference, Texas, 2013.

［16］ 钱亚军, 肖文勇, 刘理, 等. 大直径X90M管线钢的开发与试制［J］. 焊管, 2014, 37(1): 22-26.

［17］ 章传国, 郑磊, 张备, 等. X90大口径UOE焊管的开发研究［J］. 宝钢科技, 2013, (3): 30-34.

［18］ 刘生, 张志军, 李玉卓, 等. 天然气输送管件三通用X90钢板的开发［J］. 宽厚板, 2014, 20(6): 1-5.

［19］ ZHAO Wen-gui, WANG Wei, CHEN Shao-hui, et al. Effect of simulated welding thermal cycle on microstructure and mechanical properties of X90 pipeline steel［J］. Materials Science and Engineering A, 2011, 528(24): 7417-7422.

［20］ 夏佃秀, 王学林, 李秀程, 等. X90级别第三代管线钢的力学性能与组织特征［J］. 金属学报, 2013, 49(3): 271-276.

［21］ 李亮, 蔺卫平, 梁明华, 等. X90钢直缝埋弧焊管焊接接头的组织和性能［J］. 金属热处理, 2015, 40（2）：56-59.

［22］ Standard Practice for Slow Strain Rate Testing to Evaluate the Susceptibility of Metallic Materials to Environmentally Assisted Cracking ［S］. ASTM G129, 2000.

［23］ PARKINS R N. Predictive approaches to stress corrosion cracking failure ［J］. Corrosion Science, 1980, 20（2）：147-166.

本论文原发表于《天然气工业》2018年第38卷第8期。

热影响区软化的 X70 管线环焊缝应变容量分析

陈宏远[1,2,3]　张建勋[1]　池　强[2,3]　霍春勇[1,2,3]　王亚龙[2,3]

(1. 西安交通大学材料科学与工程学院；2. 石油管材及装备材料服役行为
与结构安全国家重点实验室；3. 中国石油集团石油管工程技术研究院)

摘　要：针对一种环焊接头拉伸试验断于热影响区附近母材的基于应变设计 X70 管线，通过数值仿真和韧性试验分别获得环焊接头缺陷的断裂驱动力和断裂阻力曲线，在此基础上使用切线法预测该环焊接头具有较高的拉伸应变容量，与宽板拉伸试验结果一致。研究结果表明，与现行标准中要求焊接接头强匹配，并且拉伸试验不应断于"近缝区"的准则相比，使用延性撕裂行为进行热影响区软化环焊接头的定量分析，可以避免对材料性能的过度要求，并且能科学有效地预测环焊接头拉伸应变容量。

关键词：基于应变设计；拉伸应变容量；单边缺口拉伸试验；宽板拉伸试验

　　轴向拉伸是一种较为恶劣的管线服役条件。对陆地管线，其起因一般和地层运动有关，如地震活动、滑坡失稳、不连续冻土运动、采矿沉降等。对于海洋管线，轴向拉伸多发生于管线铺设过程，例如通过盘卷方式铺设的钢管，纵向应变可能高达 2% 以上。针对上述问题，就需要引入基于应变的管线设计方法，以替代传统的基于应力的设计方法。影响管线轴向拉伸应变的关键环节是环焊缝的应变容量，其思路就是在管线钢和环焊缝的强度、韧性均能达到预期性能的前提下，存在缺陷的环焊接头可以承受一定的轴向拉伸塑性变形而不发生断裂[1]。这种设计方法需要定义不同的性能指标，这些性能指标可能会交互影响，并常常难以确定，如韧性、断后伸长率、屈服强度等[2]。这就是管线基于应变设计工作的主要难点。对于等强度匹配或者高强度匹配的无缺陷环焊缝，在轴向拉伸载荷下可以很容易地确认管线轴向应变。而存在环焊缝裂纹时，材料性能、裂纹尺寸及位置、错高低边量、内压等都会影响管线的拉伸应变容量[3]。

　　目前在管线基于应变设计的研究领域，还没有一个能广泛认可的可以进行环焊缝拉伸应变容量的通用标准。出于对焊接热过程劣化材料性能的考虑[4,5]，国内的相关标准规定对于环焊接头拉伸试验，需要确保断裂位置不发生在焊缝和热影响区及其附近区域[6]。但由于含缺陷环焊缝的拉伸应变容量对于材料性能和几何缺陷具有很高的敏感性，仅仅规定拉伸试验的断裂位置，无法满足焊接接头拉伸应变容量定量分析的要求。目前国外的趋势是通过延

基金项目：国家重点研发计划项目"油气长输管道及储运设施检验评价与安全保障技术"（2016YFC0802101）。

作者简介：陈宏远，男，1979 年 12 月出生，高级工程师，博士生，长期从事管线钢变形与断裂研究，发表论文 20 余篇，起草基于应变设计管线相关行业标准和企业标准 10 余项，研究成果获得省部级科技奖励 10 余项，E-mail：chenhongyuan@cnpc.com.cn。

通信作者：霍春勇，男，教授级高级工程师，E-mail：huochunyong@cnpc.com.cn。

性断裂行为表征的失稳扩展作为拉伸应变容量的预测基础。并且在挪威 SINTEF 研究院和埃克森美孚上游研究公司（Exxon Mobil Upstream Research Company）的工作中，大量讨论了在基于应变设计中将裂纹驱动力法作为失效准则的研究[7-9]。

针对一种 X70 管线环焊接头拉伸试验断裂于热影响区附近区域的情况，对相关区域材料性能进行研究，并基于断裂力学手段完成了特定缺陷水平下的环焊接头拉伸应变容量预测的研究，研究的方法对现行标准中的规定进行了改善。

1 试验方法

1.1 试验用管材

试验用钢管采用 JCOE 工艺制管。表1为管体材料的化学成分测试结果，化学成分测试使用的设备为 Spectro LAB 直读光谱仪。试验用的样品规格为 $\phi 813mm \times 14.7mm$，长度约为 3m。将样管截断后进行环焊对接。

表1 管体材料的化学成分[%(质量分数)]

元素	C	Si	Mn	P	S	Cr	Nb	V	碳当量
含量	0.052	0.13	1.49	0.0079	0.0021	0.036	0.051	0.0041	0.15

1.2 试验用环焊接头

由同一根钢管上截取管段，进行对接环焊。采用手工电弧焊完成根焊和热焊，半自动药芯焊丝自保护焊完成填充焊和盖面焊。焊缝表面进行了额外的盖面焊，以达到焊缝结构的强匹配。表2为焊接用材料及工艺。

表2 焊接材料及工艺

焊层	焊材型号	牌号	规格 ϕ(mm)	焊接位置	焊接工艺
根焊	AWS A5.1 E6010	BOHLER FOX CEL	4.0	5G	SMAW
填充焊	AWS A5.29 E81T8-Ni2	金桥 JC30	2.0	5G	FCAW-S
盖面焊	AWS A5.29 E81T8-Ni2	金桥 JC30	2.0	5G	FCAW-S

注：根据 NB/T 47014—2011《承压设备焊接工艺评定》，5G 为管材水平固定对接焊缝试件位置。

1.3 拉伸性能试验

在管体和环焊缝填充金属上取圆棒拉伸试样进行拉伸性能试验。试样标距长为50mm，试验机为 MTS-810。

1.4 SENT 试验

金属准静态断裂韧性测试中经常使用的单边缺口弯曲（single-edge notched bending, SENB）试验具有较高的裂尖约束度，为此要求以三点弯曲形式加载，并且试样满足一定厚度。近年来，在油气管线基于应变设计研究领域快速发展起来了单边缺口拉伸(single-edge notched tension，SENT)试验，要求试样厚度与管线壁厚接近，从而获得接近实际缺陷的裂尖约束条件。因此，单边缺口拉伸试验获得的断裂阻力曲线，能较好地表征全尺寸管线拉伸试验的断裂行为[10]，并可保证结果具有合理的保守度。此外，当 SENT 试验中的初始缺陷率 a_0/W(a_0 为初始缺陷深度；W 为壁厚)比全尺寸管线大0.1时，两者获得的断裂阻力曲线结果非常接近[11]。单边缺口拉伸试验在 MTS-810 试验机上进行。沿管体纵向取样，试样截面尺寸为14mm×14mm。如图1所示，初始缺陷深度为4.3mm，即 $a_0/W=0.3$，代表比真实管线容许缺陷的 a_0/W(3mm/14.7mm=0.2，3mm 代表一个焊道的深度，14.7mm 是壁厚)高0.1，从而获得的阻力曲线就可以代表全尺寸管线的阻力曲线。

对缺陷位于焊缝金属和热影响区的情况都进行了测试。对缺陷位于焊缝金属的试样，在外焊道焊缝金属中心线上预制裂纹；缺陷位于热影响区的试样，在外表面预制裂纹，并确保原始裂尖位于熔合线附近0.1mm内。图1描述了试样具体尺寸如下：截面厚度为14mm，宽度为14mm，夹持部分长度140mm(10倍试样宽度)，边槽深度0.7mm(5%试样宽度)，初始缺陷深度 $a_0=4.3$mm(10%原始壁厚)。使用双引伸计法测量试验过程中的裂纹尖端张开位移(crack tip opening displacement，CTOD)，如图2所示。通过三角形法则，由上、下引伸计位移外推获得裂纹尖端张开位移。

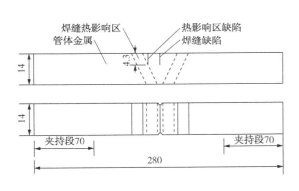

图1 试样示意图(单位：mm)

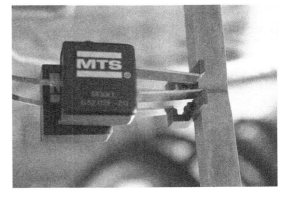

图2 使用双引伸计测量 SENT 试样的 CTOD

1.5 裂纹驱动力计算

拉伸应变预测由裂纹驱动力和极限状态两个要素构成。裂纹驱动力可表达为一定尺寸和材料参数条件下的远端应变(所谓远端应变是指焊缝两侧管体上，分别以500mm为标距测量得到的应变)的函数。

使用环焊缝含缺陷的整管拉伸变形的有限元模型计算裂纹驱动力。有限元分析中管材和焊缝金属材料模型使用拉伸试验获得的弹塑性数据。使用 ABAQUS V6.10 商业软件，采用1/2对称模型，首先建立 CAD 模型(圆柱壳体，包括焊缝和热影响区，焊接坡口特征)，然后进行网格划分，网格单元采用了 ABAQUS 的 C3D8R 单元。裂尖网格采用蛛网型，以获得大应变下的裂尖张开位移。对不同缺陷深度的模型进行有限元分析，获得对应不同裂纹扩展量下的裂纹驱动力。参考典型工艺缺陷50mm×3mm，初始缺陷的长度和深度比均为17，裂尖半径为0.075mm。图3为裂纹位于热影响区的裂尖网格。

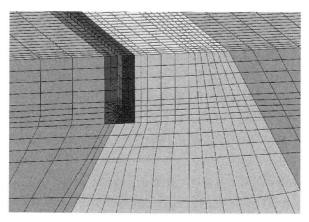

图3 有限元模型裂尖网格示意

如图4所示,使用不同裂纹深度 a_1,a_2,a_3,a_4 的管体—环焊缝缺陷的有限元模型进行仿真计算,可以计算裂纹驱动力 CTOD 与远端应变的关系。这样就能通过静态断裂模型获得裂纹驱动力曲线。进一步通过设定应变水平 ε_1,就能将驱动力—应变曲线转化为等应变驱动力—裂纹生长量曲线,并与断裂阻力曲线进行对比获得切点。

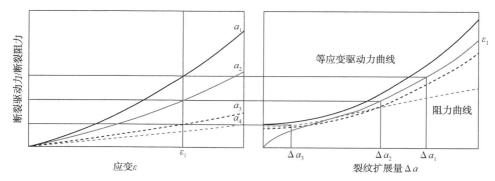

图4 建立驱动力等应变曲线示意图

1.6 宽板拉伸试验

1.6.1 试验样品

宽板拉伸试样全长约1500mm,夹持端宽度约175mm,标距长1000mm,宽300mm。环焊缝位于试样中间位置。缺陷尺寸为长度50mm,深度3mm。缺陷尖端的半径不大于0.2mm。

1.6.2 试验装置

试验在1000t通用拉伸试验系统上进行,试验系统最大拉伸载荷为15MN。试样竖直安装在加载框架内,试样上安装有位移传感器,引伸计等测试装置。

1.6.3 数据获取

试验中试验载荷由位移控制,加载速率为1.5mm/min。平均应变通过焊缝两侧管体上的位移传感器求平均值获得,代表试样的远端应变。全截面应力通过轴向载荷除以标距部分的全截面面积获得。失效应变定义为峰值载荷对应的远端应变。

2 试验结果

2.1 拉伸试验结果

拉伸性能测试的试验结果见表3。由试验结果可见，环焊缝金属材料的屈服强度达到强匹配，抗拉强度比管体材料抗拉强度略低。同时环焊缝金属拉伸屈强比为0.85，管体拉伸屈强比为0.69。环焊缝金属拉伸最大力总断后伸长率为10.5%，管体材料金属拉伸最大力总断后伸长率为9.0%。

表3 管体和环焊缝拉伸试验结果

位置	屈服强度 R_{eL}(MPa)	抗拉强度 R_m(MPa)	屈强比 R_m/R_{eL}	最大力总断后伸长率 A(%)
管体	465	672	0.69	9.0
全焊缝	557	657	0.85	10.5

2.2 SENT试验结果

通过进行SENT试验，获得预制裂纹位于焊缝金属中心线和缺陷位于热影响区试样的撕裂过程描述，由裂纹扩展长度和裂纹尖端张开位移CTOD表达的阻力曲线见图5，预制裂纹位于焊缝和热影响区试样的拟合的阻力曲线公式分别为 $\delta = 1.481\Delta a^{0.630}$（焊缝缺陷）和 $\delta = 2.251\Delta a^{0.607}$（热影响区缺陷）。这里 δ 为裂纹尖端张开位移，Δa 为裂纹生长量。阻力曲线公式描述了随着裂纹的生长，断裂扩展阻力的变化情况。

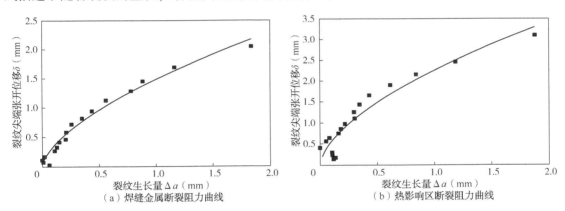

图5 SENT试样获得的焊缝和热影响区阻力曲线

2.3 裂纹驱动力计算结果

按图4所示的裂纹驱动力等应变线的建立方法，分别获得针对焊缝中心缺陷和热影响区缺陷，不同应变水平对应的等应变线。并将对应的阻力曲线与等应变驱动力曲线进行分析对比，最终确定焊缝金属缺陷的阻力曲线与3.8%的等应变线相切；对于热影响区缺陷的阻力曲线，与5.4%的等应变线相切，如图6所示。因此3.8%与5.4%即为预制缺陷分别位于焊缝金属和热影响区时拉伸应变的临界值。

2.4 宽板拉伸试验结果

由于能够更加真实地引入焊缝高低错边、焊接残余应力等因素对环焊缝撕裂行为的影响，宽板拉伸试验可以用于进行含缺陷环焊缝接头的拉伸应变容量验证。

表4为宽板拉伸试验获得的主要试验结果。结果显示，对于焊缝和热影响区存在

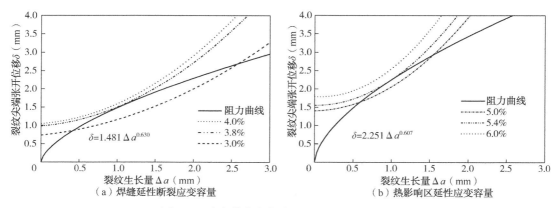

图 6 驱动力等应变曲线和裂纹生长量的关系

50mm×3mm 缺陷的管线，在宽板拉伸试验中焊缝和热影响区应变分别为 3.98% 和 4.73%。

表 4 试样和缺陷尺寸

试样编号	壁厚(mm)	缺陷位置	缺陷尺寸(mm×mm)	远端应变 ε(%)
1	15.0	—	—	5.78
2	15.2	焊缝	3×50	3.98
3	15.0	熔合线	3×50	4.73

3 结论

（1）研究的管线环焊接头，焊缝金属拉伸屈服强度达到强匹配，抗拉强度略低于管体母材纵向抗拉强度。同时，焊缝金属和管体母材金属都具有较低的屈强比，较高的最大力总断后伸长率。

（2）使用 SENT 试验，获得环焊接头不同位置缺陷在延性撕裂过程中的裂纹扩展阻力曲线公式。

（3）通过静态断裂数值仿真获得环焊接头不同位置缺陷的延性撕裂过程裂纹驱动力，并结合与 SENT 试验获得的阻力曲线，通过相切法进行管线环焊接头拉伸应变能力定量分析。结果表明，对于本研究的 X70 环焊接头，缺陷位于焊缝金属时，拉伸应变容量预测值为 3.8%，缺陷位于热影响区时，拉伸应变容量预测值为 5.4%。

（4）针对缺陷位于焊缝金属和热影响区的情况，宽板拉伸试验的结果为 3.98% 和 4.73%。考虑到宽板拉伸试验的试样未去除高低错边，并保留了残余应力的影响，认为切线法预测的 3.8% 和 5.4% 的应变容量具有合理性。

参 考 文 献

[1] Wang Y, Liu M, Song Y. Second generation models for strain-based design：PR-350-074509-R01[R]. 2011-07-31.

[2] 姚乾瑜，邓彩艳，龚宝明，等. 海底管道工程临界评估（ECA）参数敏感性分析[J]. 焊接学报，2016，37(3)：41-44.

[3] Hertelé S, O'Dowd N, Minnebruggen K, et al. Effects of pipe steel heterogeneity on the tensile strain capacity of a flawed pipeline girth weld[J]. Engineering Fracture Mechanics. 2014, 115(1)：172-189.

[4] 徐杰，李朋朋，樊宇，等. 温度对焊接热模拟X80管线钢断裂韧性的影响[J]. 焊接学报，2017，38（1）：22-26.

[5] 赵洪运，王国栋，李冬青，等. 400MPa级超级钢焊接热影响区晶粒尺寸的预测[J]. 焊接学报，2005，26(9)：1-4.

[6] 中国石油管道建设项目经理部. Q/SY GJX 137.1—2012 油气管道工程焊接技术规范 第1部分：线路焊接[S]. 2012：12.

[7] Chiesa M，Nyhus B，Skallerud B，et al. Efficient fracture assessment of pipelines. A constraint-corrected SENT specimen approach[J]. Engineering Fracture Mechanics. 2001，68(5)：527-547.

[8] Nyhus B，Østby E，Knagenhjelm H，et al. Fracture control-offshore pipelines：experimental studies on the effect of crack depth and asymmetric geometries on the ductile tearing resistance[C]. Proceeding of the 24th International Conference on Offshore Mechanics and Arctic Engineering，Halkidiki，Greece，2005：731-740.

[9] Fairchild D，Macia M，Kibey S，et al. A multi-tiered procedure for engineering critical assessment of strain-Based pipelines[C]. Proceeding of 21st international offshore and polar engineering conference，Maui，U.S.A，2011：698-705.

[10] Kibey S，Minnaar K，Chen W，et al. Development of a physics-based approach for the prediction of strain capacity of welded pipelines[C]. Proceeding of 19th International Offshore and Polar Engineering Conference，Osaka，Japan，2009：132-137.

[11] Huang T，Mario M，Minnaar K，et al. Development of the SENT test for strain-Based design of welded pipelines[C]. Proceeding of the 8th International Pipeline Conference，Calgary，Canada，2010：1-10.

本论文原发表于《焊接学报》2018年第39卷第3期。

油气输送管道用钢管标准的发展历程及趋势

李为卫[1]　谢　萍[2]　杨　明[2]　吴锦强[2]

（1. 中国石油集团石油管工程技术研究院·
石油管材及装备材料服役行为与结构安全国家重点实验室；
2. 中国石油西部管道分公司）

摘　要：为保证油气管道工程用钢管标准的先进性和适用性，跟踪研究了国内外油气输送钢管的基础标准、具体工程的管道技术条件及通用管材技术条件的最新发展，并分析了其特点、关系及差异，提出了改进建议。研究表明：基础标准发展变化快，内容系统，适用范围广，但指标低且不具体；重要的油气管道工程必须在基础标准的基础上，针对工程特点，提高或补充相关技术要求，制定具体工程的管材技术条件；通用管材技术条件是基于基础标准补充制定的，但不针对具体管道工程，需要结合具体管道工程给出钢管数据单，确定必要的技术指标并补充技术要求。

关键词：钢管；基础标准；工程技术条件；通用技术条件

为了进一步提高油气管道输送效率和降低建设成本，管道建设向着高强度、厚管壁、大口径及大输量方向发展[1,2]。提高钢管的钢级、减小壁厚可有效节约管道建设投资费用，管线钢每提高一个钢级可减少约7%的建设成本[3,4]。目前，X80管线钢在中国大输量管道上大量应用，位居世界前列，钢管技术标准对管道的安全性和经济性有非常重要的影响[5]，因而得到国内外众多学者的高度关注[6-9]。

全球应用最广泛的管线钢管的标准API SPEC 5L《管线钢管规范》已发展至第46版（2018年）。另一个国际上使用较多的管线钢管标准为ISO 3183—2012《石油和天然气工业 管道运输系统用钢管》，目前处于修订中，计划2019年发布，新版标准从格式和内容上均有很大变化。中国油气输送管道应用最广泛的管线钢管标准为GB/T 9711—2017《石油天然气工业管线输送系统用钢管》。这些标准作为管线钢管的基础标准，其特点是具有普遍适用性，但指标低且不够具体，许多方面仅给出原则性要求而没有给出具体参数。因此，国外重要管道工程的典型做法是以这些标准作为基础标准，在基础标准上增加化学成分、力学性能及缺陷验收极限等补充规定，制定适用于具体工程的补充技术条件。中国西气东输二线管道大量采用X80管线钢，为此制定了当时先进的管材技术标准。为了

基金项目：国家质检公益性行业科研专项课题"桥梁缆索用钢等23项国际标准研制"，201510205-03；中国石油天然气股份公司科学研究与技术开发项目"X80钢级在役天然气管道本质安全相关技术研究"，2016B-3106。

作者简介：李为卫，男，1965年生，高级工程师，2007年硕士毕业于西安石油大学机械工程专业，现主要从事油气输送管道材料研究及标准化工作。地址：陕西省西安市锦业路89号，电话：18191565092，E-mail：liweiwei001@cnpc.com.cn。

满足大规模管道建设需要，中国石油通过管道的标准化、模块化、信息化建设，基本构建了管道建设和运行的标准体系。该体系通过采标、制定、修订等方式，形成了一系列标准。目前，有关管线钢管的标准较多，来源不同，要求不同，跟踪国外标准的发展，不断修订完善国内标准，提高标准的科学性和适用性，对于保证中国管道工程的质量及经济性、安全性具有重要作用。

1 基础标准

1.1 国外

1.1.1 发展历史

美国石油学会(API)建于1919年，是美国第一家国家级商业协会，也是国际上最早、最成功的制定标准的商业协会之一。API的一项重要任务就是负责石油和天然气工业用设备的标准化工作，以确保该行业所用设备的安全性、可靠性及互换性[10]。1926年，API发布的第一版API SPEC 5L标准只包括3个钢级的管材(A25、A、B)，2004年(第43版)发展为11个钢级的管材(A25、A、B、X42、X46、X52、X56、X60、X65、X70、X80)。2007年该标准的第44版与ISO 3183整合，增加了X90、X100及X120共3个钢级的管材[11]。长久以来，API标准被世界石油工业普遍接受和采用，其更新快、参与制定范围广，是世界石油天然气输送钢管生产、检验及使用的重要基础标准[7]。

国际标准化组织第67技术委员会(ISO/TC67)石油与天然气工业用材料、设备和海上结构标准化技术委员会于1947年成立，但由于API的存在，最初基本没有开展工作。直到1980年，ISO首次发布了管线钢管标准ISO 3183—1980。1996年、1999年经修订又发布了3个管线钢管的系列标准：ISO 3183.1—1996、ISO 3183.2—1996、ISO 3183.3—1999。

随着世界经济一体化趋势和世界贸易组织的建立，API积极参加ISO标准的制定工作，ISO和API实行联合工作计划，双方协商制定，按照双轨程序审议，标准等同采用。借助反采政策，约45项ISO/TC67的标准被API再采用[8]。2005年API向ISO/TC 67建议，将ISO 3183的3部分系列标准整合，形成一个与API SPEC 5L类似的标准。2007年API与ISO就管线钢管标准协调一致，同年3月，ISO首先发布了ISO 3183—2007标准，同年10月API反采ISO 3183—2007，发布了API SPEC 5L—2007。

管线钢管标准ISO 3183—2007/API SPEC 5L—2007自发布以来，在内容和结构方面发生了很大变化，在国际上引起广泛关注[7]。新版ISO 3183/API SPEC 5L修订工作自2010年启动，ISO、API联合工作组共收集和处理了400多条意见，对老版标准进行修改完善，于2012年底正式出版。

API SPEC 5L—2012、ISO 3183—2012与上一版相比，主要的技术变化是增加了两个附录：欧洲陆上天然气管道PSL-2钢管订货(附录M，API SPEC 5L不包含此附录)、螺纹和接箍计算公式以及导向弯曲和夏比冲击计算公式背景(附录P)，其他在引用标准、制造工艺、性能指标及检验方法等方面也有一定的技术变化，新版标准更加完善、科学[7]。

ISO 3183—2007与API SPEC 5L—2007的一体化，促进了管线钢管标准的国际化发展，但是，由于美国提出了知识产权和贸易问题，因此，2012年API终止了与ISO的合作，声明不再共同发布协调一致的标准。但是，ISO 3183—2012/API SPEC 5L—2012管线钢管标准是ISO与API标准联合工作组共同合作的成果，内容除了API会标和欧洲陆上管线钢管订货有关附录的差异外，其他内容完全一致。2015年，ISO和API分别启动了管线钢管标准的修订工作，目前 API 已经发布 2018 版(46 版)API SPEC 5L 标准，ISO 计划于 2019 年发布 ISO 3183 新版标准。

1.1.2 API SPEC 5L—2018 的主要变化

新版标准维持了上一版标准的框架和主要内容，其主要变化体现在以下 5 个方面：

(1) 增加了一个关于基于应变设计用钢管的附录。在地震断裂带、泥石流、滑坡等地质灾害地区，需要采用具有抗变形能力的钢管，国内外对其技术指标开展了大量研究[12-20]，ISO、API 经过反复修改，终于形成了具有抗变形能力钢管的国际标准，即 API SPEC 5L—2018 附录 N(ISO 也将采用)。与普通钢管相比，附录 N 明确了具有抗变形能力钢管的订货要求，即采用 PSL-2 钢管；制造工艺要求开展外涂敷热循环对钢管性能影响的模拟试验、钢管拉伸和压缩应变能力评价试验、盘卷模拟试验、应变时效试验；化学成分 C、Mn、S、P、Cr、Mo、V、Nb、V、Ti 等元素的质量分数以及碳当量 CE_{IIW} 和 CE_{Pcm} 上限均有所降低；力学性能增加纵向拉伸性能试验要求，并控制屈服强度、抗拉强度的上限和下限，屈强比上限，均匀延伸率下限，但没有给出具体值，需要通过协议进行规定；钢级为 X65 或以下，最高硬度为 270HV10，钢级高于 X65 且低于 X80 或钢级为 X80，最高硬度为 300 HV10；协议时，可增加纵向拉伸曲线的形状、屈服强度波动范围、时效试验、断裂韧性 CTOD；无损探伤执行海洋管的相关要求，按照 API SPEC 5L 附录 K 执行；针对钢管表面质量、几何尺寸偏差的要求高于普通 PSL-2 钢管。

(2) 基于高温轧制工艺钢化学成分的变化，原计划增加 C 与 Nb 合金元素总量限制和 Ti、Al、N 元素含量的限制，但由于对此争议较大，该项修改提案未能纳入 API SPEC 5L—2018，目前仍在持续讨论中，一旦投票通过，将对 API SPEC 5L—2018 进行该内容的增补。API SPEC 5L—2018 对 X80M 钢化学成分的规定与 API SPEC 5L—2012 一致。由 API SPEC 5L—2000(42 版)以来 PSL-2 X80M 钢化学成分的变化(表 1)可知，管线钢的化学成分控制越来越严格，化学成分对管线钢的性能有重要影响，API 标准已从早期的偏重互换性向偏重技术性转变。

(3) API SPEC 5L—2018 按焊缝局部热处理和全管体整体热处理(正火或淬火+回火)，分别给出酸性和海洋用高频焊(HFW)管的硬度测试位置。对于焊缝局部热处理的 HFW 管，硬度测试位置包括焊缝、热影响区及远处的管体母材，测试点多；对于整体热处理的 HFW 管，硬度测试位置包括焊缝、热影响区及邻近的管体母材，测试点少。然而，API SPEC 5L—2012 没有区分热处理方式，硬度测试位置包括焊缝、热影响区及远处的管体母材，测试点多。

(4) API SPEC 5L—2018 总结了管端切斜的定义和距离管端的长度：外径大于等于 609.6mm 的管子，长度至少为 609.6mm；外径小于 609.6mm 的管子，长度为 609.6mm 或者一个外径。测量方法可采用任何适合的技术，如果采用固定尺测量，管体支脚的长度至少长 304.8mm，管端支脚的长度至少大于一个外径。

(5) 对接管要求的变化：旧版的附录 A 变为附录 M；要求对接的短管具有同样的壁厚和钢级；明确了工艺评定和焊工评定的标准；增加了产品性能试验要求、接头错边和咬边要求、标记要求、焊缝返修要求。

表1 API SPEC 5L 历次版本对 PSL-2 X80M 钢化学成分要求的变化情况

版本	质量分数(单个值为最大值)						碳当量(最大值)		相比上一版的变化
	C	Si	Mn	P	S	Nb+V+Ti	其他	CE_{IIW}	CE_{Pcm}	
API SPEC 5L—2000	0.22%	—	1.85%	0.025%	0.015%	Nb+V+Ti: 0.15%	—	协议	协议	—
API SPEC 5L—2004	0.22%	—	1.85%	0.025%	0.015%	Ti: 0.06% Nb+V+Ti: 0.15%	—	协议	协议	增加 Ti 质量分数限制
API SPEC 5L—2007	0.12%	0.45%	1.85%	0.025%	0.015%	Nb+V+Ti: 0.15%	Cu: 0.50% Ni: 1.00% Cr: 0.50% Mo: 0.50%	0.43%	0.25%	取消 Ti 质量分数限制；增加 Cu、Ni、Cr、Mo 质量分数限制；增加碳当量限制
API SPEC 5L—2012 API SPEC 5L—2018	0.12%	0.45%	1.85%	0.025%	0.015%	Nb+V+Ti: 0.15%	Cu: 0.50% Ni: 1.00% Cr: 0.50% Mo: 0.50% B: 0.001%	0.43%	0.25%	增加 B 质量分数限制
增补版（讨论中）	0.12%	0.45%	1.85%	0.025%	0.015%	Ti: 0.03% C+Nb: 0.20% Nb+V+Ti: 0.15%	Al: 0.07% N: 0.015% Cu: 0.50% Ni: 1.00% Cr: 0.50% Mo: 0.50% B: 0.001%	0.43%	0.25%	增加 Ti、Al、N 质量分数限制；增加 C+Nb 质量分数限制

1.1.3 ISO 3183—2019 的主要变化

计划 2019 年出版的 ISO 3183 的框架变化很大，删除了与 API SPEC 5L—2018 相同的内容，将 API SPEC 5L—2108 作为引用标准，从而保持与其一致。其核心内容只有一个关于欧洲陆上天然气订货技术要求的附录（对应 2012 版的附录 M），主要变化是针对高温轧制钢化学成分的调整。ISO 3183—2012 正文和附录 M 在钢化学成分上的规定是有区别的，在 Nb、V、Ti 总量控制的基础上，欧洲适用的附录 M 分别对钢中 Nb、V、Ti 的质量分数进行了明确限制。就 PSL-2 X80M/L555M 钢级而言，ISO 3183—2019 将 Nb 质量分数的上限由 0.07% 调整为 0.08%，增加了 C 与 Nb 合金总量质量分数最大 0.20% 的限制，Cu、Ni、Mo 的质量分数上限放宽（表2）。

表 2 欧洲陆上天然气管线钢管化学成分要求对比（PSL-2 X80M/L555M）

版本	质量分数（单个值为最大值）									碳当量（最大值）	
	C	Si	Mn	P	S	Nb	V	Ti	其他	CE_{IIW}	CE_{Pcm}
ISO 3183—2012	0.12%	0.45%	1.80%	0.025%	0.015%	Nb：0.07% V：0.11% Ti：0.07% Nb+V+Ti：0.15%			Al：0.015%~0.060% N：0.012% Al/N≥2∶1 Cu：0.25% Ni：0.30% Cr：0.30% Mo：0.10% （协议 0.35%）	0.43%	0.25%
ISO/FDIS 3183—2019（最终草案）	0.12%	0.45%	1.80%	0.025%	0.015%	Nb：0.08% V：0.11% Ti：0.07% C+Nb：0.20% Nb+V+Ti：0.15%			Al：0.015%~0.060% N：0.012% Al/N≥2∶1 Cu：0.50% Ni：0.50% Cr：0.30% Mo：0.35%	0.43%	0.25%

1.2 中国

1988 年，中国发布了第一个管线钢管国家标准 GB 9711—1988《石油天然气输送管道用螺旋缝埋弧焊钢管》。1997—2005 年，制定了 GB/T 9711.1—1997、GB/T 9711.2—1997、GB/T 9711.3—2005 系列标准。2009 年，参考 ISO 3183—2007 对这 3 个系列标准进行修订，2011 年发布 GB/T 9711—2011《石油天然气工业管线输送系统用钢管》，2012 年实施。2017 年发布了基于 ISO 3183—2012 编制的 GB/T 9711—2017《石油天然气工业管线输送系统用钢管》。

1.3 ISO 3183、API 5L、GB/T 9711 三者关系

ISO 3183—2012 比 API SPEC 5L—2012 多一个附录 M"欧洲陆上天然气管道 PSL-2 钢管订货"，少一个附录 O"API 许可证持有者的会标使用"，其他内容与 API SPEC 5L—2012 相一致。ISO 3183—2019 将 API SPEC 5L—2018 版作为引用标准，增加的内容也是附录"欧洲陆上天然气管道 PSL-2 钢管订货"。

GB/T 9711—2011 和 API SPEC 5L—2007 的技术内容相同，而 API SPEC 5L—2007 比

ISO 3183—2007 内容多，这在 API SPEC 5L—2007 附录 N 中有说明，即 GB/T 9711—2011 的内容多于 ISO 3183—2007。

GB/T 9711—2017 基于 ISO 3183—2012 进行修订，修订的内容包括：（1）增加按附录 G，当订购输气管道用钢级不低于 L485/X70 抗延性断裂扩展的 PSL-2 钢管时，采购方应该规定原材料的晶粒度、带状组织及夹杂物的具体要求。（2）增加"当落锤撕裂试验出现异常断口时，采用特殊评定要求"的附录（与 SY/T 6476—2017《管线钢管落锤撕裂试验方法》一致）。针对高强度高韧性钢落锤撕裂试验异常断口的问题，API 开展了"现代高韧管线钢抗脆性断裂试验方法"的研究[9]，但目前尚无定论。（3）删除附录 M"欧洲陆上天然气管道 PSL-2 钢管订货"、附录 O"API 许可证持有者的会标使用"、附录 P"加工有螺纹和带接箍钢管公式及导向弯曲和夏比 V 型缺口试样背景公式"、附录 L"欧洲钢级对比"，增加了相关国内外标准对比附录。

1.4 基础标准特点

以上 ISO、API、GB 管材基础标准，通用性强，其特点是：内容系统、完整；普遍适用，但指标低；针对性差，有些指标不够具体；许多要求只给出原则，不够明确。国外重要管道工程的典型做法是将其作为基础标准，加上补充技术条件后再应用于具体管道工程中。中国从西气东输管道工程开始，借鉴国外管理模式（Alliance 管道），制定了具体工程的管材技术标准，进行管材的订货、检验及验收。

2 中国管材标准现状及发展方向

2.1 西气东输管材标准

国内外管道建设经验表明：管道建设成功与否与管材的技术条件有着很大关系。以西气东输二线工程为例，介绍具体管道工程的管材技术标准。

为了制定管材技术标准，中国石油开展了管材关键技术指标研究，包括化学成分、断裂控制、应力应变行为、应变时效、基于应变地区用钢管要求、组织和夹杂物控制、无损探伤等，重点突破西气东输二线用板材、钢管、弯管、管件等关键技术指标，以及管道止裂韧性指标等重大技术瓶颈问题，在基础标准的基础上形成了西气东输二线 16 项技术标准和两项补充规定。

西气东输二线钢管标准的起草，以当时的有效版本 API SPEC 5L—2004 为框架，参考 ISO 3183—2007 的最终草案，提出补充技术要求，主要包括止裂韧性指标、晶粒度、显微组织、夹杂物级别、螺旋管张开量（残余应力控制）、管端扩径；提高了许多技术指标的要求，如化学成分、尺寸精度、水压试验压力和时间；提高了质量检测的要求，如无损探伤和尺寸检验；补充了过程质量控制要求，如首批试验、过程试验等。总体来说，这是当时国际上比较先进、严格的 X80 管道工程应用钢管技术标准，在工程中发挥了重要作用，保证了西气东输二线工程的质量和安全。

以西气东输二线为代表的具体管道工程技术标准的特点：（1）在标准框架形式上，是对基础标准的补充或修改；（2）针对具体的管型、规格，确定具体内容，条款相对简单；（3）化学成分和力学性能指标相对基础标准严格；（4）相对基础标准，提高了对外观、尺寸检验及无损探伤的要求；（5）补充了过程质量控制要求。从西气东输一线开始，陕京二线、西气东输二线等重大管道工程建设，在采标的基础上制定的工程应用管材技术标准，补充或提高了技术指标，明确了订购信息内容，增加了质量控制，在管道工程建设中发挥了重要作用。

2.2 管材通用技术规格书

中国每年都要建设、投产大量油气管道,但是由于没有统一的管材技术标准作为指导,各个天然气公司和管道设计部门在制订管材技术规范时不知如何选择标准和确定关键技术指标,有时甚至随意使不同规格、不同要求的管道技术规范。在有些情况下,采购方或设计部门为了保证管道的安全可靠性,过度提高关键技术指标,不仅增加了生产制造难度,而且造成了建设资金的浪费;有时不合理降低关键技术指标,导致安全隐患[4]。此外,编写工程技术规格书会耗费大量时间和精力,影响工程进度。

2009 年,在中油管道建设项目经理部的组织下,结合多年的研究成果和工程技术服务经验,制定和发布了 12 项管材通用技术标准,试用后效果良好,在规范管道技术规格书、降低设计工作量及加快工程进度等方面发挥了很好的作用。2010 年,针对试用中发现的问题,对其中 7 项标准进行了修订。2011 年,为了给中国石油天然气股份有限公司管道建设项目的"三化(标准化、模块化、信息化)"工作提供技术支持,将该通用标准转化为通用技术规格书,在中国石油天然气与管道分公司管辖内的油气管道工程中统一使用。

2014 年,对 12 项通用技术规格书进行了修订(表3)。

表3 中国石油天然气股份有限公司 2014 年版管材通用技术规格书

编号	名称
CDP-S-NGP-PL-006-2014-3	天然气管道工程钢管技术规格书
CDP-S-COP-PL-007-2014-3	原油管道工程钢管技术规格书
CDP-S-POP-PL-008-2014-3	成品油管道工程钢管技术规格书
CDP-S-OGP-PL-009-2014-3	油气管道工程站场钢管技术规格书
CDP-S-OGP-PL-010-2014-3	油气管道工程用 DN350 及以下管件技术规格书
CDP-S-OGP-PL-011-2014-3	油气管道工程用 DN400 及以上管件技术规格书
CDP-S-NGP-PL-012-2014-3	天然气管道工程用螺旋缝埋弧焊管热轧板卷技术规格书
CDP-S-NGP-PL-013-2014-3	天然气管道工程用直缝埋弧焊管热轧钢板技术规格书
CDP-S-OPL-PL-014-2014-3	原油成品油管道工程用螺旋缝埋弧焊管热轧板卷技术规格书
CDP-S-OPL-PL-015-2014-3	原油成品油管道工程用直缝埋弧焊管热轧钢板技术规格书
CDP-S-OGP-PL-016-2014-3	油气管道工程用感应加热弯管技术规格书
CDP-S-OGP-PL-017-2014-3	油气管道工程用感应加热弯管母管技术规格书

制定管材通用技术规格书的目的和作用是规范中国石油管道建设管材设计、订货、制造、检验及验收的技术要求,提高工程建设编制技术文件的效率,保证质量和安全。其适用范围为长输油气管道,特点是适用范围广,但参数不具体,不能直接订货,需要设计者在数据单中给出具体参数和规格,并确定必要的指标及附加要求。编制依据为:(1)多年来的科研成果和工程经验;(2)国外先进的管线钢管标准(API、ISO、CSA)和管道工程技术条件(美国、加拿大、英国、俄罗斯)。具体框架和内容是在最新版 ISO 3183/API SPEC 5L 的基础上,依据管道压力、管径、介质对管道进行分级,对不同级别管道的钢管,在材料性能指标、尺寸精度、无损检测等方面提出了不同的补充要求。

2.3 发展方向

中国油气管线钢管标准技术要求和指标比较严格,内容相对完善,多年的管道安全运行也证明了技术标准的科学性和适用性,但在对标工作中也发现,与国外著名石油公司的企业标准相比，中国的管道工程用钢管技术标准有些指标较低，有些规定不够完善，需要持续研究国外标准的技术指标，不断完善中国油气管道用钢管标准，主要包括以下3个研究方向。

2.3.1 完善钢管的力学性能指标

中国的管线钢管标准没有钢管的纵向拉伸性能指标，实际钢管强度波动范围大，对现场环焊缝的强度匹配、焊接材料的选用及焊接工艺的制定有不利影响，建议研究该指标对管道安全的影响，尤其对于X80高强度管道钢管，提出合理的控制指标。中国目前的X80钢管低温韧性指标低于俄罗斯天然气公司的要求，俄罗斯巴甫年科沃—乌恰管道要求钢管在-40℃(中国中俄东线管道为-10℃)的夏比冲击功不小于200J，-20℃(中国中俄东线管道为-5℃)的落锤撕裂试验剪切面积平均值不小于85%，建议研究低温韧性良好的钢管，标准指标相应予以提高。

2.3.2 合理提出化学成分的控制指标

大量研究表明[21-26]，化学成分对钢管的现场焊接性能有很大影响。为了提高现场焊接性，中国的中俄东线等管道工程标准对关键合金元素的上下限提出了比ISO、API等国标标准更严格的限制，这会提高钢材的制造难度和成本。国外某知名公司天然气管线钢管的化学成分没有对合金元素的上下限做出严格规定，而是在API标准基础上对合金元素目标值的波动范围提出补充要求，从而控制化学成分变化对现场焊接的影响，保证了现场环焊缝的质量稳定性。建议参考国外标准提出合理的化学成分控制指标。

2.4 完善无损检测技术要求

对比发现，中国的管道工程钢管标准无损检测有多处不完善之处[27]。如对厚壁管焊缝的超声波探伤，虽然有全壁厚检测的要求，但用于设备校准的对比标样上没有检测上下表面坡口面和中部厚度缺欠的人工反射体，对中部厚度缺欠的探测能力较差，埋弧焊管焊缝自动超声波检测使用的对比试块仅能保证检测壁厚12mm以内的钢管焊缝缺陷，12mm以上的钢管厚度中部的缺陷可能漏检[28]，而国外著名管道公司标准有详细的检测技术要求。中国目前的管道工程钢管标准焊缝射线检测的灵敏度要求采用API SPEC 5L—2000的指标，低于API SPEC 5L—2012的要求(表4、表5)。API SPEC 5L—2007已经取消荧光(电视)射线法，而中国的管道工程钢管标准一直在使用。建议参考国外先进标准的无损检测规定，完善中国的管线钢管标准。

表4 API SPEC 5L—2000射线检验典型厚度范围用ISO金属丝像质计

灵敏度	金属丝号码	规定壁厚(mm)	金属丝直径(mm)
4%	8	(12.7, 15.9]	0.63
	7	(15.9, 20.3]	0.80
	6	(20.3, 25.4]	1.00
	5	(25.4, 31.8]	1.25
	4	(31.8, 41.1]	1.60
2%	11	(12.7, 16.5]	0.32
	10	(16.5, 20.3]	0.40
	9	(20.3, 25.4]	0.50
	8	(25.4, 31.8]	0.63
	7	(31.8, 40.6]	0.80

表5 API SPEC 5L—2012 射线检验典型厚度范围用 ISO 金属丝像质计

金属丝号码	焊缝厚度(mm)	基本线径(mm)	Fe 金属丝系列编码
12	(11，14]	0.25	W10~W16 或 W6~W12
11	(14，18]	0.32	W10~W16 或 W6~W12
10	(18，25]	0.40	W10~W16 或 W6~W12
9	(25，32]	0.50	W6~W12
8	(32，41]	0.63	W6~W12

3 结论

（1）管线钢管的国际标准发展很快，应积极跟踪国际标准，不断修订完善中国油气管线钢管技术标准，以保证标准的先进性和适用性，从而确保油气管道的质量和安全。

（2）管线钢管基础标准、具体管道工程技术条件及通用技术条件各有特点，前期设计和采购时应根据其特点合理选择并使用好技术标准。

（3）钢管基础标准内容系统、完整，但有些指标低或不具体，对于重大管道工程，应在其基础上制定工程技术条件，提高或补充具体指标。

（4）管材通用技术条件对规范、统一长输管道技术规格书起到很好的作用，但不能直接用于采购，需要设计者编制数据单，确定必要的指标并补充技术要求，必要时应对有关指标进行计算或试验验证。

参 考 文 献

[1] Rosado DB, Waele W D, Vanderschueren D, et al. Latest developments in mechanical properties and metallurgical features of high strength line pipe steels[C]. Ghent：International Conference on Sustainable Construction & Design，2013：79-88.

[2] Ishikawan, Okatsu M, Endo S, et al. Mass production and in stallation of X100 linepipe for strain-based design application[C]. Calgary：International Pipeline Conference，2008：705-711.

[3] Demofonti G, Mannucci G, Hillebrand H G, et al. Evaluation of the suitability of X100 steel pipes for high pressure gas transportation pipelines by full scale tests[C]. Calgary：IPC 2004 International Pipeline Conference，2004：1-8.

[4] Yoo J Y, Ahn S S, Seo D H, et al. New development of high gradeX80 to X120 pipeline steels[J]. Materials and Manufacturing Processes，2011，26(1)：154-160.

[5] 王晓香. 关于管线钢管技术的若干热点问题[J]. 焊管，2019，42(1)：1-9.

[6] 方伟，李为卫，许晓锋，等. 油气输送管标准化最新进展[J]. 石油管材与仪器，2018，4(3)：1-6.

[7] 王慧，李狄楠，付宏强，等. API SPEC 5L《管线钢管规范》第45版主要内容变化分析[J]. 焊管，2013，36(8)：67-70.

[8] 王强，明廷宏，蔡亮，等. 美国石油学会油气管道标准研究[J]. 石油工业技术监督，2013(3)：24-29.

[9] 许晓锋，马秋荣，秦长毅，等. ISO 和 API 石油管标准最新动态(2015)[J]. 石油管材与仪器，2015，1(2)：85-88.

[10] 汪洋，徐文生，刘勇强. API 标准在企业中的应用[J]. 机械工业标准化与质量，2014(10)：28-31.

[11] 李鹤林，吉玲康，田伟. 西气东输一、二线管道工程的几项重大技术进步[J]. 天然气工业，2010，30(4)：1-9.

［12］李鹤林，李霄，吉玲康．油气管道基于应变的设计及抗大变形管线钢的开发与应用［J］．焊管，2007，30（5）：5-11．

［13］王国丽，韩景宽，赵忠德，等．基于应变设计方法在管道工程建设中的应用研究［J］．石油规划设计，2011，22（5）：1-4．

［14］臧雪瑞，顾晓婷，王秋妍，等．含腐蚀缺陷高钢级输气管道的失效压力模型［J］．油气储运，2019，38（3）：285-290，296．

［15］李睿，蔡茂林，董鹏，等．地震区油气管道的应变与位移检测技术［J］．油气储运，2019，38（1）：40-44．

［16］Wang Y Y, Liu M, Horsley D, et al. Overall framework of strain-based design and assessment of pipelines［C］. Calgary: The 2014 10th International Pipeline Conference, 2014: IPC2014.

［17］吉玲康，李鹤林，陈宏远，等．管线钢管局部屈曲应变分析与计算［J］．应用力学学报，2012，29（6）：758-762．

［18］张冬娜，戚东涛，邵晓东，等．复合材料增强管线钢管结构设计［J］．油气储运，2017，36（10）：1190-1195．

［19］Ostby E, Jayadevan K R, Thaulow C. Fracture response of pipelines subjected to large plastic deformation under bending［J］. International Journal of Pressure Vessels and Piping, 2005, 82(2): 201-215.

［20］杨忠文，张骁勇，毕宗岳，等．X100钢管落锤撕裂试验的断口表征与分析［J］．油气储运，2015，34（1）：24-32．

［21］Hattingh R J, Pienaar G. Weld HAZ embrittlement of Nb containing C-Mn steels［J］. International Journal of Pressure Vessels and Piping, 1998, 75: 661-677.

［22］Chen X, Qiao G, Han X, et al. Effects of Mo, Cr and Nb on microstructure and mechanical properties of heat affected zone for Nb bearing X80 pipeline steels［J］. Materials and Design, 2014, 53: 888-901.

［23］唐世春，李雅娴，代维，等．复杂环境下20#钢的腐蚀失效性［J］．油气储运，2018，37（8）：909-915．

［24］杜伟，李鹤林，王海涛，等．国内外高性能油气输送管的研发现状［J］．油气储运，2016，35（6）：577-582．

［25］Kirkwood P. Niobium and heat affected zone mythology［C］. Araxá: International Symposium on Microalloyed Steels for the Oil & Gas Industry, 2011: 136-143.

［26］Guagnelli M, Schino A D, Cesile M C, et al. Effect of Nb microalloying on the heat affected zone microstructure of X80 large diameter pipeline after in field girth welding［C］. Araxá: International Symposium on Microalloyed Steels for the Oil & Gas Industry, 2011: 176-187.

［27］蒋浩泽，李为卫，谢萍，等．天然气管道用埋弧焊钢管焊缝无损检测对标分析［J］．焊管，2018，41（10）：55-59．

［28］黄磊，赵新伟，李记科，等．厚壁SAWH焊缝自动超声波检测对比试块的合理性研究［J］．焊管，2017，37（1）：37-41．

本论文原发表于《油气储运》2019年第38卷第6期。

煤层气井用抽油杆腐蚀疲劳寿命的影响因素

李德君[1,2] 王 伟[3] 庞 斌[3] 林 伟[3] 李文正[3] 季 亮[3] 冯耀荣[1,2]

(1. 中国石油集团石油管工程技术研究院；2. 石油管材及装备材料服役行为与结构安全国家重点实验室；3. 中石油煤层气有限责任公司韩城分公司)

摘 要：通过显微组织分析、断口分析、有限元模拟分析、疲劳裂纹扩展速率测试和应力腐蚀试验研究制造因素、结构因素和材料强塑性对抽油杆腐蚀疲劳抗力的影响。结果表明，制造环节易导致抽油杆表面氧化脱碳以及镦粗段偏斜从而对抽油杆腐蚀疲劳寿命产生不利影响。抽油杆结构导致抽油杆镦粗段及前沿存在应力集中，且最大应力总是出现在杆体表面。表面氧化脱碳与应力集中的耦合作用使得抽油杆在腐蚀环境中快速诱发疲劳裂纹。在腐蚀环境中，提高抽油杆强度会导致疲劳裂纹扩展速率的增加，并且高强度抽油杆的腐蚀疲劳裂纹扩展曲线上会出现应力腐蚀平台。降低抽油杆的强度，提高抽油杆的韧性可以有效降低抽油杆在 H_2S 环境里的应力腐蚀开裂敏感性。

关键词：抽油杆；腐蚀疲劳；应力集中；应力腐蚀

抽油杆断裂已成为影响我国某煤层气生产企业正常生产作业的主要原因，腐蚀疲劳则是导致抽油杆断裂失效的主要原因[1,2]。据统计，2012 年至 2014 年的三年间，该企业发生抽油杆断脱事故逾 400 次，且有逐年上升的趋势，超 6 成的失效抽油杆服役寿命不足 1 年。煤层气开采不同于石油，排水采气是其主要的开采方式，工作介质为地层水，腐蚀较石油开采更为严重。且地层水中溶解有较高浓度的 CO_2（煤层气的产出气为甲烷，煤化作用产生的 CO_2 是煤层气的主要伴生气），含有一定浓度 Cl^-，部分区块含有一定量的 H_2S。

对现场使用的抽油杆进行跟踪发现，该企业近年采购的抽油杆虽然力学性能与结构尺寸满足 SY/T 5029《抽油杆》标准要求，但服役寿命差异很大。一些抽油杆服役不到半年就发生断裂，且断裂位置主要集中在镦粗区附近，如图 1 所示。再者，由于 SY/T 5029 标准对钢制抽油杆的化学成分没有明确规定，各生产厂家的抽油杆所使用的材质存在差异。送检的失效抽油杆大多采用 30Cr 钢制造，少数采用 30CrMo 和 35CrMo 钢制造。并且失效抽油杆的力学性能差异较大，有 D 级也有 H 级抽油杆，屈服强度为 644～989MPa，抗拉强度为 837～1039MPa，屈强比大多在 0.90 以上。此外，SY/T 5029 标准所规定的性能并不能反映抽油杆抵抗腐蚀疲劳的性能。虽然 SY/T 5029 标准中也有对抽油杆疲劳性能的要求与试验方法，但标准中的疲劳性能试验并没有考虑环境的影响。在腐蚀环境中，金属构件的疲劳抗力会降低，往往没有疲劳极限，并且腐蚀疲劳的条件疲劳极限同金属材料大气环境中的机械性能没

基金项目：国家科技重大专项(2011ZX05038)。
作者简介：李德君(1982—)，男，工程师，博士，从事高性能钢铁材料的研究与金属构件的失效分析，发表论文 10 余篇。电话：029-81887762，E-mail：lidejun352@163.com。

有直接关系。提高常规疲劳强度的措施，如提高材料的强度，对腐蚀疲劳作用有限，甚至会出现反作用[3]。

疲劳断裂是疲劳裂纹形成和扩展至断裂的过程，疲劳寿命取决于裂纹孕育期和裂纹扩展速率。就光滑构件而言，腐蚀疲劳裂纹扩展寿命约占总寿命的90%，而裂纹形成寿命则只占10%[3]。高周疲劳的裂纹形成寿命主要决定于材料的强度，裂纹扩展寿命则主要取决于强度与塑性配合，过高的屈强比会降低材料或构件缺陷容限[4,5]。很多研究者采用NaCl溶液作为腐蚀介质，研究结果表

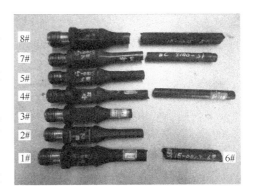

图1 失效抽油杆宏观形貌

明：材料的屈服强度越高，腐蚀疲劳裂纹扩展速率越高，应力腐蚀破裂敏感性越大[6,7]。腐蚀疲劳是一个复杂的构件失效现象，影响因素众多，包括冶金质量、材料成分、力学性能、构件结构、表面质量、应力、加载频率、介质、温度等[3,8,9]。本文以30Cr钢制抽油杆为研究对象研究抽油杆材料强塑性匹配、结构因素、制造因素对腐蚀环境中抽油杆服役性能的影响，从而为抽油杆的选用提供理论依据。

1 试验材料与方法

1.1 断口形貌与微观组织分析

使用TESCAN-VEGA Ⅱ型扫描电子显微镜和OXFORD-INCA350型能谱仪对失效抽油杆断口进行形貌观察和表面腐蚀产物成分分析。将失效抽油杆的接箍段从疲劳源区沿径向剖开，观察纵截面A-A，自断裂面连续切取金相试样直至图2中直线处，每根试样长约30mm，其取样位置如图2所示。金相试验分析内容包括：显微组织分析、氧化脱碳、表面腐蚀情况、微裂纹。

1.2 抽油杆应力有限元模拟分析

分析所用抽油杆几何模型依据SY/T 5029—2013标准建立，分析选用的抽油杆与失效抽油杆一致，杆体直径为19mm。按照标准应用ANSYS有限元模拟软件建立抽油杆三维模型，如图3所示。抽油杆的力学模型基于如下假设：

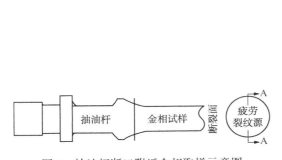

图2 抽油杆断口附近金相取样示意图

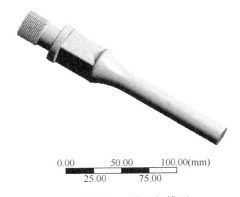

图3 抽油杆三维几何模型

(1) 抽油杆材料是均质的；
(2) 抽油杆为完全弹性；
(3) 不考虑抽油杆的振动；
(4) 抽油杆截面为圆形；
(5) 变形为小变形。

1.3 抽油杆力学性能调整

试验使用直径为19mm未经使用的抽油杆，材质为30Cr的合金钢。为了调整抽油杆的力学性能，以便研究抽油杆强塑性配合对抽油杆疲劳性能和硫化氢应力腐蚀开裂抗力的影响，将部分抽油杆在箱式炉中加热至660℃，保温100min，空冷至室温。原抽油杆和经热处理后的抽油杆的力学性能测试结果见表1。

表1 试验抽油杆的力学性能

热处理条件	屈服强度(MPa)	抗拉强度(MPa)	屈强比	A_{kv}(J)	硬度(HRC)
初始状态	910	960	0.95	86	34.0
加热至660℃	720	800	0.90	171	21.4

1.4 疲劳裂纹扩展速率测定

采用三点弯曲试样测定原抽油杆和热处理后的抽油杆在大气环境下与腐蚀环境下材料的疲劳裂纹扩展速率。试样尺寸为7mm×14mm×70mm，疲劳试验在MYS 810-100kN电液伺服疲劳试验机上进行，载荷范围$\Delta p = 1.98$ kN，载荷比$R = 0.1$，空气环境中的加载频率$f = 10Hz$。模拟腐蚀环境中的加载频率$f = 5Hz$，向3.5%NaCl溶液中通入饱和CO_2气体来模拟腐蚀环境，试验波形为正弦波，试验温度为室温。

1.5 抗H_2S应力腐蚀试验

按照GB/T 4157—2006标准A法对原始抽油杆和热处理后的抽油杆进行应力腐蚀开裂试验。将原抽油杆和热处理后的抽油杆加工成$\phi 6.35mm \times 25.4mm$标准腐蚀拉伸试样，每种抽油杆取三根平行试样。根据力学性能检测结果，分别对两种抽油杆分别施加相当于各自材料约60%屈服强度($R_{p0.2}$)的拉伸应力，浸泡于5%NaCl+0.5%冰醋酸去离子水溶液中并通入饱和H_2S气体。

2 结果及分析

2.1 抽油杆失效模式与生产制造缺陷

经SEM观察发现，失效抽油杆断口呈现明显的疲劳断口特征，断口呈现三个明显的特征区域，即疲劳源区、疲劳区和瞬断区，如图4所示。断口呈现周期性解理和准解理断裂特征，裂纹主要以穿晶方式扩展，疲劳裂纹扩展区有较多的二次裂纹，且二次裂纹多沿晶界开裂，如图5所示。研究表明，腐蚀疲劳裂纹扩展机理是由于阳极溶解与氢致开裂共存，氢致开裂加快了疲劳裂纹扩展[3,10]。进一步观察发现：多数抽油杆断口的疲劳源区和疲劳区表面有腐蚀产物覆盖。能谱分析结果表明：疲劳源区的腐蚀产物中Fe，O，C，S元素含量较高，另外还含有少量的K，Ca等元素。本次SEM能谱分析共检测了5个失效抽油杆断口，其中4个断口上的C元素含量超过了10%，最高达到15%；O元素含量超过20%，最高达到43%；少数抽油杆腐蚀产物中的S元素含量高达7.6%，如图6所示。根据煤层气井的腐蚀工况可以推断，抽油杆断口上覆盖的含C，S元素的腐蚀产物应该为CO_2和H_2S的腐蚀产物,腐蚀疲劳特征明显。

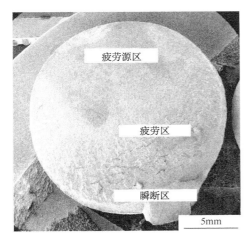

图4 抽油杆断口形貌

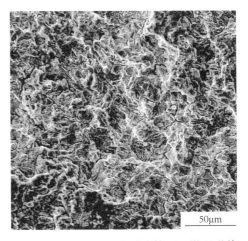

图5 失效抽油杆疲劳裂纹扩展区微观形貌

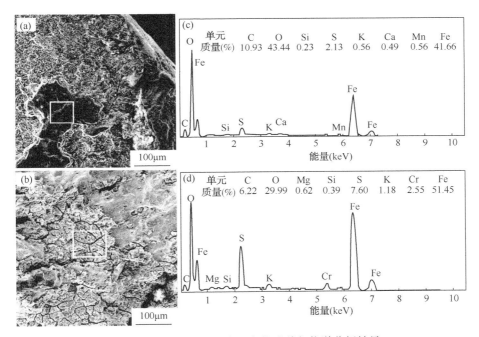

图6 抽油杆疲劳断口腐蚀产物形貌与能谱分析结果

与杆体组织一样,失效抽油杆断口区的组织也为回火索氏体 $S_{回}$,晶粒度 9~10 级,如图 7 所示。作者曾在文献[11]中对抽油杆镦粗段的组织与力学性能进行详细研究,发现抽油杆镦锻过渡区及其前沿杆体(即锻造热影响区)的组织性能和力学性能与杆体并无明显差异,主要是由于抽油杆采用了整体调质处理工艺基本消除了镦锻热影响区对抽油杆力学性能的不利影响。所有失效抽油杆,杆体表面均存在不同程度的氧化脱碳,脱碳层深 0.04~0.10mm,如图 8 所示。除氧化脱碳外,断口附近的镦锻过渡区发现较多的微裂纹,这些微裂纹垂直于杆体的轴向,并且微裂纹几乎都出现在腐蚀坑底,如图 9 所示。抽油杆同时形成多条疲劳裂纹是腐蚀疲劳区别于惰性疲劳的又一主要特征[3]。点蚀加速疲劳裂纹形成理论认为，在腐蚀疲劳初期，抽油杆表面固有的电化学不均匀性和疲劳损伤导致滑移带形成所造成的电化学不均匀，腐蚀的结果将在抽油杆表面形成点蚀坑，点蚀坑破坏了抽油杆截面的连续性产生应力集中促进疲劳裂纹萌生。应当指出，从疲劳裂纹萌生机理角度来看，表面脱碳无论对于常规疲劳还是腐蚀疲劳都是不利的，氧化脱碳层内的金属更易发生滑移造成疲劳损伤[3,5]。

 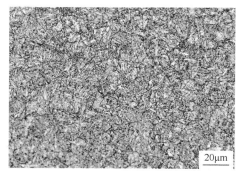

(a)抽油杆断口区　　　　　　　　　　　　　(b)杆体组织

图 7　抽油杆断口区和杆体组织

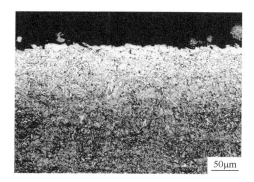 

图 8　杆体表面的氧化脱碳层　　　　　　图 9　腐蚀坑底的疲劳裂纹

此外，一些断裂抽油杆的镦锻区存在肉眼可辨的缺陷，主要表现为：杆头与杆体不同轴，镦粗段沿杆体纵向左右不对称，如图 10 所示。这些缺陷都会导致抽油杆使用过程中的应力集中，使得局部区域的应力水平提高，从而导致疲劳寿命的降低。客观上讲，采用不含 Mo 元素的 30Cr 钢替代 30CrMo 或 35CrMo 钢来制造抽油杆也可能会降低抽油杆的腐蚀疲劳抗力。Mo 元素不仅能够提高钢材的力学性能和耐蚀性，尤其是抗点蚀性能，还能够有效提高钢铁材料的氢腐蚀抗力。

2.2　抽油杆结构对疲劳性能的影响

腐蚀疲劳通常没有疲劳极限，因此在腐蚀环境中，杆体承受的应力越大，抽油杆的寿命越低。抽油杆在实际服役过程中会受到交变拉载荷与弯曲载荷共同作用。通过有限元模拟分析结果可知：在拉—弯载荷共同作用下，抽油杆应力集中区域集中在抽油杆镦锻过渡区前沿，即抽油杆经常发生断裂的部位，如图 11 所示。应力集中处横截面上的

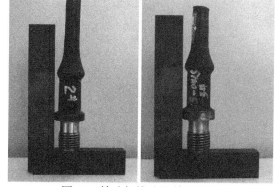

图 10　抽油杆锻造的外形缺陷

应力分布情况如图12所示。由于弯曲载荷的作用，杆体一侧表面存在高的拉应力。服役过程中，抽油杆镦锻过渡区及其前沿存在明显的应力集中，且最大应力总是出现在杆体的表面，使得该区域极易诱发疲劳裂纹从而导致抽油杆常在该区域发生断裂。

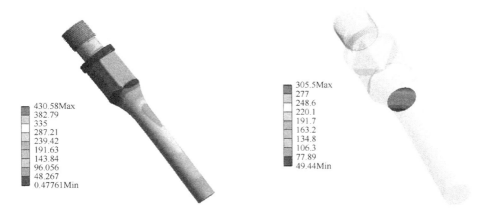

图11 拉弯载荷作用下抽油杆表面应力分布　　图12 拉弯载荷下应力集中处横截面应力分布

2.3 腐蚀环境与杆体强塑性对抽油杆疲劳裂纹扩展速率的影响

在腐蚀环境中抽油杆极易形成疲劳裂纹，采用光滑无缺陷的圆柱疲劳试样进行腐蚀环境下的疲劳试验，发现疲劳裂纹萌生寿命仅占整个疲劳寿命的10%左右[3]。腐蚀环境中服役的抽油杆，其服役寿命主要为腐蚀疲劳裂纹扩展寿命。对比原始抽油杆材料30Cr钢在大气环境与腐蚀环境中测定的疲劳裂纹扩展速率曲线可以看出，腐蚀环境会加速抽油杆疲劳裂纹的扩展，如图13所示。与大气环境中的裂纹扩展曲线相比，原始抽油杆在腐蚀环境中测得的裂纹扩展曲线上出现水平台阶，即具有应力腐蚀特征的腐蚀疲劳[3,8]。

经660℃高温回火处理的抽油杆在腐蚀环境中的疲劳裂纹扩展速率要明显低于原始抽油杆，如图14所示。合金结构钢大气环境中疲劳裂纹扩展速率的研究结果表明，在疲劳裂纹扩展B区，热处理对疲劳裂纹的扩展速率有显著影响，疲劳裂纹的扩展速率随着回火温度的升高而降低。调质状态下使用的抽油杆经过更高温度的回火处理，抽油杆有了更好的强塑性匹配和更高的韧性。韧性的提高有利于降低裂纹尖端应力水平，降低裂纹的扩展速率。

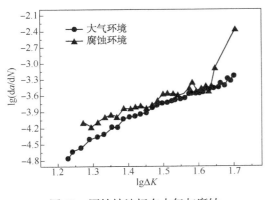

图13 原始抽油杆在大气与腐蚀
环境中裂纹扩展速率曲线

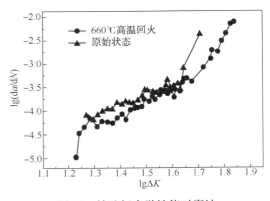

图14 抽油杆力学性能对腐蚀
疲劳裂纹扩展速率的影响

2.4 强塑性对抽油杆应力腐蚀抗力的影响

该区块煤层气井中含有 Cl^-，H_2S。钢铁材料在含 Cl^-、H_2S 环境中会有应力腐蚀开裂的风险。与腐蚀疲劳破坏相似，应力腐蚀也是腐蚀环境与应力共同作用而引起的一种局部破坏。实际上腐蚀疲劳与应力腐蚀之间的界限比较模糊，特别是处于低频率和最大应力与最小应力相差较小的交变载荷作用下时，腐蚀疲劳断口上常常呈现应力腐蚀断裂特征。有学者认为，应力腐蚀是腐蚀疲劳在应力比 $R=1$ 时的一类特殊情况[12]，两者的腐蚀机理也相类似，常用阳极溶解和氢致开裂机制来解释这两种失效现象[7,8]。

与疲劳裂纹扩展速率试验结果相似，经过 660℃ 回火的抽油杆虽然在 H_2S 环境中也发生了应力腐蚀开裂（总共 3 根试样，其中 2 根发生断裂），但是 H_2S 应力腐蚀开裂抗力明显提高，发生断裂的试样分别为 534h 和 648h。相比之下，3 根原始抽油杆试样在 H_2S 环境中不到 12h 就全部发生断裂，试验结果见表 2。取原始抽油杆 8h 发生断裂试样与 660℃ 回火处理的抽油杆 524h 发生断裂试样，使用扫描电子显微镜进行断口形貌分析。在应力腐蚀裂纹的扩展区，原始抽油杆以解理、准解理断裂为主，如图 15 所示，同时断裂面上有大量的沿晶二次裂纹，脆性断裂特征明显。相比之下，经 660℃ 回火处理抽油杆的裂纹扩展区则有明显的韧性断裂特征，出现了大量的韧窝，如图 16 所示。试验结果表明，提高回火温度，抽油杆强度/硬度的降低及韧性的提高对于降低普通抽油杆硫化物应力腐蚀断裂 SSCC 敏感性是有利的。

表 2 硫化物应力腐蚀试验条件和实验结果

热处理条件	载荷(MPa)	试样	结果
抽油杆 660℃ 回火处理	440 （约 60% $R_{p0.2}$）	1	524h 后断裂
		2	648h 后断裂
		3	720h 内无断裂
原始抽油杆	540 （约 60% $R_{p0.2}$）	1	8h 后断裂
		2	11h 后断裂
		3	9h 后断裂

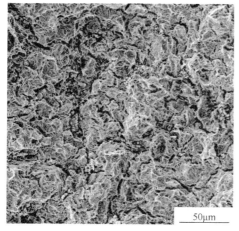

图 15 原始抽油杆 H_2S 应力腐蚀断裂裂纹扩展区形貌

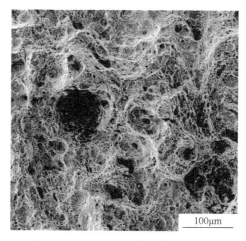

图 16 经 660℃ 回火抽油杆 H_2S 应力腐蚀断裂裂纹扩展区形貌

3 结论

(1) 由于生产制造工艺的限制,抽油杆表面均存在不同程度的氧化脱碳现象及镦粗段偏斜问题,氧化脱碳和杆体镦粗段偏斜将对抽油杆的疲劳寿命产生不利影响。

(2) 抽油杆的结构特征导致了服役过程中抽油杆镦锻区及其前沿存在明显的应力集中区,应力由心部向表面逐渐增大,最大应力总是出现在杆体表面。表面氧化脱碳层与杆体表面应力集中的耦合作用使得抽油杆在腐蚀环境中快速诱发疲劳裂纹。

(3) 高强度抽油杆在腐蚀环境里的疲劳裂纹扩展速率高于低强度抽油杆的疲劳裂纹扩展速率。

(4) 降低抽油杆的强度/硬度,提高抽油杆的韧性可以有效降低抽油杆在 H_2S 环境里的应力腐蚀开裂敏感性。

参 考 文 献

[1] 吴则中,陈强,钟永海,等. 我国29年来抽油杆研制工作回顾与展望[J]. 石油矿场机械,2012,41(1):62-67.
[2] 梁辰,邓福成,李惠子,等. 抽油杆柱疲劳断裂失效分析[J]. 石油矿场机械,2013,42(4):71-74.
[3] 王荣. 金属材料的腐蚀疲劳[M]. 西安:西北工业大学出版社,2001.
[4] 蒋祖国. 飞机结构腐蚀疲劳[M]. 北京:航空工业出版社,1991.
[5] 涂铭旌,张铁军,宋大余,等. 机械设计与材料设计[M]. 北京:化学工业出版社,2014.
[6] 郑文龙,朱国培,欧阳怀瑾,等. 强度对两种低合金钢在海水中腐蚀疲劳行为的影响[J]. 金属学报,1986,22(3):A275-282.
[7] 刘晓坤,王建军,路民旭,等. 屈服强度对40CrMnSiMoVA超高强度钢腐蚀疲劳裂纹扩展的影响[J]. 材料科学与工艺,1994,2(2):1-5.
[8] 黄永昌,张建旗. 现代材料腐蚀与防护[M]. 上海:上海交通大学出版社,2012.
[9] 付正鸿,陈辉,苟国庆,等. 载荷频率对690(TT)合金腐蚀疲劳裂纹扩展行为的影响[J]. 材料热处理学报,2016,37(2):36-41.
[10] 王新虎,邝献任,吕拴录,等. 材料性能对钻杆腐蚀疲劳寿命影响的试验研究[J]. 石油学报,2009,30(2):312-316.
[11] 白强,庞斌,林伟,等. HL型抽油杆断裂失效分析[J]. 金属热处理,2016,41(7):187-191.
[12] S. A. Shiplov. 腐蚀疲劳裂纹扩展机理[J]. 中国腐蚀与防护学报,2004,24(6):321-333.

本论文原发表于《材料热处理学报》2017年第38卷第3期。

地下储气库注采管柱气密封螺纹接头优选分析

王建军[1]　孙建华[2]　薛承文[3]　韩　军[1]　张国红[3]　王　蕊[1]

(1. 中国石油集团石油管工程技术研究院·石油管材及装备材料服役行为与结构安全国家重点实验室；2. 中国石化中原储气库有限责任公司；
3. 中国石油新疆油田公司工程技术研究院)

摘　要：为解决国内储气库井注采管柱气密封螺纹接头选用混乱且适用性不清的问题，首先通过现有标准和气密封试验数据统计比对，发现各气密封螺纹接头的拉伸效率相同，但其压缩效率差异大，且现有标准已不适用于储气库注采管柱气密封螺纹接头的选择，注采管柱气密封螺纹接头选择必须考虑管柱载荷的交变。同时结合2口储气库井注采工况，基于管柱力学理论计算获得2种规格注采管柱的最大拉伸载荷和最大压缩载荷，对比已选用的气密封螺纹接头，重点研究了接头拉伸效率、压缩效率与管柱载荷之间的关系，提出了注采管柱气密封螺纹接头优选判据，进一步利用全尺寸实物复合加载试验机对2种规格注采管柱进行多周次气密封循环试验，试验结果证实该判据的合理性。该判据不仅可以作为气密封螺纹接头优选的基本依据，还可以作为储气库井注采管柱设计依据，并纳入了中国石油天然气行业标准。

关键词：地下储气库；注采管柱；气密封螺纹；接头；压缩；优选；多周次；试验

地下储气库井因其重要的战略地位，要求其寿命周期在30年以上[1,2]。储气库注采管柱承担着注气和采气的通道，地层压力系数并不像常规气井一样逐年衰减，而是一直保持在0.9左右，管柱内运行压力、温度等载荷随注采周期交替变化[3,4]。为了确保管柱的密封性，要求储气库注采管柱和生产套管柱必须选用气密封螺纹接头[5-7]。笔者通过中国石油6座储气库的调研发现，在井深、运行压力、注采气量差异不大的情况下，各储气库选用的气密封螺纹接头各异，主要有BGT1、TPCQ、VAM TOP、3SB、BEAR等气密封螺纹。这些气密封螺纹接头的密封性能存在差异[8-10]，尤其是气密封螺纹接头的抗压缩能力最优。但油田在选用气密封螺纹接头时，仅关注接头的抗内压强度和抗拉强度[11,12]，而对接头的抗压缩能力没有任何要求。

笔者通过对现有气密封试验标准数据和储气库注采管柱载荷分析，并结合全尺寸实物模

基金项目：中国石油天然气集团公司重大项目"储气库完整性关键技术研究"（编号：2015E-4006）。
作者简介：王建军，1979年生，高级工程师，国家注册安全评价师，博士，2015年毕业于中国石油大学（华东）并获博士学位，主要从事油气井管柱安全评价技术研究。地址：陕西西安市锦业二路89号（710077），电话：029-81887677、13572281791、ORCID：0000-0003-4434-1344，E-mail：wg_j_jun@163.com。

拟试验，针对国内外气密封螺纹接头的性能差异，提出了适用于储气库工况的气密封螺纹接头优选分析方法。

1 气密封试验标准和数据

目前，对油管和套管柱接头进行气密试验检测的主要依据是 ISO 13679 标准[13]，标准试验目的是评价油管及套管螺纹连接的粘扣趋势、密封性能和结构完整性。标准规定了 4 种接头评价级别(CAL)，并指明评价级别 CAL Ⅱ 以上试验适用于气井。但在标准 B 系气密封试验，仅进行 CCW(逆时针)、CW(顺时针)、CCW 三次循环，并不能满足储气库注采 30 周次需要。此外，在标准 B 系气密封试验中，规定进行 95% 拉伸载荷、33% 和 67% 压缩载荷下的气密封试验。

统计对比国内(A、B、C 生产厂家)和国外(D、E 生产厂家)不同生产厂家的气密封螺纹接头试验结果(表1)，国内外气密封螺纹接头的拉伸效率均为 100%，但国外气密封螺纹接头在压缩载荷下的密封性能上总体优于国内气密封螺纹接头，且表1显示螺纹接头多数在 10%~40% 之间施加压缩载荷进行气密封试验，不能满足储气库井深、压力变化的多样性。

表1 国内外不同厂家和规格的气密封螺纹接头试验结果对比表

厂家	规格	循环方向	纯压缩载荷达到包络线的百分比(%)	内压下压缩载荷达到包络线的百分比(%)	接头载荷包络线VME(%)	试验结果
A	φ244.48mm×11.99mm V140	CCW、CW、CCW	30	10	95	未发生泄漏
A	φ88.90mm×6.45mm P110	CCW(逆时针)、CW(顺时针)、CCW	30	10	95	未发生泄漏
A	φ73.02mm×5.15mm P110	CCW、CW、CCW	30	10	95	未发生泄漏
B	φ177.80mm×10.36mm P110	CCW、CW、CCW	30	10	95	未发生泄漏
B	φ139.70mm×10.54mm P110	CCW、CW、CCW	60	40	95	未发生泄漏
B	φ244.48mm×11.99mm V140	CCW、CW、CCW	30	10	95	未发生泄漏
B	φ88.90mm×6.45mm Q125	CCW、CW、CCW	50	50	95	未发生泄漏
C	φ177.80mm×10.36mm P110	CCW、CW、CCW	30	10	95	未发生泄漏
C	φ244.48mm×11.99mm V140	CCW、CW、CCW	30	10	95	发生泄漏
D	φ339.7mm×12.19mm Q125	CCW、CCW、CCW	95	67	95	未发生泄漏
D	φ244.5mm×11.99mm Q125	CCW、CCW、CCW	95	67	95	未发生泄漏
D	φ250.8mm×15.88mm Q125	CCW、CCW、CCW	95	67	95	未发生泄漏
E	φ88.9mm×6.45mm P110	CCW	80	74	80	未发生泄漏
E	φ88.9mm×6.45mm P110	CCW	95	40	95	管柱失稳
E	φ88.9mm×6.45mm P110	CCW	80	60	95	管柱失稳
E	φ88.9mm×7.34mm P110	CCW、CW、CCW	57	57	90	未发生泄漏
ISO 13679 标准要求		CCW、CW、CCW	95	67	95	未发生泄漏

各生产厂家的气密封螺纹接头设计不同，性能差异明显。如气密封螺纹接头的压缩效率(耐压缩性能)参差不齐，30%~100% 都有，一旦选用不合适极易造成管柱泄漏。因储气库

注采作业的交替变化,注采管柱载荷出现拉压交变[14,15]。储气库注采管柱气密封螺纹接头选择必须考虑管柱载荷的交变,此时气密封螺纹接头的压缩效率显得尤为重要。

2 注采管柱接头选择

2.1 注采管柱载荷分析

选取国内2口储气库井的注采管柱,在各自运行工况(表2)下,按照管柱力学理论计算注采管柱轴向力。注采管柱轴向力主要由自重效应[16]、温度效应[17]、压力效应(活塞效应、膨胀效应等)[18,19]、流动效应[20]等产生的轴向力代数叠加,提取出全井段注采管柱所承受的最大拉伸载荷和最大压缩载荷(表3),并与额定抗拉强度相比,可得注采管柱在运行工况下所需要的接头拉伸效率和压缩效率(表3)。

表3中,储气库1井注采管柱气密封螺纹接头,在运行工况下的拉伸效率和压缩效率均低于额定值许多;储气库2井注采管柱气密封螺纹接头,在运行工况下的拉伸效率低于额定值,但运行工况下的压缩效率高于额定值,存在泄漏风险。而在实际运行过程中,储气库2井很快出现环空带压,储气库1井运行安全。

表2 储气库井注采管柱运行工况主要参数表

储气库井	规格	油管下深(m)	地温梯度(℃/100m)	运行压力(MPa)	注气量($10^4 m^3/d$)	采气量($10^4 m^3/d$)
1	$\phi 139.7mm \times 9.17mm$ P110	2900	2.5	15~35	60	90
2	$\phi 114.3mm \times 6.88mm$ L80	4735	2.5	29~49	45	50

表3 储气库井注采管柱接头的拉伸效率和压缩效率表

储气库井	规格	扣型	额定抗拉强度(kN)	最大拉伸载荷(kN)	最大压缩载荷(kN)	拉伸效率① 运行工况	拉伸效率① 额定	压缩效率② 运行工况	压缩效率② 额定
1	$\phi 139.7mm \times 9.17mm$ P110	T	2 853	1 035	735	36%	100%	26%	80%
2	$\phi 114.3mm \times 6.88mm$ L80	G	1 280	880	732	69%	100%	57%	40%

① 拉伸效率,在95%VME载荷包络线内,油管螺纹连接在内压+拉伸复合载荷作用下发生泄漏的临界拉伸载荷与管体拉伸屈服载荷的比值百分数;
② 压缩效率,在95%VME载荷包络线内,油管螺纹连接在内压+压缩复合载荷作用下发生泄漏的临界压缩载荷与管体压缩屈服载荷的比值百分数。

2.2 注采管柱接头优选判据

依据上述分析,所选用的气密封螺纹接头的拉伸效率和压缩效率应满足运行工况需要。因此,为了确定适用于储气库注采管柱的气密封螺纹接头选用方法,定义判据如下:

$$\frac{T_{atmax}}{T_{to}} \times 100\% \leq \frac{\delta_t}{S_{tt}} \quad (1)$$

$$\frac{T_{acmax}}{T_{co}} \times 100\% \leq \frac{\delta_c}{S_{tc}} \quad (2)$$

式中:δ_t,δ_c 为气密封螺纹接头的额定拉伸效率和额定压缩效率,%;T_{to},T_{co} 为管体拉伸屈服载荷、管体压缩屈服载荷,kN(一般均等于管体额定抗拉强度 T_o);T_{atmax},T_{acmax} 为注采管柱在注采作业过程中最大拉伸载荷和最大压缩载荷，kN；S_{tt} 为管柱接头拉伸安全系数；S_{tc} 为管柱接头压缩安全系数。

结合钻井手册和 ISO/TR 10400 标准等[21-23]，建议 S_{tt} 和 S_{tc} 取值分别不小于 1.2 和 1.1。利用式(1)和式(2)再次分析表3数据，储气库1井选用的气密封螺纹接头的安全系数均在 2.0 以上，安全余量较高。储气库2井选用的气密封螺纹接头的拉伸安全系数为 1.45，虽高于规定安全系数，但压缩安全系数为 0.70，远低于规定值，即储气库2井选用的气密封螺纹接头不适用于该运行工况，这也是造成储气库2井环空带压原因之一。

因此，选用的气密封螺纹接头必须同时满足式(1)和式(2)的要求，才能初步确定为可用的气密封螺纹接头。

3 试验验证评价

为了检验利用式(1)和式(2)选用的螺纹接头是否适用，同时验证两个公式的正确性。在 ISO 13679 标准基础上，结合表2中储气库井注采工况，改进试验程序，利用全尺寸实物复合加载试验机，进行拉压交变载荷下30周次气密封螺纹接头油管密封试验。

选取 ϕ139.7mm×9.17mm P110 T 扣形油管柱试样在 50MPa 内压下经过 2425kN（85%拉伸）→-1 141 kN（40%压缩）→-1 712 kN（60%纯压缩，无内压）→-1 141 kN（40%压缩）→2 425 kN（85%拉伸）30 个循环后，未发生泄漏（图1），且试验载荷高于运行工况载荷，保证了一定的安全余量。

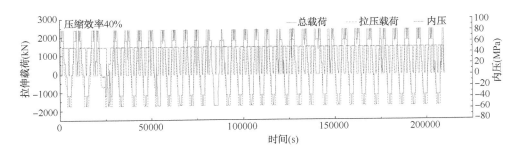

图 1　ϕ139.7mm×9.17mm P110 油管柱多周次气密封循环试验结果图

选取 ϕ114.3×6.88mm L80 G 扣形油管柱试样在 50MPa 内压下经过 1 088 kN（85%拉伸）→-448 kN（35%压缩）→-768 kN（60%纯压缩，无内压）→-448 kN（35%压缩）→1 088 kN（85%拉伸）21 个循环后，未泄漏。之后，该试样在 50MPa 内压下经过 1088kN（85%拉伸）→-512kN（40%压缩）→-768kN（60%纯压缩，无内压）→-512kN（40%压缩）→1088kN（85%拉伸）7 个循环后发生泄漏（图2），即在低于额定压缩效率下进行 21 个气密封循环试验未发生泄漏，但提高压缩载荷至额定压缩效率后仅经过 7 个循环就发生泄漏，未能完成 30 周次气密封循环试验。

图1和图2结果说明，低于接头额定拉伸和压缩效率内进行 30 周次气密封循环试验未发生泄漏。若高于或等于接头额定压缩效率，则易发生泄漏。这也进一步证实判据式(1)和式(2)的合理性。

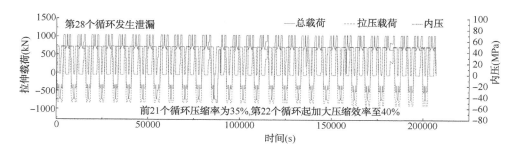

图 2 φ114.3×6.88mm L80 油管柱多周次气密封循环试验结果图

4 结论

（1）现有 ISO 13679 标准试验，不能满足储气库注采 30 周次需要。标准试验多数在 10%~40%之间施加压缩载荷进行气密封试验，不能满足储气库井深、压力变化的多样性。

（2）储气库注采管柱气密封螺纹接头选择必须考虑管柱载荷的交变，应按管柱承受的最大拉伸载荷和最大压缩载荷选择相应性能的气密封螺纹接头。

（3）依据注采管柱载荷变化和接头实际性能，提出的注采管柱接头优选判据，经多周次气密封循环试验证实了其合理性，且与现场实际相吻合。

（4）注采管柱接头优选判据，可以作为气密封螺纹接头优选的基本依据，不仅可用于储气库井，也可用于常规气井。

（5）注采管柱接头优选判据，已提交中国石油天然气行业标准，并获采纳，将作为储气库井注采管柱设计依据。

参 考 文 献

［1］丁国生，李春，王皆明，等．中国地下储气库现状及技术发展方向［J］．天然气工业，2015，35（11）：107-112.

［2］袁光杰，杨长来，王斌，等．国内地下储气库钻完井技术现状分析［J］．天然气工业，2013，33（2）：61-64.

［3］刘坤，何娜，张毅，等．相国寺储气库注采气井的安全风险及对策建议［J］．天然气工业，2013，33（9）：131-135.

［4］刘健，宋娟，张凤伟，等．多随机因素作用下储气库套管运行期安全性分析［J］．岩土力学，2012，33（12）：3721-3728.

［5］SY 6805—2010 油气藏型地下储气库安全技术规程［S］．

［6］汪雄雄，樊莲莲，刘双全，等．榆林南地下储气库注采井完井管柱的优化设计［J］．天然气工业，2014，34（1）：92-96.

［7］王建军，付太森，薛承文，等．地下储气库套管和油管腐蚀选材分析［J］．石油机械，2017，45（1）：110-113.

［8］许红林，李天雷，杨斌，等．油套管特殊螺纹球面对锥面密封性能理论分析［J］．西南石油大学学报（自然科学版），2016，38（5）：179-184.

［9］朱强，杜鹏，王建军，等．特殊螺纹套管接头柱面/球面密封结构有限元分析［J］．郑州大学学报（工学版），2016，37（5）：82-85，90.

［10］王建东，冯耀荣，林凯，等．特殊螺纹接头密封结构比对分析［J］．中国石油大学学报（自然科学版），

2010，34（5）：126-130.
［11］林勇，薛伟，李治，等．气密封检测技术在储气库注采井中的应用［J］．天然气与石油，2012，30（1）：55-58.
［12］韩志勇．关于"套管柱三轴抗拉强度公式"的讨论［J］．中国石油大学学报（自然科学版），2011，45（4）：77-80.
［13］International Organization for Standardization. Petroleum and natural gas industries—procedures for testing casing and tubing connections：ISO 13679—2002［S］. Geneva, Switzerland：International Organization for Standardization, 2002.
［14］王建军．地下储气库注采管柱密封试验研究［J］．石油机械，2014，42（11）：170-173.
［15］刘坤，何娜，张毅，等．相国寺储气库注采气井的安全风险及对策建议［J］．天然气工业，2013，33（9）：131-135.
［16］韩志勇．液压环境下的油井管柱力学［M］．北京：石油工业出版社，2011.
［17］高宝奎，高德利．高温高压井测试油管轴向力的计算方法及其应用［J］．石油大学学报（自然科学版），2002，26（2）：39-41.
［18］《海上油气田完井手册》编委会．海上油气田完井手册［M］．北京：石油工业出版社，1998.
［19］于献彬，陈晓清，纪佑军．基于数值模拟的套管受力单因素分析［J］．西南石油大学学报（自然科学版），2015，37（4）：127-134.
［20］黄春芳．油气管道设计与施工［M］．北京：中国石化出版社，2008.
［21］《钻井手册（甲方）》编写组．钻井手册（甲方）［M］．北京：石油工业出版社，1990.
［22］International Organization for Standardization. Petroleum and natural gas industries—Equations and calculations for the properties of casing, tubing, drill pipe and line pipe used as casing or tubing：ISO/TR10400—2004［S］. Geneva, Switzerland：International Organization for Standardization, 2007.
［23］SY/T 5724—2008 套管柱结构与强度设计［S］.

本论文原发表于《天然气工业》2017年第5期。

X90 超高强度输气钢管材料本构关系及断裂准则

杨锋平[1,2]　罗金恒[1,2]　李　鹤[1]　郭亚洲[3]　冯　健[4]

(1. 石油管材及装备材料服役行为与结构安全国家重点实验室；2. 西安三环科技开发总公司；
3. 西北工业大学航空学院；4. 中国石油天然气股份有限公司西部管道分公司)

摘　要：针对试制成功的 X90 输气钢管，进行 5 种不同圆棒缺口的准静态拉伸试验及应力三轴度计算，发现由于试样缺口存在，应力三轴度值增加 2.43 倍，断裂应变减少 29%，损伤应变能降低 71%。利用常规拉伸试验机和 Hopkinson 拉杆试验装置进行不同应变速率拉伸试验。发现应变速率对断裂应变的影响相对较小，准静态和高速状态下，差异最大约 10%。基于 Johnson-Cook 本构和失效模型，分别建立了考虑应变率效应的 X90 管线钢本构模型和考虑应变率、应力三轴度的失效模型。同时，基于损伤力学理论，得到了基于塑性均匀延伸率和损伤应变能的 X90 管线钢断裂准则。为数值模拟中该断裂准则的正确使用，基于材料损伤应变能密度临界值不变假设以及试验数据，得到了 X90 管线钢断裂特征长度与应力三轴度、试样直径之间的关系式。

关键词：X90 钢管；应力三轴度；应变速率；断裂应变；损伤应变能；断裂特征长度

目前，世界范围内 X90、X100 甚至 X120 钢级的管线钢已经研制成功。2007 年，API SPEC 5L 44 版[1]标准增加了这 3 个强度等级的钢管标准，但 X90 及以上钢级管材的断裂控制问题未完全解决，目前无大规模应用工程，仅有零星试验段。

设计对输气管道性能的基本要求是起裂后能迅速止裂。以夏比冲击功(CVN)为指标的止裂设计是目前国际上普遍采用的方法，如 BATTLE 双曲线法(BTC)[2]、AISI、British Gas 等模型，其中 BTC 模型应用最为广泛。但研究显示，当钢级超过 X80、压力超过 10MPa 时，BTC 方法预测的准确性急剧下降，需要加以修正[3-5]或重新建立公式，而目前缺少较为公认的预测公式。

在继续进行 CVN 指标研究基础上，寻找另一种材料指标，使其能够反映材料的止裂性能，是高钢级管道止裂研究的一个方向。如近几年欧洲正在进行以单边缺口拉伸试样代替 CVN 试样来表征管道裂纹尖端受约束情况[6-8]、意大利 CSM 使用 CTOD 准则[9]、挪威科技

基金项目：国家自然科学基金项目(No.51404294)、陕西省自然科学基金项目(2014JQ2-1004)、中国石油科技专项(2012E-2801)。

作者简介：杨锋平，男，1982 年 2 月生，2005 年获西北工业大学学士学位，2010 年获西北工业大学博士学位，现为中国石油集团石油管工程技术研究院工程师，主要从事油气长输管道风险评估与完整性评价工作。E-mail：yyffpp@163.com。

大学尝试使用 Cockcroft and Latham 准则[10]等。

随着数值模拟技术的发展，特别是有限元技术的飞速发展，流固耦合问题（气体与管道）可以解决，输气管道止裂（起裂后能否在一定长度范围内止住）问题无需单独计算气体减压波速度和裂纹扩展速度，只需在合理的管材断裂准则下，由专用或通用软件计算气体的减压行为以及裂纹尖端的应力应变情况，从而判断裂纹是否会继续扩展[9-11]。由此，可选用的断裂准则范围大大扩展[12,13]，理论上，适用金属材料的准则均可尝试在高钢级天然气管材上使用，如四大强度理论[14,15]、断裂力学 J 积分[16]、CTOD、CTOA[17]等，以及各种金属延性断裂准则[18]。

对于超高强度 X90 钢管的断裂准则及参数，由于工业需求较少，其中 Europipe 公司进行过少量的 X90 开发，并进行过 2 次全尺寸气体爆破试验。2013 年，中国在中国石油重大科技专项推动下，相关钢厂及管厂的研究工作集中在研制满足 API 基本要求的 X90 钢管上[19,20]，对断裂控制问题尚未涉及。仅有中国石油集团石油管工程技术研究院进行了先导性研究，2014 年进行了 2 次 X90 钢管实物气体爆破试验，2015 年在新疆鄯善建成了亚洲首个管道断裂控制试验场，2016 年计划继续进行 X90 管道爆破试验，以期对基于 CVN 指标的经验公式进行修正，但由于实物爆破试验次数太少，该项目尚未有公开研究成果报道。而在非 CVN 指标断裂准则研究上，尚属空白。

笔者考虑了管材小试样拉伸试验与实物钢管裂纹扩展之间的 2 个明显差异：（1）获取材料性能的小试样试验与实物钢管裂纹扩展时的应力状态不同（应力三轴度）；（2）材料的破坏速度（裂纹扩展速率）差异巨大，即实际高压天然气管道裂纹扩展时其应变速率大大高于小试样拉伸试验的应变速率。设计了含不同缺口的圆棒拉伸试验以及 hopkinson 拉杆试验，研究了国内 X90 管线钢的断裂行为，建立了考虑应变率效应的 X90 管线钢 Johnson-Cook 本构模型和考虑应变率、应力三轴度的失效模型。同时，基于损伤力学理论，得到了基于塑性均匀延伸率和损伤应变能的 X90 管线钢断裂准则。为了便于所得断裂准则在数值模拟中的应用，给出了光滑圆棒管材断裂特征长度的计算和测量方法以及其他应力状态下断裂特征长度的换算。

1 X90 管线钢拉伸试验

1.1 材料基本力学性能

试验材料取自 1219mm×16.3mm 直缝焊管，横向取样。钢管化学成分设计采用低 C、高 Mn、Nb 微合金化和 Cr、Mo 合金化的成分设计。其基本力学性能指标情况如表 1 所示，其中直径 8.9mm，标距为 35mm。

表 1 X90 棒材拉伸试验结果

编号	抗拉强度（MPa）	屈服强度（MPa）	延伸率（%）	均匀延伸率（%）	截面收缩率（%）
1	783	732	28	5.3	82.66
2	776	696	28	5.2	81.66
3	770	691	26	5.6	79.67

1.2 不同应力状态圆棒拉伸试验

研究发现，应力三轴度（平均应力/等效应力）对金属材料的断裂显著影响。为揭示应力三轴度与 X90 钢断裂行为的关系，研究设计了如图 1 所示的带圆弧缺口圆棒试样。每种类型设计 3 件试样，其中光滑圆棒试样为 1.1 节中的试样，试验结果见表 2，获得的载荷-位移曲线如图 2 所示。表 2 中断裂应变由断口的截面收缩率换算而来，损伤应变能是指从载荷-位移曲线最高点开始，到发生断裂期间，图 2 中横坐标与纵坐标围成的面积（损伤部位材料吸收的能量）。结果显示，缺口半径不同，对应的断裂应变（由截面收缩率换算真实应变所得）变化为 29%，对应的损伤应变能变化减少 71%。

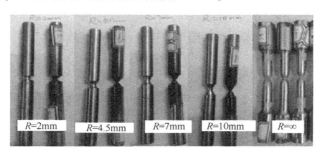

图 1　5 种不同半径缺口圆棒试样

表 2　X90 不同缺口圆棒拉伸试验结果

缺口圆弧半径(mm)	初始直径(mm)	断裂应变	损伤应变能(J)
2	4.58	1.115	12.416
4.5	4.57	1.363	17.745
7	4.59	1.396	19.813
10	4.60	1.492	22.219
∞	8.86	1.576	161.717

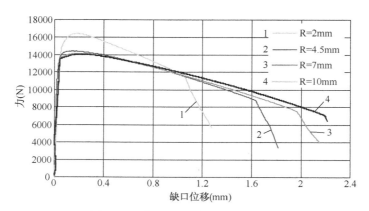

图 2　不同缺口试样载荷—位移关系

1.3　不同应变速率材料拉伸试验

试验室材料拉伸性能试验通常为准静态试验，无应变率效应。但实际管线裂纹轴向扩展中，裂纹扩展速度极快，极端情况，与声速相等，裂纹尖端材料应变速率极高。因此，必须考虑应变速率对材料断裂的影响。为此，在常规材料拉伸试验机上进行 $0.0002s^{-1}$、$0.002s^{-1}$、$0.01s^{-1}$ 等 3 个准静态或低应变速率的拉伸试验，并用 Hopkinson 拉杆试验进行高应变速率的拉伸试验，以此研究应变速率与断裂应变的关系，断裂应变以截面收缩率计算为准。试验结果见表 3，断口截面如图 3 所示（典型的断口为椭圆形，故表 3 中列出 2 个断面直径数据）。

由试验可知，材料的断裂应变随着应变速率增加而增加，但变化不大，其中在准静态（0.0002s⁻¹）与高应变速率（3500s⁻¹）情况下（相差 7 个数量级），断裂应变的变化约为 10%。工程上一般认为，应变速率越高、试样伸长量越小，断裂应变也越小。这一认识是建立在均匀伸长基础上的。实际上，应变率越高，试样其余部位来不及伸长就断裂，导致伸长量变小，但颈缩部位局部应变（真实断裂部位）并未变小，甚至略有增大。

表 3　不同应变速率下的 X90 断裂应变值

应变速率（s⁻¹）	试样直径（mm）	断面 2 个方向的直径（mm）		断裂应变
		a	b	
0.0002	8.86	4.52	3.70	1.546
	8.90	4.42	3.54	
0.002	8.84	4.50	3.60	1.590
	8.84	4.50	3.60	
0.01	8.90	4.40	3.60	1.593
	8.84	4.56	3.54	
2300	2.06	1.06	0.82	1.608
	2.02	1.00	0.80	
3500	2.02	1.02	0.78	1.715

图 3　不同应变速率试验断后照片

2　试验的应力三轴度计算

由 1.2 节中试验可知，X90 钢管断裂应变和损伤应变能与其所处的应力状态相关明显，因此在研究材料断裂准则时，需要考虑应力三轴度的影响。

应力三轴度定义为平均应力/等效应力，无法通过测量得到，故笔者通过有限元模拟得到 5 种拉伸试验的应力三轴度值，见表 4 第 3 列。需要指出的是，上述应力三轴度是指材料弹性阶段的应力三轴度，以此代表材料在不同结构中所处的应力状态。一旦材料进入塑性，即使同一个构件，其应力三轴度值也会发生变化。为了统一描述，采用弹性阶段的应力三轴度作为断裂准则拟合时的横坐标，而在断裂准则使用时，也使用结构弹性阶段的应力三轴度。

另外，缺口处不同位置的应力三轴度不同，以 $R=2$ mm 的缺口圆棒试样为例，其缺口部位静水应力（平均应力的负值）与等效应力的分布如图 4 所示。由试验断口形貌以及相关研究[21]，圆棒试样最先断裂于缺口的中心处，笔者取缺口中心处点的应力三轴度为圆棒试样应力三轴度的值。

为了验证模拟计算所得应力三轴度的准确性，参考文献[22]提出的公式：

$$h = \frac{1}{3} + \ln\left(1 + \frac{r}{2R}\right) \tag{1}$$

对有限元模拟结果进行验证，见表4第4列，发现有限元模拟结果与式(1)计算结果很接近，差异不超过5%，说明模拟结果可信。

表4 不同缺口圆棒拉伸的应力三轴度计算值

缺口圆弧半径(mm)	试样缺口处半径(mm)	模拟应力三轴度	计算应力三轴度
2	2.25	0.803	0.780
4.5	2.25	0.539	0.556
7	2.25	0.463	0.482
10	2.25	0.420	0.440
∞	4.45	0.333	0.333

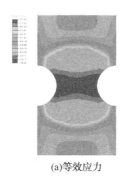 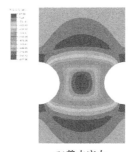

(a)等效应力　　　　　　　(b)静水应力

图4　缺口部位等效应力以及静水应力分布

3 Johnson-Cook 本构及失效模型

3.1 本构模型

根据光滑圆棒不同应变率下准静态拉伸试验和高应变率下动态拉伸所得真实应力—应变曲线，如图5中不同形状的点所示，进行拟合获取 X90 钢 Johnson-Cook(J-C)模型的各项参数。不考虑温度的影响，J-C 模型表达式为：

$$\sigma = (A + B\varepsilon^n)\left[1 + C\left(\ln\frac{\dot{\varepsilon}}{\dot{\varepsilon}_0}\right)\right] \tag{2}$$

拟合可得：$A=680\text{MPa}$，$B=468.8\text{MPa}$，$n=0.586$，$C=9.65\times10^{-3}$，拟合曲线如图5中点实线所示。拟合后，低应变速率下，J-C 模型与试验结果误差在5%以下，高应变速率下，误差最高为12.7%。

3.2 失效模型

根据 J-C 断裂失效模型，材料的断裂应变与应力三轴度和应变率之间可表示为：

$$\varepsilon_{\text{fc}} = [D_1 + D_2\exp(D_3\eta)] \times \left[1 + D_4\ln\frac{\dot{\varepsilon}}{\dot{\varepsilon}_0}\right] \tag{3}$$

将表2中的断裂应变与表4中的应力三轴度进行拟合，可得不考虑应变率时，X90管线钢的断裂应变如下：

$$\varepsilon_{fc} = -3.331 + 5.215\exp(-0.199\eta) \tag{4}$$

在考虑应力三轴度基础上,再考虑应变速率,将表 3 中的应变速率与断裂应变拟合,可得 J-C 失效模型为:

$$\varepsilon_{fc} = [-3.331 + 5.215\exp(-0.199\eta)] \times \left[1 + 0.0034\ln\left(\frac{\dot{\varepsilon}}{\dot{\varepsilon}_0}\right)\right] \tag{5}$$

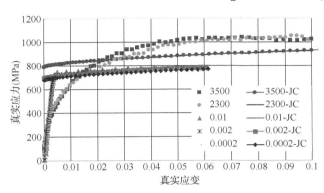

图 5 J-C 模型拟合曲线与实际应力—应变曲线对比

4 基于损伤的 X90 钢管断裂准则

4.1 损伤应变能失效模型

Johnson-Cook 失效模型以断裂应变为基础,而管线钢领域常用的 CVN 是基于能量的准则,同时断裂力学中能量密度释放率、J 积分等均以能量为原理,因此,笔者以损伤力学为基础,仍以能量为判据,给出了一个新的基于能量判据,该能量称为损伤应变能,定义为从应力-应变曲线最高点之后的面积。损伤模型的一般表达式如下[23,24]:

$$D = \begin{cases} f(\sigma, \varepsilon, K) & \varepsilon_p \geqslant \varepsilon_{pD} \\ 0 & \varepsilon_p < \varepsilon_{pD} \end{cases} \tag{6}$$

4.1.1 初始损伤

对于初始损伤门槛值 ε_{pD},一般认为,不同的受载工况下,如单轴拉伸、高周疲劳、蠕变等,ε_{pD} 取值不同。对于比例加载情况,累积等效塑性应变 ε_p 即为等效塑性应变大小 $\bar{\varepsilon}^{pl}$,根据损伤力学一般假设,载荷-位移曲线最高点对应的等效塑性应变 $\bar{\varepsilon}_0^{pl}$(此时材料承载能力达到峰值,之后由于损伤而承载能力下降)即为初始损伤临界门槛值[23]:

$$\varepsilon_{pD} = \bar{\varepsilon}_0^{pl} \tag{7}$$

4.1.2 损伤演化

损伤演化是材料的一种复杂行为,现阶段并无公认的统一模型。当材料等效塑性应变达到均匀延伸率对应的等效塑性应变以后,参考 Freudenthal 延性断裂准则[18],给出损伤累积的表达式如下:

$$D = \frac{G}{G_c} = \frac{\int_{\bar{\varepsilon}_0^{pl}}^{\bar{\varepsilon}^{pl}} \sigma d\bar{\varepsilon}^{pl}}{\int_{\bar{\varepsilon}_0^{pl}}^{\bar{\varepsilon}_f^{pl}} \sigma d\bar{\varepsilon}^{pl}} \tag{8}$$

G_c 为 G 的临界值,对于特定的材料,在特定的服役工况下(如拉伸、疲劳、蠕变等),

为一个定值,由试验确定。当 $G=G_c$ 时,材料失效。

4.2 材料参数 $\bar{\varepsilon}_0^{pl}$、G_c

损伤失效模型需要确定材料载荷-位移曲线最高点对应的等效塑性应变 $\bar{\varepsilon}_0^{pl}$(塑性均匀延伸率)和损伤应变能临界值 G_c。$\bar{\varepsilon}_0^{pl}$ 值可通过光滑圆棒拉伸试样测得,如表1所示,X90管线钢的塑性均匀延伸率为:

$$\bar{\varepsilon}_0^{pl} = 0.05 \tag{9}$$

G_c 为材料从损伤出现(开始颈缩)到断裂时期的塑性应变能,其值等于最高点之后应力-应变曲线与坐标轴所围成的面积。由于 G_c 是能量密度,因此需要测定从损伤到断裂之间累积的塑性应变能及材料的损伤体积。损伤体积为颈缩发生时的试样截面积与颈缩长度之积,该长度为断裂特征长度,因此根据 G_c 定义可推导得:

$$G_c = G_t/V_t = G_t/LS_t \tag{10}$$

G_t 值等于载荷—位移曲线最高点至断裂期间与横坐标围成的面积。由于对于拉伸等比例加载的服役工况(非疲劳、蠕变等工况),G_c 为材料常数,故可通过光滑圆棒拉伸试验确定 G_c。由表2最后一行数据可知,光滑圆棒拉伸试验 G_t 为161.717J,S_t 为61.65mm²,又根据表1中的延伸率和均匀延伸率之差可知,L 为7.68mm,从而根据式(10)可得 $G_c = 0.342\text{J}/\text{mm}^3$。

5 断裂特征长度

利用 J-C 失效模型或损伤失效模型进行工程问题的数值计算时,均需明确材料多大范围发生破坏,体现为数值计算(有限元)的网格划分尺度。理论上,破坏源于微观,尺寸越小越好,如 0.1mm,但对于至少几十米长的管道,计算规模显然不可接受。若单元划分过大,超出了实际材料断裂发生的范围,则会导致计算失真。为初步解决该问题,笔者认为在工程应用中,可以用断裂特征长度来表示材料断裂发生的范围,并将此作为网格划分的上限值。

根据式(10)中 L 的定义,作为均匀化处理,L 表征断裂特征长度。由试验结果可知,L 并非定值,而是随应力三轴度 η 和试样直径 d 变化而变化。对于X90管材,根据式(10)、G_c 值和表2的数据,利用直径4.5mm的带缺口试样,可计算得 L 与 η 的关系如图6所示。

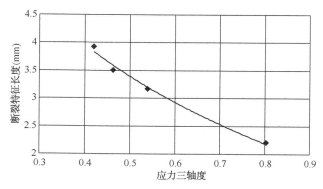

图6 断裂特征长度与应力三轴度关系

拟合公式得:

$$L = -2.56\ln h + 1.62 \tag{11}$$

L 与直径 d 的关系可由本试验 3 种不同直径(2mm、4.5mm 和 8.9mm)的光滑圆棒拉伸试验结果获得。其中 $d=2$mm 时,测量得 L 为 1.7mm,如图 7 所示。

当 $d=8.9$mm 时,前文已根据试验中延伸率和均匀延伸率之差计算得 $L=7.68$mm;当 $d=4.5$mm 时,可根据图 6 或者建立的式(11)计算得 $L=4.45$mm(光滑圆棒拉伸 η 为 0.33),由此得 L 与 d 的关系如图 8 所示。

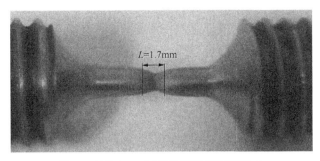

图 7　直径 2mm 试样动态拉伸断裂残余

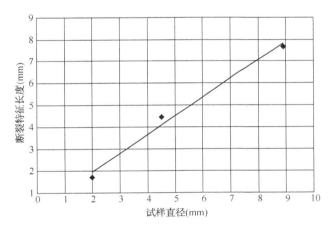

图 8　断裂特征长度与试样直径的关系

拟合 d 与 L 的公式如下:
$$L = 0.85d + 0.24 \tag{12}$$

同时考虑 L 与 η、d 的关系,将 $\ln\eta$、d 视为自变量,通过线性回归,可得
$$L = -2.21\ln\eta + 0.81d - 1.85 \tag{13}$$

由此,对某个 X90 管材的结构进行数值计算时,可首先根据预设的数值模型计算预计断裂区域或扩展路径上的应力三轴度,并根据结构的厚度或直径,根据式(13)计算 L,根据 L 重新划分断裂区网格,然后根据断裂准则进行断裂预测。由于拟合公式用试验数据量较少,不具有普遍性,笔者认为对于管壁不大于 16.3mm 的钢管,公式具有较强参考意义。

由于 L 是平均意义上的断裂区域,故 G_c 的获取也应以 L 为界限,获得平均意义上的应变能密度,而不是在一个微观或细观范围内计算取得 G_c 值。由此,试验获取断裂准则以及数值计算网格划分均以 L 为界限,两者保持一致,作为平均化判断结构断裂的方法,具有一定应用价值。

6 结论

(1) 对不同应力三轴度和不同应变速率的材料拉伸进行了试验,并拟合了 X90 管线钢的 Johnson-Cook 本构模型和断裂模型。发现应力三轴度增大 2.43 倍,断裂应变减小 29%,而应变率增加 7 个数量级,断裂应变增加 10%。断裂应变的应力三轴度效应明显而应变速率效应不明显。

(2) 基于损伤力学原理,根据不同应力三轴度的拉伸试验,得到了以材料均匀塑性延伸率、损伤应变能密度为材料指标的断裂准则,并得到 X90 的均匀塑性延伸率为 0.05,损伤应变能密度为 $0.342 J/mm^3$。

(3) 以试验为依据,提出了 X90 管材断裂特征长度的拟合公式。建议试验获取单位体积损伤应变能时,在断裂特征长度范围内获取,并且有限元数值模拟时单元网格尺度为断裂特征长度。

符 号 注 释

R——拉伸试样缺口处圆弧半径,mm;
$\dot{\varepsilon}$——拉伸试验的应变速率,s^{-1};
η——表示应力三轴度(平均应力/等效应力);
r——为拉伸试样缺口处的最小半径,mm;
σ——等效应力,MPa;
ε——表示等效塑性应变;
$\dot{\varepsilon}_0$——参考应变速率,文中取为 $2\times10^{-4} s^{-1}$;
A——J-C 本构模型材料参数,表示参考应变速率下的初始屈服应力,MPa;
B——J-C 本构模型材料应变硬化模量,MPa;
C——J-C 本构模型材料应变速率强化参数;
n——J-C 本构模型材料应变硬化指数;
ε_{fc}——J-C 失效模型中断裂应变临界值;
D_1——J-C 失效模型中材料参数;
D_2——J-C 失效模型中材料参数;
D_3——J-C 失效模型中材料参数;
D_4——J-C 失效模型中材料参数;
D——损伤力学中的损伤变量;
ε_p——累积等效塑性应变;
σ——等效应力,GPa;
ε_{pD}——初始损伤临界门槛值,ε_p 达到 ε_{pD} 时,开始发生损伤;
$\bar{\varepsilon}^{pl}$——比例加载情况下的等效塑性应变;
$\bar{\varepsilon}_0^{pl}$——拉伸试验载荷最大点对应的等效塑性应变;
G——材料超过均匀延伸率后损伤应变能密度累积量,J/mm^3;
G_t——从损伤到断裂的应变能总量,J;
G_c——G 的临界值,J/mm^3,当 $G=G_c$ 时,材料失效;
$\bar{\varepsilon}_f^{pl}$——断裂时对应的等效塑性应变;
V_t——断裂影响区域的体积,mm^3;
S_t——损伤前试样的截面积,mm^2;

L——断裂的特征长度,断裂时裂纹两侧(发生严重塑性变形)的长度之和,测量时采用损伤前的值,mm;
d——拉伸试验试样直径,mm。

参 考 文 献

[1] API SPEC 5L 44th-2007, Specification for line pipe[S]. Wsshington, D.C.: API pubilishing seevices, 2007.

[2] Maxey W A. Fracture initiation, propagation and arrest[C]//Proceedings of 5th Symposium on Line Pipe Research. Houston, USA: American Gas Association, 1974.

[3] Eiber B. Fracture propagation-1: fracture-arrest prediction requires correction factors[J]. Oil and Gas Journal, 2008, 106(39): 567-569.

[4] Eiber B. Fracture propagation-conclusion: prediction steel grade dependent[J]. Oil and Gas Journal, 2008, 106(40): 145-153.

[5] Zhu Xiankui, Leis B N. CVN and DWTT energy methods for determining Fracture arrest toughness of high strength pipeline steels[C]//Proceedings of the 2012 9th International Pipeline Conference. Calgary, Canada: ASME, 2012: 565-573.

[6] Zhou D W. Measurement and modelling of R-curves for low-constraint specimens[J]. Engineering Fracture Mechanics, 2011, 78(3): 605-622.

[7] Xu J, Zhang Z L, Østby E, et al. Constraint effect on the ductile crack growth resistance of circumferentially cracked pipes[J]. Engineering Fracture Mechanics, 2010, 77(4): 671-684.

[8] Kalyanam S, Wilkowski G M, Shim D J, et al. Why conduct SEN(T) tests and considerations in conducting/analyzing SEN(T) testing[C]//Proceedings of the 2010 8th International Pipeline Conference. Calgary, Canada: ASME, 2010: 337-346.

[9] Berardo G, Salvini P, Mannucci G, et al. On longitudinal propagation of a ductile fracture in a buried gas pipeline: numerical and experimental analysis[C]//Proceedings of the 2000 3rd International Pipeline Conference. Calgary, Canada: ASME, 2000.

[10] Nordhagen H O, Kragset S, Berstad T, et al. A new coupled fluid-structure modeling methodology for running ductile fracture[J]. Computers & Structures, 2012, 94-95: 13-21.

[11] O'Donoghue P E, Kanninen M F, Leung C P, et al. The development and validation of a dynamic fracture propagation model for gas transmission pipelines[J]. International Journal of Pressure Vessels and Piping, 1997, 70(1): 11-25.

[12] 帅健, 张宏, 王永岗, 等. 输气管道裂纹动态扩展及止裂技术研究进展[J]. 石油大学学报:自然科学版, 2004, 28(3): 129-135.

[13] 庄苗, O'Donoghue P E. 能量平衡结合有限元数值计算分析天然气管道裂纹稳定扩展问题[J]. 工程力学, 1997, 14(2): 59-67.

[14] 赵均海, 李艳, 张常光, 等. 基于统一强度理论的石油套管柱抗挤强度[J]. 石油学报, 2013, 34(5): 969-976.

[15] 崔铭伟, 曹学文. 腐蚀缺陷对中高强度油气管道失效压力的影响[J]. 石油学报, 2012, 33(6): 1086-1092.

[16] 赵新伟, 罗金恒, 路民旭, 等. 管线钢断裂和疲劳裂纹扩展特性研究[J]. 石油学报, 2003, 24(5): 108-112.

[17] 冯耀荣, 庄苗, 庄传晶, 等. 裂纹尖端张开角及在输气管线止裂预测中的应用[J]. 石油学报, 2003, 24(4): 99-102, 107.

[18] 杨锋平, 罗金恒, 张华, 等. 金属延性断裂准则精度的评价[J]. 塑性工程学报, 2011, 18(2): 103-106.

[19] 夏佃秀,王学林,李秀程,等.X90级别第三代管线钢的力学性能与组织特征[J].金属学报,2013,49(3):271-276.

[20] 钱亚军,肖文勇,刘理,等.大直径X90M管线钢的开发与试制[J].焊管,2014,37(1):22-26.

[21] Bao Yingbin, Wierzbicki T. A comparative study on various ductile crack formation criteria[J]. Journal of Engineering Materials and Technology, 2004, 126(3): 314-324.

[22] Freudenthal A M. The Inelastic Behavior of engineering materials and structures[M]. New York: John Wiley and Sons, 1950: 112-181.

[23] Lemaitre J, Desmorat R. Engineering damage mechanics[M]. Berlin Heidelberg: Springer, 2005: 13-93.

[24] 杨锋平,孙秦,罗金恒,等.一个高周疲劳损伤演化修正模型[J].力学学报,2012,44(1):140-147.

本论文原发表于《石油学报》2017年第38卷第1期。

双金属机械复合管环焊工艺及强度匹配设计研究现状及趋势

杨专钊[1,2]　王高峰[1,2]　闫　凯[1,2]　魏亚秋[3]　李安强[3]　杜中强[3]　卢卫卓[3]

（1. 中国石油集团石油管工程技术研究院；2. 石油管材及装备材料服役行为与结构安全国家重点实验室；3. 北京隆盛泰科石油管科技有限公司）

摘　要：随着油气开发逐渐向高腐蚀、高温高压、深海等苛刻环境转移，耐蚀合金衬里或内覆双金属复合管逐渐在苛刻的油气开采环境中广泛应用。本文首先介绍了双金属机械复合管典型焊接工艺，管端封焊及对焊焊接工艺的优点、应用案例、失效案例，结果显示其中封焊、过渡焊是此类工艺的质量薄弱点。文中给出了施工对焊小角度坡口和间隙、焊接工艺精确控制和焊接检验加强等建议。随后，介绍了管端堆焊及其施工对焊的研究现状、趋势和应用案例。结果显示，采用同种合金焊接材料完成焊缝打底、填充和盖面等的连接方式，可彻底避免封焊、过渡焊等形成的缺陷和应力，并成功应用于海底天然气开发用双金属复合管焊接，取得较好效果，是今后双金属复合管管端加工和施工对焊重要趋势和方向。此外，介绍了双金属复合管环焊接头强度匹配设计、接头缺陷容限设计等研究现状，结果显示目前尚无统一的焊接接头强度设计、容限设计系统理论，并且指出双金属复合管管道系统在线检测、腐蚀评价、寿命预测以及管道修复等研究方向。

关键词：耐蚀合金；衬里复合管；环焊；焊接工艺规范；强度匹配设计

　　双金属复合管是指将两种或者多种不同金属材料通过一定工艺、技术复合而成，以发挥其不同金属特性的复合管。一般双金属复合管，均由基体钢管和衬里钢管或者覆层金属组合而成，基体钢管起结构作用，为钢管承压承载，衬里钢管或者覆层金属，起耐蚀或者耐磨作用，保护基体钢管。

　　油气输送管道用双金属复合管是防止油气介质内腐蚀的新型防腐技术[1-3]，复合管作为油气田开发腐蚀预防和控制的主要措施，与碳钢相比具有耐蚀优势；与耐蚀合金纯材相比，具有价格优势，因而基于其良好的经济性和技术性。复合管相关技术开发和应用研究，在近20多年来取得长足发展[4-6]。

　　据统计，国外油气田应用双金属复合管从1990年至2005年累计达$2×10^4$t[7]。德国Butting公司生产的机械复合管已有上千公里应用于欧洲、北美以及亚洲等国的海底和陆上油气

基金项目：陕西省青年科技新星项目（编号：2015KJXX-73）。

作者简介：杨专钊，男，生于1979年07月，博士研究生，高级工程师，主要从事油气输送管道工程管材研究、设计、制造、施工技术支持、咨询及监理服务。电话：13571917003，E-mail：yangzhuanzhao@cnpc.com.cn。

管道；英国PROCLAD公司生产的冶金复合管也在全球许多国家得到应用[8-9]。Voestalpine（奥钢联）1992年开始供货轧冶金复合板，年产$3×10^4$t热轧冶金复合板。截至2011年，Technip已经累计完成13个项目超过300km的机械复合管应用业绩以及7个项目超过60km的冶金复合管应用业绩。

国内油气田双金属复合管应用始于1991年，不仅应用于苛刻腐蚀环境地面油气集输系统中，还成功应用于海底油气田开发中。塔里木油田牙哈凝析气田地面集输管道采用20G/316L复合管；长庆油田分公司靖安首站采用20/0Cr18Ni9复合管；大庆油田、长庆油田的注水管道也已开始采用双金属复合管[6]；迪那2气田地面建设工程气田集输工程计量管道和集气管道采用L245NB/316L双金属复合管[10]；新疆塔里木盆地的大涝坝凝析气田循环注气工程采用X65/316L双金属复合管[11]；新疆克深2气田应用了三种材质（L245/316L、L360/316L、L415/316L）、五种规格（直径114.3~508mm）的双金属复合管，累计108.7km[12]；普光气田集输管道安全隐患治理工程采用L360QS/N08825衬里复合管。据作者不完全统计，我国陆地油气田开发应用双金属复合管超过500km。我国首个海洋用双金属复合管项目——长约22.5km崖城13-4海管项目于2011年9月8日铺设完工，后续双金属复合管先后在南海番禺35-1/35-2气田、东海平湖HY1-1/HY1-2气田海底油气开发中应用，海洋区块应用双金属复合管累计超过100km[13-14]。

双金属复合管的连接是耐蚀管道集输系统非常重要环节。如果连接点环焊缝内外层金属同时接触腐蚀性流体，将会出现更加严重的电化学腐蚀现象，并加快腐蚀，失去原有功能，从而导致管道腐蚀、泄漏、穿孔等失效。截至目前，集输管网用双金属复合管连接时，都采用焊接连接，这就给焊接技术带来新的挑战，即不仅仅要实现复合管连接，同时要保证耐蚀层的完整性和耐蚀性。通常耐蚀层厚2~3mm，材料为不锈钢、镍基合金等，基管多为管线钢、结构钢、容器钢等碳钢。异种材料焊接本来比较困难，外加焊接对口多为单面焊双面成型，受基管椭圆度、坡口加工等影响，极易引起错边而引入残余应力。再加上异种焊丝填充、金属熔入，为焊接接头的质量保证带来了难题。

鉴于双金属复合管在我国应用尚属起步阶段，对于双金属复合管焊接技术和经验存在一定不足，本文重点介绍了双金属复合管封焊工艺、堆焊工艺、对焊工艺，对比分析各种工艺的优缺点，并且总结分析了国内外双金属复合管焊接技术研究现状，同时给出国内双金属复合管环焊工艺建议和发展趋势，以便为复合管设计、制造、施工等相关技术人员提供参考。

1 管端封焊及施工对焊工艺

1.1 工艺简介

关于双金属复合管焊接安装工艺研究文献较多。依据相关研究文献，早期工艺可以总结为封焊、根焊、填充、盖面等环节。焊接材料及其焊接工序为管端封焊、不锈钢打底焊或根焊、过渡材料过渡、基材匹配焊丝填充和盖面。

双金属复合管管端封焊就是指双金属机械复合管或衬里复合管管端坡口部分的衬层和基层之间采用焊接形式密封，用来隔离间隙内的水气、杂物等进入焊缝，从而提高焊缝质量。封焊一般在复合管制造厂进行，管端先预加工坡口，将内衬层伸出3~10mm，环形封焊，封焊后再进行坡口修磨。封焊前后外形实物如图1所示。

管端封焊后，施工对焊一般都经过根焊、过渡焊、填充、盖面等过程，最终完成双金属复合管环焊缝焊接。双金属复合管环焊缝焊接示意图如图2所示[10]。

图1 封焊前后外形图

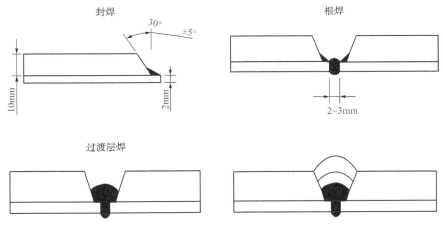

图2 封焊、根焊、过渡焊、填充、盖面等焊接示意图

1.2 国内外应用研究现状

国外关于管端封焊及施工焊接技术研究文献不多,据文献介绍,国际上复合管较早的德国Butting工厂,可提供管端封焊或管端堆焊,供用户选择。国内关于封焊及对焊技术研究、工艺开发和应用较多。

文献[15]设计了 $\phi 114mm×(12+1.5)20G/316L$ 复合管施工堆焊坡口,如图3所示,焊接工艺顺序为封焊(手工钨极氩弧焊,焊材为ER316L)、打底焊(手工钨极氩弧焊,焊材为ER316L)、过渡焊(焊条电弧焊,焊材为E309-15)、基层填充与盖面焊接(焊条电弧焊,焊材为E4215碳钢焊条),详细焊接方法、焊接材料和工艺参数见表1,过渡焊焊材E309-15化学成分表见表2。现场焊接的600多道焊口X射线检测一次合格率在96%以上。该工艺对复合管的焊接有很好的效果。

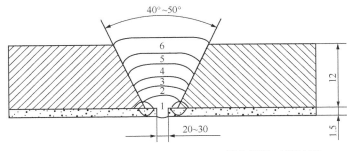

图3 $\phi 114mm×(12+1.5)316L+20G$ 复合管坡口设计图

表1 ϕ114mm×(12+1.5)316L+20G 复合管焊接工艺参数

焊道名称	焊接方法	焊材型号	规格 d(mm)	电源极性	气体流量 Q(L/min) 正面	气体流量 Q(L/min) 背面	焊接电流 I(A)	电弧电压 U(V)	焊接速度 v(cm/min)
封焊层	氩弧焊	ER316L	1.2	DC-	7~12	—	50~70	8~10	5~8
打底焊道	氩弧焊	ER316L	2.0	DC-	7~12	15~20	80~100	9~12	5~8
过渡层	焊条电弧焊	E309-15	2.5	DC+	—	12~15	50~70	22~26	8~15
填充层	焊条电弧焊	E4315	3.2	DC+	12~15		70~90	22~26	8~15
填充层	焊条电弧焊	E4315	3.2	DC+			70~90	22~26	8~15
盖面层	焊条电弧焊	E4315	4.0	DC+			90~120	22~26	8~15

表2 过渡焊焊材 E309-15 化学成分

元素	C	Si	Mn	P	S	Ni	Cr	Mo
含量(%)	≤0.08	≤1.00	1.00~2.50	≤0.030	≤0.020	14.0~16.0	19.00~22.00	2.00~3.00

西安交通大学焊接研究所与中国航天科技集团公司四院41所联合研究了薄壁不锈钢内衬复合管(20/0Cr18Ni9)的焊接工艺和焊接接头性能。第1层采用背部充氩保护的钨极氩弧焊(TIG)和超低碳奥氏体不锈钢焊丝 TGS-309L 焊接不锈钢层。第二层采用手弧焊和超低碳 CHS 062 奥氏体不锈钢焊条焊接过渡层。第三层为碳钢焊条进行填充和盖面[16-18]。文献[19]研究制订了常减压蒸馏装置工艺管道 16MnR/00Cr17Ni14Mo2 复合管 DN3000×(22+3)mm 焊接工艺,和先焊接基层,再焊接过渡层,最后焊接衬层的焊接顺序,并按照 JB 4708—2000《钢制压力容器焊接工艺评定》[20]完成焊接工艺评定。文献[21]也采用封焊(手工钨极氩弧焊)—过渡焊(手工钨极氩弧焊)—基层的焊接(焊条电弧焊)工序,试验研究了 L360QB/316L 复合管电弧焊环焊缝焊接工艺。

随着双金属机械复合管管端封焊和环焊工艺不断研究和经验总结,封焊及对焊工艺研究以及应用日趋增加。文献[10]介绍了2008年至2010年期间,封焊、根焊、过渡焊和填充焊等焊接工艺在迪那2气田地面建设工程气田集输工程计量管道和集气管道焊接过程中的应用。计量管道和集气支线管道为双金属复合管 ϕ168.3×(10+2),ϕ114.3×(8+2)L245NB/316L,并再次验证了该焊接工艺的可行性和稳定性。据文献[22],位于新疆塔里木盆地的大涝坝凝析气田循环注气工程用 X65/316L 双金属复合管,经过封焊后,采用打底焊(钨极氩弧焊)—过渡焊(钨极氩弧焊)—基层的焊接(焊条电弧焊)工序,填充金属全部采用 ER309Mo 为主材的焊丝,对焊完成的342道焊口一次合格率达到97.2%,详细焊接工艺参数见表3。

表3 大涝坝气田和雅克拉气田双金属复合管焊接工艺参数

焊接层次	焊接方法	填充金属 焊材型号及规格(mm)	焊接电流 极性	焊接电流 焊接电流(A)	电弧电压(V)	焊接速度(cm/min)	热输入(kJ/min)
打底	GTAW	TGF309,ϕ2.6	负	80~95	12~13	5~6	12.35
过渡层	GTAW	ER309MoL,ϕ2.5	负	120~135	12~14	8~9	12.6
填充	SMAW	ER309MoL-16,ϕ2.5	负	70~80	23~26	6~7	17.83
盖面	SMAW	ER309MoL-16,ϕ2.5	负	70~80	23~26	6~7	15.6

关于双金属复合管是否需要过渡焊,文献报道和研究结果不尽一致。部分文献认为过渡焊具有良好的隔离和过渡作用,也有部分文献报道过渡焊引起焊接缺陷。文献[23]对20G/316L复合管进行了TIG焊试验,复合管坡口即设计及焊道工序如图4所示。焊道分为不锈钢根焊、过渡焊和填充焊接。不锈钢焊接,试验选用与母材成分相近的1.2mm HS316L焊丝,基层选用焊丝为1.2mm ER55-G。为了防止碳层材料对基层不锈钢材料的稀释以及碳元素的扩散,在基层与复合层之间添加镍铬含量较高的过渡焊丝,试验选用规格为1.2mm的ER309L。结果表明,焊缝分为碳钢层、碳钢与过渡层间的扩散层、过渡层和不锈钢层四个区域,如图5所示。焊缝根部Ni,Cr合金元素与焊接材料相比无明显变化,采用过渡焊丝起到了保持根部焊缝合金元素含量的作用。文献[24]也证实过渡层起到了良好的隔离作用,防止合金元素稀释,保证了不锈钢焊缝的性能。

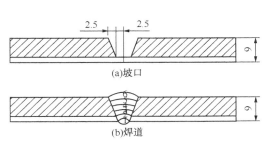

图4 复合管坡口设计及焊道工序图

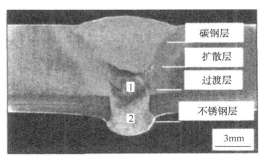

图5 焊缝整体形貌

文献[25]在上述工艺基础上,进行了工艺改进,略去了过渡焊。该文章研究了X65/Incoloy825复合管的焊接工艺试验,焊接方法是钨极氩弧焊,保护气体为氩气,焊丝为ERNiCrMo-3,焊丝直径为1.6mm,焊接工艺参数见表4。从根焊到填充,全部采用同种焊材ERNiCrMo-3完成。焊后力学性能检验、模拟腐蚀环境失重试验表明,X65/Incoloy825复合管焊接接头具有良好的性能,可应用于腐蚀性油气田开发中。焊材ERNiCrMo-3焊材化学成分见表5。

表4 Incoloy825合金+X65复合管焊接工艺参数

层次	电源极性	焊接电流(A)	焊接电压(V)	焊接速度(cm/min)	线能量(kJ/cm)	氩气流量(L/min)	
						正面	背面
1	直流正接	70~110	10~12	8~14	5~15	7~13	15~25
2~3	直流正接	100~160	10~12	6~12	5~15	7~13	10~20
4~n	直流正接	100~160	10~12	10~18	5~15	7~13	

表5 ERNiCrMo-3焊材化学成分

元素	Ni	Cr	Mo	Nb	Fe
含量(%)	≥58.0	22.00~23.00	8.0~10.0	3.0~4.0	≤1.0

复合管对焊工艺中的封焊、根焊、过渡、填充、盖面焊接工序较为成熟,但此类焊缝薄弱环节仍然为封焊和过渡焊位置。文献[26]针对某油气田地面工程用双金属复合管,基管材质L360、衬管材质316L,规格为273mm×9(7+2)mm,焊接过程包括封焊、打底、过渡、填充、盖面焊接等。在进行复合管焊接质量监督检查中,抽查焊口220道,存在超标缺陷焊

口50道,一次焊接合格率仅为77.2%,分析缺陷焊口原因主要源于封焊位置和过渡焊位置。文献[27]对L415QB+316L机械复合管焊接接头裂纹进行分析,结果说明封焊、过渡焊稀释、过渡、熔合等问题难以控制,熔合区易出现裂纹。

邝献任等[28]报到过某油田地面集输管线用DN500 L415/316L双金属复合管环焊缝开裂。其工艺设计为早期传统焊接工艺:封焊、根焊、过渡焊、填充、盖面。焊接材料见表6。从焊缝硬度测试结果可以看出,该环焊缝填充焊、根焊硬度值为180~240HV10,过渡焊局部区域硬度值超过400HV10,显微硬度测试如图6所示,环焊缝基管母材、填充焊及过渡焊组织如图7所示。

表6 L415+316L双金属复合管焊接材料

焊道	型号	牌号	验收标准	规格
封焊	R309LT1-5	ATS-F309L	AWS A5.22	2.2
根焊	R316LT1-5	ATS-F316L	AWS A5.22	2.2
过渡焊	ER309LMo	ATS-309MoL	AWS A5.9	2.4
填充、盖面	E5015	CHE507	GB/T5117-1995	3.2/2.4

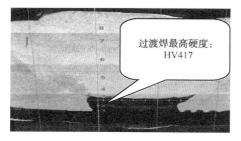

图6 硬度试验试验结果及位置示意图

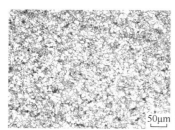
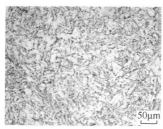

图7 环焊缝基管母材、填充焊及过渡焊组织

从失效环焊缝金相及显微硬度试验中可以看出,复合管环焊缝中过渡焊焊缝组织出现了较多马氏体,硬度值超过350HV10,是典型的硬脆组织。在现场传送过来的焊接工艺评定中可以看出,该失效环焊缝对接时采用了4种材质进行焊接,即根焊、过渡焊、封焊、填充焊,并分别采用了不同的焊材。其中根焊采用ATS-F316L焊条(不锈钢焊条),封焊采用ATS-F309L焊条(不锈钢焊条),过渡焊采用ATS-309MoL焊条(不锈钢焊条),填充焊采用CHE507焊条(碳钢焊条)。焊接顺序依次为封焊(管厂进行)、根焊、过渡焊、填充焊。由上述分析可以看出,在过渡焊(不锈钢材质)上面紧随着填充焊(碳钢材质),使得在过渡焊和填充焊相熔的区域形成了中合金焊缝。这种中合金焊缝在快速冷却下,极易产生硬脆的马氏体组织。现场焊接接头封焊和过渡焊部位硬度高,抗裂纹萌生及扩展能力低,环焊缝在封

焊处产生裂纹并迅速扩展是本次环焊缝开裂的主要原因之一。外加环焊缝受外界约束、内部载荷以及两侧壁厚差等因素，在环焊缝处形成较大的应力集中，加速焊缝扩展和开裂。

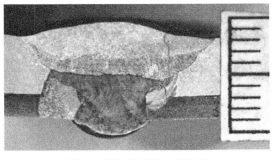

图8 环焊缝开裂典型形貌

文献［29］研究分析了新疆克深2气田双金属复合管应用中出现了焊口开裂、延迟裂纹、打压刺漏和投产后刺漏等问题，环焊缝开裂典型形貌如图8所示。分析结果显示，焊缝Ni、Cr元素稀释造成填充层、盖面层硬度超高，韧性变差，止裂能力变弱等为本次复合管焊接失效的直接原因。此外，由于基管与衬管之间为机械结合，两种材质的热膨胀系数不同，封焊处会有明显应力集中，组对焊接时会有强制应力，容易成为内部裂纹源多发区。在冬季施工后管道收缩使焊口高硬度区承受较大弯曲应力导致复合管失效。同时，文章根据调查分析结果，为消除填充层及盖面层出现高硬度区现象，实验室尝试新的焊接工艺，即采用不锈钢焊条（ER309Lmo）直接焊接填充层及盖面层，并在封焊、根焊和过渡焊时进行背面氩气保护，取得较好效果。

1.3 优缺点及趋势分析

综上，双金属机械复合管早期形成的典型焊接工艺，即封焊、根焊、过渡、填充、盖面焊接工序，相对比较成熟稳定，且具有成本低、便于施工等优点，在迪那2气田、大涝坝凝析气田、克深2气田等国内工程项目中应用，取得较好效果。但是封焊、过渡焊焊接缺陷、焊缝稀释等问题难于控制。如焊接环境、异种材料焊接、焊接工艺控制不当时，极易诱导产生焊接缺陷、焊接应力。因此，此类焊接工序的应用，需要具有稳定的焊接工艺控制手段和严格的焊接检验手段，确保封焊和过渡焊焊缝质量，从而保证环焊接头的焊缝质量。同时，坡口改进为小角度窄型坡口，小间隙对焊是主要设计趋势，不仅大量节约焊材，同时可提高焊接效率和施工进度。此外，采用同时适用于基层和耐蚀层焊接性较好的合金焊材，不仅能提高焊接效率，更能有效避免多种焊材的过渡、稀释等问题，切实提高焊缝质量。更重要的是，改进管端封焊工艺为管端堆焊工艺。

2 管端堆焊及对焊工艺研究现状及趋势

2.1 工艺简介

据API 5LD-2015，双金属机械复合管管端堆焊可以替换封焊。

双金属机械复合管管端堆焊就是指将双金属机械复合管或衬里复合管管端50~100mm内衬层去除，采用堆焊形式连接内衬层和基层，并替代管端内衬层。管端堆焊流程基本工序包括：双金属复合管内镗（内镗长50~100mm，内镗深3~4mm）、堆焊、对焊后内镗、管端坡口加工。管端堆焊坡口及堆焊工序示意图如图9所示，堆焊前后实物效果如图10所示。

管端堆焊后，经过坡口加工，按照工艺评定合格的复合管环焊工艺规范焊接。

2.2 国内外应用研究现状

伴随双金属复合管焊接经验积累和技术发展，不断研究开发出新的焊接工艺。如前所述，采用一种焊材，完成整个环焊缝焊接，取消不锈钢焊材、过渡焊材和碳钢焊材的交替使用，有效避免了焊缝稀释、过渡不良等问题，提高了环焊缝质量。此外，通过管端堆焊替代

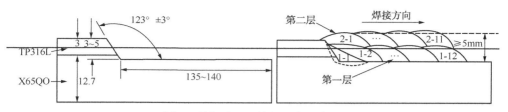

图 9 堆焊坡口及堆焊工序示意图

图 10 堆焊实物图(堆焊前、堆焊后及堆焊剖面)

封焊,坡口改进为"U"形坡口,以及焊接方法技术改进等方法可有效提高环焊质量,提高一次合格率。

文献[30]介绍了双金属复合管的焊接工艺可以单独采用 GTAW,GMAW 和 SMAW 或这些焊接工艺相结合的方法。焊接可以采用手工焊、半自动焊和自动焊。目前,国外的焊接技术较为成熟,全自动 TIG 或 MIG 焊接工艺已经开发成功,如法国 SERIMAX 公司开发的全自动 GMAW 焊接设备、AIR-LIQUIDE WELDING 公司推出的 TOP TIG 焊接设备均可用于复合管的焊接,这些工艺使复合管焊缝质量好,生产效率高,但对装配精度要求高,工艺使用范围较窄,一般用于对焊接效率要求较高的管线,如长距离集输管线的焊接。海洋石油工程股份有限公司承担的国内首条双金属复合海底管道施工中,引进了 TIP TIG 高效焊接设备,成功开发出半自动焊接工艺,克服了国产复合管椭圆度大、组对精度不高的客观条件,取得了较好的施工效果[7]。

文献[31]认为采用气保焊工艺具有电弧在保护气压缩下热量集中、影响区窄、焊接效率高、焊接材料节省且利于焊接自动化技术推广的优点,设计了 273mm×(7.1+3)mm L360QS/N08825 镍铁合金复合管气体保护焊工艺及焊接试验,坡口设计为"J"形坡口,采用单面焊双面成型的施焊工艺,依次完成耐蚀层(TIG)、过渡层 TIG 和基层 MAG 的焊缝金属填充,J 形坡口结构和焊接示意图如图 11 所示。

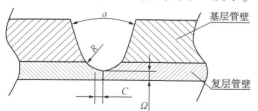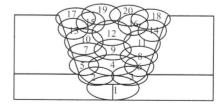

图 11 J 形坡口结构和焊接示意图

文献[32]研究了 φ168.3mm×(12.7+3)mmX65/316L 双金属复合管全自动 TIP TIG 接焊工艺,包括坡口设计、管道的组对、背面保护及全位置焊接等。整个焊接过程分为根焊、热道焊、填充焊及盖面,对接焊所用焊丝材料全部为Inconel625合金,直径为1.2mm。本试验采用奥地利焊机TIP TIG,电源为WIG500IDC,配备独立振动送丝系统,PLC控制焊接参数。对口设备采用ϕ168mm内对口器,提高了效率。采用(5±2)°的窄间隙坡口进行焊接,焊接的坡口形式如图12所示,其具体焊接工艺参数见表7。焊接完成后对焊缝进行外观检验、X射线无损检测、力学性能试验及金相组织分析。结果表明,焊缝外观成型良好,无损检测均为Ⅰ级,力学性能达到标准要求。

表7 ϕ168.3mm×(12.7+3)mm X65/316L双金属复合管全自动TIP TIG接焊工艺

焊接过程	电流(A)	送丝速度(m/min)	爬行速度(m/min)	焊道摆宽(mm)
根焊	160~180	40~50	0.20~0.24	0
热焊道	200~240	50~60	0.18~0.22	2~4
填充焊	250~280	60~70	0.24~0.30	1
盖面焊	220~240	30~40	0.10~0.18	10~12

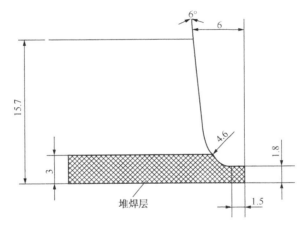

图12 X65+316L双金属复合管全自动TIP TIG接焊工艺坡口形式

2013年建造的我国南海番禺35-1/35-2海底输气工程项目用双金属复合管,在机械复合管规范设计中,管端明确要求:管端堆焊100mm,堆焊材料为ASTM B443 Alloy 625(UNS N06625),堆焊厚度为3mm。对焊后加工"U"形坡口,采用625焊丝打底、填充和盖面形成双金属复合管环焊缝接头,宏观形貌如图13所示。

图13 番禺35-1/2项目双金属复合管环焊缝接头外观

文献[33]研究了API 5LD ϕ219.1mm×(12.7+3)mm X65/Inconel 625复合管管端采用V形坡口,TIG焊接工艺,焊丝为ERCrNiMo-3,打底、填充和盖面后进行电化学腐蚀分析。试验结果表明,焊接接头具有较好的耐蚀性能。

据文献[34]介绍，PY35-1/35-2气田位于中国南海，南距香港约250km，总长度约为50km。PY35-1/35-2海上气田开发项目海底管线海管使用3LPE、3LPP双金属复合钢管(无水泥配重层)，钢管规格为ϕ168.3mm×12.7/3mm X65/316L、ϕ273.1mm×15.9/3mm X70/316L。该项目的铺设作业船为HYSY201船。主线节点焊接采用GTAW(TIP TIG)工艺焊接。管线对接焊缝无损检测方法包括X射线探伤、渗透探伤。一次合格率达到92%以上，可见该工艺在海管铺设应用取得了良好效果。番禺35-1/35-2双金属复合管焊接效率统计见表8。

表8 番禺35-1/35-2双金属复合管海上铺管焊接效率统计

管线铺设位置	钢管规格、牌号	总焊口数	返修数量	一次合格率
PY34-1 CEP—PY35-1 PLET	ϕ168.3×12.7/3、X65/316L	1572	71	95.48%
PY34-1 CEP—PY35-2 PLET3	ϕ273.1×15.9/3、X70/316L	1499	106	92.93%
PY35-2 PLET1—PY35-2 PLET2	ϕ168.3×12.7/3、X65/316L	769	18	97.66%

2015年建造的我国东海平北黄岩油气田群(一期)开发项目绍兴36-5至黄岩1-1WHPA双金属机械复合管，ϕ219.1mm×(11.1+3)mmX65/316L，规格书设计中也明确要求，管端90mm范围内进行堆焊，堆焊材料采用ASTM B443 Alloy 625(UNS N06625)，堆焊方法采用GMAW或者脉冲GTAW。

据文献[35]~[36]介绍，平北黄岩二期由两条管线即BYT至KQTA和KQTA至KQTB组成。BYT至KQTA为30.8km，由18.3km ϕ273.1×15.9mm碳钢管管线及12km ϕ273.1mm×(12.7+3)mm X65QO/316L复合管管线构成；KQTA至KQTB为8.3km ϕ219.1mm×(12.7+3)mm X65QO/316L复合管管线，由海隆106船铺设完成。该项目是该施工船首次铺管，技术和经验尚不足。在前期预制中，ϕ219.1mm×(12.7+3)mm-X65QO/316L一次合格率为86%，ϕ273.1mm×(12.7+3)mm-X65QO/316L一次合格率为88%，平北黄岩二期双金属复合管预制接长焊接效率统计见表9。在后期正式海管铺设安装过程中，通过前期技术和经验积累，焊接技术和焊接效率显著提高，ϕ219.1mm×(12.7+3)mm-X65QO/316L一次合格率达到98.52%，ϕ273.1mm×(12.7+3)mm-X65QO/316L一次合格率为91.51%，平北黄岩二期双金属复合管海上铺管焊接效率统计见表10。

表9 平北黄岩二期双金属复合管预制接长焊接效率统计

双金属复合管规格与材质	焊口数	返修数量	一次合格率
ϕ219.1mm×(12.7+3)mm-X65QO/316L	363	43	86%
ϕ273.1mm×(12.7+3)mm-X65QO/316L	506	37	88%

表10 平北黄岩二期双金属复合管海上铺管焊接效率统计

双金属复合管规格与材质	焊口数	返修数量	一次合格率
ϕ219.1mm×(12.7+3)mm-X65QO+316L	338	5	98.52%
ϕ273.1mm×(12.7+3)mm-X65QO+316L	471	40	91.51%

2.3 优缺点及趋势分析

综上，双金属机械复合管管端堆焊后采用同种合金焊接材料完成焊缝打底、填充和盖面等，是今后双金属复合管管端加工和施工对焊重要趋势和方向。

相比封焊，管端堆焊具有焊缝质量高，易于自动化，焊接高效率，工艺可控等优点，避免了多种焊材的过渡、稀释等问题，杜绝了封焊、过渡焊焊缝缺陷和焊接应力。同时，其能提高现场环焊补口和修补效率，从而提高双金属复合管利用率。但是对焊设备、对焊材料和填充用合金焊材都成本较贵，是适用单位和制造单位不可忽视的重要成本开支。

我国率先在崖城13-4项目、番禺35项目、平黄项目等海底天然气开发用双金属复合管上采用此类工艺，并取得较好效果。

3 双金属复合管环焊接头强度及容限设计研究

近年来，国内外学者针对双金属复合管环焊接头强度匹配设计、容限设计、ECA评估等课题研究逐渐增加。

2007年，TWI开展关于冶金复合管焊接工艺开发研究，包括焊接方法选择、参数设定、焊材选择，以及复合管环焊缝工程临界评估(Engineering Critical Assessment，ECA)方法等[37]。项目目标是减少安装时间和成本，增加热输入量和焊接效率，以提高焊接生产效率，同时研究焊接检验技术，并提高环焊缝的完整性。2010年，挪威Acergy Norway AS和英国TWI，联合研究了机械和冶金复合管环焊缝工程临界评估，首次给出环焊缝允许缺欠评估。ECA分析结果显示，复合管接头缺陷允许水平比碳钢管材纯材低些。建议设计复合管环焊缝的强度过匹配[38]：

（1）尽可能限制复合管屈服强度上限，如SMYS+100MPa或者更低；
（2）通过焊接工艺、焊材选择等方法获得过匹配焊缝。

2013年，DNV OS-F101-2013给出了基于BS7910:2005的ECA分析方法，包括冶金和机械复合管材料。但是该方法仅适用于等匹配或者过匹配的冶金复合管材料和焊接材料。2013年，TWI GSP联合赞助项目，利用CTOD、复合管拉伸性能、FEM有限元法等，结合弹塑性边界理论，研究机械和冶金双金属复合管环焊缝断裂评估，旨在研究并验证一种断裂评估技术以获得机械和冶金双金属复合管环焊缝缺陷ECA评估[39]。同年，文献[40]提出等效材料方法(Equivalent Material Method，EMM)，该方法遵循了标准工程临界评估设计路线，考虑了焊接接头是由一个等效的材料组成，即基材流变曲线(base metal flow curve)和焊缝流变曲线共同决定的一个等效材料流变曲线，如图14所示。

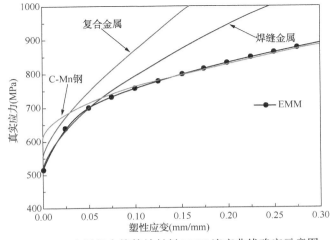

图14 双金属复合管等效材料EMM流变曲线确定示意图

2014年，塞班SAIPEM SA等[41]开展了双金属复合管环焊缝完整性评价研究，主要针对双金属复合管环焊缝ECA评估，认为Bonora N等[40]在OMAE2013中提出的等效材料方法（equivalent material method，EMM），没有考虑焊接接头中的三种材料的差异性，而是简化等效为一种材料。该等效材料具有定义的流变曲线，即为焊接接头三种材料较低流变曲线的插值流变曲线。本文通过考虑复合管内压载荷和焊接残余应力的影响，验证了采用EMM方法的可行性。研究了2个缺陷模型，并且给出多道焊接模拟。模拟结果显示，EMM方法计算的裂纹驱动力通常相对保守。依据DNV-OS-F101-2013目前的设计路线，同样也验证了EMM是一种可行有效的双金属复合管环焊缝接头结构完整性评估方法。文献[42]研究了X65+625双金属复合管环焊缝接头缺陷裂纹衍生及发展。小直径圆棒试样，试样工作直径为2mm。采用连续破坏原理来准确测定含缺陷双金属复合管环焊缝焊接接头延性裂纹萌生和扩展条件，建立了预测模型，包含复合管基材、覆材以及焊材等参数以及几何模型。Morten Hval等[43]研究了S铺管法安装的冶金复合管运行阶段的工程临界评估ECA。复合材料由于其基材为低合金碳钢，屈服强度为415MPa或者450MPa。而覆材具有相当低的强度，再加上焊接材料的过匹配，这种强度不均匀，导致复合材料具有各向异性，从而给断裂评估技术带来新的挑战。运行状态下，局部材料性能可能超过弹性极限，通常基于BS7910的评估方法可能不适用于复合管，而且应该结合更为准确详细的FEM方法。该文献相对载荷选取基于S铺管载荷，并且采用LINKpipe有限元程序来分析和验证复合管环焊缝允许临界缺陷。结果显示，即便是最保守的情况，选择其中多个组合缺陷参数，包括环向缺陷位置、尺寸、焊偏、最大屈曲应变、低边界撕裂阻力，可获得允许的最大临界焊接缺陷。敏感性分析说明得出的安全系数是比较大的。

2015年，Rodolfo F Souza[44]提出了含不同尺寸环向表面裂纹和非匹配环焊焊缝在弯曲载荷作用下，一种J积分评价程序。同时，提出了海洋用双金属复合管低匹配625环焊缝临界缺陷尺寸评价程序。采用3D数字模拟J积分方法在含裂纹非等强度匹配环焊缝进行应用，并探讨其ECA的应用。2016年，新加坡国立大学和挪威船级社DNV研究了温度对镍基冶金复合管环焊缝性能的影响评估。在自制试验设备上，采用断裂试验和拉伸试验研究分析了温度（24℃、100℃、180℃）对镍基冶金复合管环焊缝性能的影响。断裂试验为单边缺口弯曲试验和单边缺口拉伸试验两种，缺口深度为0.2。拉伸试验结果显示，在评价温度升高24℃、100℃、180℃时，基材、焊材、和覆材的拉伸强度、屈服强度、延伸率以及弹性模量都有所下降。而断裂韧性下降0.6倍，依据缺口位置和试样形状不同而不同[45]。

综上，虽然国内外学者近期已逐渐开展双金属复合管环焊接头强度匹配设计、容限设计、ECA评估等相关内容研究，但是还不多见，没有形成统一的焊接接头强度设计、容限设计系统理论。焊接接头强度（包括静强度、疲劳强度等）、焊缝缺陷类型、缺陷尺寸、腐蚀缺陷等对双金属复合管系统运行稳定性、安全性评价研究非常少见，以及双金属复合管管道系统在役检验与评价、双金属复合管腐蚀缺陷安全评价、剩余寿命预测以及双金属复合管管道修复技术等研究少之又少，所以今后这些方面研究是重要研究热点和研究趋势。

4 结语

早期形成的双金属机械复合管典型焊接工艺，即封焊、根焊、过渡、填充、盖面焊接工序较为成熟，且具有成本低、便于施工等优点，并具有较多的应用案例。但是封焊、过渡焊受焊接环境、异种材料焊接、焊接工艺控制等多种因素影响，极易出现焊接缺陷、应力集中、焊缝稀释等问题。因此，此类焊接工序的应用，需要具有成熟的焊接工艺控制手段和严格的焊接检验手段，确保封焊和过渡焊焊缝质量，从而保证双金属复合管环焊接头的焊缝质量。同时，坡口改进为小角度窄型坡口，小间隙对焊是主要设计趋势，不仅可以大量节约焊材，同时也能提高焊接效率和施工进度。

相比封焊，管端堆焊及采用同种合金焊接材料完成焊缝打底、填充和盖面等的连接方式，可避免多种焊材的过渡、稀释等问题，杜绝封焊、过渡焊焊缝缺陷和焊接应力；同时，更能提高现场环焊补口和修补效率，从而提高双金属复合管利用率。我国率先海底天然气开发用双金属复合管上采用此类工艺，并取得较好效果。管端堆焊具有焊缝质量高、易于自动化、焊接效率高、工艺可控等优点，是今后双金属复合管管端加工和施工对焊重要趋势和方向。

此外，关于双金属复合管环焊接头强度匹配设计、容限设计、ECA 评估等相关内容研究不足，没有形成统一的焊接接头强度设计、容限设计系统理论。同时，焊接接头强度（包括静强度、疲劳强度等）、焊缝缺陷类型、缺陷尺寸、腐蚀缺陷等对双金属复合管系统运行稳定性、安全性影响评价研究也是重要研究方向。伴随双金属复合管的推广应用，双金属复合管管道系统在役检验与评价、腐蚀缺陷安全评价、剩余寿命预测以及双金属复合管管道修复技术等研究也迫在眉睫。

参 考 文 献

[1] 杜轻松，曾德智，杨斌. 双金属复合管塑性成型有限元模拟[J]. 天然气工业，2008，28(9)：64-66.
[2] 曾德智，杜轻松，谷坛，等. 双金属复合管防腐技术研究进展[J]. 油气田地面工程，2008，27(12)：64-65.
[3] 曾德智，杨斌. 孙永兴，等. 双金属复合管液压成型有限元模拟与试验研究[J]. 钻采工艺，2010，33(6)：78-79.
[4] Colin Macrae. One pipe or two-manufacturing clad pipe for energy applications[J]. Tube & Pipe Journal，2008.
[5] 傅广海. 徐深气田 CO_2 防腐技术分析[J]. 油气田地面工程，2008，27(4)：66-67.
[6] 李发根，魏斌，邵晓东，等. 高腐蚀性油气田用双金属复合管[J]. 油气储运，2010，29(5)：359-362.
[7] 周声结，郭崇晓，张燕飞. 双金属复合管在海洋石油天然气工程中的应用[J]. 中国石油和化工标准与质量，2011，11：115-116.
[8] 钱乐中. 油气输送用耐腐蚀双金属复合管[J]. 特殊钢，2007，28(4)：42-44.
[9] 宋亚峰. 油气田开采用冶金复合双金属复合管开发与应用[C]//双（多）金属复合管/板材生产技术开发与应用学会研讨会文集，2008：57-63.
[10] 袁世昌，王凤仙. 双金属复合管焊接工艺及常见缺陷分析[J]. 中国石油和化工标准与质量，2011，31(5)：54-55.
[11] 杨刚. X65/316L 复合管的焊接工艺及焊接质量控制[J]. 焊接技术，2012，41(12)：56-57.
[12] 许爱华，张靖，院振刚，等. 新疆克深2气田双金属复合管失效原因[J]. 油气储运，2014，33(9)：1024-1028.
[13] 胡春红，李秀锋，熊海荣，等. 海上油气田海底管道用耐腐蚀合金复合管[J]. 石油化工设备，2015：85-91.
[14] 魏斌，李鹤林，李发根. 海底油气输送用双金属复合管研发现状与展望[J]. 油气储运，2016，35(4)：343-355.
[15] 杨胜金，肖国豪. 316L+20G 不锈钢复合管焊接技术[J]. 焊接，2006(1)：58-04.

[16] 王能利,潘希德,薛锦,等.20/0Cr18Ni9复合管焊接工艺和接头的抗腐蚀性能[J].焊接,2003(5):23-27.

[17] 王能利,柏朝晖,张希艳等.20/0Cr18Ni9复合管手工电弧焊工艺研究[J].热加工工艺,2005(9):36-38.

[18] 王能利,张希艳,潘希德,等.多层焊对A/P异种钢复合管SMAW接头组织及性能的影响[J].焊接学报,2009,28(9):51-54.

[19] 张立辉.16MnR+00Cr17Ni14Mo2复合管的焊接[J].石油工程建设,2010(1):48-50.

[20] JB 4708—2000 钢制压力容器焊接工艺评定[S].

[21] 朱丽霞,何小东,仝珂,等.L360QB/316L复合管电弧焊环焊缝接头组织性能研究[J].热加工工艺,2013,42(5):188-189.

[22] 杨刚.X65/316L复合管的焊接工艺及焊接质量控制[J].焊接技术,2012,41(12):56-57.

[23] 吕世雄,王廷,冯吉才.20G/316L双金属复合管弧焊接头组织与性能[J].焊接学报,2009,30(4):93-96.

[24] 范兆廷,张胜涛,殷林亮,等.316L-20G双金属复合管焊缝组织元素扩散分析[J].重庆大学学报,2012,35(11):99-103.

[25] 杨旭.Incoloy825合金+X65复合管焊接性和焊接工艺[J].焊管,2008,31(5):33-35.

[26] 姜汉胜,权高军.双金属复合管焊接质量控制[J].石油工业技术监督,2011,27(10):18-20.

[27] 王斌,朱洪亮,王维.L415QB+316L复合管焊接裂纹分析[J].热加工工艺,2014,43(1):193-196.

[28] 邝献任,魏友国,杨洋,等.某油田地面集输管线双金属复合管环焊缝开裂分析[C]//2015年全国失效分析学术会议论文集,2015.

[29] 许爱华,张靖,院振刚,等.新疆克深2气田双金属复合管失效原因[J].油气储运,2014,33(9):1024-1028.

[30] 赵晨光,彭清华,马宏伟,等.双金属复合管焊接技术探讨[J].焊管,2013,36(1):60-63.

[31] 胡伟 高永杰.镍铁合金复合管气体保护焊工艺研究[J].石油化工建设,2015,37(2):70-73.

[32] 王东红,郭江涛,钟炜,等.双金属复合管全自动TT对接焊工艺研究[J].热加工工艺,2014,43(19):216-217.

[33] Xu L Y, Li M, Jing H Y. Electrochemical behavior of corrosion resistance of X65/inconel 625 welded joints [J]. International Journal of Electrochemical Science, 2013, 8(2): 2069-2079.

[34] LSZJ13-366. PY35-2/ PY35-1海上气田开发项目海管铺设监造报告[R],2013.

[35] LSZJ15-S212:平北黄岩二期双金属复合管陆地预制项目复合管陆地预制驻厂监造报告[R],2015.

[36] LSZJ15-S211.平北黄岩二期双金属复合管海铺设管驻厂监造报告[R],2015.

[37] TWI. Improved Welding, Inspection and Integrity of Clad Pipeline Girth Welds[R]. TWI PROJECT Out line, PR12141, 2007.

[38] Macdonald K A, Cheaitani M. Engineering critical assessment in the complex girth welds of clad and lined linepipe materials[C]//Proceedings of the 8th International Pipeline Conference, 2010: 823-843.

[39] Eren E. Fracture assessment of clad and lined pipe girth welds. Group sponsored project outline[R]. PR21981-1, 2013.

[40] Bonora N, Carlucci A, Ruggiero A, et al. Simplified approach for fracture integrity assessment of bimetallic girth weld Joint [C]//ASME 2013. International Conference on Ocean, Offshore and Arctic Engineering, 2013.

[41] Carlucci A, Bonora N, Ruggiero A, et al. Integrity assessment of clad pipe girth welds[C]//ASME 2014 33rd International Conference on Ocean. Offshore and Arctic Engineering, 2014.

[42] Carlucci A, Bonora N, Ruggiero A, et al. Crack initiation and propagation of clad pipe girth weld flaws[C]// ASME 2014 Pressure Vessels & Piping Conference, 2014.

[43] Hval M, Lamvik T, Hoff R. Engineering critical assessment of clad pipeline installed by S-lay for the operation phase[J]. Procedia Materials Science, 2014, 3: 1216-1225.

[44] Souza R F, Ruggieri C. Fracture assessments of clad pipe girth welds incorporating improved crack driving force solutions[J]. Engineering Fracture Mechanics, 2015, 148: 383-405.

[45] Chong T V S, Kumar S B, Lai M O, et al. Effects of elevated temperatures on the mechanical properties of nickel-based alloy clad pipelines girth welds[J]. Engineering Fracture Mechanics, 2016, 152: 174-192.

本论文原发表于《油气储运》2017年第36卷第3期。

柔性管内衬高密度聚乙烯的气体渗透行为研究

张冬娜　李厚补　戚东涛　丁　楠
邵晓东　魏　斌　蔡雪华

(中国石油集团石油管工程技术研究院·
石油管材及装备材料服役行为与结构安全国家重点实验室)

摘　要：柔性管内衬材料的气体渗透行为对管材的气体输送效率及管材的使用寿命都具有非常重要的影响，为此，针对目前国内广泛使用的柔性管内衬材料——高密度乙烯(HDPE)，使用压差法对其气体渗透行为进行了研究。实验使用的渗透测试气体分别为 CO_2 及 CH_4，测试温度为 30~80℃，对不同测试条件下所获得的渗透系数、溶解度系数及扩散系数进行了分析，并使用煤油模拟油气耦合环境，研究了不同环境下 HDPE 的气体渗透行为。结果表明：(1)不同温度下 CO_2 和 CH_4 在 HDPE 中的气体渗透系数满足阿伦尼乌斯方程，两种气体透过 HDPE 的表观活化能分别为 32.61kJ/mol 及 43.84kJ/mol；(2)随着温度的升高，CO_2 的扩散系数变化不明显，溶解度系数逐渐增加；(3)随着温度的升高，CH_4 的扩散系数逐渐增加，溶解度系数变化不明显；(4)在模拟油气耦合介质的测试中，煤油起到了溶胀及增塑作用，CH_4 在煤油浸泡 HDPE 中的扩散系数及渗透系数增加。

关键词：柔性管；内衬材料；气体渗透；高密度聚乙烯；油气耦合；溶解度系数；扩散系数

1　研究背景

与传统钢管相比，以热塑性塑料为内衬的柔性管(也称 RTP 管、增强复合管等)具有耐腐蚀、柔韧性高、疲劳性能好等优点，现已普遍应用于油气地面集输管线，并且在海洋油气集输领域也开始应用[1]。作为内衬的热塑性塑料通常使用聚乙烯(PE)、聚丙烯(PP)、聚氯乙烯(PVC)和聚偏氟乙烯(PVDF)等。虽然热塑性塑料具有耐腐蚀等性能优点，但在使用过程中，气体分子的自由运动会在塑料层中发生渗透现象。气体渗透不仅会导致输送气体的浪费，还会对内衬材料造成破坏。据报道，在高温、高压、酸性环境集输状态下，与各种气体

基金项目：国家自然科学基金项目"油气耦合介质在热塑性塑料中的渗透特性及机理研究"(编号：51304236)、陕西省自然科学基金项目"酸性气体在热塑性塑料中的渗透特性及机理研究"(编号：2014JQ6227)。

作者简介：张冬娜，女，工程师，博士；2013 年毕业于西北工业大学并获博士学位；主要从事油气集输用非金属管的研发工作。地址：(710072)陕西省西安市锦业二路 89 号，电话：(029)81887597、13891849739，ORCID：0000-0001-9156-3551，E-mail：zhangdna@cnpc.com.cn。

（CH_4、H_2S、CO_2等）接触的塑料内衬容易起泡失效[2,3]。其原因是气体组分在材料表面大量吸附后渗透进入内衬层，长期作用下，材料内部积聚的气体压力与管道运行压力趋于平衡。当管线因检修等原因突然降压时，热塑性塑料内气体体积瞬间膨胀，从而在管材内表面形成较多的"气泡"。图1为国内油气集输用热塑性塑料内衬复合管，在投用约2年后发现类似起泡现象。另外，吸附在热塑性塑料内衬表面的气体介质会沿内衬厚度方向发生扩散，进而腐蚀复合管的金属增强层[3]，导致管线承压能力降低，甚至引发恶性事故，造成经济损失和生态破坏。

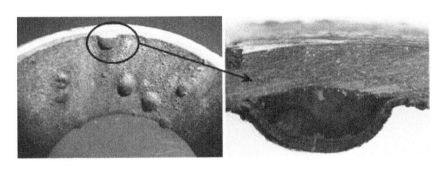

图1 油气集输管线热塑性塑料起泡失效图

内衬材料渗透规律的研究不仅能量化气体的渗透情况，更能为高阻隔热塑性内衬材料的研发建立理论基础。因此针对目前使用最广泛的内衬材料高密度聚乙烯（HDPE）进行气体渗透行为的研究。

塑料薄膜及薄片气体渗透性能测试的方法有压力法（压差法和等压法）、浓度法、体积法、气相色谱法和热传导法等。根据不同的测试方法，国内外已研制出各种类型的塑料薄膜薄片透气性测试装置，其中压差法是塑料薄膜透气性测试中使用较为广泛的一种方法，实验参考GB/T 1038—2000《塑料薄膜和薄片气体透过性实验方法 压差法》，对不同温度下CH_4和CO_2在HDPE中的渗透行为进行研究，并模拟油气耦合环境，研究输送介质对渗透行为的影响。

2 渗透实验

2.1 实验原材料和仪器设备

HDPE薄片由赢创上海分公司制备，厚度为350μm，使用的渗透测试仪器为济南兰光机电技术有限公司生产的VAC-V2型压差法气体渗透仪。

2.2 气体渗透过程

气体分子在聚合物材料中的渗透过程可分为4个步骤：（1）气体分子吸附在聚合物表面；（2）气体分子溶解在聚合物内部；（3）气体分子在聚合物内部由高压侧向低压侧扩散；（4）气体分子达到聚合物的另一侧并在其外表面解吸附[4]。气体溶解过程符合亨利定律[5]：

$$C = Sp \tag{1}$$

式中：C为单位体积聚合物材料所溶解的气体体积，cm^3；p为外界压力，Pa；S为溶解度系数。

气体扩散过程则可用菲克定律[6]来描述：

$$q = -D\frac{dc}{dx} \tag{2}$$

式中：dc/dx 为气体浓度梯度，kg/m^4；D 为扩散系数，cm^2/s；q 为单位时间、单位面积的气体透过量，$kg/(m^2 \cdot s)$。

根据式（1）、（2）可得出气体通过聚合物材料的"稳态"渗透系数（P）的计算公式为：

$$P = DS \tag{3}$$

以上溶解-扩散机理是过去几十年来描述气体在聚合物中渗透行为最为广泛接受的机理模型。

2.3 塑料薄膜薄片气体渗透测试方法

在VAC-V2型压差法渗透仪的测试过程中，使用高精度真空规测量低压侧的压力变化量，高压侧的压力为0.1MPa。当渗透过程达到稳定时，通过式（4）计算得到测试气体的气体渗透量（Q），气体渗透系数（P）由式（5）得出。

$$Q = \frac{\Delta p}{\Delta t} \times \frac{V}{S} \times \frac{T_0}{p_0 T} \times \frac{24}{p_1 - p_2} \tag{4}$$

$$P = \frac{\Delta p}{\Delta t} \times \frac{V}{S} \times \frac{T_0}{p_0 T} \times \frac{d}{p_1 - p_2} = 1.1574 \times 10^{-9} Qd \tag{5}$$

式中：Q 为材料的气体渗透量，$cm^3/(m^2 \cdot d \cdot Pa)$；$\Delta p/\Delta t$ 为稳定透过时，单位时间内低压室气体压力变化的算术平均值，Pa/h；V 为低压室体积，cm^3；S 为试样的实验面积，m^2；T 为实验温度，K；$p_1 - p_2$ 为试样两侧的压差，Pa；T_0、p_0 为标准状态下的温度（274.15K）和压力（$1.0133 \times 10^5 Pa$）；P 为材料的气体渗透系数，$(cm^3 \cdot cm)/(cm^2 \cdot s \cdot Pa)$；$d$ 为试样厚度，cm。

3 结果与讨论

分别在30℃、40℃、60℃及80℃下测试 CO_2 及 CH_4 在HDPE薄片中的气体渗透行为，并研究油气耦合介质对渗透行为的影响。

3.1 HDPE的气体渗透系数分析

气体渗透系数是以恒定温度和单位压力差下，在稳定透过时，单位时间内透过试样单位厚度、单位面积的气体体积。不同温度下 CO_2 和 CH_4 在HDPE中的渗透系数如表1所示。从表1可以看出，随着温度的升高，两种气体在HDPE中的渗透系数逐渐增大。

表1 两种测试气体在HDPE中不同温度的渗透系数表

温度(℃)	渗透系数[$(cm^3 \cdot cm)/(cm^2 \cdot s \cdot Pa)$]	
	CO_2	CH_4
30	8.918×10^{-14}	5.586×10^{-14}
40	1.764×10^{-13}	1.029×10^{-13}
60	2.683×10^{-13}	2.646×10^{-13}
80	6.430×10^{-13}	6.771×10^{-13}

气体渗透系数与材料种类、环境条件、气体的尺寸、极性、可压缩性等有关[7]。从表1可以看出，随着温度的升高，两种气体的渗透系数增加。对于任意一种气体，不同温度下的气体渗透系数满足阿伦尼乌斯方程。

$$P = P_0 \exp\left(\frac{-E_p}{RT}\right) \tag{6}$$

式中：P_0为指前因子，$(cm^3 \cdot cm)/(cm^2 \cdot s \cdot Pa)$；$E_p$为表观活化能[7]，kJ/mol。

图2为不同温度下两种气体在HDPE中的渗透系数取对数后与$1/T$的关系，根据图2及式(6)，可以计算出两种测试气体在HDPE中渗透的表观活化能及指前因子，通过计算结果，可对不同温度下的CO_2和CH_4在HDPE中的渗透系数进行预测，具体的计算结果如表2所示。

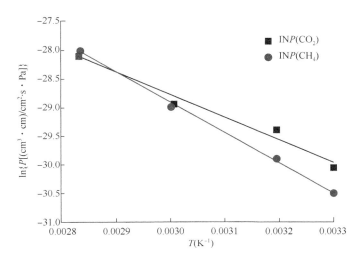

图2 两种测试气体的$\ln P$与$1/T$线性拟合直线图

表2 两种测试气体在HDPE中的表观活化能及指前因子表

测试气体	E_p(kJ/mol)	$P_0[(cm^3 \cdot cm)/(cm^2 \cdot s \cdot Pa)]$
CO_2	32.61	4.09×10^{-8}
CH_4	43.84	2.07×10^{-6}

表观活化能的值越小，气体渗透系数的温度依赖性就越低，由计算出的数据可知，CO_2的表观活化能的值稍低，两种气体活化能的值为同一数量级。

3.2 HDPE的气体渗透行为分析

通过对CO_2和CH_4在HDPE中气体渗透系数的分析，随着温度的升高，两种测试气体在HDPE中气体渗透系数都随之增加，并且与温度存在一定的线性关系。通过1.2对气体渗透过程的描述，可知气体在塑料材料中的渗透过程包括吸附、溶解、扩散和解吸4个过程，使用VAC-V2型渗透仪，分别对气体扩散系数及溶解度系数进行研究，分析气体在HDPE中的渗透行为。

渗透仪通过时间滞后法测定气体在塑料片材中的扩散系数，也称为"高真空法"，即在高真空下，通过对达到平衡状态时的"滞后时间"计算扩散系数，计算公式为：

$$D = \frac{L^2}{6\theta_0} \tag{7}$$

式中：L为试样厚度，cm；θ_0为滞后时间，s。

"滞后时间"如图3所示，在测试软件中通过测定压力与时间的关系，得到相关数据，通过式(7)计算出扩散系数(D)[8]。

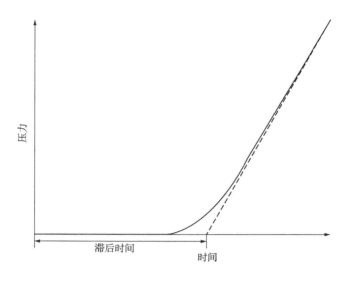

图 3 滞后时间计算方法示意图

根据式(3)可知气体渗透系数(P)是扩散系数(D)与溶解度系数(S)的乘积,在得到渗透系数和扩散系数的数据后,可计算出气体在塑料材料中的溶解度系数,实验设备提供的溶解度系数的单位为 $cm^3/(cm^2 \cdot s \cdot cmHg)$。

对于无孔的聚合物薄膜或片材,其溶解度系数与实验气体的可压缩性有关,即与气体的临界温度(T_c)相关,临界温度是指气体可以转化为液相的最高温度,即当温度高于临界温度时,无论施加多大的压力,气体都不会转变为液相[7]。通常,临界温度越高,气体在聚合物中的溶解性越好,渗透率也越高。而其扩散系数与测试气体的尺寸有关,通常表征气体尺寸的物理量为临界体积或范德华体积[9]。实验针对不同温度下气体在 HDPE 中的渗透行为进行了研究,讨论温度对溶解度系数及扩散系数的影响,图 4 为不同温度下 CO_2 在 HDPE 中的溶解度系数及扩散系数。

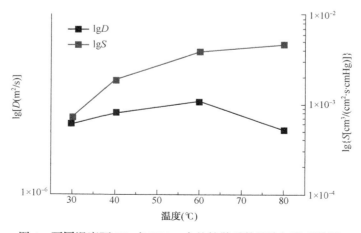

图 4 不同温度下 CO_2 在 HDPE 中的扩散系数及溶解度系数图

通常情况下，随着温度的升高，塑料材料分子链更易运动，气体分子活动能力增加，扩散系数提高，同时气体在聚合物中的溶解度降低。但在图4的数据中，随着温度的升高，CO_2在HDPE中的溶解度系数逐渐增加，扩散系数变化不明显。

气体透过塑料材料的过程是扩散和溶解共同作用的结果，在温度逐渐升高的过程中，扩散和溶解对气体渗透的影响存在相互竞争。对于CO_2，虽然其分子尺寸与N_2和CH_4接近，但是CO_2的临界温度高（CO_2的T_c为304.21K，N_2的T_c为126.20K，CH_4的T_c为191.05K），因此其溶解度高。现有渗透理论认为，气体在聚合物中的溶解过程可分为2个步骤：(1)首渗透剂压缩为液态；(2)以液相的状态与聚合物混合[10]。结合实验结果，在实验的温度范围内，升温对CO_2气体的压缩行为影响有限，但是加速了混合过程，导致溶解度系数的升高。

通常升温状态下扩散系数逐渐增加，但是由于CO_2的高溶解度，更多的CO_2溶解于塑料材料中，随压差扩散的CO_2含量减少，此现象使扩散系数降低，综合以上两种现象，升温对扩散系数的影响不大。

图5为不同温度下CH_4在HDPE中的溶解度系数及扩散系数，与CO_2的渗透过程不同的是，随着温度的升高，CH_4的扩散系数逐渐升高，而溶解度系数变化不大。随着温度的升高，渗透气体的活动能力增加，更易通过塑料材料，因此扩散系数增加。根据之前的分析，CH_4的临界温度低，可压缩性较CO_2差，因此溶解度也比CO_2低。通常情况下升温溶解度系数降低，但是由于气体扩散能力增加，单位时间内更多气体分子聚集在材料内部，两个过程的最终结果是CH_4在HDPE中的溶解度系数随温度的升高无明显的变化。

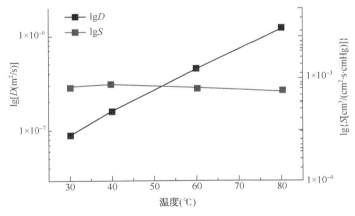

图5 不同温度下CH_4在HDPE中的扩散系数及溶解度系数图

由于实验针对HDPE薄片的气体透过行为进行了研究，因而排除了塑料材料自身结构的影响，其次是渗透气体种类的影响，包括气体分子的尺寸、可压缩性和极性等，其中CO_2和CH_4两种气体表现出了不同的渗透行为。最后是环境条件的影响，包括温度等，从扩散系数和溶解度系数中看出，不同温度下测试气体的透过行为是比较复杂的，通常升温提高扩散系数，降低溶解度系数，这两个过程是同时存在的，渗透系数也是这两个过程共同作用的结果。从实验数据中不难发现，不同温度CO_2和CH_4在HDPE中的透过过程中，两种气体的扩散系数和溶解度系数表现出不同的规律，最基础的理论并不适用于所有气体，需要对目标气源的渗透行为进行系统的分析。

3.3 油气耦合介质对HDPE气体渗透行为的影响

目前,对于热塑性塑料渗透性能主要围绕单组分气体(CH_4、O_2、H_2、N_2和CO_2等)和气体分离膜进行研究。其中单组分气体渗透行为主要为渗透机理方面的理论研究[8-9],但针对柔性管内衬材料的研究较少,而在气体分离的研究中,期望待分离的组分气体在塑料材料中的渗透系数相差较大,从而达到气体分离的效果[5-7,11-15]。在油气集输领域,柔性管输送的流体多为气相与液相的耦合物质。因此实验模拟油气耦合介质环境,分别在30℃、40℃、60℃和80℃下,将HDPE薄片在煤油中浸泡240h后,在浸泡温度下对煤油处理的HDPE试样进行渗透系数测试,测试气体为CH_4。

图6为HDPE试样及浸泡过煤油后HDPE试样的渗透系数对比,从图6可以看出,CH_4在煤油浸泡HDPE中的渗透系数随温度升高逐渐增加,总体趋势与未浸泡的试样相同,但是渗透系数更大。

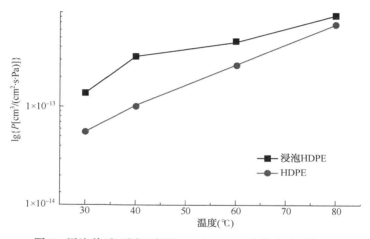

图6 浸泡前后不同温度下CH_4在HDPE中的渗透系数图

聚合物材料的渗透性能取决于材料的自由体积和链段运动能力[16]。自由体积即分子间的空隙,以大小不等的空穴无规分布在聚合物中,提供了分子的活动空间。聚合物链段的运动能力受材料不饱和度、交联度、取代基类型、结晶度和塑化程度等因素影响[17]。其中结晶度是影响气体透过性的重要因素,聚合物的气体渗透理论认为,气体是在非晶区,即无定型区域发生渗透过程的,溶解度与结晶度的关系为$S=S_a X_a$,其中S_a表示聚合物全部处在非晶态时的溶解度;X_a表示聚合物非晶区所占的体积分数[18]。高分子材料内的扩散也主要发生在非晶区,扩散系数与结晶度关系为$D=D_a/\tau$,D_a表示气体在聚合物非晶区的扩散系数,τ表示透过气体分子绕过晶区的路径弯曲度。因此,结晶度越高,有效渗透面积越小,增加了扩散路径的弯曲程度,导致气体的渗透性减小[19]。

针对煤油浸泡前后的HDPE试样进行了示差扫描量热法(DSC)分析,不同浸泡温度下的HDPE的熔融焓(H_m)及熔融峰温(T_m)如表3所示,对于浸泡前后的样品,H_m及T_m相差不大。H_m与材料的结晶度成正比,因此在煤油浸泡的试样中,结晶度最大相差5.9%,这是由于实验的最高温度为80℃,低于HDPE的熔点,很难使材料的结晶结构发生变化,溶剂诱导结晶的影响也非常有限。熔融峰温的一致表明晶片厚度也没有大的改变,因此结晶对气体渗透行为的影响可以忽略。

表3 浸泡HDPE试样的熔融焓及熔融峰温表

试样类别	H_m(J/g)	T_m(℃)
未浸泡	143.0	128.6
浸泡、测试温度为30℃	138.6	128.4
浸泡、测试温度为40℃	142.0	129.0
浸泡、测试温度为60℃	138.6	128.0
浸泡、测试温度为80℃	146.8	127.5

HDPE不存在不饱和度、交联度、取代基在浸泡后变化的问题，但是煤油渗入HDPE中，填充在分子链之间，一方面溶胀使分子间自由体积增加，另一方面煤油起到了增塑的作用，与增塑剂作用类似[16]，提高了链段活动能力，更有利于气体的通过。浸泡前后CH_4通过HDPE试样的溶解度系数及扩散系数如表4所示，煤油浸泡试样的扩散系数增加明显，但是溶解度系数变化不大。这说明在煤油浸泡试样中，气体分子的活动能力提高，更易通过聚合物材料，但是对气体分子的溶解度无明显影响，综上，气体在HDPE中的渗透系数增加。因此在油气耦合介质的集输过程中，不仅要考虑内衬层的气体渗透系数，还需考虑液相介质对内衬材料的溶胀及增塑作用。

表4 浸泡前后HDPE试样的溶解度系数及扩散系数表

实验温度(℃)	D(cm^2/s)		S[cm^3/(cm^2·s·cmHg)]	
	未浸泡试样	煤油浸泡试样	未浸泡试样	煤油浸泡试样
30	9.031×10^{-8}	2.473×10^{-7}	8.247×10^{-4}	7.354×10^{-4}
40	1.578×10^{-7}	5.286×10^{-7}	8.770×10^{-4}	8.205×10^{-4}
60	4.378×10^{-7}	6.466×10^{-7}	8.059×10^{-4}	9.237×10^{-4}
80	1.204×10^{-6}	1.237×10^{-6}	7.498×10^{-4}	8.909×10^{-4}

4 结论

（1）通过对不同温度下CO_2及CH_4在HDPE中气体渗透系数的研究，得出两种气体的渗透系数随温度升高而增加，并且通过阿伦尼乌斯方程计算出两种气体在HDPE中的表观活化能分别为32.61kJ/mol及43.84kJ/mol。

（2）研究了CO_2在HDPE中的扩散系数及溶解度系数与温度的关系，结果表明，随着温度的提高，CO_2的扩散系数变化不明显，但溶解度系数逐渐增加。

（3）研究了CH_4在HDPE中的扩散系数及溶解度系数与温度的关系，结果表明，随着温度的提高，CH_4的扩散系数逐渐增加，但溶解度系数变化不明显。

（4）模拟了油气耦合介质下CH_4在HDPE中的渗透行为，结果表明，煤油起到了溶胀及增塑作用，提高了CH_4的扩散系数及渗透系数。

参 考 文 献

[1] Laney P. Use of composite pipe materials in the transportation of natural gas[R/OL]. Bechtel BWXT Idaho: Idaho NationalEngineering and Environmental Laboratory, 2002. http://igs. nigc. ir/STANDS/BOOK/COMPOSIT-PIPE. PDF.

[2] Shamsuddoha M, Islama MM, Aravinthan T, Manalo A & Lau KT. Effectiveness of using fibre-reinforced polymer composites for underwater steel pipeline repairs[J]. Composite Structures, 2013, 100: 40-54.

[3] Scheichl R, Klopffer MH, Benjelloun-Dabaghi Z, et al. Permeation of gases in polymers: parameter identification and nonlinear regression analysis[J]. Journal of Membrane Science, 2005, 254(1/2): 275-293.

[4] 丁楠,戚东涛,魏斌,等. 海洋柔性管气体渗透机理及其防护措施的研究进展[J]. 天然气工业, 2015, 35(10): 112-116.

[5] Ren Xiaoling, Ren Jizhong, Deng Maicun. Poly(amide-6-b-ethylene oxide) membranes for sour gas separation [J]. Separation and Purification Technology, 2012, 89: 1-8.

[6] Kamaruddin HD, Koros WJ. Some observations about the application of Fick's first law for membrane separation of multicomponent mixtures[J]. Journal of Membrane Science, 1997, 135(2): 147-159.

[7] Merkel TC, Gupta RP, Turk BS, et al. Mixed-gas permeation of syngas components in poly(dimethylsiloxane) and poly(1-trimethylsilyl-1-propyne) at elevated temperatures[J]. Journal of Membrane Science, 2001, 191(1/2): 85-94.

[8] Flaconneche B, Martin J, Klopffer MH. Transport properties of gases in polymers: experimental methods[J]. Oil & Gas Science and Technology, 2001, 56(3): 245-259.

[9] Merkel TC, Bondar VI, Nagai K, et al. Gas sorption, diffusion, and permeation in poly(dimethylsiloxane) [J]. Journal of Polymer SciencePart B: Polymer Physics, 2000, 38(3): 415-434.

[10] Alentiev AY, Shantarovich VP, Merkel TC, et al. Gas and vapor sorption, permeation, and diffusion in glassy amorphous Teflon AF1600[J]. Macromolecules, 2002, 35(25): 9513-9522.

[11] Khanbabaei G, Vasheghani-Farahani E, Rahmatpour A. Pure and mixed gas CH_4 and n-C_4H_{10} permeation in PDMS-fumed silica nanocomposite membranes[J]. Chemical Engineering Journal, 2012, 191: 369-377.

[12] Sadrzadeh M, Amirilargani M, Shahidi K, et al. Pure and mixed gas permeation through a composite polydimethylsiloxane membrane[J]. Polymers for AdvancedTechmologies, 2011, 22(5): 586-597.

[13] Pinnau I, He Zhenjie. Pure- and mixed-gas permeation properties of polydimethylsiloxane for hydrocarbon/methane and hydrocarbon/hydrogen separation[J]. Journal of Membrane Science, 2004, 244(1/2): 227-233.

[14] Singh A, Freeman BD, Pinnau I. Pure and mixed gas acetone/nitrogen permeation properties of polydimethylsiloxane[PDMS][J]. Journal of Polymer SciencePart B: Polymer Physics, 1998, 36(2): 289-301.

[15] Gleason KL, Smith ZP, Liu Qiang, et al. Pure- and mixed-gas permeation of CO_2 and CH_4 in thermally rearranged polymers based on 3,3'-dihydroxy-4,4'-diamino-biphenyl(HAB) and 2,2'-bis-(3,4-dicarboxyphenyl)hexafluoropropane dianhydride(6FDA)[J]. Journal of Membrane Science, 2015, 475: 204-214.

[16] George SC, Thomas S. Transport phenomena through polymeric systems[J]. Progress in Polymer Science, 2001, 26(6): 985-1017.

[17] KatochS, SharmaV, Kundu1PP. Swelling kinetics of unsaturated polyester and their montmorillonite filled nanocomposite synthesized from glycolyzed PET[J]. DiffusionFundamentals, 2011, 15(4): 1-28.

[18] 丁运生, 张志成, 史铁钧. 阻隔性高分子材料研究进展[J]. 功能高分子学报, 2001, 14(3): 360-364.
[19] Kim JH, Lee YM. Gas permeation properties of poly(amide-6-b-ethylene oxide)-silica hybrid membranes [J]. Journal of Membrane Science, 2001, 193(2): 209-225.

本论文原发表于《天然气工业》2017年第37卷第3期。

外径 1422mm 的 X80 钢级管材技术条件研究及产品开发

张伟卫　李　鹤　池　强　赵新伟　霍春勇
齐丽华　李炎华　杨　坤

(中国石油集团石油管工程技术研究院·石油管材及装备
材料服役行为与结构安全国家重点实验室)

摘　要：本文结合中俄东线天然气管道工程用 ϕ1422mm、X80 管材技术条件的研究制定过程，对国内外管线钢管技术标准进行了对比分析，同时对制定的 ϕ1422mm X80 管材技术条件中的化学成分、止裂韧性等关键技术指标及制定过程进行了分析探讨，并对 ϕ1422mm X80 管线钢管的开发过程及产品性能进行了介绍。通过生产试制和产品检测，证明技术条件合理有效的解决了化学成分控制、断裂控制、产品焊接稳定性等技术问题，不仅满足工程要求，而且也适应生产情况，可以保障中俄东线天然气管道的本质安全。本工作可为中俄东线建设 ϕ1422mm 天然气管道提供强有力的技术支撑，同时对于其他天然气管道工程技术条件的制定具有重要的指导意义。

关键词：管线钢管；天然气管道；技术条件；技术指标；化学成分；止裂韧性；焊接

近年来，随着天然气需求的日益增加，我国油气管道特别是天然气管道建设进入了一个新的高峰期，大口径、厚壁、高钢级管线钢管成为管道建设的主要选择[1,2]。在"十一五"期间中石油正式立项开展了 ϕ1219mm X80 管线钢前期先导技术研究，取得了一些成果，制定了一系列 ϕ1219mm X80 管材技术条件，并用于西气东输二线等天然气管线建设中。就当时全球已经建成和正在建设的天然气高压长输管道而言，不论钢级、长度、管径、壁厚还是输送压力，西二线工程都堪称世界之最[3]。

2014 年 5 月，中国石油天然气集团公司与俄罗斯天然气工业股份公司正式签署了《中俄东线管道供气购销协议》，约定从 2018 年起，俄罗斯开始通过中俄东线向中国供气。为了满足中俄东线 $380\times10^8 m^3/a$ 超大输气量的要求，中国石油通过对 ϕ1422mm、X80 钢级管线钢管应用技术的攻关，形成了第三代大输量天然气管道应用配套技术，并决定在国内 737km 的中俄东线黑河—长岭段干线，首次使用 ϕ1422mm、钢级 X80 钢管。

管线钢的质量是保证管线安全的最基本也是最关键的因素之一，钢管订货技术条件是钢管生产、检验和验收的依据，确定其合理的技术要求对保证管线的安全可靠性、经济性和可

作者简介：张伟卫，男，1981 年生，高级工程师，硕士，2008 年毕业于北京科技大学并获硕士学位，主要从事输送管与管线材料及标准方面的研究工作。地址：西安市锦业二路 89 号（710077），电话：（029）81887838，E-mial：zhangweiwei@cnpc.com.cn。

行性是非常重要的。受中国石油管道项目经理部的委托，石油管工程技术研究院负责研究、制定了中俄东线天然气管道工程用 $\phi1422mm$、X80 管线钢、钢管系列技术条件。本文对中俄东线天然气管道工程用 $\phi1422mm$、X80 管线钢、钢管系列技术条件制定过程中的几个关键问题进行了论述，并对 $\phi1422mm$、X80 管线钢及钢管的开发过程及产品性能进行了介绍。

1 国内外相关管线钢标准对比分析

油气输送管道用钢管技术标准从制定方和适用范围进行划分，可分为国际标准、国家标准、行业标准、企业标准等。如 ISO 3183 属国际标准，GB/T 9711 是国家标准，API SPEC 5L 可认为是行业标准，中国石油天然气集团公司发布的 Q/SY 1513 以及中国石油管道建设项目经理部发布的 Q/SY GJX 149—2015 等是企业标准。

目前在我国使用的陆上油气输送钢管基础标准主要有 API SPEC 5L、ISO 3183、GB/T 9711 等。中国石油在吸收国内外技术标准研究成果的基础上，还形成了自己的油气输送钢管技术标准体系，经常使用的通用技术标准有 Q/SY 1513，CDP-S-NGP-PL-006 等，此外根据不同的工程需求，还制定了大量的工程技术条件，如西气东输二线天然气管道工程用管材技术条件、中俄东线天然气管道工程用管材技术条件等。

ISO 3183 是国际标准化组织制定的石油天然气工业管道输送系统用钢管标准，被 GB/T 9711 等同采用。由于 GB/T 9711 的采标修订工作受管理因素限制，更新较 ISO3183 慢，目前等同采用的 ISO3183：2007 版本。API SPEC 5L（管线管规范）是美国石油学会制定的一个被普遍采用的规范。上述标准或规范兼顾了管线钢的技术要求与制造厂实际生产的可行性，但相对管线与制管技术的发展，这些标准或规范中的技术要求显得比较宽松，许多条款仅给出了原则性能要求，具体指标不明确，因此已经很少单独用于管线项目。

目前，世界上大多数石油公司都习惯采用 API SPEC 5L 作为管线钢管采购的基础规范，在该规范基础上，根据当地实际情况或管线工程的具体要求，制订补充技术条件。中国石油的管线钢管通用技术标准 Q/SY 1513、CDP-S-NGP-PL-006 等就是以 API SPEC 5L《管线管规范》为基础，吸收了国内外管线钢和工程经验编制而成，具有很强的实用性和可行性，但由于是通用技术条件，有些指标，如化学成分、夏比冲击功（CVN）等要求比较宽松，需要根据具体工程情况进一步确定。

在具体技术指标方面，GB 9711—2011、ISO3183：2012、API 5L：2012 等标准的最大适用管径包括 1422mm，化学成分指标要求较为宽松，其技术指标要求只满足一般钢管的最基本要求，如要求 C 含量不大于 0.12%，Si 含量不大于 0.45%，其他合金元素范围也非常宽泛[4-6]，没有考虑现场焊接对化学成分的要求，缺乏工程应用指导意义。CVN 要求三个试样最小平均为 54J，仅能满足管体材料不发生启裂失效的最基本要求，不能满足钢管自身止裂要求，因而不能保证长输管道的本质安全。

中国石油企业标准 Q/SY 1513.1—2012《油气输送管道用管材通用技术条件 第 1 部分：埋弧焊管》最大适用管径为 1219mm，最高适用钢级为 X80，理化性能指标基本与 ISO3183、GB/T9711 和 API SPEC 5L：2012 相同，区别在于 Q/SY 1513.1—2012 中碳当量不要求 CE_{IIW} 指标，见表1。夏比冲击韧性只给出了确定方法（冲击功值要求应按 API Spec 5L 的附录 G 确定），没有给出具体数值，需要根据具体工程进行计算或试验验证确定。

中俄东线天然气管道工程用管材技术条件则是在 Q/SY 1513 基础上，借鉴了 API SPEC 5L：2012 的最新成果，结合中俄东线工程的具体特点，对 $\phi1422mm$ 管线钢及钢管的各项关键技术指标进行研究攻关，主要针对近年来管线工程的热点问题，确定了更为严格的化学成分指标，计算并验证了钢管的 CVN 值，规定了夹杂物评定标准等，此外，还对管材和板材的试验检验方法和要求进行了优化，并提出了更严格的制造、检验程序和更科学合理的质量控制措施。如化学成分要求 C 不大于 0.07%，Mn 不大于 1.85%，Nb、Mo、Ni 等根据螺旋缝埋弧焊管和直缝埋弧焊管管型的不同分别有不同的要求，有效的解决了现场焊接质量的稳定性问题。关于 CVN 指标，采用了 Battelle 双曲线（BTC）方法进行了理论计算，并利用近年来管道断裂控制技术研究的最新成果，采用 Leis-2、Eiber、TGRC2 等多种修正方法进行修正[7]，同时通过全尺寸气体爆破试验进行了验证。

中俄东线天然气管道工程用 ϕ1422mm X80 管材技术条件还明确了试验样品的加工要求，如板卷力学性能试验样品要求与板卷轧制方向成 20°取样，拉伸试验采用 ϕ12.7mm 的圆棒试样。

此外，通过研究，中俄东线天然气管道工程用 ϕ1422mm X80 管材技术条件还规定了严格的非金属夹杂物验收极限。针对夏比冲击试验中普遍存在的断口分离问题，增加了夏比冲击试样断口分离程度分级方法，针对钢管管端非分层缺陷检测问题，增加了非分层缺陷的检测和验收方法等。

表 1 Q/SY 1513.1—2012 对 X80 管线钢化学成分的要求

钢号	根据熔炼分析和产品分析的最大质量分数(%)									最大碳当量[①](%)	
	C[②]	Si	Mn[②]	P	S	V	Nb	Ti	其他[⑥]	CE_{IIW}	CE_{Pcm}
L555 或 X80	0.12[③]	0.45[③]	1.85[③]	0.025	0.015	[④]	[④]	[④]	[⑤]	—	0.25

① 根据产品分析。如果碳的质量分数大于 0.12%，则 CE_{IIW} 极限适用；如果碳的质量分数小于等于 0.12%，则 CE_{Pcm} 极限适用。

② 碳含量比规定最大质量分数每降低 0.01%，则允许锰含量比规定最大质量分数增加 0.05%，对于大于等于 L485 或 X70 小于等于 L555 或 X80 的钢级，最大值不应超过 2.00%。

③ 除非另有协议。

④ 除非另有协议，铌、钒和钛的总含量不应超过 0.15%。

⑤ 除非另有规定，铜的最大含量为 0.50%，镍的最大含量为 1.00%，铬的最大含量为 0.50%，钼的最大含量为 0.50%。

⑥ 除非另有规定，不得有意加入 B，残留 B 含量应≤0.001%。

2 ϕ1422mm X80 管材技术条件制定中的几个关键问题

2.1 化学成分

自西气东输二线管道工程开始，我国 X80 管线钢的生产和应用越来越多，随着钢铁冶金技术的进步，为了降低生产成本，国内各钢铁企业根据自身的特点，开发出了多种合金体系的管线钢，不同钢铁企业生产的管线钢化学成分差别很大，甚至同一企业在不同阶段生产的管线钢的化学成分也有很大的差异[8]。这种化学成分的较大差异，会降低焊接工艺和焊材的适用性，缩小现场焊接的工艺窗口，增加管线焊接的难度，造成焊缝力学性能波动加剧，从而给管道的服役安全带来隐患，对于壁厚 20mm 以上 X80 管线钢，这一问题尤为突出。为了解决这一难题，在中俄东线天然气管道工程用 ϕ1422mm X80 管材技术条件制定过程中，对化学成分指标进行了大量的试验研究工作，目标就是限定中俄东线天然气管道工程用管线钢的化学成分波动范围，制定经济、科学的化学成分指标，从而稳定管线钢质量和现场焊接工艺窗口。

2.1.1 碳、锰、铌

化学成分对管线钢的显微组织、力学性能和焊接性能有着重要的影响。通过研究，决定 φ1422mm X80 管线钢采用低 C、Mn 的成分设计，并加入适量的 Mo、Ni、Nb、V、Ti、Cu、Cr 等元素。炼钢时钢材应采用吹氧转炉或电炉冶炼，并进行炉外精炼，并采用热机械控轧工艺(TMCP)生产，最终管线钢的晶粒尺寸达到 10 级以上，从而保证生产出具有良好的强韧性、塑性和焊接性的管线钢。

对西二线等天然气管道工程用 X80 钢管的化学成分及焊接结果进行研究分析发现，管线钢中 C、Mn、Nb 的剧烈波动(图1~图3)，对焊接性能影响具有较大的影响。在管线钢中 C 是增加钢强度的有效元素，但是它对钢的韧性、塑性和焊接性有负面影响[9]。降低 C 含量可以改善管线钢的韧脆转变温度和焊接性，但 C 含量过低则需要加入更多的其他合金元素来提高管线钢的强度，使冶炼成本提高[10]。综合考虑经济和技术因素，C 含量应控制在 0.05%~0.07% 之间。

为保证管线钢中低的 C 含量，避免引起其强度损失，需要在管线钢中加入适量的合金元素，如 Mn、Nb、Mo 等。Mn 的加入引起固溶强化，从而提高管线钢的强度。Mn 在提高强度的同时，还可以提高钢的韧性，但有研究表明 Mn 含量过高会加大控轧钢板的中心偏析，对管线钢的焊接性能造成不利影响[11]。因此，根据板厚和强度的不同要求，管线钢中锰的加入量一般是 1.1%~2.0%。Nb 是管线钢中不可缺少的微合金元素，能通过晶粒细化、沉淀析出强化作用改善钢的强韧性。但有研究表明 Nb 对阻止焊接热影响区晶粒长大和改善热影响区韧性并不十分有效，这是因为在焊接峰值温度下，Nb 的碳、氮化物的热稳定性尚有不足[11]。较低的 Nb 含量，在焊接热循环过程中不能有效抑制热影响区奥氏体晶粒长大，最终导致相变时产生大尺寸的块状 M/A 和粒状贝氏体产物，使韧性恶化。过高的 Nb 含量，在焊接热循环过程中会导致较大尺寸的沉淀析出，同时使晶粒均匀性恶化，也会损害热影响区韧性[12,13]。研究表明，Nb 的加入量一般控制在 0.03%~0.075% 比较合理。

通过对 φ1422mm X80 管线钢的大量实验研究、工业试制分析和专家组研讨，认为 X80 管线钢的 Mn 含量最高不宜大于 1.85%，Nb 的含量应控制在 0.04~0.08 之间。图4和图5给出了按最新制定的技术条件工业试制的 φ1422mm X80 管线钢管的环焊缝及热影响区在 -10℃ 下的 CVN 值，可以看出其合格率高达 97% 以上。

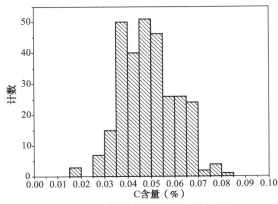

图1 X80 钢管的 C 含量分布统计

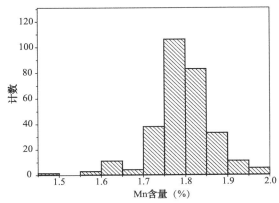

图2 X80 钢管的 Mn 含量分布统计

2.1.2 其他合金元素

Ti 是强的固 N 元素，在管线钢中可形成细小的高温稳定的 TiN 析出相。这种细小的 TiN 粒子可有效地阻碍再加热时的奥氏体晶粒长大，有助于提高 Nb 在奥氏体中的固溶度，同时对改善焊接热影响区的冲击韧性有明显作用。研究表明 Ti/N 的化学计量比为 3.42 左右，利用 0.02% 左右的 Ti 就可以固定钢中 0.006% 的 N。管线钢中的 N 含量一般不超过 0.008%，因此技术条件中 Ti 的含量规定控制在 0.025% 以下。

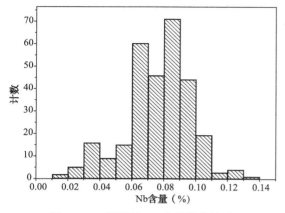

图 3 X80 钢管的 Nb 含量分布统计

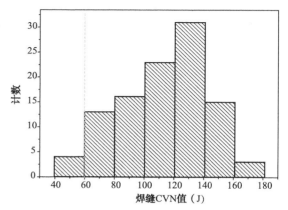

图 4 ϕ1422mm 钢管环焊缝 CVN 分布

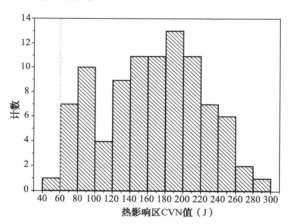

图 5 ϕ1422mm 钢管热影响区 CVN 分布

Cr、Mo 是扩大 γ 相区，推迟 α 相变时先析铁素体形成、促进针状铁素体形成的主要元素，对控制相变组织起重要作用，在一定的冷却条件和终止轧制温度下超低碳管线钢中加入 0.15%~0.35% 的 Mo 和低于 0.35% 的 Cr 就可获得明显的针状铁素体及贝氏体组织，通过组织的相变强化提高钢的强度。

Cu、Ni 可通过固溶强化作用提高钢的强度，同时 Cu 还可以改善钢的耐蚀性，Ni 的加入主要是改善 Cu 在钢中易引起的热脆性，且对韧性有益。在厚规格管线钢中还可补偿因厚度的增加而引起的强度下降。一般管线钢中铜含量低于 0.30%，镍含量低于 0.5%。

为了更好的稳定产品的理化性能，保证钢管具有良好的现场焊接性，结合国内管线钢生产中合金元素的实际控制能力，ϕ1422mm X80 管材技术条件根据钢管类型对 C、Mn、Nb、Cr、Mo、Ni 的含量进行了约定。通过试验研究，并组织冶金和焊接专家讨论协商，确定管线钢中 C 的含量目标值为 0.060%，Mn 的目标值为 1.75%，Nb 的目标值为 0.06%。直缝钢管中 Ni 目标值为 0.20%，必须加入适量的 Mo，且含量应大于 0.08%。螺旋缝钢管中 Cr、Ni、Mo 的目标值均为 0.20%。考虑到生产控制偏差、检测误差及经济性，ϕ1422mm X80 管材技术条件中规定 C 含量不大于 0.070%，Mn 含量不大于 1.80%。直缝钢管 Nb 的含量范围为 0.04%~0.08%，Mo 的含量范围为 0.08%~0.30%，Ni 的含量范围为 0.10%~0.30%；螺旋缝钢管中 Nb 的含量范围为 0.05%~0.08%，Cr 的含量范围为 0.15%~0.30%，Mo 的含量范围为 0.12%~0.27%，Ni 的含量范围为 0.15%~0.25%。表2给出了 ϕ1422mm X80 管材技术条件确定的化学成分含量要求。

表2 ϕ1422mm X80 钢管的化学成分要求

元素	产品分析 max	螺旋缝钢管成分推荐范围	直缝钢管成分推荐范围
碳	≤0.09	≤0.070	≤0.070
硅	≤0.42	≤0.30	≤0.30
锰	≤1.85	≤1.80③	≤1.80③
磷	≤0.022	≤0.015	≤0.015
硫	≤0.005	≤0.005	≤0.005
铌①	≤0.11	0.050~0.080④	0.040~0.080④
钒①	≤0.06	≤0.03	≤0.030
钛①	≤0.025	≤0.025	≤0.025
铝	≤0.06	≤0.06	≤0.06
氮	≤0.008	≤0.008	≤0.008
铜	≤0.30	≤0.30	≤0.30
铬	≤0.45	0.15~0.30	≤0.30
钼	≤0.35	0.12~0.27	0.08~0.30
镍	≤0.50	0.15~0.25	0.10~0.30
硼②	≤0.0005	≤0.0005	≤0.0005
CE_{Pcm}	≤0.23	≤0.22	≤0.22

① V+Nb+Ti≤0.15%。
② 不得有意加入硼和稀土元素。
③ 碳含量比推荐最大含量每减少 0.01% 时，锰推荐最大含量可增加 0.05%，但锰含量不得超过 1.85%。
④ 碳含量比推荐最大含量每减少 0.01% 时，铌推荐最大含量可增加 0.005%，但铌含量不得超过 0.085%。

2.2 止裂韧性

API SPEC 5L：2012 和 ISO3183：2012 中规定的四种止裂韧性计算方法中，只有对 BTC 计算结果进行修正的方法适用于 12MPa、X80、OD1422 管道的止裂韧性计算[14]，其中修正系数的确定来源于 X80 全尺寸爆破试验数据库。目前国际上通用的全尺寸爆破试验数据库如图6所示，由此确定的中俄东线管道工程止裂韧性修正方法为 TGRC2，修正系数为 1.46。

中俄东线的天然气组分如表3所示，按照中俄东线实际工况 ϕ1422mm、壁厚 21.4mm、输送压力 12MPa、运行温度 0 进行止裂韧性计算。用 BTC 方法计算其止裂韧性结果为

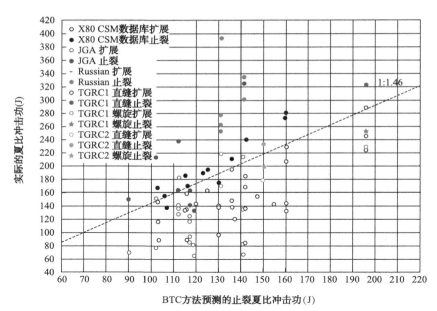

图 6 全尺寸爆破试验数据库

167.97J，按 1.46 倍修正后止裂韧性为 245J，如表 4 所示，表 4 中还给出了 Leis-2、Eiber、Wilkowski 等方法的修正结果。

表 3 中俄东线计算用气质组分

组分	C_1	C_2	C_3	C_4	C_5	N_2	CO_2	He	H_2
Mol%	91.41	4.93	0.96	0.41	0.24	1.63	0.06	0.29	0.07

表 4 中俄东线止裂韧性计算结果

计算方法	BTC(J)	BTC+Eiber 修正	BTC+Leis2 修正	BTC+Wilkowski 修正	BTC+TGRC2 修正
CVN 值(J)	167.97	251	250	286	245

由于现有的全尺寸气体爆破实验数据库无法覆盖中俄东线天然气管线 X80 钢级、ϕ1422mm、12MPa 压力下输送富气的设计参数要求。因此，2015 年 12 月在中国石油管道断裂控制试验场，针对中俄东线天然气管道具体的设计参数和服役条件，对 ϕ1422mm、X80 钢管的延性断裂止裂指标进行了全尺寸爆破试验验证，结果表明，采用 BTC 方法计算，并用 TGRC2 方法进行修正后的 ϕ1422mm、X80 钢管止裂韧性指标为 245J 是安全和经济的。

2.3 非金属夹杂物

近年来，许多管道工程使用的高钢级管线钢均在金相检测过程中发现了超尺寸大型夹杂物。管线钢中大型夹杂物的存在会对其力学，焊接，耐腐蚀等性能产生不利影响，进而给油气输送管道的安全运营带来很大的工程风险。为了有效降低管线钢中大型夹杂物的存在给管道输送系统带来的风险，石油管工程技术研究院李炎华等[15]，针对高钢级管线钢中大型夹杂物的特性进行了大量的研究工作，进而为高钢级管线钢中大型夹杂物级别判定标准的制定提供了依据。

目前，国内管线钢夹杂物评判通常采用 ASTM E 45—2005《Standard Test Methods for Determining the Inclusion Content of Steel》和 GB/T 10561—2005《钢中非金属夹杂含量的测定标准评级图显微检验法》。ASTM E45—2005 将夹杂物按形态和分布分为四类，即 A（硫化物类）、B 类（氧化铝类）、C 类（硅酸盐类）和 D（球状氧化物类）；而 GB/T 10561—2005 将夹杂物分为五类，即除上述四种外，还增加了 DS（单颗粒球类）。

李炎华等，从大型夹杂物在高钢级管线钢冶炼过程中的运动规律角度进行了分析，认为对于形态呈单颗粒球状的 DS 类夹杂物的厚度应当控制在 50μm 以下，对于形态比小于 3 的 B 类夹杂物，其厚度应当控制在 33μm 以下，即如果按照标准 GB/T 10561—2005 对管线钢中的大型夹杂物进行评定，DS 类夹杂物评级应该在 2.5 级（53μm）以下。中俄东线天然气管道工程 ϕ1422mm、X80 管材技术条件采用了这一研究成果，在非金属夹杂物级别验收极限中，定义了超标大型夹杂物的概念，并给出了验收和复验标准，见表 5。

表 5　ϕ1422mm、X80 管材技术条件中的非金属夹杂物级别限定

类型	A[①]		B[①,②,③]		C[①]		D[①]		DS[②,③]
系列	薄	厚	薄	厚	薄	厚	薄	厚	—
级别	≤2.0	≤2.0	≤2.0	≤2.0	≤2.0	≤2.0	≤2.0	≤2.0	≤2.5

注：厚度大于 50μm 的 B 类夹杂物以及评级超过 2.5 的 DS 类夹杂物均定义为超标大型夹杂物。

① 如果代表一熔炼批试样的 A、B、C、D 四类夹杂物中有一类及以上的评价不符合规定要求，则将该熔炼批判为不合格。

② 如果评价过程中发现某一视场中同时存在两个或两个以上的同类或不同类超标大型夹杂物，将该熔炼批判为不合格。

③ 如果代表一熔炼批钢管的夹杂物检验中发现某一视场中存在单个超标大型夹杂物，则需要在同一熔炼批中再随机抽取两个试样进行复验。如果两个试样的复验结果均符合 A、B、C、D 四类夹杂物规定要求且未出现超标大型夹杂物，则除原取样不合格的那根钢管外，该熔炼批合格。如果任一个试样的复验结果不符合 A、B、C、D 四类夹杂物规定要求或出现了超标大型夹杂物，则该熔炼批判为不合格。

2.4　力学性能试样取样位置

西气东输二线以来，油气管道工程用螺旋缝埋弧焊钢管的管径均小于 1219mm，为了取样方面，热轧板卷技术条件中力学性能取样位置均要求与板卷轧制方向成 30°取样。取样角度与板宽和钢管管径的关系，如公式（1）。

$$\sin\alpha = B/(\pi D)$$

式中：α 为螺旋角；B 为板宽；π 为圆周率；D 为钢管直径。

按目前主流热轧板卷产品宽度 1500~1600mm 计算，对于管径 1219mm 的螺旋缝埋弧焊管，热轧板卷的取样角度为 23.1°~24.7°，对于管径 1422mm 的螺旋缝埋弧焊管，热轧板卷的取样角度为 19.6°~21°。因此对于外径 1422mm 的螺旋缝埋弧焊管，与板卷轧制方向成 20°取力学性能样，更符合实际情况。

图 7 和图 8 给出了实际生产的热轧板卷 20°、30°位置的力学性能对比图。可以看出与轧制方向夹角 20°位置的屈服强度、抗拉强度、DWTT 剪切面积高于 30°位置，若按与轧制方向成 30°位置取样，容易低估热轧板卷的力学性能，造成不必要的浪费。因此在中俄东线天然管道工程用热轧板卷技术条件中力学性能的检测取样位置更改为与轧制方向成 20°位置。

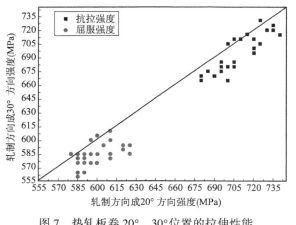

图 7 热轧板卷 20°、30°位置的拉伸性能

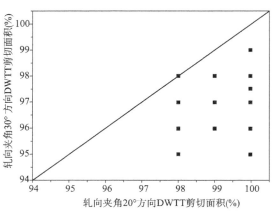

图 8 热轧板卷 20°、30°位置的 DWTT 性能

3 φ1422mm、X80 钢级大口径钢管开发

2013 年以来，中国石油组织相关科研单位和国内大型钢铁企业和制管企业，开展了 φ1422mm、X80 钢级大口径钢管的联合开发。在研发阶段，共进行单炉产品试制 3 轮，参与生产制造企业 15 家，试制产品 2000 余吨。工业应用阶段，进行了 1 次千吨级小批量试制，参与生产制造企业 8 家，试制产品 6000t。经第三方检测评价表明，试制的 φ1422mm X80 钢管的化学成分和力学性能均符合中俄东线天然气管道工程用 φ1422mm X80 管材技术条件要求。试制钢管的屈服强度为 595~668MPa，抗拉强度为 677~745MPa，母材 CVN 值为 324~486J，焊缝 CVN 值为 138~232J，热影响区 CVN 值为 172~354J。通过环焊试验证明，所试制的 φ1422mm X80 钢管的环焊缝性能均能满足标准要求。

4 结论

（1）中俄东线天然气管道工程用 φ1422mm X80 管材技术条件，借鉴了 API SPEC 5L：2012 的最新成果，结合中俄东线工程的具体特点，提出了适合外径 1422mm 管材的化学成分、夹杂物评定、CVN 值、力学性能试验方法等关键技术指标要求，技术条件具有很强的可操作性，即能满足工程要求，也适应生产情况，其研究经验在我国未来的天然气管道工程建设上推广应用。

（2）中俄东线天然气管道工程用 φ1422mm X80 管材技术条件规定管线钢采用低 C、Mn 的成分设计，并对添加的 Mo、Ni、Nb、V、Ti、Cu、Cr 等合金元素含量进行了严格的限定，是国内油气管道建设以来对化学成分要求最为严格的工程技术条件。工业试制结果表明，该技术条件制定的化学成分指标符合生产要求，并可有效解决管材的理化性能和焊接性能稳定性问题。

（3）采用 BTC 方法计算，并用 TGRC2 方法进行修正来确定 φ1422mm、X80 钢管的止裂韧性指标是安全和经济的，φ1422mm、X80 钢管的止裂韧性指标应为 245J。

（4）中俄东线天然气管道工程用 φ1422mm、X80 管材技术条件在非金属夹杂物级别验收极限中，采用最新研究成果定义了超标大型夹杂物的概念，并给出了验收和复验标准，有利于提高钢管的力学，焊接，耐腐蚀等性能。

（5）对于 φ1422mm 的螺旋缝埋弧焊管用热轧板卷，与板卷轧制方向成 20°取力学性能样，更符合生产实际情况，对板卷的力学性能评估也更为准确。因此在中俄东线天然管道工程用热轧板卷技术条件中力学性能的检测取样位置为与轧制方向成20°位置。

参 考 文 献

[1] 刘清梅，杨学梅，赵谨，等. 中国管道建设情况及管道用钢发展趋势 [J]. 上海金属，2014，36（4）：34-37.

[2] 彭涛，程时遐，吉玲康，等. X100管线钢在应变时效中的脆化 [J]. 热加工工艺，2013（20）：179-183.

[3] 李鹤林，吉玲康. 西气东输二线高强韧性焊管及保障管道安全运行的关键技术 [J]. 世界钢铁，2009，(1)：56-64.

[4] 全国石油天然气标准化技术委员会. 石油天然气工业管线输送系统用钢管：GB/T 9711 [S]. 北京：中国标准出版社，2011.

[5] International Organization for Standardization. Petroleum and natural gas industries—Steel pipe for pipeline transportation systems [S]. ISO 3183，2012.

[6] American Petroleum Institute. Specification for Line Pipe. 45th Edition [S]. API SPEC 5L，2012.

[7] 高惠临. 管线钢管韧性的设计和预测 [J]. 焊管，2010，33（12）：5-12.

[8] 尚成嘉，王晓香，刘清友，付俊岩. 低碳高铌X80管线钢焊接性及工程实践 [J]. 焊管，2012，35（12）：11-18.

[9] Bai Lu, Tong Lige, Ding Hongsheng, et al. The Influence of the Chemical Composition of Welding Material Used in Semi-Automatic Welding for Pipeline Steel on Mechanical Properties [C] //ASME 2008 International Manufacturing Science and Engineering Conference, 7-10 October 2008, Evanston, Illinois, USA. DOI：10.1115/MSEC_ICMP2008-72110.

[10] 孙磊磊，郑磊，章传国. 欧洲钢管集团管线管的发展和现状 [J]. 世界钢铁，2014，(1)：45-53.

[11] 高惠临. 管线钢与管线钢管 [M]. 北京：中国石化出版社，2012：22-27.

[12] Wang BX, Liu XH, Wang GD. Correlation of microstructures and low temperature toughness in low carbon Mn-Mo-Nb pipeline steel [J]. Materials Science and Technology, 2013, 29（12）：1522-1528. DOI：10.1179/1743284713Y.0000000326.

[13] 缪成亮，尚成嘉，王学敏，等. 高Nb X80管线钢焊接热影响区显微组织与韧性 [J]. 金属学报，2010，46（5）：541-546.

[14] 霍春勇，李鹤林. 西气东输二线延性断裂与止裂研究 [J]. 金属热处理，2011，36（增刊）：4-9.

[15] 李炎华，吉玲康，池强，等. 高钢级管线钢中大型夹杂物的特性 [J]. 管道技术与设备，2013，(1)：4-6.

本论文原发表于《天然气工业》2016年第36卷第6期。

扩散温度对 TC4 合金表面 Cu/Ni 复合镀层结构及耐蚀性能的影响

朱丽霞[1,2]　罗金恒[1,2]　武　刚[1,2]　李丽锋[1,2]
张庶鑫[1,2]　王　楠[3]　刘双双[3]　陈永楠[3]

(1. 中国石油集团石油管工程技术研究院·
2. 石油管材及装备材料服役行为与结构安全国家重点实验室；3. 长安大学)

摘　要：采用扩散热处理研究了 Cu/Ni/Ti 复合镀层不同温度下的扩散行为，分析了扩散层结构，并讨论了扩散温度对镀层结构及耐蚀性能的影响。结果表明：由于 Cu/Ni/Ti 原子之间的互扩散，形成稳定的扩散层，可以有效提高镀层表面耐蚀性能；随着热扩散温度上升到 700℃，膜层结构致密，在扩散层中形成了 Ni_xTi_y 金属间化合物及少量的 Cu_xTi_y 金属间化合物，镀层表面的耐蚀性最好；温度升高到 800℃ 时，在膜层界面处引发了 Kirkendall 效应，所形成的 Kirkendall 空位相互聚集长大，形成裂纹或孔洞，使得镀层疏松多孔，从而降低了耐蚀性。

关键词：关键词：钛合金；扩散行为；Kirkendall 效应；耐蚀性

　　钛及钛合金因其密度小、比强度高、良好的热稳定性等特点而作为结构材料被普遍应用于航空航天和石油化工等领域[1,2]。特别是在石油化工领域，由于钛合金耐蚀性较好，近年来在海油中开始广泛推广使用钛合金油管和钛合金接箍[3-5]。但是，由于钛合金较低的热导率，易于在高应力状态下形成粘扣，因此常用镀层来提高热导率。其中，金属铜由于其熔点高，化学稳定性强且具有较好的延展性、导热性，电镀在钛合金表面可改善其热导率和耐蚀性，防止粘扣事故[3,6]。Aydın[7] 等人通过对 Cu/Ti 热扩散的研究认为，采用热处理使得原子之间发生互扩散，形成扩散层，可以显著提高铜层与钛基体之间的结合力。Sabetghadam[8] 和 Kundu[9] 等人分别使用中间层 Ni，通过扩散热处理改善了扩散层的综合性能。

　　近年来，Chen[10] 通过 Ni 作为中间过渡层，采用在 TC4 合金表面电镀 Cu 制备了 Cu/Ni 复合镀层，通过扩散热处理改善了其扩散层结构，提高了力学性能。但对于石油化工行业，钛合金作为油管和接箍材料，其服役环境比较苛刻，而镀层的结构和成分将影响其耐蚀性能，从而影响膜层的服役安全性。Wei[11] 等人研究认为，在扩散过程中会形成一些析出相，使得扩散层组织得到细化，形成"Enveloping effect"，从而改善扩散层表面耐蚀性能。Salgado[12] 通过向 Ni/Ti 合金中添加 Cu 并进行热处理的研究发现，在扩散层表面结构将发生变化，形成致密钝化膜，提高扩散层耐蚀性。Qin[13] 等人研究认为，热扩散不仅可以形成有

基金项目：国家重点基础研究发展计划（"973"计划）（2017YFC0805804）；国家自然科学基金（51471136）。

作者简介：朱丽霞，女，1980 年生，硕士，高级工程师，中国石油集团石油管工程技术研究院，陕西西安（710077），电话：029-81887868，E-mail:zhulx@cnpc.com.cn。

效的保护膜,同时由于 Cu 和 Ni 的梯度分布,使得点蚀生长受到抑制,从而提高其耐蚀性。

综上所述,通过热扩散可以改变膜层结构,改善扩散层性能,但其膜层结构将对耐蚀性能产生显著的影响。基于此,本实验对 TC4 表面 Cu/Ni 复合镀层体系进行热扩散处理,研究了在不同的温度下(500℃、600℃、700℃和800℃)镀层的扩散行为,讨论了扩散温度对镀层结构以及腐蚀性能的影响。

1 实验

实验选用的基体材料为铸造 α+β 型 TC4(Ti-6Al-4V)钛合金,线切割后试样尺寸为 10mm×10mm×5mm。经除污打磨后的 TC4 基体,在氢氟酸与甲酰胺配比为 1∶10 的溶液中进行 5min 的活化处理,然后冲洗掉试样表面的活化液放置于镀镍液(180g/L $NiSO_4·6H_2O$,70g/L Na_2SO_4,30g/L $MgSO_4$,10g/L NaCl 和 30g/L H_3BO_3)中进行 20min、电压为 3.5V 的预镀镍处理,形成 2~5μm 厚且均匀的银白色镍镀层,冲洗掉试样表面的镀液,放置于镀铜液(220g/L $CuSO_4·5H_2O$,20mg/L NaCl 和 70g/L H_2SO_4)中进行 30min、电压为 0.65V 的镀铜处理,最终形成约 20μm 厚的 Cu/Ni 复合镀层,如图 1 所示。热处理采用 OTF-1200X 开启式真空管式炉进行。以氩气作为保护气,热处理温度分别为 500℃、600℃、700℃和800℃,保温 3h 后,随炉冷却至室温。

采用 Hitachi S-4800 型场发射扫描电镜,观察热扩散温度对复合镀层形貌的影响;利用能谱仪(EDS)对扩散层进行面扫描研究原子的扩散行为。采用 D/M2500 型 X 射线衍射仪分析热扩散后复合镀层物相组成,管压 40kV,管流 30mA,连续扫描模式,扫描速率 8°/min,衍射角范围 20°~80°。

采用 PARSTAT 2273 型电化学工作站在标准三电极体系下测量 TC4 钛合金复合镀层的极化曲线,溶液为 3.5%的 NaCl 溶液,参比电极为 KCl 饱和甘汞电极(SCE),辅助电极为铂电极,工作电极为试样,扫描速率 1mV/s。

2 结果与讨论

2.1 不同热扩散温度下镀层原子扩散行为

由扩散理论可知,温度越高,原子热激活能量越大,扩散速度越快,越易发生迁移[14]。图 2 为不同热扩散温度下保温 3h 的 Cu/Ni/Ti 的截面微观形貌。结合图 1 未经热处理时的 Cu/Ni 复合镀层,可以发现,未经热处理的 TC4 合金表面 Cu/Ni 镀层各界面清晰,热处理后,在 Cu/Ni 和 Ni/Ti 界面间均形成了明显的扩散层[图 2(a)(b)],并且热扩散温度越高,扩散层越厚。同时,由于 Cu/Ni/Ti 原子之间的互扩散,引发了典型的 Kirkendall 效应[15,16],使得扩散层界面变得疏松多孔。研究认为,扩散是原子朝着晶格点阵中空位处迁移,由于扩散速度不对等,将导致原子通量和空位通量向 2 个相反的方向移动[15]。同时,由于 Ni 原子具有较高的扩散速率,将朝着扩散速率较低的 Ti 侧移动[17],在扩散层中形成

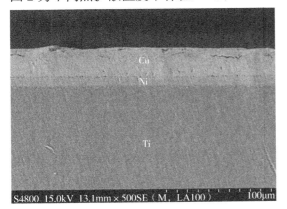

图 1 TC4 合金表面未经热处理的
Cu/Ni 复合镀层截面微观形貌

了大量的Kirkendall空位。随着温度的升高,原子的扩散速率加快,这些Kirkendall空位相互聚集并长大,形成了裂纹或孔洞[18][图2(c)(d)]。Puente[19]等人通过对Ni/Ti互扩散的研究发现,高温下的这种扩散行为,不仅形成Kirkendall孔洞,同时在一定的取向上,形成了Kirkendall扩散通道,这些均使得扩散层结构变得疏松多孔,对扩散层的性能造成不利的影响。

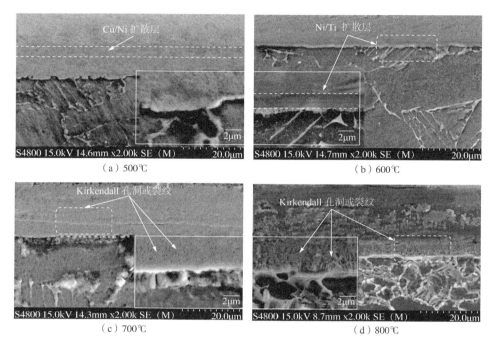

图2 不同热扩散温度下保温3h后Cu/Ni/Ti横截面微观形貌

通过对Cu、Ni、Ti元素的EDS面扫描发现,未经热处理时,Cu/Ni/Ti各界面几乎没有发生原子的互扩散[图3(a)],而经过真空热处理后,Cu/Ni和Ni/Ti界面均出现互扩散现象[图3(b)(e)];随着热扩散温度的升高,Cu、Ni和Ti元素扩散能力增强,中间Ni层的存在并不能完全阻碍Cu和Ti原子互扩散的发生。结合成分分析(图3和图4)可以发现,在500~800℃范围内,随着温度增加,扩散反应越充分,扩散层也越厚,在Cu/Ni镀层和Ni/Ti镀层界面处均发生了反应扩散,形成Ni_3Ti、$NiTi_2$、Ni_4Ti_3、$CuTi$、$CuTi_2$等金属间化合物。

根据Cu-Ni二元相图,在500~800℃的热处理范围内,只发生Cu、Ni元素间的互扩散,无明显相变发生。而在Ni/Ti扩散层中,根据Ni-Ti二元相图及扩散反应驱动力原理[20],在二元系反应扩散层中不会存在两相共存区,所以Ni/Ti扩散界面过渡区应有3种化合物层组成,即近Ni侧的过渡层(Ni_3Ti)、中间扩散层($NiTi$)和近Ti侧过渡层($NiTi_2$),这些化合物层组织结构相对独立。由于不同热处理温度引起Ni、Ti原子扩散程度改变,进而导致Ni_xTi_y金属间化合物发生改变,基于Ni-Ti二元相图,其形成自由能与温度之间关系如下式所示:

$$G(NiTi_2) = -49120 + 17.208T (\text{J/mol}) \tag{1}$$

$$G(Ni_3Ti) = -55585 + 15.962T (\text{J/mol}) \tag{2}$$

$$G(NiTi) = -54600 + 18.133T (\text{J/mol}) \tag{3}$$

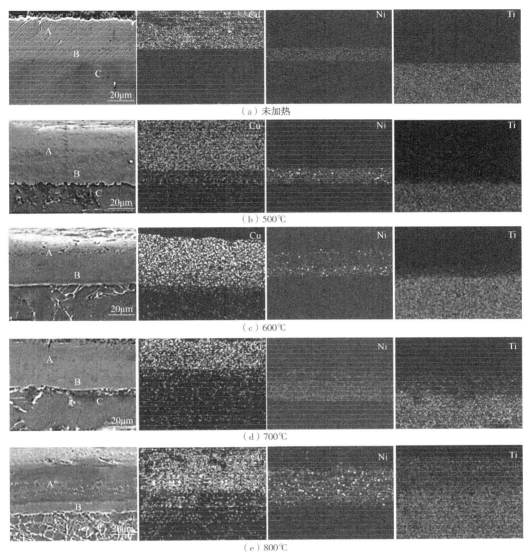

图 3 不同温度热扩散后 Cu/Ni/Ti 截面形貌及元素面分布
A—Cu； B—Ni； C—Ti

通过式(1)到式(3)，可以计算出 Ni_xTi_y 脆性金属间化合物在不同热扩散温度下的形成自由能，见表1。其中，在 500~800℃ 之间 Ni/Ti 界面 Ni_3Ti 形核概率最大；其次为 NiTi 相，最后形成 $NiTi_2$ 相。由图4可知，随着热扩散温度的上升，Ni/Ti 界面首先形成 Ni_3Ti 相，再形成 Ni_4Ti_3 相，最后形成了 $NiTi_2$ 相。在 Ti/Ni 界面经过退火处理后发生扩散反应，界面会发生相转变，如：亚稳态的 Ni_4Ti_3 相→亚稳态的 Ni_3Ti_2 相→稳态 Ni_3Ti 相。高温下亚稳态的 Ni_4Ti_3 相的出现是由于保温时间较短，亚稳态的 Ni_4Ti_3 相未能转变为稳态 Ni_3Ti。一般来说，在 Ni/Ti 界面发生扩散反应后，从 Ni 侧到 Ti 侧出现的扩散层依次为：Ni_3Ti→NiTi→$NiTi_2$ 相。XRD 图谱中并未出现 NiTi 金属间化合物，可能是由于 NiTi 相形成后又发生 $5NiTi→2NiTi_2+Ni_3Ti$ 相转变反应。这些金属间化合物的形成，可以极大提高复合镀层综合性能。

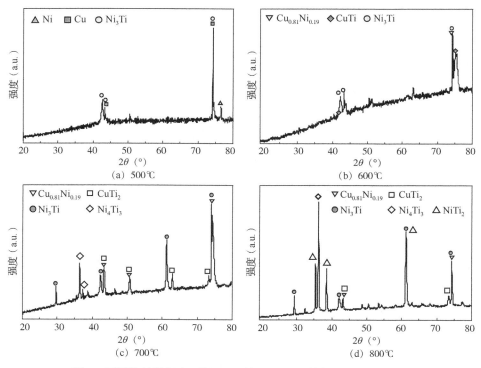

图 4 不同热扩散温度下保温 3h 的 Cu/Ni/Ti 结构表面 XRD 图谱

表 1 Ni/Ti 在不同热扩散温度下生成金属间化合物的自由能　　　单位：kJ/mol

化合物	数值			
Ni_xTi_y	500℃	600℃	700℃	800℃
$NiTi_2$	−35.82	−34.09	−32.37	−30.65
NiTi	−40.58	−38.77	−36.95	−35.14
Ni_3Ti	−43.25	−41.65	−40.05	−38.46

2.2 扩散温度对镀层腐蚀行为的影响

Cu/Ni/Ti 结构在不同热扩散温度下保温 3h 后，由于原子间的互扩散，形成了稳定的扩散层，改变了 Cu/Ni 复合镀层试样的组织结构。Salgado 等[12]研究了 Cu 含量和热处理对 NiTi 合金腐蚀行为的影响，并显示了在 NiTi 合金中增加 5% 或 10% 的 Cu 含量或在 800℃ 进行热处理后会提高 NiTi 合金的表面耐蚀性能。图 5 为不同热扩散温度下保温 3h 后 Cu/Ni/Ti 结构在 3.5%NaCl 溶液中的极化曲线。可以发现，TC4 合金经过热处理的 Cu/Ni 复合镀层比未经热处理 Cu/Ni 复合镀层具有更高的腐蚀电位和更低的腐蚀电流密度，意味着耐蚀性得到改善。经热处理后的镀层试样阳极极化曲线随着外加电压的升高，阳极电流密度逐渐上升且都显示出一个稳定钝化区，表明了 Cu/Ni/Ti 结

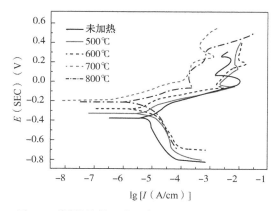

图 5 不同热扩散温度下保温 3h 的 Cu/Ni/Ti 结构在 3.5%NaCl 溶液中腐蚀的极化曲线

构表面钝化膜层的形成。Cu/Ni/Ti 结构的极化曲线通过 Tafel 外推法获得的腐蚀参数如腐蚀电位 E_{corr}、腐蚀电流密度 I_{corr} 等，见表 2。此外，随着热扩散温度从 500℃ 上升到 700℃，腐蚀电位从 $-330.87mV$ 增大到 $-201.14mV$；腐蚀电流密度从 $4.02×10^{-3}mA/cm^2$ 降低到 $0.514×10^{-3}mA/cm^2$。这是由于随着热扩散温度的上升，Ni 原子和 Ti 原子扩散到表面 Cu 镀层中，并发生扩散反应提高了耐蚀性；此外，Cu/Ni/Ti 结构表面的耐蚀性在热扩散温度 700℃ 下表现最好。Ma 等人[21]发现，Ni 以 Ni^{2+} 络合进入腐蚀产物 Cu_2O 晶格中，且在腐蚀膜层中检测到 NiO，从而提高了耐蚀性。因而，在热扩散温度 700℃ 下腐蚀电流密度较低，也可能是由于铜氧化物的形成以及 NiO、TiO 的弥散强化作用。然而，当热扩散温度达到 800℃ 时，腐蚀电位相比于 700℃ 反而降低到 $-207.21mV$，腐蚀电流密度增加到 $1.622×10^{-3}mA/cm^2$；这主要是由于高温下 Kirkendall 效应显著，使得镀层疏松多孔，进而影响其腐蚀性能。

表2 不同热扩散温度下保温 3h 的 Cu/Ni/Ti 结构在 3.5%NaCl 溶液中腐蚀的动电位极化结果

温度(℃)	E_{corr}(SCE)(mV)	I_{corr}(10^{-3}mA/cm)	b_a(mV/dec)	b_c(mV/dec)
未加热	-370.59	6.911	183.64	-142.35
500	-330.87	4.024	148.77	-105.80
600	-280.82	2.329	108.87	-91.75
700	-201.14	0.5136	96.98	-59.43
800	-207.21	1.622	76.07	-105.60

为了进一步研究扩散温度对 TC4 合金表面 Cu/Ni 复合镀层腐蚀性能的影响，绘制 Nyquist 图[图 6(a)]和 Bode 图[图 6(b)、图 6(c)]。可以看出，在测量的频率范围内，Cu/Ni/Ti 结构腐蚀表面的阻抗图谱出现 2 个重叠的相位角，表明了阻抗图谱由 2 个时间常数组成，即低频区扩散阻抗和高频区容抗弧。其中，热扩散温度 500℃ 处理的 Cu/Ni/Ti 结构在 3.5%NaCl 溶液腐蚀后 Nyquist 点的高频区显示出一个容抗弧，随之在低频区出现一条倾斜的直线。通常低频区的称为 Warburg 扩散阻抗，Warburg 扩散阻抗表明了热扩散温度 500℃ 处理的 Cu/Ni/Ti 结构在 3.5%NaCl 溶液腐蚀过程中是受扩散控制的，可能是由于在金属/溶液界面腐蚀离子和可溶的腐蚀产物转移或者是 Cu/Ni/Ti 表面溶解的氧扩散引起的[22,23]。随着热扩散温度的进一步升高，Nyquist 点低频区的 Warburg 扩散阻抗消失，只有一个较大的容抗弧出现。消失的 Warburg 扩散阻抗表明了 Ni、Ti 原子扩散到试样表面形成密实氧化物阻止了腐蚀反应的扩散过程，使得较高热扩散温度处理的 Cu/Ni/Ti 结构在 3.5%NaCl 溶液腐蚀过程中是受电荷转移控制的。此外，高频区的容抗弧经历了先缓慢增大，在热扩散温度 700℃ 达到最大，而后在热扩散温度 800℃ 又减小的过程，表明了 Cu/Ni/Ti 结构表面耐蚀性先增大后降低。从图 6(b)、6(c)可以看出 2 个相位角，出现 2 个时间常数。根据阻抗谱特征，绘制腐蚀体系的等效电路为 $R(Q(R(Q(RW))))$ 模型，表示拟合热扩散后 Cu/Ni/Ti 结构表面腐蚀反应。等效电路中[图 6(d)]，R_s 为电极系统的溶液电阻；R_f 为 Cu/Ni/Ti 结构表面形成的膜层电阻；R_{ct} 为电荷转移电阻；W 为 Warburg 阻抗，反映 Cu/Ni/Ti 结构腐蚀过程中浓差极化和扩散对电极反应影响的阻抗。不同热扩散温度 500℃、600℃、700℃ 和 800℃ 实验值在拟合时卡方值分别为 $4.03×10^{-3}$、$8.58×10^{-3}$、$2.69×10^{-3}$ 和 $1.29×10^{-3}$，表明等效电路 $R(Q(R(Q(RW))))$ 适用于实验数据值，拟合后的腐蚀参数见表 3。

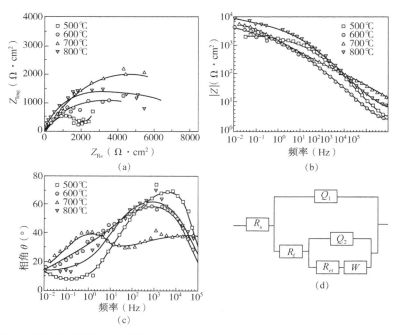

图6 不同热扩散温度下保温3h的Cu/Ni/Ti在3.5%NaCl溶液的Nyquist图和Bode图以及匹配拟合阻抗数据的等效电路图

表3 不同热扩散温度处理的Cu/Ni/Ti结构在3.5%NaCl溶液中腐蚀后的EIS结果

温度(℃)	$R_s(\Omega \cdot cm^2)$	$R_f(\Omega \cdot cm^2)$	$R_{ct}(\Omega \cdot cm^2)$	W	$Z(\Omega \cdot cm^2)$
500	2.694	15.53	2001	0.004903	2019.22
600	2.358	951	6356		7309.36
700	2.406	1066	9119		10187.41
800	2.576	1046	8252		9300.58

低频区的阻抗模量($|Z|$)可以评估腐蚀产物形成的膜层保护作用。随着热扩散温度的上升，Bode图中$f=0.01Hz$的$|Z|$缓慢增加，表明了保护膜层逐渐形成。此外，通过Chen等人[24]的研究，利用定量的Z值表征Cu/Ni/Ti结构在电化学过程中的所有电阻是非常合理的，阻抗Z可以用式(4)表示：

$$Z = R_s + R_f + R_{ct} \tag{4}$$

基于表3所示，随着热扩散温度的升高，R_{ct}和R_f的值都先增加后稍微减低。在3.5% NaCl溶液中腐蚀后，Cu/Ni/Ti结构中较高的R_{ct}值显示出较好的耐蚀性；在Cu/Ni/Ti表面形成膜层阻碍Cl^-诱发的腐蚀会提高R_f的值，进而提高腐蚀电位[24]。拟合后不同热扩散处理试样的阻抗Z值也表明了与极化曲线类似的趋势。因此，对Cu/Ni/Ti结构进行热扩散会提高Cu/Ni复合镀层的耐蚀性，并且热扩散温度为700℃时，Cu/Ni复合镀层耐蚀性能最好。

3 结论

（1）对TC4合金表面的Cu/Ni复合镀层进行热扩散后，形成Ni_3Ti、$NiTi_2$、Ni_4Ti_3、$CuTi$、$CuTi_2$金属间化合物，改善了膜层的耐蚀性能。

（2）由于热扩散使得原子之间发生互扩散，形成稳定的扩散层，可以有效提高镀层表面耐蚀性能。随着热扩散温度上升到 700℃，镀层表面的耐蚀性最好。温度进一步升高到 800℃时，发生显著的 Kirkendall 效应，使得镀层疏松多孔，从而又降低了耐蚀性。

参 考 文 献

[1] Yao Xiaofei, Xie Faqin, Han Yong et al. Rare Metal Materialsand Engineering[J]. 2012, 41(8)：1463
[2] Xia Yixiang, Li Di. Rare Metal Materials and Engineering[J]. 2001, 30(5)：388
[3] Shen Z C, Xie F Q, Wu X Q et al. China Surface Engineering[J]. 2012, 25(5)：45
[4] Khosravi G, Sohi M H, Ghasemi H M et al. International Journal of Surface Science and Engineering[J]. 2015, 9(1)：43
[5] Deillon L, Zollinger J, Daloz D et al. Materials Characterization[J]. 2014, 97：125
[6] DingY, Hao J, Chen Y et al. Hot Working Technology[J]. 2014, 67：321
[7] Aydin K, KayaY, Kahraman N. Materials & Design[J]. 2012, 37：356
[8] Sabetghadam H., Hanzaki A Z, Araee A. Materials Characterization[J]. 2010, 61(6)：626
[9] Kundu S, Chatterjee S. Materials Characterization[J]. 2008, 59(5)：631
[10] ChenY, Liu S, Zhao Y et al. Vacuum[J]. 2017, 143：150
[11] Wei H, Wei Y H, Hou L F et al. Corrosion Science[J]. 2016, 111：382
[12] Salgado F. International Journal of Electrochemical Science[J]. 2016, 11(11)：9282
[13] Qin L, Qin Z, Wu Z, et al. Corrosion Science[J]. 2018, 138：8
[14] Askeland D R, Phule P. Essentials of Materials Science and Engineering[M]. Beijing：Tsinghua University Press, 2005：22
[15] Seitz F. Acta Metallurgica[J]. 1953, 1(3)：355
[16] Fan H J, Gösele U, Zacharias M. Small[J]. 2007, 3(10)：1660
[17] Bastin G F, Rieck G D. Metallurgical Transactions[J]. 1974, 5(8)：1817
[18] Tavoosi M. Surfaces and Interfaces[J]. 2017, 9：196
[19] Puente A E P Y, Dunand D C. Intermetallics[J]. 2018, 92：42
[20] Abdul-Lettif A M. Physica B：Condensed Matter[J]. 2007, 388(1-2)：107
[21] Ma A L, Jiang S L, Zheng Y G et al. Corrosion Science[J]. 2015, 21(91)：245
[22] ZhouY, Zhang S, Guo L et al. International Journal of Electrochemical Science[J]. 2015, 13(10)：2072
[23] Ormellese M, Lazzari L, Goidanich S et al. Corrosion Science[J]. 2009, 51(12)：2959
[24] ChenY, Yang Y, Zhang T et al. Surface & Coatings Technology[J]. 2016, 307：825

本论文原发表于《稀有金属材料与工程》2019 年第 48 卷第 6 期。

Constitutive Equation for Describing True Stress–Strain Curves over a Large Range of Strains

Cao Jun[1]　Li Fuguo[2]　Ma Weifeng[1]　Li Dongfeng[1, 3]　Wang Ke[1]
Ren Junjie[1]　Nie Hailiang[1]　Dang Wei[1]

(1. State Key Laboratory of Performance and Structural Safety for Petroleum Tubular Good Goods and Equipment Materials, Tubular Goods Research Institute of CNPC; 2. State Key Laboratory of Solidification Processing, School of Materials Science and Engineering, Northwestern Polytechnical University; 3. China University of Petroleum (East China))

Abstract: Full-range strain hardening behaviour, containing the post-necking stage, is difficult to account for in detail. Hence, a Swift-Voce(S-V) model was used to describe the full-range stress-strain behaviour of 7050-T7451 aluminium alloy under uniaxial-tension, tension-with-notch and pure-shear processes with a hybrid experimental-numerical framework. Since the S-V model is not able to describe accurately the behaviour of Ti-6Al-4V alloy, a combined Swift and 4th degree polynomial (S-P4) model is proposed. Results indicate that the S-V model can successfully describe large-deformation behaviour of the 7050-T7451 aluminium alloy, and the S-P4 model can successfully describe that of the Ti-6Al-4V alloy based on careful designing process of the constitutive equation.

Keywords: Stress-strain measurements; Numerical simulation; Tensile testing; Necking

1　Introduction

True stress-strain curves are indispensable in precision forming process because the accuracy of simulation depends mainly on the constitutive relation. In the large-deformation stage, the stress and strain distributions are no longer uniform, so it is difficult to estimate the true stress-strain relation within necking zone using an engineering stress-strain relation. Theoretical work in this regards has brought significant scientific thoughts into the essential understanding of the necking phenomena[1]. In this study, large deformation means local deformation of necking phenomenon.

Post-necking strain hardening behaviour was used to predict the limiting major strain for sheet metal formability[2]. Besides, it was also applied to predict plastic strain localisation of dual-phase steels using finite-element analyses (FEA)[3]. The post-necking strain hardening behaviour was

---

Corresponding author: Cao Jun, caojun1@cnpc.com.cn, juncao1105@163.com

especially required in the prediction of ductile damage[4].

Many efforts were made to obtain full-range true stress-strain curves using the finite-element method (FEM)[5-7] and the digital image correlation (DIC) method[8,9]. A testing method employed the DIC technique and iterative FEA method to obtain stress-strain curves including post-necking strain[10]. Although this method can obtain accurate stress-strain curves, the method is too complex to tackle complicated project problems. Tu et al.[11] summarised three groups of methods to describe post-necking strain hardening behaviour of metallic materials. The first group consists of analytic solutions derived with round bar specimens[12,13], of which the Bridgman method[13] is the most well known, although suitable only for smooth round-bar specimens. The second group consists of experimental-numerical iterative methods[9,14-17], which yield trustful stress-strain behaviour in the post-necking regime. A method of this second group is applied in the present study. The third group contains an inverse method[18-20], in which the correction formula applied in an inverse way is based on predefined hardening rules. The accuracy of this latter method is a controversial issue for an actual material[11].

In this paper, a Swift-Voce (S-V) model is used to describe the large-deformation behaviour of 7050-T7451 aluminium alloy under uniaxial-tension, tension-with-notch and pure-shear processes. Since the S-V model cannot describe the large-deformation behaviour of Ti-6Al-4V alloy, a combined Swift model and 4th degree polynomial (S-P4) model is proposed to better describe this.

2 Methods

2.1 Experiments

Uniaxial-tension, tension-with-notch and pure-shear specimens of 7050-T7451 aluminium alloy plate and a cylindrical uniaxial tension specimen of Ti-6Al-4V alloy were prepared to obtain force-displacement responses, as illustrated in Figure 1. All specimens were processed along the rolling direction. Tests on the uniaxial-tension, tension-with-notch and pure-shear specimens were carried out on an Instron 3382 (Instron Inc., USA) machine. The tests were conducted at a constant crosshead velocity of 1 mm/min. The testing temperature was room temperature (18 ~ 25℃). Meanwhile, the specimens of 7050-T7451 aluminium alloy plates were tested with a DIC system to obtain displacement and local strain data. Since the specimen of Ti-6Al-4V alloy is cylindrical and so is hard to test with a DIC system, the displacement was obtained using an extensometer at small strains and a beam at large strains.

2.2 Simulations

Three-dimensional finite element (FE) models of uniaxial-tension, tension-with-notch and pure-shear specimens were performed using ABAQUS/Explicit software with a user material subroutine (VUMAT). Quarter geometric FE models were adopted for the specimens of uniaxial tension and tension with notch. All the specimens were meshed using reduced-integration eight-node solid elements (C3D8R).

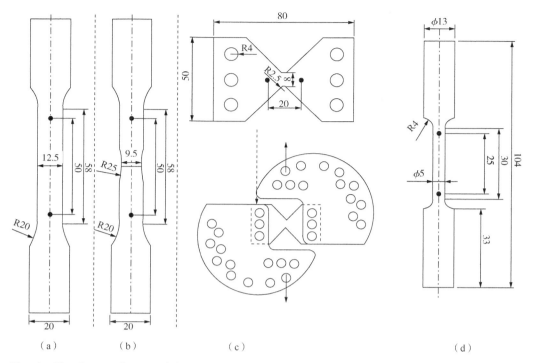

Fig. 1 The shape and size of (a) uniaxial tension (b) tension with notch (c) pure shear specimens for 7050-T7451 aluminium alloy plate and (d) uniaxial tension specimen for Ti-6Al-4Vcylindrical alloy specimen

2.3 Constitutive equations

Two common constitutive equations were applied to describe the stress-strain curves before the onset of necking. These equations can be written as

$$\sigma = K(\varepsilon_p + \varepsilon_0)^n \quad \text{(Swift equation)} \quad (1)$$
$$\sigma = \sigma_0 - \sigma_0 A \exp(-\beta \varepsilon_p) \quad \text{(Voce equation)} \quad (2)$$

where K is the strength coefficient, n is the strain hardening exponent and ε_0 is the pre-strain, σ_f and A are saturation stress and material coefficient, respectively. β is used to determine the rate at which the stress tends to reach a steady state value.

The Swift and Voce equations were used to fit the true stress-strain curves of the 7050-T7451 aluminium alloy plate and the Ti-6Al-4V alloy before the onset of necking. The constitutive parameters of K, n and ε_0 were obtained by fitting the true stress-strain curves of the two alloys with the Swift equation. σ_f, A and β were obtained by fitting the true stress-strain curves of the two alloys with the Voce equation. These constitutive parameters are listed in Table 1.

Table 1 The constitutive parameters of 7050-T7451 aluminium alloy and Ti-6Al-4V alloy

Material	K	n	ε_0	σ_f	A	β
7050-T7451	827	0.16316	0.0233	600	3.92	18.96
Ti-6Al-4V	1207	0.05638	0.0039	1082	0.1643	24.20

In the non-uniform deformation stage, the largest local deformation occurs in the post-necking region. The law of volume invariance cannot be used to transform from the engineering stress-strain relation to the true stress-strain relation in this region. However, the stress-strain relation of the post-necking stage is crucial to obtain an accurate local strain in the process of simulation. Therefore, a linear combination of the Swift and Voce models was used in the FE models. The special form of this model is

$$\sigma = q[K(\varepsilon_p+\varepsilon_0)^n] + (1-q)[\sigma_0 - \sigma_0 A\exp(-\beta\varepsilon_p)] \qquad (3)$$

where q is a weight factor, used to weight the difference of large-deformation behaviour in the post-necking stage. It relates to the deformation capacity in the post-necking stage and an increase of q indicates a decrease of the large deformation.

In the S-V model, firstly the constitutive parameters of the Swift and Voce models need to be determined by fitting the uniform true stress-strain curve of the tensile experiment. Secondly, the initial q value should be set. Thirdly, the S-V model is submitted to finite-element simulation (FES) by VUMAT, and the simulated full-range force-displacement curves are obtained to compare experimental data. Finally, the accurate stress-strain relation is obtained by optimising the value of q to make the simulated force-displacement curves agree with experimental data. Then, the accurate local strain can be obtained from the FE model with an optimised S-V model.

Can the S-V model describe full-range stress-strain curves containing large deformation for all materials? Which constitutive equation is more suitable when the full-range stress-strain curves cannot be described by the S-V model? A combination of the Swift model and a correction function is proposed to describe the full-range stress-strain curve. The correction function could be polynomial since it is easily modified. The correction process needs to be based on a necking type of material. This is an iterative process obtained by adjusting the correction function and weight factor q to make a simulated force-displacement curve that accords with experiment. The constitutive equation is expressed as

$$\sigma = qK(\varepsilon_p+\varepsilon_0)^n + (1-q)f(\varepsilon_p) \qquad (4)$$

where $f(\varepsilon_p)$ is a correction function. The power-law type constitutive equation (Swift model) is suitable for body-centred cubic (BCC) metals[21], whereas the exponential-type equation (Voce model) is suitable for most face centred cubic (FCC) metals such as aluminium and copper[22]. Some materials (for example, titanium alloys) have a close-packed hexagonal (HCP) structure. For these materials, the power-law type and exponential-type constitutive equation are not suitable for describing the full-range stress-strain curve. To better describe the full-range stress-strain curve of HCP materials, the correction function $f(\varepsilon_p)$ is used to modify the shape of the stress-strain curve in the post-necking type; this is a modification for large-deformation behaviour induced by material type.

3 Results and discussion

Figure 2(a) - Figure 2(c) exhibits a comparison of force-displacement responses and local equivalent plastic strain between experiment and simulation under uniaxialtension, tension-with-notch and pure-shear processes, respectively. These simulated force-displacement responses of 7050-T7451 aluminium alloy were obtained from the FE model with an embedded S-V model, and the local equivalent plastic strain was obtained from the surface of the specimen. It is noted that the force-displacement responses of the FE model in the post-necking stage are sensitive to the constitutive equation.

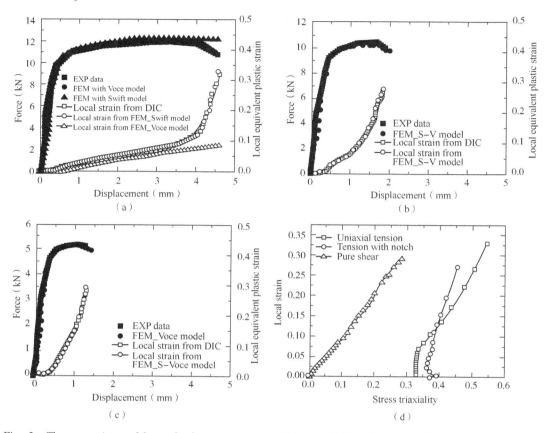

Fig. 2 The comparisons of force-displacement curve and local equivalent plastic strain between experiment and simulation with Swift model and Voce model for of 7050-T7451 aluminium alloy plate (a) uniaxial tension process, (b) tension with notch process and (c) pureshear; (d) the local strain evolution at necking region with the variation of stress triaxiality for the three kinds of specimens

In Figure 2(a), the force-displacement responses and local strain obtained from the FE model with the Voce model are consistent with those obtained from experimental data with the DIC method. The weight factor q of the S-V model is 0. Nevertheless, the FE model with the Swift model does not predict the stress-strain curves and local strain very well. It indicates that the Voce model is suitable for describing the full-range stress-strain relation of 7050-T7451 aluminium alloy (FCC structure).

In Figure 2(b), the force-displacement responses and local strain of FE model obtained with the S-V model are consistent with those of experiment data for tension-with-notch specimens, with a weight factor q is 0.7. It reflects the fact that the large-deformation capacity of the tension-with-notch process is less than that of the uniaxial-tension process. In Figure 2(c), the force-displacement responses and local strain of simulation with the Voce model are consistent with those of experimental data for pure shear; the weight factor q of the S-V model is 0. This indicates that the Voce model is suitable for describing the large-deformation behaviour of the pure shear specimen.

The three post-necking types of uniaxial-tension, tension-with-notch and pure-shear processes are different, as illustrated in Figs. 2(a)-(c). The ratio of the post-necking stage to the uniform deformation stage for pure shear is larger than for the other two kinds of specimens. The uniform deformation stage of the uniaxial tension process is longest in the three kinds of specimens. In addition, the fracture strain of the uniaxial-tension specimen is highest at 0.3272, while that of the tension-with-notch specimen is lowest at 0.2726. Comparing the relationship between the weight factors q and three fracture strains, the fracture strain is higher with a lower weight factor q. It reflects the fact that the increasing weight factor q reduces the large-deformation capacity.

Figure 2(d) shows the local strain evolution within the post-necking region of the uniaxial-tension, tension-with-notch and pure-shear specimens with the variation of stress triaxiality for 7050-T7451 aluminium alloy. As can be seen from Figure 2(d), the large variation range of stress triaxiality leads to a high local strain. By recalling the q values for describing the constitutive behaviour of the three kinds of specimen, q is 0 for the uniaxial-tension and pure-shear specimens and q is 0.7 for the tension-with-notch specimen. It is to be noted that increasing the large deformation leads to an increasingly large variation of stress state.

Figure 3(a) shows a comparison of the force-displacement responses between experimental and simulation data for Ti-6Al-4V alloy. In Figure 3(a), the force-displacement response of simulation with the Swift model is not consistent with that of experiment. As the Swift model is unsaturated, the force-displacement response of simulation in the post-necking stage is largely inconsistent with that of experiment. How can one find a suitable constitutive equation to describe this special large-deformation behaviour? A combined Swift model and a 4th degree polynomial (S-P4) were used to describe the local deformation behaviour. Since the correction function needs two inflection points from yielding point to fracture point, a 4th degree polynomial was chosen as the correction function. The S-P4 model can be expressed as:

$$\sigma = qK(\varepsilon_p + \varepsilon_0)^n + (1-q)(a\varepsilon_p^4 + b\varepsilon_p^3 + c\varepsilon_p^2 + d\varepsilon_p + e) \tag{5}$$

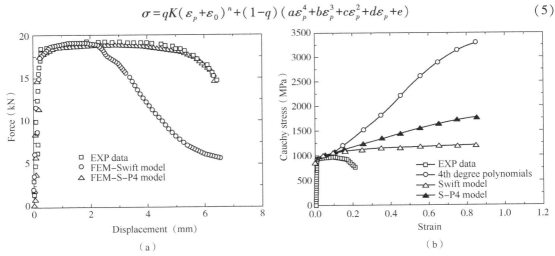

Fig. 3 (a) Comparisons of force-displacement responses between experimental data and simulation with different models for Ti-6Al-4V alloy; (b) the Cauchy stress-strain responses with different models of Ti-6Al-4V alloy

This 4th degree polynomial (P4) is necessary to better describe the local deformation of Ti-6Al-4V alloy. The procedure for determining S-P4 is as follows. Firstly, the constitutive parameters of the Swift model are determined by fitting the uniform true stress – strain curve of the tensile experiment, the Swift model being the dominant constitutive equation. Secondly, the Swift model is substituted into FES by VUMAT, and the simulated full-range force-displacement curve is obtained to compare with experimental data. Thirdly, a 4th degree polynomial needs to be designed based on the comparative result. The curve of P4 before necking needs to be consistent with the stress–strain curve of the tensile experiment. Then the concavity and convexity of the P4 curve from the beginning of necking to the end need to be changed from convexity to concavity, and an initial P4 equation should be established. Finally, the parameters of P4 and the weight factor q need to be adjusted iteratively until the experimental and numerical force-displacement curves are consistent.

As illustrated in Figure 3(b), there are two stages of post-necking, namely a slow descent stage and a rapid descent stage. The type of necking of the Ti-6Al-4V alloy is different from that of the 7050-T7451 aluminium alloy. Therefore, the full-range of the constitutive equation containing the post-necking stage needs to be designed based on the correction P4 function. Figure 3(b) shows the designed constitutive equation of S-P4 in the full-range of the deformation stage of the Ti-6Al-4V alloy. The slow-descent necking stage needs to be described by combining the Swift model and a convex P4 function. On the other hand, the rapid-descent necking stage needs to be described by combining the Swift model with a concave P4 function. Therefore, the S-P4 constitutive equation is designed as follows:

$$\sigma = 0.73 \times 1207(\varepsilon_p + 0.0039)^{0.05638} + 0.27 \times (-3282\varepsilon_p^4 + 383\varepsilon_p^3 + 3638\varepsilon_p^2 + 1452\varepsilon_p + 906) \quad (6)$$

The force-displacement response of simulation with the S-P4 constitutive equation is consistent with that of experiment, as shown in Figure 3(a). The q of the S-P4 constitutive equation is 0.73, which reflects the fact that the Swift model is the dominant constitutive equation and the 4th degree polynomial an auxiliary correction function.

Figure 4 shows a comparison of the local strain evolution within the necking region of the uniaxial tension specimens of the Ti-6Al-4V and the 7050-T7451 aluminium alloys as a function of stress triaxiality. As can be seen, the local strain with stress triaxiality has two stages for the Ti-6Al-4V alloy, while that of the 7050-T7451 aluminium alloy has only one stage in the local-deformation stage. The variation in the relationship also reflects the post – necking process. In addition, the local strain in the Ti-6Al-4V alloy is close to 1, because the large-deformation stage is a large proportion of the whole deformation stage.

4 Conclusions

An S-P4 model is proposed to describe the large-deformation behaviour of Ti-6Al-4V alloy because the S-V model cannot describe this successfully. A hybrid experimental-numerical method was performed to investigate the applicability of the S-V model and the S-P4 model in describing the full-range stress-strain behaviour of 7050-T7451 aluminium and Ti-6Al-4V alloys. The conclusions are as follows:

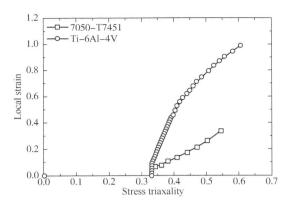

Fig. 4 The comparison of local strain evolution at necking region of uniaxial tension specimen between Ti-6Al-4V alloy and 7050-T7451 aluminium alloy with the variation of stress triaxiality

(1) The S-V model can successfully describe the full-range stress-strain behaviour of 7050-T7451 aluminium alloy containing a necking stage. Comparing the relationship between the weight factors q and three fracture strains of uniaxial-tension, tension-with-notch and pure-shear processes, the fracture strain is higher at lower weight factor q. Increasing the weight factor reduces the large-deformation capacity.

(2) The S-P4 model can successfully describe the full-range stress-strain behaviour of Ti-6Al-4V alloy under uniaxial tension whereas the S-V model does not.

(3) The S-P4 model needs to be carefully designed based on the post-necking type. The constitutive equation used to describe the full-range stress-strain curve of Ti-6Al-4V alloy in the post-necking stage needs to be designed using correction (P4) functions.

References

[1] M. K. Duszek and P. Perzyna, Int. J. Solids Struct. 27 (1991) pp. 1419–1443.
[2] L. Smith, R. Averill, J. Lucas, T. Stoughton and P. Matin, Int. J. Plast. 19 (2003) pp. 1567–1583.
[3] X. Sun, K. S. Choi, W. N. Liu and M. A. Khaleel, Int. J. Plast. 25 (2009) pp. 1888–1909.
[4] M. Luo, M. Dunand and D. Mohr, Int. J. Plast. 32 (2012) pp. 36–58.
[5] J. Isselin, A. Iost, J. Golek, D. Najjar and M. Bigerelle, J. Nucl. Mater. 352 (2006) pp. 97–106.
[6] H. D. Kweon, E. J. Heo, D. H. Lee and W. K. Jin, J. Mech. Sci. Technol. 32 (2018) pp. 3137–3143.
[7] M. Rossi, A. Lattanzi and F. Barlat, Strain 54 (2018) p. e12265.
[8] S. K. Paul, S. Roy, S. Sivaprasad and S. Tarafder, J. Mater. Eng. Perform. 27 (2018) pp. 4893–4899.
[9] L. Wang and W. Tong, Int. J. Solids Struct. 75–76 (2015) pp. 12–31.
[10] M. Kamaya and M. Kawakubo, J. Nucl. Mater. 451 (2014) pp. 264–275.
[11] S. Tu, X. Ren, J. He and Z. Zhang, Fatigue Fract. Eng. Mater. Struct. 43 (2020).
[12] W. W. Davidenkov, Proc. ASTM 46 (1946) pp. 1147–1158.
[13] P. W. Bridgman, Studies in Large Plastic Flow and Fracture, Vol. 177, McGraw-Hill, New York, 1952.
[14] S. Coppieters, S. Sumita, D. Yanaga, K. Denys, D. Debruyne and T. Kuwabara, Identification of post-necking strain hardening behavior of pure titanium sheet, in Residual Stress, Thermomechanics & Infrared Imaging, Hybrid Techniques and Inverse Problems Simon Quinn, Xavier Balandraud, ed., Springer, Orlando, 2016, p. 59–64.

[15] S. Coppieters, S. Cooreman, H. Sol, P. V. Houtte and D. Debruyne, J. Mater. Process. Tech. 211 (2011) pp. 545-552.
[16] S. Coppieters and T. Kuwabara, Exp. Mech. 54 (2014) pp. 1355-1371.
[17] J.-H. Kim, A. Serpantié, F. Barlat, F. Pierron and M.-G. Lee, Int. J. Solids Struct. 50 (2013) pp. 3829-3842.
[18] W. J. Yuan, Z. L. Zhang, Y. J. Su, L. J. Qiao and W. Y. Chu, Mater. Sci. Eng. A 532 (2012) pp. 601-605.
[19] Z. Zhang, J. Ødegård and O. Søvik, Comput. Mater. Sci. 20 (2001) pp. 77-85.
[20] K. Zhao, L. Wang, Y. Chang and J. Yan, Mech. Mater. 92 (2016) pp. 107-118.
[21] H. S. Ji, H. K. Ji and R. H. Wagoner, Int. J. Plast. 26 (2010) pp. 1746-1771.
[22] B. K. Choudhary, E. I. Samuel, K. B. S. Rao and S. L. Mannan, Met. Sci. J. 17 (2001) pp. 223-231.

本论文原发表于《Philosophical Magazine letters》2020年。

Failure Analysis of a Sucker Rod Fracture in an Oilfield

Ding Han[1,2] **Zhang Aibo**[1] **Qi Dongtao**[2] **Li Houbu**[2] **Ge Pengli**[3]
Qi Guoquan[1,2] **Ding Nan**[2] **Bai zhenquan**[2] **Fan Lei**[2]

(1. Department of Applied Chemistry, School of Science, Northwestern Polytechnical University;
2. Tubular Goods Research Institute, China National Petroleum Corporation&State Key Laboratory for Performanceand Structure Safety of Petroleum Tubular Goods and Equipment Materials;
3. Northwest Oil Field Branch Company Petroleum Engineering Technology Research Institute)

Abstract: The sucker rod was fracture after 729 days' service in an oilfield. The fracture position is located at the welding seam that is connecting the sucker rod's upsetting end and the polished rod end. The failure causes were analyzed by the nondestructive test(NDT) of penetration detection, direct-reading spectrometer, tensile strength test machine, impact test machine, vickers hardness tester, optical microscope (OM), macroscopic fracture morphology, scanning electron microscopy (SEM) in this paper. The process of the fracture is under cyclic loading and the stress concentration, the sucker rod'soutside weld was brittle cracking in the first fracture at the beginning. Then the crack in first facture is passed to the inside welding residual, which is corresponding to the source area of the second fracture. Finally, the crack in the second fracture propagation along the weld seam and connected with the first fracture surface lead the rod broke completely. The causes of sucker rod breaking for the sucker rod did not meet the requirements of the HL grade rod material and the friction welding defects.

Keywords: Sucker rod; Friction welding; Fracture; Fatigue

1 Introduction

A history of human exploitation of energy is the history of human civilization. Since thousands of years ago, man began to make use of fire, which enabled man to survive and develop in the harsh natural environment. Britain has a plenty of coal resources, the invention of steam engine prompted the British to develop and utilize a large number of coal resources, thus promoting the first industrial revolution and making a qualitative leap in human development. After a hundred years of development of modern industry, petroleum has replaced coal as the blood of modern industry, and the petroleum industry is like a huge heart, pumping continuous fresh blood for global industry.

Corresponding authors: Zhang Aibo, zhab 2003@ nwpu. edu. cn, Qi Dongtao, qidt@ cnpc. com. cn (D. T. Qi).

In the process of oil production, we will inevitably encounter all kinds of failure problems, and our job is to find out the cause of failure and avoid the recurrence of such problems. China's most of oilfield is located in the hinterland of the desert or gobi, where the natural environment is harsh and the temperature difference between day and night is huge. Many scholars have studied the failure cases of the oilfield. A. Q. Fu was studied the downhole corrosion behavior of NiW coated carbon steel in spent acid and formation water and girth weld cracking of mechanically lined pipe in the northwest oilfield in 2016[1,2], Y. Long studied the 13Cr valve cage of tubing pump failure in an oilfield in 2018[3], H. Ding analyzed the connecting rod for oil pumping unit in China western oilfield[4]. Production and transportation pipelines and equipment used in oil and gas fields can easily fail under harsh service conditions or due to material quality. D. L. Duan studied the failure of sucker rod coupling or tubing in sucker rod pumping system[5], J. An studied the effect of boronizing from solid phase on the tensile mechanical properties and corrosion and wearresistances of steel AISI 8620[6].

In the oilfield, after 729 days' of service, the sucker rod failed to break and failure. The fracture position is located at the welding seam which connecting the upsetting end of the sucker rod with the polished rod, and its macroscopic morphology was shown in Fig. 1. Polished rod model is $\phi38mm \times 11mm$, the grade is HL, material made of 35CrMoand welding methods was friction welding. In order to find out the reasons of the rod's breaking, the macro and micro morphology of the sample and the physical and chemical properties of the material were tested and analyzed, and the failure reasons of the sample were analyzed comprehensively.

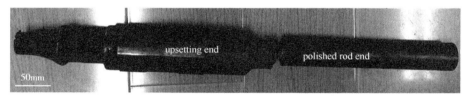

Fig. 1 Fracture of the sucker rod visual appearance

2 Experimental

In order to find out the causes of the friction welding seamfracture of a sucker rod, a series of materials characterization was carried out on the fracture sucker rod. The nondestructive test(NDT) of the Ⅱ C-d method adopted, penetration detection was carried out near the fracture. The chemical compositions were detected by ARL4460 photoelectric direct reading spectrometer. TheUTM – 5305 Mechanical testing machine was used to test the tensile strength and yield strength of the fracture sucker rod each side on room temperature. The PIT752D-2(300J) impact machine was used to test the impact energy of the polished rod end and upsetting end at room temperature respectively. TheKB30BVZ-FA vickers hardness testing machine was used to test the HV_{10} hardness of the base material, weld and the heat affected zone (HAZ). Besides, the OLS – 4100 laser confocal microscope was used to observe the microstructure of the base material, weld and the HAZ. Finally, the fracture surface and crack growth of the fracture were analyzed by macroscopic morphology and the scanning electron microscope(SEM).

3 Results and discussion

3.1 Nondestructive testing

According to NB/T 47013.5—2015 standard[7], the results of penetration detection near the fractureshowed in Fig. 2, that no cracks were found near the fracture surfaces of the sucker rod except the fracture position.

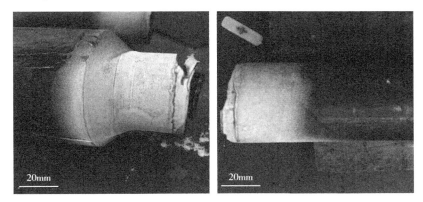

Fig. 2 Penetration detectionfor the sucker rod

3.2 Chemical composition

Spectral analysis for material composition of the sucker rod end and the upsetting end are shown in Table 1. The results of chemical composition analysis showthat the chemical composition of the sucker rod end material has some deviation from the standard requirements, mainly because the content of C element is lower than the requirements of GB/T 26075—2010 standard[8] for 35CrMo low-alloy steel.

Table 1 Chemical composition of the rod end and the upsetting end (wt. %)

Element	C	Si	Mn	P	S	Cr	Mo	Ni	Cu
GB/T 26075—2010	0.32~0.40	0.17~0.37	0.50~0.70	≤0.025	≤0.025	0.80~1.10	0.15~0.25	≤0.30	≤0.20
rod end Measured	0.28	0.23	0.64	<0.01	<0.008	0.98	0.17	0.019	<0.02
upsetting end Measured	0.38	0.23	0.61	0.022	0.0098	0.88	0.18	0.028	0.067

3.3 Mechanics property analysis

The tensile test results of the sucker rod end and upsetting end are shown in Table 2. The tensile strength of the sample on the rod end at room temperature was higher than the upper limit of SY/T 5029—2013 standard[9] for HL-grade rod, and the tensile strength and yield strength of the upsetting end sample were lower than the minimum requirements of the standard. Compared the stress-strain curves in Fig. 3, there is a vast difference in mechanical properties between the rod end and the upsetting end obviously.

The impact energy test results of the sucker rod end and the upsetting end are shown in Table 3. As to the test results, the impact performance of the sucker rod end and the upsetting end meets the requirements of SY/T 5029—2013 standard for HL-level sucker rod.

Table 2 Tensile test results at room temperature

Sample	Size(mm)	Tensile Strength(MPa)	Yield Strength(MPa)	Elongation(%)
rod end	φ6.25×25	1232	1176	17
upsetting end		1223	1145	16
		1219	1145	17
		742	632	26
		747	612	22
		749	647	20
SY/T 5029—2013		965~1195	≥793	≥10

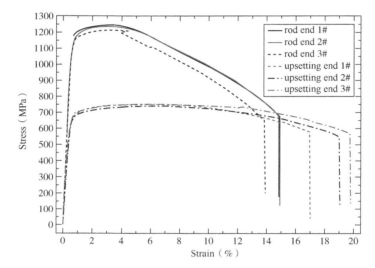

Fig. 3 Stress-strain curves of the sucker rod

Table 3 Impact test result(J)

Sample	Size(mm)	Notch	Temperature(℃)	Measured	SY/T 5029—2013
rod end	10×10×55	U	20	140	≥60J
upsetting end				140	
				147	
				144	
				136	
				143	

The HV_{10} hardness test results are shown in Table 4. As the vickers hardness test results distribution in Fig. 4, we can clearly see the hardness of the weld seam is much higher than that of the base material and the HAZ.

Table 4 Vickers hardness test results(HV_{10})

Measuring position	HAZ	Weld	Base metal
Measured	280, 256	386	286, 281

3.4 Metallographic analysis

The fracture sample was cut and prepared for the metallurgical examination. Metallurgical structure of the weld is high hardness tempered martensite as shown in Fig. 5, that's also explains why the hardness of weld is much higher thanthat of the HAZ and base metal. The structure of the HAZ is tempered sorbite with grain size of 9.5, besides the metallographic structure has the characteristics of flow line deformation as shown in Fig. 6. The base metal of the rod body is tempered sorbite with grain size of 8.0as shown in Fig. 7.

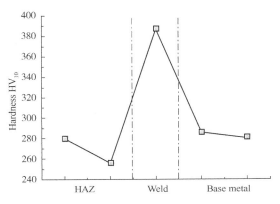

Fig. 4　Vickers hardness test results(HV_{10})

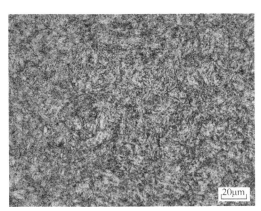

Fig. 5　Metallurgical structure of the weld

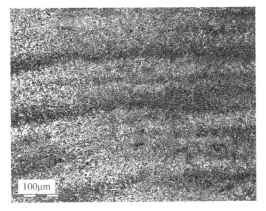

Fig. 6　Metallurgical structure of HAZ

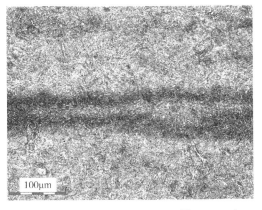

Fig. 7　Metallurgical structure of base metal

3.5 Fractography analysis

Fig. 8 shows the side morphology of the sucker rod fracture. It can be seen that the fracture cracks laterally on both sides of the weld seam. Clearly, we observe two fracture surface on different plane, besidesthe residual height of the weld is 0.5mm higher than the rod body. The sucker rod fractured on both sides of the weld, crack growth along the two cross section one after another. The crack in the first fracture starts from the outside of the sucker rod, and stoppedwhen the crack extends to its 1/2 circle. The first fracture' scrack end corresponds to the the second fracture' ssource area. After the the second fracture' scrack continues to expand thoroughly, the sucker rod breaks completely.

Fig. 9 shows the macroscopic morphology of the second fracture surface. It can be seen that the second fracture surface is divided into two characteristic sections: the flat section and the shear lip section. The flat zone occupies about 1/3 of the circumference of the fracture and has brittle cracking characteristics. The cowrie pattern lines points to the source area where inner side of the weld seam. The herring bone pattern points to the direction of the crack source along the rod circle. The shear lip area

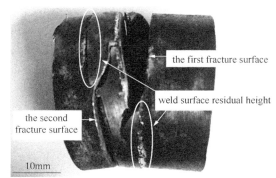

Fig. 8   Macroscopic morphology of fracture side

occupies about 2/3 of the circumference of the fracture, which is the instantaneous fracture area formed by the final fracture of the rod body. Brittle cracking occurred in the source region of the second fracture after the first fracture's crack extended to its end. The second fracture's crack continuous growth lead the sucker rod fractured completely.

Fig. 10 shows the axial section of the fracture on the sucker rod. The fracture surface is located at the joint of the residual height of the inner weld seam and the base material. The crack originates from the inner surface of the interface between the weld seam and the base material, which is consistent with the cowrie pattern lines of the fracture surface points direction in Fig. 8, and the residual height of the weld inner side is curled and itas high as 4mm.

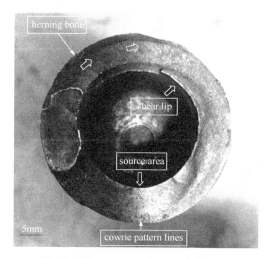

Fig. 9   The macroscopic morphology of the second fracture surface

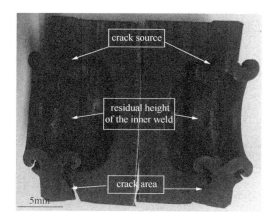

Fig. 10   The macroscopic morphology of the fracture was cut along the axial direction

The fracture source area SEM photo shown as in Fig. 11, the surface is relatively flat. Fig. 12 and Fig 13 show the micromorphology of the fracture propagation zone. A large number of fatigue striations can be seen at the micro level of the fracture in Fig. 12, extrusion marks were shown in Fig. 13 as the white arrow pointed, indicating that the fracture has fatigue propagation after repeated folding and extrusion.

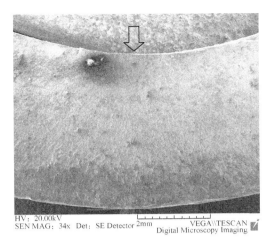
Fig. 11　Source area SEM photo

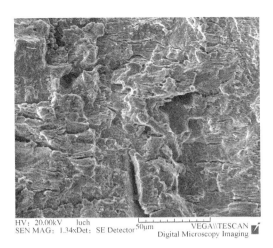
Fig. 12　Propagation area fatigue striations

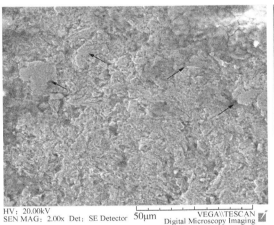
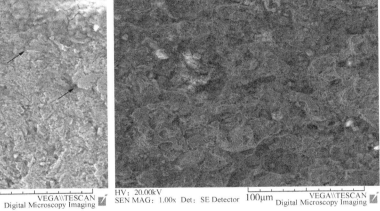
Fig. 13　Propagation area extrusion trace

4　Root cause analysis

Based on the test results and analysis above, the chemical composition of the sample has some deviation from the standard requirements, the metallographic structure of the weld is high hardness tempered martensite as shown in Fig. 5, the tensile properties at room temperature do not meet the standard requirements of SY/T 5029—2013, the tensile strength and yield strength of the upsetting end samples are lower than the minimum requirement of the standard. It can be concluded that the poor mechanical properties of the material is one of the main reasons for the fracture of the rod.

The main reasons and type of the rod's fracture can be determined by combining the macroscopic morphology and service stress of the sample. First of all, because the rod is a welding structure member, it is prone to dimensional change at the welding seam. According to the macroscopic morphology of the fracture in Fig. 10 above, the weld reinforcement of the inner and outer sides weld reaches 4mm and 0.5mm respectively, and the stress concentration in this part was caused by dimensional change change. Secondly, in the mechanical test, the hardness value of the weld seam is significantly higher than the base material and the heat affected area, because of high hardness tempered martensite was formed. So the weld became more prone to crack than the base material. Finally, during the service period, the sucker rod is subjected to alternating load. When such cyclic load exists for a long time, it is easy to cause fatigue damage of the structure. Such fatigue damage will often give priority to the formation of fatigue cracks in the material dimension where is not uniform or the dimensional change site. From the macroscopic morphology of the fracture in Fig. 8-10, it can be seen that the crack originated from the weld residual height part, and it expanded laterally along the lateral residual height edge until complete fracture. The fatigue striation and extrusion marks of the fracture can be clearly seen from the fracture micromorphology in Fig. 12 and Fig. 13. In addition, the metallographic analysis also shown that the microstructure of the heat-affected zone of the weld has the characteristics of streamline deformation in Fig. 6. It can be inferred that, under the action of cyclic loading, fatigue and fracture occurred in the high weld residual part after repeated folding and extrusion, which eventually led the weld of the sucker rod to fatigue fracture.

5 Conclusions and recommendations

5.1 Conclusions

(1) The reason for the fracture of the sucker rod is fatigue fracture. The fracture process is as follows: under cyclic loading, the residual height of the sucker rod in the inner side of the weld forms a crack source because of stress concentration at first, then the crack propagates along the heat affected zone, after further extends along the residual height of the outer weld lead the sucker rod complete fracture.

(2) The content of C in the chemical composition of polished rod end material is lower than the requirement of GB/T 26075—2010 standard for 35CrMo low alloy steel, and its room temperature tensile property does not meet the requirement of SY/T 5029—2013 standard.

5.2 Recommendations

(1) Improving the structure of sucker rod, especially the excess weld height in the welded part, needs to be effectively handled.

(2) Improve welding quality, control welding temperature, nondestructive testing of weld seam after welding, etc.

(3) The quality of sucker rod products should be effectively controlled, and materials that meet the standards and specifications should be selected.

References

[1] Fu A Q, Feng Y R, Cai R, et al. Downhole corrosion behavior of NiW coated carbon steel in spent acid & formation water and its application in full-scale tubing Engineering Failure Analysis, Volume 66, August 2016, Pages 566-576.

[2] Fu A Q, Kuang X R, Han Y, et al. Failure analysis of girth weld cracking of mechanically lined pipe used in gasfield gathering system. Engineering Failure Analysis, 2016, 68: 64-75.

[3] Y. Long, G. Wu, A. Q. FuFailure analysis of the 13Cr valve cage of tubing pump used in an oilfieldEngineering

Failure Analysis, Volume 93, November 2018, Pages 330-339.
［4］ H. Ding, J. F. Xie, Z. Q. BaiFracture analysis of a connecting rod for oil pumping unit in China western oilfieldEngineering Failure Analysis, Volume 105, November 2019, Pages 313-320.
［5］ D. L. Duan, Z. Geng, S. L. Jiang, S. Li Failure mechanism of sucker rod coupling Engineering Failure Analysis 36(2014)166-172.
［6］ J. An, C. Li, Z. Wen A study of boronizing of steel AISI 8620 for sucker rods Metal Science and Heat Treatment, Vol. 53, Nos. 11-12, March, 2012.
［7］ NB/T 47013.5—2015, Nondestructive testing of pressure equipments, Part 5: Penetant testing.
［8］ GB/T 26075—2010, Steel bars for sucker rods, 2010.
［9］ SY/T 5029—2013, Sucker rods, 2013.

本论文原发表于《Engineering Failure Analysis》2020年第109卷。

Effect of Streaming Water Vapor on the Corrosion Behavior of Ti60 Alloy under a Solid NaCl Deposit in Water Vapor at 600℃

Fan Lei[1,2] Liu Li[1] Cui Yu[4] Cao Min[4] Yu Zhongfen[1,2]
Emeka E. Oguzie[3] Li Ying[4] Wang Fuhui[1]

(1. Shenyang National Laboratory for Materials Science, Northeastern University, Shenyang; 2. State Key Laboratory for Performance and Structure Safety of Petroleum Tubular Goods and Equipment Materials, CNPC Tubular Goods Research Institute; 3. Africa Centre of Excellence in Future Energies and Electrochemical Systems (ACE-FUELS), Federal University of Technology Owerri; 4. Institute of Metal Research, Chinese Academy of Sciences)

Abstract: The corrosion behavior of Ti60 alloy coated with solid NaCl layer in dry O_2 and humid O_2 (30.8 vol.% H_2O) environments at 600℃ have been investigated. The isotopic tracing of ^{18}O showed involvement of H_2O in both the outer corrosion layer (OCL) and the inner corrosion layer (ICL). Theoretical calculations revealed that H_2O provided 14% and 21% of the oxygen required for the formation of the outer and inner corrosion layers, respectively. Detailed analysis of corrosion products revealed that H_2O enhanced the reaction of solid NaCl and TiO_2 to form $Na_4Ti_5O_{12}$ and promoted the crystallinity of corrosion products.

Keywords: Titanium; Alloy; High temperature corrosion

1 Introduction

Titanium alloys are considered as appropriate materials for compressor blades due to their low specific density in addition to high strength and creep resistance at elevated temperatures[1-4]. Indeed, some titanium alloys like Ti-1100[5], IMI 834 and Ti60[6], have found useful applications even at the highest operational temperatures for compressor blades (600℃).

When deployed in service in marine environments, compressors blades are surrounded by moist air which specifically contains abundant salt (especially NaCl) and water vapor. The compressor

---

Corresponding author: Liu Li, liuli@mail.neu.edu.cn.

blades generally operate at temperatures (300~600℃) wherein NaCl exists as a solid deposit on the surface of the blades. Titanium alloys suffer severe active corrosion under solid NaCl deposits at 300~600℃[7-12], due to generation chlorides, like Cl_2 or HCl as reaction products. These chlorides diffuse inward to the substrate and react with substrate metal cyclically[7-12]. This kind of active corrosion has also been reported by Wang et al. for several traditional alloys considered as candidate materials for compressors blades, including Fe-Cr alloy and Ni-based alloy, under a solid NaCl deposit in water vapor at 500~700℃[13-18].

H_2O can significantly influence on the corrosion of Ti alloys[19-23]. For instance, H_2O promotes formation of defective multilayered oxides on Ti-6Al-4V alloy[19], which could imply existence of different oxidation mechanisms[20]. On the other hand, H_2O can increase oxides thickness of pure Ti at 700℃[21]. The detrimental effect of H_2O vapour on the oxidation resistance of Ti alloys results from the tendency of H_2O to dissociate into free H atoms and OH groups at defect sites of TiO_2 at high temperature[24-26]. The generated hydrogen dissolves into TiO_2 forming hydrogen defects, which increases the concentration of crystal defects[26], thereby greatly increasing the diffusion of ions in the oxide scale (both the inward diffusion of oxygen anions and the outward diffusion of metallic cation) as well as the oxidation rate[22-25]. When Ti alloys are coated with a solid NaCl deposit, the H_2O also significantly enhances the corrosion rate compared with that exposed in dry O_2[13-17], which is similar to the behaviour of Fe-xCr alloy. A possible explanation for these phenomena is that the HCl generated under wet conditions is more active and more corrosive than Cl_2 produced in the dry O_2 condition. This active corrosion mechanism of Cl has been studied and reported[8,27].

The corrosion scale on Ti alloy under the combined effects of solid NaCl deposit and humid O_2 is complex, and the effect of H_2O on the corrosion behaviour not yet well understood. Accordingly, this study is focused on the corrosion behaviour of Ti60 alloy coated with a solid NaCl layer, exposed in dry O_2 and H_2O+O_2 (30.8 vol.% H_2O) at 600℃. The involvement of H_2O in the corrosion products was investigated by stable isotopic (^{18}O) labelling methods. The chemical and structural information on the corrosion products were studied by scanning electron microscopy (SEM) equipped with an energy dispersive spectrometer (EDS), transmission electron microscopy (TEM), time of flight-secondary ion mass spectrometry (ToF-SIMS) and X-ray diffraction (XRD).

2 Experiments

2.1 Materials preparation

The chemical composition of the test Ti60 alloy is shown in Table 1. The size of the samples was 10mm × 15mm × 2mm. Prior to experiments, all the samples were first mechanically ground to 800 grid with SiC paper, ultrasonically degreased in alcohol for about 20 min, and then dried in air. This surficial treatment is for helping solid NaCl deposit on the surface of samples. Preheated specimen surfaces were covered with NaCl deposit by repeatedly brushing and drying (surface of samples is preheated at 60~80℃) with saturated NaCl solution[9,10,15], that about (4±0.2)mg/cm² solid NaCl was deposited.

Table 1　The chemical composition of Ti60 alloy studied in this work (wt%)

Alloy	Al	Zr	Sn	Mo	Nb	Ta	Si	Ti
Ti60	5.62	2.98	3.85	0.9	0.4	1.05	0.35	Bal.

2.2　Corrosion experiments

All corrosion tests were carried out in a thermo-balance[9,10]. The continuous mass gain during the corrosion experiment was monitored by thermo-gravimetric analysis (TGA).

The (humid O_2) test atmosphere was obtained by bubbling pure O_2 into distilled water in a glass bubbler (in this study, the flow rate of pure O_2 was about 140mL/min when the inner diameter of the tube was about 3.2cm). The amount of the water vapor was controlled by precisely setting up the temperature of the distilled water in the glass bubbler according to the relationship between vapor pressure of water and temperature. In this study, the temperature of the distilled water was about 70℃, producing about 30.8 vol.% water vapor. This amount of water vapor is for happening strong corrosion reactions. In the old work, this amount of water vapor accelerated the corrosion of Ti60 alloy[10]. To avoid the water vapor condensing inside the thermobalance, a counter flow of pure N_2 was passed through the thermobalance, and its flow rate was about 400mL/min. After the furnace reached the desired temperature (600℃) and the gas flows of humid O_2 and pure N_2 stabilized, the sample was quickly lowered into the constant temperature zone of the furnace tube.

In this work, the corrosion experiments were conducted under the following corrosion environments: solid NaCl deposit layer in a humid O_2 flow at 600℃ (NHO600), solid NaCl deposit layer in a dry O_2 flow at 600℃ (NO600), humid O_2 flow at 600℃ (HO600), and dry O_2 flow at 600℃ (O600). The specific environmental parameters are presented in detail in Table 2.

Table 2　Experimental parameters

Environment	Mass of NaCl (mg/cm^2)	Amount of water vapor (vol.%)	Flow rate of O_2 (mL/min)	Temperature (℃)
NHO600	4.0	30.8	140	600
NO600	4.0	0	140	600
HO600	0	30.8	140	600
O600	0	0	140	600

2.3　Morphologies and chemical composition analysis

After corrosion, the surface and cross-sectional morphologies of the samples were analyzed by SEM and TEM. The chemical composition of the surface corrosion products and the inner corrosion layer was identified by XRD and XPS.

XPS analysis was performed on the powders of the corrosion products scratched from the corroded samples. The XPS measurements were carried out with ESCALAB250 X-ray photoelectron spectrometer, whose X-ray source was monochromatic Al Ka 1486.6eV. The energy scale was calibrated by placing the C 1s peak at 284.6eV. The identification of peaks was performed by reference to an XPS database.

The samples after corrosion were wrapped into a thin nickel foil which protected the oxide scale from fracture and spall during the metallographic preparation. Then, they were embedded into the epoxy resin, ground to 5000 grit with SiC paper and finally polished with diamond paste for cross-sectional observation by SEM.

The detailed microstructures of the corrosion products were studied by TEM analysis performed on a JEOL 2100 TEM operating at 200kV. The cross-section samples for TEM were prepared as follows[9,10]: (1) after corrosion, the outer corrosion layer was removed by mechanical polishing and then the samples with the inner corrosion layer were cut into rectangles (3mm × 1.5mm × 0.7mm); (2) the rectangular samples were glued together with topcoat face to face with an adhesive and mechanically polished to a thickness of approximately 50μm; (3) the thin samples were glued onto a 3mm copper ring, dimpled to a thickness less than 20μm and thinned by argon ion beam; (4) TEM observation.

2.4 Time of flight-secondary ion mass spectrometry, ToF-SIMS

After corrosion, the samples were studied using a ToF-SIMS5 instrument (ION-TOF GmbH), which allowed parallel mass registration with high sensitivity and high mass resolution. A cesium liquid-metal ion (LMI) gun at 20keV beam energy was used for spatially resolved ToF-SIMS analysis. Before ToF-SIMS analysis, the outer corrosion layer was removed by mechanical polishing after 20h exposure in NHO600, and the samples with the inner corrosion layer were ultrasonically degreased in alcohol for about 5min and dried in the air.

2.5 Oxygen isotope analysis

Stable isotope analyses were performed at the environmental stable isotope laboratories, Chinese Academy of Agricultural Sciences (Beijing, China) with elementar PyroCube in contact with ISOPRIME-100. The samples for isotope analysis were 0.35mg powders of the corrosion products, scratched from the sample after 20h exposure in NHO600.

Two kinds of samples were analyzed in this study. The normal sample, scratched from Ti60 alloy exposed in the normal environment (NaCl + normal H_2O + normal O_2) for 20h. The labeled sample, scratched from Ti60 alloy exposed in the labeled environment (NaCl + labeled $H_2^{18}O$ + normal O_2) for 20h. The labeled $H_2^{18}O$ means that the water in the glass bubbler composes of normal distilled water (500mL) and 98% ^{18}O enriched water (5mL).

3 Results

3.1 Corrosion kinetics

The mass gain curves for Ti60 alloy under different conditions as shown in Fig.1(a) indicate that the mass gain of the Ti60 alloy without NaCl (O600 and HO600) was relatively low during the testing time (20h), when compared with the solid NaCl deposit. The mass gain of the Ti60 alloy in NHO600 was significantly larger than that in NO600, which implies that water vapor further accelerates the corrosion of the Ti60 alloy under the solid NaCl deposit layer. Also, Fig.1(b) shows the mass gain in HO600 did not significantly differ from that in O600 after 20h of exposure, but increased gradually and became markedly higher on prolonged exposure (up to 100h). It is thus clear that H_2O promotes both the corrosion of Ti60 alloy under solid NaCl deposit, as well as the oxidation of Ti60 alloy at 600℃.

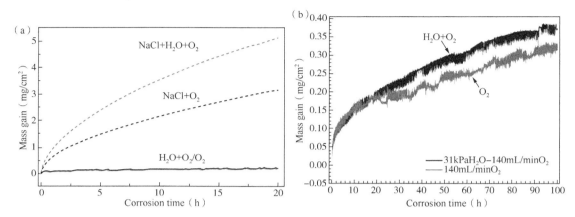

Fig. 1 Mass gain curves of Ti60 alloy: (a) under different corrosion environments for 20h at 600℃:
(I) dry O_2, (II) H_2O+O_2, (III) $NaCl+O_2$, (IV) $NaCl+H_2O+O_2$;
(b) in dry O_2 and H_2O+O_2 for 100 h at 600℃

3.2 Morphologies and structure characterization of the corrosion products

The surface and the cross section morphologies of Ti60 alloy after the samples were exposed in O600 and HO600 for 20h are shown in Fig. 2. There is no clear difference between the surface and cross section morphologies both environments within this time interval. As shown in Fig. 2(a) and (b), a very thin, compact and continuous scale was formed on Ti60 alloy after 20h exposure in the dry O_2 as well as in the H_2O+O_2 stream at 600℃. This oxide scale is very thin, with visible scratch lines (produced by mechanical abrasion during the preparation of specimen). It is thus obvious that the effect of H_2O vapor on the corrosion and oxidation behaviors of Ti60 alloy did not manifest after 20h of exposure at 600℃. However, after 100h of exposure, the surface morphologies of Ti60 alloy became significantly altered by H_2O vapor, as shown in Fig. 3. The oxide scale on Ti60 alloy though remained thin with visible scratch lines, but with obvious granular oxides piles on the surface of samples in HO600, especially inside the scratch lines (Fig. 3c). Therefore, H_2O enhanced the oxidation of Ti60 alloy at 600℃.

The surface and the cross section morphologies of Ti60 alloy after 20h exposure in NO600 and WNO600 are presented in Figs. 4 and 5 respectively. A compact, thick and multilayered corrosion product scale (about 45μm), is formed when Ti60 alloy, coated with a solid NaCl deposit, is exposed in dry O_2 flow at 600℃ for 20h, as shown in Fig. 4. The morphology of the outer corrosion products is nodular, with some small pores (Fig. 4a). From cross section morphologies (Fig. 4b), this complex corrosion scale comprises a 25μm outer corrosion layer (OCL) and a 20μm inner corrosion layer (ICL) from the original surface of the substrate. The OCL is a compact layer without any defects, whereas the ICL consists of lamellar corrosion products with grain orientation, which is similar with results obtained previously[9,10].

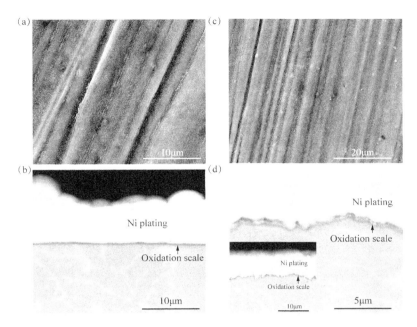

Fig. 2 Surface and cross-sectional morphologies of Ti60 alloy oxidized in dry O_2 and H_2O+O_2 at 600℃ for 20h: (a) surface morphologies in dry O_2; (b) cross section morphologies dry O_2; (c) surface morphologies in H_2O+O_2; (d) cross section morphologies in H_2O+O_2

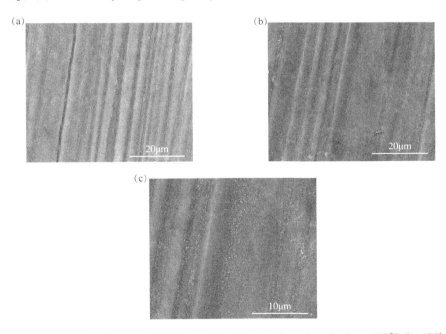

Fig. 3 Surface morphologies of Ti60 alloy oxidized in dry O_2 and H_2O+O_2 at 600℃ for 100h: (a) in dry O_2; (b) in H_2O+O_2; (c) in H_2O+O_2(large magnification)

Fig. 5 shows the surface and cross section morphologies of Ti60 alloy after exposure in NHO600 for 20h. A lot of holes and pores are observed on the surface as shown in Fig. 5(a), and these deep holes are also observed in the cross-sectional morphologies, which are filled with Ni plating as shown in Fig. 5(b). This porous morphology of surface can testify the formation of some volatile species, which also agree with old work and public papers[9,10]. As the corrosion of Fe−Cr alloy under solid NaCl and water vapor at 600℃, formed volatile species such as $TiCl_2(g)$[7], $TiCl_4(g)$[8], $HCl(g)$[7,8], $Cl_2(g)$[7,8], will produce the porous corrosion products on Ti alloy. The cross section morphologies show the corrosion product scales to be thick (about 80μm) and double layered, which also is divided into the outer corrosion layer (OCL, 55μm) and the inner corrosion layer (ICL, 25μm) from the original surface of the substrate. Asides the deep holes, there are also lots of defects formed within the OCL and on the interface between the OCL and ICL. The ICL corrosion products are also lamellar, with grain orientation, like the results previously obtained for NO600[9,10].

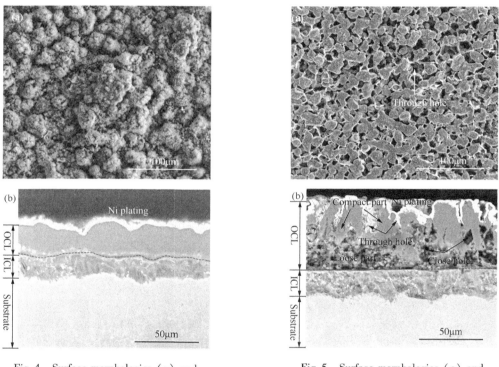

Fig. 4 Surface morphologies (a) and cross-sectional morphologies (b) of Ti60 alloy under NaCl deposit in an atmosphere of dry O_2 at 600℃ for 20h

Fig. 5 Surface morphologies (a) and cross-sectional morphologies (b) of Ti60 alloy under NaCl deposit in an atmosphere of H_2O+O_2 at 600℃ for 20h

Figs. 6 and 7 show the detailed TEM microstructure information of the inner corrosion layer (ICL) of corrosion products formed after exposure for 20h in NO600 and NHO600, respectively. The TEM samples of the ICL were produced by polishing off the OCL. Fig. 6(a) shows the TEM image of the bilayer ICL in NO600, comprising an outer compact region near the interface between the ICL and the OCL (labeled blue) and an inner lamellar region near the base metal (labeled red). The thickness of the compact region formed in NO600 is approximately 0.5μm. The selected area electron diffraction (SAED) technique was employed to determine the crystal structure of the lamellar region. The ringed SAED patterns, as shown in Fig. 6(b), indicate that the lamellar region is almost amorphous with slight polycrystalline character. The white contrast points in the figure should be the defects like nanovoids in the corrosion products. These nanoscale defects in the corrosion products may be produced by corrosion reactions. It can be seen that the grains of the corrosion products in the lamellar structure zone are long rice-like grains (the red dotted lines), and many grains are arranged one by one in the direction of the longer grains. Statistics show that the grain sizes to be 5~10nm wide and 10~30nm long.

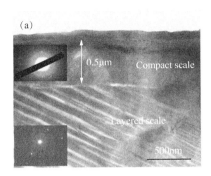
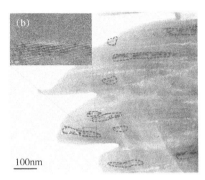

Fig. 6 STEM image of the inner corrosion layer of Ti60 alloy after being exposed under NaCl solid deposit in dry O_2 at 600℃ for 20h (a), and TEM image of the layered scale (b)
(For interpretation of the references to colour in the text, the reader is referred to the web version of this article.)

The TEM image of ICL in NHO600 shown in Fig. 7 also reveals a bilayer structure, as with NO600, though with much thicker scales. The compact region is mainly amorphous, containing a small amount of grains [Fig. 7(b)], and mainly enrich in Ti and O, with a small amount of Al [Fig. 7(d)]. Therefore, the compact region is mainly amorphous Ti oxide. The thickness of the compact region formed in the presence of H_2O was about 1μm. Moreover, in the diffraction of the lamellar region near the metal matrix [Fig. 7(c)], reveals a very obvious amorphous diffraction in the middle, as well as a very clear polycrystalline diffraction ring. From these results, it may be concluded that H_2O vapor increased the crystallinity of the lamellar region near the metal matrix.

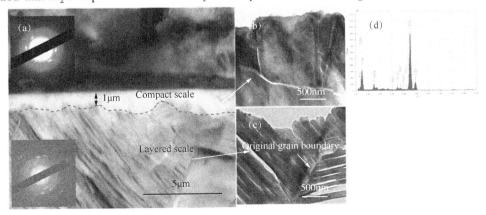

Fig. 7 Detailed cross-sectional morphologies of the inner corrosion layer of Ti60 alloy after being exposed under NaCl solid deposit in H_2O+O_2 for 20h (a), TEM image of the compact scale (b) and the layered scale (c) and the EDS patterns of the compact scale (d)

3.3 Compositions of the corrosion products

TiO_2 is the main corrosion products of Ti60 alloy after 20h corrosion in O600 and HO600. When solid NaCl is deposited on the samples, chemical compositions of corrosion products changed as shown in Fig. 8(a) shows the XRD patterns of the surface corrosion products of Ti60 alloy after 20h exposure in NHO600 and NO600. In NHO600, the surface corrosion products consist mainly of TiO_2 and $Na_4Ti_5O_{12}$, with the residual NaCl on the surface. While, in NO600, TiO_2 is the main corrosion product and $Na_4Ti_5O_{12}$ has not been identified in the XRD result. Fig. 8(b) shows that the ICL of Ti60 alloy in NHO600 and NO600 comprises of TiO_2 and Ti_2O. However, in the absence of H_2O (in NO600), the distinct diffraction peaks of NaCl were detected in the ICL, while those of TiO_2 were relatively subdued. In the presence of H_2O (in NHO600), the diffraction peaks of TiO_2 were very distinct, but NaCl could not be detected. Therefore, it was concluded that H_2O promoted the formation of TiO_2 in the ICL. On the other hand, O can dissolute into based Ti alloy[28,29]. In this work, authors also found O can dissolute into Ti alloy after corrosion 20h under this experiment.

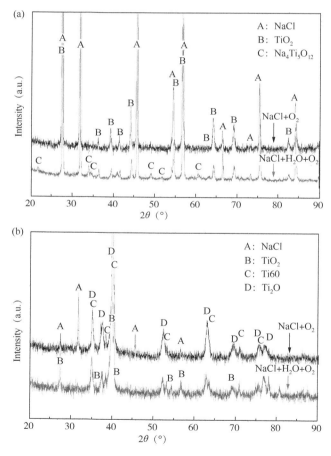

Fig. 8 XRD patterns of the surface corrosion products (a) and the corrosion products of the inner corrosion layer (b) of Ti60 alloy under NaCl deposit in an atmosphere of dry O_2 and wet O_2 at 600℃ for 20h

The high resolution XPS spectra was used to identify chemical composition of the corrosion products. The samples used in XPS testing, are corrosion products powder scratched from samples. The outer layer is relative loose, and scratched easily. After scratching the loose outer layer, the compact inner layer is then scratched from the sample. Figs. 9 and 10 show detailed XPS spectra for the powder corrosion products of OCL and ICL of Ti60 alloy in NO600, respectively. The results show the corrosion products of the OCL to be TiO_2, Ti_2O, $Na_4Ti_5O_{12}$, Al_2O_3, ZrO_2 and SnO_2, with the residual NaCl. Similarly, the ICL contains Ti_2O, TiO_2, Al_2O_3, ZrO_2 and SnO, with NaCl diffused inward. Ti oxides are the main corrosion products in both layers, with Ti_2O occurring mainly in the inner corrosion layer, while the outer corrosion layer had more of TiO_2. Al_2O_3, ZrO_2 and SnO_2 were detected in both the internal and external corrosion layers, though ZrO_2 occurred predominantly in the outer corrosion layer and SnO_2 mainly occurred in the inner corrosion layer. $Na_4Ti_5O_{12}$ was detected only in the outer corrosion layer. In addition, the Ti XPS spectra revealed an additional chemical state (Ti $2P_{3/2}$ at 458.57eV) in the internal and external corrosion layers, which is similar to those of $TiCl_4$ ($2P_{3/2}$ at 459eV[30,31]) and $TiCl_3$ ($2P_{3/2}$ at 458.50eV[32]). The Cl XPS spectra, showed a Cl chemical state (Cl $2P_{3/2}$ at 198.48eV) in addition to NaCl, which is similar to $TiCl_4$ ($2P_{3/2}$ at 198.4eV[30,31]) and a-$TiCl_3$ ($2P_{3/2}$ at 198.0eV[33]). Therefore, it is reasonable to suggest that there are Ti–Cl bonds in the corrosion product layer, predominantly in the inner corrosion layer. This result agrees with the old work[10]: Cl penetrates into the inner layer of corrosion products of Ti60 alloy, which will help the transport of me-tallic ions and promote corrosion.

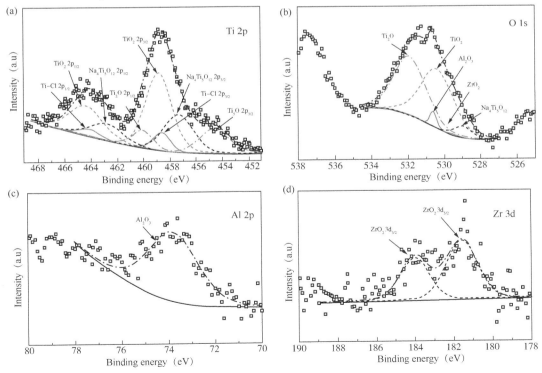

Fig. 9 XPS spectra for Ti (a), O (b), Al (c), Zr (d), Sn (e), Na (f) and Cl (g) of the powder corrosion products of the OCL of Ti60 alloy under NaCl deposit in dry O_2 at 600℃ for 20h

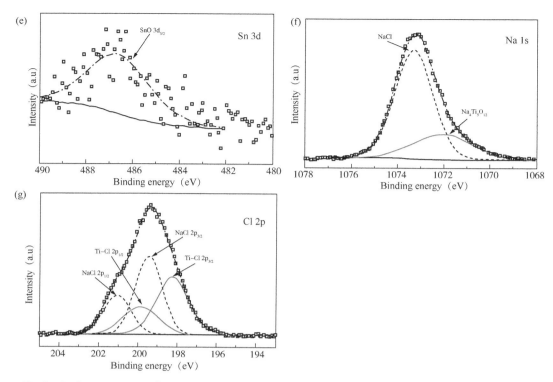

Fig. 9　XPS spectra for Ti (a), O (b), Al (c), Zr (d), Sn (e), Na (f) and Cl (g) of the powder corrosion products of the OCL of Ti60 alloy under NaCl deposit in dry O_2 at 600℃ for 20h (continued)

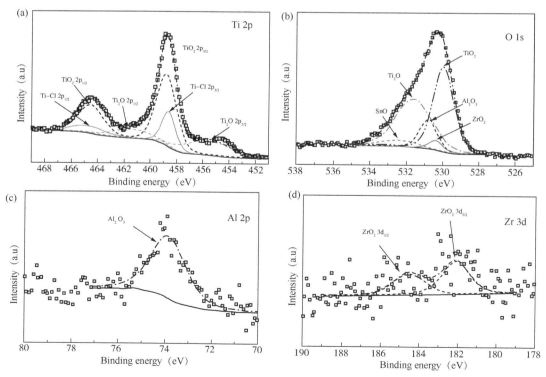

Fig. 10　XPS spectra for Ti (a), O (b), Al (c), Zr (d), Sn (e), Na (f) and Cl (g), of the powder corrosion products of the ICL of Ti60 alloy under NaCl deposit in dry O_2 at 600℃ for 20h

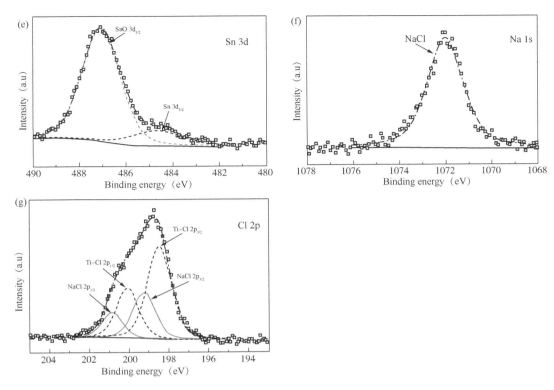

Fig. 10　XPS spectra for Ti (a), O (b), Al (c), Zr (d), Sn (e), Na (f) and Cl (g), of the powder corrosion products of the ICL of Ti60 alloy under NaCl deposit in dry O_2 at 600℃ for 20h (continued)

XPS spectra of the corrosion products of Ti60 alloy in NHO600 show similar composition to that of NO600, though the relative amount of $Na_4Ti_5O_{12}$ in the OCL in NHO600 is much larger than that in NO600, as shown in Fig. 11.

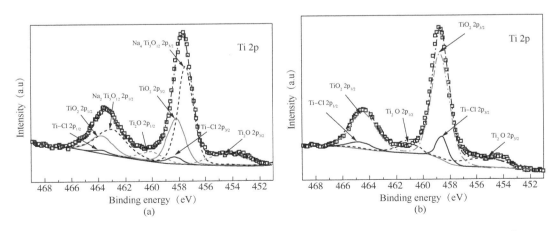

Fig. 11　XPS spectra for Ti from the powder corrosion products of the OCL (a), Ti from the powder corrosion products of the ICL (b), and Na from the powder corrosion products of the OCL (c) of Ti60 alloy under NaCl deposit in H_2O+O_2 at 600℃ for 20h (d)

· 237 ·

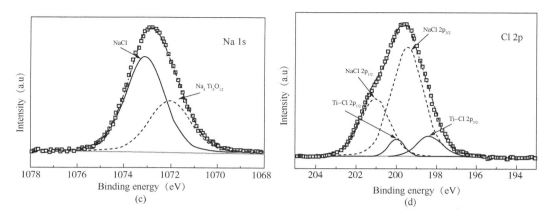

Fig. 11 XPS spectra for Ti from the powder corrosion products of the OCL (a), Ti from the powder corrosion products of the ICL (b), and Na from the powder corrosion products of the OCL (c) of Ti60 alloy under NaCl deposit in H_2O+O_2 at 600℃ for 20h (d) (continued)

3.4 Stable isotopes of the corrosion products

In order to isolate the effect of H_2O on the corrosion behavior of Ti60 alloy under solid NaCl deposit at 600℃, the labeled $H_2^{18}O$ heavy water was used for all corrosion studies. The amount of labeled ^{18}O in $H_2^{18}O$ heavy water can be calculated by following function:

$$C^{18O} = {}^{18}O/({}^{18}O+{}^{16}O) \times 100\% \qquad (1)$$

The testing samples are also corrosion products powder scratched from samples, like XPS. Stable isotope analyses were performed to test the amount of ^{18}O in the powder samples. And then the ratio of ^{18}O in OCL and the ICL of Ti60 alloy exposed for 20h in NHO600 has been calculated, as shown in Fig. 12. The results indicate that when Ti60 alloy is exposed in the ^{18}O labeled $H_2^{18}O$, C^{18O} of the OCL are significantly larger than that exposed in the normal H_2O, which means that H_2O participates in the corrosion reactions to form the OCL. Similarly, by the results, H_2O also participates in the corrosion reactions to form the ICL since C^{18O} of the ICL are larger in the labeled $H_2^{18}O$ than that in the normal environment. Thus, these results demonstrate

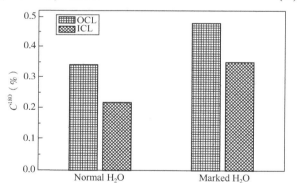

Fig. 12 The concentration of ^{18}O, namely $C^{18} = {}^{18}O/({}^{18}O+{}^{16}O)$ 100%, of the OCL and the ICL of Ti60 alloy under NaCl deposit in H_2O+O_2 at 600℃ for 20h. The normal one is exposed in the normal environment (NaCl+normal H_2O+normal O_2) for 20h. The marked one is exposed in the marked environment (NaCl+labeled H_2O+ normal O_2) for 20h

that H_2O participates in the chemical reaction of Ti60 alloy and NaCl at 600℃, and also participates in the both formation of OCL and ICL.

4 Discussion

According to our obtained experimental results, H_2O significantly influences the corrosion behavior of Ti60 alloy with or without solid NaCl deposit at 600℃. Fe-Cr alloys, Ni-based super alloy and Ti alloys undergo pronounced corrosion in the presence of NaCl at 500 ~ 700℃[7-10,13-18]. According to previous findings, Ti60 alloy also under-goes accelerated corrosion with solid NaCl deposit in $H_2O + O_2$ at 600℃[9,10]. Generally, Ti is sensitive to NaCl above 400℃[8,10]. Ti and Ti oxides undergo chemical reactions with NaCl at 500~700℃[8,9,10].

The mechanism of Ti alloy corrosion in the presence of NaCl has been studied[8]. The schematic diagram of corrosion mechanism with and without H_2O has shown in Fig. 13. The main chemical reactions have been shown to proceed as follows:

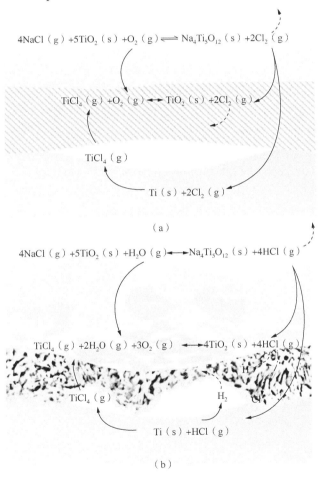

Fig. 13 The schematic diagram of corrosion mechanism of Ti60 alloy under solid NaCl deposit without (a) and with (b) H_2O at 600℃

Ti alloys under solid NaCl deposit at 600℃ undergoes a series of chemical Reactions(2) ~ (4).

$$4NaCl_{(g)} + 5TiO_{2(s)} + O_{2(g)} \rightleftharpoons Na_4Ti_5O_{12(s)} + 2Cl_{2(g)} \qquad (2)$$

$$Ti_{(s)} + 2Cl_{2(g)} \rightleftharpoons TiCl_{4(g)} \qquad (3)$$

$$TiCl_{4(g)} + O_{2(g)} \rightleftharpoons TiO_{2(s)} + 2Cl_{2(g)} \qquad (4)$$

Ti alloys under solid NaCl deposit in H_2O at 600℃ also undergoes a series of chemical Reactions (5) ~ (7).

$$4NaCl_{(g)} + 5TiO_{2(s)} + H_2O_{(g)} \rightleftharpoons Na_4Ti_5O_{12(s)} + 4HCl_{(g)} \qquad (5)$$

$$Ti_{(s)} + 4HCl_{(g)} \rightleftharpoons 2TiCl_{4(g)} + 2H_{2(g)} \qquad (6)$$

$$TiCl_{4(g)} + 2H_2O_{(g)} + 3O_{2(g)} \rightleftharpoons 4TiO_{2(s)} + 4HCl_{(g)} \qquad (7)$$

From these chemical reactions, it is clear that NaCl reacts with TiO_2 first, not Ti. The chemical reactions of Ti occur at the interface between the metal and corrosion products in NO600 and NHO600. Also the in-volvement of Cl in the ICL is observed clearly in NO600 and NHO600, and Cl replaces O in the Ti oxides lattice in the ICL to form Ti-Cl bond (as shown in XPS results), resulting in the increase in the electro-chemical properties of ICL and the fast diffusion of ions in the ICL[9,10]. Furthermore, the dopied Cl diffuses into the metal/ICL interface quickly through the interfaces of the layered ICL as well as grain boundaries and reacts with the substrate metal cyclically, resulting in the fast corrosion of Ti60 alloy[9,10].

The involvement of H_2O accelerates the corrosion of Ti alloys[9,10], and the corrosion products changed, with higher amounts of volatiles, leading to formation of corrosion products with more porous structure (from SEM results). Also, H_2O promotes the formation of $Na_4Ti_5O_{12}$, which implies that H_2O promotes the chemical reactions between NaCl and Ti oxides. It is thus of great interest to analyze and categorize the involvement of H_2O in the formation of the OCL or ICL. Consequently, the ratio of H_2O in the OCL and ICL have been studied in detail.

4.1 Ratio of O supplied by H_2O in the outer layer and inner layer corrosion products

The testing of labelled ^{18}O in the two layers (OCL and ICL) of corrosion products scale formed after 20h in NHO600 formed the basis for this calculation. In this part, the involvement of H_2O to form two layers oxides is tried to study by quantitative analysis. The first step is to select a right calculation way to do this quantitative analysis according to the both formed oxides information.

The first part is outer layer. Since the outer corrosion layer (OCL) formed in the $NaCl + H_2O + O_2$ environment is very porous, the outer corrosion layer has no barrier protecting it from the corrosion medium. Hence, both H_2O and O_2 can rapidly diffuse inward, react with the metal ions that diffuse outward, and further form the outer corrosion layer. Therefore, the ratio of $H_2^{18}O$ can be calculated directly based the measured amount of concentration of ^{18}O.

The ^{18}O of the corrosion products in the OCL is supplied by both the $H_2^{18}O$ and O_2, as follows:

$$C_{OCL}^{18O} = \alpha_{OCL} \cdot C_{H_2O}^{18O} \cdot \beta_{OCL}^{H_2O} + (1 - \alpha_{OCL}) \cdot C_{O_2}^{18O} \cdot \beta_{OCL}^{O_2} \qquad (8)$$

Where C_{OCL}^{18O} is the concentration of ^{18}O in the OCL, α_{OCL} is the percent of O supplied by H_2O in the OCL (in this case, the percent of oxygen supplied by O_2 gas is $1 - \alpha_{OCL}$), $C_{H_2O}^{18O}$ and $C_{O_2}^{18O}$ are the concentrations of ^{18}O for $H_2^{18}O$ and O_2 (^{18}O exists in general O_2, so it also can be detected in corrosion products) respectively, and $\beta_{OCL}^{H_2O}$ and $\beta_{OCL}^{O_2}$ are the isotope effect during the corrosion reaction for $H_2^{18}O$ and O_2 gas respectively.

The experimental H_2O solution is composed of 500mL normal H_2O and 5mL labeled $H_2^{18}O$. We should calculate the ratio of labeled $H_2^{18}O$. Total volume is 505 mL. Considering that the normal H_2O also contains some $H_2^{18}O$, the measured concentration C^{18O} consists of $C^{18O}_{H_2O-normal}$ and labeled $H_2^{18}O$, which means the $C^{18O}_{H_2O} = \dfrac{500 \times C^{18O}_{H_2O-normal} + 5 \times 98\%}{500+5}$ in this study. Therefore, two equations for the OCL have been set up as follows:

$$C^{18O}_{OCL-normal} = \alpha_{OCL} \cdot C^{18O}_{H_2O-normal} \cdot \beta^{H_2O}_{OCL} + (1-\alpha_{OCL}) \cdot C^{18O}_{O_2} \cdot \beta^{O_2}_{OCL} \tag{9}$$

$$C^{18O}_{OCL-marked} = \alpha_{OCL} \cdot \dfrac{500 \times C^{18O}_{H_2O-normal} + 5 \times 98\%}{500+5} \cdot \beta^{H_2O}_{OCL} + (1-\alpha_{OCL}) \cdot C^{18O}_{O_2} \cdot \beta^{O_2}_{OCL} \tag{10}$$

Then, from Eqs. (9) and (10), Eq. (11) has been obtained:

$$\alpha_{OCL} = \dfrac{101 \times (C^{18O}_{OCL-marked} - C^{18O}_{OCL-normal})}{(98\% - C^{18O}_{H_2O-normal}) \cdot \beta^{H_2O}_{OCL}} \tag{11}$$

Generally, the isotope effect of O is minimal during the chemical reaction; thus, we assume $\beta^{H_2O}_{OCL} \approx 1$. Since the changes of α_{OCL} are minor by $C^{18O}_{H_2O-normal}$ and we assume it is same with the standard mean ocean water (SMOW), in which case $C^{18O}_{H_2O-normal} \approx 0.20012\%$ [29]. Additionally, $C^{18O}_{OCL-normal}$ and $C^{18O}_{OCL-marked}$ are the results of oxygen isotope analysis as shown in Fig. 12, $C^{18O}_{OCL-normal} = 34.016\%$ and $C^{18O}_{OCL-marked} = 47.801\%$. The ratio of O supplied by H_2O in the OCL can be calculated: $\alpha_{OCL} = 14.24\%$. This means that for the OCL, 14.24% O of the corrosion products comes from H_2O.

Secondly, we try to calculate the ratio of labeled $H_2^{18}O$ in the inner layer, which is different from the outer layer. Because the oxides are different between two layers, the calculation ways are different too. When Ti60 alloy is exposed in NHO600 for 20h, a compact scale forms in the ICL, around the interface of the OCL and the ICL as shown in Fig. 7a and b. Therefore, we think the H_2O cannot diffuse into the ICL as molecules. And then O, not H_2O, should diffuse into the oxides and react with the substrate metal. The calculation way of O has been changed. During the diffusion, the isotope fractionation of O occurs and the mean concentration of ^{18}O in the ICL (C^{18O}_{ICL}) could be displayed as follows:

$$C^{18O}_{ICL} = \dfrac{\int_0^h m_O \cdot C^{18O}_x \cdot s \, dx}{\int_0^h m_O \cdot s \, dx} \tag{12}$$

where x is the distance in the ICL from the interface of the OCL and ICL; h is the thinness of the ICL; m_O is the O concentration as function of x; C^{18O}_x is the concentration of ^{18}O at x and s is the area of the sample. m_O was analyzed by SIMS and the results presented in Fig. 13 indicates that the amount of O^- did not noticeably change with the sputter time. We then assume that m_O is a constant ($m_O = 9.5 \times 10^4$). Also, s is a constant. If we assume $C^{18O}_x = C^{18O}_{x=0} \cdot f(x)$, where $f(x)$ is the function of x, Eq. (12) accordingly yields Eq. (13).

$$C^{18O}_{ICL} = C^{18O}_{x=0} \times \dfrac{\int_0^h f(x) \, dx}{\int_0^h 1 \, dx} \tag{13}$$

$C_{x=0}^{18O}$ is determined by the reaction of H_2O and O_2 at the interface of the OCL and the ICL. Thus, $C_{x=0}^{18O}$ is given as follows, similar with the OCL:

$$C_{x=0}^{18O} = \alpha_{ICL} \cdot C_{H_2O}^{18O} \cdot \beta_{ICL}^{H_2O} + (1-\alpha_{ICL}) \cdot C_{O_2}^{18O} \cdot \beta_{ICL}^{O_2} \quad (14)$$

where α_{ICL} is the percentage of O supplied by H_2O in the ICL, $C_{H_2O}^{18O}$ and $C_{O_2}^{18O}$ are the concentrations of ^{18}O for H_2O and O_2 respectively, $\beta_{ICL}^{H_2O}$ and $\beta_{ICL}^{O_2}$ are the isotope effects during the corrosion reaction for H_2O and O_2 respectively. If we identified the factor of isotope fractionation $A = \dfrac{\int_0^h f(x)\,dx}{\int_0^h 1\,dx}$, we can obtain Eq. (15) according to Eqs. (13) and (14).

$$C_{ICL}^{18O} = A \cdot [\alpha_{ICL} \cdot C_{H_2O}^{18O} \cdot \beta_{ICL}^{H_2O} + (1-\alpha_{ICL}) \cdot C_{O_2}^{18O} \cdot \beta_{ICL}^{O_2}] \quad (15)$$

Similarly, two equations for the ICL have been set up when considering the normal solution ($C_{H_2O}^{18O} = C_{H_2O-normal}^{18O}$) and the labeled solution ($C_{H_2O}^{18O} = \dfrac{500 \times C_{H_2O-normal}^{18O} + 5 \times 98\%}{505}$).

$$C_{ICL-normal}^{18O} = A \cdot [\alpha_{ICL} \cdot C_{H_2O-normal}^{18O} \cdot \beta_{ICL}^{H_2O} + (1-\alpha_{ICL}) \cdot C_{O_2}^{18O} \cdot \beta_{ICL}^{O_2}] \quad (16)$$

$$C_{ICL-marked}^{18O} = A \cdot \left[\alpha_{ICL} \cdot \frac{500 \times C_{H_2O-normal}^{18O} + 5 \times 98\%}{505} \cdot \beta_{ICL}^{H_2O} + (1-\alpha_{ICL}) \cdot C_{O_2}^{18O} \cdot \beta_{ICL}^{O_2}\right] \quad (17)$$

Then, from Eqs. (16) and (17), Eq. (18) is obtained:

$$\alpha_{ICL} = \frac{101 \times (C_{ICL-marked}^{18O} - C_{ICL-normal}^{18O})}{(98\% - C_{H_2O-normal}^{18O}) \cdot A \cdot \beta_{ICL}^{H_2O}} \quad (18)$$

Further permutations yield Eq. (19) as follows:

$$C_{O_2}^{18O} = \frac{C_{OCL-normal}^{18O} \cdot (0.98 - C_{H_2O-normal}^{18O}) - 101 \times (C_{OCL-marked}^{18O} - C_{OCL-normal}^{18O}) \cdot C_{H_2O-normal}^{18O}}{[(0.98 - C_{H_2O-normal}^{18O}) \cdot \beta_{OCL}^{H_2O} - 101 \times (C_{OCL-marked}^{18O} - C_{OCL-normal}^{18O})] \cdot \beta_{OCL}^{O_2}/\beta_{OCL}^{H_2O}} \quad (19)$$

Generally, the isotope effect of O is small during the chemical rereaction; thus we assume $\beta_{OCL}^{H_2O} \approx 1$, $\beta_{OCL}^{O_2} \approx 1$, $\beta_{ICL}^{H_2O} \approx 1$, $\beta_{ICL}^{O_2} \approx 1$. Then, from Eqs. (16), (18) and (19), we can obtain:

$$A = \frac{C_{ICL-normal}^{18O} - \dfrac{101 \times (C_{ICL-marked}^{18O} - C_{ICL-normal}^{18O})}{0.98 - C_{H_2O-normal}^{18O}} \cdot C_{H_2O-normal}^{18O}}{\dfrac{C_{OCL-normal}^{18O} \cdot (0.98 - C_{H_2O-normal}^{18O}) - 101 \times (C_{OCL-marked}^{18O} - C_{OCL-normal}^{18O}) \cdot C_{H_2O-normal}^{18O}}{0.98 - C_{H_2O-normal}^{18O} - 101 \times (C_{OCL-marked}^{18O} - C_{OCL-normal}^{18O})}} + \frac{101 \times (C_{ICL-marked}^{18O} - C_{ICL-normal}^{18O})}{0.98 - C_{H_2O-normal}^{18O}} \quad (20)$$

$C_{OCL-normal}^{18O}$, $C_{OCL-marked}^{18O}$, $C_{ICL-normal}^{18O}$ and $C_{ICL-marked}^{18O}$ are the results of O isotope analysis (as shown in Fig. 12, $C_{OCL-normal}^{18O} = 34.016\%$, $C_{OCL-marked}^{18O} = 47.801\%$, $C_{ICL-normal}^{18O} = 21.700\%$ and $C_{ICL-marked}^{18O} = 34.931\%$). Eq. (20), then, yields Eq. (21):

$$A = \frac{\left(0.2170\% - \dfrac{0.1336 \times C_{H_2O-normal}^{18O}}{0.98 - C_{H_2O-normal}^{18O}}\right) \cdot (0.8408 - C_{H_2O-normal}^{18O})}{0.3402\% \times (0.98 - C_{H_2O-normal}^{18O}) - 0.1392 \times C_{H_2O-normal}^{18O}} + \frac{0.1336}{0.98 - C_{H_2O-normal}^{18O}} \quad (21)$$

Thus, from Eqs. (18) and (21), α_{ICL} could be given as a function of $C_{H_2O-normal}^{18O}$:

$$\alpha_{\text{ICL}} = \frac{0.1336/(0.98 - C_{H_2O\text{-normal}}^{18O})}{\dfrac{\left(0.2170\% - \dfrac{0.1336 \times C_{H_2O\text{-normal}}^{18O}}{0.98 - C_{H_2O\text{-normal}}^{18O}}\right) \cdot (0.8408 - C_{H_2O\text{-normal}}^{18O})}{0.3402\% \times (0.98 - C_{H_2O\text{-normal}}^{18O}) - 0.1392 \times C_{H_2O\text{-normal}}^{18O}} + \dfrac{0.1336}{0.98 - C_{H_2O\text{-normal}}^{18O}}} \quad (22)$$

From Eq. (22), the value of α_{ICL} is presented in Fig. 14 when $C_{H_2O\text{-normal}}^{18O}$ ranges from 0.00% to 0.50%. The results show that α_{ICL} has no obvious change and its value is around 21%. Consequently, it can be proposed that the percentage of O supplied by H_2O in the ICL is about 21%.

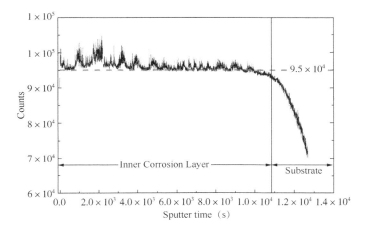

Fig. 14 ToF-SIMS sputter depth profiles in the corrosion products scales formed on Ti60 alloy under NaCl deposit in an atmosphere of H_2O+ O_2 at 600℃ for 20h

Through ^{18}O isotope test and model calculation, it can be concluded that H_2O provides 14% and 21% O for the formation of outer and inner corrosion layers, respectively. It can be seen that the effect of H_2O on the inner corrosion layer is more obvious, and the proportion of O supplied by H_2O in the inner corrosion layer is larger (Fig. 15).

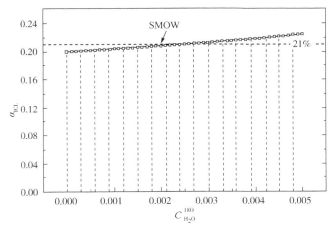

Fig. 15 Percent fraction of O supplied by H_2O in the ICL (α_{ICL}) when the concentration of ^{18}O for the normal H_2O ($C_{H_2O\text{-normal}}^{18O}$) ranged from 0.00% to 0.50%

4.2 Effect of water vapor on the corrosion behaviour

H_2O tends to dissociate into free H atoms and OH groups in TiO_2[24-26] and promotes chemical reactions between NaCl and Ti60 alloy, including the formation of corrosion products. Asides this, the H released by H_2O during the chemical reactions also significantly influences the corrosion process. As the chemical reactions also produce H, the generated H are more prone to diffuse into TiO_2 through the crystal channels in the c-axis direction[24,25]. This dissolution of H in TiO_2 can increase the concentration of crystal defects by forming H defects, thereby increasing the outward diffusion paths metal ions [26], which react with O to form the OCL. This fast diffusion of metal ions results in more active reactions and a higher corrosion rate. On the other hand, H can combine with the doped Cl in the corrosion products to form the gaseous product HCl at the interface between the outer corrosion layer and the inner corrosion layer, thus consuming the doped Cl. The addition of H_2O increases the consumption of doped Cl and as more NaCl become decomposed into doped Cl, more Na will combine with Ti oxides on the surface to form $Na_4Ti_5O_{12}$ (Fig. 11).

H_2O enters the inner corrosion layer and participates in the formation of the inner corrosion layer. Under the NaCl+O_2 condition, the peak value of Ti in XPS shows that there is a large amount of TiO_2 in the inner corrosion layer (Fig. 9), but the structure of TiO_2 is not detected in the XRD diffraction results of the inner corrosion layer (Fig. 8). At the same time, TEM results show that the structure of the inner corrosion layer is mainly amorphous, with very few grains (Fig. 6b). Under the NaCl+H_2O+O_2 condition, a large amount of TiO_2 was detected in the results of Ti peak of XPS (Fig. 11) and XRD diffractogram (Fig. 8), so the main structure of TiO_2 in the inner corrosion layer is the crystal structure. The crystal structure was also observed in TEM, and the inner corrosion layer was mainly nano-polycrystalline (Fig. 7a). Previous studies have found that at temperatures over 400℃, H_2O will promote the transformation of amorphous TiO_2 into the crystalline form[35]. In this corrosion experiment, 21% of the O content of the oxide in the inner corrosion layer comes from H_2O, so a large amount of H_2O takes part in the corrosion reaction during the corrosion process, which greatly promotes the crystallization process of the corrosion products in the inner corrosion layer[35-40]. Therefore, H_2O may promote the transformation of amorphous TiO_2 into the crystalline form.

5 Conclusion

H_2O plays an important part during the corrosion of Ti60 alloy under solid NaCl deposit at 600℃. H_2O was found to promote chemical reactions between NaCl and the Ti60 alloy, and was as well involved in the formation of corrosion products. Through ^{18}O isotope test and model calculation, we determined that H_2O provides 14% and 21% of the O content for the formation of the outer and inner corrosion layers, respectively. H_2O increased the consumption of doped Cl to form HCl, which promoted formation of porous corrosion products, and also promoted the transformation of amorphous TiO_2 into the crystal form, both of which are beneficial to the outward diffusion of metal ions and hence promote corrosion.

Acknowledgment

This investigation was supported by the National Natural Science Foundation of China under the

Contract No. 51622106 and No. 51871049.

References

[1] G. Lutjering, Influence of processing on microstructure and mechanical properties of (alpha+beta) titanium alloys, Mater. Sci. Eng. A: Struct. Mater. Prop. Microstruct. Process. 243 (1998) 32–45.

[2] W. J. Evans, Optimising mechanical properties in alpha + beta titanium alloys, Mater. Sci. Eng. A: Struct. Mater. Prop. Microstruct. Process. 243 (1998) 89–96.

[3] D. Eylon, S. R. Seagle, Titanium technology in the USA—an overview, J. Mater. Sci. Technol. 17 (2001) 439–443.

[4] S. L. Semiatin, V. Seetharaman, I. Weiss, Hot workability of titanium and titanium aluminide alloys—an overview, Mater. Sci. Eng. A: Struct. Mater. Prop. Microstruct. Process. 243 (1998) 1–24.

[5] D. H. Lee, S. W. Nam, High temperature fatigue behavior in tensile hold LCF of nearalpha Ti–1100 with lamellar structure, J. Mater. Sci. 34 (1999) 2843–2849.

[6] H. Guleryuz, H. Cimenoglu, Oxidation of Ti-6Al-4V alloy, J. Alloys Compd. 472(2009) 241–246.

[7] Y. Shu, F. Wang, W. Wu, Corrosion behavior of Ti60 alloy coated with a solid NaCl deposit in O_2 plus water vapor at 500–700℃, Oxid. Met. (52) (1999) 463–473.

[8] C. Ciszak, I. Popa, J. Brossard, D. Monceu, S. Chevalier, NaCl induced corrosion of Ti–6Al–4V alloy at high temperature, Corros. Sci. 110 (2016) 91–104.

[9] L. Fan, L. Liu, M. Cao, Z. Yu, Y. Li, M. Chen, F. Wang, Corrosion behavior of pure Ti under a solid NaCl deposit in a wet oxygen flow at 600℃, Metals (6) (2016) 72–83.

[10] L. Fan, L. Liu, M. Cao, Z. Yu, Y. Li, F. Wang, The corrosion behavior of Ti60 alloy coated with a solid NaCl deposit in wet oxygen flow at 600℃, Sci. Rep. (2016)29019.

[11] P. Dumas, C. St. John, NaCl–induced accelerated oxidation of a Titanium alloy, Oxid. Met. 10 (1976) 127–134.

[12] C. Ciszak, I. Popa, J. Brossard, D. Monceu, S. Chevalier, NaCl–induced high–tem–perature corrosion of 21S Ti alloy, Oxid. Met. 87 (2017) 729–740.

[13] Y. Shu, F. Wang, W. Wu, Synergistic effect of NaCl and water vapor on the corrosion of 1Cr–11Ni–2W–2Mo-V steel at 500–700℃, Oxid. Met. 51 (1999) 97–110.

[14] Y. Shu, F. Wang, W. Wu, Corrosion behavior of pure Cr with a solid NaCl deposit in O^2 plus water vapor, Oxid. Met. 54 (2000) 457–471.

[15] C. Wang, F. Jiang, F. Wang, Corrosion inhibition of 304 stainless steel by nano-sized Ti/silicone coatings in an environment containing NaCl and water vapor at 400–600℃, Oxid. Met. 62 (2004) 1–13.

[16] F. Wang, Y. Shu, Influence of Cr content on the corrosion of Fe–Cr alloys: the synergistic effect of NaCl and water vapor, Oxid. Met. 59 (2003) 201–214.

[17] L. Liu, Y. Li, C. Zeng, F. Wang, Electrochemical impedance spectroscopy (EIS) studies of the corrosion of pure Fe and Cr at 600℃ under solid NaCl deposit in water vapor, Electrochim. Acta 51 (2006) 4736–4743.

[18] F. Wang, S. Geng, S. Zhu, Corrosion behavior of a sputtered K38G nanocrystalline coating with a solid NaCl deposit in wet oxygen at 600 to 700℃, Oxid. Met. 58 (2002) 185–195.

[19] F. Motte, C. Coddet, P. Sarrazin, M. Azzopardi, J. Besson, A comparative study of the oxidation with water vapor of pre titanium and of Ti–6Al–4V, Oxid. Met. 10 (1976) 113–126.

[20] S. R. J. Saunders, M. Monteiro, F. Rizzo, The oxidation behaviour of metals and alloys at high temperatures in atmospheres containing water vapour: a review, Prog. Mater. Sci. 53 (2008) 775–837.

[21] Y. Wouters, A. Galerie, J. P. Petit, Thermal oxidation of titanium by water vapour, Solid State Ionics 104 (1997) 89–96.

[22] S. Taniguchi, N. Hongawara, T. Shibata, Influence of water vapour on the isothermal oxidation behaviour of TiAl at high temperatures, Mater. Sci. Eng. A: Struct. Mater. Prop. Microstruct. Process. 307 (2001) 107–112.

[23] A. Zeller, F. Dettenwanger, M. Schutze, Influence of water vapour on the oxidation behaviour of titanium aluminides, Intermetallics 10 (2002) 59–72.

[24] O. W. Johnson, S. H. Paek, J. W. Deford, Diffusion of H and D in TiO_2 suppression of internal fields by Isotope-exchange, J. Appl. Phys. 46 (1975) 1026–1033.

[25] J. D. Fowler, D. Chandra, T. S. Elleman, A. W. Payne, K. Verghese, The diffusion on Al_2O_3 and BeO, J. Am. Ceram. Soc. 60 (1977) 155–161.

[26] D. L. Douglass, P. Kofstad, A. Rahmel, G. C. Wood, International workshop on high-temperature corrosion, Oxid. Met. 45 (1996) 529–620.

[27] H. J. Grabke, E. Reese, M. Spiegel, The effects of chlorides, hydrogen chloride, and sulfur dioxide in the oxidation of steels below deposits, Corros. Sci. 37 (1995) 1023–1043.

[28] H. Ding, G. Nie, R. Chen, J. Guo, H. Fu, Infulence of oxygen on microstructure and mechanical properties of directionally solidified Ti-47Al-2Cr-2Nb alloy, Mater. Des. (41) (2012) 108–113.

[29] X. Ji, S. Emura, T. Liu, K. Suzuta, X. Min, K. Tsuchiya, Effect of oxygen addition on microstructures and mechanical properties of Ti-7.5Mo alloy, J. Alloys Compd. (737) (2018) 221–229.

[30] C. Moustydesbuquoit, J. Riga, J. J. Verbist, Solid state effects in the electronic structure of TiCl4 studied by XPS, J. Chem. Phys. 79 (1983) 26–32.

[31] W. Chen, J. T. Roberts, Surface chemistry of $TiCl_4$ on W(100), Surf. Sci. 359 (1996) 93–106.

[32] C. Sleigh, A. P. Pijpers, A. Jaspers, B. Coussens, R. J. Meier, On the determination of atomic charge via ESCA including application to organometallics, J. Electron Spectros. Relat. Phenom. 77 (1996) 41–57.

[33] M. Ohno, P. Decleva, Satellites of the 2s and 2p XPS spectra of $TiCl_4$, Phys. Rev. B: Condens. Matter 49 (1994) 818–825.

[34] Yun-Mo Sung, Yong-Ji Lee, Seong-Min Lee, Anatase crystal growth and photo-catalytic characteristics of hot water-treated polyethylene oxide-titania nanohy-brids, J. Cryst. Growth 267 (2004) 312–316.

[35] A. H. Yuwono, B. H. Liu, J. M. Xue, J. Wang, H. I. Elim, W. Ji, Y. Li, T. J. White, Controlling the crystallinity and nonlinear optical properties of transparent TiO_2-PMMA nanohybrids, J. Mater. Chem. 14 (2004) 2978–2987.

[36] Y. Yunzhi, K. Kyo-Han, C. M. Agrawal, J. L. Ong, Effect of post-deposition heating temperature and the presence of water vapor during heat treatment on crystallinity of calcium phosphate coatings, Biomaterials 24 (2003) 5131–5137.

[37] C. H. Chen, S. W. Yang, M. C. Chuang, W. Y. Woon, C. Y. Su, Towards the continuous production of high crystallinity graphene via electrochemical exfoliation with molecular in situ encapsulation, Nanoscale 7 (2015) 15362–15373.

[38] Q. Jiang, J. Zhao, X. X. Li, W. J. Ji, Z. B. Zhang, C. T. Au, Water modification of PEG-derived VPO for the partial oxidation of propane, Appl. Catal. A: Gen. 341 (2008) 70–76.

[39] R. Muydinov, A. Steigert, S. Schonau, F. Ruske, R. Kraehnert, B. Eckhardt, I. Lauermann, B. Szyszka, Water-assisted nitrogen mediated crystallisation of ZnO films, Thin Solid Films 590 (2015) 177–183.

本论文原发表于《Corrosion Science》2019 年第 160 期。

Corrosion Behavior of Reduced-Graphene-Oxide-Modified Epoxy Coatings on N80 Steel in 10.0wt% NaCl Solution

Feng Chun[1,2] Cao Yaqiong[1,2] Zhu Lijuan[1,2] Yu Zongxue[3]
Gao Guhui[4] Song Yacong[1,2] Ge Hongjiang[5] Liu Yaxu[1,2]

(1. CNPC Tubular Goods Research Institute;
2. State Key Laboratory for Performance and Structure Safety of Petroleum Tubular Goods and Equipment Materials; 3. Southwest Petroleum University; 4. TsinghuaRedbud Innovation Institute;
5. Oil Production Technology Research Institute of Petrochina Dagang Oilfield Company)

Abstract: This study is examined the effect of modified epoxy (EP) coatings with various contents of reduced graphene oxide (RGO) in high temperature and high salinity for long time on the corrosion resistance of N80 steel. The fracture surfaces of RGO-modified EP coatings were characterized by scanning electron microscopy (SEM). The corrosion resistance of the coatings after inmersion in 10.0wt% NaCl solution for 60 days at 50℃ was characterized via electrochemical impedance spectroscopy (EIS) and Tafelcurves. It indicated that the addition of 1.0 wt% RGO nanosheets not only improve the strength and toughness of EP coatings, but also effectively decreased the size of pores by two to three times and the quantity of pores decreased by two magnitudes in the coatings than that of neat EP coatings after immersed in the 10.0 wt% NaCl for 60 days, which bring in excellent corrosion resistance.

Keywords: Reduced Graphene Oxide; Epoxy; High temperature; High salinity; Corrosion resistant

1 Introduction

Epoxy (EP) coatings have been widely used in the corrosion protection of tubing due to the excellent adhesion properties, low shrinkage low price, and outstanding chemical stability of epoxy[1-6]. However, with the development of ultra-deep oil fields, the problem of tubing corrosion is more severe due to the high-temperature and high-salinity exploitation environment[7,8]. In onshore oilfields with high sodium chloride content of the for mation brine, such as Pucheng oilfield, in which the Na^+ content is $(8.5 \sim 14.5) \times 10^4$ mg/L and the Cl^- content is $(19.15 \sim 22.50) \times 10^4$

Corresponding author: Feng Chun, fengchun003@ cnpc.com.cn; Cao Yaqiong, caoyaqiong1992@ 163.com; Zhu Lijuan, zhulijuan1986@ cnpc.com.cn.

mg/L[9], the corrosion of N80 tubing is extremely serious[10,11]. In these high temperature and high salinity exploitation environment, the EP coatingsis exposed its limitations: the high degree of crosslinking density makes it brittle, thereby reducing their fracture toughness[12,13], and they exhibit poor resistance to crack propagation[1]. Defects, such as pin holes, easily occur after curing[13]. Corrosive electrolytes, such as small molecules especially Cl⁻ in high-salinity and high-temperature environmens, will penetrate these cracks and pores to form corrosion products under the coatings. Therefore, the EP coatings can't provide long-term corrosion protection[14,15]. Fillers are typically added to improve the corrosion resistance of EP coatings[16-21].

In recent years, graphene has attracted extensive attention due to its substantial potential in improving the properties of resin-based materials and excellent physical properties. Graphene that is well-dispersed in the voids of the coatings can improve the compactness of the coatings, and the lamellar structure of graphene can effectively prevent the permeation of corrosive media, such as H_2O, O_2 and Cl^-, thereby resulting in excellent corrosion resistance of the graphene – modified coatings[19-22]. The effects of graphene addition on the corrosion resistance of EP coatings in simulated seawater and normal atmospheric temperature environment have been widely reported[23-29]. Feng.[30] prepared graphene solid lubrication coating by using graphene water dispersion and water-based EP, and found thatgraphene solid lubrication coatingsshowed good corrosion resistanceafter 48 hours immersion in seawater with 3.5wt% NaCl aqueous solution at room tempareture. Wang Yuqiong et al.[31] added 0.5 wt% graphene dispersion to two-component waterborne EP to prepare EP coatings, and soaked in 3.5wt% NaCl simulated seawater solution for 48 hours at room temperature. The results show that the coatings has good corrosion resistance. However, the corrosion resistances of graphene modified EP coatings in high salinity and high temperature oil production environments are seldom reported.

In our previous study[32], RGO-reinforced EP composite coatings were prepared on N80 steel. These composite coatings showed improved adhesion, toughness, and corrosion resistance after immersion in simulated oil and gas production environments with high-temperature and high-salinity at 80℃ for 10 hours. The addition of 1.0 wt% RGO nanosheets effectively reduced the number and size of the pores in the as-prepared EP composite coatings. However, quantitative expressions and variation rules of the sizes of the pores in the EP composite coatings after corrosion testing with the contents of RGO have not been defined, and the long-term service performances of RGO reinforced EP composite coatings in oil and gas production environments remain unclear. The working temperature of most water injection tubings is approximately 50℃. However, the corrosion resistances of RGO-modified EP coatings in high-salinity and high-temperature oil production environments at 50℃ for long times are rarely reported.

Therefore, the corrosion resistance of RGO-modified EP coatings that are immersed in 10.0 wt% NaCl solution for 60 days at 50℃ was studied via EIS and potentiodynamic polarization curves in this study. In addition, the morphology and microstructure of the RGO was analyzed via FTIR and TEM. The fracture surfaces of RGO – modified EP coatings were characterized via SEM, and quantitative expressions and variation rules for the sizes of pores in the EP composite coatings as functions of the RGO content were defined.

2 Experimental

2.1 Preparation of modified EP coatings

N80 tubing samples (200mm×20mm×3mm) are selected as coating spraying substrates. The specimens were polished with silicon carbide paper and cleaned in an ultrasonic cleaner. The cleaning solution consisted of acetone and ethanol, and the steel substrates were reserved after drying. RGO was mixed into an organic solvent (0.8g/L), and the mixture was dispersed in an ultrasonic disperser. The prepared dispersions were added into EP coatings (solid content of 61.0 wt%) for magnetic stirring, and the stirring time was 15 min. The mixed slurries of RGO with various contents (0 wt%, 0.5 wt%, 1.0 wt%, 2.5 wt% and 4.0 wt%) were sprayed on the cleaned N80 tubing steel substrates, which were subsequently dried at 55 °C for 15 hours. The thickness of the coatings was approximately 300μm.

2.2 Characterization of RGO and RGO modified EP coatings

The morphology and microstructure of the RGO was investigated via FTIR (WQF-520, Beijing, China) and transmission electron microscopy (TEM, Tecnai G2 F20). The fracture surfaces of RGO-modified EP coatings after liquid nitrogen embrittlement were characterized via SEM (Inspect F50, FEI).

2.3 Electrochemical anti-corrosion test

Potentiodynamic polarization curves were plotted and electrochemical impedance spectroscopy (EIS) was conducted to evaluate the corrosion performance of coatings via potentiostat/galvanostat/ZRA (CS-350H, Wuhan Corrtest Instruments Co. Ltd., China). The frequency range was 10^5 Hz to 10^{-2} Hz, and the signal was a 10mV sinusoidal wave. Tafel analysis was conducted to measure the corrosion rates in the range of -0.25V to 0.25V (vs. OCP), and the scanning rate was 1mV/s. A three-electrode system was adopted: a platinum electrode was used as the counter electrode, a saturated calomel electrode (SCE) was used as the reference electrode, and an N80 steel substrate with a coating was used as the working electrode. The working electrode was immersed in 10.0 wt% NaCl solution at 50°C. The impedance and polarization curves of the coatings were measured after 60 days of immersion. Then, Tafel fitting of the potentiodynamic polarization curves and equivalent electrical circuit (EEC) fitting of the EIS results were conducted using the ZSimpWin software.

3 Results and Discussion

3.1 Characterization of RGO

FTIR spectra of GO and RGO are presented in Fig. 1. The characteristic absorption bands of GO and RGO at 3447, 1741, 1625, 1200 and 1120 cm-1 correspond to -OH, C=O, C=C, C-O and C-O-O, respectively. For GO, many oxygen groups are distributed on the surface, and the carboxyl and carbonyl groups are distributed on the edges of the sheets[33]. In contrast to the groups of GO, the C=O, C-O and C-O-O groups in RGO almost disappeared. Therefore, the RGO that is used in this study has a high degree of graphitization.

The surface morphologies of RGO were identified via TEM. Fig. 2 presents a TEM image of dispersed RGO. The color of the RGO sheets is lighter; hence, the dispersion is satisfactory and there are no large aggregates. Due to the large surface energy of a single layer or several layers of RGO, curling and folding occur at the edges of the RGO sheet[34]. The thermodynamic stability of the two-dimensional structure is maintained by this curling and folding, which results in unique structural characteristics.

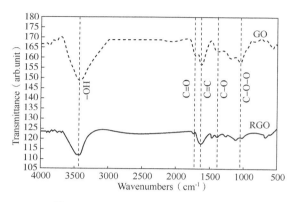

Fig. 1 FTIR spectra for GO and RGO

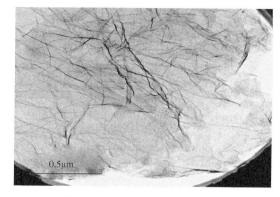

Fig. 2 TEM photograph of RGO

3.2 Characterization of RGO-modified EP coatings

In our previous work, we found that the addition of 1.0 wt% RGO nanosheets effectively reduced the numbers and sizes of the pores in the as-prepared EP composite coatings. To define the variation rules of the sizes of the pores in the EP composite coatings after long-term service, brittle fracture experiments of coatings with various RGO contents were conducted after immersing them at 50 ℃ in 10.0 wt% NaCl solution for 60 days. The fracture surfaces of 0 wt%, 0.5 wt%, 1.0 wt%, 2.5 wt%, and 4.0 wt% RGO-modified EP coatings are shown in Fig. 3. The cross-sectional surface of the 0 wt% RGO-modified EP coating shows a clean fracture surface, thereby implying a brittle fracture. However, in the ductile fracture morphology in Fig. 3[(b)-(e)], tearing ridges can be readily observed from the RGO-modified EP coatings, which show RGO that is warped without peeling off the resin matrix. Hence, the interfacial adhesion between the RGO and the EP was strong, and RGO can improve the strength and toughness of EP coatings. In the process of deformation, RGO, which is the stress concentration center, can cause the resin matrix around it to yield and absorb a substantial amount of energy. The interface between RGO and the EP is separated to form holes, which can passivate the cracks and prevent the generation of destructive cracks. Moreover, the specific surface area of RGO is large, which increases the contact area with the EP. If the material is under external force, it can produce more microcracks and absorb a substantial amount of stress. However, as the RGO content was increased to 4.0 wt%, the tearing ridge decreased significantly, which was likely due to the aggregation of the RGO nanosheets.

The pore sizes of the samples were randomly measured, and the top five sizes of pores for each sample are listed in Tab. 1. According to Tab. 1, the maximum pore size that was measured on the

EP coatings with 0 wt% is 1305 nm, with an average size of 826.5 nm. The addition of 0.5 wt% RGO caused no significant change in the quantity or size of the pores [Fig.3(a) and Fig.3(b)]. However, when the RGO content was 1.0 wt%, the size of the pores decreased by two to three times and the quantity of the pores decreased by two orders of magnitude in the coatings compared to the neat EP coatings [Fig.3(c)]. When the RGO contents were increased to 2.5 wt% and 4.0 wt%, the quantity and size of the pores increased substantially, and the average size of the pores in these coatings even exceeds that in the neat EP coating.

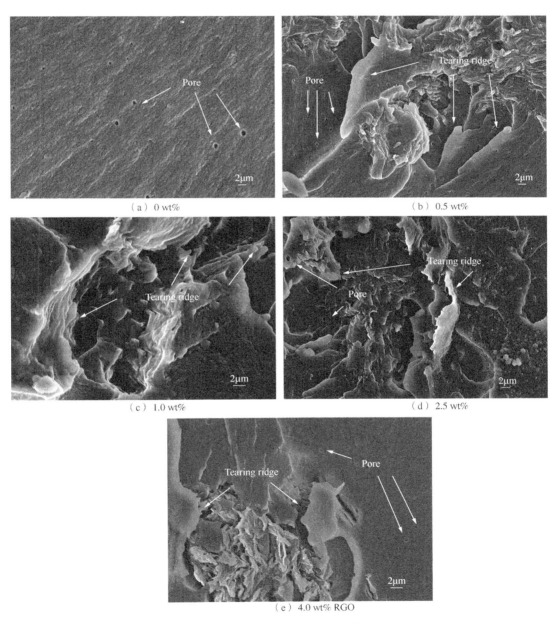

Fig. 3 SEM images of fractures in the EP coatings

Table 1 Pore size on the cross-sectional surfaces of coatings with various RGO contents

Content of RGO	Pore size (nm)					Average size (nm)
0 wt%	435.0	870.0	652.5	870.0	1305.0	826.5
0.5 wt%	431.0	431.0	862.0	1293.0	1115.0	826.4
1.0 wt%	652.5	261.0	211.0	196.0	356.0	335.3
2.5 wt%	862.0	1115.0	1115.0	1293.0	1293.0	1135.6
4.0 wt%	431.0	862.0	1293.0	1293.0	1293.0	1034.4

If the coating is subjected to external stress, large pore defects will lead to poor stress transfer and stress concentration, and corrosive media such as H_2O, O_2 and Cl^- will enter the coating more easily through the defects and reach the metal substrate surface, thereby forming severe corrosion, which will eventually cause the coating to fall off easily. Many studies have shown that the number of pores in the coating can be decreased by adding graphene into the EP. For example, nano oxide can be used to modify GO and blend with EP to form composite materials, which can fill pores in the coating[35-38]. Zhang[39] prepared a GO-modified EP coating, in which the GO modification effectively reduced the porosity of the coating. However, these studies use graphene to fill the pores, to decrease the number of pores, and the mechanism of graphene in the process of bubble generation is not discussed.

In this study, a possible mechanism is proposed: When RGO is dispersed by ultrasonic waves, the high temperature and high pressure of the generated bubbles through the cavitation of the rupture process promotes the uniform dispersion of RGO in EP coatings. Therefore, when the content of RGO is 0.5 wt%, many pores remain in the coatings. As the content of RGO increases, RGO will destroy large bubbles, thereby resulting in small bubbles. This facilitates bubble overflow; hence, the number of pores in the coatings will decrease. However, when the RGO content was increased to 2.5 wt% and 4.0 wt%, aggregation of the RGO nanosheets occurred, which significantly lowered the bubble-breaking effect, thereby resulting in the formation of pores of increased quantity and size in the coatings. Therefore, there is a balance between the pore reduction that is induced by the bubble-breaking effect of RGO nanosheets and the pores increase that is induced by the aggregation of the RGO nanosheets. This balance is realized when the optimal amount of RGO has been added into the coatings. The bubble-breaking effect and the mechanism of RGO will be investigated in our future work. Therefore, the addition of a suitable amount of RGO not only increases the toughness but also decreases the quantity and size of the pores of these coatings in high-temperature and high-salinity environments.

3.3 EIS analysis of RGO-modified EP coatings

EIS and potentiodynamic polarization curves were used to evaluate the corrosion performances of EP coatings with various contents of RGO. Many electrochemical studies on the coating have been conducted by immersing the coatings in 3.5% NaCl solution at room temperature in accordance with the marine simulation environment[23]. For example, GO was introduced into a polypyrrole (PPY) matrix, and PPY-GO composite coatings with various GO contents were electrodeposited in situ on 304 stainless steel (SS) bipolar plates to protect them from corrosion in an aggressive working environment[40]. N. N. Taheri[27] grafted GO particles onto zinc doped polyaniline (PANI), added it to the EP, and immersed it in 3.5% NaCl solution for electrochemical testing. To more closely approximate the conditions of the underground environment of onshore oil wells, prior to the test, the coatings were immersed in 10.0 wt% NaCl solution at 50℃ for 60 days, and the results are presented in Fig. 4. The higher the impedance in the Nyquist plot, the better the anti-corrosion effect of the coatings. The addition of RGO increases the radius of the capacitance; hence, the addition of RGO effectively enhances the anticorrosion performance of the EP coating. The effect depends on the chemical stability of RGO. When RGO is added to the EP coating, a retardant layer can be formed, which prevents direct contact between the corrosive media and the substrate, and increases the tortuosity of the diffusion path. Moreover, the RGO is a nanometer-scale material, which can be used to fill the curing defects of the EP coating. Moreover, it is hydrophobic; hence, it can delay the pervasion of water and the corrosion. The TEM image in Section 3.1 shows that there are many folds on the surface of RGO. When it is added to EP as a filler, the folds on the surface can have a larger contact area with the EP, thereby making the diffusion path of active media more tortuous and, thus, substantially improving the corrosion resistance of the coating. Therefore, according to the Nyquist plots and Bode diagrams in Fig. 4, with the increase of the RGO content, the radius of capacitance initially increases and subsequently decreases. The modified EP coating with 1.0 wt% RGO shows excellent corrosion resistance, which is due to its high impedance and large radius of capacitance. However, when the content of RGO is high, agglomeration occurs in the coating, thereby resulting in many defects, such as pores in the EP coating, which provide channels for the diffusion of corrosive media; thus, the anti-corrosion performance of the coating is decreased.

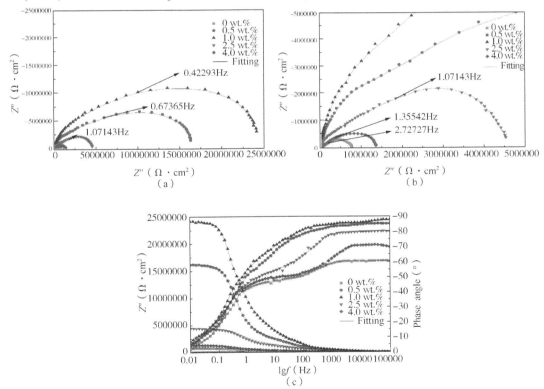

Fig.4 (a) Nyquist plots, (b) locally enlarged Nyquist plots, and (c) Bode diagrams of the EP composite coatings after immersion in 10.0 wt% NaCl solution at 50℃ for 60 days

To further characterize the anti-corrosion effect of the coatings, the impedance data are analyzed and fitted with the ZSimpWin simulation software. The equivalent circuit and the analysis results are presented in Fig. 5 and Tab. 2. The electronic components R_s, C_c, R_c, R_{ct} and CPE_{dl} in the equivalent circuit represent the electrolyte

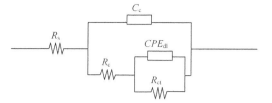

Fig. 5　Equivalent circuit of the sample

resistance, coating capacitance, coating resistance, charge transfer resistance and double layer capacitance, respectively. C_c represents the amount of corrosive media that penetrate into the coating, R_c can be used to represent the number and area of the pores on the coating surface, CPE_{dl} represents the failure area of the coating, and R_{ct} represents the resistance value of the charge transfer on the metal surface, which can directly reflect the corrosion rate of the interface between the coating and N80 steel.

Table 2　EIS analysis of electrochemical parameters of the coating samples after immersion in 10.0 wt% NaCl solution for 60 days

Content of RGO	0 wt%	0.5wt%	1.0 wt%	2.5 wt%	4.0 wt%
$R_{ct}(10^6 \Omega \cdot cm^2)$	0.11	2.81	10.16	1.02	0.18
$C_c(10^{-10} F \cdot cm^{-2})$	314.50	12.31	1.13	15.17	17.48
$R_c(10^5 \Omega \cdot cm^2)$	1.11	28.09	101.60	10.18	1.79
$CPE_{dl}(10^{-9} F \cdot cm^{-2})$	2.64	21290.00	1.67	36.87	77.35

As summarized above, $R_c(1.016 \times 10^7 \Omega \cdot cm^2)$ and $R_{ct}(1.016 \times 10^7 \Omega \cdot cm^2)$ attained their maximum values and $C_c(1.127 \times 10^{-10} F \cdot cm^{-2})$ and $CPE_{dl}(1.67 \times 10^{-9} F \cdot cm^{-2})$ reached their minimum values when the RGO content was up to 1.0 wt% in the modified EP coatings. Therefore, the optimal addition of RGO is 1.0 wt%. RGO can greatly improve the anti-corrosion performance of EP coatings. However, with the increase of graphene content, graphene will agglomerate, which will increase the defects of EP coatings and reduce the corrosion resistance.

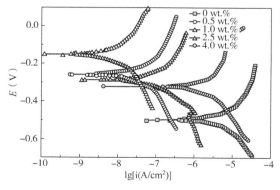

Fig. 6　Potentiodynamic polarization curves of the EP composite coatings after immersion in 10.0 wt% NaCl solution at 50 ℃ for 60 days

3.4　Tafel analysisof RGO-modified EP coatings

The potentiodynamic polarization curves are plotted in Fig. 6, and Tab. 3 lists the kinetic parameters that were calculated from the potentiodynamic polarization curves, which include I_{corr} and E_{corr}. The corrosion current (I_{corr}) decreased significantly and the corrosion potential (E_{corr}) exhibited a positive shift when graphene was added. I_{corr} initially decreased and subsequently increased with the increase of the RGO content, while E_{corr} initially exhibited a positive shift and subsequently exhibited a negative shift.

Table 3 Electrochemical parameters that were obtained from potentiodynamic polarization curves via Tafel extrapolation for the coated samples after immersion in 10.0 wt% NaCl solution for 60 days

Content of RGO	0 wt%	0.5 wt%	1.0 wt%	2.5 wt%	4.0 wt%
$I_{corr}(\mu A/cm^2)$	9.190	0.074	0.009	0.705	5.415
$E_{corr}(V)$	−0.501	−0.262	−0.151	−0.287	−0.325

This is due to the small-size effect of RGO filling in the pores of the coating. The corrosive media can not penetrate the pore paths that are blocked by RGO t via natural convection mass transfer to the reaction area of the electrode interface corrosion. Due to the consumption of depolarization, the concentrations of H_2O, O_2 and Cl^- at the bottom of the pores can not be supplemented timely, thereby resulting in a concentration difference with the corrosive media in the solution of the coating surface at the top of the pore. Instead, Faradaic processes of the electrode reaction are controlled by the tangent diffusion of the corrosive media. The electrochemical polarization of the corrosion system is gradually weakened, and concentration polarization occurs.

Compared with neat EP coatings, the corrosion resistance of the coatings with RGO is substantially improved. The modified EP coatings with 1.0 wt% RGO had the lowest corrosion current density (9.5172×10^{-9} A/cm²), the highest corrosion potential (−0.15107V), and the maximum polarization resistance ($6.04 \times 10^9 \Omega \cdot cm^2$). Therefore, the optimum addition of RGO is 1.0 wt%, which is in accordance with the results of EIS.

3.5 Anti-corrosion mechanism

In this study, after immersion in 10.0 wt% NaCl solution for 60 days, the modified EP coatings with 1.0 wt% RGO still showed superior corrosion resistance to that of the neat EP coatings. Researchers proposed possible anti-corrosion mechanisms of graphene-modified coatings: (1) graphene can decrease the coating porosity[30-32]; (2) graphene causes a barrier effect[30-32]; (3) graphene can block the cathodic reaction between the coating and the metal interface[41]. As described in the above investigation, the addition of 1.0 wt% RGO nanosheets not only improved the strength and toughness of the coatings but also effectively decreased the size of the pores by two to three times and decreased the quantity of the pores by two orders of magnitude in the coatings compared to neat EP coatings. Therefore, the added RGO nanosheets effectively block the penetration of the corrosive media via the penetration effect and decrease the quantity and size of the pores in the coatings, which results in excellent anticorrosion performance. However, with the increase of the RGO content, RGO nanosheets will agglomerate, the balance between the pores reduction induced by the bubble-breaking effect of the RGO nanosheets and the pore increase that is induced by the aggregation of the RGO nanosheets will be broken, which will increase the severity of the defects in the EP coatings and decrease the corrosion resistance of the coatings.

4 Conclusions

(1) After immersion in 10.0 wt% NaCl solution at 50℃ for 60 days, the EP composite coatings with 1.0 wt% RGO nanosheets still showed improved strength and toughness, with pores that were two to three times smaller and two orders of magnitude less numerous compared to the neat

EP coatings.

(2) The modified EP coatings with 1.0 wt% RGO showed excellent corrosion resistance with the lowest corrosion current density (9.5172×10^{-9} A/cm^2), the highest corrosion potential (−0.15107V) and the maximum polarization resistance (6.04×10^9 Ω·cm^2) after immersion in the 10.0 wt% NaCl solution for 60 days at 50℃.

Acknowledgement

This work was supported by the National Natural Science Foundation of China (NO.51804335): Corrosion resistance and mechanism of graphene-modified epoxy coatings in a coupled oil and water multi-phase medium; Major science and technology projects of Petro China Co Ltd (2018E-11-06): Key technologies for improving the development level in complex fault block oilfields at the high water cut stage; and The National Key Research and Development Program of China(2019YFF0217504): Research and application of common technology in the national quality foundation-The complete process inspection and quality control technology of 13Cr/110SS products in service.

References

[1] X. M. Shi, T. A. Nguyen, Z. Y. Suo, Y. J. Liu and R. Avci, Surface and Coatings Technology, 204 (2009) 237.

[2] I. Zaman, T. T. Phan, H. C. Kuan, Q. S. Meng, L. T. B. La, L. Luong, O. Youssf and J. Ma, Polymer, 52 (2011) 1603.

[3] S. Chatterjee, J. W. Wang, W. S. Kuo and N. H. Tai, Chemical Physics Letters, 531 (2012) 6.

[4] J. P. Pascault, R. J. J. Williams, Polymer Bulletin, 24 (1990) 115.

[5] D. Chmielewska, T. Sterzyński, and B. Dudziec, Journal of Applied Polymer Science, 131 (2014) 8444.

[6] S. Hichem, E. M. M. Lassaad, H. Guermazi, S. Agnel and A. Toureille, Journal of Alloys and Compounds, 477 (2009) 316.

[7] Y. G. Liu, Corrosion & Protection, 8 (2003) 361 (In Chinese).

[8] Y. Y. Ge, C. J. Han, Y. Q. Yuan, X. Y. Ma, S. F. Sheng and Y. C. Deng, Henan Petroleum, 19 (2005) 76 (In Chinese).

[9] G. P. He, Q. Y. Yang, Journal of Jianghan Petroleum University of Staff and Workers, 22 (2009) 56 (In Chinese).

[10] J. Z. Li, Inner Mongolia Petrochemical Industry, 20 (2013) 41 (In Chinese).

[11] A. Bisht, K. Dasgupta, and D. Lahiri, Journal of Applied Polymer Science, 135 (2018) 46101.

[12] A. Montazeri, J. Javadpour, A. Khavandi, A. Tcharkhtchi and A. Mohajeri, Materials & Design, 31 (2010) 4202.

[13] S. Chhetri, N. C. Adak, P. Samanta, P. K. Mallisetty, N. C. Murmu and T. Kuila, Journal of Applied Polymer Science, 135 (2018) 46124.

[14] H. Feng, X. D. Wang, and D. Z. Wu, Industrial & Engineering Chemistry Research, 52 (2013) 10160.

[15] R. J. Day, P. A. Lovell, and A. A. Wazzan, Composites Science and Technology, 61 (2001) 41.

[16] R. Bagheri, R. A. Pearson, Polymer, 41 (2000) 269.

[17] T. Kawaguchi, R. A. Pearson, Polymer, 15 (2003) 4239.

[18] S. H. Lee, D. R. Dreyer, J. An, A. Velamakanni, R. D. Piner, S. J. Park, Y. W. Zhu, S. O. Kim, C.

W. Bielawski and R. S. Ruoff, Macromolecular Rapid Communications, 31 (2010) 281.
[19] G. X. Wang, X. P. Shen, B. Wang, J. Yao and J. Park, Carbon, 47 (2009) 1359.
[20] Q. F. Jing, W. S. Liu, Y. Z. Pan, V. V. Silberschmidt, L. Li and Z. L. Dong, Materials & Design, 85 (2015) 808.
[21] T. Kuilla, S. Bhadra, D. H. Yao, N. H. Kim, S. Bose and J. H. Lee, Progress in Polymer Science, 35 (2010) 1350.
[22] M. Naderi, M. Hoseinabadi, M. Najafi, S. Motahari and M. Shokri, Journal of Applied Polymer Science, 135 (2018) 46201.
[23] H. H. Di, Z. X. Yu, Y. Ma, Y. Pan, H. Shi, L. Lv, F. Li, C. Wang, T. Long and Y. He, Polymers Advanced Technologies, 27 (2016) 915.
[24] Y. H. Tang, D. Xiang, D. H. Li, L. Wang, P. Wang and P. D. Han, Surface Technology, 47 (2018) 203.
[25] J. A. Q. Rentería, L. F. C. Ruiz, and J. R. R. Mendez, Carbon, 122 (2017) 266.
[26] Y. H. Wu, X. Y. Zhu, W. J. Zhao, Y. J. Wang, C. T. Wang and Q. J. Xue, Journal of Alloys and Compounds, 777 (2019) 135.
[27] N. N. Taheri, B. Ramezanzadeh, and M. Mahdavian, Journal of Alloys and Compounds, 800 (2019) 532.
[28] M. S. Selim, S. A. El-Saftya, N. A. Fatthallahd and M. A. Shenashen, Progress in Organic Coatings, 121 (2018) 160.
[29] F. D. Meng, T. Zhang, L. Liu, Y. Cui and F. H. Wang, Surface and Coatings Technology, 361 (2019) 188.
[30] C. Feng, L. J. Zhu, Y. Q. Cao, Y. Di, Z. X. Yu and G. H. Gao, International Journal of Electrochemical Science, 14 (2019) 1855.
[31] C. Feng, L. J. Zhu, Y. Q. Cao, Y. Di, Z. X. Yu and G. H. Gao, International Journal of Electrochemical Science, 13 (2018) 8827.
[32] L. J. Zhu, C. Feng, and Y. Q. Cao, Applied Surface Science, 493 (2019) 889.
[33] A. Dimiev, D. V. Kosynkin, L. B. Alemany, P. Chaguine and J. M. Tour, Journal of the American Chemical Society, 134 (2012) 2815.
[34] D. R. Nelson, T. Piran, and S. Weinberg, World Scientific, 2004, Singapore.
[35] Z. X. Yu, H. H. Di, Y. Ma, Y. He, L. Liang, L. Lv, X. Ran, Y. Pan and Z. Luo, Surface and Coatings Technology, 276 (2015) 471.
[36] Z. X. Yu, H. H. Di, Y. Ma, L. Lv, Y. Pan, C. L. Zhang, Y. He, Applied surface science, 351 (2015) 986.
[37] Y. Ma, H. H. Di, Z. X. Yu, L. Liang, L. Lv, Y. Pan, Y. Y. Zhang and D. Yin, Applied surface science, 360 (2016) 936.
[38] H. H. Di, Z. X. Yu, Y. Ma, F. Li, L. Lv, Y. Pan, Y. Lin, Y. Liu and Y. He, Journal of the Taiwan Institute of Chemical Engineers, 64 (2016) 244.
[39] Z. Y. Zhang, W. H. Zhang, D. S. Li, Y. Y. Sun, Z. Wang, C. L. Hou, L. Chen, Y. Cao and Y. Q. Liu, International journal of molecular sciences, 16 (2015) 2239.
[40] L. Jiang, J. A. Syed, H. B. Lu and X. K. Meng, Journal of Alloys and Compounds, 770 (2019) 35.
[41] S.S.Chen, L. Brown, M. Levendorf, W. W. Ca, S. Y. Ju, J. Edgeworth, X. S. Li, C. W. Magnuson, A. Velamakanni, R. D. Piner, J. Y. Kang, J. Park and R. S. Ruoff, ACS Nano, 5(2011)1321.

本论文原发表于《International Journal of Electrochemical Science》2020 年第 7 卷。

Downhole Corrosion Behavior of Ni-W Coated Carbon Steel in Spent Acid & Formation Water and Its Application in Full-Scale Tubing

Fu Anqing[1,2]　Feng Yaorong[1,2]　Cai Rui[1,2]　Yuan Juntao[1,2]
Yin Chengxian[1,2]　Yang Dongming[3]　Long Yan[1,2]　Bai Zhenquan[1,2]

(1. CNPC Tubular Goods Research Institute;
2. State Key Laboratory of Performance and Structural Safety for Petroleum Tubular Goods and Equipment Materials;
3. Northwest Oilfield Company, China Petroleum & Chemical Corporation)

Abstract: Downhole corrosionbehavior of Ni-W coated carbon steel tubing was investigated in spent acid (dilute HCl) solution and formation water by using autoclave and electrochemical techniques, in which dilute HCl with different pH is used to simulate the spent acid during flowback in acidizing process before well production, and the formation water is used to simulate the fluid during well production. Weight loss test in autoclave and electrochemical measurement indicated that Ni-W coated carbon steel exhibited higher corrosion resistance than carbon steel in spent acid and formation water, especially in spent acid with low pH and formation water at high temperature. According to weight loss test, corrosion rate ratio of carbon steel and Ni-W coated carbon steel in spent acid with pH=1, pH=2, and pH=4 is 13.7, 14.5, and 6.9, respectively, while corrosion rate ratio of carbon steel and Ni-W coated carbon steel in formation water at 30℃, 60℃, and 90℃ is 85.5, 73.5, and 125.9, respectively. Moreover, the sealing performance of full-scale Ni-W coated carbon steel tubing was evaluated by using make-and-break test, hydraulic bursting test, and extreme downhole condition corrosion test. No sticky thread was found after 4 times of make up and three times of break out by using maximum recommended makeup torque of 4258 lbf·ft, no leakage was detected after hydraulic bursting test at 95.0 MPa for 30 mins, and it is observed very slight corrosion on the tubing shoulder experienced make-and-break test under extreme downhole condition.

Keywords: Ni-W coated steel tubing; Downhole corrosion; Sealing; Spent acid; Formation water; Make-and-break test

1 Introduction

With the continuously growing demand in oil and gas energy globally, the search for new

---

Corresponding author: Fu Anqing, fuanqing@cnpc.com.cn.

sources of oil and gas makes the operation condition became more and more severe. High temperature high pressure (HTHP) gas well was developed in northwest China increasingly with years. The average reservoir temperature is 90℃ and the initial reservoir pressure is nearly 100MPa, the downhole partial pressure of CO_2 and H_2S is up to 4MPa and 2MPa, respectively, and chloride concentration is as high as 150,000 mg/L. Acidizing is employed to enhance the productivity of HTHP gas well, the acid system is HCl-based including 15% HCl, 1.5% HF, 3% HAc, and inhibitor. Super 13Cr martensitic stainless steels tubing was used for well completion and production, field statistics shows several gas well failure after 1~3 years of production due to pitting corrosion of 13 Cr tubing, moreover, over 70% failed gas well was acidized before production. Besides the corrosion attack by acid during acidizing process, in the long-term production process, the presence of CO_2, H_2S, and high chloride concentration in formation water plays a key role in the corrosion failure of downhole tubing. The corrosivity of downhole environment is further complicated and enhanced by the high temperature and high pressure, it is well acknowledged that temperature accelerated the kinetics of corrosion reactions, as well as pressure contributes to internal corrosion in terms of higher downhole pressure increasing the partial pressure or solubility of naturally occurring corrosive acid gases, such as CO_2 and H_2S[1,2].

Both field data and lab research indicated that 13Cr martensitic stainless steel tubing is no longer the best choice for HTHP well, especially in spent HCl solution during flowback and high chloride concentration at high temperature[3-5]. The correct strategy in the choice of tubing material is becoming increasingly significant, safety and cost have to be preferentially considered for tubing material selection, in the petroleum exploration and production industry, the major corrosion resistant alloys used fall into three categories: martensitic stainless steels, duplex stainless steels, and Ni-based alloys[6]. It is well-known that the Ni-based alloys are the best choice with overall desirable properties, but the biggest disadvantage is that the Ni-based alloys are prohibitively expensive used as downhole tubing. However, development of Ni-based alloy coating on carbon steel is an economical alternative choice, providing excellent corrosion resistance and wear resistance. Extensive studies have been carried out to investigate the mechanical and corrosion performance of Ni-based alloy coating, such as Ni-P[7], Ni-P-W[8], Ni-SiC[9], Ni-TiN[10], under various environmental conditions.

In this work, downhole corrosion behavior of Ni-W coated carbon steel tubing was studied in HCl solution with different pH and formation water by using autoclave, HCl solution with different pH is used to simulate the spent acid during flowback in acidizing process before well production, and the formation water is used to simulate the fluid during well production. Meanwhile, corrosion mechanism in terms of electrochemical corrosion behavior was characterized by potentiodynamic polarization technique, electrochemical impedance spectroscopy (EIS). For all the tests, the carbon steel sample without coating was used for comparison. Moreover, Ni-W coating was applied to full-scale carbon steel tubing, the full-scale tubing sealing performance was evaluated by using make-and-break test, hydraulic bursting test, and extreme downhole condition corrosion test. It is expected that this work would provide technical support for potential application of Ni-W coated carbon steel tubing in HTHP gas well.

2 Experimental

2.1 Electrode and solution

Carbon steel tubing and Ni-W coated carbon steel tubing (Hunan Nanofilm New Material Technology Co., Ltd.) were used in this work, in which Ni-W coated carbon steel is being planned to use for gas well tubing in western China. The chemical composition is given in Table 1, and microstructure is shown in Fig. 1. The specimens for weight loss test in autoclave were machined into pieces with a dimension of 40mm×10mm×3mm. The specimens for electrochemical measurement were machined into pieces with a dimension of 10mm×10mm×2mm, and then embedded in epoxy resin with an exposed working area of 1 cm^2. For Ni-W coated carbon steel sample, part of the machined coupons, including autoclave test and electrochemical measurement sample, were electrodeposited with Ni-W coating. Prior to experiment, the working surface of carbon steel specimen was sequentially grounded with 320 grit, 600 grit, 800 grit, 1000 grit and 1200 grit SiC papers, polished with 0.1μm alumina polishing powder. Then the surface-treated carbon steel sample and as-prepared Ni-W coated carbon steel sample degreased with alcohol, cleaned in water, and finally dried in air.

Table 1 Chemical composition analysis of Ni-W coated carbon steel and carbon steel (wt%)

Element	C	Fe	Si	P	Ni	W	Cr	Mn
Ni-W coated carbon steel	—	2.02	—	3.29	73.37	21.32	—	—
Carbon steel	0.62	97.78	0.21	—	—	—	0.88	0.51

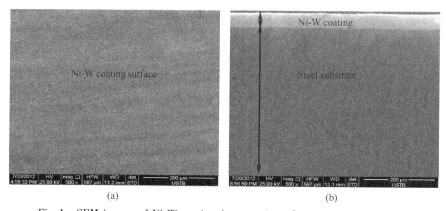

Fig. 1 SEM images of Ni-W coating (a. top view; b. cross-sectional view)

Two types of solution including spent acidizing fluid (dilute HCl) and formation water were used to study the corrosion performance during acidizing process and after acidizing in formation water, respectively. Detailed composition of formation water is given in Table 2. The solution was made from analytic grade reagents and ultra-pure water (18 MΩ cm in resistivity).

Table 2 Chemical composition of oilfield formation water (mg/L)

Na$^+$	Mg^{2+}	Ca^{2+}	Ba^{2+}	Cl$^-$	HCO$_3^-$	SO$_4^{2-}$	pH	Total Mineralization
71560	912.6	8037	18.48	127000	466.6	537.6	5.86	209000

2.2 Weight loss test in autoclave

In order to simulate the downhole corrosion of high temperature high pressure gas well, weight loss test was performed in autoclave to characterize the corrosion behavior of carbon steel and Ni-W coated carbon steel tubing. With consideration of reproducibility, three equivalent coupons were used for each test condition. Before the weight loss test, the specimens were cleaned with distilled water and acetone, dried, and then weighed using METTLER TOLEDO 4-digit electronic balance with a precision of 0.1mg, and the weight before test was recorded as the original weight (W_0). After completion of test, the corroded samples were rinsed with distilled water, and then cleaned in corrosion film removing solution to remove the corrosion product formed on the sample surface, and then rinsed and dried again, finally reweighed to obtain the final weight (W_f). The corrosion rate (CR) was reported in mm/y according to the obtained weight loss via Eq. (1). The average corrosion rate of three specimens for each test condition was used.

$$CR = \frac{(W_0 - W_f) \times 1000 \times 365 \times 24}{t \times \rho \times S} \tag{1}$$

where W_0 and W_f are the original weight and final weight of specimen, g, respectively; S is the exposed surface area of specimen, mm^2; t represents the immersion time, h; and ρ is the density.

2.3 Electrochemical measurements

Electrochemical measurements were performed using a Princeton Applied Research 273 electrochemical workstation on a three-electrode cell, where the X60 pipeline steel specimen was used as working electrode, a platinum foil as counter electrode, and a saturated calomel electrode (SCE) as reference electrode. Prior to electrochemical measurement, the working electrode was cathodically pre-polarized at potential of −800 mV to remove the oxide film formed on steel electrode surface. Finally, the corrosion potential of the working electrode was monitored for 30 min to ensure that a steady state was reached.

Potentiodynamic polarization curve measurement was conducted at a potential scanning rate of 0.5 mV/s. EIS was measured under a sinusoidal excitation potential of 10 mV in the frequency range from 100 kHz to 10 mHz. The obtained EIS results were fitted by using commercial software ZSimpWin 3.2.

2.4 Ni-W coated full-scale tubing thread sealing test

Due to thread acts as sealing and connecting part for each individual tubing string, therefore, the surface of pin and box have to be coated with Ni-W coating as well. For all the full-scale tests, ϕ88.90mm×6.45mm P110 tube was used. The sealing performance of Ni-W coated pin and box coupling was evaluated by using make-and-break test system, hydraulic bursting test system, and complex loading system. (1) Make-and-break test: the maximum recommended makeup torque of 4258 lbf·ft was used, the speed for makeup torque is 25 rpm, for each thread, 4 times of make up and 3 times of break out were performed, the high pressure API thread sealant SHELL TYPE3 was used. (2) Hydraulic bursting test: the test medium for hydraulic bursting test is water, which is conducted at 95.0 MPa for 30 mins.

3 Results and Discussion

3.1 Corrosion behavior of Ni-W coated carbon steel in spent acid solution

Fig. 2 shows the corrosion rate of Ni-W coated carbon steel and carbon steel in spent acid with different pH at 90℃. The corrosion rate ratio of carbon steel and Ni-W coated carbon steel is 13.7, 14.5 and 6.9 in spent acid with pH=1, pH=2, and pH=4, respectively. Compared to carbon steel, Ni-W coating exhibits higher corrosion resistance in acid solution, especially in lower pH solution. Fig. 3 shows the micro-morphology of corroded Ni-W coated carbon steel and composition of corroded Ni-W coating surface. It is seen that uniform corrosion occurred in spent acid solution, according to the comparison of EDS analysis results of corroded Ni-W coated carbon steel surface and chemical composition of Ni-W coating in Table 1, few Fe originally contained in coating was totally dissolved in spent acid solution, and relative content of Ni decreases from 73.37% to 54.21%, while the relative content of W increases from 21.32% to 35.57%. It is indicated that Ni was partially dissolved in spent acid solution, and W was oxidized resulting in formation of WO_3, several previous papers have reported the influence of tungsten in enhancing corrosion resistance of Ni-P based coating, it is reported that tungsten preferentially migrated to the coating surface and formed W-riched oxide film during corrosion process, which inhibited the further corrosion[11-13]. Ni-W coating was reported more corrosion resistant than stainless steel 304 in acidic medium[14].

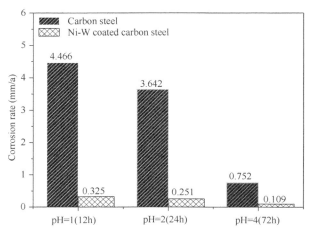

Fig. 2 Corrosion rate of Ni-W coated carbon steel and carbon steel in spent acid solution with different pH at 90℃

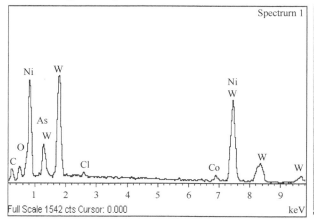 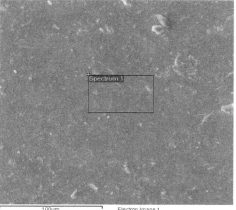

Element	C	O	Cl	Ni	W	Total
wt%	2.65	6.88	0.69	54.21	35.57	100

Fig. 3 SEM images and EDS results of corroded Ni-W coated carbon steel in spent acid solution at 90℃

Electrochemical measurements were carried out to investigate corrosion mechanism of Ni-W coated carbon steel. Fig. 4 shows the corrosion potential of Ni-W coated carbon steel and carbon steel in spent acid solution with different pH at 90 ℃, corrosion potential shifted negatively with the increase of solution pH. Correspondingly, Fig. 5 shows polarization curves of Ni-W coated carbon steel and carbon steel in spent acid solution with different pH at 90 ℃. It is seen that the polarization curves of Ni-W coated carbon steel are characterized by passivation when the solution pH is equal to 1 and 2, the passive current density

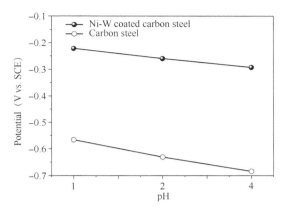

Fig. 4 Corrosion potential of Ni-W coated carbon steel and carbon steel in spent acid solution with different pH at 90 ℃

decreased with the increasing of pH, while no passive region was observed when the pH increased up to 4, which indicated that corrosion mechanism of Ni-W coating varied with acid pH, passivity dominated the anodic reaction if spent acid solution pH is less than 2, while active dissolution dominated the anodic reaction if spent acid solution pH is greater than 2. In contrast, all the polarization curves of carbon steel measured in different pH spent acid solutions have similar characteristics, active dissolution instead of passivation dominated the anodic reaction.

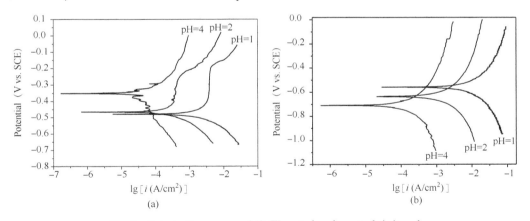

Fig. 5 Polarization curves of Ni-W coated carbon steel (a) and carbon steel (b) in spent acid solution with different pH at 90 ℃

EIS diagrams of Ni-W coated carbon steel and carbon steel measured in different pH spent acid solutions are shown in Fig. 6, EIS diagrams of Ni-W coated carbon steel and carbon steel are characterized by one semicircle. Electrochemical equivalent circuits shown in Fig. 7 (a) are proposed to fit the EIS data, where R_s is solution resistance, R_{ct} is charge-transfer resistance, Q_{dl} is double-charge layer capacitance. The fitted electrochemical impedance parameters are given in Table 3. The charge-transfer resistance and solution resistance of Ni-W coated carbon steel increases with solution pH, similar trends observed on carbon steel. Since corrosion rate is inversely proportional to the charge-transfer resistance, consequently, the corrosion rate ratio of carbon steel and Ni-W coated carbon steel calculated based on charge-transfer resistance is 97, 62, and 12 in spent acid with pH = 1, pH = 2 and pH = 4, respectively. The corrosion rate ratio obtained both from autoclave weight loss test and electrochemical test indicated that Ni-W coated carbon steel exhibited higher corrosion resistance than carbon steel in spent acid, especially in low pH condition. The reason why Ni-W coated carbon steel exhibits high corrosion resistance than carbon steel in low pH solution (pH=1, 2) than that in high pH solution, which is mainly due to passive nature of Ni-W coating in low pH solution, as shown in Fig. 5.

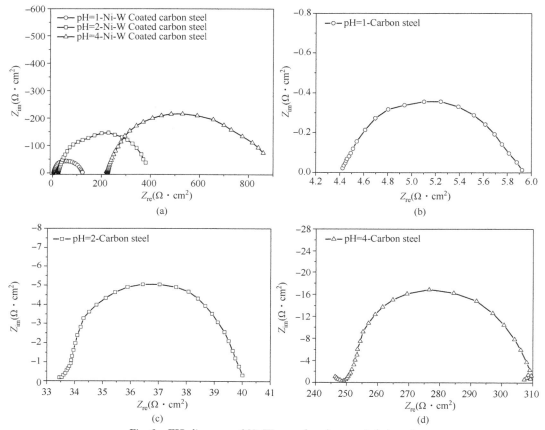

Fig. 6 EIS diagrams of Ni-W coated carbon steel (a) and carbon steel (b, c, d) in spent acid solution with different pH at 90℃

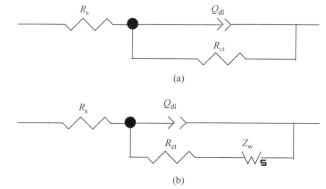

Fig. 7 Electrochemical equivalent circuit (EEC) for EIS data fitting

Table 3 EIS fitted data of Ni-W coated carbon steel and carbon steel in spent acid solution

	pH	$R_s(\Omega \cdot cm^2)$	$Q_{dl}(F/cm^2)$	n	$R_{ct}(\Omega \cdot cm^2)$
Ni-W coated carbon steel	1	4.535	7.32×10^{-4}	0.8079	138.5
Ni-W coated carbon steel	2	23.71	3.125×10^{-4}	0.8338	396.5
Ni-W coated carbon steel	4	242.2	2.688×10^{-4}	0.8612	722.8
Carbon steel	1	4.42	3.20×10^{-3}	0.6589	1.43
Carbon steel	2	33.86	1.15×10^{-3}	0.8077	6.37
Carbon steel	4	254.9	5.87×10^{-4}	0.7469	58.45

3.2 Corrosion behavior of Ni-W coated carbon steel in oilfield formation water

Fig. 8 shows the corrosion rate of Ni-W coated carbon steel and carbon steel in oilfield formation water at different temperatures. The corrosion rate ratio of carbon steel and Ni-W coated carbon steel is 85.5, 73.5 and 125.9 at 30℃, 60℃, and 90℃, respectively. Compared to the corrosion rate ratio measured in spent acid solution, Ni-W coating has higher corrosion resistance in oilfield formation water. Fig. 9 shows the micro-morphology of corroded Ni-W coated carbon steel and composition of corrosion products. It is seen that uniform

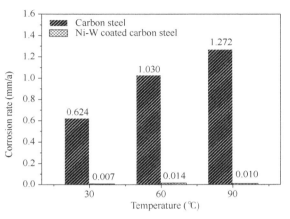

Fig. 8 Corrosion rate of Ni-W coated carbon steel and carbon steel in oilfield formation water with CO_2(5MPa), H_2S (2MPa), and Cl^-(100000mg/L)

corrosion occurred in oilfield formation water as well, the corrosion product pieces scattered over the Ni-W coated carbon steel substrate surface. EDS analysis results of corrosion products and corroded Ni-W coated carbon steel are shown in the table of Fig. 9, for the corrosion products (Spectrum 2), it is observed that most Fe contained in coating was dissolved and a few Ni was dissolved, resulting in formation of FeS and NiS, no W was found in corrosion products, this result confirms that W is preferentially to form oxide film during corrosion process[11-13]; for the corroded Ni-W coated carbon steel surface (Spectrum 3), except the content of Ni, it is seen that the content of Fe and W is very close to the original chemical composition of Ni-W coating in Table 1. According to the content in two tables, it is deduced that some Ni and W oxides and sulfides covered on surface.

Fig. 10 shows the corrosionpotential of Ni-W coated carbon steel and carbon steel in oilfield formation water at different temperatures, corrosion potential shifted negatively with temperature. Correspondingly, Fig. 11 shows the polarization curves of Ni-W coated carbon steel and carbon steel in oilfield formation water at different temperatures, it is seen that all the polarization curves of Ni-W coated carbon steel are characterized by passivation. Critical passive potential, start and end of passive potential in passive region are almost independent of temperature, while the critical passive current density and passive current density are dependent of temperature, moreover, it is observed that the increase of passive current density in logarithm is proportionally to temperature increase. It is concluded that thermodynamic parameters are temperature-independent and kinetic parameters are temperature-dependent. The passivity nature becomes more and more indistinct with temperature. In contrast, all the polarization curves of carbon steel measured at different temperatures have similar characteristics, active dissolution instead of passivation dominated the anodic reaction.

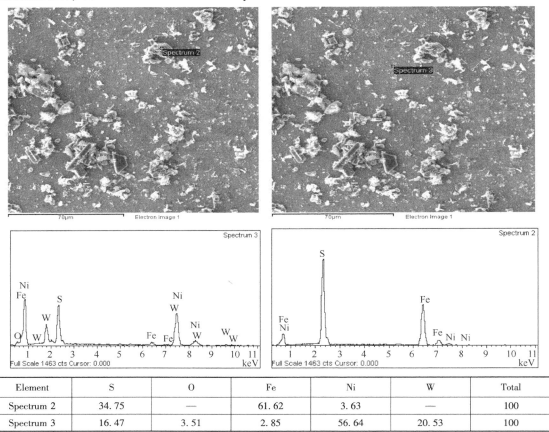

Element	S	O	Fe	Ni	W	Total
Spectrum 2	34.75	—	61.62	3.63	—	100
Spectrum 3	16.47	3.51	2.85	56.64	20.53	100

Fig. 9 SEM images and EDS results of corroded Ni-W coated carbon steel in oilfield formation water with CO_2(5MPa), H_2S (2MPa), and Cl^-(100000mg/L)

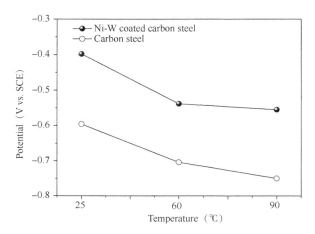

Fig. 10 Corrosion potential of Ni-W coated carbon steel and carbon steel in formation water at different temperatures

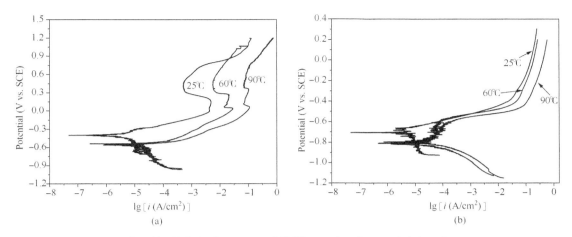

Fig. 11 Polarization curves of Ni-W coated carbon steel (a) and carbon steel (b) in oilfield formation water at different temperatures

EIS diagrams of Ni-W coated carbon steel and carbon steel measured in oilfield formation water at different temperatures are shown in Fig. 12, all the EIS diagrams are characterized by one semicircle, while diffusion control is observed on carbon steel at 60℃ and 90℃. EIS diagrams in Fig. 12 (a, c, d) are fitted by electrochemical equivalent circuits shown in Fig. 6(a), and EIS

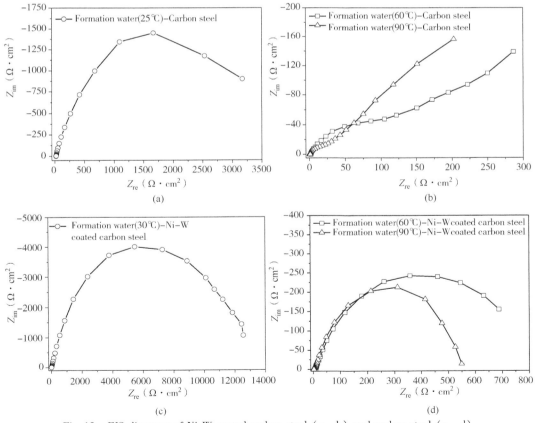

Fig. 12 EIS diagrams of Ni-W coated carbon steel (a, b) and carbon steel (c, d) in oilfield formation water at different temperatures

diagrams in Fig. 12 (b) with diffusion are fitted by electrochemical equivalent circuits shown in Fig. 6(b). The fitted electrochemical impedance parameters are given in Table 4. The charge-transfer resistance of Ni-W coated carbon steel and carbon steel decreases significantly with temperature increasing, the corrosion rate ratio of carbon steel and Ni-W coated carbon steel calculated based on charge-transfer resistance is 3.2, 5.0, and 29.2 in oilfield formation water at 30℃, 60℃, and 90℃, respectively. The corrosion rate ratio obtained both from autoclave weight loss test and electrochemical test indicated that Ni-W coated carbon steel exhibited higher corrosion resistance than carbon steel, especially in high temperature.

Table 4 EIS fitted data of Ni-W coated carbon steel and carbon steel in oilfield formation water

Parameter	$T(℃)$	$R_s(\Omega \cdot cm^2)$	$Q_{dl}(F/cm^2)$	n	$R_{ct}(\Omega \cdot cm^2)$	$Y_w(\Omega^{-1} \cdot cm^{-2} s^{1/2})$
Ni-W coated carbon steel	30	2.887	2.622×10^{-5}	0.7616	12320	—
Ni-W coated carbon steel	60	8.069	4.505×10^{-4}	0.7193	800.8	—
Ni-W coated carbon steel	90	1.532	5.17×10^{-4}	0.7203	555.6	—
Carbon steel	30	21.91	1.077×10^{-3}	0.8025	3886	—
Carbon steel	60	1.764	1.827×10^{-3}	0.6928	160	0.01825
Carbon steel	90	0.7646	5.59×10^{-4}	0.8532	19.01	0.01605

3.3 Thread sealing test of full-scale Ni-W coated carbon steel tubing

Table 5 shows the make-and-break test results of full-scale Ni-W coated carbon steel tubing thread, Fig. 13 shows that the largest makeup torque (4220 lbf · ft) was reached after 2.75 turns of screw. It is seen that after 4 times of make up and three times of break out, the coating on pin and box still exhibits good appearance and adhesion, no sticky thread and damage was observed, as shown in Fig. 14. There is no leakage was found after hydraulic bursting test at 95.0 MPa for 30 mins. Since the shoulder plays a key role in tubing sealing performance, therefore, the tubing shoulder was cut from pin [Fig. 14 (b)] after make-and-break test for corrosion evaluation, prior to corrosion test, the Ni-W coating was mechanically scratched by knife, it is visually seen that coating and steel substrate still has good adhesion even after 4 times of make up and three times of break out. Corrosion test was carried out in an extreme downhole condition in terms of high temperature, high H_2S, high CO_2, and high Cl^- containing solution, the specific content of each species is given in Table 6. Both macro-image and micro-image in Fig. 15 shows that Ni-W coating is highly corrosion resistant in extreme downhole condition, no pitting corrosion was observed, moreover, the macro-image shows that the tubing shoulder still has metallic luster. The results of 4 times of make up and three times of break out test, hydraulic bursting test at 95.0 MPa, and extreme downhole condition corrosion test demonstrated that tubing thread sealing performance was maintained with application of Ni-W coating layer, while the corrosion resistance was greatly enhanced.

Table 5 Make-and-break test results of full-scale Ni-W coated carbon steel tubing thread

Numbers of make up/break out	Makeup torque (lbf · ft)	Break out torque (lbf · ft)	Results
1	4476	4435	Not sticky
2	4347	4178	Not sticky
3	4384	4126	Not sticky
4	4578	—	—

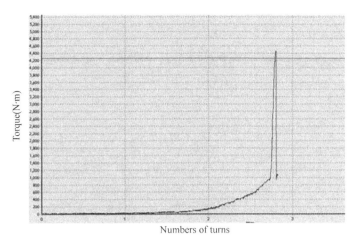

Fig. 13 Torque vs. numbers of turns for full-scale Ni-W coated carbon steel tubing thread

(a)   (b)

Fig. 14 Images of full-scale Ni-W coated carbon steel tubing thread (a. box; b. pin) after make-and-break test

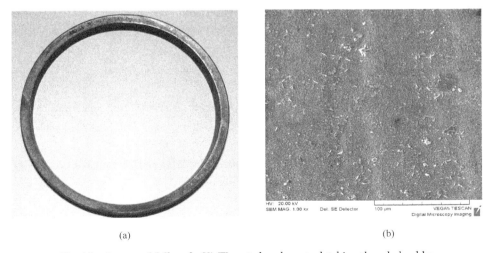

(a)   (b)

Fig. 15 Images of full-scale Ni-W coated carbon steel tubing thread shoulder
after make-and-break test and corrosion test

Table 6 Parameters for tubing shoulder corrosion test

T(℃)	P_{H_2S}(MPa)	P_{CO_2}(MPa)	Cl⁻(mg/L)	Flow rate (m/s)	Duration (d)
120	2.07	5.96	100000	5	7

4 Conclusions

(1) Ni-W coated carbon steel exhibited higher corrosion resistance than carbon steel in spent acid, especially in low pH spent acid. The corrosion rate ratio of carbon steel and Ni-W coated carbon steel in spent acid with pH=1, pH=2, and pH=4 is 13.7, 14.5, and 6.9, respectively.

(2) Ni-W coated carbon steel exhibited higher corrosion resistance than carbon steel in formation water, especially at high temperature. The corrosion rate ratio of carbon steel and Ni-W coated carbon steel in formation water at 30℃, 60℃, and 90℃ is 85.5, 73.5, and 125.9, respectively.

(3) No sticky thread was found after 4 times of make up and three times of break out by using maximum recommended makeup torque of 4258 lbf · ft, no leakage was detected after hydraulic bursting test at 95.0 MPa for 30 mins, and it is observed very slight corrosion on the tubing shoulder experienced make-and-break test under extreme downhole condition.

Acknowledgement

This work was supported by the Key Laboratory for Mechanical & Environment Behavior of Tubular Goods, China National Petroleum Corporation.

References

[1] G. M. Abriam, Controlling corrosion of carbon steel in sweet high temperature and pressure downhole environments with the use of corrosion inhibitors. NACE Northern Area Western Conference, Calgary, Alberta, 2010.

[2] J. A. Carew, A. Al-Sayegh, A. Al-Hashem, The effect of water-cut on the corrosion behaviour L80 carbon steel under downhole conditions. Corrosion'2000, Paper No. 00061, NACE International, Houston, 2000.

[3] Q. J. Meng, B. Chambers, R. Kane, J. Skogsberg, M. Kimura, K. Shimamoto, Evaluation of localized corrosion resistance of high strength 15Cr steel in sour well environments. Corrosion'2010, Paper No. 10320, NACE International, Houston, 2010.

[4] H. A. Nasr-EI-Din, S. M. Driweesh, G. A. Muntasherr, Field application of HCl-formic acid system to acid fracture deep gas wells completed with super Cr-13 Tubing in Saudi Arabia, SPE international, Paper No. 84925, Kuala Lumpur, Malaysia, 2003.

[5] H. Marchbois, H. EL Alami, J. Leyer, A. Gateaud, Sour service limits of 13% Cr and super 13% Cr stainless steels for OCTG: effect of environmental factors. Corrosion'2009, Paper No. 09084, NACE International, Houston, 2009.

[6] T. Henke, J. Carpenter, Cracking tendencies of two martensitic stainless alloys in common heavy compeltion brine systems at down-hole conditions: A laboratory investigation. Corrosion'2004, Paper No. 04128, NACE International, Houston, 2004.

[7] H. Liu, R. X. Guo, Z. Yun, B. Q. He, Z. Liu, Comparative study of microstructure and corrosion resistance of electroless Ni-W-P coatings treated by laser and furnace-annealing. Transactions of Nonferrous Metals Society of

China 20 (2010) 1024-1031.
[8] Y. W. Yao, S. W. Yao, L. Zhang, H. Z. Wang, Electrodeposition and mechanical and corrosion resistance properties of Ni-W/SiC nanocomposite coatings. Materials Letters 61(2007) 67-70.
[9] M. R. Vaezi, S. K. Sadrnezhaad, L. Nikzad, Electrodeposition of Ni-SiC nano-composite coatings and evaluation of wear and corrosion resistance and electroplating characteristics. Colloids and Surfaces A: Physicochem. Eng. Aspects 315(2008)176-182.
[10] F. F. Xia, C. Liu, F. Wang, M. H. Wu, J. D. Wang, H. L. Fu, J. X. Wang, Preparation and characterization of Nano Ni-TiN coatings deposited by ultrasonic electrodeposition. Journal of Alloys and Compounds 490(2010) 431-435.
[11] W. H. Hui, J. J. Liu, Y. S. Chaug, Surf. Coat. Technol. 68/69 (1994) 546-551.
[12] M. Obradovic, J. Stevanovic, A. Despic, R. Stevanovic, J. Stocii, Journal of Serbian Chemical Society 66 (2001) 899-912.
[13] R. Z. Valiev, R. K. Islamgaliev, I. V. Alexandrov, Bulk nanostructured materials from severe plastic deformation. Progress in Material Science 45(2000) 103-89.
[14] S. Yao, S. Zhao, H. Guo, M. Kowaka, A new amorphous alloy deposit with high corrosion resistance. Corrosion 52(1996) 183-186.

本论文原发表于《Engineering Failure Analysis》2016年第66卷。

Fatigue and Corrosion Fatigue Behaviors of G105 and S135 High-Strength Drill Pipe Steels in Air and H$_2$S Environment

Han Lihong[1]　Liu Ming[2, 3]　Luo Sheji[4]　Lu Tianjian[2, 3, 5]

(1. State Key Laboratory of Service Behavior and Structure Safety for Petroleum Tubular Goods and Equipment Material, CNPC Tubular Goods Research Institute;
2. State Key Laboratory for Strength and Vibration of Mechanical Structures;
3. MOE Key Laboratory for Multifunctional Materials and Structures, Xi'an Jiaotong University;
4. School of Materials Science and Engineering, Xi'an Shiyou University; 5. State Key Laboratory of Mechanics and Control of Mechanical Structures, Nanjing University of Aeronautics and Astronautics)

Abstract: Fatigue and corrosion fatigue(CF) tests were carried out to investigate the behaviors of G105 and S135 low carbon high-strength drill pipe steels under different stress amplitudes in air as well as simulated H$_2$S contained drilling environment. The regression analysis method was applied to obtain empirical equations governing the fatigue and CF lives of drill pipe steels in different environmental conditions. Results revealed that there exist fatigue limits for G105 and S135 drill pipe steels in air, and the fatigue life equations for G105 and S135 are $N_f = 3.28 \times 10^8 (S_{eqv} - 406.1)^{-2}$ and $N_f = 3.81 \times 10^8 (S_{eqv} - 472.5)^{-2}$ respectively. For both types of pipe steels, quasi-cleavage and cleavage fracture was identified as the main feature in the fatigue source zone of the two steels, while fatigue striations were the main feature in the stable crack growth zone. However, in H$_2$S solution, no obvious fatigue limits were found for G105 and S135, and the corresponding CF life equations are $N_f = 3.58 \times 10^8 (S_{eqv} - 143.7)^{-2}$ and $N_f = 2.91 \times 10^8 (S_{eqv} - 119.6)^{-2}$. The CF sensitivity levels of G105 and S135 in H$_2$S solution are high (64.6% and 74.7%, respectively), but S135 displays a higher sensitivity (74.7%) than G105 (64.6%). Further, no apparent plastic deformation appeared on the fracture surface in H$_2$S solution, and the fatigue cracks sprout from the surface and expand into the specimen with radiation pattern.

Keywords: Drill pipe steel; Hydrogen sulfide; Corrosion fatigue; Elastic fracture mechanics

1　Introduction

Corrosion fatigue(CF) is a serious failure mode for materials subjected to cyclic mechanical

---

Corresponding author: Han Lihong, hanlihong@cnpc.com.cn; Liu Ming, liuming0313@xjtu.edu.cn.

loading and corrosive media. It has been found that failures resulting from corrosion fatigue can even occur on metal materials of good corrosionresistance. Thatis, CF is different from stress corrosion and not selective to materials and the environments(Zhao et al., 2018; Cheng and Chen, 2018; Zenget al., 2018; Zamani et al., 2016; Luo et al., 2018). Under the interaction of cyclic loading and corrosive environment, so long as the critical mechanicalconditions of CF aremet, CF damage/failurewill inevitably appear and eventually lead to low stress brittle fracture of the material (Liu et al., 2015; Tian et al., 2014; Zhao et al., 2017; Sabelkin et al., 2016).

As an important part of the drill string system, drill pipe is often used as the circulation channel of drilling fluid. It is also used to transfer torque in rotary drilling, and hence bears the complex effect of a variety of loads, such as tension, bending and torsion. During drilling operations, some additives and dissolvedoxygen(O_2), hydrogen sulfide(H_2S) and carbon dioxide (CO_2) in oil and gas reservoirs, various salts(such as NaCl), bacteria, and metabolites will cause serious corrosion to the drilling tools(Donahue and Burns, 2016; Thodla et al., 2012), resulting in severe CF failures. Under the combined action of corrosion and alternating stressing, the drill pipe steel is prone to CF failures, such as sudden(brittle) fracture. Statistically, CF is the main form of drill pipe failure, accounting for almost 80% of drill pipe failure in all kinds of failure accidents, causing huge economic losses (Hansford and Lubinski, 1966; Lin et al., 2013; Chang et al., 2018; May et al., 2018). The CF failure not only accelerates the scrap of drill pipes and increases the cost of drilling production but also causes drilling accidents and even casualties due to sudden breaking.

Inrecentyears, with the gradual depletion of shallow resources, the petroleum industry is developing towards ultra-deep wells and high-speed drilling. As increasingly more deep and ultra-deep wells in abominable conditions (such as the oil and gas fields of elevated H_2S and CO_2 concentrations)are drilled, CF damage of drillpipe has become one of the harsh problems restricting its development (II man, 2014; Slifka et al., 2014; Cheng and Chen, 2017). In our previous work, the fatigue crack propagation (FCP) behavior of S135 drill pipe steel in air was characterized(Luo et al., 2013). It was found that with the increase of stress ratio R, the fracture near the threshold zone is a cleavage step, while the fracture in the stable propagation zone is dominated by fatigue striation propagation. Although the corrosion fatigue(CF), stress corrosion cracking(SCC) and other service behaviors of petroleum engineering materials in H_2S containing environment have been studied(Vosikovsky and Cooke, 1978; Tsukadaetal., 1983; Casarini, 1993; Revie et al., 1993), corrosion fatigue behavior and fracture mechanisms of low-carbon high-strength sulfur-resistant drill pipes in H_2S containing environment need further investigations. To address this deficiency, in the current study, we aimed to investigate the CF behavior of G105 and S135 drill pipe steels in H_2S containing solution, establish empirical equations governing CF life, analyze fracture morphologies, and explore physical mechanisms underlying CF failure. The results could be of great the oretical significance and engineering applicationvalues.

2 Experimental

2.1 Materials andsolution

G105 and S135 carbon high strength drill pipe steels were used for experiments. Table 1 shows their chemical compositions(in % by mass) determined by analytical analysis. The specimen shape and sizes are shown in Fig. 1. To ensure the specimen fractures within the gauge range, paint was sprayed outside the gauge to insulate the corrosive media. Prior to experimentation, all the samples were polished with silicon carbide water polishingpaper down to 2000 #, after polishing, the samples were ultrasonically cleaned in acetone and rinsed in distilled water, then dried in cold air and put in a drying dish foruse.

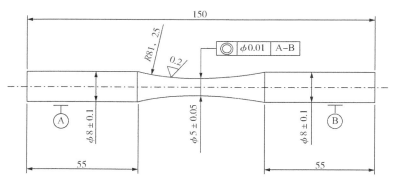

Fig. 1 Schematic illustration of fatigue specimen(all units in mm). The 5.0wt.% NaCl + 0.5wt.% CH_3COOH solution was chosen to simulate the drilling environment with H_2S according NACE TM 0177(Veloz and Martínez, 2006). The NaCl and CH_3COOH were dissolved in deionized water with a pH of approximately 2.6-2.8

Table 1 Chemical compositions(wt.%) of G105 and S135 drill pipe steels

Steel	C	Si	Mn	P	S	Ni	Cr	Mo	Ti	Cu	Fe
G105	0.23	0.25	0.47	0.006	0.002	0.011	0.888	0.658	0.060	0.06	Balance
S135	0.20	0.24	0.54	0.007	0.004	0.403	0.861	0.858	0.007	0.06	Balance

2.2 CF test

To investigate the CF behavior and mechanisms of G105 and S135 in H_2S environment, the tests were conducted in a purposely-designed device that simulated the environment (Fig. 2). CF tests were performed on the PLD-100 microcomputer controlled electro-hydraulic servo fatigue test machine. During the actual drilling process, the speed of drill pipe is usually between 60~120r/min (12Hz). In order to accelerate the fatigue test, 5Hz was used to carry out the experiment in air. 1 Hz was adopted to simulated the actual working condition in H_2S environment. Triangular wave was adopted, with a fixed stress ratio of 0.1. In the fatigue tests, the normal stress S_{max} applied to the specimen was calculatedas:

$$S_{max} = \frac{P}{S} \tag{1}$$

where P is the applied load and S is the cross sectional area of the specimen.

With reference to Fig. 2, a gas cylinder was used to generate H_2S passing into the self-designed device that contained glacial aceticacid. The partial pressure of H_2S was 0.83kPa and the amount of H_2S inflow was several bubbles per minute. Before CF test, the pH value of the solution after bubbling H_2S was tested, and the pH value of the solution was approximately 2.7, which was constant after bubbling. The H_2S was continuously introduced until the specimen was fractured. In the process of CF test, in order to maintain the consistency of solution and remove the influence of corrosion products, the solution was replaced every 6h during testing. A SIN pH 3.0 industrial pH meter was used to test the pH value of the solution, and it was measured per 1h duringt he test. The pH value of the test solution was between 2.6 and 2.8. After the experiment, the morphologies of fracture surfaces were examined under SEM (JSM 6390 A). All the tests were conducted at ambient temperature, 25℃.

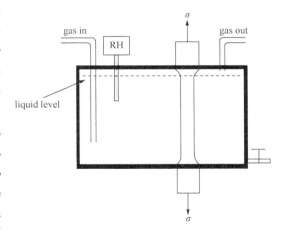

Fig. 2 Schematic of purposely-designed device for CF testing

2.3 S-N curve fitting

Inair, the fatigue life of drill pipe steels are not affected by corrosive environment, the fatigue strength of drill pipe steels is defined as 10^7 cycles. However, the fatigue life of steels will be significantly reduced by the H_2S corrosive environment, the 10^6 cycles is defined as fatigue strength in H_2S environment. Under cyclic loading, the fatigue life of a material is closely related to the applied stress amplitude and the cyclic stress ratio. The effects of stress amplitude and stress ratio on fatigue life can be normalized by the equivalent stress amplitude. The fatigue life can be expressed as a function of equivalent stress amplitude, as (Zheng et al., 1995; Wang, 2001):

$$N_f = S_f [S_{eqv} - (S_{eqv})_c]^{-2} \tag{2}$$

Where S_f is the resistan cecoefficient for fatigue, $(S_{eqv})_c$ is the fatigue limit expressed as an equivalentstress; S_{eqv} is the equivalent stress applied to the specimen during the fatigue process which depends on the nominal stress range and the stress ratio (Zheng et al., 1995).

$$S_{eqv} = -\frac{1}{2(1-R)} AS = -\frac{(1-R)}{2} S_{max} \tag{3}$$

Here, AS is the nominal stress amplitude, S_{max} is the maximum stress during cyclic loading and R is the stress ratio.

Performing logarithm operation on both sides of the Eq. (1), we arrived at:

$$\lg N_f = \lg S - 2\lg[S_{eqv} - (S_{eqv})_c] \tag{4}$$

The above equation indicates that, in double logarithmic coordinates, $\lg N_f \sim \lg[S_{eqv} - (S_{eqv})_c]$ is straighe line, with a slope equal to 2. Using the tail difference method, we obtained the CF resistance coefficient S_f and CF limit $(S_{eqv})_c$ values of G105 and S135 drill pipe steels via regression analysis at slope 2±0.002.

3 Results anddiscussion

3.1 Fatigue behavior

3.1.1 Fatigue life and S-N curve

The fatigue test results and S-N curves of G105 and S135 drill pipe steels under different cyclic stresses in air are shown in Table 2 and Fig. 3. In air, the fatigue life (N_f) of the two steels are prolonged as the equivalent stress amplitude (S_{eqv}) is decreased, and the N_f of G105 is lower than that of S135. For G105, the N_f is outnumbering more than 10^5 cycles when the S_{eqv} is reduced to less than 444.4MPa. Correspondingly, with the decrease of S_{eqv}, the N_f is increased significantly, showing obvious fatigue limit characteristic. When the S_{eqv} drops below 410.2MPa, the N_f reaches 10^7 cycles, tending to infinite life. In comparison, for S135, the N_f is more than 10^5 cycles when the S_{eqv} is reduced to less than 546MPa and, with the decrease of the S_{eqv}, the N_f is increased significantly, showing obvious fatigue limit characteristic. When the S_{eqv} is reduced to 464.8MPa, the test specimen has gone through 8.3×10^6 cycles without breaking, indicating the S_{eqv} is close to the fatigue limit of the steel.

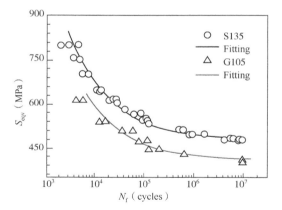

Fig. 3　S-N curves of drill pipe steels in air

Table 2　Fatigue test results of G105 and S135 drill pipe steels in air

Steel	No.	P_{max}(kN)	S_{max}(MPa)	S_{eqv}(MPa)	N_f
G105	1	18	917.1	615.3	6012
	2	18	917.1	615.3	4013
	3	16	815.2	546.8	12123
	4	16	815.2	546.8	17018
	5	15	746.3	512.7	37325
	6	15	746.3	512.7	59059
	7	14	713.3	478.5	73214
	8	14	713.3	478.5	108329
	9	13	662.4	444.4	119047
	10	13	662.4	444.4	203457
	11	12.5	636.8	427.2	613245
	12	12	614.2	410.2	1×10^7
	13	11.7	597.0	400.5	1×10^7

Continued

Steel	No.	P_{max}(kN)	S_{max}(MPa)	S_{eqv}(MPa)	N_f
S135	1	23	1171.9	786.1	2025
	2	23	1171.9	786.1	3023
	3	23	1171.9	786.1	2201
	4	22	1121	752.0	2941
	5	22	1121	752.0	3645
	6	20.5	1044.5	700.7	4740
	7	20.5	1044.5	700.7	5550
	8	19	968.1	649.4	7530
	9	19	968.1	649.4	9200
	10	19	968.1	649.4	11096
	11	18	917.1	615.3	17810
	12	18	917.1	615.3	23429
	13	18	917.1	615.3	27522
	14	17.5	891.7	598.2	29927
	15	17	866.2	581.1	42790
	16	16.5	840.7	564.8	60772
	17	16.5	840.7	564.8	84120
	18	16	815.2	546.8	125300
	19	16	815.2	546.8	97106
	20	15	746.3	512.7	412340
	21	15	746.3	512.7	632578
	22	14.5	738.8	495.5	1.1×10^6
	23	14.5	738.8	495.5	1.8×10^6
	24	14	713.3	478.5	6.1×10^6
	25	14	713.3	478.5	7.2×10^6
	26	14	713.3	478.5	8.2×10^6
	27	13.6	692.9	464.8	8.3×10^6

Based on Eqs. (2) and (4), the fatigue test results of G105 and S135 in air were analyzed by regression, and the resultsare listed in Table 3.

Table 3 Fatigue test fitting results of drill pipe steels in air

Steel	S_f(MPa^2)	$(S_{eqv})_c$(MPa)	r	s
G105	3.28×10^8	406.1	−0.9473	0.2173
S135	3.81×10^8	472.5	−0.9791	0.2053

(r, s are linear correlation coefficients and mean square variance, respectively) The linear correlation coefficients of G105 and S135 in air are $|r| = 0.9473$ and $|r| = 0.9791$, respectively, which are greater than the minimum values of 99.9% confidence. Therefore, Eq. (2) can be used to describe the fatigue life of G105 and S135 in air. The refore, using the results of Table 3, we obtained the equations governing the fatigue life of G105 and S135 in air as:

$$G105 \quad N_f = 3.28 \times 10^8 (S_{eqv} - 406.1)^{-2} \quad (5)$$

$$S135 \quad N_f = 3.81 \times 10^8 (S_{eqv} - 472.5)^{-2} \quad (6)$$

Fig. 3 also presents the fitting curves represented by Eqs. (5) and (6), which agree well with the experimental results. Therefore, these equations can be used to characterize the general fatigue rule of G105 and S135. So long as the external load is known, the fatigue life of the drill pipe steels in air can be calculated using Eqs. (5) and (6).

3.1.2 Fatigue fracturemechanisms

Fatigue fracture morphologies of G105 and S135 drill pipe steels under different values of S_{eqv} in air are displayed in Figs. 4 and 5. There fracture is brittle as no apparent plastic deformation is found on the fracture surface, and the cross section of which is roughly perpendicular to the axial direction. Cracks are originated from or near the surface of the specimen, and it is a single fatigue source even though S_{eqv} is varied. The fatigue fracture is composed of three typical zones: fatigue source zone, fatigue crack propagation zone and fatigue instantaneous fracture zone. The proportion of each zone on the fracture surface depends on the stress state, the specimen shape, the loading mode, and the mechanical properties of steel. Statistical analysis of the three fracture zones under different S_{eqv} indicates that the fatigue source zone is relatively small, and the fatigue source zone and crack propagation zone combine to occupyless than 1/2 of the fracture surface. In contrast, the instantaneous fracture zone is the largest, accounting for more than 1/2 of the fracture section. Further, with the increasing of S_{eqv}, the fatigue source zone and the fatigue crack propagation zone both decrease, while the instantaneous fracture area increases.

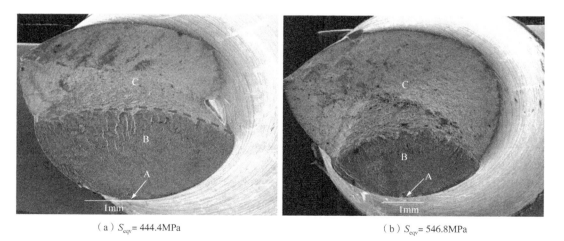

(a) S_{eqv} = 444.4MPa　　　　　　　　(b) S_{eqv} = 546.8MPa

Fig. 4　Fracture morphology of G105 drill pipe steel in air
"A" represents fatigue source zone, "B" represents fatigue crack
propagation zone and "C" represents fatigue instantaneous fracture zone

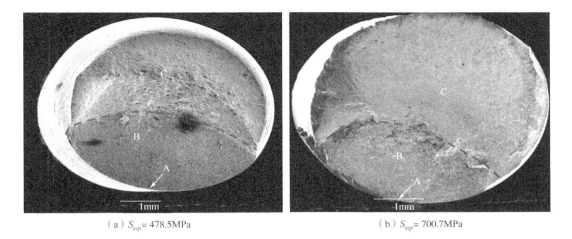

(a) S_{eqv}= 478.5MPa (b) S_{eqv}= 700.7MPa

Fig. 5 Fracture morphology of S135 drill pipe steel in air
"A" represents fatigue source zone, "B" represents fatigue crack
propagation zone and "C" represents fatigue instantaneous fracture zone

Figs. 6 and 7 display the microscopic fracture morphologies of crack source sites of G105 and S135 under different S_{eqv} in air. The fatigue crack sprouts on or near the surface of the specimen. The crack starts from the initiation point and expands in fluvial pattern on different crystallographic planes. The tip of the crack deviates in the extension direction due to different expansion resistance, and then expands on the respective plane. Two different fracture planes are mutually cutto form steps, with radial ray sformed on the fracture surface. Fatigue source zone is the place where cracks are initiated, wherein the crack propagation rate is slow (usually less than 10^{-3} mm/cycle). The fracture surface is constantly squeezed due to repeated opening and closing. As a result, the fracture surface of the fatigue source region is smooth compared to the other two zones.

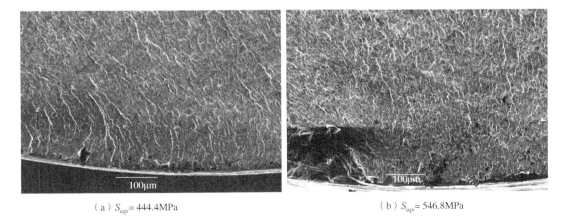

(a) S_{eqv}= 444.4MPa (b) S_{eqv}= 546.8MPa

Fig. 6 Fracture morphologies of crack initiation region of G105 drill pipe steelin air

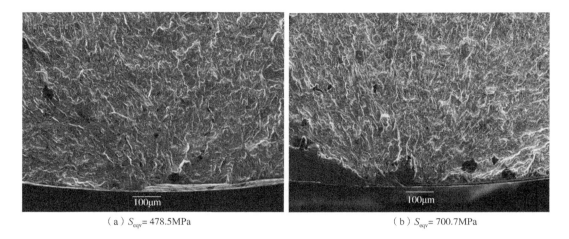

(a) S_{eqv}= 478.5MPa　　　　　　　　(b) S_{eqv}= 700.7MPa

Fig. 7　Fracture morphologies of crack initiation region of S135 drill pipe steel in air

As no obvious fatigue striations are observed on the crack source zone, the radiating source with convergence characteristics exists around the crack source zone, which is the mark left on the fracture surface after fatigue crack initiation. The fracture exhibits quasi cleavage mode with characteristics of tiny cleavage planes and tearridges, more obvious as S_{eqv} is increased. The typical quasi cleavage fracture behavior on the crack propagation region near crack source zone, which is formed by blunting/resharpening at crack tips by cyclic deformation. The tiny cleavage planes become smaller and the tear ridges become denser with the increasing of S_{eqv}, as shown in Fig. 6 (b) and 7(b). The higher the loading stress is, the more ductile the fracture modetends to be. Thus it should be of more tear ridges at the crack source under higher loading stress, which makes the radial ridges denser.

Fracture morphologies in the crack propagation zone of G105 and S135 under various S_{eqv} in air are presented in Figs. 8 and 9. Fatigue striations now become the main characteristics. As S_{eqv} is increased, the number of striations decreases(i. e., the spacing between striations increases), and secondary cracks appear on the surface of the fracture. For G105, the fracture is mainly fatigue striations, which have small spacing when S_{eqv} = 478. 5MPa; when S_{eqv} = 700. 7MPa, the fracture is characterized by cleavage fracture and fatigue striations. As for S135, the fracture is mainly via fatigue striations and the spacing between the striations is small when S_{eqv} = 478. 5MPa. When S_{eqv} = 546. 8MPa, the fracture is again characterized by fatigue striations but the spacing is greater. As S_{eqv} is increased to 700. 7MPa, the cracks are mainly characterized by transgranular cleavage and fatigue striations, and the number of striations is obviously reduced relative to the case of S_{eqv} = 546. 8MPa. At S_{eqv} = 752MPa, the cracks are mainly transgranular cleavage fracture but a small amount of striations can be observed on the cleavage surface; also, a small number of secondary cracks appear on the fracture surface, which show a certain angle with the maincrack.

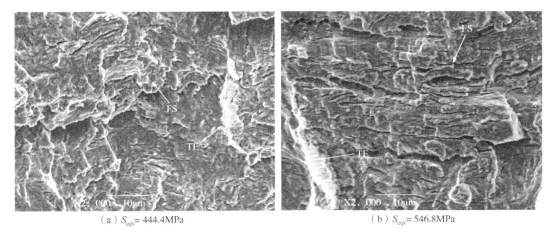

(a) S_{eqv}= 444.4MPa (b) S_{eqv}= 546.8MPa

Fig. 8 Fracture morphology of crack propagation zone of G105 drill pipe steel in air
"FS" represents fatigue striation and "TE" represents tear ridge

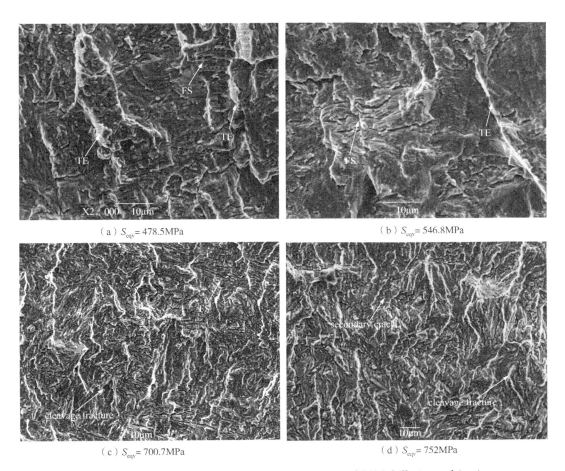

(a) S_{eqv}= 478.5MPa (b) S_{eqv}= 546.8MPa

(c) S_{eqv}= 700.7MPa (d) S_{eqv}= 752MPa

Fig. 9 Fracture morphology of crack propagation zone of S135 drill pipesteel in air
"FS" represents fatigue striation and "TE" represents tearridge

3.2 CF behavior

3.2.1 Corrosion fatigue life and S-N curve

Table 4 and Fig. 10 present the fatigue test results and $S-N$ curves of G05 and S135 drill pipe steels in H_2S solution for selected values of S_{eqv}. During CF testing, the maximum stress corresponding to 1 million cycle life is very low. For S135 drill pipe steel, the maximum stress is only about 1/5 of the yield strength; for G105 drill pipe steel, the maximum stress is only about 1/3 of the yield strength. Therefore, the CF test were carried out at 10^6 cycles. The results show that the fatigue life N_f of both steels is prolonged as S_{eqv} is decreased, and the N_f of G105 is higherthan that of S135 nearly in the whole stress range. For G105, the N_f becomes larger than 10^5 cycles when S_{eqv} drops below 205.1MPa. In comparison, the N_f of S135 exceeds 10^5 cycles only when S_{eqv} is reduced to 170.9MPa and, as S_{eqv} is further decreased, it is significantly prolonged. For instance, when S_{eqv} is reduced to 160MPa, the specimen has gone through 4.71×10^5 cycles without breaking.

Table 4 CF test results of G105 and S135 drill pipe steels in H_2S solution

Steel	No.	P_{max}(kN)	S_{max}(MPa)	S_{eqv}(MPa)	N_f
G105	1	13	662.4	444.4	2912
	2	13	662.4	444.4	3726
	3	12	611.5	410.2	6356
	4	11	560.5	376.0	8123
	5	11	560.5	376.0	7579
	6	10	509.5	341.8	10749
	7	9	458.5	307.6	20158
	8	8	407.6	273.5	40220
	9	7	356.7	239.3	113891
	10	6	305.7	205.1	198000
	11	5	254.8	170.9	361792
	12	5	254.8	170.9	499548
	13	4.5	229.2	153.8	921453
S135	1	15	764.3	512.7	1698
	2	15	764.3	512.7	2245
	3	15	764.3	512.7	1425
	4	13	662.4	444.4	2165
	5	13	662.4	444.4	3414
	6	13	662.4	444.4	5018
	7	11	560.5	376.0	6257
	8	11	560.5	376.0	4523
	9	11	560.5	376.0	7161
	10	9	458.6	307.6	8251
	11	9	458.6	307.6	10839
	12	9	458.6	307.6	14441
	13	7	356.7	239.3	30343
	14	7	356.7	239.3	20123
	15	7	356.7	239.3	43814
	16	5	254.8	170.9	223461
	17	5	254.8	170.9	174024
	18	4.8	244.5	164.0	323318
	19	4.7	239.4	160.0	471246

Based on Eqs. (2) and (4), we analyzed the CF results of G105 and S135 drill steels in H_2S solution by regression and presented the results in Table 5. As shown in Table 5, the linear correlation coefficients of G105 and S135 in H_2S solution are $|r| = 0.9579$ and $|r| = 0.9706$, respectively, both greater than the minimum values of 99.9% confidence. Therefore, in the presence of corrosive environment (H_2S solution), Eq. (2) can still be used to describe the fatigue life of G105 and S135. It follows that, for the case considered, the empirical equations for the fatigue life of G105 and S135 are:

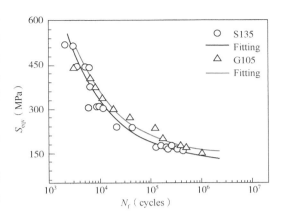

Fig. 10 S–N curves of drill pipe steels in H_2S solution

$$G105 \; N_f = 3.58 \times 10^8 (S_{eqv} - 143.7)^{-2} \tag{7}$$
$$S135 \; N_f = 2.91 \times 10^8 (S_{eqv} - 119.6)^{-2} \tag{8}$$

The fitting curves of (7) and (8) are in good agreement with the experimental results; Fig. 10, the fitting model has been proved by many scholars in corrosive media (Zheng and Wang, 1999, 1997; Zheng and Lu, 1987). Therefore, the fatigue life of the two drill pipe steels in H_2S solution can be predicted using Eqs. (7) and (8) if the external load is known.

Table 5 Fitting results of fatigue test for drill pipe steels in H_2S solution

Steel	$S_f(MPa^2)$	$(S_{eqv})_c$(MPa)	r	s
G105	3.58×10^8	143.7	-0.9579	0.2271
S135	2.91×10^8	119.6	-0.9706	0.1543

3.2.2 Corrosion fatigue fracture mechanisms

The macroscopic fracture morphologies of G105 and S135 in H_2S solution are presented in Figs. 11 and 12 show for selected values of S_{eqv}. In the presence of H_2S solution, the specimens exhibited brittle fracture, with no apparent plastic deformation observed on fracture surface. The fatigue cracks sprout from specimen surface and expand into the specimen with radiation pattern. Again, the CF fracture is composed of three distinct zones. With the increasing of S_{eqv}, the area of instantaneous fracture increases while the other two areas decrease. For G105, thefatigue source zone and the crack propagation zone combine to account for about 4/5 and 2/5 of the total fracture surface when $S_{eqv} = 170.9$ and $S_{eqv} = 410.2$ MPa, respectively. For S135, incomparison, the fatigue source and crack propagation zones account for about 1/2 and 2/5 of the fracture surface when $S_{eqv} = 307.6$ and $S_{eqv} = 444.4$ MPa, respectively. Remarkably, as S_{eqv} is increased, multiple cracks replace single crack in the fatigue source zone: single fatigue sources are shown in Fig. 12 (a) and (b), while multiple fatigue sources are shown in Fig. 12(c) and (d). Note, however, although both main crack and secondary crack sources appear on the test specimens, the propagation and fracture of the specimens are usually caused by the main crack.

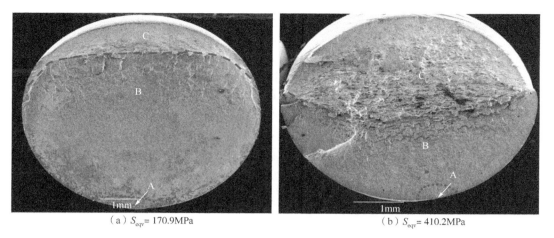

(a) S_{eqv}= 170.9MPa (b) S_{eqv}= 410.2MPa

Fig. 11 Macroscopic CF fracture morphology of G105 drill pipe steel in H_2S solution

"A" represents fatigue source zone, "B" represents fatigue crack propagation zone and
"C" represents fatigue instantaneous fracture zone

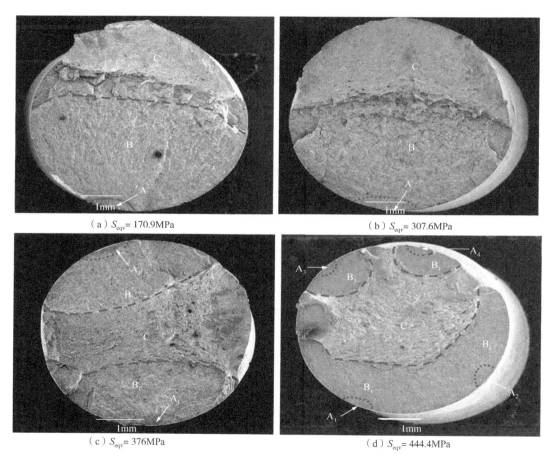

(a) S_{eqv}= 170.9MPa (b) S_{eqv}= 307.6MPa

(c) S_{eqv}= 376MPa (d) S_{eqv}= 444.4MPa

Fig. 12 Macroscopic CF fracture morphology of S135 drill pipe steel in H_2S solution

"A" represents fatigue source zone, "B" represents fatigue crack propagation zone and
"C" represents fatigue instantaneous fracture zone

Corresponding microscopic CF fracture morphologies of the crack source zone are displayed in Figs. 13 and 14. It can be seen that, in the presence of H_2S solution, more cleavage characteristics are observed in this zone and, with the increase of the S_{eqv}, the cleavage characteristics become more obvious. As the stress level is conducive to the opening of CF cracks, the ability of steel to resist the initiation of CF cracks is reduced. Due to the slow expansion and corrosion effect of H_2S, the color of the CF crack source zone is relatively dark compared to that observed in air. For G105, the CF crack source zone is covered by a layer of corrosion product, which is a typical cleavage plane.

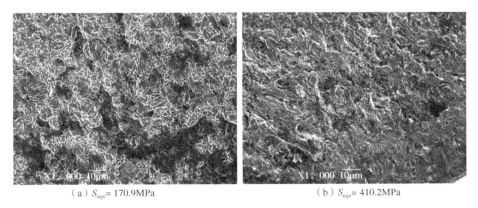

(a) S_{eqv} = 170.9MPa   (b) S_{eqv} = 410.2MPa

Fig. 13 Fracture morphologies of fatigue crack source zone of G105 drill pipe steel in H_2S solution

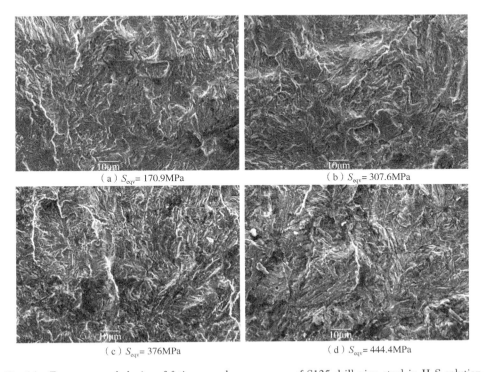

(a) S_{eqv} = 170.9MPa   (b) S_{eqv} = 307.6MPa

(c) S_{eqv} = 376MPa   (d) S_{eqv} = 444.4MPa

Fig. 14 Fracture morphologies of fatigue crack source zone of S135 drill pipe steel in H_2S solution

Figs. 15 and 16 present the microscopic fracture morphologies of fatigue crack propagation zone for G105 and S135 in H_2S solution. Fatigue striations are absent in crack propagation zone for both of steels. In Fig. 15, quasicleavage fracture is dominant in crack propagation zone for G105 steel. When S_{eqv} = 170.9MPa, there are more tear edges and cleavage characteristics on the fracture surface. When S_{eqv} = 410.2MPa, the cracking surface is dominant with a few tear edges and secondary cracking. In Fig. 16, cleavage fracture is dominant in crack propagation zone for G135 steel. When S_{eqv} = 170.9MPa, cleavage fracture is dominant on the fracture surface with a few tear edges. When S_{eqv} = 307.6MPa, the fracture surface is characterized by cleavage fracture, a small amount of tearing edge and intergranular fracture. When S_{eqv} = 376MPa, the fracture surface is characterized by cleavage fracture and intergranular fracture. When S_{eqv} = 376MPa, the fracture surface is characterized by cleavage step and intergranular fracture.

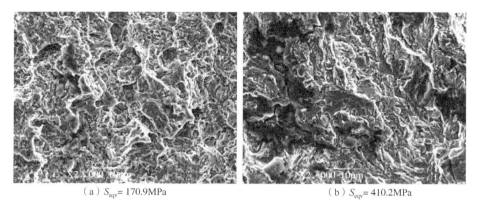

(a) S_{eqv}= 170.9MPa   (b) S_{eqv}= 410.2MPa

Fig. 15 Fracture morphologies of fatigue crack propagation zone of G105 drill pipe steel in H_2S solution

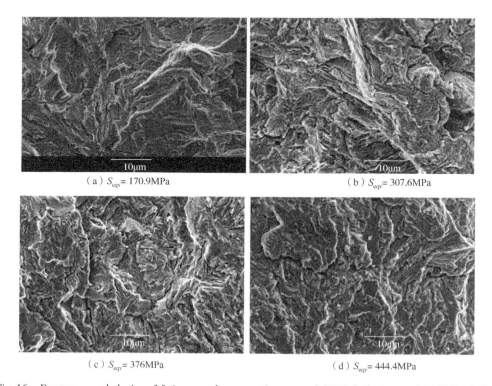

(a) S_{eqv}= 170.9MPa   (b) S_{eqv}= 307.6MPa

(c) S_{eqv}= 376MPa   (d) S_{eqv}= 444.4MPa

Fig. 16 Fracture morphologies of fatigue crack propagation zone of G105 drill pipe steel in H_2S solution

3.2.3 Comparison of fatigue properties in air and corrosive environment

Fig. 17 compares the $S-N$ curves of G105 and S135 in air and H_2S solution. The fatigue behaviors of the steels in air are significantly different from those in H_2S solution. For G105 [Fig. 17 (a)], under the same stress level, the N_f in H_2S solution is much lower than that in air; for $S_{eqv} <$ 500MPa, the difference is usually one order of magnitude, increasing as S_{eqv} is reduced. While there is obvious fatigue limit for G105 in air, no such limit is observed in H_2S solution. For S135 [Fig. 17 (b)], the N_f in H_2S solution is significantly lower than in air, and the difference increases with decreasing S_{eqv}. In air, as S_{eqv} is reduced, the N_f of S135 gradually decreases in the range of $10^6 \sim 10^7$ cycles, approaching progressively towards the fatigue limit. However, in H_2S solution, the slope of the $S-N$ curve for S135 varies little in the range of $10^5 \sim 10^6$ cycles and there is no fatigue limit.

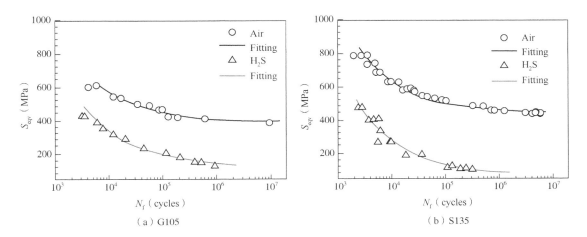

Fig. 17 Comparison of $S-N$ curves of drill pipe steel in air environment and H_2S solution

In the current study, the formation of cracks and expansion are under the interaction of cyclic load and corrosive medium. The presence of corrosive effect reduces the fatigue resistance of steel. In the current study, fatigue damage first occurs on the surface of steel under cyclic loading, and fracture eventually occurs under the continuous action of corrosive agent. The interaction between corrosive medium and cyclic stress accelerates the corrosion process, while corrosion also accelerates the fatigue process. CF failure is governed by both the corrosion environment and fatigue load, but it is not simple superposition of the two. The susceptibility index of CF may be defined by the loss of steel fatigue strength. The equivalent fatigue limit with clear physical meaning is selected (Távara et al., 2001; Locke, 2016; Arunachalam et al., 2018; Wang, 1998), as:

$$I_f = \frac{(S_{eqv})_{c,F} - (S_{eqv})_{c,CF}}{(S_{eqv})_{c,F}} \times 100\% \tag{9}$$

where $(S_{eqv})_{c,F}$ represents the fatigue limit in air and $(S_{eqv})_{c,CF}$ represents the fitting results from Table 4. For G105 and S135 in H_2S solution, the I_f values calculated by Eq. (9) are 64.6% and 74.7% in H_2S solution, S135 is more prone to corrosion fatigue failure.

Table 6 CF sensitivities of drill pipe steels in H_2S solution

Steel	$(S_{eqv})_{c,F}$(MPa)	$(S_{eqv})_{c,CF}$(MPa)	I_f(%)
G105	406.1	143.7	64.6
S135	472.5	119.6	74.7

3.2.4 CF mechanisms

CF is caused by a combination of corrosion and cyclic loading. Typically, CF damage gradually accumulates in the components, and the CF crack begins to grow when the crack reached a critical value. After the initiation, the crack grows continuously under the effect of alternating loading and corrosion. When the crack length reaches a critical level, the crack propagates rapidly to break the whole specimen. Therefore, for smooth specimens, the CF process includes fatigue crack initiation, subcritical propagation and ultimate unstable fracture. The CF process can be considered from two aspects: one is how the corrosive medium accelerates the initiation and expansion process; the other is how the cyclic load promotes the corrosion process. The interaction of corrosion medium and cyclic stressing on steel is different at each stage, and the physical mechanisms governing CF are complicated. At present, there are four main viewpoints about the mechanisms of CF: pitting accelerated crack initiation, selective electrochemical corrosion, protective film rupture, media absorption, and hydrogen induced cracking (Huetal., 2016; Wangetal., 2016; Kimetal., 2013; Chenet al., 2018).

In H_2S solution, H_2S reacts with Fe to form FeS, it is cathode after drilling pipe surface is attached by H_2S, and the drill pipesteel is anode. They form micro batteries, causing different degrees of corrosion plaque on the surface of drill pipes. H_2S reacts with the steel matrix, which will release hydrogen atoms, penetrate into the grain boundaries of the steel, form hydrogen molecules to generate bubbles, resulting in hydrogen embrittlement of drill pipes. The corrosion mechanism is expressed as follows(Yang et al., 2018):

$$H_2S \longrightarrow HS^- + H^+ \quad (10)$$

$$HS^- \longrightarrow H^+ + S^{2-} \quad (11)$$

Oneside, the dissociation products of H_2S, HS^-, S^{2-} are adsorbed on the surface of themetal forming the adsorption complex of ions $FeHS^-$. The reaction is as follows(Dong et al., 2009):

$$Fe + H_2S + H_2O \longrightarrow FeHS^-_{ads} + H_3O^+ \quad (12)$$

$$FeHS^-_{ads} \longrightarrow FeHS^+ + 2e^- \quad (13)$$

$$FeHS^+ + H_3O^+ \longrightarrow Fe^{2+} + H_2S + H_2O \quad (14)$$

$$Fe^{2+} + HS^- \longrightarrow FeS + H^+ \quad (15)$$

The corrosion products mainly include Fe_9S_8, Fe_3S_4, and FeS etc, which reduce the plasticity of the steel and accelerate the initiation of fatiguecracks.

H_2S reacts with the steel to produce hydrogen, which is adsorbed on the surface of the metal and diffused into the metal. The hydrogen atoms diffuse into the lattice or follow the dislocation movement to the stress concentration region under the action of stress gradient. The interaction between hydrogen and metal atoms weakens the binding force of metal atoms, therefore, cracks and expansion will occur in the hydrogen enriched region, resulting in the brittle fracture. In addition, the diffusion hydrogen atoms will enrich into grain boundaries, phase boundaries, dislocations etc and become trappedhy drogen. As the number of trapped hydrogen atoms increases, the hy drogen atoms bind to become hydrogen molecule and form high pressure at the trap. The atomic bonds in some regions break up and form microcracks when the hydrogen pressure is equal to the bonding force of metal atoms. Through a certain period of time, with the increase of diffused hydrogen trap, the crack pressure will rise further, leading to a new round of crack propagation(Jain et al., 2013; Thodal et al., 2012).

Fig. 18 displays the CF crack initiation location of G105 and S135 drill pipe steels in H_2S solution. The fatigue cracks of the two steels sprout on the surface of each specimen, and the cracks initiation point is smooth, with no corrosion pit observed. In the H_2S drilling environment, drill pipes are easily to react with H_2S to release H, which is adsorbed by the surface of steel(Huanget al., 2017; Li et al., 2017; Mayén et al., 2017). Under the action of cyclic loading, persistent slip striations(PSBs) are formed on the surface of steel, and the dislocation density in the extrusion ridge or squeezing groove is higher. Simultaneously, the adsorbed H will diffuse into the metal and promote plastic deformation, causing micro cracks to form(Mayén et al., 2017). In due course, these micro cracks will connect to the front end of the main crack and expand forward continuously, leading eventually to brittle fracture(Dong et al., 2009; Sangid et al., 2011), as shown in Figs. 15 and 16.

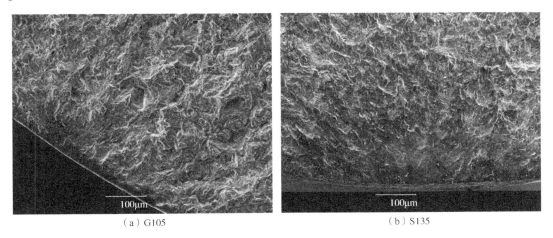

(a) G105   (b) S135

Fig. 18 Crack initiation site of drill pipe steels in H_2S solution

4 Conclusions

The fatigue behavior of G105 and S135 drill pipe steels in air and H_2S solution were studied based on the actual working conditions. Regression analysis was employed to obtain empirical equations governing the CF life of both steels in different environmental conditions. Physical mechanisms underlying CF fracture were explored. The main conclusions are summarized as follows:

(1) There exist fatigue limits for G105 and S135 in air. In sharp contrast, in H_2S solution, there is no fatigue limit for both G105 and S135. While both G105 and S135 exhibithigh CF sensitivity in H_2S solution, the high grade drill pipe steel S135 displays a higher CF sensitivity

(74.7%) than that (64.6%) of G105.

(2) In air, The fracture exhibits quasi cleavage mode with characteristics of tiny cleavage planes and tear ridges. The typical quasi cleavage fracture behavior on the crack propagation region near crack source zone, which is formed by blunting/resharpening at crack tips by cyclic deformation.

(3) In H_2S solution, fatigue striations are absent in crack propagation zone for both of steels. Quasi cleavage fracture is dominant in crack propagation zone for G105 steel, cleavage fracture is dominant in crack propagation zone for G135 steel.

Acknowledgment

This work was supported by the China National Science Foundation Project (No. 51574278, No. U1762211, No. 51801149), China postdoctoral Science Foundation (No. 2017M620448), Shaanxi Province Excellent Youth Foundation Project (No. 2018JC 030) and China State Oil & Gas Project (No. 2015QN014).

References

[1] Arunachalam, S. R., Dorman, S. E. G., Buckley, R. T., Conrad, N. A., Fawaz, S. A., 2018. Effect of electrical discharge machining on corrosion and corrosion fatigue behavior of aluminum alloys. Int. J. Fatigue 111, 44–53.

[2] Casarini, G., 1993. Problems related to safety and reliability of materials in environments polluted by hydrogen sulphide. Int. J. Pres. Ves. Pip. 55(2), 313–322.

[3] Chang, Y., Wu, X., Chen, G., Ye, J., Chen, B., Xu, L., Zhou, J., Yin, Z., Ren, K., 2018. Comprehensive risk assessment of deepwater drilling riser using fuzzy Petrinet model. Process. Saf. Environ. Prot. 117, 483–497.

[4] Chen, J., Diao, B., He, J., Pang, S., Guan, X., 2018. Equivalent surface defect model for fatigue life prediction of steel reinforcing bars with pitting corrosion. Int. J. Fatigue 110, 153–161.

[5] Cheng, A., Chen, N., 2017. Fatigue Crack growth modelling for pipeline carbon steels under gaseous hydrogen conditions. Int. J. Fatigue 96, 152–161.

[6] Cheng, A., Chen, N., 2018. An extended engineering critical assessment for corrosion fatigue of subsea pipeline steels. Eng. Fail. Anal. 84, 262–275.

[7] Donahue, J. R., Burns, J. T., 2016. Effect of chloride concentration on the corrosionfatigue crack behavior of an age hardenable martensitic stainless steel. Int. J. Fatigue 91, 79–99.

[8] Dong, C., Liu, Z., Li, G., Cheng, Y., 2009. Effects of hydrogen charging on the susceptibility of x100 pipeline steel to hydrogen induced cracking. Int. J. Hydrogen Energy 34(24), 9879–9884.

[9] Hansford, J. E., Lubinski, A., 1966. Cumulative fatigue damage of drill pipe in doglegs. J. Pet. Technol. 18(3), 359–363.

[10] Hu, P., Meng, Q., Hu, W., Shen, F., Zhan, Z., Sun, L., 2016. A continuum damage mechanics approach coupled with an improved pit evolution model for the corrosion fatigue of aluminum alloy. Corros. Sci. 113, 78–90.

[11] Huang, B., Yang, J., Zhang, H., Liu, G., Chen, Y., Li, J., 2017. Influence of H_2S corrosion on rotating bending fatigue properties of S135 drillpi pesteel. T. IndianI. Metals71(1), 1–9.

[12] IIman, M. N., 2014. Chromate in hibition of environmentally assisted fatigue crack propagation of aluminium

alloy AA2024-T3 in 3.5% NaClsolution. Int. J. Fatigue 62(7), 228-235.

[13] Jain, S., Thodal, R., Sridhar, N., 2013. Role of hydrogen in fatigue crack growth rate (FCGR) of X65 alloys: modeling study. Corrosion 13, 1-8.

[14] Kim, T. W., Baek, J., Lee, H. J., Choi, J. Y., 2013. Fatigue performance evaluation of SBS modified mastic asphalt mixtures. Constr. Build. Mater. 48(11), 908-916.

[15] Li, J., Wu, J., Wang, Z., Zhang, S., Wu, X., Huang, Y., Li, X., 2017. The effect of nanosized NbC precipitates on electrochemical corrosion behavior of high-strength low alloy steel in 3.5% NaCl solution. Int. J. Hydrogen Energy 42(34), 22175-22184.

[16] Lin, Y., Qi, X., Zhu, D., Zeng, D., Zhu, H., Deng, K., Shi, T., 2013. Failurean alysis and appropriate design of drill pipe upset transition area. Eng. Fail. Anal. 31(31), 255-267.

[17] Liu, W., Liu, Y., Chen, W., Shi, T., Singh, A., Lu, Q., 2015. Longitudinal crack failure analysis of box of S135 tool joint in ultra-deep well. Eng. Fail. Anal. 48, 283-296. Locke, J., 2016. Comparison of age hardenable Al alloy corrosion fatigue crack growth susceptibility and the effect of testing environment. Corrosion 72(7), 1941.

[18] Luo, S., Zhao, K., Wang, R., 2013. Fatigue cracks propagation behavior of S135 drill pipe steel at different stress ratios. Mater. Mech. Eng. 37(7), 72-76.

[19] Luo, S., Liu, M., Lin, X., 2018. Corrosion fatigue behavior of S135 high-strength drill pipe steel in a simulated marine environment. Mater. Corros., 1-10, http://dx.doi.org/10.1002/maco.201810542.

[20] May, M. E., Saintier, N., Palin Luc, T., Devos, O., Brucelle, O., 2018. Modelling of corrosion fatigue crack initiation on martensitic stainless steel in high cycle fatigue regime. Corros. Sci. 133, 397-405.

[21] Mayén, J., Abúndez, A., Pereyra, I., Colín, J., Blanco, A., Serna, S., 2017. Comparative analysis of the fatigue short crack growth on Al 6061-T6 alloy by the exponential crack growth equation and a proposed empirical model. Eng. Fract. Mech. 177, 203-217.

[22] Revie, R. W., Sastri, V. S., Hoey, G. R., Ramsingh, R. R., Mak, D. K., Shehata, M. T., 1993. Hydrogen-induced cracking of linepipe steels part 1 threshold hydrogen concentration and pH. Corrosion 49(1), 17-23.

[23] Sabelkin, V., Misak, H., Mall, S., 2016. Fatigue behavior of Zn-Ni and Cdcoated AISI 4340 steel with scribed damage in saltwater environment. Int. J. Fatigue 90, 158-165.

[24] Sangid, M. D., Maier, H. J., Sehitoglu, H., 2011. Aphysically based fatigue model for prediction of crack initiation from persistent slip striations in polycrystals. ActaMater. 59(1), 328-341.

[25] Slifka, A. J., Drexler, E. S., Nanninga, N. E., Levy, Y. S., DavidMcColskey, J., Amaro, R. L., Stevenson, A. E., 2014. Fatigue crack growth of two pipeline steels in a pressurized hydrogen environment. Corros. Sci. 78, 313-321.

[26] Távara, S. A., Chapetti, M. D., Otegui, J. L., Manfredi, C., 2001. Influence of nickel on the susceptibility to corrosion fatigue of duplex stainless steel welds. Int. J. Fatigue 23(7), 619-626.

[27] Thodal, R., Gui, F., Mueller, M., Gordon, R., 2012. Critical factors affecting the fatigue crack growth rate behavior of X65 steels in sourenvironments. Corrosion 5, 3808-3827.

[28] Thodla, R., Gui, F., Mueller, M. G., Gordon, R., 2012. Critical factors affecting the fatigue crack growth rate behavior of X65 steels in sourenvironments. NACE Corros. 6, 4652-4671.

[29] Tian, Y., Wang, J., Liu, Y., 2014. Modeling of corrosion fatigue crack growth of X70 pipeline steel and numericals simulation of crack propagation. Appl. Mech. Mater. 685, 43-47.

[30] Tsukada, K., Minakawa, K., McEvily, A. J., 1983. On the corrosion fatigue behavior of a modified SAE 4135 type steel in a H_2S environment. Metall. Mater. Trans. A14(8), 1737-1742.

[31] Veloz, M. A., Martínez, I. G., 2006. Effect of some pyridine derivatives on the corrosion behavior of carbon

steel in an environment like NACE TM0177. Corrosion 62(4), 283-292.

[32] Vosikovsky, O., Cooke, R. J., 1978. Ananalysis of crack extension by corrosion fatigue in a crude oil pipeline. Int. J. Pres. Ves. Pip. 6(2), 113-129.

[33] Wang, R., 1998. Fracture model of corrosion fatigue crack growth. J. Chin. Soc. Corros. Prot. 18(2), 88-94.

[34] Wang, R., 2001. 2001 Corrosion Fatigue of Metallic Materials [M]. Xi'an: Northwestern Polytechnical Universitypress.

[35] Wang, C., Xiong, J., Shenoi, R. A., Liu, M., Liu, J., 2016. Amodified model to depictcorrosion fatigue crack growth behavior for evaluating residual lives of aluminum alloys. Int. J. Fatigue 83, 280-287.

[36] Yang, X. J., Du, C. W., Wan, H. X., Liu, Z. Y., Li, X. G., 2018. Influence of sulfides on thepassivation behavior of titanium alloy TA2 in simulated seawater environments. Appl. Surf. Sci. 458, 198-209.

[37] Zamani, S. M., Hassanzadeh Tabrizi, S. A., Sharifi, H., 2016. Failure analysis of drillpipe: A review. Eng. Fail. Anal. 59, 605-623.

[38] Zeng, D., Li, H., Tian, G., Liu, F., Li, B., Yu, S., Ouyang, Z., Shi, T., 2018. Fatigue behavior of high-strength steel S135 under coupling multi-factor in complex environments. Mat. Sci. Eng. AStruct. 724, 385-402.

[39] Zhao, T., Liu, Z., Du, C., Dai, C., Li, X., Zhang, B., 2017. Corrosion fatigue crack initiation and initial propagation mechanism of E690 steel in simulated seawater. Mat. Sci. Eng. AStruct. 708, 181-192.

[40] Zhao, T., Liu, Z., Du, C., Sun, M., Li, X., 2018. Effects of cathodic polarizati on oncorrosion fatigue life of E690 steel in simulated seawater. Int. J. Fatigue 110, 105-114.

[41] Zheng, X. L., Lu, B. T., 1987. On a fatigue formula under stress cycling. Int. J. Fatigue 9(3), 169-174.

[42] Zheng, X. L., Wang, R., 1997. On the corrosion fatigue crack initiation model and expression of metallic notched elements. Eng. Fract. Mech. 57(6), 617-624.

[43] Zheng, X. L., Wang, R., 1999. Overload effects on corrosion fatigue crack initiation life and life prediction of aluminum notched elements under variable amplitude loading. Eng. Fract. Mech. 57(5), 557-572.

[44] Zheng, X., Lü, B., Jiang, H., 1995. Determination of probability distribution of fatigue strength and expressions of P-S-N curves. Eng. Fract. Mech. 50(4), 483-491.

本论文原发表于《Process Safety and Environmental Protection》2019 年第 124 卷。

Analysis of Corrosion Behavior on External Surface of 110S Tubing

Han Yan[1] Li Chengzheng[2] Zhang Huali[3] Li Yufei[3] Zhu Dajiang[3]

(1. CNPC Tubular Goods Research Institute, State Key Laboratory for Performance and Structure Safety of Petroleum Tubular Goods and Equipment Materials;
2. Oilfield Development Division, Changqing Oilfield Branch of Petrochina;
3. Petro China Southwest Oil and Gas Field Company Engineering Technology Research Institute)

Abstract: The failure analysis of 110S tubing during acidizing process was addressed. Results showed that serious pitting corrosion occurred on the outer wall of tubing, and there was no obvious pitting on the inner wall. The maximum pitting depth on the outer wall was 1019 μm. According to the results of simulation corrosion test, needle-shaped pitting appeared on the sample surface in the test without inhibitor, the maximum depth of pitting was 158 μm; and no pitting was found on the sample surface in the test within 1.5% TG501 inhibitor; the original pitting were deepened after spent acid test, and the sample with no pitting at the beginning also showed deep pitting corrosion after 96 hours spent acid test. It was indicated that the spent acid accelerated the development of pitting significantly. The external surface corrosion of the 110S tubing was caused by the chemical reaction between the high-concentration acidifying liquid and the outer wall of the tubing. There is a gap between the tubing and coupling threaded connection, which caused the acid solution entered into the thread position, and hence the severe corrosion of the thread and pin end of the tubing happened, the joint strength was continuously reduced with corrosion development till the tripping of the coupling, and then the lower string dropped. Some suggestions were proposed for avoiding or slowing down this kind of failure based on this study.

Keywords: 110S tubing; Acidizing; Spent acid; Pitting corrosion; Inhibitor; Threaded connection

1 Introduction

Acid-fracture technology is the most effective oil extraction process to increase production of carbonatite. It expands, extends, presses and communicates cracks through the chemical dissolution of acid and the hydraulic action when acid is squeezed into the formation. The oil and gas seepage channel with high

---

Corresponding author: Han Yan, hanyan@cnpc.com.cn.

circulation capacity can effectively stabilize the oil production and ensure the stability during the oil exploitation process [1]. Therefore, acid-fracture technology is a very important and necessary construction technique in the oil exploitation process [1-6]. However, the acid solution will cause various degree corrosion to casing, tubing and construction equipment during the acidification process. Although the corrosion of the wellbore can be alleviated by adding corrosion inhibitor to the acid solution, there is still much oil casing corrosion failure occurs caused by acidification [7-11]. The long residence time of fresh acid and residual acid in the string, the high temperature in the downhole, and the failure of the corrosion inhibitor to play an effective role are the main causes of casing failure [10,11]. In this paper, the corrosion failure of external wall 110S tubing after acid-fracture process was systematically analyzed. The reason of failure was verified by the simulation experiment, while the relevant preventive measures and suggestions were proposed, in order to prevent or slowdown the corrosion failures that may occur during oilfield acidification design and construction.

2 Sample preparation and experimental methods

The tubing was made of 110S steel with specifications of $\phi 73$ mm × 5.5 mm. The well was approximately 3472 m deep. It was found that the packer could not be sealed tightly during the acid-fracture construction, and the tubing string was found to be tripped at 1894 m after string lifting, which resulted in the 158 lower tubing falling into the well, and the fish was tubing collar. After salvage, there were 62 tubing with seriously corroded on outer wall, included 29 tubing which not-falling into the well and 33 tubing which falling into the well. Some of the tubing also had obvious pitting corrosion on pin thread and sealing face. The severely corroded tubing was located near the tripping position of tubing string, and there was no significant corrosion was found elsewhere.

To study the cause of the external corrosion, the following series of tests were conducted: (1) visual examination of corrosion characteristics; (2) chemical composition analysis; (3) metallographic structural characterization; (4) scanning electron microscopy (SEM) observation of corrosion morphology and energy-dispersive spectrometry (EDS) analysis of corrosion products; (5) simulated condition corrosion test.

In order to clarify the corrosion resistance of the 110S tubing under acidification solution, simulated corrosion tests were carried out under three conditions. The specific test conditions are shown in Table 1. Coupon specimens with the length 40 mm, width 10 mm and thickness 3 mm, were used.

The acidification solution and duration time in simulated test were determined based on the on-site acid-fracture process. The test temperature was similar with the temperature at tripping position. Since the inhibitor used in the field cannot be obtained, 1.5% TG201 inhibitor was used in fresh acid with inhibitor test (Test No. 2 in Table 1). The spent acid was fetched from site.

Table 1 Test condition under simulated environmental

Test No.	Temperature (℃)	Solution	Duration time (h)
1	80	20% HCl	4
2	80	20% HCl + inhibitor (1.5% TG201)	4
3*	80	Spent acid	96

* Note: The specimens used in test No. 3 are the specimens used in the test No. 1 and test No. 2.

3 Results and discussion

3.1 Visual examination

The outer diameter of the corroded tubing is 73 mm, the wall thickness is 5~6 mm, and the outer wall presents a dense pitting morphology. The corrosion products are rust and black colors, as shown in Fig. 1(a). The thread of the tubing is also seriously corroded. There are many corrosion pits on the outer wall, inner wall and end face of the thread, as shown in Fig. 1(b) and Fig. 1(c). The thread was damaged too severely to connect with coupling.

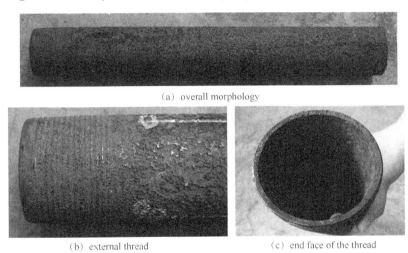

(a) overall morphology

(b) external thread   (c) end face of the thread

Fig. 1 Corrosion morphology of the tubing

Low magnification observation in Fig. 2 shows that the corrosion pits on the outer wall of the tubing are mostly circular shaped, and two or more corrosion pits are connected and superimposed to aggravate corrosion. Sampling was carried out at the severe pitting corrosion area. The maximum depth of the corrosion pit on samples fetched from outer surface and thread end were 1019 μm and 936 μm separately, as shown in Fig. 2.

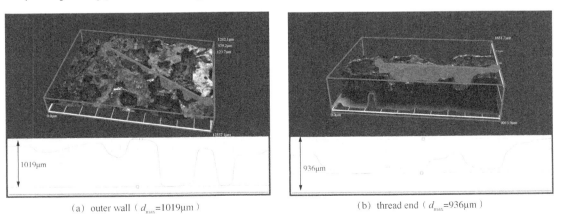

(a) outer wall (d_{max}=1019μm)   (b) thread end (d_{max}=936μm)

Fig. 2 Characteristic and maximum depth of pitting

3.2 Chemical composition analysis

The chemical composition of the 110S tubing is shown in Table 2. The elements content accorded with the requirement of SY/T 6857.1—2012 standard [12].

Table 2 Chemical composition of the tubing (wt. %)

Elements	C	Si	Mn	P	S	Cr	Mo	Ni	Nb	V	Ti	Cu
110S tubing	0.26	0.22	0.48	0.0080	0.0030	1.01	0.45	0.014	0.0012	0.0052	0.003	0.024
SY/T 6857.1—2012	≤0.35	≤0.35	≤1.00	≤0.010	≤0.005	0.40~1.20	0.15~1.00	≤0.15	≤0.040	≤0.050	≤0.040	≤0.15

3.3 Metallographic structural characterization

The metallographic structural of the 110S tubing is shown in Fig. 3, and the microstructure is tempered sorbite. No oversized nonmetallic inclusions were found, and the grain size of tubing is ASTM 9.0 grade, which is smaller than SY/T 6857.1—2012 requirement: ASTM 7.0 grade.

Fig. 3 Microstructure of the tubing

The characteristics of localized corrosion are shown in Fig. 4. The typical corrosion pits on the tubing surface are in round or elliptical, which are all defined as the primary pits, as shown in Fig. 4 (a). The gray corrosion products, as marked by arrow in Fig. 4 (a), attach to the inside of pits. In addition, some new pits continue to stack along the depth direction of the bottoms of primary pits, as shown in Fig. 4(b).

(a) microstructure and corrosion products in the pitting

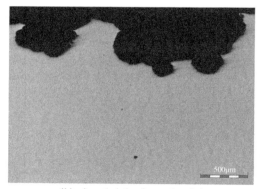

(b) characteristic of corrosion pitting

Fig. 4 Morphology and the microstructure of pitting on outer-surface of the tubing

3.4 SEM observation and EDS analysis

The micro-morphology and energy spectrum analysis specimen were taken from the pipe body and the thread end. It was found that various degree of corrosion occurred in all parts of the tubing. Among them, pitting corrosion mainly occurred on the outer wall and the thread area, the inner wall of the tubing undergoes slight uniform corrosion, as shown in Fig. 5. The outer wall of the tubing was covered with circular corrosion pitting, and the corrosion product film exhibits loose and uneven characteristic, which has no protection against further corrosion.

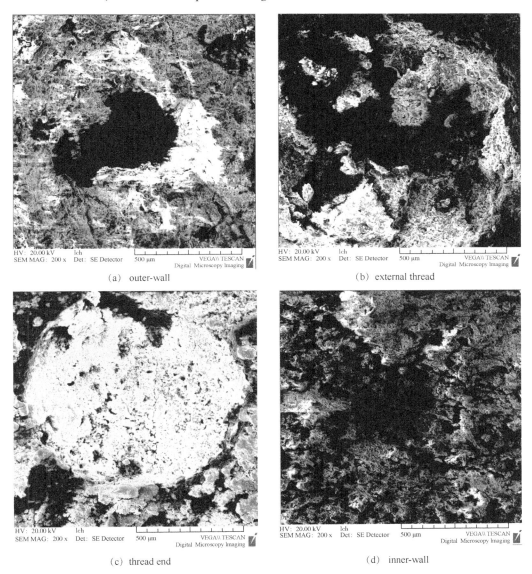

(a) outer-wall   (b) external thread

(c) thread end   (d) inner-wall

Fig. 5　Micro-morphology of corrosion

The EDS results show that the corrosion products mainly contain Fe, C, O and S elements, and the Fe and Cr content in corrosion products in outer wall is significantly higher than that in inner wall, while the content of C and O elements in corrosion products in inner wall is higher than that in outer wall, as shown in Fig. 6. This suggest that the outer wall corrosion pit exhibits the composition of the metal matrix, while the inner wall corrosion product is mainly CO_2 corrosion product.

· 297 ·

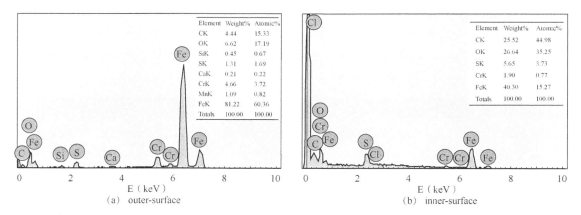

Fig. 6 Energy spectrum of the corrosion products in the pitting

3.5 Simulated condition corrosion test

Photographs of the specimens after simulated test are shown in Fig. 7, and the micro-morphology was shown in Fig. 8. It was found that the surface of the specimens were covered with pin-like pitting after 20% HCl fresh acid test, and no obvious pitting was observed on the surface of the specimens, which the corrosion inhibitor was added. After 96 hours spent acid immersion test, the pitting density of specimens in test No. 1 significantly increased, and the specimens in test No. 2, which had no obvious pitting and pitting corrosion.

Ten pitting pits were selected for depth measurement randomly, and the result was shown in Table 3. After the spent acid test, the original pitting pits were deepened, and the specimens with no pitting pits at the beginning also showed deep pitting corrosion. After 4 hours' fresh acid test without inhibitor, the maximum depth of the pitting is 158μm, and the maximum depth of the pitting after 96 hours' spent acid experiment are 226μm and 178.2 μm, respectively. The pitting morphology of the specimens is consistent with the pitting morphology of the failure tubing. It is distributed in a plurality of circular pit-like shapes, and has traces of superposition and development in the depth direction. The profiles of the pitting were shown in Fig. 9 to Fig. 10.

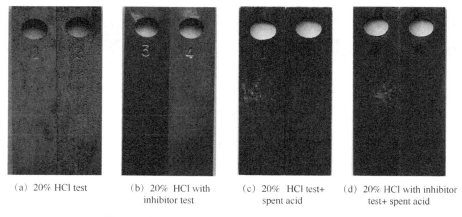

(a) 20% HCl test (b) 20% HCl with inhibitor test (c) 20% HCl test+ spent acid (d) 20% HCl with inhibitor test+ spent acid

Fig. 7 Photographs of specimens after simulated test

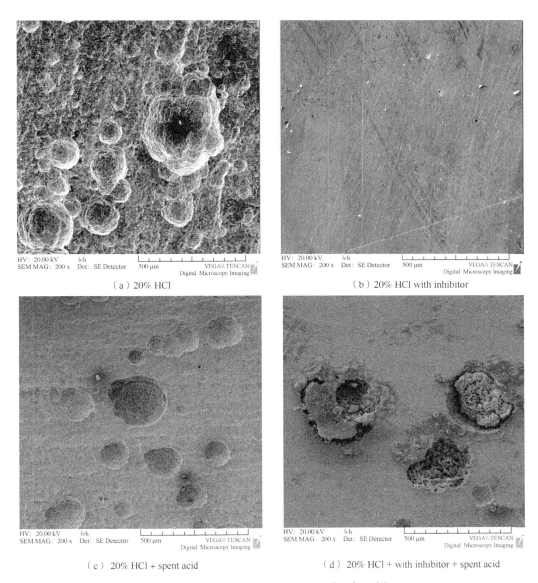

(a) 20% HCl (b) 20% HCl with inhibitor
(c) 20% HCl + spent acid (d) 20% HCl + with inhibitor + spent acid

Fig. 8 Micro-morphology of sample after simulated acidification test

Table 3 Results of the pitting depth after simulated corrosion test (μm)

Test sample		Points										Average
		1	2	3	4	5	6	7	8	9	10	
Test No. 1	1-1#	147.7	132.8	139	107	132	134	151	137	151	158	138.95
	1-2#	122	128.8	140	125	122	128	131	110	126	131	126.38
Test No. 3	3-1#	226	93	93	142	224	118	153	159	124	124	145.6
	3-2#	171.6	127.1	115	125.4	141	157	135	90	153	113	132.81
	3-3#	139.4	57	114.2	149	58	161.8	49	93	54	116	99.14
	3-4#	143.4	57	131	85	73	178.2	74	156.4	63	90	105.1

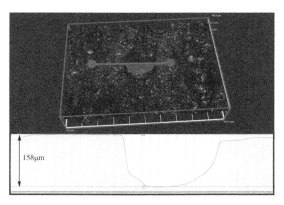

Fig. 9　The profile of the pitting after 20% HCl ($d_{max} = 158\mu m$)

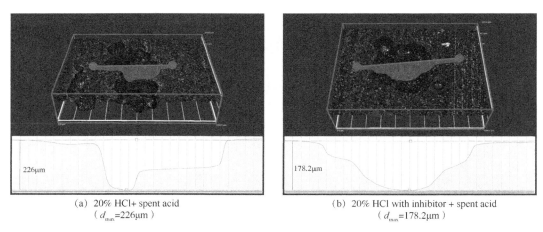

(a) 20% HCl+ spent acid
(d_{max}=226μm)

(b) 20% HCl with inhibitor + spent acid
(d_{max}=178.2μm)

Fig. 10　The profile of the pitting after test No. 3

According to the construction record, theacidizing process as follows: Firstly, acidification of the 3414~3416m section was carried out for 4 hours. Secondly, the packer is set after pressure test of the column to 68MPa, the sealing pressure of the packer was 19 MPa. 204.8m³ acid was injected into the tubing with 61MPa average pressure, and then 27m³ common acid was injected with 54MPa average pressure, the bursting pressure is 63.5MPa, the pump pressure is 36MPa when stop, the highest applied pressure is 64MPa, the balance pressure is between 14MPa to 26MPa. Subsequently, the fracturing pump car casts the ball and to raise the pressure of column, but it was found that the packer cannot be sealed tightly and the upper section acidified fracturing cannot be implemented.

Therefore, the process of spray, drain, kill well, and wash off with clean water were executed until the water quality of the import and export is consistent, and then salvage. The upper 198 tubing, have been subjected to acidification for about 4 days, and the lower 158 tubing have been salvaged out 9 days after acidification. From the observation of more than 300 tubing, it was found that only 62 tubing corroded seriously on their outer surface, and the corrosion range is about 300m above and below the tripping position. Some of the threads and sealing surfaces were corroded heavily, and no serious corrosion was found on other tubing. It can be seen from the corrosion morphology of the corroded tubing that the corrosion pits are deep circular hole, and there are several small corrosion pits at the bottom of the large corrosion pit, which are superimposed pitting, which is consistent with the acid corrosion characteristics [9,10], and the pits has the same shape with the corrosion pit after the simulated corrosion test in the laboratory. The pitting corrosion of the tubing is caused by the corrosion of high concentration acid.

Based on the serious corrosion at the external thread of the tubing, it can be judged thatat field end of the tubing, there exist a gap between the buckle and the coupling. The injection of acid with high pressure caused the acid solution penetrate into the thread connection gap, then severe localized corrosion occurred at this area due to the poor fluidity at the threaded joints. As the thread corrosion strengthen, the acid solution entered into the casing annulus and caused corrosion on the outer wall of the tubing near the leakage location. At last, the coupling tripping as the continuously deteriorate of the sealing performance of threaded, and the lower tubing string dropped eventually.

Hydrochloric acid is an acid commonly used in acid-fracture processes, but it will cause severe corrosion to the tubing string. Therefore, a certain corrosion inhibitor must be added in acid-fracture processes to reduce the corrosion. The rate of uniform corrosion caused by acid can reach extreme high (>100mm/a) in absence of corrosion inhibitors or inhibitor do not play an effective role, and it will increase exponentially with increasing temperature and acid concentration[13]. SHI Zhi-ying et al.[14] shown that, the corrosion inhibitor in the acid is absorbed by the rock clay and minerals as the acid squeezes into the formation, so the corrosive of spent acid from the formation will increase higher. Considered the long processing cycle of spent acid (3~7 days), it will bring more severe corrosion than fresh acid with inhibitor. FU An-qing's studies[15] shown that, P110 steel showed different pitting degrees after the acid and spent acid tests, and the pitting density of the sample in the spent acid increased significantly. When steel material is corroded in the electrolyte solution, the anode process is the dissolution of Fe as shown in Eq.(1), and the cathode process is different due to the change of the medium. The hydrogen depolarization reaction of the cathode is prone to occur in the reflux liquid with pH value of 1.5, as shown in the Eq.(2). The reaction rate increases with the concentration increase of the H^+. The corrosion rate of P110 tubing steel in the reflux fluid can reach 47mm/a [16].

Anode process $\qquad Fe \longrightarrow Fe^{2+} + 2e \qquad$ (1)

Cathode process $\qquad 2H^+ + 2e \longrightarrow H_2 \qquad$ (2)

The corroded tubing undergoes 4 hours acidification time for fresh acid with inhibitor and 4 days in spent acid before lifted from the well, it can be inferred that the spent acid is more likely to cause serious corrosion on the outer surface of the 110S tubing.

4 Conclusions and suggestions

The chemical composition and microstructure of the 110S tubing meet the requirements of SY/T 6857.1—2012 standard. According to the results of macroscopic and microscopic morphology observation, the pitting corrosion on the outer wall is circular and shows the characteristics of superimposed corrosion, which is consistent with the acid corrosion characteristics. Combined with the results of simulated corrosion test, the pitting corrosion of the tubing is caused by the corrosion of high concentration acid. The pitting morphology of specimens in simulated test is consistent with the pitting morphology of the failure tubing. The depth of pitting is deepened after the spent acid immerse.

There is a gap between the tubing and coupling threaded connection, which caused the acid medium entered into the thread position, it brings severe corrosion of the thread and pin end of the tubing, and the joint strength continuously reduced with the development of corrosion till the tripping of the coupling, then the lower string dropped.

In order to avoid this kind of failure from happening, construction management on-site should be strengthened, and to ensure the reasonable torque we mustcarry out the torque according to the requirements of the technical specification strictly. Meanwhile, it is necessary to verify the effectiveness of the corrosion inhibitor before use.

Acknowledgement

This work was supported by the National Key R&D Program of China (2017YFB0304905), Key project of shanxi key research and development plan (2018ZDXM-GY-171) and Major science and technology project of CNPC (2016E-0606).

References

[1] Matjaž Finšgar, Jennifer Jackson. Application of corrosion inhibitors for steels in acidic media for the oil and gas industry: A review [J]. Corrosion Science, 2014, 86: 17 – 41.

[2] Luo Jing-qi. Acidizing fundamentals [M]. Beijing: petroleum industry press, 1983.

[3] M. U. Shafiq, H. B. Mahmud, MOHSEN Ghasemi. Integrated mineral analysis of sandstone and dolomite formations using different chelating agents during matrix acidizing[J]. Petroleum, 2018, in press: 1–10.

[4] A. Q. Fu, Y. R. Feng, R. Cai, et al. Downhole corrosion behavior of Ni-W coated carbon steel in spent acid & formation water and its application in full-scale tubing[J]. Engineering Failure Analysis, 2016, 66: 566–576.

[5] R. Abdollahi, S. R. Shadizadeh. Effect of acid additives on anticorrosive property of henna in regular mud acid [J]. Scientia Iranica Transactions C: Chemistry and Chemical Engineering, 2012, 19: 1665–1671.

[6] Lv Zhi-feng, Zhan Feng-tao, Wang Xiao-na, et al. Synthesis of ethyl chloroacetate quinolinium salt derivatives and evaluation of its acidification inhibition corrosion properties[J]. Materials protection, 2018, 51 (5): 25–29.

[7] X. W. Lei, Y. R. Feng, A. Q. Fu, et al. Investigation of stress corrosion cracking behavior of super 13Cr tubing by full-scale tubular goods corrosion test system[J]. Engineering Failure Analysis, 2015, 50: 62–70.

[8] StefanBachu, Theresa L. Watson. Review of failures for wells used for CO_2 and acid gas injection in Alberta, Canada[J]. Energy Procedia, 2009, 1(1): 3531–3537.

[9] Xie Junfeng, Fu An-qing, Qin Hong-de, et al. Influence of Surface Imperfection on Corrosion Behavior of 13Cr Tubing in Gas Well Acidizing Process [J]. Surface technology, 2018, 47(6): 51–56.

[10] Li Yan, Xie Jun-feng, Chang Ze-liang, et al. Reasons for Tube Corrosion in a Gas Well Due to Acidizing in Tarim Oilfield [J]. Corrosion and protection, 2016, 37(10): 861–864.

[11] Zhang Shuang-shuang, Zhao Guo-xian, Lv Xiang-hong. Corrosion Behavior of TP140 in Simulated Oilfield Fresh Acid and Spent Acid Environments [J]. Corrosion and protection, 2014, 35(1): 28–32.

[12] SY/T 6857.1-2012, Petroleum and natural gas Industries – OCTG used for special environment–Part 1: Recommended practice on selection of casing and tubing of carbon and low alloy steels for use in sour service

[S]. Beijing：Beijing：petroleum industry press，2012.

[13] E. Barmatov，J. Geddes，T. Hughes，et al. Research on corrosion inhibitors for acid stimulation[C]，NACE，2012，C2012-0001573.

[14] Shi Zhi-ying，Tian Zhen-yu，Chen Li. The study of spent acids' corrosiveness and its control[J]. Oil drilling technology，1999，6(3)：52-53.

[15] Fu An-qing，Geng Li-yuan，Li Guang，et al. Corrosionfailure analysis of an oil tube used in a western oilfield [J]. Corrosion and protection，2013，34(7)：645-648.

[16] Liu Zongzhao，Wang Yu，Wang Jun. Research on corrosion behavior of release sewage of acidizing in the exploration of LD10-1 oilfield[J]. Total corrosion control，2013，27(6)：35-39.

本论文原发表于《Materials Science Forum》2020年第993卷。

The Microstructure Evolution of Dual-Phase Ferrite-Bainite X70 Pipeline Steel with Plastic Deformation at Different Strain Rates

Ji Lingkang[1, 2] Xu Tong[3] Wang Haitao[1, 2] Zhang Jiming[1, 2]
Tong Mingxin[3] Zhu Ruihua[3] Zhou Genshu[3]

(1. Tubular Goods Research Institute of CNPC; 2. State Key Laboratory of Performance and Structural Safety for Petroleum Tubular Goods and Equipment Materials; 3. State Key Laboratory for Mechanical Behavior of Materials, Xi'an Jiaotong University)

Abstract: Tensile properties of dual-phase ferrite-bainite X70 pipeline steel have been investigated at room temperature under the strain rates of $2.5\times10^{-5}\mathrm{s}^{-1}$、$1.25\times10^{-4}\mathrm{s}^{-1}$、$2.5\times10^{-3}\mathrm{s}^{-1}$ and $1.25\times10^{-2}\mathrm{s}^{-1}$. The microstructures at different amount of deformation were examined by using scanning and transmission electron microscopy. Generally, the ductility of typical body-centered cubic (bcc) steels is reduced when its stain rate increases. However, we observed a different ductility dependence on strain rates in the dual-phase X70 pipeline steel. The uniform elongation (UEL%) and elongation to fracture (EL%) at the strain rate of $2.5\times10^{-3}\mathrm{s}^{-1}$ increase about 54% and 74% respectively compared to those at 2.5×10^{-5} s^{-1}. The UEL% and EL% reach to their maximum at the strain rate of 2.5×10^{-3} s^{-1}. Whether or not the ductility can be enhanced with increasing strain rates depends on the competition between the homogenization of plastic deformation among the microconsituents (ultra-fine ferrite grains, relatively coarse ferrite grains as well as bainite) and the progress of cracks formed as a consequence of localized inconsistent plastic deformation. This phenomenon was explained by the observed grain structures and dislocation configurations.

Keywords: Microstructure; Deformation; Uniform elongation; Strain rate, Pipeline; Steel

1 Introduction

Currently there is a growing demand for higher grade pipeline steels in petroleum industry. Increasing the strength of the steels enables a significant reduction in the pipe wall thickness or the pipe to transport crude oil and gas at a higher operating pressure (Shanmugam et

---

Corresponding author: Xu Tong, Tel.: +86 2982668614; fax: +86 2982663453; E-mail address: xutong@mail.xjtu.edu.cn.

al., 2008). At the same time, the ductility and toughness of the steels have to be kept high enough to prevent the pipe from local buckling and girth weld fracture, especially in harsh environment such as seismic regions where ground movements can be expected to impose large and fast strains. Therefore high strength, high toughness and formability are important requirements of the pipeline steels (Shanmugam et al., 2008).

Rashid (1977) has reported the dual-phase steels with optimized microstructures were selected for modern pipelines due to their fine combinations of strength and ductility. Nakagawa et al. (1985) described the dual-phase steel as a "composite" which might contain three or more phases despite the generic name "dual-phase". Generally, the composite microstructure of the "dual-phase" steels contains soft ductile ferrite matrix and the strengthening phase(s). Nakagawa et al. (1985) have reported this composite microstructure exhibited typical mechanical properties including low yield strength, continuous yielding, high initial work hardening rate, high tensile strength, good ductility and formability, among which the uniform elongation in uniaxial tension tests was an important parameter. The onset of necking in uniaxial tension tests is suppressed to yield high uniform elongation value (Nakagawa et al. 1985).

API X70 steel has been used worldwide for pipelines. Both microstructure and mechanical properties have been well characterized. Several attempts have been made by Nakagawa et al. (1985), Okatsu et al. (2005), and Ishikawa et al. (2008) to relate the mechanical properties with the microstructures. However, the detailed microstructures related to deformability and ductility in terms of the uniform elongation and post-necking elongation at different strain rates have not been clarified yet. Especially, few researches have been concerned with the deformation behavior of the dual-phase pipeline steels. Moreover, strain-based design have been developed which is a new concept that line pipes require higher resistance against larger compressive and tensile strains (Ishikawa et al. 2008). In the present work, the tensile properties of dual-phase X70 containing polygonal ferrite and bainite have been investigated at different amount of deformation and strain rates. The corresponding microstructure changes have been examined with scanning electron microscope (SEM) and transmission electron microscope (TEM). These results would provide industry with an applicable approach to the improvement of the reliability and economic usefulness of the pipeline steels.

2 Experimental procedures

The dual-phase ferrite-bainite X70 pipeline steel (the X70 steel) was supplied by Tubular Goods Research Institute of CNPC, China. The alloy was hot rolled to a minimum of 17.5mm gage and subsequently fabricated to 150mm diameter pipes. The Chemical composition is listed in Table 1.

Cylindrical dog-bone specimens with the gage length of 40 mm and the diameter of 8 mm according to ASTM E8 specification were machined from one of the pipes. The cylinder axis of the specimens is parallel to the longitudinal direction of the pipe. Standard tensile tests were conducted at room temperature at a constant crosshead speed of 0.001mm/s, 0.005mm/s, 0.1mm/s and 0.5mm/s respectively, employing INSTRON 1195 computerized tensile testing system. The specimens were drawn up to an engineering strain of 5%、10%、15%、20% and to fracture respectively.

Table 1 Chemical composition of the experimental X70 steel

element	C	Si	Mn	P	S	Cr	Mo	Ni	Nb	V
mass fraction(%)	0.046	0.19	1.5	0.007	0.0018	0.14	0.16	0.02	0.051	0.005
element	Ti		Cu		B		Al		N	
mass fraction(%)	0.01		0.032		0.0003		0.026		0.004	

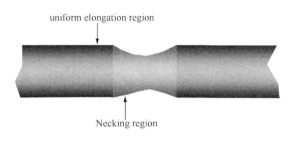

Fig. 1 Schematic of the observed regions in dog-bone specimen

The samples for SEM and TEM were cut from the un‑deformed and deformed specimens, in parallel tension direction at the middle‑thickness portion. The observed areas are schematically shown in Figure 1 with the labels of uniform elongation region and necked region. The samples for SEM were grinded, mechanically polished, and etched in 4% of nital solution. Hitachi S‑2700 SEM was used for the examination of the microstructures.

Thin foil samples for TEM were prepared by cutting 0.3mm thin wafers from the specimens by using electro-spark method and ground to about 0.08~0.1mm in thickness. Three-millimeter discs were punched from the wafers and electro‑polished using a solution of 5% perchloric acid in ethanol. The polishing voltage is 50V and current is 20 mA. The examination of the microstructure is performed with a JEM2000FX transmission electron microscope equipped with a double‑titled specimen holder.

3 Results and Discussion

3.1 Mechanical properties

Typical example of the engineering stress–strain curves at different strain rates are shown in Figure 2. At least five tensile specimens were tested at every strain rate to increase the statistics for the uniform elongation and elongation at fracture measurements. The curves show the typical round-house type. Unlike the conventional relationship between flow stress and strain rate, the yield strength (YS) and ultimate strengths (UTS) stay at the same level as the strain rate increases from 2.5×10^{-5} to $1.25 \times 10^{-2} s^{-1}$. However, the uniform elongation (UEL%) as well as the elongation to facture (EL%) is enhanced at high strain rates, especially at the strain rate of 1.25×10^{-4} and $2.5 \times 10^{-3} s^{-1}$, as seen in Figure 2. Zhang et al. (2010) have reported similar results previously for the X80 pipeline steel.

The tensile properties of the X70 steel are summarized in Table 2. The steel clearly exhibits a low yield strength (YS) and YS/UTS ratio (<0.8), but moderate high ultimate tensile strengths, sufficient uniform elongation and elongation to fracture. All the properties are representative of the high-deformability pipeline steel. The results showed that the YS, UTS and YS/UTS ratio exhibit

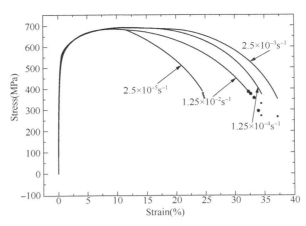

Fig. 2 Engineering stress-strain curves of the X70 steel under different strain rate

essentially the same level as the strain rate increases from 2.5×10^{-5} to $1.25 \times 10^{-2} s^{-1}$. However, the UEL% and the EL% increase about 54% and 74% respectively as the strain rate increases from 2.5×10^{-5} to $2.5 \times 10^{-3} s^{-1}$, and then decrease at the strain rate of $1.25 \times 10^{-2} s^{-1}$. Nevertheless, the UEL% and the EL% at $1.25 \times 10^{-2} s^{-1}$ are still higher than those at the strain rate of $2.5 \times 10^{-5} s^{-1}$. This result is reproducible from the repeated experiments.

Table 2 Mechanical properties of the experimental X70 pipeline steel

Strain rate	Yield strength YS(MPa)	Tensile strength UTS(MPa)	YS/UTS ratio	Elongation EL(%)	Uniform elongation UEL(%)
2.5×10^{-5}	487±3	678±4	0.71	21.1±2.0	8.2±0.7
1.25×10^{-5}	492±2	683±5	0.72	31.9±1.0	13.9±0.4
2.5×10^{-3}	491±2	686±3	0.71	32.4±1.3	14.3±0.5
1.25×10^{-2}	490±4	681±4	0.72	23.7±2.1	9.0±1.0

Yield strength: flow stress at 0.2% plastic strain; Uniform elongation was designated as the elongation at which the load began to decrease

3.2 Microstructure characterization of the un-deformed specimen

Polygonal ferritic/bainitic "dual-phase" microstructures are evident in the un-deformed specimen, as shown in the SEM micrographs in Figure 3. Ferrite grains are mainly polygonal and their sizes are not homogeneous, ranging from about $2 \sim 5 \mu m$ to dozens of micrometer (over $20 \mu m$). Anyway, the ferrite phase is relatively fine enough and with an adequate volume fraction that higher uniform elongation and lower YS/UTS can be obtained in the steel, which was reported previously by researchers, such as Ishikawa et al. (2008).

Detailed characteristics of ferrites were examined under TEM, as shown in Figure 4. A large number of fine ferrite grains with the size of several micrometers and polygonal shape were observed. The sizes of the polygonal ferrite grains are rather inhomogeneous and their differences can reach up to ten times. The distribution of the grain sizes can be described as a bimodal distribution where the fine grains are embedded inside the coarse ferrite matrix. The fine grains impart high strength, as expected in the Hall-Petch relationship. Meanwhile, Andrade et al. (1994), Asgari

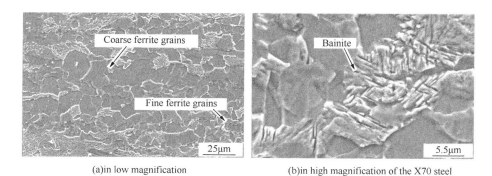

(a) in low magnification (b) in high magnification of the X70 steel

Fig. 3　The SEM image

et al. (1997), Gao et al. (1999), Youngdahl et al. (2001) and Wang et al. (2002) have reported that this inhomogeneous microstructure stabilized the tensile deformation, leading to a high uniform elongation and elongation to fracture. The improvement of the ductility with increasing strain rates in terms of uniform elongation (UEL%) and the elongation (EL%) to fracture can be mainly ascribed to the ferrite phase with the bimodal grain size distribution in the experimental X70 steel.

(a) fine ferrite grains in the microstructure　　(b) irregular tangled dislocations and fine scale precipitation on dislocations in ferrite matrix

Fig. 4　Bright field transmission electron micrographs showing

Irregular tangled dislocations with no preferred orientation were observed in ferrite phase. In some ferrite grains these tangles had formed incipient cell structures. And fine cementite or alloy carbide precipitates were scattered among the dislocations in the ferrite matrix, as shown in Figure 4 (b). The dispersion of the precipitates may contribute to the strengthening of the ferrite matrix or the strength of the steel without lessening its ductility and toughness, which was reported previously by Maloney et al. (2005) and Kimura et al. (2006).

Besides the ferrite grains, the so called "bainite" was observed under TEM, which had several phases or microconstituents, such as M/A constituent in ferrite grain boundary、granular bainite with different orientations、ferrite bainite、low bainite/degenerated pearlite、sublaths in bainite laths and upper bainite/ degenerated pearlite. The sublaths in bainite laths are shown in Figure 5(a) and the upper bainite/ degenerated pearlite is shown in Figure 5(b). The body-centered cubic (bcc) structure was identified with micro-diffraction in the strips between bainite laths [Figure 5(a), (b)]. There also should be some martensite and retained austenite (M/

A). The transformation of retained austenite to martensite may introduce a high density of mobile dislocations in the surrounding ferrite matrix, which allows the steels to be deformed at low stresses with continuous yielding.

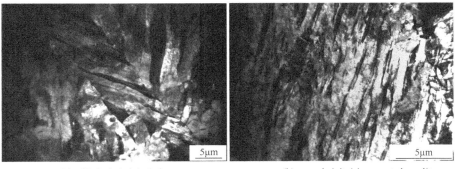

(a) sublaths in bainite laths   (b) upper bainite/ degenerated pearlite

Fig. 5 Bright field TEM micrographs

The microstructures described above are complicatedly mixed and it is unnecessary to distinguish them from each other. Meanwhile, the mixture of above phases has a very characteristic appearance (the light etching microconstituent labelled as "bainite" in Figure 3(b)). So it may be simplified and classified into one microconstituent as "bainite". The microstructure of the X70 steel can be described as the ferrite-bainite dual-phase microstructure (Ishikawa et al. 2008), although the term dual-phase is a misnomer that is the prevalent usage. The sizes of the bainite colony are also extremely inhomogenous.

3.3 Microstructure characterization of the deformed ferrite-bainite X70 samples

The microstructure of the deformed specimens was examined on the longitudinal-cross section of the specimens. A large number of SEM and TEM observations were carried out on the specimens under different strained and strain rate conditions. The representative changes in the grain structures at relatively lower strain rates were present in Figure 6. When the specimens were elongated to 8.2% at $2.5 \times 10^{-5} s^{-1}$, many coarse grains are elongated in the direction of the stress applied. Nevertheless most of the fine grains, especially the fine grains assembled together, still keep their polygonal shapes, as indicated by arrows in Figure 6(a). This phenomenon is more obvious when the specimens were drawn to 13.9% at $1.25 \times 10^{-4} s^{-1}$. Most of coarse grains are elongated and some of fine grains are elongated or distorted. However, many fine grains that still keep polygonal shapes can be found in Figure 6(b). The aspect ratios of the deformed coarse grains marked as a, b and c in Figure 6(b) have reached 2.9, 3.2, 3.4, respectively. Figure 6(c) is the micrograph with high magnification of the specimen drawn to 13.9% at $1.25 \times 10^{-4} s^{-1}$ and shows more clearly the undeformed assembled polygonal fine grains.

Figure 7 with the indicated strain rates and the amount of plastic strain shows the representative changes in the grain structures as a result of deformation at relatively higher strain rates. When the specimen was elongated to the plastic strain of 14.3% at the strain rate of $2.5 \times 10^{-3} s^{-1}$, as well as elongated to 9.0% at $1.25 \times 10^{-2} s^{-1}$, almost all the fine and coarse polygonal grains were deformed and elongated. In contrast to those at the lower strain rates in Figure 6, it is obvious that the fine

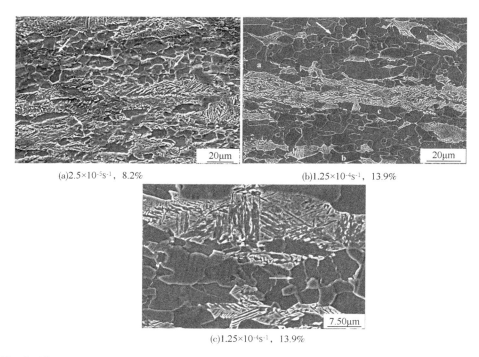

Fig. 6 The representative SEM micrograph for the samples deformed at relatively lower strain rates

polygonal grains begin to undertake deformation at more small plastic strains at the relatively higher strain rate. This result is more explicit as shown in Figure 7(b). Even the fine grains assembled together were elongated and distorted when the specimen was drawn to 9% at $1.25 \times 10^{-2} s^{-1}$. Most of the fine grains were severely deformed and the aspect ratios of some deformed fine grains have reached more than 4, as shown in Figure 7(b). However, the aspect ratios of the coarse grains in Figure 7(b) are relatively low. With increasing strain rates, the fine ferrite grains begin to be deformed and undertake deformation at relative more small plastic strains.

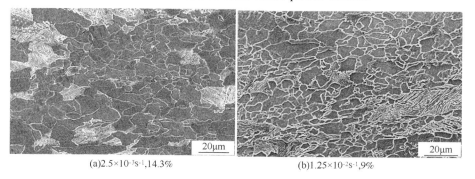

Fig. 7 The representative SEM micrograph of the samples deformed at relatively higher strain rates

At the beginning, the plastic deformation of the bainite micro-constituents could be neglected. The plastic deformation can be ascribed mainly to the soft ferrite phase with the bimodal distribution. Plastic deformation starts first in the coarse grains with low yield strength. With increasing strain hardening as the deformation of the coarse grains proceeds, the stress concentration at the boundaries between the coarse and neighboring fine ferrite grains increases. The fine ferrite grains begin to deform once the stress concentration is beyond their critical resolved shear stress. The contribution to the total deformation can be described approximately as following:

$$\varepsilon_t = \varepsilon_c \times f + \varepsilon_f \times (1-f) \quad (1)$$

ε_t is the total plastic strain, ε_c and ε_f are the strain in coarse and fine ferrite grains respectively, while f is the volume fraction of coarse ferrite grain. The corresponding strain rate is

$$\dot{\varphi}_t = \dot{\varphi}_c \times f + \dot{\varphi}_f \times (1-f) \quad (2)$$

$\dot{\varphi}_t$ is the total strain rate, $\dot{\varphi}_c$ and $\dot{\varphi}_f$ are the strain rate in coarse and fine ferrite grains respectively. According to the theory of the intrinsic yield phenomena proposed by Johmston and Gilman in 1959 (Cahn et al. 1996), the kinematic equation of shear strain rate is:

$$\dot{\gamma} \propto b\,\bar{v}\rho_M \quad (3)$$

Where

$$\bar{v} = b\left(\frac{\tau}{\tau_0}\right)^m \quad (4)$$

b is Burgers vector, $\bar{v}$ is the average velocity of dislocation movement, ρ_M is the mobile dislocation density, the exponent m is typically in the range of 40, τ is the effective shear stress on dislocations and τ_0 is the stress for unit speed of dislocation motion. Assuming the shear strain rate $\dot{\gamma}$ is proportional to the applied strain rate $\dot{\varphi}_t$. In the initiation of plastic deformation, as $\dot{\varphi}_t$ increases, $\dot{\gamma}$ is increased. A higher ρ_M or τ is needed. From the TEM analysis above, we can confirm that enough mobile dislocations exist in the ferrite grains. Besides, ferrite phase itself has a high plastic deformation capacity. It is reasonable that the yield strengths (YS) exhibit rate-independent behavior and keep essentially at the same level at the different experimental strain rates, as shown in Table 2.

In the early plastic deformation stage, each infinitesimal plastic deformation can be regarded as an intrinsic yielding of materials. As $\dot{\varphi}_t$ increases, the effective shear stress on dislocations (τ), the average velocity of dislocation movement ($\bar{v}$) or the mobile dislocation density (ρ_M) has to be enhanced. This will create a higher stress concentration at grain boundaries and facilitate the dislocation movement in neighboring grains. As a result, the fine ferrite grains begin to undertake the plastic deformation earlier and more at higher strain rates. The contributions from coarse and fine ferrite grains to the total deformation are therefore rate dependent. The contribution from the fine ferrite grains is increased as the strain rate increases. At the same time the straining in the coarse ferrite grains is alleviated. Distribution of the plastic deformation should be more homogeneous among fine and coarse ferrite grains as well as in the bainite colonies as the strain rate increases.

The length and shape of the necking regions of the specimens at the nominal strain of 20% were carefully measured and examined. No obvious differences have been found at the strain rates of 1.25×10^{-4} and 2.5×10^{-3} s^{-1}. The amount of plastic deformation in the necking part is comparable at the approximately same positions. The necking region of specimens at the nominal strain of 20% at strain rates of 1.25×10^{-4} and 2.5×10^{-3} s^{-1} were examined carefully by SEM. The representative SEM micrographs are shown in Figure 8. Almost all ferrite grains and some large bainite colonies were deformed severely at 1.25×10^{-4} s^{-1}. A large number of fibrous ferrite grains were produced, as shown in Figure 8 (a). The bainite laths in large bainite colonies are aligned along tensile direction, indicated by arrow in Figure 8 (a) and (b). However, at 2.5×10^{-3} s^{-1}, no obvious fibrous ferrite grains with severe plastic deformation were observed in the whole necked region, as shown in Figure 8(b). Such differences in the microstructures can be attributed to more homogenous deformation distributed in fine and coarse grains and bainite colonies at the relative high strain rate.

A void was observed at the interface of severely deformed ferrite and bainite, indicated by arrows in Figure 8 (a). A large number of SEM observations of the voids, especially in the necking regions were carried out and showed that the void nucleation is either from the fracture of severe plastic deformed bainite, especially the large-sized bainite colonies, or as a consequence of localized inconsistent plastic deformation. As the strain rate increases, the fine ferrite grains begin to undertake the plastic deformation earlier and more in the early plastic deformation stage. The distribution of the plastic deformation is more homogeneous among fine and coarse ferrite grains as well as in the bainite colonies. The formation and coalescence of voids would be postponed. Therefore, the uniform elongation and the elongation to fracture are increased. However, at higher strain rates the homogenization of plastic deformation makes the hard "phase" bainite to undertake more plastic deformation in which the voids could be formed easily along the interfaces of ferrite laths. The increase of the stress concentration at phase boundaries also would facilitate formation and propagation of cracks. Whether or not the uniform elongation can be enhanced with increasing strain rates depends on the competition between triggering plastic deformation in the fine ferrite grains as well as bainite, and the progress of cracks. At even high strain rates, the fracture already happens before the deformation develops intensively in the fine ferrite grains. The uniform elongation is decreased. Reducing the sizes of the bainite colonies may restrain the decrease in uniform elongation to a certain extent.

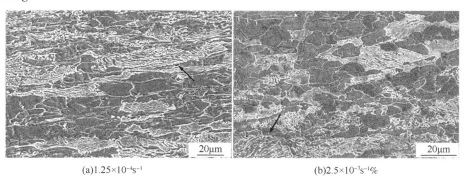

(a)1.25×10^{-4}s^{-1} (b)2.5×10^{-3}s^{-1}‰

Fig. 8 The representative SEM micrograph in the necking region for the samples at the nominal strain of 20% at the strain rate

To facilitate understanding the detailed information of microstructure changed during deformation, TEM examinations were carried out. At the small plastic strain of 5%, the microstructure at both the strain rates (1.25×10^{-4} s^{-1} and 2.5×10^{-3} s^{-1}) is almost similar to that of the undeformed specimens. Dislocations in ferrite grains are only slightly inhomogeneous. With increasing the plastic strain to 13.9%, the fine grains basically keep the polygonal shape at strain rate 1.25×10^{-4} s^{-1}. Only tangled dislocations are formed in these grains, indicating slight deformation, as shown in Figure 9(a). Well developed cell structures are formed in the coarse ferrite grains [Figure 9(b)], where the deformation seems to be more intensive as compared to that in the fine grains. With increasing the strain hardening in the coarse ferrite grains, the plastic deformation gradually transfers from the coarse ferrite grains to the fine ferrite grains and bainite. Deformed bainite with higher dislocation density and incipient cell structure could be observed. At high magnification, a high dislocation density near martensite/austenite microconstituent which lies between ferrite laths appears and the interfaces between martensite and ferrite are not clear [Figure 9(c)]. According to Korzekwa et al. (1984), the high dislocation density near the martensite has been attributed to the martensite transformation shear and the incompatibility of deformation between ferrite and martensite. The voids located in bainite colonies along ferrite laths can be explained as the result of the strain incompatibility between ferrite and martensite. This indicates that the bainite microconstituent undertaking more plastic deformation at even high strain rates would reduce the uniform elongation and the elongation to fracture.

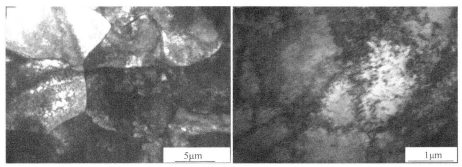

(a) Polygonal fine ferrite grains   (b) Dislocation cell in coarse grain

(c) Deformed bainite at high magnification

Fig. 9 The TEM micrograph of deformed ferrite-bainite samples deformed to 13.9% at the strain rate of 1.25×10^{-4} s^{-1}

In contrast, most of the fine ferrite grains are elongated, twisted and out of polygonal shape when deformed to 10% at 2.5×10^{-3} s^{-1}. Incipient and well developed cell substructure has already been formed in these grains, as shown in Figure 10(a) and Figure 10(b). The deformation in fine ferrite grain is more severe as compared to that at the low strain rate. The TEM observations also confirmed that the fine ferrite grains begin to undertake the plastic deformation earlier and more in the early plastic deformation stage at relatively higher strain rates.

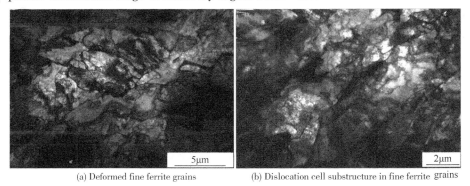

(a) Deformed fine ferrite grains　　(b) Dislocation cell substructure in fine ferrite grains

Fig. 10　The TEM micrograph of deformed ferrite-bainite samples deformed to 10%
at the strain rate of $2.5 \times 10^{-3} s^{-1}$

4　Conclusions

The main results in the present investigation can be summarized as following:

(1) The yield strength, ultimate tensile strengths and YS/UTS ratio (<0.8) of the dual-phase ferrite-bainite X70 pipeline steel are independent of the applied strain rates as the strain rate increases from $2.5 \times 10^{-5} s^{-1}$ to $1.25 \times 10^{-2} s^{-1}$. However, the uniform elongation and elongation at fracture are enhanced as the strain rates increase from $2.5 \times 10^{-5} s^{-1}$ to $2.5 \times 10^{-3} s^{-1}$, and then decrease at the strain rate of $1.25 \times 10^{-2} s^{-1}$. Nevertheless, as compared to that at $2.5 \times 10^{-5} s^{-1}$, the uniform elongation and elongation to fracture at the strain rate $1.25 \times 10^{-2} s^{-1}$ is still higher by 12% and 9% respectively.

(2) The microstructures of the X70 pipeline steel observed are composed of ferrite and bainite microconstituents. The ferrite grains are mainly polygonal and have a bimodal size distribution with fine grains ($2 \sim 5\mu m$) embedded in coarse ferrite grains (over $20\mu m$). The grain structures and dislocation configurations examined under SEM and TEM show that the fine ferrite grains were deformed early and undertook more plastic strain at high strain rates. The distribution of the plastic deformation was more homogeneous among fine and coarse ferrite grains as well as in the bainite colonies, as compared to that at low strain rates. The ferrite-bainite dual-phase steel made of the ferrite phase with a bimodal distribution of grain sizes, i.e., fine grains embedded inside a relatively coarse ferrite matrix, and moderate size of bainite colonies can provide a higher ductility in terms of uniform elongation and elongation to fracture at relatively higher strain rates. Such a grain structure is applicable in the production of the high-deformability pipeline steel without consideration of uniform fine grains in the steels that are not always easily achieved with thermo-mechanical processing of massive productions.

Acknowledgment

The work was financially supported by Tubular Goods Research Institute of CNPC. The authors thank Mr. J. Q Fa in Xi'an Jiaotong University for his help with experiments.

References

[1] Andrade, U., Meyers, M. A., Vecchio, K. S., Chokshi, A. H., 1994. Dynamic recrystallization in high-strain, high-strain-rate plastic deformation of copper. Acta Metall. Mater. 42, 3183–3195.

[2] Asgari, S., El-Danaf, E., Kalidindi, S. R., Doherty, R. D., 1997. Strain hardening regimes and microstructural evolution during large strain compression of low stacking fault energy fcc alloys that form deformation twins. Metall. Mater. Trans. A 28, 1781–1795.

[3] Cahn, R. W., Haasen, P., 1996. Physical metallurgy, 4th ed, Elsevier Science B V. VOLUME III, Part I: 1938.

[4] Gao, H., Huang, Y., Nix, W. D., Hutchinson, J. W., 1999. Mechanism-based strain gradient plasticity-I. Theory. J. Mech. Phys. Solids. 47, 1239–1263.

[5] Ishikawa, N., Shikanai, N., Kondo, J., 2008. Development of ultra-High strength linepipes with dual-phase microstructure for high strain application. JFE TECHNICAL REPORT No. 12: 15–19.

[6] Kimura, Y., Inoue, T., Yin, F. X., Tsuzaki, K., 2008. Inverse Temperature Dependence of Toughness in an Ultrafine Grain-Structure Steel. Science 320, 1057–1060.

[7] Korzekwa, D. A., Matlock, D. K., Krauss, G., 1984. Dislocation Substructure as a Function of Strain in a Dual-Phase Steel. Metall. Trans. A 15A, 1984–1221.

[8] Maloney, J. L., Garrison Jr., W. M., 2005. The effect of sulfide type on the fracture behavior of HY180 steel. Acta Mater. 53, 533–551.

[9] Nakagawa, A. H., Thomas, G., 1985. Microstructure-mechanical property relationships of dual-phase steel wire. Metall. Trans. A 16A, 831–840.

[10] Okatsu, M., Shinmiya, T., Ishikawa, N., Endo, S., Kondo, J., 2005. Development of high strength linepipe with excellent deformability. Proc. of OMAE2005, OMAE2005-67149.

[11] Rashid, M. S., 1977. Relationship between steel microstructure and formability, formable HSLA and dual-phase steels. A. T. Davenport, Ed., AIME, New York, 1–24.

[12] Shanmugam, S., Ramisetti, N. K,, Misra, R. D. K., Hartmann, J., Jansto, S. G., 2008. Microstructure and high strength-toughness combination of a new 700MPa Nb-microalloyed pipeline steel. Mater. Sci. Eng. A 478: 26–37.

[13] Wang, Y. M., Chen, M. W., Zhou, F. H. and Ma, E., 2002. High tensile ductility in a nanostructured metal. Nature. Vol. 419, 912–915.

[14] Youngdahl, C. J., Weertman, J. R., Hugo, R. C., Kung, H. H., 2001. Deformation behavior in nanocrystalline copper. Scripta Mater. 44, 1475–1478.

[15] Zhang, X. L., Zhang, Z. G., Fan, J. W., Mu, Y. C., Wang, L. L., 2010. Study on the Effect of Loading Rate on Fracture Mode of High Grade Pipeline Steels. Journal of Zhongyuan Institute of Technology. Vol. 21, No. 6, 1–4.

本论文发表于《Journal of material science engineering and performance》2017年第7期。

Deformation Stability of a Low-Cost Titanium Alloy Used for Petroleum Drilling Pipe

Jiang Long[1,2,3]　Feng Chun[1,2,3]　Liu Huiqun[4]　Wang Le[4]　Han Lihong[1,2,3]
Feng Yaorong[1,2,3]　Li Fangpo[1,2,3]　Lu Caihong[1,2,3]　Zhu Lijuan[1,2,3]
Wang Hang[1,2,3]　Yang Shangyu[1,2,3]

(1. State Key Laboratory for Performance and Structure Safety of Petroleum Tubular Goods and Equipment Materials, CNPC Tubular Goods Research Institute; 2. Shaanxi key Laboratory for Performance and Structure Safety of Petroleum Tubular Goods and Equipment Materials; 3. Key Laboratory of Petroleum Tubular Goods Engineering, CNPC; 4. School of Materials Science and Engineering, Central South University)

Abstract: A new modified low-cost titanium alloy, Ti-Al-X, was designed for petroleum drilling applications. The alloy ingots were prepared by combination of vacuum consumable electrode arc melting, forging/hot rolling, homogenization, and solid-solution/aging treatments. The hot deformation behavior of Ti-Al-X alloy was investigated by a thermal simulation machine Gleeble 1500 in temperature range of 850~1000℃ with the strain rate range of 0.001~1s^{-1}. The results showed that the deformation resistance significantly decreased with the increase of deformation temperature and the strain rate. The alloy exhibited flow instability under the deformation conditions of strain rates about 0.001 s^{-1} and temperature above 1000℃, which should be avoided during hot working. In addition, the instability area enlarged in processing map with the increasing of true strain.

Keywords: Titanium Alloy; Hot Deformation; Flow Stress; Processing Map

1 Introduction

Withthe development of the petroleum industry, the number of deep wells, ultra deep wells and horizontal wells is increasing[1]. Compared with titanium alloys, the specific strength of traditional steel drilling pipe is lower. In complex drilling /working conditions, failure accidents such as stress corrosion cracking, fatigue fracture, leakage and overload have caused great threat to the safety production of oil and gas well[1,2]. The titanium alloy drilling pipe has the characteristic of high specific strength, good resistance to H_2S stress corrosion and excellent fatigue performance[3-6]. It can significantly reduce the weight of the drilling string, relieve the stress concentration, improve the maximum drilling depth and corrosion-fatigue resistance, prolong the

---

Corresponding author: Jiang Long, jianglong 003@ cnpc. com. cn.

service life, and reduce the safety accidents caused by the failure of the drilling tools[3-6].

In addition, the optimum combination of performance and cost of corrosion-resistant titanium alloy drilling pipe has attracted growing attention. At present, most of the widely used titanium alloys are designed with precious metal elementssuch as V, Zr, Pd and Ru[6,7]. The wide application of titanium alloy has been restricted due to the high cost, poor formability, and low yield rate.

In this study, a low cost Ti-Al-X alloy without any precious alloying additions was designed to possess excellent mechanical characteristics and corrosion resistance. Hot deformation in various industrial processes such as hot forging or rolling is extensively used for manufacturing of this alloy. As a result, the aim of present research is to study the behavior of the Ti-Al-X alloy under straining by hot compression. Particularly, the attention was paid to apparent features of flow behavior and characteristic points to achieve more exact and reliable results.

2 Experimental procedures

(1) Material Preparation and Heat Treatments. The used materials are Ti-5wt.% Al-X (abbreviated as Ti-Al-X) alloys. The alloy ingots were prepared by combination of vacuum consumable electrode arc melting, forging/hot rolling, homogenization, and solid-solution/aging treatments. The billet was heated in a furnace and homogenized at 970~990℃ for 1 hour. After homogenization, the pipe billet is extruded at 850~950℃, with extrusion ratio of ~16, and extrusion speed of 4~6mm/s. Finally, the pipe was pre-stretched ~3%.

(2) Thermal Simulation Test. The thermal simulation was carried out on the Gleeble 1500 with single pass compression. The experimental temperature were 850℃, 900℃, 950℃, and 1000℃, and the strain rate were $0.001s^{-1}$, $0.01s^{-1}$, $0.1s^{-1}$ and $1s^{-1}$, respectively.

3 Results and discussions

(1) High Temperature Deformation. Fig. 1 shows the true stress-strain curve of Ti-Al-X alloy obtained at different temperatures/strain rates by using the principle of volume invariance. At 850℃, the stress increased with the increase of strain and strain rate, decreased with the increase of temperature, and showed dynamic softening after reaching the peak stress. The dynamic recrystallization occurred when the alloy was deformed at 850℃. With the temperature increasing to more than 900℃, the stress level tended to be stable after reaching the peak stress. However, the stress level of the alloy had a certain degree of strain hardening after a period of steady flow deformation, which makes the stress level rise and appear periodic fluctuation. This is the result of alternation of softening caused by partial dynamic recrystallization and deformation/re-hardening of recrystallized grains.

(2) Hot Processing Map. The relationship between flow stress and strain was fitted by three spline function.

$$\ln\sigma = k_1 + k_2\ln\dot{\varepsilon} + k_3(\ln\dot{\varepsilon})^2 + k_4(\ln\dot{\varepsilon})^3 \tag{1}$$

The strain rate sensitivity m can be calculated by,

$$m = \frac{\partial\sigma}{\partial\dot{\varepsilon}}\bigg|_{\varepsilon,T} = k_2 + 2k_3\ln\dot{\varepsilon} + 3k_4(\ln\dot{\varepsilon})^2 \tag{2}$$

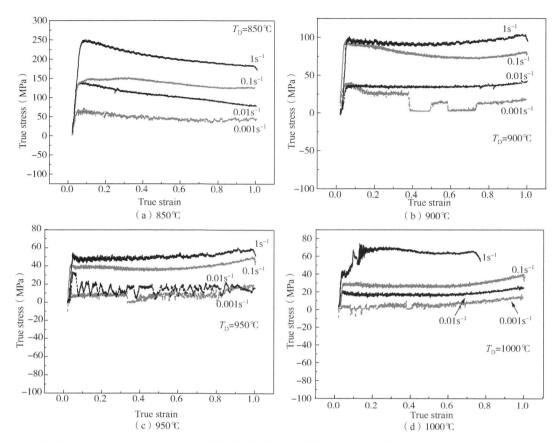

Fig. 1 True stress-strain curves of Ti-Al-X alloys at different deformation temperatures strain rates

Where k_1, k_2, k_3, and k_4 are coefficients. σ is stress. $\dot{\varepsilon}$ is the strain rate. T is temperature.

In the case of a given strain and deformation temperature, the coefficients k_1, k_2, k_3, and k_4 were obtained by fitting the Eq. 1, and then the function relation between the strain rate sensitivity m and the strain rate $\dot{\varepsilon}$ can be given. Next, the various expression of the Eq. 2 at different temperatures was calculated, and then the m values at different temperatures and strain rates can be calculated. The m values under peak stress conditions are listed in Table 1.

Table 1 Strain rate sensitivity m of Ti-Al-X alloy under different deformation conditions

Alloys	$\dot{\varepsilon}$	850℃	900℃	950℃	1000℃	1050℃
Ti-Al-X	0.001s^{-1}	0.42957	0.15017	0.41387	1.07534	1.08677
	0.01s^{-1}	0.17444	0.16094	0.20658	0.33338	0.39173
	0.1s^{-1}	0.10701	0.17903	0.13641	0.08002	0.11056
	1s^{-1}	0.22726	0.20443	0.20334	0.31529	0.24326

As shown in Table 1, the relationships among strain rate sensitivity m, strain rate $\dot{\varepsilon}$ and deformation temperature T are very complicated. Generally speaking, when the temperature is certain, for Ti-Al-X alloy, at lower strain rate, the value of m decreases with the increase of $\dot{\varepsilon}$ at various temperature except at 900℃, while at the higher strain rate, the m value increases with the increase of $\dot{\varepsilon}$. With the increase of T, the m value is increased.

At the same time, the m values of Ti−Al−X alloy at conditions of 1000℃/0.001s^{-1} and 1050℃/0.001s^{-1} are slightly more than 1, which are 1.07534 and 1.08677, respectively. It is known from the flow deformation instability criterion that the Ti−Al−X alloy had undertook the flow deformation instability at these conditions.

The power dissipation efficiency η under each deformation condition can be calculated according to the Eq. 3, and the contour lines of the equal η are drawn in the Fig. 2 in the surface of T and $\ln\dot{\varepsilon}$.

$$\eta = \frac{2m}{m+1} \qquad (3)$$

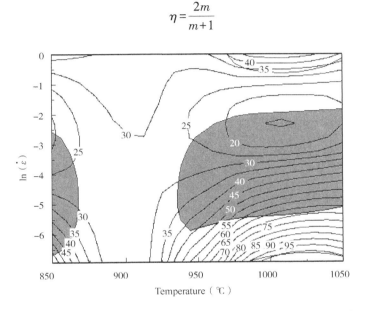

Fig. 2 Hot working diagram of Ti−Al−X alloy at strains of ~0.05
(The shadow part is the flow instability region, and the number on the contour line represents the percentage of power dissipation efficiency)

As a function of the deformation temperature and the strain rate, the region of the flow deformation instability on the power dissipation map can be defined by Eq. 4.

$$\xi(\dot{\varepsilon}) = \frac{\partial \ln\left(\frac{m}{m+1}\right)}{\partial \ln \dot{\varepsilon}} + m < 0 \qquad (4)$$

The criterion for stable flow deformation is $0 < \eta < 2m$ and $0 < m \leq 1$.

Table 2 Power dissipation efficiency of Ti−Al−X alloy under different deformation conditions(%)

Aalloys	$\dot{\varepsilon}$	850℃	900℃	950℃	1000℃	1050℃
Ti−Al−X	0.001s^{-1}	60.09779	26.11266	58.54428	103.6302	104.1581
	0.01s^{-1}	29.70607	27.72581	34.24224	50.00525	56.29397
	0.1s^{-1}	19.33316	30.36903	24.00718	14.81824	19.91068
	1s^{-1}	37.03535	33.94635	33.79593	47.94228	39.1326

Table 3 Deformation instability parameters of Ti-Al-X under different deformation conditions(%)

Alloys	$\dot{\varepsilon}$	850℃	900℃	950℃	1000℃	1050℃
Ti-Al-X	0.001s^{-1}	9.026	18.067	21.608	128.138	122.118
	0.01s^{-1}	-13.002	19.873	-2.436	-29.727	-12.24
	0.1s^{-1}	9.668	21.951	9.629	-13.745	-12.32
	1s^{-1}	86.034	24.298	57.798	176.087	121.876

The hot working diagram of the alloy at strains of ~0.05 (under peak stress) is shown in Fig. 2.

4 Conclusions

The apparent differences of flow stress curves obtained in the Ti-Al-X alloys have been analyzed in term of different dependence of flow stress to temperature and strain rate. The deformation resistance significantly decreases with the increase of deformation temperature and the strain rate. The alloy exhibits flow instability under the deformation conditions of strain rates about 0.001s^{-1} and temperatures above 1000℃, which should be avoided during hot working. In addition, the instability area enlarges in processing map with the increasing of true strain.

Acknowledgement

This work was supported by the National Key Basic Research Program of China (Grant No. 2016YFB0300904), National Science and Technology Major Project of China (2016ZX05020-002, 2016ZX05022-005 and 2017ZX05009-003) and the National Natural Science Foundation of China (U1762211), CNPC Program on Basic Research Project (2016A-3905) and Shaanxi Natural Science Basic Research Project.

References

[1] S. A. Holditch and R. R. Chianelli, Factors that will Influence Oil and Gas Supply in 21st Century, MRS Bull., 33(2008) 317-323.

[2] R. Zeringue, HPHT Completion Challenges, SPE High Pressure/High Temperature Sour Well Design Applied Technology Workshop, May 17-19, 2005 (The Woodlands, TX), SPE 97589.

[3] R. W. Schutz and H. B. Watkins, Recent Developments in Titanium Alloy Application in the Energy Industry, Materials Science and Engineering A, 243(1998) 305-315.

[4] R. W. Schutz, Performance of ruthenium-enhanced α-β Titanium alloys in aggressive sour gas and geothermal well produced fluid brines, Corrosion, 1997, 32, NACE, Houston, TX, 1997.

[5] M. Ziomek-Moroz. Environmentally Assisted Cracking of Drill Pipes in Deep Drilling Oil and Natural Gas Wells, Journal of Materials Engineering and Performance, 21(2012) 1061-1069.

[6] X. H. Lü, Y. Shu, G. X. Zhao, Research and application progress of Titanium alloy petroleum pipe, Rare Metal Materials and Engineering, 43(2014) 1518-1524.

[7] Z. H. Chen. Titanium and Titanium Alloys, Chemical Industry Press, Beijing, 2006.

本论文原发表于《Materials Science Forum》2019年第944卷。

Experimental and Simulation Investigation on Failure Mechanism of a Polyethylene Elbow Liner Used in an Oilfield Environment

Kong Lushi[1] Fan Xin[2] Ding Nan[3] Ding Han[1] Shao Xiaodong[1]
Li Houbu[1] Qi Dongtao[1] Liu Qingshan[3] Xu Yanyan[3] Ge Pengli[3]

(1. State Key Laboratory for Performance and Structure Safety of Petroleum Tubular Goods and Equipment Materials, Tubular Goods Research Institute, China National Petroleum Corporation; 2. Natural Food Macromolecule Research Center, School of Food and Biological Engineering, Shaanxi University of Science&Technology; 3. Key Laboratory of Enhanced Oil Recovery in Carbonate Fractured-vuggy Reservoirs, CNPC and SINOPEC Northwes Company of China Petroleum and Chemical Corporation)

Abstract: Internal collapse of a polyethylene (PE) liner in the steel elbow between furnace and manifold of second-stage separator occurred during the operation of the crude oil treatment system in a certain factory. To analyze the causes and prevent such cases from happening again, macroscopic and microscopic observation, FTIR, density and hardness measurements, thermogravimetric analysis (TGA), tensile test, and thermal stress simulation were conducted. The results show that the PE liner was swelled by the conveying medium, resulting in a decrease in yield strength and module. Moreover, high operating temperature leads to softening and expansion of the swelled liner pipe. Finally, the stress caused by the PE liner's inability to expand freely due to the external constraints of steel elbow increased dramatically and induced the liner pipe collapse inward, resulting in pipe blockage. Failure mechanism is further verified by computer simulation.

Keywords: Polyethylene; Elbow liner; Swelling; collapse

1 Introduction

In recent years, polymeric pipes are being considered as suitable candidates to substitute for metallic pipes in the transportation of oil and oil derivatives. Carbon steel is still the main material used to manufacture pipelines, but corrosion and a fairly high internal roughness are two main drawbacks of steel pipelines[1,2]. Therefore, the substitution of old steel grids by new ones using polymers or steel tubes with an inner polymeric layer, instead of an all-steel pipe is a common trend, since polymers are corrosion resistant and can also be manufactured with very small surface

Corresponding author: Kong Lushi, kongls@cnpc.com.cn.

roughness[3]. Among the many available polymeric materials, polyethylene (PE) is a promising choice due to its good properties, availability, and cost[4-6].

PE pipes are mainly used in the protection of steel pipes in two fields: corrosion protection in the transportation of corrosive chemical reagents and repair of damaged steel pipes.[7] Trenchless interpenetration technology for repairing damaged pipelines was originally developed in Europe and North America for repairing water pipes.[8] During trenchless internal repairing, the diameter of PE pipe is first reduced by a special reducing tool, which reduces the cross-section area of the lining pipe by about 40%. The deformed PE pipe is pulled into the damaged pipe under the effect of traction. When PE liner pipe is in place, the deformed pipe will bounce back with the help of its memory characteristics or the effect of pressure and temperature, and the outer wall of the thermoplastic pipe will stick to the inner wall of the damaged steel pipe with an interference fit, forming a pipe-in-pipe structure. The joint will be connected by flange, ensuring good integrity and sealing of the pipe[9, 10]. The repaired steel pipe with PE liner combines the advantages of corrosion resistance of PE pipe and high strength and good pressure-bearing performance of the steel pipe. The conveying medium flows in the inner liner pipe without contact with the steel pipe, which prevents from corrosion of steel pipe and extends the service life of the steel pipe. The cost, using the trenchless interpenetration technology to repair the damaged pipelines, is only about 60% of the new pipeline comprehensive cost. In the engineering of the ground system construction, it can greatly shorten the construction period, achieve early start-up, and save the cost of investment[11].

Repairing steel pipe by trenchless interpenetration technology with thermoplastic plastics has drawn more and more attention. However, some problems have been exposed during the service life, such as collapse or rupture of the liner pipe[12-14]. In this work, an accident about the collapse of the elbow liner was reported. To avoid the recurrence of similar accidents, failure analysis was conducted on the polymeric liner pipe, and countermeasures were offered.

2 Background of the incident

The processing pipeline lined with a polymeric pipe in the crude oil treatment system station of a certain plant was put into operation in October 2013 with an operating pressure of 0.4~0.5 MPa. The conveying medium was crude oil, in which hydrogen sulfide content was less than 10%, carbon dioxide content was 4%~6%, water content was 20%~30%. The amount of daily treatment liquid was about 15000 tons. On May 15, 2019, a polymeric lined elbow located between the heating furnace and the manifold of the secondary separator was blocked. According to the record of the incident, the outlet pipeline temperature exceeded 90 ℃ when the failure occurred. Figure 1 shows the failed polymeric lined elbow. The inside diameter of the steel elbow is about 350 mm. The bend centerline radius and bend angle are 400 mm and 65°, respectively. The collapse appeared in the middle part of the polymeric liner, as shown and yellow-marked in Fig. 1(a) and (c). Besides, at one end of this elbow, the liner ruptured obviously between the liner body and the flanging zone.

3 Failure description

To conduct a complete visual examination, the polymeric liner was taken out from the steel

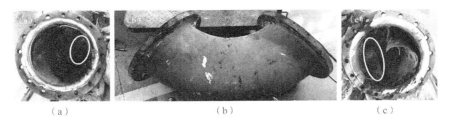

Fig. 1　Photographs of failed polymeric lined elbow

elbow. Fig. 2 shows the failed polymeric liner. It can be seen from Fig. 2(a)(b), and 2(c) that collapse occurred in three zones of the liner body, locating at the intrados side of the liner and the area of 90° and −90° from the intrados side, respectively. Among them, the most serious one happened in the area shown in Fig. 2(a) and the collapse runs almost through the liner body. In Fig. 2(d), the extrados side of the liner body remained intact. Besides, the three collapses are all close to the same one end of the liner at which more than half of the circle ruptured between the liner body and the flanging zone, as shown in Fig. 2(e). Fig. 2(f) shows that the other end of the liner had a crack at the inside edge of the flanging zone.

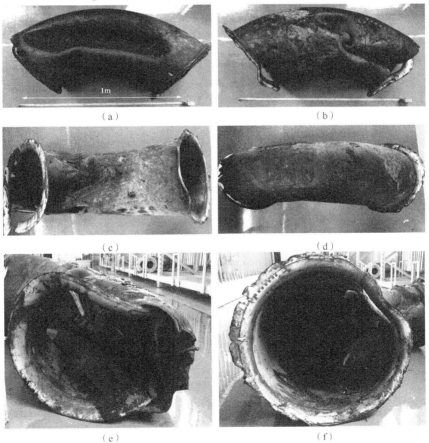

Fig. 2　Photographs of failed polymeric liner

4 Methodology

In this work, the failure analysis contains three parts. First, based on the background information and failure description, the causes that might lead to the failures of the liner was proposed. Then, probable causes of the failure were systematically analyzed by kinds of measurements, including microscopic observation, Fourier Transform infrared spectroscopy (FTIR), density and hardness measurements, thermogravimetric analysis (TGA), tensile test, and Thermal stress simulation. Finally, a comprehensive analysis based on the above results was conducted and countermeasures are offered.

To analyze the causes of the failure, the samples used for chemical, physical, and mechanical tests were cut from the area close to the collapse zone and flanging zone, respectively, as shown and red-marked in Fig. 2(b) and 2(f). In theory, the material at the flanging zone was not in contact with the medium, which means that properties of the material are close to that of the liner material before service. So the sample obtained from the flanging zone was used as the reference sample. Before tests, all samples were firstly washed by kerosene to remove the oil pollution at the surface of the samples. Then, the samples were washed by ethanol and deionized water under ultrasonic treatment. Finally, the samples were dried in an oven at 60℃ for 4 h.

FTIR spectra of the samples via the attenuated total reflection model were collected by using a Nicolet iS 50 IR spectrometer. The density of the samples was measured by an ET-12SL electronic densitometer. The hardness of the samples was obtained by a TIME5410 Shore A durometer. Thermo gravimetric analysis (TGA) was conducted from 40℃ to 650℃ at 30K/min at the N_2 atmosphere on a TGA-2 analyzer. The tensile tests were evaluated by an AGS-X10kN large stretch tensile testing machine according to ISO 6259-3: 2015[15]. An S261TR microscope was employed to observe the samples. Thermal stress simulation of elbow liner was performed by using Fusion 360 software. Polyethelene was selected as the liner material from the material database of the software. Because of the constraint of flange connection, the two end faces of the elbow liner were fixed. According to the record of the incident, the outlet pipeline temperature exceeded 90 ℃ when the failure occurred. Based on this information, the temperature of internal surface of the elbow liner was set as 90 ℃. The direction of deformation was chosen as inward deformation due to the external constraints of steel elbow. Based on 10% of the model size, The mesh was generated automatically by the software. There are 33575 nodes and 16399 elements in this model.

5 Results and discussions

5.1 Probable cause of the collapse

Based on the background information and the Failure description (Fig. 2), it can be predicted that the polymeric liner may suffer a swelling induced by conveying medium and thermal stress caused by the liner's inability to expand freely due to the external constraints of steel elbow under high working temperature. According to references[5, 6, 16], swelling can cause a severe decrease in the mechanical properties of polymeric materials. It means that the swelling could weaken the structural stability of this polymeric liner. Glock[17] derived an analytical expression relating buckling pressure with pipe geometry (w/D) valid within the elastic regime as shown in Eq. (1).

$$P_c = \frac{E}{1-v^2}\left(\frac{w}{D}\right)^{2.2} \tag{1}$$

Where P_c is the bucking collapse critical pressure expressed in MPa, E is the module of liner materials, v is the Poisson's ratio of liner materials, and w/D is the ratio of wall thickness and diameter.

According to Eq. (1), the decrease in E will result in a decrease in Pc, which means that the liner became easier to collapse. On the one hand, contact with a high-temperature medium can soften the materials, resulting in lower yield strength and module. On the other hand, at high temperature, thermal stress increased dramatically, posed a high risk of collapse of PE liner. Based on the above-mentioned analysis, it can be speculated that the collapse of the polymetric liner was caused by a combination of swelling and thermal stress. To verify this speculation, tests, including microscopic observation, FTIR spectra, density and hardness tests, TGA, tensile test, and thermal stress simulation, were conducted.

5.2 Microscopic observation

Fig. 3 shows the photographs of the surface and cross-section of samples at the flanging and failed zone of the liner. It can be seen in Fig. 3(a) that the surface of the liner in the flanging zone is relatively smooth. The surface of the liner in the failed zone became rough and some pittingis distributed at the surface, as shown in Fig. 3(b). The color changed from white to black. Fig. 3(c) shows that the materials at the flanging zone are white and homogeneous, with flat edges and no cracks. Because of no contact with the conveying medium, there is no interaction between liner and medium and the liner material in the flanging zone remains in its original state. Fig. 3(d) shows that the color of the material at the failure zone changed significantly. The part near the upper and lower surfaces is black, the middle part is brown, and cracks appeared at the upper and lower surfaces, indicating that the conveying medium has infiltrated into and swelled the materials in the failure zone.

5.3 Density and hardness

The density and hardness of samples from the failed zone and flanging zone were measured respectively for comparison. The density of the sample from the ruptured zone (0.9377g/cm^3) is higher than that of the sample from the flanging place (0.9245g/cm^3). Compared with the sample from the flanging zone, the hardness of the sample from the ruptured zone shows an obvious decrease from 45.6 to 25.7. Based on the changes of color, density, and hardness, it can be inferred that the conveying medium with high density diffused into the liner, causing color change and density to increase. The medium in the liner may act as the plasticizer and soften the liner, causing hardness to decrease.

5.4 FTIR analysis

To confirm the chemical structure of the liner, samples from the ruptured zone and flanging place were characterized by FTIR. The FTIR spectra of the samples from the flanging zone and ruptured zone are shown in Fig. 4. The structure of PE can be identified from the characteristic peaks at 2915cm^{-1}, 2847cm^{-1}, 1462cm^{-1}, and 718cm^{-1}, corresponding to C-H stretching, C-H stretching, -CH$_2$ twisting and -CH$_2$ rocking, respectively. It implies that the chemical structure of the liner in the failed zone did not change obviously. Compared with the spectrum of the sample from the flanging zone, three new peaks at 1605cm^{-1}, 538cm^{-1}, and 450cm^{-1} appeared in the spectrum

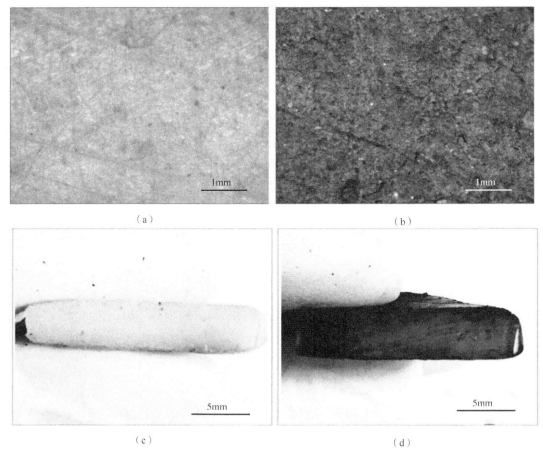

Fig. 3　Surface and cross-section photographs of samples (a, c) flanging zone and (b, d) ruptured zone

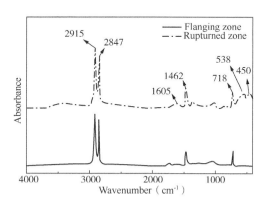

Fig. 4　FTIR spectra of the samples
(a) flanging place and (b) ruptured zone

of the sample from the ruptured zone. Considering the changes of color, density, and hardness, these three peaks could be attributed to characteristic absorption of the medium diffused into the matrix of the liner.

5.5　TGA analysis

To confirm the swelling of PE liner caused by the conveying medium, TGA was used to characterize the liner. TGA curves of the samples from the flanging place and ruptured zone are shown in Fig. 5. In Fig. 5(a), there is only one 100% mass loss step in the TGA curve, which could be attributed to the mass loss of thermal decomposition of PE in the flanging place. Being different from the TGA curve of PE in the flanging place, two mass loss steps appear in the TGA curve of PE in the failed zone [Fig. 5(b)]. The temperature of the first step is lower than the PE thermal decomposition temperature. This part of mass loss was

34.87%, which could be attributed to the evaporation of medium absorbed by the PE liner. The second step of mass loss (60.86%) belongs to the thermal decomposition of PE, being like the features of the TGA curve in Fig. 5(a). It should be noted that the residue is 4.27%, which may be attributed to some inorganic composition absorbed by PE during its service. The results of TGA confirm the swelling of PE by conveying the medium further. More importantly, the medium with a high weight percentage (34.87%) plays a key role in swelling and plasticizing of the PE liner. According to references[5, 6, 16], plastination can cause a severe decrease in mechanical properties of PE materials. It means that the plastination could weaken the structural stability of PE liner.

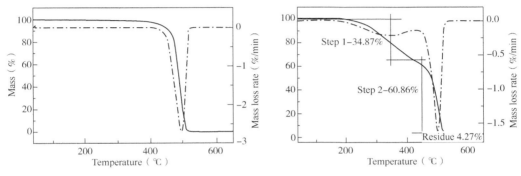

Fig. 5　TGA curves of the samples from (a) flanging place and (b) ruptured zone

5.6　Mechanical property of the PE liner

Fig. 6 shows the stress–strain curves of the sample in the failed zone and pristine PE. Compared with the pristine PE, yield strength, module, and breaking elongation of the sample in the failed zone exhibits a significant decrease. The yield strength of PE liner decreased from 20.4 MPa to 8.1 MPa and the module decreased from 822 MPa to 86.8 MPa. It verifies the plastination of medium absorbed by PE liner. In this case, the breaking elongation of the plasticized sample in the failed zone shows an obvious decrease instead of increase. The reason

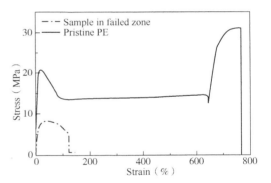

Fig. 6　Stress-strain curves of the sample in the failed zone and pristine PE

may be explained that pitting and cracks distributed at the surface, as shown in Fig. 3(b), will become stress concentrations during tensile, lead to the material break and low breaking elongation. It could be used to explain the rupture and crack of the liner in Fig. 2(e) and 2(f).

5.7　Thermal stress simulation

The results of FTIR, TGA, and tensile tests indicate that the structural stability of PE liner decreased significantly due to the swelling and plastination induced by the conveying medium exited in the PE matrix. Additionally, according to the incident records, the working temperature exceeded 90 ℃. On the one hand, contact with a high-temperature medium can soften the PE materials, resulting in lower yield strength and module. On the other hand, at high temperature, the stress caused by the PE liner's inability to expand freely due to the external constraints of steel elbow increased dramatically and posed a high risk of collapse of PE liner. The results of the thermal stress simulation conducted by using Autodesk's Fusion 360 software confirmed the speculation. Fig. 7 shows the results of the thermal stress simulation at 90℃. Because of high stress, the intrados collapsed. The maximum stress is 10 MPa, which is higher than that of yield strength of plasticized PE liner at room temperature. With increasing temperature, the yield strength and module of PE material will decrease[18]. At 90℃, the yield strength of PE is lower than that at room temperature. So, this high stress is likely to cause the yield or even break of PE liner. The position of maximum stress matches well with the ruptured zone shown in Fig. 2(a) and 2(b). Additionally, the stress in the transition part between flanging and interior is relatively high, which may cause the rupture of PE liner shown in Fig. 2(e).

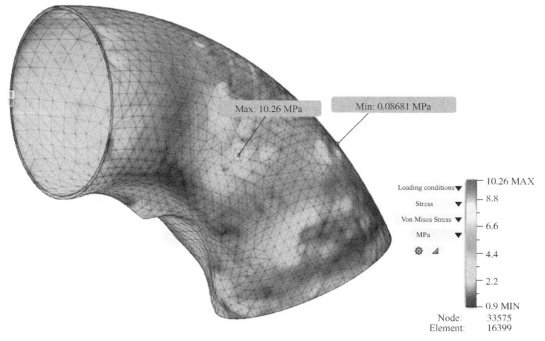

Fig. 7 Von Misses stress contours of deformed geometry obtained by FEM simulation
(The legend shows color references of Von Mises stress values) Color figure online

6 Conclusions and recommendations

6.1 Conclusions

Based on the background information of the failed PE lined elbow and results of tests, including macroscopic and microscopic observation, FTIR, density and hardness measurement, Thermogravimetric analysis (TGA), tensile test, and thermal stress simulation, it is deduced that the failure of this liner pipe was caused by a combination of swelling and heat. The intrinsic reason for the failure of the liner is that the PE liner was swelled by the conveying medium, resulting in a decrease of mechanical properties and poor structure stability. Meanwhile, high operating temperature (90 ℃) accelerated the decrease of the performance of the swelled liner pipe. At high temperature, the stress caused by the PE liner's inability to expand freely due to the external constraints of steel elbow increased dramatically and induced the liner pipe collapse inward, resulting in pipe blockage.

6.2 Recommendations for failure prevention

According to working conditions, suitable materials and structure should be selected and designed to produce the liner pipe, such as high module materials and high w/D value to ensure the structural stability of liner. During its service, the working conditions, including the composition of medium, pressure, and temperature, should be continuously monitored to ensure that its change does not go beyond the capability of the material.

Acknowledgement

This work was supported by the SINOPEC Northwest Company of China Petroleum and Chemical Corporation (No. 34400007-19-ZC0607-0058), the Key Research and Development Program Shaanxi Province (No. 2018ZDXM-GY-171), and Shaanxi Province Innovation Capability Foundation (No. 2019KJXX-092).

References

[1] A. H. U. Torres, J. R. M. d´Almeida, J. -P. Habas, Aging of HDPE Pipes Exposed to Diesel Lubricant, Polymer-Plastics Technology and Engineering 50(15) (2011) 1594-1599.

[2] L. Fan, F. Tang, S. T. Reis, G. Chen, M. L. Koenigstein, Corrosion Resistances of Steel Pipes Internally Coated with Enamel, Corrosion 73(11) (2017) 1335-1345.

[3] A. Habas-Ulloa, J. -R. M. D´Almeida, J. -P. Habas, Creep behavior of high density polyethylene after aging in contact with different oil derivates, Polymer Engineering & Science 50(11) (2010) 2122-2130.

[4] J. P. Manaia, F. A. Pires, A. M. P. de Jesus, S. Wu, Yield behaviour of high-density polyethylene: Experimental and numericalcharacterization, Engineering Failure Analysis 97 (2019) 331-353.

[5] M. Erdmann, M. Böhning, U. Niebergall, Physical and chemical effects of biodiesel storage on high-density polyethylene: Evidence of co-oxidation, Polymer Degradation and Stability 161 (2019) 139-149.

[6] W. Ghabeche, K. Chaoui, N. Zeghib, Mechanical properties and surface roughness assessment of outer and inner HDPE pipe layers after exposure to toluene methanol mixture, The International Journal of Advanced Manufacturing Technology 103(5-8) (2019) 2207-2225.

[7] E. Engle, Pipe Rehabilitation with Polyethylene Pipe Liners, Iowa. Dept. of Transportation, 2003.

[8] M. Kim, D. -G. Lee, H. Jeong, Y. -T. Lee, H. Jang, High Temperature Mechanical Properties of HK40-type Heat-resistant Cast Austenitic Stainless Steels, Journal of Materials Engineering and Performance 19 (2010) 700-704.

[9] A. Morris, T. Grafenauer, A. Ambler, City of Houston 30-Inch Water Transmission Main Replaced by Compressed Fit HDPE Pipe Lining, Amer Soc Civil Engineers, New York, 2017.

[10] V. A. Orlov, I. O. Bogomolova, I. S. Gureeva, Trenchless renovation of worn-out pipelines through their prior destruction and dragging new polymer pipes in place of the old, Vestnik MGSU / Proceedings of Moscow State University of Civil Engineering (7) (2014) 101-109.

[11] Z. H. Deng, Y. S. Fu, P. Ning, The application case of high density polyethylene pipe linings for the renovation of water mains, in: M. J. Chu, X. G. Li, J. Z. Lu, X. M. Hou, X. Wang (Eds.), Progress in Civil Engineering, Pts 1-4, Elsevier Science Bv, Amsterdam, 2012, pp. 2424-2427.

[12] F. Rueda, A. Marquez, J. L. Otegui, P. M. Frontini, Buckling collapse of HDPE liners: Experimental set-up and FEM simulations, Thin-Walled Structures 109 (2016) 103-112.

[13] S. R. Frost, A. M. Korsunsky, Y. Wu, A. G. Gibson, Collapse of polymer and composite liners constrained within tubular conduits, Plastics, Rubber and Composites 29(10) (2013) 566-572.

[14] F. Rueda, J. P. Torres, M. Machado, P. M. Frontini, J. L. Otegui, External pressure induced buckling collapse of high density polyethylene (HDPE) liners: FEM modeling and predictions. Thin-Walled Structures 96 (2015) 56-63.

[15] ISO 6259-3: 2015 Thermoplastics pipes-Determination of tensile properties-Part 3: Polyolefin pipes.

[16] W. Ghabeche, L. Alimi, K. Chaoui, Degradation of Plastic Pipe Surfaces in Contact with an Aggressive Acidic Environment. Energy Procedia 74 (2015) 351-364.

[17] D. Glock, Behavior of Liners for Rigid Pipeline Under External Water Pressure and Thermal Expansion, Der Stahlban (English translation) 7 (1997) 212-217.

[18] H. W. Mu Leijin, Tao Junlin, Experimental study on compressive property of polyethylene pipe material within wide temperature and wide strain rate range, New Building Materials 10 (2017) 132-138.

本论文原发表于《Journal of Failure Analysis and Prevention》2020 年第 20 卷第 6 期。

Failure Analysis and Solution to Bimetallic Lined Pipe

Li Fagen[1]　Li Xunji[2]　Li Weiwei[1]　Wang Fushan[2]　Li Xianming[2]

(1. CNPC Tubular Goods Research Institute, State Key Laboratory of Performance and Structural Safety for Petroleum Tubular Goods and Equipment Materials;
2. PetroChina Tarim Oilfield Company)

Abstract: Owing to low cost, excellent pressure and corrosion resistance, bimetallic lined pipes were regarded as one of the most important methods to resolve corrosion of traditional steel pipes used for oilfield. Nowadays, the pipes were widely used in the projects of oil and gas gathering and transportation. However, there were some failure cases in succession in recent years. In this paper, the failure causes were excavated from multiple perspectives, and on this basis further countermeasures were put forward. Firstly, three typical failure accidents, including CRA layer collapse, CRA layer corrosion and weld joint failure, were listed throughout the whole life cycle from product ordering, construction technology to later operation. Secondly, failure analysis was carried out from five aspects: product quality, test technology, welding process, standard specification and application threshold, and a serial of comprehensive views were proposed. (1) Manufacturing period: Water seepage and tightness between CRA layer and backing pipe could not be effectively monitored. The ratio of collapse test was low so that proposed relevant risk could not be eliminated. (2) Welding period: the process has high risk failure in theory and lack of process assessment and construction acceptance standards. Potential danger could not be effectively assessed, and weld quality could not be guaranteed. (3) Application period: The CRA application range remained unclear, and bimetallic lined pipes were used in the environment beyond the threshold sometimes. The paper also summarized further improvement measures and research directions about these five aspects, including process quality, inspect technique, welding process, standard specification and application range. Finally, solution suggestions were proposed for the whole chain from manufacturers, testing institutions, construction units to oilfield users.

Keywords: Bimetallic lined pipe; CRA layer collapse; CRA layer corrosion; Weld joint failure; Failure analysis

---

Corresponding author: Li Fagen, lifg@ cnpc. com. cn.

1 Introduction

Owing to low cost, excellent pressure and corrosion resistance, bimetallic lined pipes were regarded as one of the most important methods to resolve corrosion of traditional steel pipes used for oilfield. Nowadays, the pipes were widely used in the projects of oil and gas gathering and transportation. Preliminary statistics showed that there were more than 2500 km long bimetallic lined pipes used in oilfields in China[1-4].

Bimetallic lined pipes could be equally allocated pressure to backing pipe and corrosion to CRA layer. Material combinations dramatically reduced the cost. Furthermore, cost effectiveness increased as lengths extended, and operational expenses were lower than the corrosion inhibitor in a long period. Regarding corrosion environment, bimetallic lined pipe selected suitable CRA as the layer, totally owned the performance like CRA materials. Comparing with corrosion inhibitors, this solution avoided the risk resulted in the complicated management process, and it guaranteed operational safety. Due to special structure with two layers, there were higher requirements in manufacturing technology, performance testing, welding construction and field application about bimetallic lined pipe than the single pipe. Therefore, despite the development of manufacturing technology and application technology for many years, there were some failure cases, including CRA layer collapse, CRA layer corrosion, welding joint failure in succession in recent years. Frequent failure seriously disturbed the normal production order in oil fields[5-10].

At present, the scholars have done a lot of work to analyze the failure causes of bimetallic lined pipes, but the understanding was still lack of systematic, and no comprehensive solution had been put forward[5-13]. Based on that, this paper would focus on typical failure cases, analyze failure reasons from five aspects (product quality, inspect technology, welding process, standard specification and application range), and then put forward five solution suggestions.

2 Typical failure problems

2.1 CRA layer collapse

The CRA layer was affixed or tightly fitted to the external pipe full length by expansion, or some other means. There was inevitable gap between CRA layer and backing pipe. Therefore, the CRA layer might collapse once it was subjected to external pressure. In practice, CRA Layer collapse shown in Fig.1 had been one of the main failure modes. For example, a wide range of collapse with the highest proportion even reaching 28% happened in the early stage of external anticorrosion in one project, while in the later stage of operation in another project more than 250 similar failure cases were found.

2.2 CRA layer corrosion

The corrosion resistance was one of the key factors to determine service life. However, transportation environment in oilfields was complex and severe, and even residual acid was sometimes mixed into the oil/gas medium. It was important to ensure CRA material used in appropriate conditions. Once the operation was improper, the risk of pipe corrosion was high. Fig.2 showed that corrosion case about 316L liner pipe occurred.

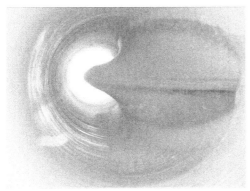 

Fig. 1　Failure morphology of CRA layer collapse　　Fig. 2　Failure morphology of CRA layer corrosion

2.3　Weld joint failure

The welded joint structure and process about 316L lined pipes were complex so that failure cases occurred frequently. The structure and process were often consisted of five steps: seal weld, root weld, transition weld, filler weld and cap weld. Root pass and transition weld pass were welded by 309Mo or 309MoL electrode, while carbon steel electrode was often selected to weld fill pass and cap pass in China. However, the first success rate of welding operation was low. Moreover, welding quality was not easy to guarantee. Weld joint failure, such as the crack and corrosion shown in Fig. 3 had taken place for many times.

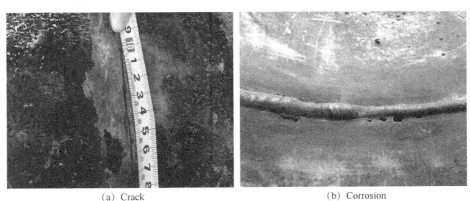

(a) Crack　　　　　　　　　　　　　　　(b) Corrosion
Fig. 3　Failure morphology of weld joint

3　Failure factor analysis

3.1　Product quality

CRA layer collapse was essentially one style of stiffness instability. in bimetallic lined pipes, there must be an instability threshold whose calculation formula was given in the literature[14]. Once external pressure was more than CRA's instability threshold, layer collapse would happen. As a matter of fact, not only 316L's stiffness was very low, but also diameter–thickness ratio and tightness that effected instability threshold were also difficult to control. The instability threshold about 316L liner pipes was not high, and it was easy to collapse. Additionally, impurities, such as water and air between backing pipe and CRA layer, could not be completely removed during early manufacture stage. External pressure was introduced while impurities were heated during external anticorrosive stage. Similarly, once oil and gas medium was entered into the gap between backing pipe and CRA layer due to weld corrosion and other reasons, stress load would also be introduced at the moment of shutdown during operation stage. Once external pressure went beyond 316L's instability threshold, the collapse phenomenon would occur.

3.2 Inspect technique

The quality control of bimetallic lined pipes depended not only on the standard requirements and manufacturing process, but also on a series of inspect techniques. However, there were still many problems to resolve about inspect technology, such as how to monitor tightness in the manufacturing stage and variation of corrosion resistance elements in the seal weld stage. If tightness could be accurately tested and controlled by real-time, product performance would be more reliable. In recent years, although monitoring modal parameters had been proposed to control the tightness by the relationship between modal parameters and tightness timely, the practical results need still to be proved by further application [15]. Similarly, there was also lack of on-line monitoring methods for the seal weld quality. If appropriate test could be used to find out the change of corrosion resistance elements, and then weld current was dynamically adjusted and the corrosion resistance might not turn worse.

3.3 Welding process

The early weld process was not mature and it was closely related to frequent weld joints failures. Previous process would easily result in weld defects in the region of seal weld, promote crack initiation and rapid propagation both in seal weld and weld joint. Weld defects became the failure source, while the high hardness martensite area was transformed into the crack propagation pass further, and then cracks passed through weld joint. Thermal stress between backing pipe and CRA layer promoted weld defects formation. The martensite would form during the welding process by carbon steel electrode under stainless steel transition pass. Additionally, previous process also resulted in weld corrosion. Once weld current was slightly larger during seal weld stage, local ablation was more serious when CRA was needed to repeat heating. The joint welding by the cored wire without gas protective measures would be oxidized and the protective effect couldn't meet the demand. Therefore, the two together led to poor corrosion resistance in the heat affected zone and promoted to weld corrosion[16].

3.4 Standard specification

API Spec 5LD standard was used to control the product quality, but its performance requirement was too loose to meet the application requirements. The pipe end sizes could not make sure CRA layer to be effective welded. In addition, the standard also has the problem of insufficient testing items. The tightness mentioned above not only had no acceptance index, but also had no test mean to prevent CRA layer collapse. In weld quality control, weld joints must meet both the strength requirement of backing pipe and the corrosion resistance of CRA layer. Nowadays, the assessment and construction specification still remained unsolved, and the welding quality was difficult to guarantee. The existing standards, which mainly aimed at carbon steel or stainless steel, were not obviously suitable for bimetallic lined pipes.

3.5 Application range

Corrosion resistance of CRA was limited, and each material had its usage threshold in severe corrosion conditions. The corrosion resistance range about 316L was given[17,18]. However, the oilfield environment was too complex to take into account all of effect factors, and there was one consensus that 316L could not be used in the low pH value condition, high chloride ion and high temperature environment. In the process of production allocation, it was easy to overlook CRA's threshold by users. For example, one pipe was used in 90℃ environment for a long time despite its design temperature was 60℃. In addition, high corrosion media, such as residual acid, was mixed in the pipe mentioned above without any protective measures. It was obvious that service condition beyond the threshold would lead to CRA corrosion.

4 Application recommendations

Following suggestion were address for manufacturers, testing institutions, construction units and oilfield users.

(1) The manufacturer should improve the manufacturing process to reduce or eliminate the problem of CRA Layer Collapse. A new process might be tried to solve the problem [19]. Meanwhile, structure optimization, such as improving dimension accuracy, has been considered to reduce the collapse probability. Finally, manufacturers could also make efforts to develop cheap bimetallic clad pipes and remove the trouble of CRA layer collapse.

(2) Corrosion environment applicability and collapse sensitivity, which were not mentioned in API Spec 5LD standard, were the vital factors of service life about bimetallic lined pipes. It was necessary to develop relevant test technology including dynamic monitoring tightness, non-destructive testing weld joint etc.

(3) The new process that the whole joint was welded by CRA materials was not only widely used abroad but also has been applied in China[20]. The new process had more simple operation and more reliable quality than previous process mentioned above in China. Later large-scale promotion was carried out about the new process. It was suggested that large-scale welding application should be carried out around the new process in the future.

(4) It was very urgent to compile technical specifications applicable to the site and welding process qualification and construction specifications. At present, the main work was to optimize or form the evaluation methods and acceptance requirements based on the key issues of tightness, dimensions, and collapse of pipes, mechanical properties and corrosion resistance of welded joints, and so forth.

(5) It was suggested that design units, oil fields, manufacturers and relevant research departments worked together to screen corrosion factors, and clarify the scope of materials applicability further. Finally, bimetallic lined pipes were used in the suitable environment so that it could reduce unnecessary corrosion damage.

Acknowledgement

The authors acknowledge the financial support of the Key Research and Development Program of Shaanxi Province (No. 2019KJXX-091, No. 2018ZDXM-GY-171).

References

[1] LI Fagen, WEI Bin, SHAO Xiaodong, et. al. Bimetal Lined/Clad Pipe Used in Highly Corrosive Oil and Gas Fields. Oil & Gas Storage and Transportation, 2010, 32(12): 92-96.

[2] Russell D, Wilhelm S M. Analysis of Bimetallic Pipe for Sour Service. SPE Production Engineering, 1991, 6(3): 291-296.

[3] ROMMERSKIRCHEN I. New Progress Caps 10 Years of Work with Bubi Pipes. World Oil, 2005, 226(7): 69-70.

[4] Fu, G. H. Technical Solutions Analysis of CO_2 Corrosion Problems in Xushen Gas Field. J. Oil-Gasfield Surface Engineering, (2008) 66-67.

[5] PAN Xu, ZHOU Yongliang, FENG Zhigang, et al. Problems and Suggestions on Collapse of Double Metal Composite Pipe Lining. Petroleum Engineering Construction, 2017, 43(1): 57-59.

[6] GUO Chongxiao, JIANG Qinrong, ZHANG Yanfei, et al. Stress Corrosion Cracking Failure Analysis on Bimetal Composite Pipe Lining Layer. Welded Pipe and Tube, 2016, 39(02): 33-38.

[7] CHEN Hao, GU Yuanguo, JIANG Shengfei. Failure Reasons for 20G/316L Double Metal Composite Pipe. Corrosion & Protection, 2015, 36(12): 1194-1197.

[8] Fu A. Q, Kuang X. R, Han Y. Failure Analysis of Girth weld Cracking of Mechanically Lined Pipe sed in Gas field Gathering System. Engineering Failure Analysis, 2016, (68): 64-75.

[9] CHEN Haiyun, CAO Zhixi. Influence of Heat Load to the Reisdue Pressure of Bimetal Composite Pipe. Journal of Plasticity Enginnering, 2007, 14(2): 86-89.

[10] WEI Fan, JIANG Yi, WU Ze, et al. Mechanism Analysis and Testing Research on the Buckling of the Bimetal Lined Pipe. Natural Gas and Oil, 2017, 35(5): 06-11.

[11] LIN Yuan, KYRIAKIDS Stelios. Wrinking Failure of Lined Pipe under Bending. 32nd International Conference on Ocean, Offshore and Arctic Engineering, Nantes, France, 2013: 1: 7.

[12] CHANG Zeliang, JIN Wei, CHEN Bo, et al. Effect of Welding Process on Pitting Potential of Welded Joints of 316L Lining Composite Pipe. Total Corrosion Control, 2017, 30(11): 18-22.

[13] VASILIKIS D, Karamanos S A. Mechanical Behavior and Wrinkling of Lined Pipes. International Journal of Solids and Structures, 2012, (49): 3432-3446.

[14] DANIEL Vasilikis, Spyros A. Karamanos. Mechanics of Confined Thin-Walled Cylinders Subjected to External Pressure. Applied Mechanics Reviews, 2014, 66: 010801-1: 010801-15.

[15] WEI Fan, ZHANG Yanfei, GUO, et al. Inspection and Control of Bonding Strength for Mechanical Composite Pipe, Welded Pipe and Tube, 2015, 38(2): 32-36.

[16] LI Fagen, MENG Fanyin, GUO Lin, et al. Analysis of Welding Technology about Bimetal-Lined Pipe. Welded Pipe and Tube, 2014, 37(6): 40-43.

[17] Li Weiwei, Liu Yaxu, Xu Xiaofeng, et al. One girth weld process of bimetallic lined pipe, China, ZL201310202717. X. 2013-10-02.

[18] LI Ke, SHI Daiyan, et al. Application Boundary Conditions for 316L Austenite Stainless Steel Used in Gas Field Containing CO_2. Materials for Mechanical Engineering, 2012, 36(11): 26-28.

[19] Bruce D Craig. Selection Guidelines for Corrosion Resistant Alloys in the Oil and Gas Industry. https://stainless-steel-world.net/pdf/10073.pdf.

[20] YANG Gang. Welding Technology and Quality Control about X65/316L Bimetallic Lined Pipe. Welding Technology, 2012, 41(12): 56-57.

本论文原发表于《Materials Science Forum》2020 年第 993 卷。

Development and Application of Sour Service Pipeline Steel with Low Manganese Content

Li Yanhua[1,2]　Chi Qiang[1,2]　Li Weiwei[1,2]　Zou Bin[3]　Xie Ping[3]
Hu Meijuan[1,2]　Yang Ming[3]

(1. Tubular Goods Research Institute, State Key Laboratory of Performance and Structural Safety for Petroleum Tubular Goods and Equipment Materials; 2. CNPC Key Lab for Petroleum Tubular Goods Engineering; 3. Petrochina West Pipeline Company)

Abstract: Comprehensive analysis on the requirement and evaluation result of pipeline steel for sour service was carried out in this paper in order to provide technical support for the development of the high toughness sour service pipeline steel with low manganese content. The evaluation results showed that the high toughness sour service pipeline steel with low manganese content had excellent comprehensive properties, and the excellent comprehensive mechanical properties and corrosion resistance properties of trial products achieved through reasonable composition and micro-structure design. The HIC and SSC test result complied with pipeline steel requirement for sour service drafted by major international manufacturers. In order to guarantee the application prospect of the trial products, the weldability of the sour service pipeline steel was analyzed finally, and the test result showed that HAZ softening could be improved when the cooling rate was higher than 20 ℃/s.

Keywords: Pipeline steel; Corrosion Resistance properties; Sour service; Mechanical properties; Weldability

1　Key Technical Parameters of Pipeline Steel for Sour Service of Major Oil and Gas Companies

HIC is presumed to occur in the following manner: First, hydrogen atoms, generated at the surface through corrosion reaction, penetrate and diffuse in the steel, and are trapped at the interface between the steel matrix and nonmetallic inclusions. When the internal pressure due to molecular hydrogen gas precipitated at the interface exceeds a critical value, HIC is initiated. HIC tends to propagate easily along the pearlite or bainite band which is sensitive to hydrogen embitterment[1]. Therefore, the following three measures should be considered to increase the HIC resistance of steel.

Corresponding author: Li Yanhua, E-mail: liyanhua003@cnpc.com.cn.

(1) Suppress hydrogen penetration into the steel;
(2) Reduce HIC initiation sites;
(3) Increase the resistance of the steel matrix to HIC propagation.

As a result, the pipes furnished to specifications shall be made from basic-oxygen steel or electric arc furnace steel. The steel shall be fully killed and made with fine grain structure with a grain size of ASTM 7 or finer as per ASTM E 112. Steel shall be vacuum degassed or alternative processes shall be applied. The material shall be treated for inclusion shape control to increase resistance to hydrogen-induced (blistering and stepwise) cracking[2-5]. In 2020D of India oil and gas company, Steel shall be made by continuous casting only. Manufacturing procedure specification of this specification shall be prepared and submitted to Company for approval prior to start of production.

HIC tends to propagate a low temperature microstructure consisting of bainite or martensite, because a harder microstructure is generally more susceptible to hydrogen embitterment. Such hard microstructures are attributed to enrichment of C, Mn, P and other elements and sometimes appear in segregation zones of CR plates. An addition of alloying elements is necessary to ensure strength of steel. But it may cause HIC even in super-low sulfur Ca-treated steel[6-8]. HIC test results can be divided into several groups with respect to effects of manganese content. As-rolled steels containing high carbon show a sudden increase in HIC susceptibility when the manganese content exceeds about 1.0%, while as-rolled low-carbon steels with low carbon contents exhibit excellent HIC resistance. Quenched and tempered steels also show excellent HIC resistance, regardless of their chemical compositions.

As a result, the chemical requirements of pipes for sour service are strictly controlled as shown in Table 1.

Table 1 Chemical requirements of pipeline steel for sour service

Standard name	C	Mn	P	N	Ca/S
2020D of India oil and gas company	≤0.14	≤1.45	≤0.015	≤0.010	For 0.002%<S≤0.003%, 2.0≤ Ca/S≤3.0
DEP 31.40.20.31-Gen. of Shell	≤0.16	≤1.30 (For X70 or above, 1.40)	≤0.015	≤0.012	For 0.0015%<S≤ 0.003%, Ca/S≥2.0
Annex H of API 5L	≤0.16	≤1.65	≤0.020	≤0.012	For 0.0015%<S≤ 0.003%, Ca/S≥1.5

Resistance to hydrogen-induced cracking and sulfide stress-corrosion cracking properties of pipeline steel is usually evaluated in accordance with the method specified in NACE TM 0284 standard and NACE TM 0177[9-12]. But there is significant difference between the acceptance criteria of pipeline steel for sour service. The difference between the acceptance criteria of pipeline steel for sour service is shown in Table 2.

These standards has been implemented for several years, as a result, they has important reference value to the requirements for the development of high grade pipeline steel for sour service.

Table 2 The HIC acceptance criteria of pipeline steel for sour service

Standard name/requirement difference	2020D	DEP 31.40.20.31-Gen.	Annex H of API 5L
HIC requirement	CSR ≤ 1.00%	CLR ≤ 15%, CSR ≤ 1.5%, CTR ≤ 5%, The maximum individual crack length on any section shall not exceed 5 mm	CLR ≤ 15%, CSR ≤ 2%, CTR ≤ 5%

It can be obtained from Table 1 and Table 2 that requirements in Annex H of API 5L is adopted as the basic requirement for pipeline steel for sour service, the relevant technical requirements and acceptance criteria of major international manufacturers are more stringent based on the actual service condition, such as chemical requirements.

2 Evaluation Result of X65 Pipeline Steel for Sour Service in TGRI

During the past few years, comprehensive analysis on the characters of pipeline steel for sour service is carried out in TGRI, such as the corrosion behavior of pipeline steel for sour service, the main factors affecting corrosion resistant properties of pipeline steel, characterization and test methods of corrosion resistant properties of pipeline steel for sour service, the acceptance criteria of pipeline steel for sour service, etc.. The HIC test results of X65 pipeline steel for sour service in china are shown in Figs. 1-3.

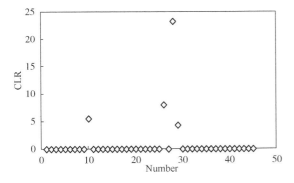

Fig. 1 Statistic figure of CLR for solution A

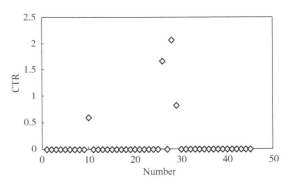

Fig. 2 Statistic figure of CTR for solution A

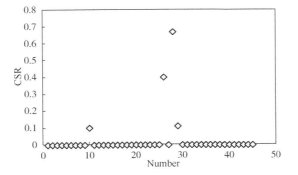

Fig. 3 Statistic figure of CSR for solution A

It can be obtained from the statistics figures that the resistance to hydrogen-induced cracking properties of X65 pipeline steel for sour service in china is excellent generally, but there are some cracks has appeared in the samples casually, and the cracking probability of pipeline steel for sour service would increase with the corrosion increase of corrosive medium[13].

3 Development and Evaluation Results of the High Toughness Sour Service Pipeline Steel with Low Manganese Content

The high toughness sour service pipeline steel with low manganese content is developed based on the research above, the chemical composition of the pipeline steel is shown in Table 3.

Table 3 Chemical composition analysis result ($Wt \times 10^{-2}$)

Chemical composition	Pipe body	Pipe body requirement
C	0.043	≤0.10
Mn	0.27	≤1.60
Al/N	11.2	≥2.0

Comprehensive property evaluation is carried out at the same time, such as Vicker hardness, metallographic analysis and HIC test. The results are shown in Table 4 and Table 5.

Table 4 Vicker hardness test result (HV10)

Sample Location	Vicker hardness test result (HV_{10})								
	1	2	3	4	5	6	7	8	9
Pipe body	229	226	222	238	237	237	241	226	227
Weld seam	182	186	207	184	195	202	175	183	204

Note: Q/SY 1513.2 and Q/SY-TGRC 76-2014 requirement: ≤250HV_{10}

Table 5 HIC test result

Sample number	Section number									Average		
	I			II			III					
	CLR (%)	CTR (%)	CSR (%)	CLR (%)	CTR (%)	CSR (%)	CLR (%)	CTR (%)	CSR (%)	CLR (%)	CTR (%)	CSR (%)
Pipe body, centred = 90° from weld seam	1	0	0	0	0	0	0	0	0	0	0	0
	2	0	0	0	0	0	0	0	0	0	0	0
	3	0	0	0	0	0	0	0	0	0	0	0

The SSC test result shows that the tension surface of the specimen was examined under a low-power microscope at ×10 magnification (Fig. 4, Fig. 5). There are no surface breaking fissures or cracks on the tension surface of the test specimen.

4 Weldability of the Sour Service Pipeline Steel

The weld thermal cycle results in significant changes in microstructure and, consequently, mechanical properties of the weld heat affected zone (HAZ). Simulated heat affected zone continuous cooling transformation curve was measured by means of Gleeble 3500 thermal simulator. According to the results of microstructure observation and hardness measurement, influence of cooling rates on the microstructure and hardness in coarse grain zone of the sour service pipeline steel was studied[14,15].

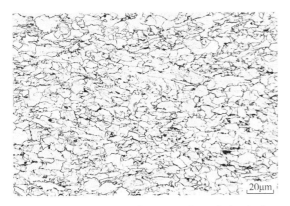 

Fig. 4 Microstructure close to surface of pipe body   Fig. 5 Banded orientation of pipe body

The microstructure of X65MS pipeline steel is polygonal ferrite, a small amount of bainite and pearlite. The critical phase transition temperatures are very important for the research and development of new kind of steel.

According to the thermal expansion curve of X65MS, there are mainly three kinds of phase change:

827.54℃ –882.77℃, bainite ⟶ austenite;
882.77℃ –916.64℃, pearlite ⟶ austenite;
916.64℃ –957.61℃, ferrite ⟶ austenite;

As shown in Fig. 6, there are mainly two types of microstructure in coarse grain zone of X65MS during the cooling process: (1) polygonal ferrite and a small quantity of pearlite(PF+P); (2) granular bainite(B).

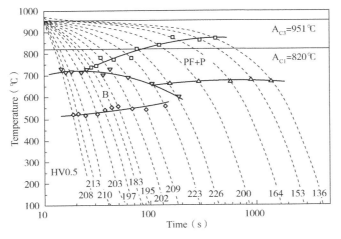

Fig. 6 SH-CCT of X65MS

When the cooling rate is lower than 2℃/s, PF+P was obtained in coarse grain zone.

When the cooling rate is higher than 12℃/s, B was obtained in coarse grain zone.

When the cooling rate is between 2℃/s and 12℃/s, the mixture of PF+P and B was obtained, and the content of B increased with the cooling rate.

Different microstructures can be obtained at different cooling rates of X65MS pipeline steel HAZ as shown in Fig. 7.

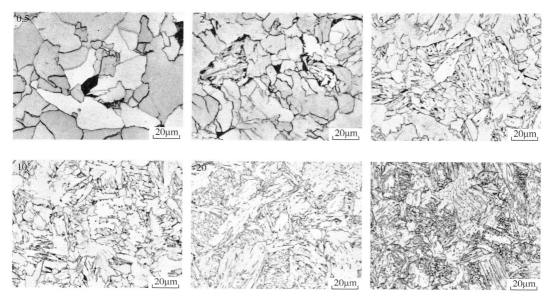

Fig. 7 Microstructures at different cooling rates of X65MS pipeline steel HAZ

It can be obtained that the cooling rate of sub-merged arc welding during pipe manufacture is between 5℃/s and 15℃/s, and according to the microhardness and impact energy curve, HAZ softening was appeared during the sub-merged arc welding of X65MS. The HAZ softening can be improved when the cooling rate was higher than 20℃/s as shown in Fig. 8.

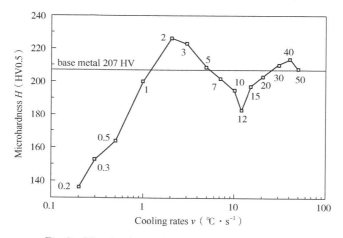

Fig. 8 Microhardness of X65MS pipeline steel HAZ

5 Conclusions

The excellent comprehensive mechanical properties and corrosion resistance properties of trial products are achieved through reasonable composition and microstructure design. The HIC and SSC test result complies with pipeline steel requirement for sour service drafted by major international manufacturers.

Acknowledgement

The work is supported by The National Key Research and Development Program of China (2017YFB0304900, 2017YFB0305000) and China National Petroleum Corporation Projects (2016B-3106, 2016B-3002).

References

[1] Gao Jian-zhong, Wang Chun-fang, Wang Changan, et al. Strain aging behavior of high strength linepipe steel [J]. Development and Application of Materials, 2009, 24(3): 86-90.

[2] Allen T, Busby J, Meyer M, et al. Materials challenges for nuclear systems[J]. Mater Today, 2010, 13: 14-23.

[3] Deodeshmukh VP, Srivastava SK. Effects of short and long-term thermal exposures on the stability of a Ni-Co-Cr-Si alloy[J]. Mater. Des, 2010, 31: 9-2501.

[4] Panait CG, Zielińska-Lipiec A, Koziel T, et al. Evolution of dislocation density, size of subgrains and MX-type precipitates in a P91 steel during creep and during thermal ageing at 600℃ for more than 100000h [J]. Mater. Sci. Eng. A, 2010, 527: 4062-4069.

[5] Sanderson N, Ohm RK, Jacobs M. Study of X100 linepipe costs points to potential savings [J]. Oli&Gas journal, 1999, 3(15): 54-57.

[6] Suzuki N, Kato A., Yoshikawa M. Line pipes having excellent earthquake resistance[J]. NKK Giho, 1999, 167: 44-49.

[7] M. K. Gräf, H. G. Hillenbrand, C. J. Heckmann, K. A. Niederhoff, Europipe, 2003, 35: 1-9.

[8] S. Shanmugam, R. D. K. Misra, J. Hartmann, S. G. Jansto, Mater. Sci. Eng. 2006, A441: 215-229.

[9] Y. B. Xu, Y. M. Y, B. L. Xiao, et al. Mater. Sci. 2010, 45: 2580-2590.

[10] H. Nakagawa, T. Miyazaki, H. Yokota. J. Mater. Sci. 2000, 35: 2245-2253.

[11] C. Z. Wang, R. B. Li, J. Mater. Sci. 2004, 39: 2593-2595.

[12] M. Ozawa, J. Mater. Sci. 2004, 39: 4035-4036.

[13] W. G. Zhao, M. Chen, S. H. Chen, et al. Static strain aging behavior of an X100 pipeline steel [J]. Mater. Sci. Eng., 2012, A550: 418-420.

[14] N. Guermazi, K. Elleuch, H. F. Ayedi. The effect of time and aging temperature on structural and mechanical properties of pipeline coating [J]. Mater. Des. 2009, 30: 2006-2012.

[15] M. S. Rashid. Strain-aging of vanadium, niobiumor titanium-strengthened high-strength low-alloy steels[J], Metal. Trans., 1975, 6A: 1265-1267.

Investigation on Leakage Cause of Oil Pipeline in the West Oilfield of China

Liu Qiang[1]　Yu Haoyu[2]　Zhu Guochuan[1]　Wang Pengbo[1]　Song Shengyin[1]

(1. State Key Laboratory for Performance and Structural Safety of Petroleum Tubular Goods and Equipment Materials, CNPC Tubular Goods Research Institute;
2. School of Mechano-Electronic Engineering, Xidian University)

Abstract: Acrude oil pipeline used in an oilfield in the west oilfield of China leaked many times in a short time. Serious corrosion was found near the leaking position of the failed pipe, the wall thickness of pipeline also decreased near the failure section. To determine the reason for this, the failed pipeline was investigated and analysed by macroscopic analysis, chemical composition tests, metallurgical analysis, mechanical property analysis, scanning electronic microscopy (SEM) analysis, X-ray diffraction (XRD) analysis and corrosion simulation tests. Based on the systematic analysis, it can be concluded that the mechanical properties of the failed pipeline met the related standards. The crude oil conveyed in the pipeline contained a large amount of formation water; because the formation water was acidic (the pH value was 5.0~6.5) and had high salinity, the materials of the pipeline had a very high corrosion rate when exposed tothe crude oil, which was confirmed by the corrosion simulation test. At the same time, because the fluid flow changed suddenly in the pipeline lifting location, fouling deposition occurred inside the pipeline, which led to serious localized corrosion caused by thesmall anode/big cathode corrosion galvanic cell reaction, and the wall thickness decreased quickly in a short time. Finally, the pipeline leaked due to fluid erosion and the high speed impact of solid particles and grit in the crude oil. Improvement suggestions are put forward to the pipeline user at the end of this paper, such as improving the pipeline design, reducing the salinity in the conveying fluid, adding a corrosion inhibitor and replacing the pipeline lifting segment with an anticorrosive coating inside the tube; the crude oil pipeline did not leak again after followingthese suggestions.

Keywords: Pipeline; Leak failure; Localized corrosion; SEM; XRD; Corrosion simulation test

---

Corresponding author: Liu Qiang, liuqiang030@ cnpc. com. cn.

1　Introduction

In China's central and western oilfields, crude oil pipelines are prone to leakage and failure, often resulting in serious economic losses and environmental pollution. In recent years, the crude oil pipelines of Changqing oilfield in west of China had been leaking and failing frequently; one crude oil pipeline lifting segment used in the Changqing oilfield leaked many times in a short period. The specification of the pipeline in the lifting segment is 114mm×4.5mm, and the material is 20-steel (according to the Chinese GB/T 699—1999 standard[1]). The working pressure and output oil amount of this pipeline are 1.4 MPa and 13 m^3/h, respectively.

Many studies have been conducted to investigate similar leakage failures of oil and gas pipelines in the oilfield. Hu et al.[2] analysed the corrosion failure of an oil pipeline by chemical composition analysis, metallographic examination, mechanical testing and corrosion product analysis methods and found that under-scale corrosion was the direct cause of the failure. Ding et al.[3] studied the cause of gathering pipeline perforation in one western oil field through physical and chemical inspection as well as corrosion products analysis; the results showed that high CO_2 content in the pipes and a high chlorine ion environment caused local corrosion. Qin et al.[4] investigated the influences of various corrosion defects and the interaction of multiple corrosion defects on pipeline failure with full-scale pressure burst tests and the finite element method. Physical and chemical inspection, microstructure analysis, finite element analysis, corrosion product composition and structure analysis are the main methods to conduct this kind of pipeline failure analysis, but corrosion process research and simulation tests are rare. The lack of corrosion simulation tests under real conditions has a negative influence on the accurate analysis of the corrosion failure mechanism and the discovery of reasons for failure.

Fig. 1　Operation leak morphology of crude oil pipeline

To determine the reason for the failure of this pipeline lifting segment, the failed part of the pipeline lifting segment was dug out and examined by a nondestructive test. The leakage position was at the bottom of pipeline near the corner of the lifting segment, as shown in Fig. 1. An obvious wall thickness decrease was also found near the failure position, as shown in Fig. 2. To determine the cause of the leak failure in the pipeline, macroscopic analysis, tensile tests, flattening tests, metallurgical analysis and chemical composition tests were used for the pipeline. Material physical-chemical examination was conducted, and corrosion products near the leak position were investigated by X-ray diffraction (XRD) analysis and scanning electronic microscopy (SEM) analysis. Finally, corrosion simulation tests were designed for this study to simulate the corrosion process of the pipeline for the analysis of leakage causes.

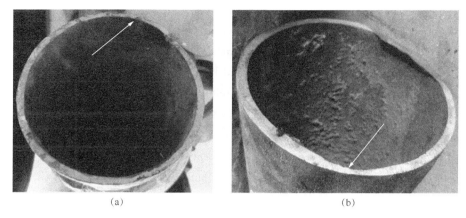

Fig. 2 Wall thickness decreasemorphology of crude oil pipeline(arrow position)
(a)position 1, (b) position 2

2 Experiments

The chemical composition sample of the failed pipeline was taken from the longitudinal pipe and analysed by a Baird Spectrovac 2000 direct reading spectrometer and a LECO CS-444 IR Carbon-Sulfur Spectrometer. To prepare specimens for metallurgical analysis, the samples were cut near the leak position after being ground and mechanically polished with diamond pastes of 6 μm and 1 μm. Specimens were etched and observed by MEF3A and MEF4M metallographic microscopes.

The mechanical properties of the pipeline were determined by tensile testing and flattening testing. Tensile test samples were cut along the pipe longitudinal direction with a section gauge of 50 mm×11.2mm, and the tests were carried out by MTS810-15 machines according to the GB/T 228.1-2010 standard[5]. The tensile properties were determined at room temperature. The leak surface morphology and corrosion products were analysed by visual examination, scanning electronic microscopy (SEM) with a TESCAN-VEGA II system and measurements with an OXFORD-INCA350 Energy Dispersive Spectrometer (EDS). The compositions of the corrosion products were examined by X-ray diffraction with a D8 Advance instrument.

To simulate the corrosion process of the pipeline, corrosion simulation tests were carried out in a high-temperature autoclave made by Cortest Co. Ltd. The hanging corrosion samples with dimensions of 50mm×10mm×3mm were prepared from a longitudinally failed pipeline and polished by silicon carbide papers of 240 to 1200 grit; the surface roughness was less than 1.6μm. Corrosion samples were loaded into the autoclave before the corrosion test, and high purity nitrogen was injected into the autoclave for more than 10 h for deoxygenation; then the samples were loaded, and the autoclave was sealed. The high purity nitrogen was continuously injected for deoxygenation, and the medium gas was injected. The test pressure was set to 1.8MPa, and the autoclave was heated to 40℃, which is the same as the pipeline working temperature. When the temperature in the autoclave was raised to the required temperature, the test timing was started. Corrosion simulation tests were carried out for 120h. The samples were weighed with a balance before and after testing.

3 Results

3.1 Visual examination

Macroscopic examination was carried out on the failed pipeline. As shown in Fig. 3, the leak failure occurred at the bottom of the pipeline lifting segment, which was 40cm above the welding seam. The wall thickness inspection showed that the wall thickness of the pipeline bottom was not uniform. The AB section wall thickness of the pipeline bottom 20cm away from the welding seam was normal (4.5cm). In the BC section, 20~40cm above the welding seam, the wall thickness of the pipeline gradually became thinner at the bottom; the thickest portion of the BC section was only 1.79mm. The wall thickness of the CD section, which was 20~140cm above the welding seam, gradually became thicker; the thickest wall of the CD section was 3.75cm thick. The wall thickness of the pipeline above the CD section was normal (4.5cm).

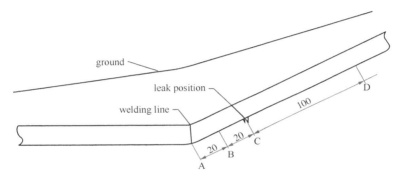

Fig. 3 Measurement Position of external diameter of the burst steam injection pipe

Fig. 4 Leak position morphology of failure crude oil pipeline

The external diameters of the failed pipeline were also measured. The results showed that there was no outer diameter distortion in the pipeline; the leak position morphology of the pipeline is shown in Fig. 4.

3.2 Chemical composition

The chemical composition tests of the failed pipeline were carried out, and the results are shown in Tab. 1. The pipeline materials were qualified and complied according to the requirements of national standard GB/T 8163—1999[6].

Table 1 Chemical composition results of leak failure pipeline (wt., %)

Sample	Elements								
	C	Si	Mn	P	S	Cr	Mo	Ni	Cu
2#	0.19	0.23	0.43	0.022	0.0085	0.014	—	0.0078	0.010
GB/T 8163—2008 requirement	0.17~0.23	0.17~0.37	0.35~0.65	≤0.030	≤0.030	≤0.25	—	≤0.30	≤0.25

3.3 Microstructure examination

The microstructure of failure pipeline material is ferrite and pearlite, as shown in Fig. 5. According to standard GB/T 6394—2002[7], the grain fineness of ferrite and pearlite is 10 grade. The non-metallic inclusion grade are A0.5, B0.5 and D1.0. It is indicated that metallographic test results were qualified and accorded with the standard GB/T 8163—1999 requirement.

The inner surface of leak failure pipeline was analyzed by metallographic microscope, it can be seen that there were several corrosion

Fig. 5 Microstructure of the material of the crude oil pipeline with leak failure

pits at the bottom of pipeline BC section, no crack was found around the leak position, the microstructure of the material under the corrosion pits is still ferrite and pearlite, no microstructure deformation was occurred, as shown in Fig. 6.

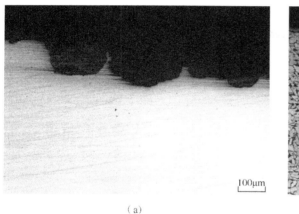
(a)

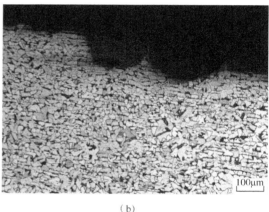
(b)

Fig. 6 Inner surface corrosion pitsmorphology(a) and microstructure of pipeline(b)

3.4 Mechanical properties

The tensile and flattening properties of the failed pipeline at room temperature are shown in Tab. 2 and 3, respectively. The mechanical properties of the burst pipe met the requirement of standard GB/T 8163—1999.

Table 2 Results of tensile test

Items	Ultimate tensile strength(MPa)	Yield tensile strength(MPa)	Elongation(%)
Results	448	330	22
	486	334	29
	464	309	36
Average	466	324	29
GB/T 8163—2008 requirement	410~530	≥245	≥20

Table 3 Results of flattening test

Sample		Plate distance	Results
Item	Diameter(mm)		
pipe	Φ114×100	D/3(38mm)	NO crack
GB/T 8163—2008 Requirement		D/3(38mm)	NO crack

3.5 Corrosion products XRD analysis

The leaking section (BC section) of the failed pipeline was selected and opened in the middle plane along the longitudinal direction. The internal corrosion morphology of the failed pipeline is shown in Fig. 7. The inner surface of the pipeline has corrosion products attached; significant wall thinning at the bottom of the tube and corrosion perforation of local areas were both observed. After the corrosion products were cleaned through physical methods, many corrosion pits and grooves were visible around the leak position (Fig. 8).

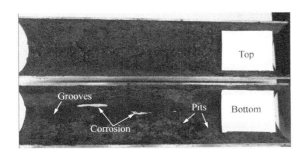
Fig. 7 Internal corrosion morphology of the failed pipeline

Fig. 8 Corrosion pits and groovesin the failed pipe

Corrosion products collected from the inner surface of the pipeline were examined by X-ray diffraction (XRD) analysis, as shown in Fig. 9. The XRD results for the corrosion products are shown in Fig. 10. The results revealed that these corrosion products were composed of ferric oxide (Fe_2O_3), alkali-type ferrous oxide (FeO(OH)), silicon dioxide (SiO_2) and a small amount of ferrous salt carbonate ($FeCO_3$). The ferric oxideis produced by the pipeline material reacting with oxygen dissolved in the oil-water mixture, while FeO(OH) was the reaction product of oxygen in the atmosphere of the pipeline operation and/oroxygen dissolvedin the liquid. $FeCO_3$ was the corrosion product of the pipeline material and CO_2 dissolved in the liquid. SiO_2 in the corrosion products was derived from impurities in the oil-water mixture pipe adhered to the pipeline wall.

Fig. 9　Corrosion products from the failed pipeline

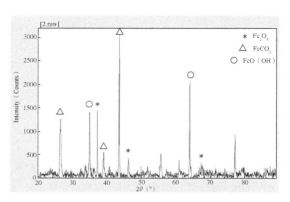

Fig. 10　XRD analysis results of the corrosion products

3.6　SEM and EDS analysis

The inner corroded surface of the failed pipeline was further examined by SEM and EDS, as shown in Fig. 11. The results revealed that the inner surface of the pipeline presented an obvious corrosion morphology; many etched pits and grooves were found near the leakage. The EDS analysis revealed that there were obvious thick corrosion products covering the inner surface of the pipeline, and the major elements of the corrosion products were Fe, O, C and Si. Based on the analysis of the atomic ratio, the corrosion products may be Fe_2O_3 and $FeCO_3$. Small amounts of S, Ca, Si and other elements were also found in the inner surface, as shown in Fig. 12. According to the information provided, sulfates were contained in the oil-water mixture, so the products may be calcium sulfate salt and silica, which are consistent with the XRD analysis.

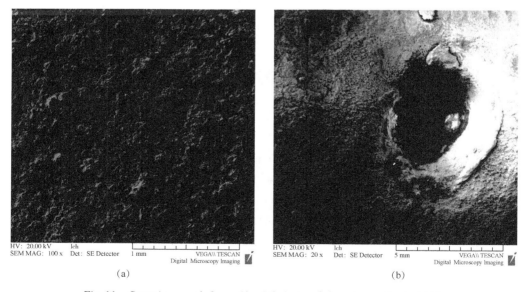

Fig. 11　Corrosion morphology of the failed pipe (a) and near the leak (b)

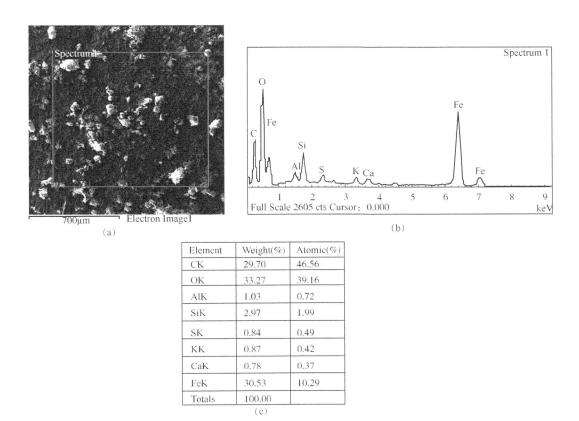

Fig. 12 EDS analysis results(b) and the failed pipe corrosion products(a)

3.7 Corrosion simulation tests

The diameter and weight of the samples were measured before and after the corrosion simulation tests, as shown in Fig. 13. The corrosion rate of the pipeline material was calculated according to the NACE RP0775 − 99 standard[8], as shown in Tab. 4. The results showed that under the field conditions of the working medium, temperature and pressure, the samples obtained after corrosion simulation tests presented apparent corrosion characteristics according to the corrosion degree classification rules in NACE standard RP0775−99, as shown in Tab. 5. The corrosion rate of the pipeline materials in this working condition was calculated to be 0.174357mm/a, which is classified as a serious corrosion level.

Fig. 13 Sample surface morphology of the failed pipe before (a) and after (b) the corrosion simulation test

Table 4 Results of the samples corrosion rates calculation

Items		Weight before test (g)	Weight after test (g)	Weight loss (g)	The sample surface area (mm^2)	corrosion rates (mm/a)
samples	4-1	11.7577	11.7298	0.0279	14.2909	0.181551
	4-2	11.6316	11.609	0.0226	14.22141	0.147781
	4-3	11.5057	11.4762	0.0295	14.15972	0.193741
Average corrosion rates						0.174357

Table 5 Classification of the corrosion rates in NACE RP-0775-99

Classification	Corrosion rate (mm/a)
Slight corrosion	<0.025
Moderate corrosion	0.025~0.125
Serious corrosion	0.126~0.254
Extremely Serious corrosion	>0.254

4 Discussion

The results obtained indicate that the chemical composition, mechanical properties and microstructure of the failed pipelinewere all in accordance with the relevant technical requirements of the standards. Macromorphological observation of the pipeline indicates that the outer surface of the pipeline is intact with no obvious corrosion. This illustrates that the outer soil corrosion of the pipeline was very slight, so this failure was caused by internal corrosion perforation of the pipeline from the inside to the outside.

The leaking section of the pipeline was opened longitudinally, showing that there were thinning areas all around the lower part of the pipeline above the corner; this is a typical local corrosion phenomenon. There was a mass of corrosion products covering the surface of the corrosion and perforation site; these corrosion products were thicker and easily peeled off the pipeline. The inner structure of the corrosion products was loose and porous. There were many corrosion pits and grooves under the corrosion product, so local corrosion very easily occurred under these loose corrosion products.

The SEM and XRD analysis results revealed that the corrosion products were composed of ferric oxide, alkali-type ferrous oxide, silicon dioxide, a small amount of ferrous salt carbonate and calcium sulfate. The ferric oxide was produced by the pipeline material reacting with oxygen dissolved in the oil-water mixture. FeO(OH) is a reaction product from the corrosion products reacting with oxygen in the atmosphere of the pipeline operation and/or oxygen dissolved in the liquid[9]. $FeCO_3$ is obtained by Fe^{2+} reacting with CO_3^{2-} generated from CO_2 dissolved in the liquid[10-12]. Silicon dioxide and calcium sulfate werederived from mineral salts and gravel conveyed with the oil-water mixture. From the analysis results of the oil-water mixture conveyed in the pipeline, the liquid contained a large amount of formation water, which was high in Cl^-, Ca^{2+} and salinity. Previous studies have reported that the existence of Ca^{2+} ions could significantly improve the ionic strength of the liquid, the electrical conductivity of the medium and the scaling tendency, which accelerated the local corrosion of the pipeline[13]. A large number of Cl⁻ ions penetrated into the pores and defects of the corrosion product film, which covered the steel surface and resulted in a microrupture of the corrosion product film. The nucleation of pitting corrosion also occurred. As a result of the constant accumulation of Cl⁻ ions, the pitting corrosion of the pipeline was also accelerated under the action of the occluded cell[14-17]. Because the formation water was acidic (the pH value was 5.0~6.5) and had high salinity, the oil-water mixture conveyed in the pipeline was strongly corrosive, which was the internal cause of corrosion and perforation of the pipeline[18].

According to the information provided by the oil field, corrosion perforations of the pipeline could only be found in positions with a large drop height. Moreover, the leakages often occurred approximately 200~400 mm from the upstream welding seam of pipeline elbows, as shown in Fig. 1. The velocity of the transmission fluid decreased at these climbing sections of the pipeline, and then the oil-water separation, sediment deposition and deposition of corrosive media took place in these positions. The corrosion product film and $CaCO_3$ incrustation easily formed beneath the sediment. These corrosion product films and incrustations did not completely prevent corrosion of the pipeline due to the poor uniformity and compactness of the film, which is composed of corrosion products and incrustation. Some weak spots existing in the film became anodes for the corrosion reaction, and the residual surface protected by the film became the cathode for the corrosion reaction. It is common knowledge that a large cathode coupled with a small anode has an important effect on acceleration of corrosion[19]. Localized corrosion is very serious near these weak spots; corrosive pitting can form on the inner surface and reduce the wall thickness of the pipeline. In the worst cases, the corrosive pitting will lead to leakage of the pipeline[15].

At the same time, the bottom of the pipeline was covered by scale and corrosion products, which were scoured by many small solid particles and gravel, resulting in the corrosion product film being destroyed and corrosion occurring at the bottom of the pipeline[20]. With increased accumulation of a high concentration of Cl⁻ ions in the corrosion pits, the dissolution rate of anodic Fe in the corrosion pits was accelerated, so the wall thickness of the pipeline continuously decreased over a short time. The pipeline leaked because the remaining wall thickness was not enough to bear the pipeline working pressure.

To prevent such leak failures, the crude oil pipeline lifting segment should be upgraded with an internal anti-corrosion coating. A corrosion inhibitor should also be added to the conveyed liquid to decrease scaling and reduce the corrosion rate. Deoxygenation should be carried out regularly in the pipeline in order to reduce the salinity of the conveyed crude oil, especially the contents of Cl⁻ and Ca^{2+} ions, which could decrease the corrosion rate. At the same time, the design of the pipeline structure should be optimized when the pipeline crosses ravines; a sudden large height deviation in the pipeline should be avoided in the pipeline design, and large radius elbows are proposed for use in the corners of the pipeline. Regular pigging operation is also recommended. The crude oil pipeline in this oilfield did not leak again after following these suggestions.

5 Conclusions

The chemical composition, mechanical properties and microstructure of thefailed pipeline are all in accordance with the relevanttechnical requirements of the standards. The corrosion rate of the pipeline in the working conditions is 0.174357 mm/a, which is a serious corrosion level. The crude oil conveyed in the pipeline contained a large amount of formation water and was very corrosive. Because the fluid flow changed suddenly at the pipeline lifting location, the fouling deposition occurred at the inner bottom of the pipeline, which resulted in serious localized corrosion caused by the small anode/big cathode corrosion galvanic cell reaction. The wall thickness of the pipeline decreased significantly in a short time. Finally, the pipeline lifting segment leaked due to fluid erosion and the high speed impact of solid particles and grit in the crude oil.

Acknowledgment

The author thanks the financial supports from the Essential Research and Strategic Reserve Technology Research Fund Program of China national petroleum corporation with Grand No. 2015Z-05, and the Shaanxi Innovation Capability Support Program with Grand No. 2018KJXX-046.

References

[1] GB/T 699-1999 Quality Carbon Structure Steel[S]. National Standard of the People Republic of China.

[2] J.-GHu, H.-j Luo, Z.-h Zhang, et. al, Analysis on corrosion failure of an oil pipeline in Changqing oilfield [J], Corrosion & Protection, 2018, 39(12): 962-970.

[3] H. Ding, G.-f Lin, Y. Long, et. al, Failure Analysis of Ground Pipeline in the Western Oil Field [J], Petroleum Tubular Goods & Instruments, 2017, 3(1): 56-63.

[4] P.-c Qin, C.-b Xiong, Z. LI, et. al, Failure Pressure Assessment of Submarine Pipelines Considering theEffects of Multiple Corrosion Defects Interaction [J], Surface Technology, 2020, 49(1): 237-244.

[5] GB/T 228.1-2010 Metal materials-Tensile testing-Part 1: Method of test at room temperature [S]. National Standard of the People Republic of China.

[6] GB/T 8163-1999 Seamless Steel Tubes for Liquid Service[S]. National Standard of the People Republic of China.

[7] GB/T 6394-2002, Metal-methods for estimating the average grain size [S]. National Standard of the People Republic of China.

[8] NACE Standard RP0775-99 Preparation, Installation, Analysis, and Interpretation of Corrosion Coupons in Oilfield Operations [S]. NACE International, 1999.

[9] Y.-P. Nin. Study of the Oil Transportation Pipeline Corrosion and Corrosion Rate Prediction [J]. Pipeline Technique and Equipment. 2011, 32(5): 45-47.

[10] Nesic S. Key Issues Related to Modeling of Internal Corrosion of Oil and Gas Pipelines A Review [J]. Corrosion Science, 2007, 49(12): 4308-4338.

[11] Ogundele G. I., White W. E., Some Observation on Corrosion of Carbon Steel in Aqueous Environment Containing Carbon Dioxide[J]. Corrosion, 1986, 42(2): 71~77.

[12] G.-x ZHAO, H.-x LV. Analysis of the Reason on Tubing Corrosion Failure [J]. Journal of Materials Engineering, 2010, 3: 51-55.

[13] C.-z ZHANG. Metal corrosion and protection[M]. Metallurgical industry press, 2000. 105.

[14] Fierro G, Ingo G M, Mancla Fi. XPS-investigation on the corrosion behavior of 13Cr martensitic stainless steel in CO_2-H_2S-Cl^- environment[J]. Corrosion, 1989 (10): 814.

[15] W.-j ZHOU, D. GUO, Y. ZHANG, H.-p ZHU, Comprehensive Research to the Corrosion Failure Behavior of

X52 Pipeline Steel in Acidic Corrosive Medium[J]. Pipeline, 2009, 23(9): 13-18.

[16] L.-p WAN, Y.-f MENG, C.-x WANG, L. YANG, Mechanism of Corrosion and Scale Deposit of Tubes in Western Oilfield[J]. Journal of Chinese Society for Corrosion and Protection, 2007, 27(8): 247-251.

[17] Z.-c LIAO, B.-q LIN, J.-c ZHANG, The corrosion and protection of oil and gas well[J]. Storage, Transportation & Preservation of Commodities, 2008, 30(6): 115-118.

[18] Bassam G. N. M., MounaA., Mohammed H. M., et. al. Inspection of internal erosion-corrosion of elbow pipe in the desalination station[J]. Engineering Failure Analysis, 2019, 102(8): 293-302.

[19] A. Colombo, L. Oldani, S. P. Trasatti, Corrosion failure analysis of galvanized steel pipes in a closed water cooling system[J]. Engineering Failure Analysis, 2018, 84(2): 46-58.

[20] S. Arielya, A. Khentovb, Erosion corrosion of pump impeller of cyclic cooling water system[J]. Engineering Failure Analysis, 2006, 13(6): 925-932.

本论文原发表于《Engineering Failure Analysis》2020 年第 113 卷。

Failure Analysis of the 13Cr Valve Cage of Tubing Pump Used in an Oilfield

Long Yan[1, 2] Wu Gang[1, 2] Fu Anqing[1,2] Xie Junfeng[3] Zhao Mifeng[3]
Bai Zhenquan[1,2] Luo Jinheng[1,2] Feng Yaorong[1,2]

(1. Tubular Goods Research Institute, China National Petroleum Corporation;
2. State Key Laboratory for Performance and Structure Safety of Petroleum
Tubular Goods and Equipment Materials; 3. Tarim Oilfield Company,
PetroChina Company Limited)

Abstract: The 13Cr valve cage of tubing pump was fractured after 191 days service in an oilfield in China. The failure causes were analyzed by direct-reading spectrometer, Rockwell hardness tester, optical microscope, scanning electron microscopy, energy dispersive spectrometer and finite element method in this paper. The results showed that the failure reason was typical corrosion fatigue fracture initiating at the root of the last thread next to the thread relief groove, where lots of corrosion pits and cracks were observed. In general, co-effect of two factors occurred to cause the valve cage to corrosion fatigue fracture. The downhole medium with high CO_2 and Cl^- concentrations was the environmental factor of the failure. And alternating stress caused by pumping cyclic motion was the main stress factor of the failure, while the high stress concentration on thread connection further induced cracking.

Keywords: Tubing pump; 13Cr Martensitic stainless steel; Corrosion fatigue; Stress concentration

1 Introduction

In recent years, the continuous increase in energy demand has caused the decline of the availability of petroleum resource in the world. Artificial lift systems to produce oil are the most effective means to enhance production capacity of oil fields[1,2]. As a consequence, a lot of subsurface sucker rob pumps including tubing pumps and insert pumps are used in world's oil fields. The American Petroleum Institute provides the requirements and guidelines for the design of subsurface sucker rod pumps in API Spec 11AX[3]. Tubing pumps are the oldest type of sucker rod pumps with a simple construction, including working barrel, plunger, travelling valve and standing valve, as shown in Fig. 1. Due to the relatively larger barrels to allow more fluids to pass than any

---

Corresponding author: Long Yan, longyanol@cnpc.com.cn.

other type of pumps, tubing pumps are highly efficient and widely used in petroleum industry[4]. However, the high impact of the ball in travelling valve and the reciprocating motion of the rod string could cause pump components to fracture easily[5].

With the development of the deeper wells, the downhole temperature is increasing significantly, and the downhole medium usually contains substantial amounts of salt water, carbon dioxide, and hydrogen sulfide (sour wells)[6]. The aggressive environment may cause cracking to speed up, such as corrosion fatigue characterized by a unexpected and sudden failure[7,8]. The corrosion fatigue failure of tubing pumps is catastrophic for oilfield production, causing huge economic losses. It is very important to evaluate and study the affect of materials, environment, stresses, etc. on the corrosion fatigue behavior of tubing pumps.

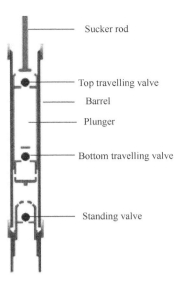

Fig. 1 Structural schematic diagram of a tubing pump with two travelling valves

Due to the better corrosion resistance and mechanical property, 13Cr stainless steel has been widely used in petroleum equipment[9]. In the current work, the failure reason of the 13Cr valve cage of tubing pump was studied by chemical analysis, metallographic analysis, fracture morphology analysis, corrosion products analysis and stresses simulating, which could provide some suggestions for the material applicability and manufacture of tubing pump to avoid the similar failure in corrosive environment.

2 Background of the Failure

The ϕ38.10mm tubing pump used in an oilfield was radically fractured on the top travelling valve in 3801 m depth downhole. The environment medium is oil-water mixture with some natural gas. The CO_2 content of natural gas is up to 23.21 mol%. The formation water with 128500mg/L chloride content is mainly calcium chloride type.

The top travelling vale cage composed of a top open valve cage and a bottom closed valve cage which were connected through thread. The fracture surface is located near the last pin thread of the open valve cage, as shown in Fig. 2(a), and another part of the fracture pin thread remains in the box thread of adjacent closed valve cage. Fig. 2(b) shows the inner morphology of the closed valve cage by section. It can be seen the valve ball and valve seat in closed valve cage are undamaged without obvious general corrosion.

3 Experimental Method

All the failure analysis specimens were cut by wire electrical discharge machining. The chemical composition of the failed open valve cage was analyzed using the direct-reading spectrometer. The hardness of the failed open valve cage was measured using Rockwell hardness tester. The microstructure of the failed open valve cage was analyzed using the metallographic

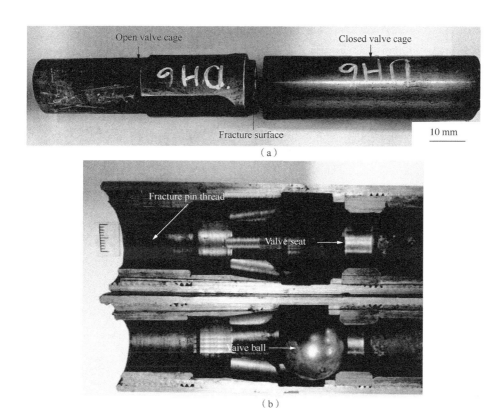

Fig. 2 The failed valve cages(a) and inner morphology of the closed valve cage by section(b)

microscope, meanwhile, the morphology of the cracks near the fracture surface was also observed. The morphology of the fracture surface of the failed open valve cage was analyzed using the optical camera and the scanning electron microscope (SEM). The element distribution of the corrosion products on the fracture surface and the cracks was analyzed using the energy dispersive spectrometer (EDS). Additionally, the finite element modeling (FEM) of the thread connection was created in order to calculate the stresses distribution, which could further determine the cause of the failure.

4 Experimental Results

4.1 Chemical composition and mechanical property

The chemical composition of the failed open valve cage is listed in Table 1. The result shows that the failed open valve cage is compliant with GB/T 20878—2007[10] for 3Cr13 stainless steel. The Rockwell hardness test result of the failed open valve cage is listed in Table 2. The test result shows the failed valve cage has high hardness, and the average hardness is about 38.7 HRC.

Table 1 Chemical composition of the failed open valve cage (wt, %)

Element	C	Si	Mn	P	S	Cr	Mo	Ni	Nb	V	Ti	Cu	Al
Measured	0.30	0.30	0.59	0.016	0.0086	12.11	<0.06	<0.10	<0.03	<0.04	<0.03	<0.04	<0.02

续表

Element	C	Si	Mn	P	S	Cr	Mo	Ni	Nb	V	Ti	Cu	Al
GB/T 20878—2007	0.26~0.35	≤1.00	≤1.00	≤0.040	≤0.030	12.00~14.00	—	≤0.60	—	—	—	—	—

Table 2 Rockwell hardness of the failed open valve cage (HRC)

Position	A	B	C
Measured	38.7	38.3	39.0

4.2 Fracture morphologies

In general, the fracture surface of the failed open valve cage is perpendicular to the axial direction and consists of three typical regions of crack initiation, crack propagation and final fracture, as shown in Fig. 3(a). The crack initiation region is relative flat and bright. On the contrary, the crack propagation and final fracture regions show dark color, indicating the corrosion feature. The surface of final fracture region is rougher and the size is smaller, just less than 1/10 of the area of the whole fracture, which implies the failure is caused by brittle fracture under low stress. It is clear that the fracture initiates at the root of the last thread next to the thread relief groove, as shown in Fig. 3(b). In addition, some beach arcs could be observed on the innermost portion of the fracture surface, as shown in Fig. 3(c). Beach arc is the typical macro characteristic of fatigue crack propagation[11].

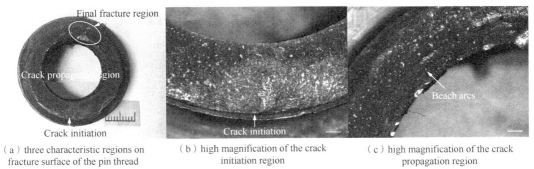

(a) three characteristic regions on fracture surface of the pin thread (b) high magnification of the crack initiation region (c) high magnification of the crack propagation region

Fig. 3 Macro morphologies of the fracture surface

According to the SEM analysis, the micro morphologies of the fracture surface are shown in Fig. 4. The fracture surface presents mainly transgranular feature without secondary cracks. According to Fig. 4(a), fatigue striations can be found near the crack initiation region at high magnification, further confirming the fatigue feature of the fracture surface. Meanwhile, there are lots of corrosion products observed on the fracture surface. The main morphology of the corrosion product scales is compact and crystal like, as shown in Fig. 4(b). From above, it can be deduced that the corrosion may play an important role in the failure of the open valve cage.

4.3 Cracks morphology and microstructure

After cutting the fracture into two parts, the cross section analysis is shown in Fig. 5(a). Many corrosion pits and parallel cracks can be observed on the surface of the thread relief groove near the fracture surface where structure and dimension change most greatly. It is also seen the cracks are transgranular and perpendicular to the axial, initiating at the bottom of the corrosion pits. All the cracks look relatively straight and have no branches. The microstructure of the failed open valve cage is tempered martensite which is the normal microstructure of 13Cr martensitic stainless steel as shown in Fig. 5(b).

(a) the fatigue striations near the crack initiation (b) the corrosions products on the crack propagation region

Fig. 4 Micro morphologies of the fracture surface

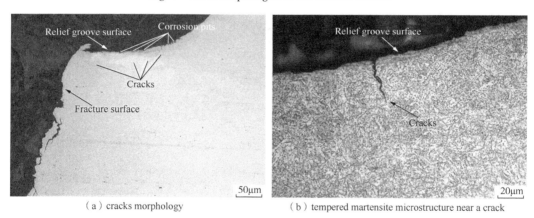

(a) cracks morphology (b) tempered martensite microstructure near a crack

Fig. 5 Cross section of the failed open valve cage near fracture surface

4.4 EDS analysis

Fig. 6 shows the compositions of the corrosion products on the fracture surface and in the crack of the threadrelief groove based on EDS analysis results. As shown in the table of Fig. 6, there are significant differences in the composition among the two test positions. The corrosion products on fracture surface contain higher Ca, C, O contents, but the contents of Cr, Fe are below the original chemical composition given in Table 1. By contrast, the corrosion products in the crack contain higher Cr, Fe contents and a few Cl contents, especially the content of Cr is up to 25.38

wt. %, about ten times than that on the fracture surface. The composition variation of the corrosion products on the fracture surface is shown in Fig. 7, based on EDS line scanning analysis results. It is clear that the element distribution varies with the thickness of the corrosion products on the fracture surface. The line scanning graphs reveal that there are sudden rises in the contents of Fe, O, Cr at about 15 μm thickness position. On the contrary, there is a sudden dropping in the content of Ca at in same thickness position. Meanwhile, C content peaks on the outermost surface and reaches the minimum on the matrix surface. From above results, it can be deduced that corrosion products on the fracture surface may have a multilayer structure and element distribution of the scale nearest the fracture surface is similar with element distribution in crack.

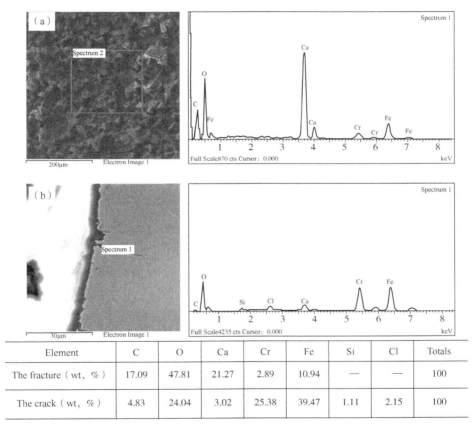

Element	C	O	Ca	Cr	Fe	Si	Cl	Totals
The fracture (wt, %)	17.09	47.81	21.27	2.89	10.94	—	—	100
The crack (wt, %)	4.83	24.04	3.02	25.38	39.47	1.11	2.15	100

Fig. 6 EDS analysis results:
(a) the fracture; (b) the crack

5 Discussion

As mentioned above, the failure characteristics of the open valve cage are discussed as follows. First, the fracture surface is flat with beach arcs, where obvious fatigue striations and corrosion products could be observed through SEM analysis. Second, some localized corrosion such as pitting is found on the surface of the relief groove. Third, most of cracks are transgranular and flat without bifurcation, initiating at the bottom of the corrosion pits. According to above characteristics, it can be inferred that the failure type is the corrosion fatigue, caused by the combined effect of

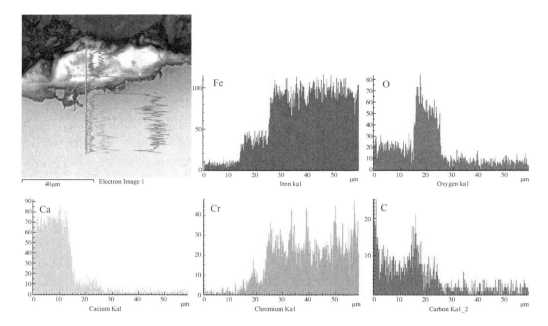

Fig. 7　EDS line scanning analysis results of the cross section of the fracture specimen

corrosive medium and stresses.

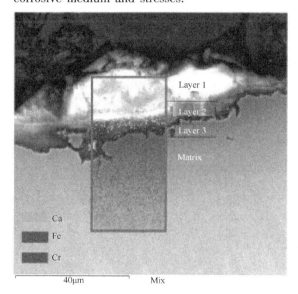

Fig. 8　EDS mapping analysis results of the cross section of the fracture specimen

5.1　Corrosion mechanism of the 13Cr valve cage

The main element distribution such as Ca, Cr and Fe elements on the surface of the failed valve cage fracture is shown in Fig. 8, based on EDS mapping analysis results. The innermost layer (layer 3) is thin and inhomogeneous, and composes of high Cr content and a few Fe content. The high Fe concentrates in outer layer of the scales (layer 2), which indicates 13Cr stainless steel had appeared obvious anodic reaction. And the scales are eventually covered by salt sediment containing Ca element (layer 1) from formation water. It is seen that the scales on the surface has a complex multilayer structure. In CO_2 corrosion environment, two main types of the scales on the surface of 13Cr stainless steel are passive film and corrosion product scale, respectively. The main passive film of 13Cr stainless steel is $Cr(OH)_3$ which has non-crystalline structure and could prevent aggressive anions such as Cl^- to transfer to the metal surface. And the formation process of $Cr(OH)_3$ can be expressed as Eq(1)[12]:

$$Cr + 3H_2O \longrightarrow Cr(OH)_3 + 3H^+ + 3e \qquad (1)$$

The corrosion product scale of 13Cr stainless steel mainly composed of $FeCO_3$ followed by the anodic reaction of Fe at high temperature, the process can be expressed as following equations[13,14]:

$$Fe \longrightarrow Fe^{2+} + 2e \qquad (2)$$

$$Fe^{2+} + CO_3^{2+} \longrightarrow FeCO_3 \qquad (3)$$

$$Fe^{2+} + 2HCO_{3+} \longrightarrow Fe(HCO_3)_2 \longrightarrow FeCO_3 + CO_2 + H_2O \qquad (4)$$

In some cases, the formation and deposition of $FeCO_3$ could repair the breakdown of passive film to prevent 13Cr stainless steel from further corrosion.

Corrosion resistance of 13Cr stainless steel greatly depends on the structure of the scales, and protective scales on the surface of the steel should be stable and compact, as shown in Fig. 9(a). However, the thicker scale is not always better for corrosion resistance. If a thick scale is easy to be broken and can not be timely repaired or repassivated, the aggressive Cl^- will easily penetrate the passive film and the anodic reaction of Fe will be accelerated, both of which may result in the serious localized corrosion, even the initiation of cracks, as shown in Fig. 9(b). The Characteristics of scales on surface of the failed open valve cage are similar with the broken-type scale. A lot of corrosion pits are observed on the surface of the failed valve cage, which shows that the scales have no sufficient protective ability to restrain localized corrosion. In addition, Cl element concentrates in the crack(Fig. 6), which also explains Cl^- had transferred to the metal through the scales. From above, it can be inferred that conventional 13Cr stainless steel may be likely to fail due to the downhole medium containing high CO_2 and Cl^- concentrations at high temperature. There are some environment limits for selection of the corrosion resistance alloys commonly used in oil and gas production, including temperature, pH value and corrosion medium concentration[15]. In severe environment, modified stainless steels such as super 13Cr with more Ni, Mo, Cu alloy elements are considered a suitable choice of material and have a better performance on corrosion pitting resistance than conventional 13Cr[16-18].

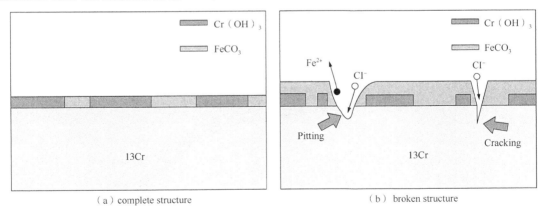

(a) complete structure　　　　　　　　　　　(b) broken structure

Fig. 9　Schematic diagrams of the scales on the surface of 13Cr stainless steel

5.2　stresses analysis of the 13Cr valve cage

The partial structure and force of the tubing pump are shown in Fig. 10. The failed open valve cage upon the closed valve cage is connected to the lower end of the rod string, while it is directly moved by the rod string. In general, the main force on open valve cage varies with pumping cycle. In the stage of upstroke, the high hydrostatic pressure(P_1) acts on the travelling valve ball due to the closing of the ball and the decreasing of plunger pressure(P_p), which causes the open valve cage to withstand the high pulling force, as F_1 shown in Fig. 10(a). When the downstroke stage begins, the plunger pressure(P_p) increases, causing the travelling valve ball to open and arising the pushing force, as F_2 shown in Fig. 10 (b), which will further cause the open valve cage to be compressed. Due to alternating loads on the valve cage with pumping cyclic motion, the stresses of the thread connection are always varying, making the stresses analysis very complicated. For simplicity, FEM analysis method is performed to assess the stresses distribution on the cross section of the thread connection of valve cages before fracturing. The simulated serviced loads are 16.9 kN pulling force F_1 for upstroke and 32.3kN pushing force F_2 for downstroke, respectively, based on severer condition. The dimensions of the pin and box threads model are the same with failed valve cage, belonging to straight threads. The stresses distribution results given by FEM analysis are shown in Fig. 11. It is shown that the stresses in two pumping cycle stages concentrate at the root of the last thread and adjacent relief groove for both pin and box threads. Under alternating loading, high stress concentration is also a potential critical location to induce fatigue failure, which can be proved by the distribution characteristics of the fracture and cracks of the failed open valve cage. And stresses distribution can be influenced by structure parameters, pumping depth, pressure variation and fluid friction force, etc. As changing other downhole factors can not be easily achieved, adjusting the structure of the thread connection is the best way to avoid stress concentration. For instance, the maximum stress on the relief groove could significantly decrease through adding the length L, as shown in Fig. 12.

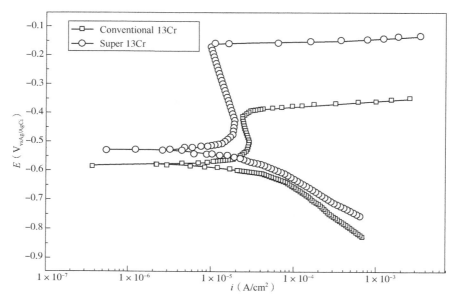

Fig. 10 Potentiodynamic polarization curves of conventional 13Cr stainless steel and super 13Cr stainless steel in CO_2 saturated formation water at 90℃

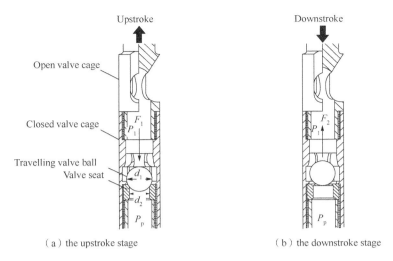

(a) the upstroke stage (b) the downstroke stage

Fig. 11 Schematic diagrams of partial structure and force with pumping cycle (d_1 is the diameter for valve ball, 23.83mm; d_2 is the diameter for valve seat, 18.02mm)

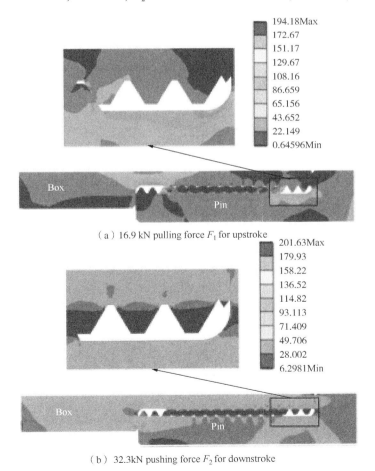

(a) 16.9 kN pulling force F_1 for upstroke

(b) 32.3kN pushing force F_2 for downstroke

Fig. 12 Mechanical stresses (von Mises stress, MPa) contours on the cross section of the threads

6 Conclusions and suggestions

In this work, the fracture reason of a 13Cr open valve cage used in tubing pump had been investigated. By analyzing test results, following conclusions and suggestions are made:

(1) The fracture reason of the failed valve cage is corrosion fatigue. The fractureinitiates at the root of the last thread next to the thread relief groove, where many corrosion pits and cracks can be observed.

(2) The downhole medium with high CO_2 and Cl^- concentrations is the environmental factor of the failure. The scales on the surface of the failed valve cage are not protective enough to avoid corrosion pitting and penetration of aggressive Cl^-.

(3) Alternating stress caused by pumping cyclic motion is the main stress factor of the failure, and the high stress concentration on thread connection can further induce cracking.

(4) It is suggested to select stainless steels with higher corrosion resistance to restrain localized corrosion and optimize the structure of the thread connection such as relief groove to reduce the stress concentration.

References

[1] S. M. Bucaram, J. D. Clegg, N. W. Hein. Recommendations and Comparisons for Selecting Artificial-Lift Methods[J]. Journal of Petroleum Technology, 1993, 45(12): 1163-1167.

[2] J. A. Villasante, L. O. Mantovano, H. A. Ernst, et al. Development of a new hollow sucker rod family for rotating pumping(progressive cavity pump systems)[J]. Journal of Petroleum Science & Engineering, 2015, 134: 277-289.

[3] API Specification 11AX-2015, Specification for Subsurface Sucker Rod Pump Assemblies, Components, and Fittings, Thirteenth edition.

[4] G. Takacs. Chapter 3-Sucker-Rod Pumping System Components and Their Operation[M]. Sucker-Rod Pumping Handbook. Elsevier Inc. 2015: 57-246.

[5] Y. S. Zhang, N. Y. Song. Improvement Design of Floating Valve in Down Hole Pump[J]. Oil Field Equipment, 2008, 38(10): 57-59.

[6] R. M. Schroeder, I. L. Müller. Fatigue and corrosion fatigue behavior of 13Cr and duplex stainless steel and a welded nickel alloy employed in oil and gas production[J]. Materials & Corrosion, 2015, 60(5): 365-371.

[7] J. Sun, S. Y. Chen, Y. P. Qu, et al. Review on Stress Corrosion and Corrosion Fatigue Failure of Centrifugal Compressor Impeller[J]. Chinese Journal of Mechanical Engineering, 2015, 28(2): 217-225.

[8] X. Ren, F. Wu, F. Xiao, et al. Corrosion induced fatigue failure of railway wheels[J]. Engineering Failure Analysis, 2015, 55: 300-316.

[9] M. Cayard, R. Kane. Serviceability of 13Cr Tubulars in Oil and Gas Production Environments [J]. Corrosion, 1998.

[10] GB/T 20878—2007, Stainless and heat-resisting steels-Designation and Chemical composition, 2007.

[11] F. V. Ellis. Fatigue and Corrosion Fatigue Failures[C]. ASME 2005 Pressure Vessels and Piping Conference, 2005: 191-197.

[12] G. X. Zhao, M. Zheng, X. H. Lv, et al. Effect of temperature on anodic behavior of 13Cr martensitic steel in CO_2 environment[J]. Metals & Materials International, 2005, 11(2): 135-140.

[13] Y. D. Cai, P. C. Guo, D. M. Liu, et al. Comparative study on CO_2 corrosion behavior of N80, P110, X52

and 13Cr pipe lines in simulated stratum water[J]. Science China Technological Sciences, 2010, 53(9): 2342-2349.

[14] H. Zhang, Y. L. Zhao, Z. D. Jiang. Effects of temperature on the corrosion behavior of 13Cr martensitic stainless steel during exposure to CO_2 and Cl^-, environment[J]. Materials Letters, 2005, 59(27): 3370-3374.

[15] L. Smith. Control of corrosion in oil and gas production tubing[J]. British Corrosion Journal, 2013, 34(4): 247-253.

[16] R. M. Moreira, C. V. Franco, C. J. B. MJoia, et al. The effects of temperature and hydrodynamics on the CO_2 corrosion of 13Cr and 13Cr5Ni2Mo stainless steels in the presence of free acetic acid[J]. Corrosion Science, 2004, 46(12): 2987-3003.

[17] X. W. Lei, Y. R. Feng, A. Q. Fu, et al. Investigation of stress corrosion cracking behavior of super 13Cr tubing by full-scale tubular goods corrosion test system[J]. Engineering Failure Analysis, 2015, 50: 62-70.

[18] G. O. Ilevbare, G. T. Burstein. The role of alloyed molybdenum in the inhibition of pitting corrosion in stainless steels[J]. Corrosion Science, 2001, 43(3): 485-513.

本论文原发表于《Engineering Failure Analysis》2018年第93卷。

Corrosion Behavior Research of Aluminized N80 Tubing in Water Injection Well

Lu Caihong[1,2]　Feng Chun[1]　Han Lihong[1]　Feng Jie[3]
Zhu Lijuan[1,2]　Jiang long[1,2]　Feng Yaorong[1,2]

(1. Tubular Goods Research Center of China National Petroleum Corporation;
2. State Key Laboratory of Performance and Structure Safety of Petroleum Tubular Goods and Equipment Materials; 3. Tianjin Dagang Oilfield New Century Machinery Manufacturing Co. (LTD))

Abstract: In this paper, in view of the problem of tubing corrosion in the water injection well, it is proposed to use the aluminizing N80 tubing to improve the corrosion problem. An aluminized layer with a thickness of about 150μm was prepared on the surface of commonly used N80 tubing. The electrochemical polarization test at room temperature and the high temperature autoclave test for simulated water injection well environment were carried out on both of the aluminized and non-aluminized N80 tubing. It is intended to provide a feasibility test basis for the application of aluminized tubing in water injection wells by comparing the corrosion resistance performance of the aluminized and non-aluminized N80 tubing. The results showed that the aluminized layer on the surface of N80 tubing was combined with matrix by metallurgical diffusion, and the thickness of the aluminized layer could reach 150μm. Compared with the non-aluminized N80 tubing, the self-corrosion current density of the aluminized N80 tubing polarization curve decreased obviously in the water injection well environment, and the corrosion rate in the high temperature autoclave test was also reduced to one quarter of that of the non-aluminized N80tubing.

Keywords: Aluminized N80 tubing; Aluminizing layer; Water injection well environment; Electrochemical polarization test; High Temperature autoclave test

1 Introduction

With the oilfield exploratioin development, the number of water injection wells is increasing, and the internal corrosion problem of water injection tubing is becoming more and more serious. Because the water quality of the injection well contains a large amount of dissolved oxygen, microorganisms (sulfate-reducing bacteria, iron bacteria,), mineral ions (Cl$^-$, Ca^{2+}, Mg^{2+}, etc.), and corrosive gases such as CO_2 and H_2S, and the high formation temperature, the inner

---

Corresponding author: Lu Caihong, lucaihong@cnpc.com.cn.

wall of the tubing that is directly in contact with the water is prone to dissolved oxygen corrosion, scale corrosion, microbial corrosion, electrochemical corrosion and temperature corrosion. In order to reduce the corrosion inside the tubing, the domestic injection well oil field mostly adopts epoxy coating or Ni-P coating to prevent the inner wall of the tubing from being treated. However, due to the bonding problem between the coating layer and the substrate, the coating layer bubbles and falls off often occurs, which in turn exacerbates the corrosion of the substrate tubing.

Aluminized surface treatment can improve the corrosion resistance and heat resistance of steel components, and the aluminized layer are prepared by metallurgical diffusion with the matrix. The interface bonding force obtained by this method is the most ideal, and the related research has received extensive attention from scholars from all over the world[1-3]. In this paper, it is proposed to use aluminized N80 tubing to improve the corrosion resistance of tubing in water injection wells and to evaluate the corrosion resistance of aluminized tubing by electrochemical test and simulated high temperature autoclave test.

2　Experimental materials and processes

The test material selected the commonly used N80 tubing, and the samples were cut from the tubing for surface aluminizing treatment. The aluminizing process was degreasing→water washing→rust removal→assisting→drying→aluminizing→air cooling to room temperature. The aluminizing process parameters were as follows: the aluminizing temperature was selected to be 740℃, and the aluminizing time was 5 minutes.

VEGAII scanning electron microscope was used to analyze the microstructure of the aluminized sample, and the thickness of the layer was measured. The ϕ15mm disk was cut from the tubing, and after aluminizing, the copper wire was welded on one side, and the non-measured area was sealed with epoxy resin, leaving only the measured surface to be tested, conducting electrochemical polarization test. The test used a three-electrode system, the reference electrode was saturated calomel, the auxiliary electrode was platinum electrode, and the working electrode was aluminized sample to be tested. The test solution was the harsh water of the injection well. The composition is shown in Table 1. After the ablation of the coupon sample from the tubing and aluminizing, the high temperature autoclave test was carried out. The solution still selected the harsh water of the injection well, the temperature was 80℃, and the partial pressure of CO_2 gas was 0.2MPa, test time was 7 days. All the above tests were carried out by parallel test with non-aluminized tubing.

Table 1　Harsh water component of an oil field

Temperature (℃)	Mineralization degree (mg/L)	Cl^- (mg/L)	HCO_3^- (mg/L)	CO_3^{2-} (mg/L)	SO_4^{2-} (mg/L)	Ca^{2+} (mg/L)	Mg^{2+} (mg/L)	K^+Na^+ (mg/L)
104	38218	11001	3853	156	2491	1023	194	12150

SRB(CFU/mL)		Iron bacteria(CFU/mL)		Saprophytic Bacteria(CFU/mL)		Dissolved oxygen(CFU/mL)		
25000		60		1100000		None		

3　Experimental results and discussion

(1) Analysis of the Microstructure of the Layer. As the Al concentration increases, the

intermetallic compound between the Fe-Al solid solution may be, in order, β1 (Fe₃Al) and β2(FeAl) FeAl₂ (ζ phase), Fe₂Al₅ (η phase), and FeAl₃ (θ phase)[4]. The composition of the aluminized layer is also closely related to the temperature and time of aluminizing. It is generally believed that in the aluminizing process, the infiltration and diffusion between the liquid Al and the solid iron, thereby forming a metallurgical bond, is the key to obtaining a dense and effective aluminized layer. Therefore, the aluminized layer is inwardly directed from the surface layer and is generally divided into a pure Al layer and a Fe – Al intermetallic compound layer, as shown in Fig. 1. The total thickness of the layer is 150μm, wherein the thickness of the pure aluminum layer is 30μm, and the layer and the matrix are metallurgically bonded, and the bonding force is strong.

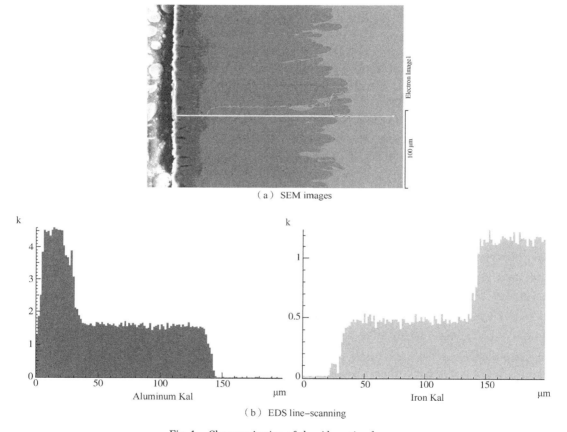

Fig. 1 Characterization of the Al-coating layer

(2) Electrochemical Test Results and Analysis. The self-corrosion potential of N80 steel after aluminizing is lower than that of non-aluminized N80 steel, as shown in Fig. 2. Because the corrosion potential of aluminum is much lower than that of iron, self-corrosion potential of electrode surface after aluminizing was decreased. Although the tendency of corrosion on the surface of the electrode is enhanced from the viewpoint of thermodynamics, the reaction rate on the surface of the electrode is controlled by the electrochemical corrosion kinetics of the electrode surface, that is, the chemical reaction rate of the anode and the cathode. After the surface of the electrode is aluminized,

the resistance of the anode-anode chemical reaction were significantly enlarged[5]. The dissolution rate of the metal decreased, and the self-corrosion current density of the aluminized N80 steel on the polarization curve was significantly reduced. It is thus shown that the aluminizing treatment significantly increases the corrosion resistance of the steel and reduces the corrosion rate, effectively improving the corrosion resistance of the N80 tubing. It can be seen from the polarization curve that there is no obvious Tafel region in the cathode strengthening region of different materials, because the cathode reaction is mainly controlled by diffusion; there is obvious Tafel region in the strong polarization region of the anode, which is due to The anodic dissolution reaction of the sample is completely controlled by activation.

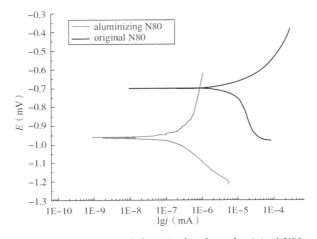

Fig. 2 Polarization curves of aluminized and un-aluminized N80 steel

(3) Experimental Results and Discussion of High Temperature Autoclave in Simulated Water Injection Well Environment. The aluminized and non-aluminized N80 tubing were tested in a high temperature autoclave for simulating the injection well environment. The samples were cleaned and weighed. The uniform corrosion rate was calculated according to Eq. 1. The calculation results are shown in Table 2. The average corrosion rate of the aluminized N80 steel is reduced to about one quarter of the non-aluminized N80 steel.

$$V = g \cdot 365000/\gamma \cdot t \cdot S \tag{1}$$

In the Equation:

g is the weight loss of the sample, g;

γ is the specific gravity of the material, g/cm^3, and γ is 7.8 g/cm^3 for N80 steel;

t is the experimental time, day;

S is the sample area, mm^2;

V is the average corrosion rate, mm/a.

Table 2 Table of corrosion loss of aluminized N80 steel and non-aluminized N80 steel

Sample	Area(mm^2)	Weight loss(g)	Corrosion rate(mm·a^{-1})
Non-aluminized N80	1105.9	0.0607	0.367
Aluminized N80	1198.7	0.0156	0.087

After the aluminized N80 steel sample and the non-aluminized N80 steel sample were subjected to a 7-day high temperature autoclave test, the sample was taken out from the autoclave and analyzed by a scanning electron microscope. The results are shown in Fig. 3. Fig. 3(a) shows the aluminized N80 sample. There is no obvious corrosion product on the surface of the sample, and white fine particles appear locally. The white particles are still dense and intact under electron microscope enlarged to 1000 times. See Figure 3(b), no holes or cracks, no signs of local corrosion. Such a aluminizing layer has a strong bonding force with the surface, and can better block the transfer of the corrosive medium to the substrate, so the uniform corrosion rate is low. Fig. 3(c) shows the non-aluminized N80 sample, the surface of which is a loose corrosion product film grain, and the local product film is peeled off to expose the substrate[6], showing obvious local corrosion characteristics as shown in Fig. 3(d).

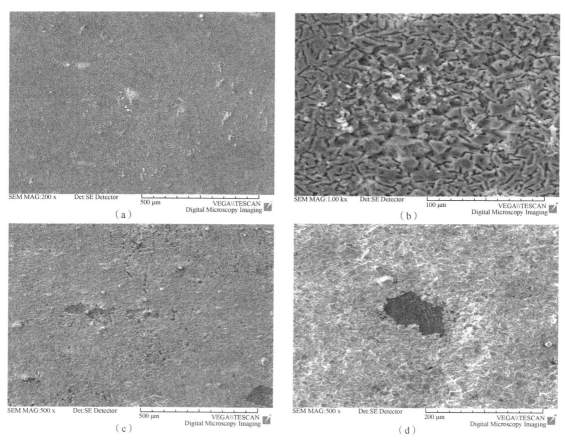

Fig. 3 Surface morphology of the sample after high temperature autoclave test of aluminized and non-aluminized N80 steel

4 Conclusion

(1) Aluminizing layer of aluminized N80 tubing surface is combined with matrix by metallurgical diffusion, the thickness can reach 150μm;

(2) Electrochemical test results under normal pressure at 25℃ show that aluminized treatment significantly increases the corrosion resistance of the steel and reduces its corrosion rate, thereby effectively improving the corrosion resistance of the N80 tubing;

(3) Under the harsh environment of water injection well with 0.2MPa CO_2 partial pressure and temperature of 80℃, the uniform corrosion rate of aluminized steel is 0.087mm/a, which is reduced to one quarter of the average corrosion rate of N80 steel, which shows superior uniform corrosion resistance.

References

[1] Xiang Z D, Datta P K. Pack aluminisation of low alloy steels at temperatures below 700℃ [J]. Surface and Coatings Technology, 2004, 184(1): 108-115.

[2] Xiang Z D, Datta P K. Relationship between pack chemistry and aluminized coating formation for low-temperature aluminisation of alloy steels [J]. Acta. Materialia, 2006, 54(17): 4453-4463.

[3] Bates BL, Wang Y Q, Zhang Y, et al. Formation and oxidation performance of low-temperature pack aluminized coatings on ferritic-martensitic steels [J]. Surface & Coatings Technology, 2009, 204(6-7): 766-770.

[4] Akdeniz M V, Mekhrabov A O. The effect of substitutional impurities on the evolution of Fe-Al diffusion layer [J]. Acta. Material, 1998, 46(4): 1185-1192.

[5] Kinsella B, Tan Y J, Bailey S. Electrochemical impedance spectroscopy and surface characterization techniques to study carbon dioxide corrosion product scales [J]. Corrosion, 1998, 54(10): 835-842.

[6] Mishra B, Al-Hassan S, Olson D L, et al. Development of a predictive model for activation-controlled corrosion of steel in solutions containing carbon dioxide [J]. Corrosion, 1997, 53(11): 852-859.

本论文原发表于《Materials Science Forum》2019年第944卷。

Tribological Properties of Ni/Cu/Ni Coating on the Ti-6Al-4V Alloy after Annealing at Various Temperatures

Luo Jinheng[1,2] **Wang Nan**[3] **Zhu Lixia**[1,2] **Wu Gang**[1,2]
Li Lifeng[1,2] **Yang Miao**[3] **Zhang Long**[3] **Chen Yongnan**[3]

(1. CNPC Tubular Goods Research Institute; 2. CNPC Key Laboratory for PetroChina Tubular Goods Engineering; 3. School of Materials Science and Engineering, Chang'an University)

Abstract: Diffusion reaction was a crucial route to enhance the wear resistance of Ti-6Al-4V alloys surface. In this work, the Ni/Cu/Ni composite layers were fabricated on the surface of Ti-6Al-4V alloy by electroplate craft, and then different annealing temperatures were applied to further optimize its tribological properties. The diffusion behaviors at various temperatures were systematically analyzed to reveal the physical mechanism of the enhanced tribological properties of the coatings. It was demonstrated that Cu_xTi_y and Ni_xTi_y intermetallic compounds with high hardness and strength were produced in the Ni/Cu/Ni coating, which acted as the reinforcing phases and improved the microhardness, reduced the friction coefficient, and lessened the wear rate. Specially, this effect reached the maximum when the annealing temperature was 800℃, showing excellent wear resistance. This work revealed the relationship between annealing temperatures andtribological properties of the Ni/Cu/Ni coating, and proposed wear mechanism, aiming to improve the surface performance of Ti-6Al-4V alloy by appropriately diffusion behavior.

Keywords: Ti-6Al-4V alloy; Ni/Cu/Ni coating; Phases transitions; Tribological properties; Wear mechanism

1 Introduction

Ti-6Al-4V alloys have the advantages of high specific strength, strong corrosion resistance and good biocompatibility, which have been widely used in aerospace, marine development and biomedical fields[1-3]. However, in some engineering applications, titanium alloys usually exhibit poor tribological properties, such as high friction coefficient, severe adhesive wear, low plastic shear resistance, weak work hardening ability and brittle oxide film on the surface[4,5], which greatly limit their application as friction components. Aydin et al.[6] suggested that the performance

---

Corresponding author: Chen Yongnan, chenyongnan@chd.edu.cn.

of matrix materials can be effectively reinforced by the thermal diffusion behaviour between different metals. Yao et al.[7] prepared the copper layer on Ti-6Al-4V alloy, exploiting diffusion between copper and titanium atoms, which can greatly improve the surface hardness and wear resistance. In addition, multilayer diffusion had been designed to form a variety of intermetallic compounds in coating by adding interlayer materials, optimizing the mechanical performance of the coating[8,9].

As well known, the atoms in surface coating will occur diffusion reaction and form the diffusion layer with a certain thickness during annealing[6,10,11]. Hu et al.[10] insisted that the Ni-Ti diffusion layer was mainly composed of the $NiTi_2$, Ni_3Ti and NiTiphases and hence it showed high hardness. Aydın et al.[6] provedthat the formed Cu_xTi_y intermetallic compounds during annealing process could apparently improve surface properties because of its high hardness. It can be concluded that the improvement of hardness in the diffusion layerwas strictly related to the intermetallic compounds, which can cause the wear mechanism transform to micro-adhesive wear from the adhesive wear, thereby improving the surface tribological properties. More importantly, the diffusion behavior of Ni and Ti atoms could be remarkably adjusted by the annealing temperatures[11]. Xia et al.[12] accurately analyzed the relationshipbetween annealing temperatures and the properties of Ti-6Al-4V alloy coatings, and the results confirmed the metallurgical bonding could be easily formed to strengthen its surface properties. Shen et al.[13] found that the depth of diffusion layer and the type of intermetallic compounds will be changed with the rise of annealing temperatures, which could observably improve the surface capability of the coatings. The above studies showed that the surface tribological properties of Ti-6Al-4V alloys can be improved by appropriate annealing of dissimilar metals.

However, the relationship between multi-layer diffusion behaviors, surface tribological properties and annealing temperatures were still indistinct. More importantly, a variety of intermetallic compounds may be dispersed in the diffusion layer without damaging the plasticity of the coating, which was essential for the study of wear mechanisms. Relevant research[2] had been done on Cu/Ni coating before, and a diffusion layer with certain properties has been obtained on the surface of Ti-6Al-4V alloy. But it was still not satisfactory in terms of the tribological properties. In thiswork, the diffusion behaviors and tribological properties of the Ni/Cu/Ni coating after annealing at various temperatures 600℃, 700℃ and 800℃ were studied. The formation mechanism of intermetallic compounds and the wear mechanism of diffusion layer were analyzed in detail. The purpose of this work was to study the diffusion layer evolution and wear resistance of the Ni/Cu/Ni coatings, obtaining the improved coating with a certain anti-wear effect by adjusting the annealing temperatures.

2 Experimental procedure

2.1 Materials synthesis

The commercial Ti-6Al-4V alloy was adopted for this study, and the chemical composition (wt.%) was 88.2Ti, 6Al, 4V, 0.6Fe, 0.5Mn, 0.4Si and 0.3Zn. The electroplated Ni/Cu/Ni composite coatings were synthesized from traditional electroplate method, the plating sequence of the Ni/Cu/Ni composite coating was first nickel plating for 10 minutes, then copper plating for 20 minutes, finally nickel plating for 10 minutes and the detail electrodeposition conditions and compositions of the bath were summarized in Table 1. The flow diagram of sample preparation process was shown in Figure 1(a), and the macro picture of sample obtained at each step of the preparation process was shown in Figure 1(b), showing the macro change process of sample. The schematic diagram of the cross section and surface of the prepared sample were shown in Figure 1(c). It could be evidenced the structure of the Ni/Cu/Ni coating on the Ti−6Al−4Vsubstrate and the surface was uniform and dense. And then, annealing process was performed in a vacuum furnace(OTF−1200X), setting the vacuum level of 1×10^{-2} Pa. The plating samples were annealing at 600℃, 700℃ and 800℃ for 3 h, respectively, after that, cooled down to 25℃ in the furnace.

Table1 The electrolyte and the operating condition

Composition and condition	Electrolyte composition	Operating condition
nickel plating	nickel sulfate hexahydrate($NiSO_4 \cdot 6H_2O$): 180g/L sodium sulfate(Na_2SO_4): 70g/L magnesium sulfate($MgSO_4$): 30g/L sodium chloride(NaCl): 30g/L boric acid(H_3BO_3): 30g/L	voltage: 3V temperature: 25℃ duration: 10min
copper plating	copper sulfate pentahydrate($CuSO_4 \cdot 5H_2O$): 210g/L sodium chloride(NaCl): 20mg/L sulfuric acid(H_2SO_4): 70g/L	voltage: 0.65V temperature: 25℃ duration: 20min

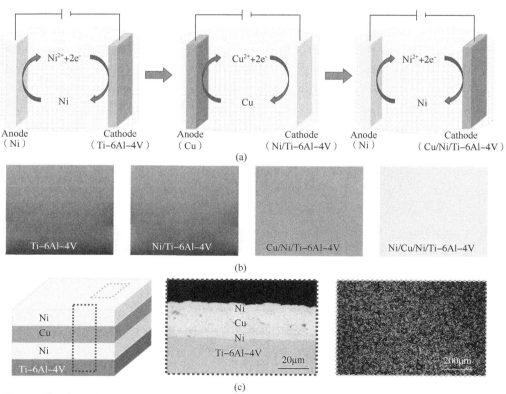

Fig. 1 The Ti−6Al−4V electroplated Ni/Cu/Ni coating(a) schematic diagram of preparation technology (b) macroscopic photographs of samples during preparation (c) schematic diagram and optical microscopy images and cross−sectional microstructure

2.2 Characterization after the thermaldiffusion

The microstructure analysis of the annealing sampleswas detected using a scanning electron microscopy(SEM, Hitachi-S4800). The thermal diffusion behaviors of Cu, Ni and Ti atoms were interpreted using an energy dispersive spectroscopy(EDS, Hitachi-S4800). The phases structures of the coatings were identified by X-ray diffraction(XRD, XRD D/M2500) with CuKa radiation (0.154 nm wavelength)at tube voltage of 40 kV, current of 40 mA, an 8.0°/min scanning speed, a 20°~80° of 2θ range.

2.3 Tribological properties

The surface roughness and microhardness of the coating surface after annealing were measured by roughness tester (SJ – 210) and microhardness tester (HV – 1000A), respectively. For data validity, five tests were performed and then averaged. The wear resistance of the coatings was assessed with a ball-on-plate wear tester (MMQ – 02G) using a 6mm GCr15 ball counterpart. A constant load of 3 N was applied normally to the sample under non-lubricated condition atroom temperature. The abrasion resistance testswere performed on with a circular track of 3mm in diameter, a rotational speed of 100 r/min and a total sliding distance of 37.70 m. The worn morphologies of the coatings were explained using scanning electron microscopy (SEM, Hitachi-S4800)and Laser Confocal Microscope(LCM, Olympus OLS5000)to show the worn mechanism.

3 Results and discussion

3.1 The diffusion behaviorsand phases transitions

As shown in Figure 2(a), the morphology and composition of Ni/Cu/Ni layers on Ti-6Al-4V alloy after annealing were researched by cross-sectional observation with EDS examination. The element distribution of Cu, Ni and Ti in diffusion layers were clearly displayed. Obviously, the Cu, Ni and Ti atoms hardly occurred diffusion behavior without annealing, where the interfacesin the Ni/Cu/Ni layers were obvious with a clear outline. However, these interfaces gradually became blurred and disappeared with the increasing of annealing temperatures, which indicated that diffusion behaviorsbetween Cu, Ni and Ti atoms intensified step by step. It was noteworthy that the diffusion layer with a certain thickness was formed between Ni/Ti interfaces, which was distinctly different from Ni layer and Ti-6Al-4Vsubstrate. Moreover, the higher the annealing temperature was, the thicker the diffusion layer was[Figure 2(b)]. Actually, Ti atom had a higher diffusion rate than Ni atom during the annealing process, which resulted in a large number of Kirkendall voids on the Ti side[14]. Furthermore, as the annealing temperatures increased, the voids will connect with each other and then coalesce into large gaps, forming the Kirkendall diffusion channels[15,16]. Finally, they would become a fast diffusion channel for all atoms[Figure 2(c-d)].

Figure 3 displayedthe elements line scan diagrams across the coatings from its substrate to the top after annealing. It can be seen that without annealing, the interface of each layer was clear, and no atomic diffusion occurred [Figure 3(a)]. With the increase of annealing temperatures, the interdiffusion between Cu, Ni, and Ti atoms intensified, gradually forming a continuous one. In addition, the interdiffusion ability of the element in the Ni layer was better than that in the Cu layer. As shown in Figure 3(b)-(d), there were fewer Ni and Ti atoms in the Cu layer. In the Ni layer, Cu and Ti atoms were fully diffused and the diffusion reaction occurs, forming a series of intermetallic compounds and solid solutions.

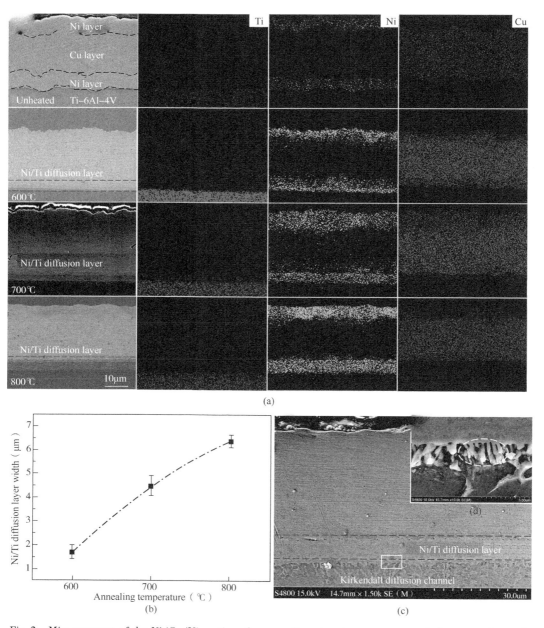

Fig. 2 Microstructure of the Ni/Cu/Ni coatingsafter annealing at various temperatures(a) cross section and corresponding element composition(b) Ni/Ti diffusion layer width(c) the diffusion layer structure of the Ni/ Ti interface for 800℃. The obvious Ni/Ti diffusion layers was observed in the red line. (d) showed the Kirkendall diffusion channel, which was showed in the white area

Figure 4(a) showed the Ni-Cu solid solution and the intermetallic compounds of Ni_xTi_y and Cu_xTi_y in the coatings when the annealing temperatures increased from 600℃ to 800℃. According to the

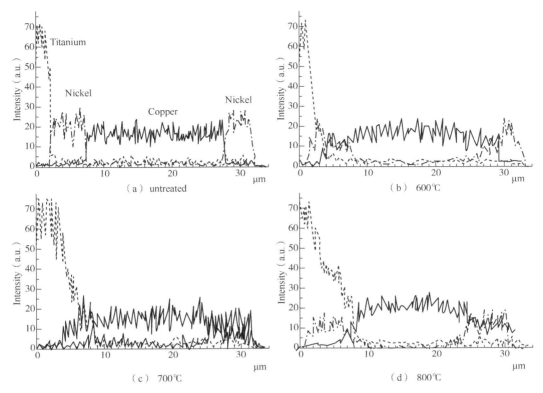

Fig. 3 The elements line scan diagrams across the coatings from its substrate to the top after annealing

Ni−Cu equilibrium diagram, the α(Cu, Ni) solid solution (Cu$_{3.8}$Ni) was mainly formed in the annealed coatings. For the Ni−Ti binary system, the following reactions (1)−(3) occurred after the diffusion of Ni and Ti atoms, and the Ni$_3$Ti, NiTi$_2$ and NiTi phases were formed, respectively[17,18]:

$$3Ni + Ti = Ni_3Ti \quad (1)$$

$$Ni + 2Ti = NiTi_2 \quad (2)$$

$$Ni + Ti = NiTi \quad (3)$$

In order to reveal the formation mechanism of Ni$_x$Ti$_y$ intermetallic compounds, the Gibbs free energies of the above reaction were summarized in Figure 4(b). The Gibbs free energies of Ni$_3$Ti and NiTi$_2$ phases were substantially lower than that of NiTi phase, which illustrated that the reaction (1) and reaction (2) were more likely to occurthan reaction (3). Bastin et al. [14] found that the nucleation and growth of NiTi phase caused a sharp decrease in NiTi$_2$ phase and Zhou et al. [19] also confirmed that the growth of NiTi phase considerably consume Ni$_3$Ti and NiTi$_2$ phases. According to previous analysis, the nucleation and growth of NiTi phase mainly depend on the Ni$_3$Ti and NiTi$_2$ phases. That is, the nucleation of the Ni$_3$Ti and NiTi$_2$ phases was earlier than that of NiTi phase, which was consistent with the order of the phase's formation calculated. The Ni$_3$Ti and NiTi$_2$ phases in this study were gradually formed in the Ni/Ti interfacial diffusion layer with the thermaldiffusion temperature increased, the absence of NiTi phase may be due to insufficient diffusion time or a small amount of formation[20,21].

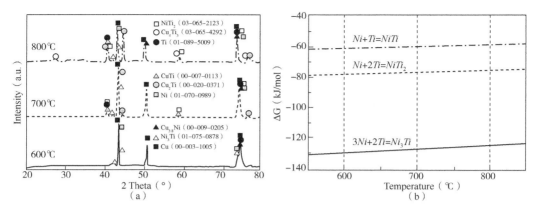

Fig. 4 Phases transitions of the Ni/Cu/Ni coatings after annealing at various temperatures (a) surface XRD images (b) Gibbs free energy of the Ni-Ti reaction

The atom diffusion direction in the Cu – Ti binary system could be theoretically analyzed by chemical potential. It is well known that the atom always spontaneously transfers from a high chemical potential to a low chemical potential, and the following theory is summarized by Wu et al.[22]:

$$\mu_i = \frac{\partial G}{\partial n_i} \tag{4}$$

Where μ_i is the chemical potential of component i, G is the Gibbs free energy and n_i is the atomic number of component i. It can be calculated that the chemical potential of Ti atoms was higher than that of Cu atoms. Under the driving force of the diffusion, the Ti atoms diffused toward the Cu side while the diffusion reaction occurred. Previous researches proved that the Ti atoms with greater diffusion capacity formed Cu_xTi_y intermetallic compounds on the Cu side[6,22]. Therefore, under the above synergistic effect, the Ti atoms rapidly diffused toward the Cu side, gradually forming the CuTi, Cu_2Ti and Cu_4Ti_3 intermetallic compounds, which can be expressed as follows[22,24]:

$$2Cu + Ti = Cu_2Ti \tag{5}$$

$$Cu + Ti = CuTi \tag{6}$$

$$4Cu + 3Ti = Cu_4Ti_3 \tag{7}$$

Apparently, the intermetallic compounds of Ni_xTi_y and Cu_xTi_y were gradually formed in the coating with the annealing temperatures increased. These intermetallic compounds with strong atomic bonds and high hardness served as the reinforcing phases in the diffusion layer, which can effectively improve the tribological properties of the coating[25,26].

3.2 Tribological properties of the Ni/Cu/Ni coating on Ti-6Al-4V alloy

It is well known that surface topography is an important factor in describing tribological properties of the coating[27]. As depicted in Figure 5(a), the surface of untreated coating was characterized of continuous uniform grain size. As the annealing temperature increased, the coating surface gradually transformed to granulated and densified status [Figure 5(b-d)]. It showed that a large number of fine particles like the Ni_xTi_y and Cu_xTi_y phases existed on the surface of the Ni/Cu/Ni coating after the thermal diffusion at 700℃ and 800℃, which was resulted from a stable diffusion layer with the diffusion reaction between Ni, Cu and Ti[26].

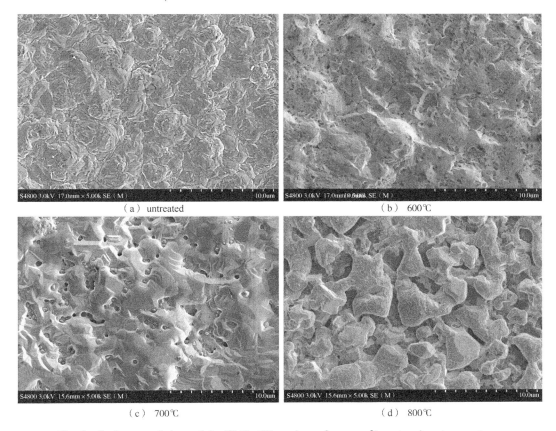

(a) untreated   (b) 600℃

(c) 700℃   (d) 800℃

Fig. 5 Surface morphology of the Ni/Cu/Ni coatings after annealing at various temperatures

The microhardness of the annealed coatings was shown in Figure 6(a). With the annealing temperatures up to 800℃, the microhardness progressively increased from 155HV to 357HV. The increased microhardness was ascribed to the formation of the intermetallic compounds and the strengthening effect of solid solution[25,28]. The Cu_xTi_y and Ni_xTi_y intermetallic compounds with high hardness can substantially enhance the surface microhardness of the Ni/Cu/Ni coating. At the same time, the annealed coatings had a lower amount of mass loss, and can even be reduced to half of the Ti-6Al-4V substrate. This was mainly because the hard phases dispersed on the coating surface can effectively enhance the coating's resistance to high contact stresses, thereby optimizing the coating surface wear behavior.

Figure 6(b) displayed the friction coefficient of the annealed Ni/Cu/Ni coatings. Clearly, the friction coefficient of the Ti-6Al-4V substrate was around 0.6, and it gradually decreased and stabilized at around 0.4 after annealing, which were all composed of the running-in stage and stabilization stage. It was noteworthy that some anomalous peaks appear during the stabilization stageafter annealing at 600℃. This may be related to severe plastic deformation and adhesive wear caused by low microhardness. The process of adhesion, delamination, debonding and re - stabilization resulted in these anomalous peaks, which implied poor surface tribological properties of the coating. As for the Ni/Cu/Ni coating after annealing at 800℃, there was no obvious running-in stage and reached quickly to a relatively stabilization stage, which implied the increase of microhardness and the uniform dense diffusion layer can improve the surface tribological properties of the coating. Meanwhile, the plastic deformation ability of the Ni/Cu/Ni coating under stress concentration was also significantly reduced due to the increase of yield strength and microhardness[29]. Thus, the increase of thermal diffusion temperature was conductive to reducing the friction coefficient of the coating, which was put down to the strengthening effect of the hard particle during the diffusion process.

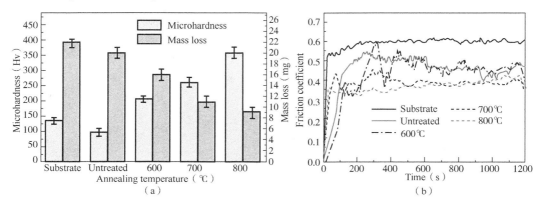

Fig. 6 The surface tribological properties of the Ni/Cu/Ni coatings after annealing at various temperatures (a)microhardness and mass loss(b)the change curves of friction coefficient with sliding time

3.3 Wear mechanism

The Ni/Cu/Ni coating without annealing was susceptible to plastic deformation and adhered to the contact surface to form adhesive wear, which caused continuous cutting penetrated on the coating and wore the Ti-6Al-4V alloysubstrate. In this process, a large amount of grinding debris was formed on both sides of the grinding mark, as shown in Figure 7(a), which indicated poor wear performance. After annealing, the coating was tightly bonded to the Ti-6Al-4V substrate and formed an effective diffusion layer and distributed a mass of intermetallic compounds, resulting in a prominent improvement in wear performance. The wear marks on the surface of the samples gradually became shallower and flatter [Figure 7(b-d)], indicating that the adhesive wear gradually weakened. Especially, as the annealing temperatures raised up to 800℃, only slight wear was caused on the surface of the coating, showing the best tribological properties[Figure 7(d)].

Obviously, the tribological properties of the coatings on Ti-6Al-4V alloy were improved after annealing, which could be stated using Archard's law[30,31]:

$$Q = KW/H \tag{8}$$

where Q is the wear rate, K is the friction coefficient, W is the applied load, and H is the hardness. A low friction coefficient and a high hardness can result in a low wear rate under the same wear

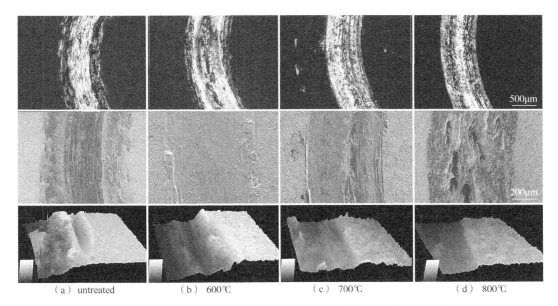

Fig. 7 The surface wear morphology of the Ni/Cu/Ni coatings after annealing at various temperatures

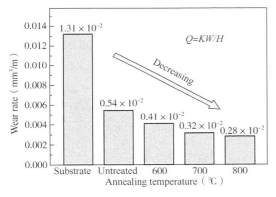

Figure 8 The wear rare of the Ni/Cu/Ni coatings after annealing at various temperatures

conditions. As shown in Figure 8, the wear rate of the Ni/Cu/Ni coating on the Ti-6Al-4V alloy was effectively reduced after annealing. Compared with Ti-6Al-4V substrate, the wear rate was reduced from 1.31×10^{-2} mm^3/m to 0.28×10^{-2} mm^3/m after annealing 800℃. It was mainly related to changes in the surface state of the coating. Wang et al.[11] studied that intermetallic compounds can increase the resistance of coatings to high contact stress due to their high microhardness. In addition, the Ni layer and the Cu layer had excellent ductility, improving the brittle fracture ability of the coating. During the adhesive wear process, the Ni layer and the Cu layer with lower hardness first underwent a certain degree of deformation, and the adhesive wear contact area was continuously increased. On the other hand, the Ni_xTi_y and Cu_xTi_y intermetallic compounds with high hardness could resist more wear loads and prevented the Ni layer and the Cu layer with low hardness from further deformation. Meanwhile, the good ductility of Ni layer and the Cu layer can alleviate cracks caused by the loading of the hard – intermetallic compounds. Therefore, the intermetallic compounds with the high hardness and the Ni layer and the Cu layer with good ductility can cooperate with each other to jointly improve the wear resistance of the Ni/Cu/Ni coating.

When the annealing temperature was up to 800℃, the $NiTi_2$ and Cu_4Ti_3 phases could be formed in the coating, which had a higher microhardness than the Ni_3Ti, $CuTi$ and Cu_2Ti phases[19,32]. Therefore, they would act as a supporting load to reduce the furrow effect of the counter ball to the coating during the wear test. The diffusion layer with a certain thickness reduced the area of adhesion in the friction surface, weakening the adhesion effect. Especially, the hard Ni-Ti diffusion layer of the Ni/Cu/Ni coating increased the yield strength of the contact surface and effectively reduces the tangential stresses and the interfacial stresses[33,34]. As a result, the adhesion effect and the furrow effect on the surface of the coating can be effectively suppressed by each otherafterannealing at 800℃.

4 Conclusions

The Ni/Cu/Ni coating were prepared on the surface of Ti-6Al-4V alloy by theelectrodeposition and subsequentlyannealing at 600℃, 700℃ and 800℃ for 3 h, respectively. The microstructure and tribological properties of the diffusion layer were systematically investigated. Base on achieve results the following conclusions can be made:

The Cu, Ni, Ti atoms in coating displayed complex diffusion behaviors and formed the diffusion layers with a certain thickness. As the annealing temperature increased, the Kirkendall diffusion channel appeared, accelerating the diffusion behaviors of the atoms. In addition, the diffusion layers were mainly composed of the Ni_3Ti, $NiTi_2$, $CuTi$, Cu_2Ti and Cu_4Ti_3 intermetallic compounds and the α(Cu, Ni) solid solutionwith high hardness and strength, which acted as the reinforcing phases of wear resistant materials. The Ni_xTi_y and Cu_xTi_y intermetallic compounds were continuously and uniformly distributed in the Ni/Cu/Ni coating, which can tremendously strengthen the surface hardness, reduce the friction coefficient and lessen the wear rate. In wear process, the intermetallic compounds with the high hardness and the Cu layer and the Ni layer with good ductility can cooperate with each other to jointly improve the wear resistance of the coating. The synergistic effect was particularly remarkable at the annealing temperature of 800℃, showing excellent wear resistance.

References

[1] Wang, F.; Wang, D. Study on the Effects of Heat-treatment on Mechanical Properties of TC4 Titanium Alloy Sheets for Aviation Application. Titan. Ind. Progress. 2017, 45, 56-60.

[2] Chen, Y.; Liu, S.; Zhao, Y.; Liu, Q.; Zhu, L.; Song, X.; Zhang, Y.; Hao, J. Diffusion behavior and mechanical properties of Cu/Ni coating on TC4 alloy. Vacuum. 2017, 143, 150-157.

[3] Shen, Z. C.; Xie, F. Q.; Wu, X. Q.; Yao, X. F. Properties of Coating on TC4 Titanium Alloy by Copper Electroplating. China Surf. Eng. 2012, 25, 45-49.

[4] Zhang, Y.; Zhang, H. L.; Wu, J. H.; Wang, X. T. Enhanced thermal conductivity in copper matrix composites reinforced with titanium-coated diamond particles. Scripta. Mater. 2011, 65, 1097-1100.

[5] Jiang, P.; He, X. L.; Li, X. X.; Wang, H. M. Wear resistance of a laser surface alloyed Ti-6Al-4V alloy. Surf. Coat. Tech. 2000, 130, 24-28.

[6] Aydın, K.; Kaya, Y.; Kahraman, N. Experimental study of diffusion welding/bonding of titanium to copper. Mater. Design2012, 37, 356-368.

[7] Yao, X.; Xie, F.; Wang, Y.; Wu, X. Research on Tribological and Wear Properties of Cu Coating on TC4 Alloy. Rare. Metal. Mat. Eng. 2012, 41, 2135-2138.

[8] Petrovic, S.; Peruško, D.; Mitric, M.; Kovac, J.; Dražić, G.; Gaković, B.; Homewoodc, K. P.;

Milosavljević a, M. Formation of intermetallic phase in Ni/Ti multilayer structure by ion implantation and thermal annealing. Intermetallics2012, 25(3), 27-33.

[9] Wang, F. L.; Sheng, G. M.; Deng, Y. Q. Impulse pressuring diffusion bonding of titanium to 304 stainless steel using pure Ni interlayer, Rare Metals. 2016, 35(4), 1-6.

[10] Hu, L.; Xue, Y.; Shi, F. Intermetallic formation and mechanical properties of Ni-Ti diffusion couples. Mater. Design2017, 130, 175-182.

[11] Wang, Z.; He, Z.; Wang, Y.; Liu, X.; Tang, B. Microstructure and tribological behaviors of Ti6Al4V alloy treated by plasma Ni alloying. Appl. Surf. Sci. 2011, 257(23), 10267-10272.

[12] Xia, Y. X.; Li, D. Effects of Post Heat-treatment of Electroplating on Plating Adhesion for TC4 Titanium Alloys. Rare Metal Mat. Eng. 2001, 30(5), 390-391.

[13] Shen, Q.; Xiang, H.; Luo, G.; Wang, C.; Li, M.; Zhang, L. Microstructure and mechanical properties of TC4/oxygen-free copper joint with silver interlayer prepared by diffusion bonding. Mater. Sci. Eng. A. 2014, 596, 45-51.

[14] Bastin, G. F.; Rieck, G. D. Diffusion in the titanium-nickel system: I. occurrence and growth of the various intermetallic compounds. Metall. Trans. 1974, 5, 1817-1826.

[15] Puente, A. E. P. Y.; Dunand, D. C. Synthesis of NiTi microtubes via the Kirkendall effect during interdiffusion of Ti-coated Ni wires. Intermetallics. 2018, 92, 42-48.

[16] Fan, H. J.; Gosele, U.; Zacharias, M. Formation of nanotubes and hollow nanoparticles based on Kirkendall and diffusion processes: a review. Small. 2007, 3, 1660-1671.

[17] He, P.; Liu, D. Mechanism of forming interfacial intermetallic compounds at interface for solid state diffusion bonding of dissimilar materials. Mater. Sci. Eng. A. 2006, 437(2), 430-435.

[18] Vandal, M. J. H.; Pleumeekers, M. C. L. P.; Kodentsov, A. A.; Vanloo, F. J. J. Intrinsic diffusion and kirkendall effect in Ni-Pd and Fe-Pd solid solutions. Acta Mater. 2000, 48, 385-396.

[19] Zhou, Y.; Wang, Q.; Sun, D. L.; Han, X. L. Co-effect of heat and direct current on growth of intermetallic layers at the interface of Ti-Ni diffusion couples. J. Alloy. Compd. 2011, 509, 1201-1205.

[20] Lin, C. M.; Kai, W. Y.; Su, C. Y.; Tsai, C. N.; Chen, Y. C. Microstructure and mechanical properties of Ti-6Al-4V alloy diffused with molybdenum and nickel by double glow plasma surface alloying technique. J. Alloy. Compd. 2017, 717, 197-204.

[21] Simoes, S.; Viana, F.; Ramos, A. S.; Vieira, M. T.; Vieira, M. F. Reaction zone formed during diffusion bonding of TiNi to Ti6Al4V using Ni/Ti nanolayers. J. Mater. Sci. 2013, 48, 7718-7727.

[22] Akbarpour, M. R.; Moniri, J. S. Wear performance of novel nanostructured Ti-Cu intermetallic alloy as a potential material for biomedical applications. J. Alloy. Compd. 2017, 699, 882-886.

[23] Wu, M. F.; Yang, M.; Zhang, C.; Yang, P. Research on the liquid phase spreading and microstructure of Ti/Cu eutectic reaction. Trans. China Weld. Inst. 2005, 26(10), 68-71.

[24] Campo, K. N.; Lima, D. D. D.; Lopes, E. S. N.; Caram, R. Erratum to: On the selection of Ti-Cu alloys for thixoforming processes: phase diagram and microstructural evaluation. J. Mater. Sci. 2016, 51, 9912-9913.

[25] Semboshi, S.; Iwase, A.; Takasugi, T. Surface hardening of age-hardenable Cu-Ti alloy by plasma carburization. Surf. Coat. Tech. 2015, 283, 262-267.

[26] Zhang, X.; Ma, Y.; Lin, N.; Huang, X.; Hang, R.; Fan, A.; Tang, B. Microstructure, antibacterial properties and wear resistance of plasma Cu-Ni surface modified titanium. Surf. Coat. Tech. 2013, 232, 515-520.

[27] Zhou, X.; Shen, Y. Surface morphologies, tribological properties, and formation mechanism of the Ni-CeO_2 nanocrystalline coatings on the modified surface of TA2 substrate. Surf. Coat. Tech. 2014, 249, 6-18.

[28] Fan, D.; Liu, X.; Huang, J.; Fu, R.; Chen, S.; Zhao, X. An ultra-hard and thick composite coating metallurgically bonded to Ti-6Al-4V. Surf. Coat. Tech. 2015, 278, 157-162.

[29] Chen, H.; Zheng, L. J.; Zhang, F. X.; Zhang, H. X. Thermal stability and hardening behavior in super elastic Ni-rich Nitinol alloys with Al addition. Mater. Sci. Eng. A. 2017, 708, 514-522.

[30] Wang, L.; Gao, Y.; Xue, Q.; Liu, H.; Xu, T. Microstructure and tribological properties of electro deposited Ni-Co alloy deposits. Appl. Surf. Sci. 2005, 242, 326-332.

[31] Ranganatha, S.; Venkatesha, T. V.; Vathsala, K. Development of electroless Ni-Zn-P/nano-TiO_2, composite coatings and their properties. Appl. Surf. Sci. 2010, 256, 7377-7383.

[32] Cai, Q.; Liu, W.; Ma, Y.; Zhu, W.; Pang, X. Effect of joining temperature on the microstructure and strength of W-steel HIP joints with Ti/Cu composite interlayer. J. Nucl. Mater. 2018, 507, 198-207.

[33] Wang, N.; Chen, Y. N.; Zhang, L.; Li, Y.; Liu, S. S.; Zhan, H. F.; Zhu. L. X.; Zhu, S. D.; Zhao, Y. Q. Isothermal diffusion behavior and surface performance of Cu/Ni coating on TC4 alloy. Materials 2019, 12, 3884.

[34] Dipak, T. W.; Chinmaya, K. P.; Ritesh, P. NiTi coating on Ti-6Al-4V alloy by TIG cladding process for improvement of wear resistance: Microstructure evolution and mechanical performances. J. Mater. Process. Tech. 2018, 262, 551-561.

本论文原发表于《Materials》2020年第13卷。

Effect of Type B Steel Sleeve Rehabilitate Girth Weld Defect on the Microstructure and Property of X80 Pipeline

Ma Weifeng[1,2]　**Ren Junjie**[1,2]　**Zhou Huiping**[3]　**Wang Ke**[1,2]
Luo Jinheng[1,2]　**Zhao Xinwei**[1,2]　**Huo Chunyong**[1,2]

(1. State Key Laboratory for Performance and Structure Safety of Petroleum Tubular Goods and Equipment Materials, CNPC Tubular Goods Research Institute;
2. Shanxi Key Laboratory for Performance and Structure Safety of Petroleum Tubular Goods and Equipment Materials;
3. PetroChina West Pipeline Company)

Abstract: In recent years, accident caused by girth weld defects is relatively common. The failure mode of girth weld is mostly fracture. It is an effective way to avoid the failure accident of girth weld to find the hidden trouble of girth weld and repair it in time before the failure. The type B steel sleeve is an effective method to repair the girth weld defects. However, quite few works have been carried out on the reliability verification and engineering practice of repairing the X80 pipeline girth welds. Therefore, it is necessary to carry out full-size physical test for engineering and obtain the influence law of welding operation on main pipeline performance, with the aim of reliability of repair quality and service after repair. In the present work, the reliability study of repairing with type B steel sleeve on X80 pipe girth weld defects was verified. The results show that the bearing capacity of the type B sleeve can effectively guarantee the safe operation of the pipeline in the case of leakage from the girth weld. But there are obvious influence on the material properties of the outer surface of the main pipe due to the welding heat. Effects from longitudinal weld seam is smaller than effects from fillet weld. Grain roughening occurred in heat affected zone of fillet weld which results in that the hardness of the main pipe under the fillet weld increased and the yield strength was reduced by 5.5%. Surface cracking occurs during bending test. There is a certain service risk and it should be used under more frequent monitoring than usual.

Keywords: X80 pipeline; Girth weld defect; Type B steel sleeve; Rehabilitation

After the long service of oil and gas pipeline, it is common that the failure of oil and gas pipeline is caused by the defect of girth weld. According to the statistical result of U.S. office of

Corresponding author: Ma Weifeng, mawf@cnpc.com.cn.

pipeline safety, From 1985 to 1996, The accident rate of the dangerous liquid and gas pipeline caused by pipeline weld problems were 12% and 8% respectively. In our country, the problem of pipeline failure caused by girth weld fracture is particularly prominent. With the construction of high steel grade (X70, X80) and large caliber pipeline (ϕ1016mm, 1219mm ϕ), As the difficulty of field welding increases, the undetected rate of defects increases. In recent years, there have been more than 30 cracking and leakage accidents in girth welds during the initial test and operation, within them more than 70% are caused by welding defects. Therefore, the repair of high steel pipeline girth welds have become the engineering technical problems to ensure the safe and economic operation of pipelines.

The failure mode of girth weld is mostly fracture. It is an effective way to avoid the failure accident of girth weld to find the hidden trouble of girth weld and repair it in time before the failure. The type B steel sleeve is an effective method to repair the girth weld defects. However, there are some problems exist in the actual implementation process. First, the main standards for on-site welding repair are SY/T 6150.1—2017 *Technical standard of tapping plugging on steel pipeline* and Q/SY 1592 *Repair technical specification for oil & gas pipeline*. There is no calculation model of bearing capacity of type B sleeve (fillet weld) after welding repair. Secondly, the welding heat input may reduce the strength and toughness of the main pipe and affect the bearing capacity, but it is lacking in verification with test. In addition, the flowing medium in the pipeline will take away a lot of heat and accelerate the cooling of the weld, whick increases the possibility of cracks in the weld heat affected zone and posed a certain risk to the quality of the welding. Thirdly, the material of type B steel sleeve is different from the main pipe in steel grade, wall thickness and diameter, which may cause various aspects of repair risk and quality reliability problems. Especially with the construction of high steel grade pipeline, discrepancy between repair materials (low steel grade generally, such as Q345R, etc) and main pipe (high steel grade, such as X80, etc). The problems of welding matching and difficulty of field welding are becoming more and more prominent, together with the grown of risk.

Aimed at the service security technology needs of inservice X80 high strength pipeline girth weld, full-size verification test of the type B steel sleeve on repairing X80 pipeline girth weld defects was carried out, assisting with full-scale hydraulic bursting test. Effects of welding operation on the microstructure and mechanical properties of main pipe were investigated. Service safety evaluation of the main pipe was studied to ensure the essential safety of main pipeline service.

1 Sample preparation and test method

1.1 Sample preparation

The basic information of steel tube and sleeve for test is shown in Table 1. X80 straight seam submerged-arc welding pipes in the second-level area of west-east gas transmission was adopted as the main pipeline. According to Q/SY - GJX 0110—2007 *Technical specification of mainline construction for second west-east natural gas transmission pipeline project*, the butt girth weld groove is machined and welded. A penetration crack defect of 300mm was prepared using electrospark machining (Fig.1).

Table 1 The basic information of steel tube and sleeve

Samples	Material type	Diameter (mm)	Wall thickness (mm)	Length (m)	Welding specification
Main pipe	X80	1219	22	10	Girth weld: *Q/SY-GJX* 0110—2007 *Technical specification of mainline construction for second west - east natural gas transmission pipeline project*
Type B steel sleeve	Q345R	1219	50	0.53	(1) *Evaluation report on maintenance and repair welding technology of X80 grade steel pipeline in service of west-east gas transmission*; (2) *Welding type repair sleeve sleeve sleeve repair operation instructions*

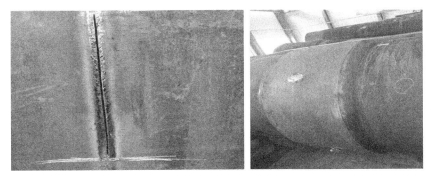

Fig. 1 Macroscopic picture of prefabrication defect of girth weld

The type B sleeve used in the test is a common specification and material in the market. Width of the groove is 40mm. The structure diagram is shown in Fig. 2. Firstly, two longitudinal welds were welded at the same time. After the longitudinal welds, two ends of the sleeve were welded with circular fillet weld one by on. Nondestructive testing was executed using layered testing method. Root welding, 50% fill welding and 100% fill welding were tested using a South Korea CCZ-212 magnetic particle detector. No surface crack defect were detected which meet the requirement of the operation instruction.

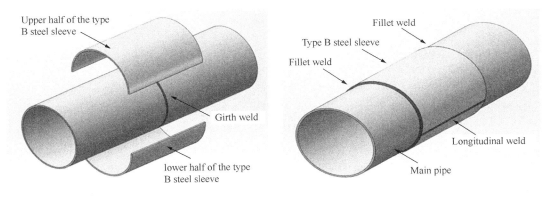

Fig. 2 Schematic diagram of type B steel sleeve structure

1.2 Test method

Full-scale hydraulic bursting test was executed according to SY/T 5992—2012 *Method of hydrostatic burst test for line pipes*. Theoretical yield pressure and theoretical blasting pressure of X80 pipe were evaluated according to Eq. 1 and Eq. 2.

$$p_s = \frac{2R_{t0.5}t}{D} = \frac{2 \times 555 \times 22}{1219}\text{MPa} \approx 20\text{MPa} \quad (1)$$

$$p_b = \frac{2R_m t}{D} = \frac{2 \times 625 \times 22}{1219}\text{MPa} \approx 22.6\text{MPa} \quad (2)$$

Here, p_s is the theoretical yield pressure (MPa), p_b is the theoretical yield pressure (MPa), t is wall thickness of the sample (mm), $R_{t0.5}$ is the yield strength (MPa), R_m is the yield strength (MPa), D is the external diameter (mm).

The physical and chemical properties sampling scheme was designed carefully after Full-scale hydraulic bursting test for clarifying the influence degree and influence range of welding repair of type B sleeve on main pipe material and performance. Sampling plan is shown in Fig. 3. Physical and chemical properties of the main pipe base material (M), the main pipe material under the longitudinal welding seam (LW) of the sleeve, and the main material under the fillet weld seams (1# FW and 2# FW) of the sleeve were tested, including chemical composition, metallographic structure, hardness, tensile properties, impact properties and bending properties.

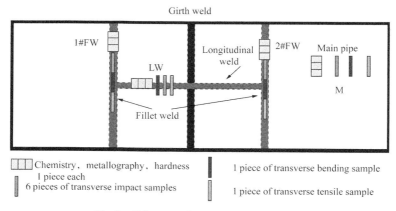

Fig. 3 Schematic diagram of sampling plan

2 Results and discussion

2.1 Full-scale hydraulic bursting test

Continuous pressure data is shown in Fig. 4. Due to the penetrating crack, the load caused by internal pressure was passed to the fillet welds and longitudinal weld of type B steel sleeve. It can be seen from Fig. 3 that piercing-caused leakage occurred at the pressure of 18.1MPa, which indicates that the type B sleeve can meet the requirements of girth weld repairation compared with the pipeline design pressure of 10MPa. The leakage point is located at the cross point of longitudinal welding and fillet weld, which means that the position is weak point of the type B sleeve.

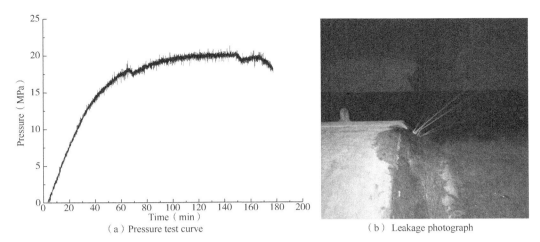

(a) Pressure test curve (b) Leakage photograph

Fig. 4 Full-scale hydraulic bursting test data and leakage photograph

2.2 Physical and chemical properties

Tensile properties of M, LW, 1#FW and 2#FW was tested according to GB/T 228.1—2010, compared with the requirement for X80 pipeline steel in Q/SY GJX 0104—2007 *Technical Specification of LSAW Line Pipe for the 2nd West-East Pipeline Project*. Results are listed in Table 2. As can be seen, tensile properties of several positions are all meet the standard requirements. Yield strength of LW is slightly higher than M position, which means that good effects were caused by heat input under welding. On the contrary, yield strength of 1#FW and 2#FW reduced by 5.5% compared with M position. Tensile strength does not change much, and the tensile property of the main pipe parent material meets the standard.

Table 2 Results of tensile properties

Sample	Width (mm)	Gauge length (mm)	Tensile strength (R_m)(MPa)	Yield strength ($R_{t0.5}$)(MPa)	Elongation(A) (%)	Reduction of area(Z)(%)
M	38.1	50	723	621	39.5	61
LW	38.1	50	724	640	41.0	63
1#FW	38.1	50	710	587	43.0	62
2#FW	38.1	50	727	587	40.5	59
Q/SY GJX 0104			625~825	555~690	—	—

Impact properties of M, LW, 1#FW and 2#FW was tested according to GB/T 229—2007, compared with the requirement for X80 pipeline steel in Q/SY GJX 0104—2007 *Technical Specification of LSAW Line Pipe for the 2nd West-East Pipeline Project*. Results are listed in Table 3. As can be seen, impact properties of several positions are all meet the standard requirements, and much more greater than the standard requirements. The impact energy of M, LW, 1#FW and 2#FW is 294.7J, 322J, 305J and 332J respectively. The impact properties of the main pipe after weld repairation increase slightly.

Table 3 Results of charp impact properties

Sample	Size(mm)	Notch geometry	Temperature (℃)	KV_2 (J)			Shear section rate (%)		
M	10×10×55	V	0	282	291	311	100	100	100
	10×10×55	V	−20	298	285	296	100	100	100
LW	10×10×55	V	0	323	319	325	100	100	100
	10×10×55	V	−20	318	325	315	100	100	100
1#FW	10×10×55	V	0	278	308	330	100	100	100
	10×10×55	V	−20	322	301	295	100	100	100
2#FW	10×10×55	V	0	333	331	333	100	100	100
	10×10×55	V	−20	318	329	294	100	100	100
Q/SY GJX 0104			−10	180(140)			90(80)		

Bending properties of M, LW, 1#FW and 2#FW was tested according to GB/T 2653—2008, compared with the requirement for X80 pipeline steel in Q/SY GJX 0104—2007 Technical Specification of LSAW Line Pipe for the 2nd West-East Pipeline Project. Results are listed in Table 4. As can be seen, bending properties of M, LW and 1#FW are all meet the standard requirements, no cracks can be observed after bending 180°. Face bending of 2#FW show crack with the length of 16mm [Fig. 5(b)], which indicates that the weld has great influence on the structure and properties of the main pipe under fillet weld.

Table 4 Results of bending properties

Sample	Size (mm×mm×mm)	Bend shaft diameter (mm)	Results
M	300×38×22	65	No crack
LW	300×38×22	65	No crack
1#FW	300×38×22	65	No crack
2#FW	300×38×22	65	Two craks with the length of 16mm
Q/SY GJX 0104			No crack

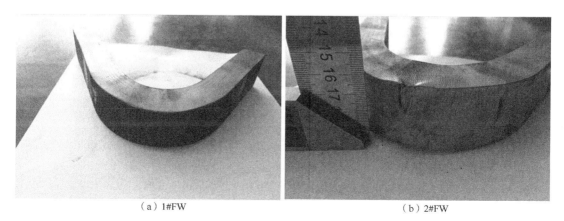

(a) 1#FW (b) 2#FW

Fig. 5 Photographs of 2#FW after bending test

Hardness of M, LW, 1#FW and 2#FW was tested according to GB/T 4340.1—2009. Schematic diagram of pointing position is shown in Fig. 6, and the results are listed in Table 5. As can be seen, average hardness of M is 223HV10. The hardness of LW is 234HV10. But the hardness of 1#FW and 2#FW increase obviously (about 260HV10). Overall analysis, hardness of main pipe surface incerase after weld heat affect. Change of fillet weld is more greater. The risk of brittle cracking on the outer surface is increased, which is consistent with the test results of bending performance.

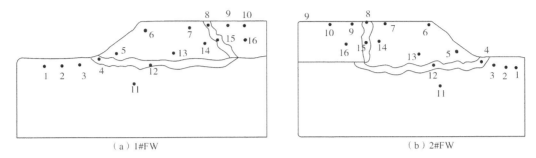

(a) 1#FW　　　　　　　　　　　　　　(b) 2#FW

Fig. 6　Schematic diagram of hardness test pointing position

Table 5　Results of hardness test (HV10)

Sample	Position															
	1	2	3	4	5	6	7	8	9	10	11	12	13	14	15	16
M	223	228	230	219	217	210	224	229	228	—	—	—	—	—	—	—
LW	239	242	235	237	229	226	236	230	236	—	—	—	—	—	—	—
1#FW	268	268	228	236	257	261	229	220	193	197	215	206	226	216	260	201
2#FW	251	253	259	213	217	240	279	212	178	174	243	217	235	227	216	180
Average	M									223HV10						
	LW									234HV10						
	1#FW and 2#FW									260HV10						

The macroscopic metallogra phicpictures of welding joints between the fillet weld and the main pipe are shown in Fig. 7. As shown in Fig. 7, Both samples contain the sleeve, the main pipe and the fillet weld. Effects can be identified in Wall thickness and axial direction. The affect depth in wall thickness is among 4.15 to 4.85mm, which is 18.9 to 22% of the wall thickness. The affect length is about 31.40 to 36.61mm.

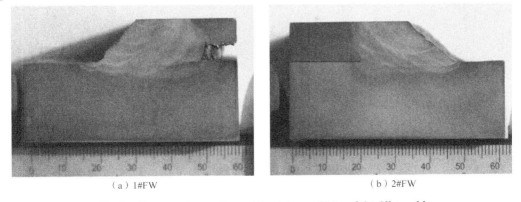

(a) 1#FW　　　　　　　　　　　　　　(b) 2#FW

Fig. 7　Macroscopic metallographic pictures of 1# and 2# fillet weld

The metallographic microstructures of M, LW, 1#FW and 2#FW regions were compared, and the results are shown in Fig. 8 and Table 6. Microstructure of M position is granular bainite with grain size is 10.5 grade. The microstructure of the main pipe under longitudinal weld seam is almost granular bainite which means that effect caused by longitudinal weld seam is quite slight. However, effects caused by fillet weld is quite conspicuous, which can be evolved from the appearance of pearlite and a small amount of polygonal ferrite. At the same time, the grain size decrease from 10.5 grade to about 9.79~9.9 grade, which signify that the grain size is coarsened.

Table 6 Microstructure and grain size

Sample	Microstructure	Average grain size (grade)
M	granular bainite	10.50
LW	granular bainite+ pearlite	10.50
1#FW	granular bainite+ pearlite+ polygonal ferrite	9.90
2#FW	granular bainite+ pearlite+ polygonal ferrite	9.79

Note: (1) Magnification: 500X; (2) Grain size detection method: cut-off method.

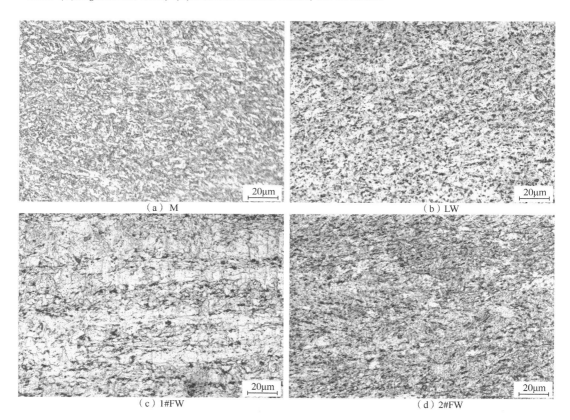

Fig. 8 Metallographic microstructures of M, LW, 1#FW and 2#FW regions

3 Conclusion

(1) The type B sleeve made of Q345R can effectively repair the girth weld defects of X80 pipe. Repair effect meets the requirements of pipeline design pressure. When the pressure is up to

18.1MPa, puncture leakage occurs at the cross joint of the longitudinal weld of type B sleeve and the circumferential fillet weld, where the crack occurs. The position is the weak point of the whole type B sleeve.

(2) Microstructue and properties of the main pipe are affected obviously by welding of longitudinal welds and fillet welds. Influence caused by fillet welds are greater than those caused by longitudinal welds. The affect depth is 18.9 to 22% of the wall thickness. Pearlite and a small amount of polygonal ferrite are formed in the microstructure of the weld thermal affect zone, and the grain size was coarsened, which results in that hardness of the main pipe under fillet weld increase and the yield strength is reduced by 5.5%. Surface cracking occurs during bending test. There is a certain service risk and it should be used under more frequent monitoring than usual.

Acknowledgement

This work was supported by the National Key R&D Program of China (2016YFC0802101).

References

[1] China petroleum pipeline corporation. World pipeline overview. 2014[M]. Beijing: Scientific and Technical Documentation Press, 2015: 232-234.

[2] Chen Anqi, Ma Weifeng, Ren Junjie, et al. Study on rehabilitation of high steel pipeline girth weld defects[J]. Natural Gas and Oil, 2017, 35(5): 12-17.

[3] Li Rongguang, Du Juan, Zhao Guoxing, et al. Overview of long distance oil/gas pipeline defects and repair technology[J]. Petroleum Engineering Construction, 2016(1): 10-13.

[4] Shi Renwei, et al. Oil and gas pipeline maintenance and repair technology[M]. Beijing: China Petrochemical Press Co. ltd, 2017: 66-79.

[5] Yin Changhua, Xue Zhenkui, Liu Wenhu. Overview of welding technology commonly used in domestic foreign minister pipeline[J]. Petroleum Engineering Construction, 2010(1): 42-47.

[6] China petroleum pipeline corporation. Oil and gas pipeline inspection and repair technology[M]. Beijing: Petroleum Industry Press, 2010: 138-145.

[7] Wang Hongju, Qian Chengwen, Wang Yumei, et al. Current status of gas pipeline pressure testing at home and abroad[J]. Petroleum Engineering Construction, 2007(1): 8-10.

[8] Sui Yongli. Welding technology research for girth of domestic X80 grade line pipe[D]. Doctoral Dissertation of Tianjin University, 2008.

本论文原发表于《Materials Science Forum》2019 年第 944 卷。

A Review of Dynamic Multiaxial Experimental Techniques

Nie Hailiang[1, 2]　Ma Weifeng[1]　Wang Ke[1]　Ren Junjie[1]
Cao Jun[1]　Dang Wei[1]

(1. Institute of Safety Assessment and Integrity, State Key Laboratory for Performance and Structure Safety of Petroleum Tubular Goods and Equipment Materials, Tubular Goods Research Center of CNPC; 2. Northwestern Polytechnical University)

Abstract: Structures and materials are usually exposed to exploding and attacking loading, experiencing large multiaxial plastic deformation under high strain rate, thus experimental techniques under dynamic multiaxial loading are significant and practical. In this paper, the history and progress of dynamic multiaxial loading experiment techniques are reviewed, the key technical problems in the dynamic multiaxial testing are analyzed, and a feasible solution is proposed.

Keywords: multiaxial loading; dynamic; Hopkinson bar; material property

1 Introduction

In many cases, structures and materials are subjected to explosive and impact loadings, For example: military weapons platform and the important military facilities may be impacted by the ammunition explosion and penetration loadings, the civilian infrastructure, such as bridge, nuclear power plants, large hydropower station are likely to be influenced by the impact loadings from ship collision or earthquakes, civilian aircraft, engine and so on may be subjected to abnormal landing impacts and discrete hit. The common characteristic of the above phenomena is that the materials or structures bear the transient strong impact loadings and experience the large multiaxial plastic deformation under high deformation rate. Fig. 1 shows the stress state of the armor and the armor-piercing projectile during penetration[1]. It can be seen that multiaxial stress states, large plastic deformation and transient state are their main characteristics. In order to design the impact protection of these structures, it is necessary to fully understand the mechanical behavior of plastic yield, plastic flow, plastic failure and dynamic fracture of materials under the action of multiaxial stress. The key is to test the dynamic mechanical properties of the materials under the condition of high deformation rate and multiaxial stresses.

Corresponding author: Nie Hailiang, niehll@ cnpc. com. cn.

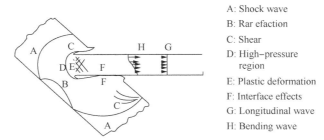

A: Shock wave
B: Rarefaction
C: Shear
D: High-pressure region
E: Plastic deformation
F: Interface effects
G: Longitudinal wave
H: Bending wave

Fig. 1 Schematic diagram of complex stress state in armor penetration process

At present, in the field of quasi-static mechanics, the multiaxial mechanical test is a relative proven technique, the multiaxial loading experiments such as tensile-torsion tests, compression-torsion tests and biaxial tensile/compression tests can almost be tested under quasi-static conditions by using the servo hydraulic testing machines[2]. The characteristics of some materials, such as unidirectional tensile compression curve of plastic materials, loading and unloading principles of plastic states, isotropic and follow-up reinforcement, Mises yield criterion and Tresca yield criterion, which were derived theoretically under quasi-static conditions, were also verified by various experimental means[3-6]. In terms of structural design, the four strength theories are also put forward under quasi-static conditions[3,5]. As for the failure mechanism of materials and the propagation of cracks in structures, some conclusions and rules have been obtained under quasi-static uniaxial loading and in-plane multiaxial loadings[7-10]. However, does this also apply to dynamic multiaxial stress states?

Is the yield surface affected by the strain rate?

Is the flow model also affected by the strain rate?

The answers or proofs to these scientific questions depend on dynamic multiaxial loading experiments.

The split Hopkinson bar technique is a commonly used technique to measure the dynamic properties of materials. Currently, most experimental methods for testing the dynamic properties of materials are developed on the basis of the split Hopkinson bar.

The development of dynamic testing equipment can be traced back to Kolsky. In 1949, professor Kolsky developed the split Hopkinson pressure bar device based on the studies of Hopkinson father and son[11,12], and the dynamic properties of various metal materials under impact compression load were tested[13]. In the book of "Split Hopkinson (Kolsky) bar: design, testing and applications"[14], Chen and Song introduces the history and development of Hopkinson bar in detail, so it will not be repeated here. Fig. 2 shows the schematic diagram of the split Hopkinson pressure bar device.

The split Hopkinson bar technique was originally used to measure the dynamic properties of ductile materials such as metals, later, some researchers extended this testing techniques to brittle materials. A recent review by Zhang and Zhao[15] has provided a detailed summary of the split Hopkinson bar in testings of rock, including the pulse shaping technique, end friction effects, inertia effects, dispersion and strain rate limit, etc. In a review of Xia and Yao[16], the key loading

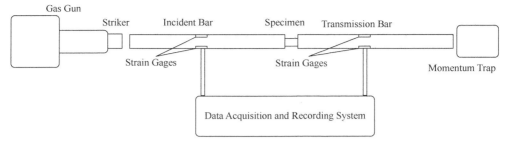

Fig. 2 Schematic diagram of split Hopkinson pressure bar device

techniques that are useful for dynamic rock tests with SHPB including the multiaxial loading techniques are introduced in detail, and various measurement techniques for rock tests in SHPB are fully discussed. In addition to the loading technology, data measurement technology is also a key point of dynamic testing. In 2017, Xing et al.[17] summarized the high-speed photography and digital optical measurement techniques for geomaterials in detail, which is worth learning and studying for readersIn order to study the dynamic tensile properties of the material, the Hopkinson tensile bar was developed based on the pressure version. Various Hopkinson tensile bars have been proposed by many researchers[18-20]. At present, the most widely used is the direct drawing Hopkinson tensile bar, which were improved by Qgawa based on J. Harding's device in the 1980s[21].

The torsional test has no three-dimensional problem, lateral inertia effect, and frictional effect on the ends of the specimen, thus the dynamic torsional experiment is paid more and more attentions. At present, the main loading methods of Hopkinson torsional bar equipment include prestored energy loading, explosive loading, direct impact loading, flywheel loading, and electromagnetic loading[14]. In a recent review of Yu[22], these five typical types of Hopkinson torsional bar were systematically reviewed, and interested readers can refer to it.

The successful development of the Hopkinson bar device indicates that the dynamic compressive, tensile and shear mechanical properties of the material at the strain rate of $10^2 \sim 10^4 s^{-1}$ can be measured separately. The loading process during the experiment is usually within tens to hundreds of microseconds. However, the single stress state is only an ideal condition, further study of the structure and material under the actual working conditions still requires multiaxial loading tests. The development of dynamic multiaxial testing equipment and experimental platform under high strain rate has always been the forefront of solid mechanics research and the dream of dynamists, especially in the high speed collision, penetration, protection engineering, fracture dynamics, plastic dynamics, material plastic molding and processing, explosion and impact and other fields of important basic issues. In recent years, with the efforts of some researchers, considerable progress has been made in the dynamic multiaxial testing, but there is still a lot of room for growth.

2 Dynamic multi-axial loading technique

2.1 Compression-shear dynamic biaxial loading devices

There are few researches about compression-shear dynamic biaxial loading devices. Most of the existing compression-shear dynamic biaxial loading techniques are based on falling weight experiment, direct impact experiment and high speed flying disc impact experiment.

Chung et al.[23,24] introduced biaxial loadings by limiting the lateral displacement of specimens in the traditional falling weight experiments. In the device, the specimen deformation is obtained by time integration of the acceleration signal of the falling weight. However, this data processing method may cause a large error.

Hong et al.[25] developed a set of direct impact dynamic compression-shear composite loading device on the basis of static oblique loading techniques. In this device, the air gun fires the plate to hit the loading conversion unit, and then hits the specimen. The specimen is fixed on the target plate, and the loading information is obtained through the multiaxial loading sensor. The defect of the direct impact device is that the impulse generated by impact loading is too short, and the deformation rate of specimen is not constant. In their device, Hong et al. designed a loading transfer unit, which applied the impact kinetic energy to the specimen through the inertia movement of the massive block, so as to correct and make up for the above defects.

The dynamic compression-shear test based on high speed flying disc appeared in the 1960s and 1970s, and the shear and high hydrostatic mechanical properties of materials can be tested at a very high strain rate ($10^5 s^{-1}$)[26]. A high speed air gun is used to shoot the elastic flying piece at a certain angle, and the specimen sheet is attached to it. The elastic flying piece carries the specimen to hit a thick elastic cutting plate, so as to produce the dynamic shear deformation. The flying strip shear experiment is applicable to the ultra-high strain rate, however due to the limitation of sample size, the experiment can only be conducted on materials less than the thickness of the specimen, which requires high requirements on the specimen, long preparation period and complex operation. At the beginning of the 21st century, many researchers are devoted to the research of compression-shear composite loading techniques. However, due to the limitation of loading equipment, the existing compression-shear composite loading is to use a uniaxial Hopkinson pressure bar to load a specially designed specimen, so that the specimen can be subjected to local compression-shear composite loading, or to modify the Hopkinson pressure bar to realize compression-shear composite loading.

Rittle et al.[27] proposed a compression-shear composite loading specimen, it has a groove in the middle of the specimen with a certain angle with the axial direction. In the experiment, when the split Hopkinson pressure bar is used to load along the axial direction of the specimen, the compression-shear composite load will be generated in the groove. However, the disadvantage of this method is that there is a large stress concentration in the groove and the specimen is easy to be destroyed along the root of the groove.

On the basis of the Hopkinson torsional Bar (TSHB, Torsional Split Hopkinson Bar), Huang et al.[28] established a compression-torsion composite loading device (Fig. 3). However, due to the propagation velocity difference between the compression wave and the torsional wave, the compression and torsional loadings cannot be applied to the specimen synchronously.

In order to realize synchronous loading of compression and shear, Zhao et al.[29] designed a Hopkinson compression-shear composite loading system with double transmission bars (Fig. 4). In the device, there are two symmetrical inclined planes at the end of the incident bar, the two

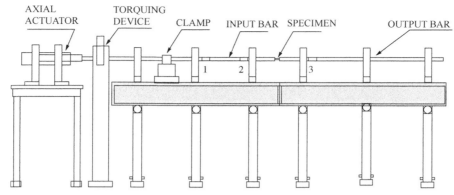

Fig. 3 The schematic diagram of dynamic compression-torsion composite bar by Huang et al. [28]
[Reproduced with permission from Int J Solids Struct. 41(11), 2821 (2004). Copyright 2004 Elsevies]

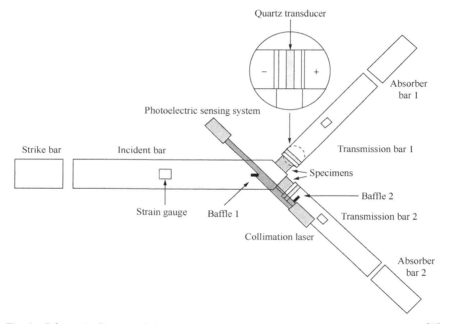

Fig. 4 Schematic diagram of dynamic compression-shear composite bar by Zhao et al. [29]
{Reproduced from [Zhao PD, Lu FY, Chen R, Lin YL, Li JL, Lu L, "A technique for combined dynamic compression-shear test," Rev Sci Instrum. 82(3): 35 (2011).], with the permission of AIP Publishing}

transmission bars form a certain included angle, and are opposite to the two inclined planes of the incident bar respectively. The two specimens are symmetrically placed between the two inclined planes of the incident bar and the end planes of the two transmission bars. Since the axial direction of the incident bar is at a certain angle to that of the transmission bar, when the stress wave is propagated into the transmission bar, the specimen will be subjected to compression-shear composite loading at the same time. In this experiment, the shear and compression stress wave signals on the incident and transmission bars were measured by piezoelectric sensors respectively. Similarly, Hou et al. [30,31] realized compression-shear composite loading by changing the end of the contact ends of the incident and transmission bars into inclined planes and installing the specimen between the two inclined planes (Fig. 5). However, in the experiment the shear force completely depends on the friction transfer between the specimen and the contact surface of the loading bars, while in order to ensure the uniform deformation of the specimen, the Hopkinson bar experiment requires the end friction to be as small as possible, so the friction between the specimen and the loading bars is limited, and the amplitude of the shear load will be limited.

Fig. 5 Dynamic compression-shear composite loading device of Hou et al. [30]
[Reproduced with permission from Int J Solids Struct. 48(5), 687 (2011). Copyright 2011 Elsevier]

2.2 In-plane dynamic biaxial loading device

Compared with the dynamic compression-shear loading, the in-plane biaxial loading is more difficult, because the loads from both directions are hard to synchronously control within hundreds of microseconds.

Grolleau[32] developed a dynamic bulge test using a split Hopkinson bar to obtain an equal-biaxial tension. In this device, a movable "bulge cell" composed of a thick-walled steel cylinder and a die ring was designed to perform dynamic bulge tests in a SHPB system. During the test, a round sheet specimen is clamped between the cylinder and the die ring. The input bar is sealed insert the cylindrical cell, and a fluid is filled into the cell to transmit the pressure. An output bar with tubular cross-section is contact with the end of the die ring. When the incident stress pulse propagates to the input bar/fluid interface, the pressure wave is transmitted through the fluid and ultimately causes the bulging of the sheet specimen to form an equal-biaxial tension stress condition. This experimental procedure is more reliable than direct tension tests because it circumvents the inherent gripping and force measurement issues associated with direct tension tests. However, the sealing of the "bulge cell" is required to be very high, and the diameter of the incident bar and transmission bar is large, waveform dispersion effect is serious. Sealed fixtures also have safety hazards under high pressure loading.

Shimamoto et al. [33] developed and validated a biaxial testing rig to generate universal biaxial impact loading in the cruciform specimen. The testing device consisted of four actuators, which were orientated at 90° to each other. A programmable controller was used to control the circuits. Four hydraulic actuators operated independently generate dynamic loads on the four arms of the specimen and the center point was always maintained at the home position without movement. Tests under unequal biaxial stress (load ratio of 1 : 1 to 1 : 4) were also possible. However, due to the power limitation of the hydraulic actuators, the strain rate obtained in the experiment is much lower than that in the traditional split Hopkinson bar experiment.

Hummeltenberg et al.[34] invented a biaxial in-plane tensile split Hopkinson bar system. This system consists of two sets of separated Hopkinson bar systems which are perpendicular to each other. A cruciform sample is clamped between four elastic bars, and two strikers are driven by the electromagnetic driving force. A circuit is designed to control the synchronization of the two launchers. However, the two strikers need to move a certain distance before impacting the incident bar, and there are too many influencing factors in this process, as a result, the stress waves of the two systems cannot get completely synchronized in time, which leads to the bending load in the specimen arms. In addition, the cruciform specimen is loaded asymmetrically, which makes the center of the specimen unable to remain immobile.

2.3 Dynamic triaxial loading device

Triaxial loading experiments are mainly used to study the dynamic responses of underground rocks. There are two main types of methods to realize triaxial loading, i.e. displacement boundary conditions and pressure boundary conditions.

2.3.1 Displacement boundary conditions

Displacement boundary conditions are typically achieved through jacketing the cylindrical surface of the specimen using a shrink-fit metal sleeve or a passive thick vessel. This is a passive confining technique, the boundary conditions on the specimen lateral surface include both stress and displacement, thus the key to this technology is the material of the jackets. The boundary can be treated as nearly rigid if the jacket is too harder than the specimen, while the effect is closer to pressure boundary conditions if the jacket is plastically deformable during the experiment.

Chen and Ravichandran[35] firstly employed a metal jackets in the axial loading systerm to radially confine cylindrical brittle specimens, which was further performed by Rome et al.[36], Forquin et al.[37] and Nemat-Nasser et al.[38].

Gong and Malvern[39] provides another inexpensive passive confining jacket system to study experimentally the multiaxial compressive response of rock-like specimens. In their device, jackets made of steel and aluminum and a 76.2-mm-diameter split Hopkinson pressure bar system was used.

In order to explore the damage and failure processes of a transparent polycrystalline aluminum oxynitride (AlON), Paliwal et al.[40] developed an experimental technique, which enabled a controlled and homogeneous stress state with high lateral compressive stresses. In the device, a prismatic specimen (with a rectangular cross section) was statically precompress from two perpendicular directions and then subjected to axial dynamic compressive loading using a modified compression Hopkinson bar setup.

2.3.2 The pressure boundary conditions

Pressure boundary conditions are achieved through hydrostatic pressure in a triaxial test. In such a test, a specimen placed inside a pressure chamber is isotropically loaded by hydrostatic pressure. When an additional axial load is applied to the specimen which is in the constant hydrostatic pressure, the lateral deformation of the specimen results in the change of hydrostatic pressure and an additional shear stress will apply to the specimen. The specimen is isolated from the confining fluid through a soft seal membrane that is placed over the specimen. In such an experiment, the boundary condition on the lateral surface of the specimen is pressure only, making the stress state in the specimen clearly defined.

Christensen et al.[41] and Lindholm et al.[42] performed some of the most pioneering work in the dynamic tests of rocks under hydrostatic confinement in the early 1970s. The device is composed of an SHPB system with two hydraulic cylinders, and the sample is enclosed in the lateral confining cylinder. The transverse confining stresses and the axial confining stress will be generated by the action of the lateral confining cylinder and the confining cylinder applies.

Li[43] improved the devices of Christensen et al. and Lindholm et al. In this experimental design, the two pressure cylinders was connected with two tie-rods. The claimed that they can generate triaxial confinement, while the results they showed were only axial confinement in their work.

Frew et al.[44] used a very similar idea to improve the device of Christensen et al. and Lindholm et al. In their device, four tie-rods were connected with the two cylinders to apply hydrostatic confinement. The method to achieve such a confining state is first expose the cylindrical rock sample to the confining fluid and then to maintain the same fluid pressure in both cylinders.

Cadoni and Albertini[45] designed a truetriaxial loading apparatus, this setup adopts hydraulic servo devices to apply 0 ~ 100 MPa static load to the cube sample from three-directions independently, and then the split Hopkinson pressure bar is used to apply impact dynamic load on the specimen from one incident bar. However, the very short loading times do not enable one to carry out multiaxial dynamic loading in synchronicity.

3 Electromagnetic solution of synchronization problem in dynamic multiaxial loading

The traditional split Hopkinson pressure bar generates stress waves by means of striker impact. The striker is launched through the air valve switch. The striker needs to travel a distance before impact, so the time synchronization of multiple stress waves cannot be accurately controlled. In order to solve the synchronization problem, the traditional mechanical energy conversion method should be abandoned and the energy conversion method that can accurately control the time should be developed. For this purpose, some researchers are trying to find new mechanisms of stress waves. Silva et al.[46] proposed a new design for the compressive split Hopkinson bar that makes use of the intense pressure created in a transient magnetic field formed by the passage of a pulse of electric current through a series of coils. In this device, the striker is driven and accelerated by an instantaneous magnetic field, and finally impact the incident bar to produce a stress wave. The system has some advantages over the traditional Hopkinson bar, while the principle of the stress wave generation is still the impact method, the time synchronization problem is still not solved.

A new stress wave generation principle that is helpful to solve the stress wave synchronization problem was proposed by Nie et al.[47] in 2008. They developed a novel electromagnetic split Hopkinson pressure bar (ESHPB), which employs the electromagnetic energy conversion technique of LC circuit to generate directly the incident stress pulse (Fig. 6). This technique can generate easily compressive as well as tensile incident pulses. Compared with traditional pulse generation techniques by the impact of a projectile or by a sudden release of a pre-stressed section, the electromagnetic energy conversion technique can be accurately triggered within several microseconds. It is, therefore, a good candidate to supply the symmetrical and synchronous loads in bidirectional or biaxial split Hopkinson bar systems in the future.

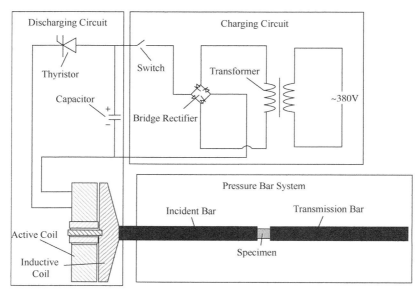

Fig. 6 Schematic diagram of the Electromagnetic SHPB setup[47] [Reproduced with permission from Int J of Impact Eng. 116, 94 (2018). Copyright 2018 Elsevier]

In fact, the electromagnetic split Hopkinson bar does show unique advantages in the problem of stress wave synchronization. At present, Nie et al. have successfully made use of electromagnetic split Hopkinson bar to develop a symmetrically loading Hopkinson bar[48] (Fig. 7). In order to obtain two synchronized stress pulses, two identical electromagnetic stress pulse generators connected to the same LC discharge circuit are used in this device. This symmetric impact loading configuration might be easily interchanged into a compressive as well as a tensile version because of the versatility of the electromagnetic stress pulse generators. They measured the stress waves in the two incident bars, and the results showed that the two incident waves basically reached the end faces of the specimen at the same time, with an error of less than 3 microseconds (Fig. 8).

The biggest disadvantage of the electromagnetic Hopkinson bar is the electromagnetic interference problem. Since the device will generate an instant electromagnetic pulse in the process of stress wave producing, which will interfere with the nearby data acquisition equipment, and the interference and stress wave signals will be both captured by the data collector almost at the same time, so it will affect the rising edge of the collected stress waves.

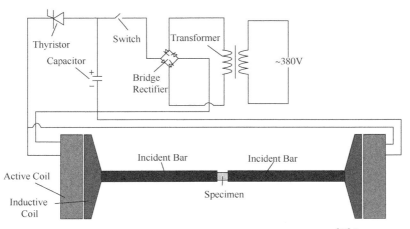

Fig. 7 Schematic diagram of the symmetric split Hopkinson compression bar[48] [Reproduced with permission from Int J of Impact Eng. 122, 73 (2018). Copyright 2018 Elsevier]

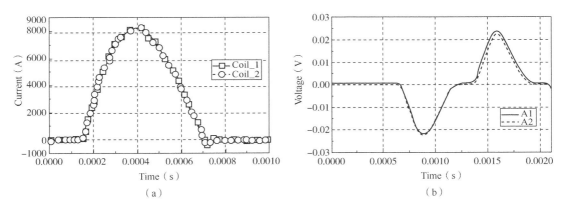

Fig. 8 Comparison of experimental results: (a) discharge currents of two active coils, (b) signals on the bars measured by the data collector[48] [Reproduced with permission from Int J of Impact Eng. 122, 73 (2018). Copyright 2018 Elsevier]

The distribution of pulse electromagnetic fields produced by unidirectional and symmetrical loading equipment are relatively simple and have a regular to follow, and the electromagnetic interference can be eliminated through some methods, such as the location optimization of data acquisition equipment, the special design of the wires and so on. In fact, we have eliminated electromagnetic interference in unidirectional and symmetric loading devices by the above method and obtained stress-strain curves of some common materials. For example, Fig. 9(a) shows a typical stress pulses obtained in symmetric compression experiment of copper. It is necessary to point out that the negative values in reflected waves are not due to an un-perfect contact between the specimen and the bars, they are the natural result of the superposition of waves. A detailed explanation can be found in reference[48]. The forces at the two sides [Fig. 9(b)] of the specimen are in a good equilibrium state. The results of copper symmetrical compression test are compared in Fig. 10(a) with that of traditional striking SHPB test and the single electromagnetic SHPB (ESHPB) test under the similar average strain rate of about 1200 1/s, since the same material was tested by using in the previous paper, the result is also added for comparison. The stress–strain curve obtained in the symmetric test is in a good agreement with that of the traditional SHPB test. It is noticed from Fig. 10(b) that the strain rate–strain curve of the symmetric test is not constant, which is a drawback of this device because the stress wave is almost half-sin shape unless the pulse shaper method is applied. However, if four devices discharge at the same time, the distribution of the electromagnetic field will be more complex, and the interference cannot be offset by the above method. Therefore, electromagnetic interference is still a great challenge to realize the dynamic biaxial or even triaxial loading device by using the electromagnetic Hopkinson bar.

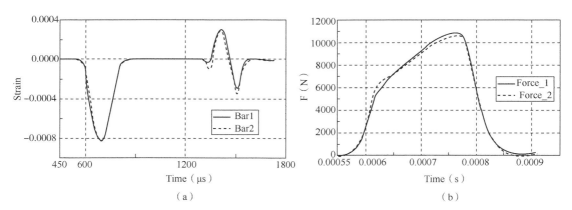

Fig. 9 Pulses recorded in a copper test using symmetric SHPB apparatus: (a) the strain signal in bars, (b) the forces at the sides of the specimen[48] [Reproduced with permission from Int J of Impact Eng. 122, 73 (2018). Copyright 2018 Elsevier]

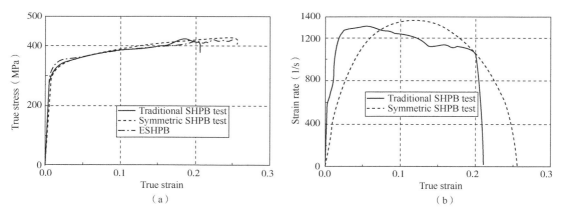

Fig. 10 Compression results of copper by using symmetric loadings, electromagnetic SHPB apparatus (ESHPB) and traditional SHPB apparatus: (a) stress-strain curves, (b) strain rate–strain curves[48] [Reproduced with permission from Int J of Impact Eng. 122, 73 (2018). Copyright 2018 Elsevier]

4 Summary and looking forward

After more than 60 years of efforts, the single-axial Hopkinson bar devices have played an important role in promoting the development of many disciplines. However, either by way of mechanical collision to produce compression and tensile pulse, or adopt the way of energy storage to release a loading impulse, its synchronicity is unable to effectively controlled within microseconds time scale, as a result, the development of tension – compression and tension – torsion coupled multiaxial Hopkinson bar devices have not made substantial progress.

Although many researchers want to develop a tension-torsion or compression-torsion multiaxial loading device by using the principle of stored mechanical energy release, satisfactory results have not been achieved due to the complexity of the release mechanism and the large difference in wave velocities between the compressive/tensile and shear waves.

Some researchers also used the uniaxial compression Hopkinson bar with inclined plane to conduct compression – shear coupling experiments. The friction between the specimen and the inclined plane of the Hopkinson bar can impose shear load on the specimen, but the compressive stress is not evenly distributed in the sample.

The development of dynamic multiaxial loading device under has not made a breakthrough, so far it is still in blank. The main reason is that the duration in dynamic tests is very short compared with that of quasi – static tests, which requires to precisely control the generating time of stress waves. However, the stress pulse generation principle of traditional Hopkinson bar is the conversion between one mechanical energy and another mechanical energy, that is, the use of impact method or energy storage method to generate stress wave, while the time control of mechanical method is unable to achieve the accuracy required by multiaxial loading. Therefore, in order to solve the time control problem of dynamic multiaxial loading technique, it is necessary to abandon the traditional stress pulse producing method in Hopkinson bar, so that the initial times of stress waves can be controlled precisely.

With the invention of the electromagnetic Hopkinson bar device, the synchronization problem of stress waves has been technically solved, and the dynamic symmetrical loading of the specimen has been successfully realized. It is believed that in the near future, it will be possible to achieve dynamic biaxial loading even more triaxial loading.

5 Data availability statement

The data that support the findings of this study are available from the corresponding author upon reasonable request.

Acknowledgement

This work was supported by the China Postdoctoral Science Fund (2019M653785).

References

[1] Wright TW, "A survey of penetration mechanics for long rods," Springer. (1983).

[2] Hoferlin E, Bael AV, Houtte P V, Steyaert G, Maré2 C, "The design of a biaxial tensile test and its use for the validation of crystallographic yield loci," Model Simul Mater Sc. 8(4), 423 (2000).

[3] Rotter, Julian B, "Generalized expectancies for internal versus external control of reinforcement," Psychol Monogr. 80(1), 1 (1966).

[4] Sutton R, Barto A, "Reinforcement Learning: An Introduction," MIT Press. (1998)

[5] Donovan PE, "A yield criterion for Pd40Ni40P20 metallic glass," Acta Metall, 37(2), 445 (1989).

[6] Shrivastava HP, Mr ÃZ, Dubey RN, "Yield Criterion and the Hardening Rule for a Plastic Solid," ZAMM-J Appl Math Mec. 53(9), 625 (1973).

[7] Jia P, Tang CA, "Numerical study on failure mechanism of tunnel in jointed rock mass," Tunn Undergr Sp Tech. 23(5), 500 (2008).

[8] Krebs F C, Norrman K, "Analysis of the failure mechanism for a stable organic photovoltaic during 10000 h of testing," Prostate. 15(8), 697 (2007).

[9] Apostolico F, Gammaitoni L, Marchesoni F, Santucci S, "Resonant trapping: A failure mechanism in switch transitions," Phis Rev E. 55(1), 36 (1997).

[10] Belov D, Yang MH, "Failure mechanism of Li-ion battery at overcharge conditions," J Solid State Electr. 12 (7-8), 885 (2008).

[11] Hopkinson J, "Further experiments on the rupture of iron wire," Original Papers-by the late John Hopkinson, Vol II, Scientific Papers. (1872).

[12] Hopkinson B, "A method of measuring the pressure produced in the detonation of high explosives or by the impact of bullet," Philos T R Soc A. 437, (1914).

[13] Kolsky H, "An investigation of the mechanical properties of materials at very high rates of loading," Proc Phy Soc B. 62(11), 676 (1949).

[14] Chen WW, Song B, "Split Hopkinson (Kolsky) bar: design, testing and applications," Springer Science & Business Media. (2010).

[15] Zhang Q, Zhao J, "A Review of Dynamic Experimental Techniques and Mechanical Behaviour of Rock Materials," Rock Mech Rock Eng. 47, 1411 (2014).

[16] Xia K, Yao W, "Dynamic rock tests using split Hopkinson (Kolsky) bar system-A Review," J Rock Mech Geotechnical Eng. 7(1), 27(2015).

[17] Xing H, Zhang Q, Braithwaite C, Pan B, Zhao J, "High-speed photography and igital optical measurement techniques for geomaterials: fundamentals and applications," Rock Mech Rock Eng. 50, 1611 (2017).

[18] Lindholm U, Yeakley L, "High strain-rate testing: tension and compression," Exp Mech. 8(1), 1(1968).

[19] Nicholas T, "Tensile testing of materials at high rates of strain," Exp Mech. 21(5), 177 (1981).

[20] Harding J, Wood E, Campbell J, "Tensile testing of materials at impact rates of strain," J Mech Eng. 2(2), 88 (1960).

[21] Ogawa K, "Impact-tension compression test by using a split-Hopkinson bar," Exp Mech. 24(2), 81 (1984).

[22] Yu X, Chen L, Fang Q, Jiang X, Zhou Y, "A Review of the Torsional Split Hopkinson Bar," Advances in Civil Engineering, 2018(PT9), 17(2018).

[23] Chung J, Waas AM, "Compressive response of circular cell polycarbonate honeycombs under inplane biaxial static and dynamic loading. Part I: experiments," Int J Impact Eng. 27(7), 729(2002).

[24] Chung J, Waas AM, "Compressive response of circular cell polycarbonate honeycombs under inplane biaxial static and dynamic loading—Part II: simulations," Int J Impact Eng. 27(10), 1015(2002).

[25] Hong ST, Pan J, Tyan T, Prasad P, "Dynamic crush behaviors of aluminum honeycomb specimens under compression dominant inclined loads," Int J plasticit. 24(1), 89 (2008).

[26] Espinosa H, Patanella A, Xu Y, "Dynamic compression-shear response of brittle materials with specimen recovery," Exp Mech. 40(3), 321 (2000).

[27] Rittel D, Lee S, Ravichandran G, "A shear-compression specimen for large strain testing," Exp Mech. 42 (1), 58 (2002).

[28] Huang H, Feng R, "A study of the dynamic tribological response of closed fracture surface pairs by Kolsky-bar compression-shear experiment," Int J Solids Struct. 41(11), 2821 (2004).

[29] Zhao PD, Lu FY, Chen R, Lin YL, Li JL, Lu L, "A technique for combined dynamic compression-shear test," Rev Sci Instrum. 82(3): 35 (2011).

[30] Hou B, Ono A, Abdennadher S, Pattofatto S, Li Y, Zhao H, "Impact behavior of honeycombs under combined shear-compression. Part I: Experiments," Int J Solids Struct. 48(5), 687 (2011).

[31] Hou B, Pattofatto S, Li Y, Zhao H, "Impact behavior of honeycombs under combined shear-compression. Part II: Analysis," Int J Solids Struct. 48(5), 698 (2011).

[32] Grolleau V, Gary G, Mohr D, "Biaxial testing of sheet materials at high strain rates using viscoelastic bars," Exp Mech. 48, 293 (2008).

[33] Shimamoto A, Shimomura T, Nam, "The development of a servo dynamic loading device," Key Eng Mater. 243-244, 99 (2003).

[34] Hummeltenberg A, Curbach M, "Entwurf und Aufbau eines zweiaxialen Split-Hopkinson-Bars," Beton-Und Stahlbetonbau. 107, 394 (2012).

[35] Chen W, Ravichandran G, "Dynamic compressive failure of a glass ceramic under lateral confinement," J Mech Phys Sol, 45(8), 1303 (1997).

[36] Rome J, Isaacs J, Nemat-Nasser S, "Hopkinson techniques for dynamic triaxial compression tests," In: Gdoutos E editor. Redent Advances in Experimental Mechanics. Netherlands Springer, 2004.

[37] Forquin P, Gary G, Gatuingt F, "A testing technique for concrete under confinement at high rates of strain," Int J Impact Eng. 35(6), 425 (2008).

[38] Nemat-Nasser S, "Introduction to high strain rate testing," In: Kuhn H, Medlin D editors. ASM Handbook: Volume 8: Mechanical Testing and Evaluation. Ohio, USA: ASM International. 427 (2000).

[39] Gong JC, Malvern LE, "Passively confined tests of axial dynamic compressive strength of concrete," Exp Mech. 30(1), 55 (1990).

[40] Paliwal B, Ramesh KT, McCauley JW, Chen M, "Dynamic compressive failure of AlON under controlled planar confinement," J Am Ceram Soc. 91(11), 3619 (2008).

[41] Christensen RJ, Swanson SR, Brown WS, "Split-Hopkinson-bar tests on rock under confining pressure," Exp Mech. 12(11), 508 (1972).

[42] Lindholm US, Yeakley LM, Nagy A, "The dynamic strength and fracture properties of Dresser basalt," Int J Rock Mech Min. 11(5), 181 (1974).

[43] Li XB, Zhou ZL, Lok TS, Hong L, Yin TB, "Innovative testing technique of rock subjected to coupled static and dynamic loads." Int J Rock Mech Min. 45(5), 739 (2008).

[44] Frew DJ, Akers SA, Chen W, Green ML, "Development of a dynamic triaxial Kolsky bar," Measurement Science and Technology. 21(10), 105 (2010).

[45] Cadoni E, Albertini C, "Modified Hopkinson bar technologies applied to the high strain rate rock tests," Advances in Rock Dynamics and Applications, New York. USA: CRC Press, 79 (2011).

[46] Silva C, Rosa P, Martins P, "An innovative electromagnetic compressive split Hopkinson bar," Int J Mech Mater Des. 5(3), 281 (2009).

[47] Nie H, Suo T, Wu B, Li Y, Zhao H, "A versatile split Hopkinson pressure bar using electromagnetic loading," Int J of Impact Eng. 116, 94 (2018).

[48] Nie H, Suo T, Shi X, Liu H, Li Y, Zhao H, "Symmetric split Hopkinson compression and tension tests using synchronized electromagnetic stress pulse generators," Int J of Impact Eng. 122, 73 (2018).

本论文原发表于《Review of Scientific Instruments》2020 年。

Analysis of Cracks in Polyvinylidene Fluoride Lined Reinforced Thermoplastic Pipe Used in Acidic Gas Fields

Qi Guoquan[1,2] Yan Hongxia[1] Qi Dongtao[2] Wei Bin[2] Li Houbu[2]

(1. Northwestern Polytechnical University; 2. State Key Laboratory for Performance and Structure Safety of Petroleum Tubular Goods and Equipment Materials, CNPC Tubular Goods Research Institute)

Abstract: Cracking of polyvinylidene fluoride(PVDF) pipe as lining layer used in reinforced thermoplastic pipe(RTP) was occurred after running for half a year operation in gathering and transportation in acidic gas fields. For investigating the cause of cracking, the characteristics of the PVDF pipe with crack, such as composition, thermal mechanical and structural, have been compared with the new one and analyzed. Results showed that the cracks have been probably caused by the weld lines inside the liner PVDF pipe, which were caused by un-proper merging of PVDF melt during extrusion after passing through the six legged spider mandrel. It is suggested to assess the extrusion conditions of PVDF. Moreover, it is recommended to use a different mandrel geometry, such as spiral mandrel, in order to avoid weld lines in the final item.

Keywords: Crack; Pipeline failures; Weld line; Microstructures

1 Introduction

With the large-scale developing, the gas field contained H_2S and CO_2 has become an important part of the natural gas resources for exploitation in China nowadays. Ordinary carbon steel pipe corrosion is very serious in such a high acid environment, such as the hydrogen sulfide contained in the medium, can cause the sulfide stress corrosion cracking(SSCC) and hydrogen induced cracking(HIC) of the pipe. Huge economic losses, casualties and ecological damage would be caused by serious corrosion problem with H_2S and CO_2, finally. In recent years, researchers take a variety of solutions to solve the problem of oil and gas field corrosion, such as using anti-H_2S alloy pipe, adding corrosion inhibitor, and so on. Due to the high cost and difficult management of the technologies mentioned above, application effect is not good. However, thermoplastics are broadly employed in the field of the oil and gas transporting which contains H_2S

Corresponding author: Qi Dongtao; E-mail: qidt@cnpc.com.cn; Yan Hongxia: hongxiayan@nwpu.edu.cn; Wei Bin: weibin006@cnpc.com.cn; Li Houbu: lihoubu@cnpc.com.cn.

and CO_2 due to its good corrosion resistance, and it is turned into an important solution for oil and gas field corrosion[1-3].

In the process of transmission, due to the product quality, processing technology, high temperature, high pressure and corrosive medium, the thermoplastics pipe as the lining layer often fail, which is caused by performance degradation. In the service environment with H_2S/CO_2, accidents such as perforation or fracture often happen, usually result in oil and gas leak, and even serious accidents may lead to huge economic losses, casualties and ecological damage[4,5]. In this study, failure analysis of polyvinylidene fluoride (PVDF) pipe as lining layer used in reinforced thermoplastic pipe (RTP) is investigated. For reliable operation and efficient design, it is urgent to investigate the failure reasons of the PVDF pipe and to further analyze the factors which affect the pipeline's serving ability. It will be of great benefits to prevent events which could trigger disastrous incidents, thus can reduce much loses in terms of service life and economics.

2 Material and methodologies

2.1 Background of the Failure

Reinforced thermoplastic pipe (DN80 PN16MPa), PVDF lining reinforced with fiber, as is shown in Fig. 1, was used for oil gathering and transportation in sour oil and gas field. The transmission medium is mostly gas, with a small amount of oil and water. The pressure and temperature of transmission medium are 9MPa and 37℃, respectively. The content of H_2S is 22900mg/m^3 while CO_2 of 2.1mol%. The whole construction project had been completed on April 2015, while the pipeline had been started to use on May 2015. After running for half a year, the inspection of pipeline application effect was carried out.

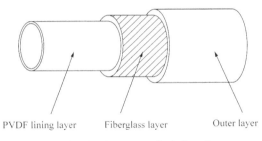

Fig. 1　Structure of reinforced thermoplastic pipe

2.2 Failure description

In order to study the performance of PVDF extruded liner pipe service performance, reinforced thermoplastic pipe is cut longitudinally. It can be seen from Fig. 2 that the PVDF extruded liner pipe shows two cracks with the color of brown along the extrusion direction at inner surface of the pipe.

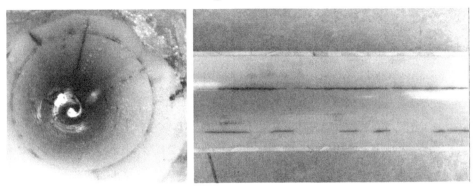

Fig. 2　Macroscopic topography of the failure sample

2.3 Material

The specimens of PVDF pipe were cut from the same failed one (Sample 1) which was taken from the pipeline after running for half a year. By comparison, two types of specimens for below tests were also investigated. One (Sample 2) is PVDF pipe extruded by same batch but never be used. The other (Sample 3) is PVDF pipe extruded by Solvay out of Solef® 60512, used as a standard reference. At least three specimens taken along the circumference of the pipe were tested for each measurement to evaluate the results statistically.

2.4 Methodologies

Many factors may cause the failure of PVDF pipe and more studies are needed. Firstly, the background information and the operation conditions that might lead to failures of the pipe were investigated in details. Secondly, probable causes for failure of the pipe were systematically analyzed by various measurements.

To check possible degradation of PVDF material, Fourier transform infrared spectroscopy (FT-IR, Wisconsin, USA) and X-ray photoelectron spectroscopy (XPS, PE Corp., USA) have been performed with the aim to find traces of degradation of PVDF and at the same time to find possible traces of foreign contaminants. XPS analysis for microanalysis is implemented on the discolored (brownish) part in the nearby of the cracks present in the inner surface of the sample, and also on areas far from the cracks for reference.

In order to evaluate the heat resistance of the PVDF pipe, Vicat softening temperature has been tested by Vicat softening point tester according to the standard of GB/T 1633 "Plastics – Thermoplastic materials – Determination of Vicat softening temperature (VST)." The test uses B50 method, that is, heating rate of 50℃/h, the application load of 50N.

To check possible alteration of the crystallinity due to un-proper processing, Dynamic Scanning Calorimetry (DSC) has been performed to check crystallinity of the two samples (the failed one and the other batch) in comparison. DSC measurements were made under N_2 atmosphere from 70℃ to 240℃ with heating and cooling rates of 10℃/min. Samples were taken by scraping away the material from the pipes' surface.

To check whether the cracks were produced by un-properly merged weld lines which were caused by the legs of the spider mandrel (Fig. 3), measurements were carried out. The Scanning Electron Microscope (SEM, JEOL - 6700F, Japan) and Optical Microscopy (OM, Leica DM LB2, Germany) with reflected polarized light have been performed in order to highlight the presence of weld-lines in the pipe.

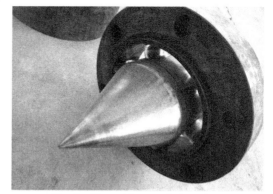

Fig. 3 The spider mandrel used for PVDF extrusion

3 Results anddiscussion

Analysis contain morphology and structure will be conducted as follows to study the actual reason for such gradual failure of PVDF pipe.

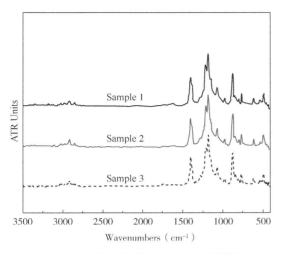

Fig. 4 The FT-IR spectrum of PVDF

3.1 FT-IR analysis

Due to service in acidic gas fields, along with the time increasing, the PVDF pipe physical properties deteriorate, some cracks appear, which may caused by molecular breakage or cross linking. To understand what compositions of the PVDF pipe with crack were changed, the molecular structure and chemical composition of the pipe were characterized by Fourier transform infrared spectroscopy in Fig. 4. The FT-IR spectrum contains three samples, Sample 1 (the used pipe), Sample 2 (the new one) and Sample 3 (the standard reference).

The results show that the peaks at around $2984cm^{-1}$, $1403cm^{-1}$ and $1182cm^{-1}$ are assigned to =CH, $-CH_2$ and CF_2 vibration, respectively. The peak at $1180cm^{-1}$ and $880cm^{-1}$ are C–C skeleton vibration, and the sharp absorption at $976cm^{-1}$, $840cm^{-1}$, $796cm^{-1}$, $763cm^{-1}$, $614cm^{-1}$ and $509cm^{-1}$ are the vibration absorption peak of the crystalline phase. Compared with the results of sample 2 and sample 3, the peak of sample 1 is close to the other two. It can be seen that the functional group of the PVDF did not change significantly, that is, the material was not modified by environmental medium corrosion.

3.2 XPS analysis

In order to further find traces of degradation of PVDF and at the same time possible traces of foreign contaminants, XPS analysis is implemented on the discolored (brownish) part in the nearby of the cracks present in the inner surface of the sample [Fig. 5(a)], and also onto areas far from the cracks [Fig. 5(b)] for reference.

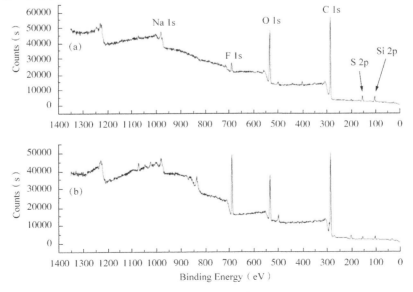

Fig. 5 XPS analysis on the PVDF

Spectra obtained in areas close to the cracks show the presence of intense signals of F, C, O and Si, and negligible signals of N, Zn, Na and Ca. Due to the little influence on the research result, the trace additives such as N, Zn, Na and Ca are not investigated in this study. By comparing the elemental quantification of two rings(Graph 1), the content of F element decrease, with the percent content of atomic(at.%) from 9.39 to 1.89, while the content of C, F, Si does not change basically. Both of the areas close to the cracks and far from the cracks show the presence of intense signals of Fluorine and Carbon(typical of PVDF polymer). The results show that PVDF polymer seems not to be degraded even in correspondence of the cracks; the secondary signals of O and Si possibly result from the media transported by the pipe while in use.

3.3 VST analysis

It can be seen from Fig. 6 that the deformation of the sample increases with increasing temperature during the test of VST. For comparison, the specimens are taken from Sample 1 and Sample 2. At higher temperatures (greater than 80℃), the increase in the shape of the sample begins to accelerate, reaching the maximum growth rate at the time of reaching the VST. When the deformation reaches 1mm, the test equipment will automatically stop. At this time, the corresponding temperature is the VST of the PVDF pipes under this test condition.

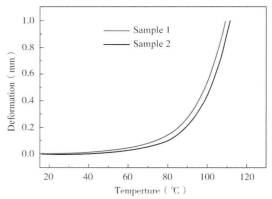

Fig. 6 Vicat softening temperature test of PVDF pipe

With the increase of temperature, the deformation of thermoplastics lining is getting bigger, indicating that the hardness decreases gradually. When the temperature is higher than 60℃, the amount of deformation at the same temperature is larger, which indicates that the mechanical properties of the lining material are weakened with the influence of service time. However, the value decreases little with the drop rate of 2%. Taking into account the actual service temperature of the pipe less than 40℃, the heat resistance of the PVDF pipe will not be reduced significantly. That is to say, the reason of cracks of the pipe is not decline of heat resistance mechanical during the service time.

3.4 DSC analysis

Thermographs of the two samples (sample 1 and sample 3) are depicted in Fig. 7 and Fig. 8. The thermograph of Sample 1 shows thermal values (first and second melting peaks, crystallization peak, melting enthalpy and crystallization) that are comparable to Sample 3. The generated crystalline morphology during extrusion molding directly affects final performance of products. The crystallinity is related to the process parameters in the forming stage, such as extrusion temperature, pressure and time. In normal circumstances, the brittleness of the material increases with the increase of the polymer's crystallinity, which will lead to the toughness decrease and the ductility deterioration, eventually. From DSC results, there should be no reason to think that the pipe of Sample 1 has suffered an un-proper extrusion process that affected the crystallinity and consequently the overall mechanical properties.

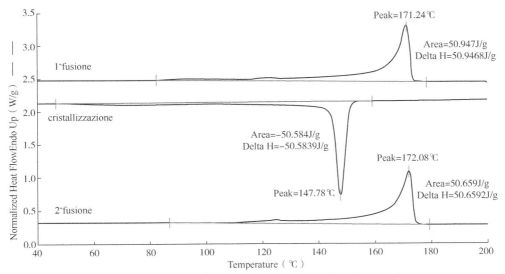

Fig. 7 Heating-cooling-reheating thermographs of Sample 1

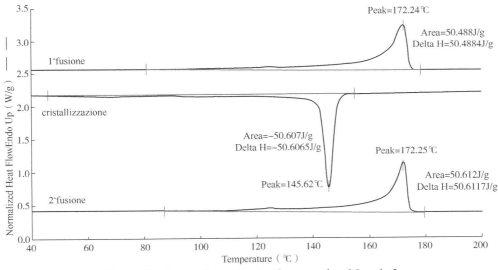

Fig. 8 Heating-cooling-reheating thermographs of Sample 3

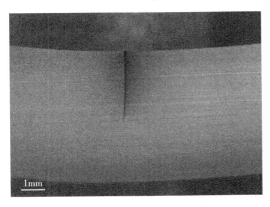

Fig. 9 Microstructure of the PVDF pipe with crack

3.5 Analysis of microstructure

As can be seen from Fig. 9, the cracks initiation area exists in the inner surface of PVDF pipe, and expands outward gradually. The crack width reduces along the inner wall toward the outer one, and the crack depth is 2mm. In the course of service time, the position with weak pressure capacity causes crack due to internal pressure combined with the media action[6].

In order to further identify the weak pressure capacity position, OM with reflected polarized light is used to view the structure of the Sample 2. In this study, the magnification of OM is 25. The OM image clearly shows a weld line exists in the inner surface of the pipe with the extrusion direction, which is consistent with the cracking direction[7-9]. Moreover, the two cracks present along the inner surface of the liner pipe in service form two lines that are at an angle of approximatively 60°, which is the same distance between the six legs of the spider mandrel.

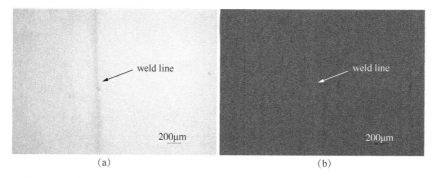

Fig. 10　25x images in reflected polarized light of the inner surface of Sample 2

3.6　Discussion

In order to identify the cause of cracking, the characteristics of the PVDF pipe with crack, such as composition, thermal mechanical and structural have been compared with the new one and analyzed. According to the composition investigation (FT-IR and XPS), the service conditions seem not to have affected PVDF material of the liner pipe, which does not show signs of degradation. The result of thermo mechanical properties (VST test) shows that cracks of the pipe are not result from the decline of heat resistance mechanical during the service time. The thermal characteristics (DSC results) of sample 2 are not evidences severe anomalies due to un-proper processing, when compared to thermal characteristics of sample 3. According to the optical microscopy investigation, the two cracks present along the inner surface of sample 1 are at a distance of approximatively 60°, which exactly matches the angle of the spider mandrel legs. The weld lines present on sample 2 further demonstrate that the cracking is caused by the processing.

In the extrusion process, when the melt flow before the peak and some subsequent parts of the separation or confluence, the weld line (or welding surface) will be formed in that position[10,11]. The weld line is a common plastic parts defect, its existence not only affects the appearance of the product quality, but also has a great impact on the mechanical properties of the product.

Any factor that affects the molecular chain entanglement, crystallization orientation, or molecular thermal can affect the strength of the weld line. These factors are related to the molecular activity during the formation of the weld line, depending on the material characteristics and process conditions[12,13]. Material characteristics include relaxation time, crystal morphology and so on; and process conditions include extrusion speed, mold temperature, melt temperature, pressure and so on[14,15].

4　Conclusions and mitigation measures

4.1　Conclusions

During the service time, the crack occurred in the internal surface of PVDF pipe as lining layer of RTP under the acidic gas service conditions. The cracks appeared on the weak point inside the liner PVDF pipe given by the weld lines, which were caused by un-proper merging of PVDF pipe melt during extrusion after passing through the six legged spider mandrel. It is indicated that this mandrel is not suitable for extrusion of PVDF pipe and the mandrel structure should be changed to improve the quality of pipe.

Table 1 The elemental comparison of areas close to the cracks and far from them

Sample No.	C1s	F1s	O1s	Si2p
1	70.89	1.89	15.55	6.87
2	67.41	9.39	12.86	3.65

4.2 Mitigation measures

Some possible suggestions are given below: It is suggested to review the extrusion conditions of PVDF, with specific care on the temperature of the melt in proximity of the die head, and also on the processing speed. Moreover, it is recommended to use a different mandrel geometry, in order to avoid weld lines in the final item. For instance, a spiral mandrel could help in this specific case.

Acknowledgement

This work was supported by the Key Laboratory for Petroleum Tubular Engineering (2016D-1604), China National Petroleum Corporation.

References

[1] D. T. Qi, H. B. Li, X. H. Cai, et al., Application and Qualification of Reinforced Thermoplastic Pipes in Chinese Oilfields. ICPTT, Beijing, 2011.

[2] G. Q. Qi, Y. Wu, D. T. Qi, et al., Experimental study on the thermostable property of aramid fiber reinforced PE-RT pipes. Natural Gas Industry B, 2(5)(2015)461-466.

[3] B. Wei, D. T. Qi, H. B. Li, et al., Corrosion resistance of reinforced thermoplastic pipe in the sour environment. Natural Gas Industry, 35(06)(2015)87-92.

[4] H. B. Li, M. L. Yan, D. T. Qi, et al., Failure analysis of steel wire reinforced thermoplastics composite pipe. ENG. FAIL. ANAL. 20(2012)88-96.

[5] P. K. Kumar, N. V. Raghavendra, B. K. Sridhara. Development of infrared radiation curing system for fiber reinforced polymer composites: An experimental investigation. Indian J, Eng. Mater, S, 18(1)(2011) 24-30.

[6] Y. J. Zhao, B. H. Choi, C. Alexander. Characterization of the fatigue crack behavior of pipe grade polyethylene using circular notched specimens. Int. J. Fatigue 51(2013)26-35.

[7] S. Yarlagadda, B. K. Fink, J. W. Gillespie. Resistive susceptor design for uniform heating during induction bonding of composites. J. Thermoplast. Compos., 11(1998)321-337.

[8] X. Y. Nie, D. S. Hou, J. Y. Zheng, et al., Eigen-line in welded structures of thermoplastic polymers. Polym. Test. 57(2017)209-218.

[9] H. Shi, I. F. Villegas, M. -A. Octeau, H. E. N. Bersee, A. Yousefpour. Continuous resistance welding of thermoplastic composites: modelling of heat generation and heat transfer. Compos. A: Appl. Sci. Manuf., 70(2015): 16-26.

[10] T. Ge, G. S. Grest, M. O. Robbins. Tensile fracture of welded polymer interfaces: miscibility, entanglements,

and crazing. Macromolecules47(19)(2014)6982-6989.

[11] H. J. Li, B. J. Gao, J. H. Dong, et al. Welding effect on crack growth behavior and lifetime assessment of PE pipes. Polym. Test. 52(2016)24-32.

[12] I. F. Villegas, P. V. Rubio. On avoiding thermal degradation during welding of high performance thermoplastic composites to thermoset composites, Compos. A: Appl. Sci. Manuf., 77(2015)172-180.

[13] T. J. Ahmed, D. Stavrov, H. E. N. Bersee, et al., Induction welding of thermoplastic composites – an overview. Compos. A: Appl. Sci. Manuf. 36(1)(2006)39-54.

[14] Merah N., M. Irfan-Ul-Haq, Z. Khan. Temperature and Weld-Line Effects on Mechanical Properties of CPVC. J. Mater. Process. Tech. 142(1)(2003)247-255.

[15] C. Lu, S. Y. Guo, L. Wen, et al., Weld Line morphology and strength of polystyrene/polyamide-6/poly (styrene-co-maleic anhydride)blends. Eur. Polym. J. 40(11)(2004)2565-2572.

本论文原发表于《Engineering Failure Analysis》2019年第99期。

Influence Factors of X80 Pipeline Steel Girth Welding with Self-Shielded Flux-Cored Wire

Qi Lihua [1,2] Ji Zuoliang [3] Zhang Jiming [1,2] Wang Yulei [3]
Hu Meijuan [1,2] Wang Zhenchuan [3]

(1. Tubular Goods ResearchCenter of CNPC; 2. State Key Laboratory of Performance and Structural Safety for Petroleum Tubular Goods and Equipment Mariteals; 3. Pipeline Construction Project Manager Department of CNPC)

Abstract: For girth weld of high pressure oil and gas transmission pipeline, there is impact toughness values deviation phenomenon with self-shielded flux-cored semi-automatic welding technology. The macro-images, microstructure and mechanical performance of girth welding joint have been investigated by OM, SEM, TEM. The results show that there are several factors of impact toughness unqualified values of weld joints, such as welding heat input, coarse grain zone and a chain of MA organizations, Al_2O_3 and Zr Precipitates particle sizes and distribution et. al. , which are the main unqualified reasons of welding impact toughness of the semi-automatic self-shielded core butt welding process of X80 pipeline steel.

Keywords: Girth welded joint; Microstructure; Mechanical properties; Coarse grain and MA mixing zone; Heat input; Precipitates

Introduction

The continuously growing demand in natural gas globally over the past twenty years promoted the high speed construction of pipeline in using high strength and large diameter pipeline steel[1-3]. Girth welding faces several new challenges when high strength and toughness pipeline steel was used, such as, X70, X80 and X90, et. al. [4,5]. It is reported that over 80% high strength pipeline girth weld was adopting self-shielded flux-cored wire semi-automatic welding method[2,4-6] in china. Various factors have significant influence on the impact toughness of girth weld[6-13], such as heat input, microstructure of weld joint especially for M-A organizations distribution in the coarse grain zone, and precipitates distribution such as AlN, Al_2O_3 and Zr. Besides the factors

---

Corresponding author: Qi Lihua, Ph.D., Senior Engineer, State Key Laboratory of Performance and Structural Safety for Petroleum Tubular Goods and Equipment Mariteals, Xi'an 710077, P.R.China, Tel: +086-029-81887719; E-mail address: qlh1973@163.com.

mentioned above, technical level of welder, filling pass of welding beam, fluctuation of welding current and voltage, the thickness of each layer also affect the impact toughness of girth weld. Generally, smaller welding heat input results in better impact toughness for girth weld of X80 grade pipeline steel. However, the present work demonstrated that the impact toughness of small heat input is far lower than that of large heat input. Therefore, it is necessary to investigate the influencing factors of girth welding process, it is expected that this work would significantly improve the girth weld quality during pipeline construction, and therefore, reducing the failure risk during pipeline operation [11-13].

1 Experimental

1.1 Two groups test of girth weld

The base material is X80 steel pipe with specification of 1219mm×18.4mm. Metal powder cored wire with 1.2 mm in diameter was used for root welding. E81T8-Ni2 wire with 2.0 mm in diameter was used for filling and cover welding.

Two groups' tests were carried out. Group-I, three base material samples, such as 1$^{\#}$ sample, 2$^{\#}$ samples and 3$^{\#}$ sample, were adopted by different manufacturers, and welding wire was adopted by the same factory; Group-II, base material was adopted by the same manufacture, and welding wire were fabricated by different manufacturers, Factory - A and Factory - B. Same welding parameters were employed for two groups, as shown in Table 1.

Table 1 welding process parameters of FCAW-S filling cover face with RMD root welding

Weld bead	Welding grade	diameter(mm)	welding current(A)	Voltage(V)	welding speed(cm/min)
Root welding	AWSA5.28 E80C-Ni	1.2	160~170	16~17	25~30
Hot welding	AWSA5.29 E81T8	2.0	200~220	18~20	25
Fill welding	AWSA5.29 E81T8	2.0	200~240	18~22	20~25
cap welding	AWSA5.29 E81T8	2.0	200~230	18~21	18~20

1.2 Chemical element analysis

Chemical element of steel pipe and weld joint were analyzed by spectrometer. The results of Group-I and Group-II are given in Table 2 and Table 3.

Table 2 the chemical composition of the deposited metal of X80 steel pipe and welding material(wt%)

element	C	Si	Mn	P	S	Cr	Mo	Ni	Nb	V	Ti	Cu	B	Al
1$^{\#}$sample	0.05	0.19	1.77	0.010	0.001	0.30	0.20	0.13	0.08	0.001	0.010	0.1	0.0001	0.030
2$^{\#}$sample	0.052	0.22	1.89	0.010	0.0018	0.31	0.18	0.048	0.084	0.003	0.014	0.057	0.0003	0.038
3$^{\#}$sample	0.054	0.19	1.82	0.082	0.0025	0.33	0.18	0.05	0.082	0.002	0.015	0.01	0.0002	0.035
E81T8-Ni2	0.042	0.06	1.46	0.0091	0.0035	0.025	0.006	2.43	0.005	0.003	0.0002	0.020	0.0013	1.1294

Table 3 the chemical composition of the deposited metal of X80 steel pipe and welding material(wt%)

element	C	Si	Mn	P	S	Cr	Mo	Ni	Nb	V	Ti	Cu	B	Al
4# and 5# sample	0.05	0.19	1.77	0.010	0.001	0.30	0.20	0.13	0.08	0.001	0.010	0.1	0.0001	0.030
Welding Factory A	0.042	0.06	1.46	0.0091	0.0035	0.025	0.006	2.43	0.005	0.003	0.0002	0.020	0.0013	1.1294
Welding Factory B	0.034	0.18	1.37	0.0107	0.0032	0.027	0.005	1.66	0.003	0.003	0.0015	0.006	0.0014	1.0908

1.3 Mechanical properties

All thesamples for mechanical tests of the weld joint were carried out. Tensile test results of Group-I and Group-II are shown in Table 4 and Table 5, respectively. Charpy test results with 3 point and 12 point position of Group-I and Group-II are shown in Figure 1(d) and Figure 2(c).

1.4 Metallurgical properties

Three samples of Group-I and two samples of Group-II for metallurgical properties of the weld joint were observed, as illustrated in Figure 1 and Figure 2.

1.5 Microstructure analysis

In order to investigate the influencing factors of girth welding process, the microstructure of coarse grain zone and M-A constitute of the weld joint was observed by scanning electron microscope (SEM) and transmission electron microscopy (TEM). The precipitate particles of Group-II samples were conducted by the method of carbon membrane extraction, observed by high resolution transmission electron microscopy (HRTEM).

1.6 Thermal Simulation Test

Computer finite element modeling of welding temperature field of the weld joint was established according to the girth welding process. According to temperature cycling curve of finite modeling result, weld thermal simulation test was carried out using Gleeble3500 testing machine, and then mechanical tests was done.

2 Results

2.1 Mechanical properties of Group-I

Macro-images of three girth weld joints prepared at 3 o'clock positions are shown in Figure 1. Except for root welding, heat welding and cap welding, there are different filling pass in three samples. Samples 1# and 3# are both with 11 filling passes, while Sample 3# is not even, the earlier deposited metal layer was melted partially by the latter deposited metal, the deposited layer boundaries are very hard to identify. Unlike Sample 1# and 3#, Sample 2# only has 3 filling passes. The tensile strength of 3 samples is given in Table 4, which is between 714~740MPa, meeting the API RP 5L standard requirement. However, the values of impact toughness of 3 samples is big different, as shown in Figure 1(d).

Table 4 mechanical performance test results

sample	Fill welding pass	tensile strength(MPa)		Standard requirements
1#	11pass	740	728	≥625MPa
2#	3pass	714	732	
3#	11pass	738	727	

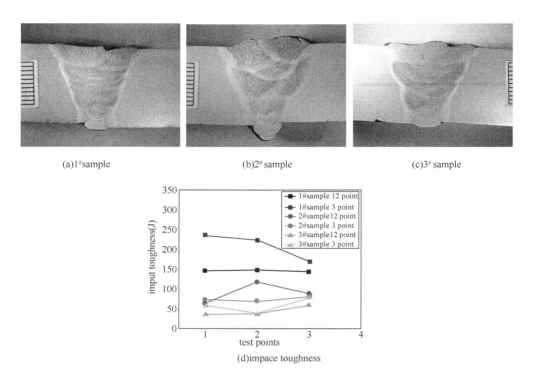

Fig. 1 Comparison of the macro and impact toughness of the different filling pass of the weld seam

The impact toughness of Sample 1[#] is the highest, mean values of which are 146J and 209J, respectively. That of Sample 3[#] is the lowest, which are 44 J and 58J, respectively. Welding current and voltage of Samples 1[#] and 3[#] are basically the same, each filling pass thickness is also close to each other. But there is big difference of impact toughness values between Samples 1[#] and 3[#]. Compared with Sample 2[#], the welding heat input of Sample 2[#] is the largest, and filling pass thickness is the most thick. However, the impact toughness values of Sample 2[#] are 90J and 74J, respectively, which is higher than that of Sample 3[#], but is lower than that of Sample 1[#]. Therefore, except for welding current and voltage, it is necessary to analysis the other factors of welding impact toughness by semi-automatic self-protective core welding method.

2.2 Mechanical properties of Group-II

Macro-images of three girth weld joints prepared at 3 o'clock positions are shown in Figure 2. Except for root welding, heat welding and cap welding, Samples 4[#] and 5[#] are both with filling passes. The tensile strength of 2 samples is given in Tab. 5, which is between 708~742MPa[14]. The mean values of impact toughness of Sample 4[#] are 91J and 215J, respectively, and thats of Sample 5[#] are 65J and 111J. The highest single value of Sample 5[#] is 162J, the lowest single value is only 25J, and several single values do not meet the standard requirements. The welding current and voltage of both samples are basically the same, the deposited metal thickness of each pass of the two samples is relatively close, but the impact toughness values of the two samples are quite different.

Table 5 mechanical performance test results

sample	Fill welding pass	tensile strength (MPa)		Standard requirements
4#	12 pass	708	719	≥625MPa
5#	12 pass	731	742	

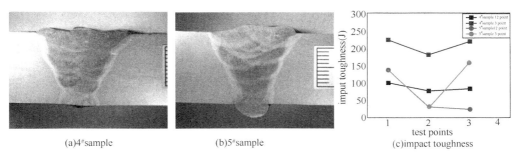

(a) 4# sample (b) 5# sample (c) impact toughness

Fig. 2 Comparison of the macro and impact toughness of the same filling pass and different wire manufacturer of the weld seam

3 Influencing Factors Analysis

3.1 Welding heat input

Macro-images and the microstructure schematic diagram of the various regions of girth welding joint are shown in Figure 3. The regions with arrows in the diagram are the columnar crystal region, the coarse grain region, the coarse grain and the M-A mixed zone respectively. Because of the great influence of coarse grain region and M-A mixed zone on the impact toughness of the welding joint, it will be investigated further.

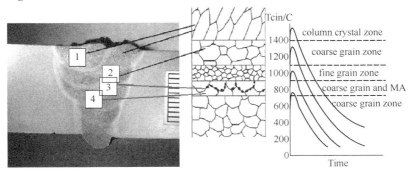

Fig. 3 macro-images and microstructure schematic diagram of the grain shape of each region

Macro-images of coarse grain area and enlarge images of filling weld are shown in Figure 4. Grain size of Sample 1# is about 50~100μm, which is composed of ferrite laths and M-A constituent. That Grain size of Sample 2# is coarser than that of 1# sample, which is mainly composed of lath ferrite and M-A constituent. Compared with Sample 1#, due to the welding heat input increase and filling passes decrease, the deposited metal thickness of each filling pass of Sample 2# is about 4~5mm (Figure 1), and the grain size is significantly coarser than that of Sample 1# and ferrite laths in austenite grain interiors are bunched structure, as shown in Figure 4 (c). Impact toughness of coarse grain microstructure of Sample 2[#] is lower.

(a)1[#] sample100X　　　(b)enlarged images 500X　　　(c)2[#] sample 100X

Fig. 4　Macro-image of coarse grain area of Samples 1[#] and 2[#] packed layers

Figure 5 is the SEM images of filling weld coarse grain of Sample 1[#], and the grain size is about 50~100μm. It is composed of ferrite lath and M-A organization distributed in the grain, but a chain of M-A constituent distributed at the grain boundary. It is well known that M-A chain weakened interfacial energy at grain boundaries and reduced the impact toughness of the weld. Tiny bright white particles with size of 1~2μm distributed around grain boundaries are the M-A constituent chain. As we all know, a single tiny distributed in grain or grain boundary is beneficial for grain nucleation, inhibiting grain growth and enhanced microstructure toughness. But for a chain M-A constitution distribution on the grain boundaries, it will be weakened the binding force between the grains, and bring the disadvantages to the toughness.

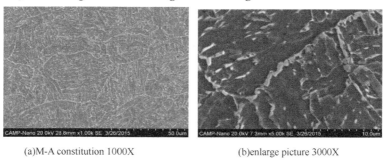

(a)M-A constitution 1000X　　　(b)enlarge picture 3000X

Fig. 5　picture of M-A constitution at coarse grain zone of the filling welding SEM

M-A constitution and it's diffraction pattern of Sample 1[#] were observed by HRTEM, as shown in Figure 6 (a). M-A constituent dark field picture is shown in Figure 6 (b), which the bright white part is retained austenite of M-A constituent. Obviously, there is more retained austenite found at internal grains and at grain boundaries of Specimen 1[#]. It is conducive to improve the impact toughness of the weld. According to retained austenite in the M-A constituent, it is beneficial for the high strength of the martensite and the deformation behavior of retaining austenite.

M-A microstructure at the grain boundaries of Sample 2[#] was investigated by HRTEM, as shown in Figure 6 (c) and (d), and martensitic diffraction pattern in Figure 6 (d). Obviously, the dislocation density of the ferrite laths around M-A constituent at grain boundaries is much lower than that of the ferrite grains around martensite interior grains, see Figure 6 (d). It means that the interface energy of Sample 2[#] is lower, and the cracks around M-A constituent at grain boundaries are easier to form and expand than that of internal grains. The retained austenite in M-A constitution of sample 2# was not observed both the grains boundary and internal grains.

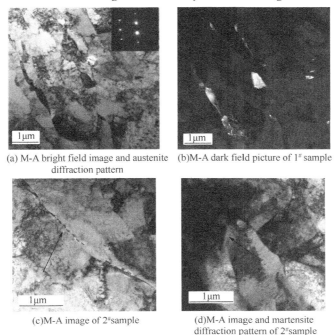

(a) M-A bright field image and austenite diffraction pattern
(b) M-A dark field picture of 1# sample
(c) M-A image of 2# sample
(d) M-A image and martensite diffraction pattern of 2# sample

Fig. 6 picture of microstructure and diffraction pattern of M-A constituent of 1# and 2# samples TEM

3.2 M-A constitution morphology

Macro-images pictures of coarse grain and M-A mixed area of Samples 1# and 3# are given in Figure 7. The black boxes in Figure 7 (c) and (d) are the sampling position of the impact toughness, there are 7 layers coarse grain and M-A mixed zone of Sample 3# and 5 layers coarse grain with M-A mixed zone of Sample 1#, respectively. Although the welding heat input, grain sizes and distribution of two samples are similar, the impact toughness of Sample 1# is about 146~209J, but that of Sample 3# just only 44~58J, it is significantly different, see Figure 1.

SEM and enlarged images of coarse grain and M-A mixed area at grain boundaries were shown in Figure 8. Ferrite laths have no obvious direction around M-A constitution due to the high temperature remelting, and lath boundary passivation around the deposited metal. The grain sizes of M-A constitution near the grain boundaries were slightly aggregated and grown. The size of M-A constitution at the boundaries is coarse, 4~5μm. By further magnification observation, there are many light fine shape material on the laths as shown in Figure 8 (c).

Coarse grain zone and M-A constitution at the grain boundaries were observed by HRTEM of directional sectioning observation, as shown in Figure 9 (a) and (b), and diffraction pattern of M-A constitution is in Figure 9 (c). The size of martensite is about 2~4μm, high carbon martensite, and the distribution is more concentrated. As we all know, martensite with high hardness and low impact toughness shows a brittle phase in the microstructure of welded joint. If it is distributed aggregated, which is easy to form stress concentration, and impact toughness decrease. It will lead to crack formation and extension.

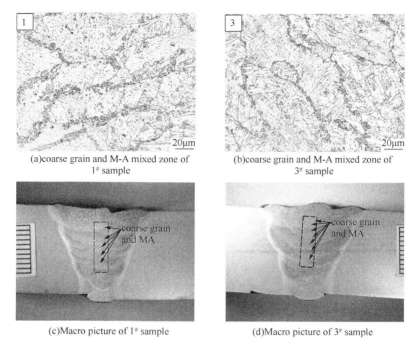

(a) coarse grain and M-A mixed zone of 1# sample

(b) coarse grain and M-A mixed zone of 3# sample

(c) Macro picture of 1# sample

(d) Macro picture of 3# sample

Fig. 7 Metallographic images of comparison of 1# and 3# samples

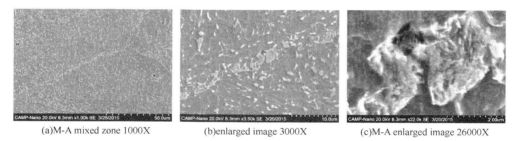

(a) M-A mixed zone 1000X

(b) enlarged image 3000X

(c) M-A enlarged image 26000X

Fig. 8 The M−A chain organization at grain boundary of coarse grain zone and its magnification SEM

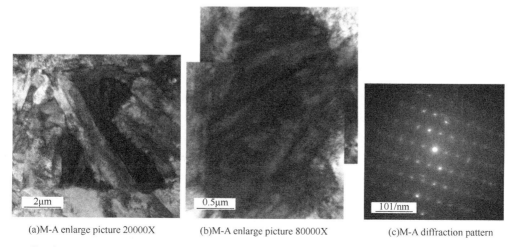

(a) M-A enlarge picture 20000X

(b) M-A enlarge picture 80000X

(c) M-A diffraction pattern

Fig. 9 TEM picture of the M−A chain organization and diffraction pattern at grain boundaries

3.3 Precipitates effect

In order to observe the type and shape of precipitates, the pipe body and the weld heat affected zoneat two sides of weld joints were separated. The weld area was retained, and the precipitation of the whole weld zone was observed by the method of carbon membrane extraction. Precipitates distributed in the weld area have been shown in Figure 10. It was obvious, there are a large number of precipitates in the whole weld area, and the particle size of the precipitates is about 100nm~1μm.

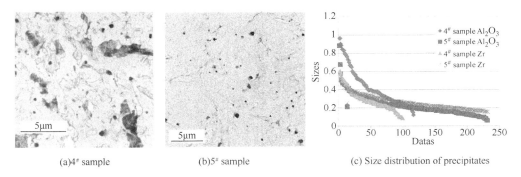

(a)4# sample   (b)5# sample   (c) Size distribution of precipitates

Fig. 10 TEM picture of precipitates distribution of two samples in the weld area

The size distribution of precipitates in the weld area is in Figure 10 (c). In the same region, the number of precipitates of Sample 4# is much more than that of Sample 5#. The precipitates sizes of Sample 4# are mostly less than 400 nm, however, nearly 50% of precipitate size of Sample 4# are greater than 400nm. It is well know that, it is beneficial to the nucleation and grain refinement when the precipitates size is smaller than 200nm, and dispersed in the matrix. It can not only enhance the strength of the matrix, but also increase the toughness. However, the existence of a large number of excessive sizes of precipitates (greater than 400 nm) weakens the matrix toughness and easy to stress concentration under the external force. Furthermore, it will lead to micro crack form, propagation and penetration of each and the formation of macroscopic crack fracture failure.

Morphology and energy spectra of Al_2O_3 precipitates were shown in Figure 11. The particle surface is approximately spherical, and the result of energy spectrum analysis is main components of Al_2O_3. The particles sizes of Samples 4# are about 200nm and smaller particle sizes of Al_2O_3 precipitates. The size of the precipitate of Sample 5# is about 580nm, and the sizes of all Al_2O_3 in the whole observation field are larger than 200nm. The residual Al_2O_3 inclusions formed by flux cored wire welding are in the interior of the welding seam.

Particles morphology of trace alloy elements Zr were shown in Figure 12, which the number of Zr particles of Sample 4# is larger, additionally, a small amount of Zr particles in the aggregate state are above 500nm, the rest of the size of the majority is about 200nm. The number of Zr particles in the Sample5# sample is less, and the most in the single crystal form, see Figure 12(c).

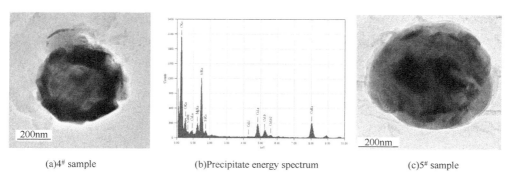

(a)4# sample (b)Precipitate energy spectrum (c)5# sample

Fig. 11 Morphology and energy spectrum of the Al_2O_3 precipitates in two samples weld zone

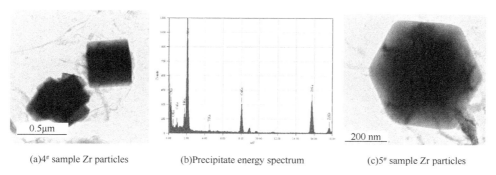

(a)4# sample Zr particles (b)Precipitate energy spectrum (c)5# sample Zr particles

Fig. 12 Morphology and energy spectrum of the Zr precipitates in the weld zone

In a word, the larger particles of Sample 4# are the aggregation state of Zr particles, a small amount of Al_2O_3 mixture. On the contrary, the larger particles of Sample 5# are mainly Al_2O_3, and a small amount of Zr precipitates. In order to improve the toughness of the weld, the amount of high temperature-micro alloying element Zr was added to the core. However, excessive residual Al_2O_3 inclusions in the weld are not conducive to the improvement of the overall toughness of the weld. Therefore, in order to improve the toughness of weld, the formula the ratio of flux cored elementary in the wire should be adjusted properly. It is favorable for the welding process of gas-slag protection, increase the fluidity of molten pool on the one hand, on the other hand to avoid excessive large particles of Al_2O_3 inclusions and Zr element residual.

4 Thermal Simulation Test

According to the analysis of Figs. 1, 7 and 8, the number and morphology of M-A constituent at grain boundaries of coarse grain zone have great influence on impact toughness. Therefore, computer finite element modeling of welding temperature field is established according to the girth welding process. Steel pipe specifications for OD1219mm × 18.4mm × 200mm, environment temperature is 0℃, interlayer temperature is 80℃. Finite element mesh selected 3 o'clock position of girth weld, butt welding groove form and the second layer of filling welding when the row welding form were selected. According to the sampling position of impact specimen, the design in the weld center every 0.5mm points to calculate the temperature in the grid, as shown by the yellow line to take the position.

According to temperature cycling curve of M-A constituent at coarse grain zone finite modeling results, weld thermal simulation tests was carried out using Gleeble3500 testing machine, and then mechanical tests was carried out. Impact toughness results after thermal simulation test were shown in Figure 13 (b). The impact toughness of the weld microstructure after a thermal simulation cycle was significantly lower than that of the welding state. The mean reduction value is 35J, the maximum reduction value is 52J. The mixed microstructure of coarse grain and M-A can significantly reduce the impact toughness of the weld.

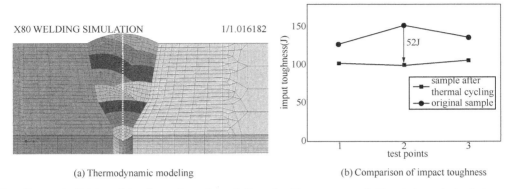

(a) Thermodynamic modeling (b) Comparison of impact toughness

Fig. 13 mesh division of the thermodynamic modeling of weld temperature field and thermal simulation test

5 conclusions

Based on the laboratory test of self-shielded flux-cored wire semi-automatic welding method, the microstructure and mechanical results of girth weld of X80 pipeline were analyzed. The conclusions are as follows:

(1) The tensile strength of the girth weld has little change with different welding heat input and flux cored wire produced by different manufacturer, and the results can meet the standard requirement.

(2) The microstructure of the welded joint is similar to each other when obtained by the same welding heat input, which is composed of ferrite lath and M-A constitution.

(3) The grain sizes of weld joint microstructure were coarser significantly and the deposited metal thickness of filling pass increased due to the heat input increase. Furthermore, there is no retained austenite in M-A constitution observed both in the grain boundaries and internal grains. Therefore, the impact toughness of the weld joints is low relatively.

(4) Under the similar low heat input condition, the microstructure of girth weld with more retained austenite is conducive to improve the impact toughness of the weld joint. Moreover, except for heat input effect, the more M-A constituent exist at grain boundary of coarse grain zone, the lower impact toughness value would be obtained.

(5) Many large sizes residual Al_2O_3 inclusions formed during flux-cored wire welding is adverse to improve the toughness of the girth weld.

Acknowledgment

This work was supported by the Science research and technology development project of CNPC (No. 2014B-3416-0501) and Project of Natural Science Foundation of Shaanxi Province in China (2012GY2-23)

References

[1] LI He-lin. Developing pulse and prospect of oil and gas transmission pipe [J]. Welded Pipe, 2004, 27 (6): 111-120.

[2] Zhao, M., Wei, F., Huang, W. Q., et. al. Experimental and numerical investigation on combined girth welding of API X80 pipeline steel [J]. Science and Technology of Welding and Joining, 2015, 20 (7): 622-630.

[3] Andia, J. L. M., de Souza, L. F. G. 2, et. al. Microstructural and mechanical properties of the intercritically reheated coarse grained heat affected zone (ICCGHAZ) of an API 5L X80 pipeline steel [J]. Advanced Materials Research, 2014, 12(2): 657-662.

[4] Wang, Xiaoyan, He, Xiaodong, Han, Xinli, et. al. Study on FCAW semi-automatic welding procedure of girth weld joint of line pipes[J]. Advanced Materials Research, 2012, 415-417: 2078-2084.

[5] YANG Liuqing, WANG Hong, SUI Yongli, et. al. Research on weld microstructure and properties of self-shiekded flux cored wire[J]. Welded pipe, 2012, 36(12): 15-19.

[6] CHEN Cuixin, LI Wushen, WANG Qingpeng, et. al. Microstructure and properties of X80 pipeline steel welded coarse grain zone[J]. Transactions of the china welding institution, 2005, 26(6): 77-80.

[7] CHEN Cuixin, LI Wushen, WANG Qingpeng, et. al. Research on influence factor of impact toughness in coarse grain heat-affected zone[J]. Material engineering, 2005, 5: 22-26.

[8] Gianetto, J. A., Fazeli, F., Chen, Y., et. al. Microstructure and toughness of simulated grain coarsened heat affected zones in X80 pipe steels [C]. Proceedings of the Biennial International Pipeline Conference, IPC2014-33254.

[9] Zhenglong, Lei, Caiwang, Tan, Yanbin, Chen, et. al. Microstructure and mechanical properties of fiber laser-metal active gas hybrid weld of X80 pipeline steel [S]. Journal of Pressure Vessel Technology, Transactions of the ASME, 2013, 135, 1.

[10] LI Yajuan, LI Wushen, XIE Qi. Research and prediction on cold cracking susceptibility of Nb-Mo X80 pipeline steel[J]. Transactions of the china welding institution, 2010, 31(5): 105-108.

[11] XU Xueli, XIN Xixian, SHI Kai, et. al. Influence of welding thermal cycle on toughness and microstructure in grain-coarsening region of X80 pipeline steel[J]. Transactions of the china welding institution, 2005, 26(8): 69-72.

[12] MIAO Chengliang, SHANG Chengjia, WANG Xuemin, et. al. Microstructure and toughness of HAZ in X80 pipeline steel with high Nb content[J]. ACTA metallugrica sinica, 2010, 46(5): 541-546.

[13] Technical specification of welding for oil and gas pipeline project Part 1: Mainline welding Q/SY GJX137.1-2012 [S].

[14] LI Yajuan, LI Wushen, XIE Qi. Research and prediction on coldcracking susceptibility of Nb-Mo X80 pipeline steel[J]. Transactions of the china welding institution, 2010, 31(5): 105-108.

本论文发表于《Materials Science and Technology》2017年第5期。

Mechanical Properties of Girth Weld with Different Butt Materials Severed for Natural Gas Station

Ren Junjie[1,2] **Ma Weifeng**[1,2] **He Xueliang**[3] **Chen Anqi**[4]
Luo Jinheng[1,2] **Wang Ke**[1,2] **Ma Qiurong**[1,2] **Huo Chunyong**[1,2]

(1. State Key Laboratory for Performance and Structure Safety of Petroleum Tubular Goods and Equipment Materials, CNPC Tubular Goods Research Institute;
2. Shaanxi Key Laboratory for Performance and Structure Safety of Petroleum Tubular Goods and Equipment Materials;
3. PetroChina Beijing Natural Gas Pipeline Co., Ltd.;
4. Northwest Institute for Non-ferrous Metal Research)

Abstract: Weld samples imitating the inservice girth welds in station (L245 straight pipe jointed to WPHY-70 tee joint and L415MB straight pipe jointed to WPHY-80 tee joint) were prepared. Tensile, bending, impact toughness and hardness of the joints were investigated. Results show that under tensile or bending load, failure occurred from the side with lower grade and smaller wall thickness. Relatived to the lower grade side, the weld seam is strong match. Significant change of impact toughness can be found in weld seam center and the heat affected zones (HAZ). The impact energy of seam center is the lowest in the weld joint. The impact energy show a trend of increase from seam center to base metal. In HAZ zone, impact toughness of the fusion line is the lowest. Impact toughness of higher grade side is higher than that of the lower grade side. Hardness of positions in HAZ zones are different distinctly. From coarse grained region to fine grained region, the hardness decrease. For the in-station girth welds jointed with different materials, lower grade and samller wall thickness side should be intensive monitored.

Keywords: High grade; Girth weld; Impact toughness

1 Introduction

In resent years, accidents caused by failure of girth welds often happened in oil & gas transportation station[1-3]. In 2011 year, a burst accident occurred in a gas compressor station because the failure of the girth weld between the compressor outlet pipeline and flange. In another gas station, flaws were examined in a girth weld between a in-serviced flange and pipeline elbow

---

Corresponding author: Ren Junjie, renjunjie@cnpc.com.cn.

while a routine maintenance was carried out. The facilities in oil & gas transportation station are concentrated, once an accident occurs, the damage is extremely high. Therefore, failure of girth welds serviced in gas station has attracted extensive attention.

Compared to the girth welds serviced in long-distance transmission pipeline, those serviced in stations have two different joints with the characteristics of different materials, different wall thickness and different components. According to investigation, a greater materials difference exists in the girth welds between reducer tee and straight pipe, where WPHY-70 joint to L245 or WPHY-80 joint to L415MB. In addition, the difference of wall thickness is almost double[4]. Few research about high grade pipeline girth welds was focused on the gas station, compared with long-distance transmission pipeline.

Fracture is the main failure type of high grade pipeline girth welds, and the fracture behavior is affected by the structure and properties of the material[5]. Impact toughness is a significant property which can reflect the material property characteristic, defect and microstructure sensitively. In the present work, girth welds with two different joints were prepared according to field investigations. The distribution of impact toughness and hardness of the girth welds were investigated.

2 Experimental procedure

Materials preparation. The 1# girth weld was prepared by welding a X80 pipe to a X60 pipe, and the 2# girth weld was prepared by welding X70 pipe to a L245 pipe. The details specifications are listed in Table 1. The welding process was carried out according to Q/SY GJX 0221—2012 "*Welding specification for station pipes of West - east gas transmission third line*"[6]. The groove angle is 60°. The root welding was done by manual welding using ER50-6 welding rod with the diameter of 2.5mm, the filling welding and cover welding were done by manual welding using E6015-Ga welding rod with the diameter of 2.5mm.

Table 1 Material specifications for weld joints

Materials	1# Girth weld		2# Girth weld	
	WPHY-80/X80	X60/L415	WPHY-70/X70	L245
Pipe diameter	914mm	914mm	1016mm	1016mm
Wall thickness	33mm	22mm	21mm	12mm
Manufacturing standards	MSS SP-75—2004 API Spec 5L	API Spec 5L GB/T 9711	MSS SP-75—2004 API Spec 5L	GB/T 9711

The tensile strength of the girth weld was tested in a general testing machine (MTS SHT4106) using whole wall thickness samples where the wide is 38mm (1#) and 50.8mm (2#). The Charpy impact test was carried out in WANCE PIT302D machine using samples 3.3mm×10mm×55mm at 0℃ sampling from three positions of near outer surface, middle of pipe wall and near inner surface. The bending test was carried out in a WZW-1000 bending test machine. The microstructure was observed using a Leica MeF3a and a Olympus microscope. A FUTURE-TECH FM-700 digital vickers hardness tester was hired to obtain HV1 hardness with a loading time of 15 s.

3 Results and discussion

Tensile properties. Tensile properties of 1# (X80-X60) and 2# (X70-L245) girth welds are shown in Table 2. Tengsile strength of 1# and 2# weld joints is 603MPa and 472MPa respectively. In addition, the fracture positions of both weld joints are all in the lower grade base metal with obvious necking effect which indicates plastic deformation before fracture occurred (as shown in Fig. 1), while obvious deformation can not be found in the position of welding seam.

Table 2 Tensile properties of the weld joints

Sample	Wide (mm)	Gauge length (mm)	R_m(MPa)	Fracture position
1#	25	50	603	X60 side base metal
2#	38	50	472	L245 side base metal

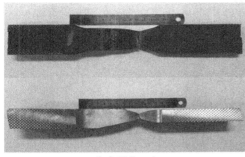 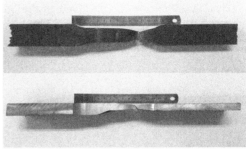

(a) 1# Sample (b) 2# Sample

Fig. 1 Photographs of 1# and 2# samples after fracture

According to GB/T 31032—2014 *Welding and acceptance standard for steel pipings and pipelines*, tensile strength of the girth weld sample should be no less than the minimum tensile strength of the pipe. If the fracture occurred in the position of the pipe base metal, and the tensile strength is no less than the nominal tensile strength of the pipe base matal, then the sample is qualified[7]. Therefore, the 1# and 2# girth welds are both qualified. The tensile strength of the 1# or 2# weld seams is more higher compared to the lower grade base metal of them which indicates a over match.

When the tensile sample of the girth welds surfered drawing force, the stress in the base metal of a girth welds with greater wall thickness is lower than the other base metal with smaller wall thickness. At the same time, the materials with greater wall thickness have higher tensile strength. As a result, the lower grade materials with a smaller wall thickness of both 1# and 2# girth welds went to failure firstly under tensile stress.

The macroscopic fracture morphology of 1# and 2# tensile samples are shown in Fig. 2. It can be seen that both samples have similar characteristics. Cup and cone morphology can be observed obviously. The tensile load led to the necking. The fracture surface contain shear lip, radial zone and fiber zone[8]. The shear lip is located in the outer edge of the fracture surface, the fiber zone is located in the midde with rough surface and the radial zone is between the other two zones. The fracture surface has dark color, no crystalline particle and no metallic lustre. A typical ductile fracture can be identified after tensile failure.

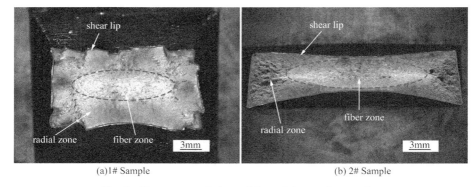

Fig. 2　Fracture morphology of 1# and 2# tensile samples

The microcosmic fracture morphology of 1# and 2# samples are shown in Fig. 3. It can be seen that dimple and tearing ridge with unequal size and deepness exist in the fracture surface. Some second phase particles can be observed in the bottom of dimples in fiber zone. The fracture surfaces have typical equalaxis dimple features, which is the trace of drawing. Both samples show ductile fracture features. The characteristics of microcosmic fracture morphology match up with the macroscopic survey.

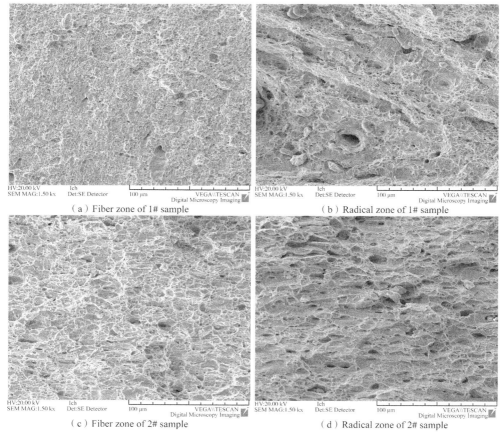

Fig. 3　Microcosmic fracture morphology of 1# and 2# samples

Side bending. Samples after side bending of 1#(X80-X60) and 2# (X70-L245) girth welds joints are shown in Fig. 4. No flaws can be observed in both samples after side bending which signify fine bending performance. The crook diverged from the weld seams although the load was pushed aligned with the weld seams. It is because that the performance of the weld seams is good enough, but the lower grade base matal have both lower strength and smaller sectional area, which is the weakest area. Compared to the lower grade base metal, the weld joints is strong match.

Impact toughness. Charp impact test was carried out to investigate the impact toughness of different positions in weld seam and heat affected zone (HAZ). Schematic diagram of sampling and V gap notching position is shown in Fig. 5. Because the two joints of the girth welds were V-groove, the transverse section of the HAZ and fusion line is not perpendicular to pipe wall surface, with a 30° intersection angle. So the V gap may across more than two zones of weld seam, coarse grained region, fine grain region, two phase region and base metal. Therefore, for obtain the properties of a certain region from charp impact samples, the size of charp impact sample was intended to be 3.3mm thickness according to ASTM A370-17 *Standard Test Methods and Definitions for Mechanical Testing of Steel Products*[9]. As can be seen in Fig. 5. L2/R2 notching is from the crossing point of fusion line and inner surface, reflecting properties of the fusion line. M position is the middle of the weld seam. L1/R1 notching is between M and L2/R2. L3/R3 position is from the crossing point of fusion line and outer surface, reflecting properties of HAZ. L4/R4 position is from the crossing point of the outer surfac and the line between HAZ and base metal.

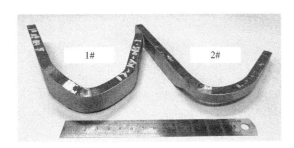

Fig. 4　Samples after side bending

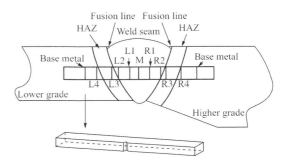

Fig. 5　Schematic diagram of sampling and V gap notching position

Fig. 6(a) is the Charp impact energy of 1# girth weld. The impact energy was changed into that of the sample with size of 10mm×10mm×55mm. It can be seen that the impact energy of seam center is the lowest in all notching positions. The impact energy show a trend of increase from seam center to base metal. Compare X60 side to X80 side, of the fusion line in X60 side (L2) is greater than the that of the seam center, but the impact toughness of R2 is almost the same to the seam center. Impact energy of L3/R3 is higher than L2/R2, but the increasing range of R3 is higher than L3. Impact energy of the lines (L4/R4) between HAZ and base metal is comparative, and are near that of the base metal. In both side, impact energy of L2/R2, L3/R3 and L4/R4 show obvious increase, which indicates the heterogeneity of impact properties in HAZ.

Fig. 6(b) is the Charp impact energy of 2# girth weld. The impact energy of seam center is the lowest. In X70 side, feature of impact toughness is similar to 1# weld joint. But in L245 side, impact energy of the fusion line (L2) is the highest.

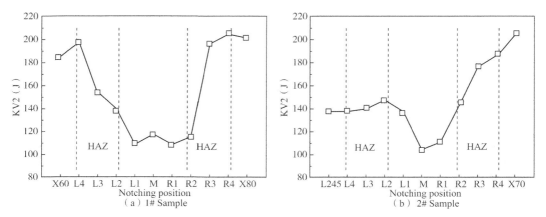

Fig. 6　Charp impact energy of 1# and 2# samples

For survey the impact fracture characteristics of different positions, fracture photographs of five notching positions from X80 side of 1# sample were observed, as shown in Fig. 7. As can be seen in the Fig. 7, cup and cone morphology can be observed in the fracture surfaces, and the surface is gray, which indicates that typical ductile fracture occurred. Fracture surfaces of L1 and L2 are smooth comparatively. And in L3 and L4, obvious buckling can be observed in the fracture surface. This means resistance of crack propagation increases and more impact energy was consumed away, which demonstrate that the impact toughness increase. In the fracture surface of X80 base metal, less buckling appear but roughness increase, which also signify good toughness. From L1 to X80 base metal, routhness of the impact surface increase, percentage of brittle fracture surface (show white in the bottom) decrease, which indicates absorption work of plastic deformation increase.

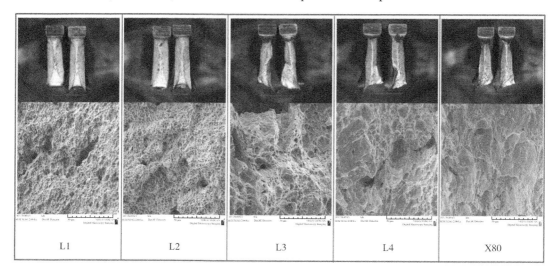

Fig. 7　Macrocosmic and microcosmic fracture morphology of impact samples
from different position of 1# sample X80 side

As can be seen from microcosmic fracture morphology of Fig. 7, dimple can be recognized in the middle of the fracture surface. Lighter and homogeneous equalaxis dimples appear spread all over the ductile fracture surface of L1 and L2. In contrast, dimples in L3 and L4 are more bulky. For the X80 base metal impact sample, elongated dimple feature seems to be exist. Characteristics reflected from the microcosmic fracture morphology agree with the impact toughness results.

Hardness. Along a line which pass trough HAZ zone, HV1 hardness was tested. Dotting position and hardness are shown in Fig. 8. It can be seen that hardness of HAZ zone, include coarse grained region (CG), fine grain region (FG) and two phase region (TP) are different distinctly. Hardness of coarse grained region is the highest in HAZ. From coarse grained region to fine grained region, the hardness decrease. The hardness is related to the grains size[10]. For the same structure phase, hardness of bulky grain is higher than that of fine grain. Moreover, dispersion of hardness in coarse grained region is higher than the fine grain region because homogeneity of the second phase in coarse grained region is far below the fine grain region. In 1# weld joint, hardness of X60 side HAZ is close to that of X80 side. But in 2# weld joint, hardness of L245 side is significantly less than that of X70 side, which is due to the main component of L245 is ferrite.

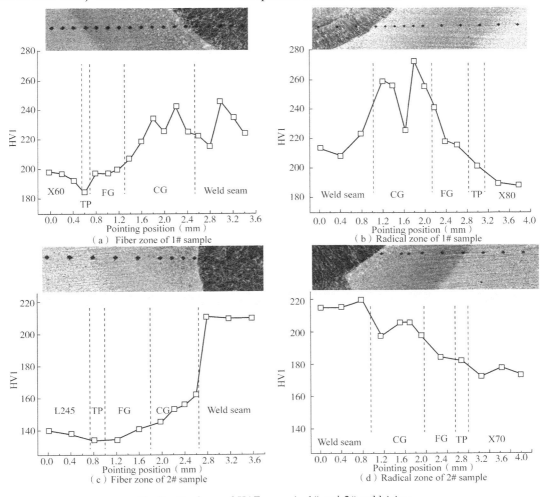

Fig. 8 Hardness of HAZ zones in 1# and 2# weld joints

4　Summary

Imitating the inservice girth welds in station, girth weld joints (X80-X60、X70-L245) were prepared according to practical welding specification. Compared to the lower grade and thin base metal side, the welds joints show strong match. Corresponding to the actual girth welds in station, the straight pipe side of the joints would failure first when surfering tensile or bending load. Impact toughness of the seam center is the lowest among positions including HAZ and base metal. The impact energy show a trend of increase from seam center to base metal. Impact fracture is typical ductile fracture, HAZ and base metal have higher impact toughness. Hardness of coarse grained region is the highest in HAZ. From coarse grained region to fine grained region, the hardness decrease.

Acknowledgement

This work was supported by the National Key R&D Program of China (2016YFC0802101).

References

[1] H. L. Li, X. W. Zhao, L. K. Ji, Failure analysis and integrity management of oil and gas pipeline, Oil & Gas Storage and Transportation. z1 (2005) 1-7.

[2] Z. Q. Huang, The Integrality and Its Management of Pipeline, Welded Pipe and Tube. 03 (2004) 1-8.

[3] S. H. Dong, Z. P. Yang, The world oil & gas pipeline integrity management and technology latest development and chinese pipeline countermeasure, Oil & Gas Storage and Transportation. 02 (2007) 1-17.

[4] A. Q Chen, W. F. Ma, J. J. Ren, K. Wang, K. Cai, J. H. Luo, C. Y. Huo, Y. R. Feng, Study on rehabilitation of high steel pipelinegirth weld defects, Natural Gas and Oil. 05 (2017) 12-17.

[5] H. L. Li, Operation safety and integrity management of oil and gas pipeline, Oil Forum. 02 (2007) 18-25.

[6] Q/SY GJX 0221—2012 Welding specification for station pipes of West - east gas transmission third line.

[7] GB/T 31032 Welding and acceptance standard for steel pipings and pipelines.

[8] Y. J. Li, Q. L. Jiang, Y. P. Bao, J. Wang, L. Zhang, Effect of heat input on the microstructure and toughness of heat affect zone of Q690 high strength steel, Sciencepaper Online. 02 (2011) 98-102.

[9] ASTM A370 Standard Test Methods and Definitions for Mechanical Testing of Steel Products.

[10] Y. L. Sui, Welding technology research for girth of domestic X80 grade line pipe. Doctoral Dissertation of Tianjin University, 2008.

本论文原发表于《Materials Science Forum》2019年第944卷。

Research on the Steel for Oil and Gas Pipelines in Sour Environment

Shao Xiaodong

(State Key Laboratory for Performance and Structure Safety of Petroleum Tubular Goods and Equipment Materials; CNPC Tubular Goods Research Institute)

Abstract: In recent years, there are more and more oil and gas fields containing H_2S corrosive media, and it is imperative to develop oil and gas transmission pipes in acidic environment. The research progress of steel for oil and gas pipelines in acidic environment is reviewed from the aspects of H_2S corrosion mechanism and influencing factors in pipeline service under acidic service conditions. At the same time, the development status of sour-resistance corrosion oil and gas transmission steel tubes at home and abroad is introduced. Pipeline steel is further explored in terms of SSCC-resistance research and improvement of corresponding standards.

Keywords: Pipeline steel; H_2S corrosion; HIC; SSCC

1 Introduction

Oil and gas pipeline transportation is characterized by high efficiency, economy and safety, and is the main form of long-distance transportation of oil and gas[1]. The development of pipelines tends to be large-caliber and high-pressure, the service conditions of pipelines are becoming more and more demanding, the transportation medium is complex, and many pipelines need to cross densely populated areas or deserts, cold regions, etc., which puts forward the performance of pipelines. High technical requirements, especially the requirements of corrosion resistance, Academician Li Helin pointed out that anti-hydrogen induced cracking(HIC) pipeline steel and steel pipe is one of the development directions of high performance oil and gas transmission steel pipe. Corrosion is a key factor affecting the reliability and service life of pipeline transportation systems. It can not only cause perforation, but also cause oil and gas leakage. It may also cause an explosion, threaten personal safety, pollute the environment, and the consequences are extremely serious. According to the National Transportation Safety Administration, 74% of the leaks in the US gas pipelines and gas gathering pipelines are caused by corrosion, and H_2S corrosion is one of the main forms of pipeline corrosion. According to the American Society of Corrosion Engineers (NACE), in natural gas containing water and H_2S, when the partial pressure of H_2S in the gas is

---

Corresponding author: Shao Xiaodong, shaoxd@ 163. com.

equal to or greater than 0.0003 MPa, it is called an acidic environment.

Since the 1970s, the development conditions of oil and natural gas have changed significantly in various countries. At present, although the purification process has been carried out before the natural gas transmission, the natural gas to be transported into the gas pipeline does not meet the requirements of dry and purified natural gas. H_2S in the pipeline. In-pipe corrosion caused by the presence of water is still inevitable. In recent years, accidents caused by pipe H_2S corrosion have occurred, resulting in huge economic losses and environmental pollution. Therefore, solving or mitigating H_2S corrosion of oil and gas pipelines has very important economic and social benefits. According to the development needs of Chinese oil and gas industry, it is imperative to develop Chinese own acidic service conditions to transport steel pipes. Foreign countries have started research in this area earlier, and there are specialized agencies to study the transportation of steel pipes in an acidic environment. Although domestic work in this area started late, it also carried out a series of research work. The author mainly reviews the research progress of steel for acid and oil pipelines from the corrosion mechanism and influencing factors of H_2S in acidic environment pipelines, and introduces the development of oil and gas pipelines in acidic environment at home and abroad. Research direction.

2 H_2S corrosion mechanism of pipeline steel

The corrosion of pipeline steel in acidic environment mainly includes hydrogen induced cracking and sulfide stress corrosion cracking(SSCC). At present, the research on H_2S corrosion of pipeline steel is mainly focused on HIC. Pipeline steel in the oil-rich environment rich in H_2S in the state of no stress or non-tensile stress, the hydrogen generated by the corrosion into the steel to form a crack called HIC, generally refers to hydrogen-induced bubbling (surface crack) and hydrogen-induced step cracking (Internal crack). SSCC means that the hydrogen atoms generated by H_2S penetrate into the interior of the steel, dissolve in the crystal lattice, cause brittleness, and form cracks under the action of external tensile stress or residual stress. SSCC is in sensitive materials, acidic environment and stretching. The three conditions of stress occur together[2]. When transporting acidic oil and gas, the inside of the pipeline is exposed to H_2S. Under the joint action of stress and H_2S and other corrosive media, SSCC often occurs in the pipeline.

3 Research on the HIC of pipeline steel

3.1 Influence of the environment on HIC

The results show that when the partial pressure of H_2S is 0.1~0.5MPa, the corresponding hydrogen permeability is higher, which means that the sensitivity of HIC is greater[3]. As the pH decreases, the HIC sensitivity of the pipeline steel also increases. When CO_2 is dissolved in water to form carbonic acid, hydrogen ions are released, which lowers the pH of the environment and increases the sensitivity of HIC. Cl^- accelerates the corrosion rate in the range of pH 3.5~4.5, and the HIC sensitivity increases[4]. When the temperature is around 30°C, the additive effect of hydrogen adsorption and diffusion is the largest, that is, the HIC sensitivity is the strongest, below 30°C, the hydrogen diffusion rate and activity decrease, and it is difficult to aggregate active hydrogen above 30°C.

3.2 Influence of composition and structure on HIC

The influence of pipeline steel composition and structure on HIC is the most in-depth study, and the results are also the most, which can be summarized into three aspects.

(1) Chemical composition. The elements that are more sensitive to HIC are carbon, manganese, sulfur, phosphorus, calcium, copper, and molybdenum. As the carbon content increases, the HIC sensitivity increases, and the increase in carbon content and carbon equivalent in the pipeline steel causes the steel to form the martensite structure most sensitive to hydrogen bubbling during hot rolling, thus reducing the carbon content and carbon equivalent. It can improve the H_2S-resistance corrosion performance of pipeline steel. Sulfur can promote the occurrence of HIC, which is the most easily nucleated site of HIC with manganese-generated MnS. Baosteel's research indicates that the addition of calcium can change the form of sulfide inclusions, making it a dispersed spheroid, thereby improving the HIC-resistance ability of steel. The test found that the best effect is when the calcium-sulfur content ratio is greater than 1.5. Adding an appropriate amount of manganese to the pipeline steel can improve the hardenability of the steel, cause solid solution strengthening, and compensate for the decrease in strength caused by low carbon. The segregation of manganese and phosphorus can cause the formation of band-like structures sensitive to HIC, thus increasing the manganese content. , will lead to more banded tissue generation, which increases HIC sensitivity. The effect of copper on HIC is obvious. In NACE B solution, HIC sensitivity decreases significantly with increasing copper content. However, in the H_2S environment with pH<4.5, the passivation film of copper is no longer formed, and at this time, the copper prevents the effect of HIC from disappearing. The addition of molybdenum can lower the phase transition temperature, inhibit the formation of massive ferrite, promote the transformation of acicular ferrite, improve the strength of steel, reduce the ductile-brittle transition temperature, and improve the HIC-resistance ability. The addition of microalloying elements such as bismuth, vanadium and titanium to pipeline steel can effectively prevent the growth of austenite grains, refine grains and enhance the corrosion resistance of pipeline steels[5,6]. Due to the influence of chemical composition on the HIC resistance of pipeline steel, the chemical composition of different steel grade HIC-resistance oil and gas transmission steel pipes has clear technical requirements in ISO 3183.

(2) Microstructure. Sodium thiocyanate was selected as the chromogenic reagent in the proposed method. The effect of the chromogenic reagent was examined by measuring the absorbance of solution containing certain amounts of molybdenum and variable amounts of sodium thiocyanate. It was found that 10ml of 100.0mg · mL^{-1} sodium thiocyanate solution sufficed to complex the amounts of molybdenum taken, with higher concentrations the absorbance was essentially constant. Ten millilitres of 100.0mg · mL^{-1} sodium thiocyanate solution were recommended as a suitable amount of chromogenic reagent. The thermodynamically balanced and stable fine grain structure is the ideal structure for HIC-resistance. For medium and low-strength pipeline steels, HIC is prone to low temperature conversion of manganese, phosphorus and other elements in the banded pearlite structure and the center of the plate thickness. In particular, in order to improve the strength of pipeline steel, the addition of many alloying elements promotes the formation of low temperature

conversion microstructure, which provides a convenient place for the occurrence of HIC[7]. With the increase of the microstructure of steel, the HIC-resistance performance decreases gradually, and the HIC - resistance performance decreases with the increase of martensite structure. With the increase of the band structure level, the HIC-resistance performance of pipeline steel is gradually reduced. Yin Guanghong et al found that the HIC-resistance performance of pipeline steel will be significantly enhanced after normalizing heat treatment.

(3) Non-metallic inclusions. The morphology and distribution of non-metallic inclusions affect the HIC resistance of pipeline steel. The non - metallic inclusion interface is a strong hydrogen trap[8]. On the interface between the pipeline steel and H_2S, the hydrogen ions generated by the electrochemical reaction are trapped in the hydrogen trap after the electrons are converted into hydrogen atoms at the anode. When the hydrogen pressure rises to a certain value, it is in the non-metal. Cracks occur at the interface of the inclusions, and plastic deformation occurs at the crack tip. In pipeline steel, the higher the content of non-metallic inclusions, the smaller the threshold value of hydrogen concentration required for HIC, that is, the stronger the HIC sensitivity[9]. In addition, in the same pipeline steel, the coarser the grain, the worse the HIC - resistance performance[10]. Some scholars have pointed out that the hardness of the material also has a great influence on HIC. In the medium and low strength steel, the higher the strength and the greater the hardness, the greater the HIC sensitivity.

4 Research on the SSCC of pipeline steel

4.1 Influence of the environment on SSCC

The lower the pH of the environment, the greater the hydrogen concentration and the stronger the sensitivity of the SSCC. At the same time, the amount of hydrogen permeation in the pipeline steel increases with the increase of H_2S concentration, and the SSCC sensitivity of the pipeline steel is greater[11], the high - strength steel SSCC has a most sensitive temperature, near this temperature, occurs The SSCC is the most likely.

4.2 Influence of composition and structure on SSCC

The composition and organization of pipeline steel is another major factor affecting pipeline steel SSCC.

(1) Chemical composition. Regarding the influence of carbon, different scholars have different opinions, but regardless of the explanation, the lower the carbon content, the lower the sensitivity of the SSCC of pipeline steel. Du Shiyu and other research through the domestic X70 pipeline steel, pointed out that the main chemical elements affecting its SSCC are manganese and phosphorus content, not carbon content. Molybdenum is a beneficial element, especially in the case where high strength of the material is required, a small amount of molybdenum is added to form carbides, so that the amount of solid solution carbon is reduced, and thus the performance against SSCC is improved. The role of chromium has two different perspectives, but it is beneficial for improving general corrosion resistance. The addition of titanium and boron improves the hardenability and also has a certain effect on improving the resistance to SSCC[12,13]. Aluminum is a beneficial element, especially for materials that add both chromium and aluminum.

(2) Microstructure. The microstructure of pipeline steel has a great influence on the sensitivity of SSCC. Martensite is generally considered to be more sensitive than pearlite and austenite. Xie Guangyu et al studied the SSCC behavior of X70 grade pipeline steel with different microstructures by three-point bending method. The results show that the uniform fine acicular ferrite microstructure has excellent SSCC resistance. Zhao Mingchun et al studied the SSCC-resistance behavior of different microstructure pipeline steels. The results showed that the acicular ferrite microstructure had the best performance against SSCC, the ultrafine ferrite structure was the second, and the microstructure of ferrite pearlite. Worst. Mansour et al[14]. studied the SSCC behavior of X100 pipeline steel and pointed out that with the increase of the microstructure of steel, the performance of SSCC decreased. In general, the higher the strength level of pipeline steel, the greater the sensitivity to SSCC, the highest hardness value of SSCC does not occur between 20~27HRC, the higher the hardness, the lower the critical stress value and the fracture time. In addition, due to a series of physical and chemical changes in the weld during the welding process and some weld defects, the ability of the welded joint to resist SSCC is reduced relative to the base metal. The SSCC resistance of the pipeline steel can be improved by eliminating the residual stress annealing treatment. Tao Yongqi et al. showed that different weld matching in H_2S-containing media resulted in different corrosion resistance of pipeline steels, with the increase of cold deformation, the resistance of pipeline steel to SSCC was reduced.

5 Development of H_2S-resistance corrosion pipeline steel

Some scholars believe that there is a difference and a relationship between the HIC and SSCC of pipeline steel. It can be considered that the SSCC crack of pipeline steel is a hydrogen-induced cracking stress corrosion. If the HIC of the pipeline steel can be controlled, the necessary conditions for the occurrence of SSCC H_2S corrosion or material sensitivity are not available, so that the SSCC crack of the pipeline steel can be controlled. Therefore, the development and evaluation of H_2S-resistance corrosion transport steel pipes at home and abroad mainly focus on the HIC-resistance performance of steel. Pipeline steels used in acidic environments are characterized by high toughness and resistance to HIC properties that depend on the high purity of the steel.

In order to meet the strength, weldability and manufacturability, it is very important to control the process parameters of the HIC pipeline steel, such as composition design, smelting, continuous casting, rolling, etc., in order to ensure a uniform microstructure. Through the research on the corrosion mechanism and influencing factors of H_2S, the measures to improve the HIC resistance of pipeline steel have formed a consensus, using concentrate and high-efficiency iron pretreatment and re-refining outside the furnace to improve the purity of steel. Calcium treatment is carried out while reducing sulfur content, electromagnetic stirring is carried out during steel smelting and continuous casting, light reduction technology is adopted in continuous casting process, multi-stage controlled rolling and forced accelerated cooling process are used to limit band structure, reduce carbon content, control manganese content, add copper and nickel, fully refine the grains by microalloying and controlled rolling processes.

At present, the general method adopted for the evaluation of steel HIC-resistance ability is the standard test method NACE TM0284 issued by NACE, including three evaluations of crack sensitivity rate(CSR), crack length ratio(CLR) and crack thickness ratio(CTR). parameter. Since the 1990s, the HIC-resistance steel pipes supplied by foreign countries are mainly concentrated in the X65 steel grade. The HIC-resistance oil and gas transmission steel pipes produced by the European steel pipe companies have accounted for more than 30% of their sales, the X70 steel pipes with HIC-resistance have also been successfully developed. It is used on a pipeline in Mexico. Some countries have developed X80 steel pipes resistant to HIC, but they have not been supplied in bulk and are still under further study. Sumitomo Metal Industries Co., Ltd. has developed X80 steel grade pipe steel with excellent HIC resistance and steel pipe made of this steel. For X70 and below steel grade HIC-resistance steel pipes, there are clear requirements for HIC resistance evaluation in the world: CSR ≤ 2%, CLR ≤ 15%, CTR ≤ 5%. Although the research and development of domestic HIC-resistance pipeline steel has just started, it has also achieved some results. According to the above design principles, China has developed HIC-resistance pipeline steel including X80 steel grade. Pangang Group Chengdu Iron and Steel Co., Ltd. is at the domestic leading level in the research and development of anti-acid corrosion seamless steel tubes, with mass production capacity. The submarine pipelines developed by the company are resistant to acid corrosion X56, X60 and X65 steel grade series seamless steel tubes. The microalloyed component design and the corresponding process technology route solve the technical problems of smelting, rolling and heat treatment processes, and the product indicators meet the requirements of the corresponding standards. Baoji Petroleum Pipe Co., Ltd. also developed a spiral submerged arc welded pipe for X52 H_2S-resistance corrosion gas gathering pipeline. Laiwu Iron & Steel Co., Ltd. has developed a manufacturing method for HIC-resistance pipeline steel. The physical and chemical properties of the product have reached X46 and X52 steel grades, which can realize multi-purpose steel. The X52, X60 and X65 grade HIC pipeline steels developed by Baosteel have high strength and high toughness and good HIC resistance. The main properties are listed in Table 1. X65 grade pipeline steel also has better resistance to dynamic tearing. Performance and higher impact toughness, which helps to ensure the safety of high pressure gas pipeline operation. Wugang developed a large-thickness and high-performance HIC-resistance steel for the Sichuan Puguang gas field, and formed mass production, which successfully replaced imported products. Baosteel is at the domestic leading level in the development of X80 steel grade HIC-resistance pipeline steel. It adopts lower carbon, phosphorus and sulfur content in the composition design, and at the same time, in order to make up for the strength reduction caused by the decrease of carbon content, proper addition is beneficial to the HIC-resistance performance. Alloying elements such as copper and nickel. In order to obtain a microstructure with less segregation, increase the niobium content and low carbon content, NbC particles are obtained, the microstructure is refined to increase the recrystallization termination temperature, and the tissue segregation is reduced, thereby ensuring strength and toughness and ensuring HIC-resistance performance. The method also adopts the pipeline steel composition and process design with "ultra-low carbon acicular ferrite" as the structural feature to eliminate the banded pearlite structure in the steel to ensure the HIC-resistance performance of the pipeline steel. The HIC-resistance performance of the product was evaluated as CSR≤0.01%, CLR≤0.09%, and CTR≤0.02%.

Table 1 Properties of HIC-resistance pipeline steel produced by Baosteel

Grade	Specification (mm)	$R_{t0.5}$ (MPa)	R_m (MPa)	$A(\%)$	$A_{KV}(-30℃)/J$			NACE TM0284 A Solution
					Tube	Heat affected zone	Weld	
X52	Φ508 ×11.9	425	546	44	375	239	198	No crack
X60	Φ711 ×8.0	476	598	41	208	180	178	No crack
X65	Φ508 ×8.7	479	599	39	388	196	192	No crack

6 Conclusions

With the advancement of metallurgical technology, after years of development and production, the output of pipeline steel in China has increased year by year.

The products have been applied to a series of major long-distance oil and gas pipeline projects such as the West-East Gas Pipeline Project. It has been able to mass produce X65 pipeline steel resistant to HIC, X80 steel grade HIC-resistance pipeline steel has also been developed. However, it must be recognized that the research and development of Chinese HIC-resistance pipeline steel has just started, and there is still a certain gap with the advanced level of foreign countries.

With Chinese energy structure adjustment and environmental protection demand, the demand for natural gas has been increasing, and natural gas development conditions have also undergone significant changes. Acidic oil and gas fields have been rapidly developed. Therefore, in the future, it is necessary to further strengthen the research and development of steel for oil and gas pipelines in acidic environments. First of all, although there is a relationship between HIC and SSCC of pipeline steel, the research on pipeline steel HIC can not completely replace the research on SSCC performance of pipeline steel. At present, there are some researches on the influencing factors of SSCC, but the lack of SSCC is unique. The study of the key factors affecting tensile stress should strengthen the research work in this area and further develop the development of targeted SSCC-resistance pipeline steel. Secondly, in order to better study the HIC generated by pipeline steel in the actual acidic environment, the standard evaluation test of NACE TM0284 should be extended and improved through relevant experimental research. NACE has published the corresponding standard test method NACE TM0177 for evaluation. The material's ability to resist SSCC, but there is no clear technical parameter requirement in the world to measure the anti-SSCC ability of steel, and should be further explored and improved. Finally, there are few studies on the influence of H_2S corrosion products on the stress corrosion rate of hydrogen sulfide, and the H_2S corrosion products have some influence on the HIC crack and SSCC crack growth rate, and then the crack growth rate model can be predicted. And the establishment of difficulties.

Acknowledgment

The authors gratefully acknowledge the State Key Laboratory for Performance and Structure

Safety of Petroleum Tubular Goods and Equipment Materials, CNPC Tubular Goods Research Institute, and China National Quality Supervision, Testing and Inspection Center of Oil Tubular Goods.

References

[1] E. J. Omonbude. The transit oil and gas pipeline and the role of bargaining: a non-technical discussion, Energy Policy, 35, 6188(2007).

[2] L. Zhang, X. Li, C. Du. Effect of applied potentials on stress corrosion cracking of X70 pipeline steel in alkali solution, Materials & Design, 30, 2259(2009).

[3] W. K. Kim, S. U. Koh, B. Y. Yang. Effect of environmental and metallurgical factors on hydrogen induced cracking of HSLA steels, Corrosion Science, 50, 3336(2008).

[4] J. L. Gonzalez, R. Ramirez, J. M. Hallen. Hydrogen-induced crack growth rate in steel plates exposed to sour environments, Corrosion, 53, 935(1997).

[5] B. Beidokhti, A. H. Koukabi, A. Dolati. Influences of titanium and manganese on high strength low alloy SAW weld metal properties, Materials Characterization, 60, 225(2009).

[6] S. S. Nayak, R. D. K. Misra, J. Hartmann. Microstructure and properties of low manganese and niobium containing HIC pipeline steel, Materials Science and Engineering: A, 494, 456(2008).

[7] G. T. Park, S. U. Koh, H. G. Jung. Effect of microstructure on the hydrogen trapping efficiency and hydrogen induced cracking of linepipe steel, Corrosion Science, 50, 1865(2008).

[8] X. Ren, W. Chu, J. Li. The effects of inclusions and second phase particles on hydrogen-induced blistering in iron, Materials Chemistry and Physics, 107, 231(2008).

[9] M. Elboujdaini, V. S. Sastri, J. R. Perumareddi. Studies on inhibition of hydrogen-induced cracking of linepipe steels, Corrosion, 62, 29(2006).

[10] V. Venegas, F. Caleyo, T. Baudin. Role of microtexture in the interaction and coalescence of hydrogen-induced cracks, Corrosion Science, 51, 1140(2009).

[11] C. Natividad, M. Salazar, A. Contreras. Sulfide stress cracking susceptibility of welded X-60 and X-65 pipeline steels, Corrosion, 62, 375(2006).

[12] B. Beidokhti, A. H. Koukabi, A. Dolati. Effect of titanium addition on the microstructure and inclusion formation in submerged arc welded HSLA pipeline steel, Journal of Materials Processing Technology, 209, 4027(2009).

[13] B. Beidokhti, A. Dolati, A. H. Koukabi. Effects of alloying elements and microstructure on the susceptibility of the welded HSLA steel to hydrogen-induced cracking and sulfide stress cracking, Materials Science and Engineering: A, 507, 167(2009).

[14] M. A. Mansour, A. M. Alfantazi, M. El-boujdaini. Sulfide stress cracking resistance of API-X100 high strength low alloy steel, Materials & Design, 30, 4088(2009).

本论文原收录于 MATEC Wed of Conferences 2018 年。

Assessment of Hydrogen Embrittlement via Image-Based Techniques in Al-Zn-Mg-Cu Aluminum Alloys

Su Hang[1,2]　Hiroyuki Toda[2]　Kazuyuki Shimizu[2]　Kentaro Uesugi[3]
Akihisa Takeuchi[3]　Yoshio Watanabe[4]

(1. State Key Laboratory for Performance and Structure Safety of Petroleum Tubular Goods and Equipment Materials; 2. Department of Mechanical Engineering, Kyushu University; 3. Japan Synchrotron Radiation Institute (JASRI); 4. UACJ Corporation)

Abstract: Hydrogen repartitioning and the related embrittlement behavior were characterized by studying Al-Zn-Mg-Cu aluminum alloys with different intermetallic particle contents. Using high-resolution X-ray tomography and related microstructural tracking techniques, hydrogen-induced quasi-cleavage cracks and the related strain localization were observed regardless of the content of the intermetallic particles. The area of quasi-cleavage cracks on the fracture surface increased and the strain localization became more intense with a decrease in the content of intermetallic particles, thereby revealing that trapped hydrogen at intermetallic particles increases the resistance to hydrogen embrittlement. In addition, a quantitative assessment of the hydrogen repartitioning taking into account vacancy production and dislocation multiplication during deformation, was applied to characterize the hydrogen embrittlement behavior. Because of the thermal equilibrium among various hydrogen trap sites, internal hydrogen atoms are mainly repartitioned to vacancies and precipitates in the strain localization region during deformation because of their high trap site densities and high hydrogen trap binding energies. Since the concentration of hydrogen trapped at dislocations is extremely limited, it can be assumed that hydrogen repartitioned to precipitates induces decohesion of precipitates along specific crystallographic planes, where quasi-cleavage cracking may originate.

Keywords: X-ray tomography; Hydrogen embrittlement; Strain localization; Hydrogen repartitioning; Al-Zn-Mg-Cu aluminum alloy

1　Introduction

Aluminum alloys can absorb hydrogen during high-temperature heat treatment (e.g. a

---

Corresponding author: Su Hang, suhang12@cmpc.com.cn.

homogenization process) and/or services in aggressive environments due to the destruction of a surface oxide layer[1-3]. The dissolved hydrogen atoms are trapped among various trap sites (e.g. dislocations, vacancies and grain boundaries) and considerably degrade the mechanical properties of aluminum alloys[4,5]. In particular, hydrogen decreases the ductility and fracture toughness, leading to an unexpected failure at applied strain levels that are far lower than a hydrogen-free material can sustain[6,7].

In addition to inducing adrastic reduction in ductility and related premature fracture, hydrogen can cause a transition of the fracture mode from ductile (i.e. micro void coalescence) into brittle fractures that appear as either intergranular fractures or transgranular quasi-cleavage fractures[8-10]. Numerous mechanisms have been proposed to explain the effects of hydrogen trapping and concentration on the initiation and propagation of hydrogen-induced cracks. According to Gerberich, et al., hydrogen trapped at grain boundaries reduces the cohesive bonding strength (i.e., the hydrogen enhanced decohesion (HEDE) mechanism), promoting the initiation and propagation of intergranular cracks[10]. Using in-situ transmission electron microscopy (TEM), Birnbaum, et al. proposed the hydrogen enhanced localized plasticity (HELP) mechanism [11], in which hydrogen trapped at dislocations promotes planar slips of dislocations[12]. Martin, et al. revealed that fracture origins are generated at the intersections of dislocation planar slip lines, leading to the initiation and propagation of quasi-cleavage cracks in the strain localization region through the expansion of fracture origins along intense slip lines in X60 pipeline steels[13]. In addition, Nagumo, et al. revealed that hydrogen trapped at vacancies accelerated the agglomeration of vacancies to form nano voids, resulting in the propagation of hydrogen-induced quasi-cleavage cracks via the coalescence of nano voids (i.e. hydrogen-enhanced strain-induced vacancies (HESIV) mechanism)[14,15]. In previous research, Su, et al. revealed that hydrogen-induced cracks appear as quasi-cleavage cracks initiated on the surface of specimens in Al-Zn-Mg-Cu aluminum alloys[16]. However, the proposed mechanisms for hydrogen-induced quasi-cleavage cracks (e.g. HELP and/or HESIV mechanisms) can not be associated with the hydrogen trapping behavior in Al-Zn-Mg-Cu aluminum alloys since the concentration of hydrogen at the dislocations or vacancies is lower compared with that at other hydrogen trap sites due to its low hydrogen trap binding energy, as determined by Bhuiyan, et al. via performing thermal desorption spectroscopy (TDS) experiments[17].

Additionally, experimental studies and finite element method (FEM) simulations revealed that the initiation and propagation of hydrogen-induced cracks are attributed to the accumulation of dissolved hydrogen atoms in regions with high internal stress and strain during deformation[17-19]. For examples, Albrecht, et al. revealed that hydrogen tends to migrate with dislocations to grain boundaries with high normal stress, leading to the intergranular fracture of commercial 7075 aluminum alloys[19]. Liang, et al. revealed that hydrogen atoms tend to be accumulated in the regions with high effective plastic strain, and the content of dissolved hydrogen in strain concentration regions is approximately 2 to 3 times higher compared to the other regions[20]. In addition, Lee, et al. revealed that hydrogen diffusion and uptake in the strain localization region during deformation increases the growth rate to 0.5μm/s and decreases the threshold stress intensity

(i. e. K_{TH} to as low as 10% of K_{IC}) of hydrogen-induced quasi-cleavage cracks in ultrahigh-strength AERMET 100 steel[21]. With the help of FEM simulations, Pañeda, et al. proposed that a high geometrically necessary dislocation (GND) density in terms of strain gradient plasticity and an elevated hydrostatic stress results in the accumulation of hydrogen ahead of the crack tip, leading to the propagation of hydrogen-induced quasi-cleavage cracks in Ni-Cu superalloys[22]. By applying in-situ tensile tests under high-resolution X-ray tomography, Su, et al. showed that the initiation and propagation of hydrogen-induced quasi-cleavage cracks in Al-Zn-Mg-Cu aluminum alloys are time-dependent due to the accumulation of hydrogen in the strain localization region[16,23]. In addition, the majority of hydrogen is repartitioned to nano voids in strain localization regions during deformation due to their high trap site density[16], indicating that in-situ hydrogen repartitioning in the strain localization is necessary for hydrogen embrittlement to occur.

With the help of the high-resolution X-ray tomography and the related 4D strain mapping technique, vacancy production and dislocation multiplication in terms of microscopic strain distribution were characterized to precisely estimate the density values for all the hydrogen trap sites during in-situ tensile tests[24]. In addition, Yamaguchi, et al. calculated the hydrogen trap binding energies of dislocations, grain boundaries, vacancies and precipitates in Al-Zn-Mg-Cu aluminum alloys by applying the first-principle simulations[25-27]. By combining the issues mentioned above, the hydrogen repartitioning behavior among various trap sites during plastic deformation and its effect on the initiation and propagation of hydrogen-induced quasi-cleavage were studied in this research.

2 Materials and experimental methods

2.1 Materials

High Zn (10% mass Zn) Al-Zn-Mg-Cu aluminum alloys were used in this study. To identify the influences of intermetallic particles on the hydrogen partitioning and embrittlement behaviors, two alloys with different Fe and Si contents were applied. The chemical compositions are summarized in Table 1. The ingots were homogenized at 743K for 24 hours, hot rolled at 673K prior to solution treatment at 748K for 2 hours, and artificially aged at 393K for 6 hours and then at 423K for 5 hours. All the specimens for the in-situ tensile tests were sampled along the rolling direction by an electro discharge machine (EDM) wire eroder. The dimensions of the in-situ tensile test specimen are shown in Fig. 1. To determine the influence of the precharged hydrogen content on the embrittlement behaviors in Al-Zn-Mg-Cu aluminum alloys, specimens with different hydrogen contents were prepared. High hydrogen content specimens were charged via EDM cutting in distilled water due to the absorption of hydrogen that was decomposed from distilled water during machining. In contrast, low hydrogen content specimens were prepared by EDM cutting in DAPHNE CUT HL-25 produced by Idemitsu Kosan Co. Ltd. to prevent the absorption of hydrogen during the cutting process. The vacuum fusion method was used to measure the total hydrogen content of specimens after EDM cutting[16], and the content of precharged hydrogen was 6.97 mass ppm for the high hydrogen content specimen and 3.26 mass ppm for the low hydrogen content specimen, respectively. Details of the vacuum fusion method used for analyzing the total hydrogen content in aluminum are shown elsewhere[28].

Table 1 Chemical compositions of the three alloys (wt, %)

Composition	Si	Fe	Cu	Mg	Zn	Ti	Zr	Al
High Fe Si	0.30	0.30	1.50	2.40	10.00	0.04	0.15	Bal
Low Fe Si	0.01	0.01	1.50	2.40	10.00	0.04	0.15	Bal

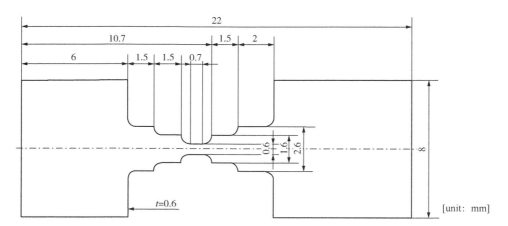

Fig. 1 Geometry of an in-situ tensile test specimen

2.2 In-situ tensile test via high resolution X-ray tomography

The in-situ tensile tests were performed at the BL20XU undulator beamline in SPring-8. A monochromatic X-ray beam with a 20keV photon energy generated by a liquid nitrogen-cooled Si (111) double crystal monochromator was applied for the present research. The image detector used included a 2048 (H) × 2048 (V) element digital CMOS camera (ORCA Flash 4.0, Hamamatsu Photonics K. K.), a single crystal scintillator (Pr: $Lu_3Al_5O_{12}$) and lens (20×). The effective pixel size of the detector was 0.5μm, and it was positioned 20mm behind the specimen. A total of 1800 radiographs with scanning 180 degrees in 0.1 degree increments, were obtained along the loading axis in each scan. Image slices were reconstructed from the 1800 radiographs using the conventional filtered backprojection algorithm, followed by a conversion process from 16-bit to 8-bit. During this conversion, the gray value of the 8-bit images was calibrated so that the linear absorption coefficients of $-30\sim40cm^{-1}$ fell within the 8-bit grayscale of $0\sim255$. The isotropic voxels in the reconstructed images were $(0.5\mu m)^3$ in size.

Al-Zn-Mg-Cu aluminum alloys with different contents of intermetallic particles and hydrogen, were employed for the in-situ tensile tests. A DEBEN CT 500 in-situ testing stage was used for the in-situ tensile tests, and the strain rate was approximately 5×10^{-4} s^{-1}. The influence of the intermetallic particles on the hydrogen accumulation and repartitioning behaviors in the Al-Zn-Mg-Cu aluminum alloys under loading was clarified by holding an applied displacement for 3.34ks at each step (the step size was approximately 0.02mm between each step). Specimens for the in-situ tensile tests were named as High Fe Si-low H_2, High Fe Si-high H_2 and Low Fe Si-high H_2 specimen, respectively, in the present research.

2.3 Image processing and analysis

The Marching Cubes algorithm was applied to calculate parameters including the volume, surface area and diameter of each particle and hydrogen micro pore at subvoxel accuracy. To suppress the inaccuracies generated from image noise, only hydrogen micro pores and particles that were over 9 voxels in volume were counted as microstructural features in the present research. Precise image registration was performed to minimize the distances of identical particles captured at neighboring loading steps. The microstructural tracking technique (MTT), which enables the visualization of deformation behaviors, was applied by tracking all the particles throughout the deformation. In addition, high-density 4D strain mapping which enables a direct understanding of strain localization mechanisms in the presence of hydrogen, was also achieved by calculating the physical displacement of the identical particles under different loading steps. Further details of the MTT technique and 3D strain mapping are available elsewhere[24,29,30].

3 Results and discussion

3.1 Influence of intermetallic particles on hydrogen embrittlement behaviors of Al-Zn-Mg-Cu aluminum alloys

Nominal stress-strain curves for the in-situ tensile tests of High Fe Si-low H_2, High Fe Si-high H_2, Mid Fe Si-high H_2 [also shown as 10 Zn (T_h = 3.34ks) specimens in Fig. 4 in Ref[16]] and Low Fe Si-high H_2 specimens are shown in Fig. 2. Due to the stress relaxation behavior of the

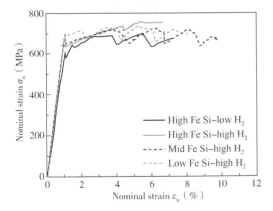

Fig. 2 Nominal stress-strain curves for the in-situ tensile tests. It is worth noting that nominal stress-strain curve of Mid Fe Si-high H_2 has already been applied in Ref[16]

material during the CT scans, vertical drops in the load are observed in Fig. 2. Due to the existence of brittle intermetallic particles, the fracture strain decreases with an increase in Fe and Si content in the presence of hydrogen. In addition, the fracture strain of the high Fe and Si content specimens decreases from approximately 10%~6% with an increase in the precharged hydrogen content, revealing a high susceptibility to hydrogen embrittlement in Fig. 2.

The fracture surface morphologies of High Fe Si-low H_2, High Fe Si-high H_2, Mid Fe Si-high H_2[16] and Low Fe Si-high H_2 are shown in Fig. 3. Fracture surfaces of all the specimens are composed of quasi-cleavage cracks and dimple fracture features. Quasi-cleavage cracks are initiated near the surface of the specimen, and gradually transform into dimple patterns with an increase in the distance from the surface to the center of the specimen. Since the precharged hydrogen content in high hydrogen content specimens (6.97 mass ppm) is approximately fifty times higher compared with that before EDM cutting (0.13 mass ppm), it can be inferred that the initiation of quasi-cleavage cracks is attributed to a high hydrogen concentration near the surface of the specimens precharged through the EDM cutting process in distilled water. In addition, the fractional area of quasi-cleavage cracks (A_{Quasi}/A_{Total}) increases with an increase in the precharged hydrogen content and decreases with an increase in the content of intermetallic particles, as summarized in Table. 2.

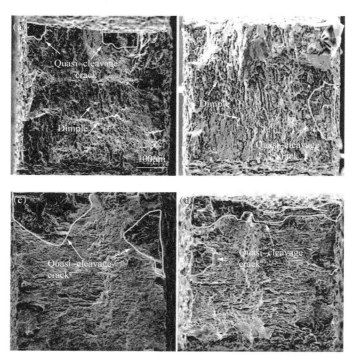

Fig. 3 Fracture surfaces after the in-situ tensile tests; (a) High Fe Si-low H_2 specimen, (b) High Fe Si-high H_2 specimen, (c) Mid Fe Si-high H_2 specimen[16] and (d) Low Fe Si-high H_2 specimen; the quasi-cleavage crack region is identified by yellow solid lines

Table 2 Areal fractions of the quasi-cleavage cracks

Materials	Areal fraction (%)	Materials	Areal fraction (%)
High Fe Si-low H_2	5.7	Mid Fe Si-high H_2	18.8[16]
High Fe Si-high H_2	8.1	Low Fe Si-high H_2	22.4

A series of 3D perspective views of the hydrogen micro pores and voids near the quasi-cleavage cracks for the High Fe Si-low H_2, High Fe Si-high H_2 and Low Fe Si-high H_2 specimen under different applied strains are shown in Fig. 4. The underlying intermetallic particles and aluminum matrix are not displayed. Hydrogen micro pores are distributed uniformly in the matrix at the unloading state, as shown in Fig. 4(a), (d) and (h). For the High Fe Si-low H_2 specimen, two quasi-cleavage cracks initiate near the surface at an applied strain of 6.3%, propagate and tend to be coalesced with an increase in applied strain levels as shown in Fig. 4(b) and (c). Voids due to the fracture of intermetallic particles are observed at an applied strain of 6.3% as shown in Fig. 4(b). In addition, micro cracks due to the growth and coalescence of voids are observed ahead of the crack tip with an increase in the applied strain from 6.3% ~ 8.6%, as shown in Fig. 4(c). In contrast, a quasi-cleavage crack initiates just after the yield strain (i.e. the applied strain, $\varepsilon_a = 1.0\%$) and propagates at relatively low applied strain levels ranging from 1.0 to 4.4% in the High Fe Si-high H_2 material, as shown in Fig. 4(e) to (f). Since intermetallic particles are the heterogeneous nucleation site for hydrogen micro pores during the homogenization process in aluminum alloys[31], the number density of hydrogen micro pores in the Low Fe Si-high H_2 material is lower than that in the High Fe Si-high H_2 material at the unloading state, compared in Fig. 4 (d) and (h). In contrast, the diameter of the hydrogen micro pores in High Fe Si-high H_2 is smaller than that in the Low Fe Si-high H_2 material, which is attributed to the restrained effects of intermetallic particles on the growth of hydrogen micro pores during the homogenization process[3,31] as shown in Fig. 4 (d) and (h). Quasi-cleavage cracks initiate at an applied strain of 2.4% and 4.4%, respectively, and gradually propagate with an increase in applied strain levels, as shown in Fig. 4 (h)-(i). With an increase in the content of pre-charged hydrogen, almost no growth of pre-existing hydrogen micro pores nor the nucleation of voids was observed during the entire in-situ tensile tests, as shown in Fig. 4 (d)-(i).

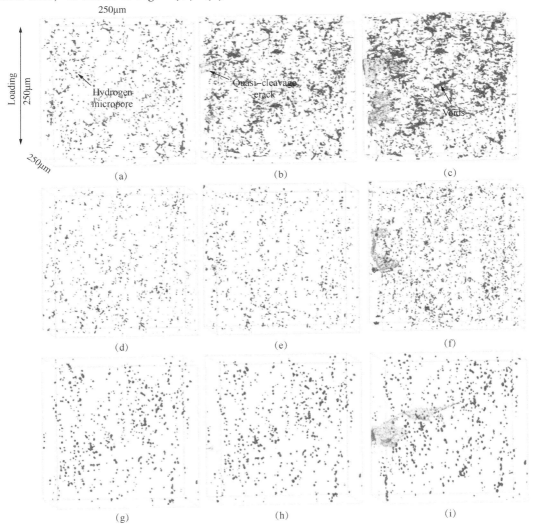

Fig. 4 4D observations of the initiation and propagation of the quasi-cleavage crack in High Fe Si-low H_2 under different applied strains; (a) ε_a = 0.0%, (b) ε_a = 6.3% and (c) ε_a = 8.6%; in High Fe Si-high H_2 under different applied strains; (d) ε_a = 0.0%, (e) ε_a = 1.0% and (f) ε_a = 4.4% and that in Low Fe Si-high H_2 under different applied strains; (g) ε_a = 0.0%, (h) ε_a = 2.4% and (i) ε_a = 6.9% Hydrogen micro pores are shown in red and the quasi-cleavage crack is shown in yellow

Fig. 5 shows the equivalent strain (ε_{eq}) mapping of the RD-ND virtual cross-section of the High Fe Si-low H_2, High Fe Si-high H_2 and Low Fe Si-high H_2 materials under similar macroscopic applied strain levels (the applied strain, $\Delta\varepsilon_a$, was approximately 4.5%). Fracture surfaces along the RD-ND direction obtained via the fracture trajectory analysis[32] are also shown as a solid black line in Fig. 5 to show the influence of strain localization on the entire fracture of Al-Zn-Mg-Cu aluminum alloys. For the high Fe and Si content materials, the strain localization region is located at the center of the specimen regardless of the pre-charged hydrogen content, as shown in Fig. 5 (a) and (b). Isolated, small regions with a high strain concentration that affect the propagation route

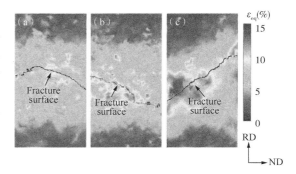

Fig. 5 The equivalent strain (ε_{eq}) distribution under different applied strains, viewed on the y-z (RD-ND) cross-section; (a) equivalent strain map calculated between ε_a of 2.1 and 6.3% in High Fe Si-low H_2, (b) equivalent strain map calculated between ε_a of 0 and 4.4% in High Fe Si-high H_2 and (c) equivalent strain map calculated between ε_a of 2.4 and 6.9% in Low Fe Si-high H_2. Fracture surface is shown as the black line

of the crack are observed in the strain localization region in the High Fe Si-high H_2 specimen, as shown in Fig. 5 (b). For the Low Fe Si-high H_2 specimen, the strain localization region appears as a band oriented in an oblique direction (approximately 45°) which is more intense compared with that in the High Fe-Si high H_2 specimen. The microscopic strain distribution also indicates the movement and related concentration of dislocations in the presence of hydrogen, as shown in Fig. 5. According to Ferreira, et al.[33], the formation of the hydrogen atmosphere around an edge component decreases the energy of the edge dislocation-hydrogen system to a certain extent which is lower than that of an edge dislocation and higher than that of a screw dislocation. In addition, hydrogen from the atmosphere revert to the solid solution and increase the total system energy when the edge dislocations are transformed to screw components during cross slip. Due to the decrease in the energy of edge dislocations and the stabilization of the edge components in the presence of hydrogen, it can be inferred that hydrogen is primarily trapped at edge dislocations instead of screw dislocations, promoting planar slip as cross slip is prohibited. Liang, et al revealed that scuh dislocation planar slip in the presence of hydrogen result in the microscopic strain localization along the shear direction by applying FEM simulations[34]. As a result, hydrogen trapped at edge dislocations leads to the strain localization region appears as a band along approximately 45° in Low Fe-Si high H_2 specimen, as shown in Fig. 5 (c). In contrast, the hydrogen distribution behavior among various hydrogen trap sites affects both localized strain concentration behavior and the degree of the strain localization region in High Fe Si-high H_2 specimen, revealing that the hydrogen-induced strain localization behavior is suppressed when there is an increase in the content of intermetallic particles, as compared in Fig. 5 (b) and (c).

Based on the issues mentioned above, influences of intermetallic particles on the hydrogen embrittlement and related premature fracture behavior in Al−Zn−Mg−Cu aluminum alloys has been studied. The introduction of intermetallic particles was found to decrease the fractional area of hydrogen−induced quasi−cleavage cracks and retard the hydrogen−induced strain localization behavior at low applied strain levels during deformation, indicating that the role of intermetallic particles is to enhance the resistance to hydrogen embrittlement in Al−Zn−Mg−Cu aluminum alloys, as shown in Fig. 3~5.

According to previous research, particles in aluminum alloys are mainly divided into three types including intermetallic particles, dispersoids and precipitates based on their formation temperature ranges. Intermetallic particles aremicrometer in size and are incoherent particles that form through a casting process and remain during the subsequent heat treatment such as homogenization and aging processes[35]. In contrast, dispersoids form during the homogenization process and the precipitates are nanometer−sized particles or clusters that form during the aging process[36,37]. With respect to intermetallic particles, Yamabe, et al. revealed that hydrogen is mainly trapped at the interface between $(Fe, Mn, Cr)_2SiAl_{12}$ intermetallic particles and the aluminum matrix in Al−Mg−Si aluminum alloys by applying secondary ion mass spectrometry (SIMS), and they proposed that an extremely low effective hydrogen diffusivity (i.e., on the order of 10^{-14} m^2/s) and the related high resistance to hydrogen embrittlement is mainly attributed to the hydrogen trapping behavior at intermetallic particles near the surface of the specimen[38]. In addition, Su, et al. studied the effects of the shape and size of intermetallic particles (e.g. Al_7Cu_2Fe particles) on the hydrogen accumulation and related hydrogen micro pore initiation behavior in Al−Zn−Mg−Cu aluminum alloys[3]. It is proposed that hydrogen is prior to accumulate at intermetallic particles with irregular shapes and large sizes due to a high hydrostatic tension concentration at the angular corners and sharp edges of the irregular intermetallic particles, promoting a higher concentration of hydrogen at more complex and irregular particles compared with that at spherical particles. With an increase in both the volume fraction and the size of intermetallic particles, it can be inferred that hydrogen trapped at intermetallic particles leads to a decrease in both the accumulated hydrogen content ahead of the quasi−cleavage crack tip[39] and the trapped hydrogen content at specific hydrogen trap sites (e.g., dislocations[40,41] and vacancies[14]) due to the thermal equilibrium among various trap sites[42]. This result in a higher resistance to hydrogen embrittlement in the High Fe Si−high H_2 specimen compared with that in the Low Fe Si−high H_2 specimen. Although numerous hydrogen atoms are trapped at the interface between intermetallic particles and the matrix in the High Fe Si−high H_2 specimen, almost no voids due to the decohesion of intermetallic particles was observed ahead of the quasi−cleavage cracks during the in−situ tensile test, as shown in Fig. 4 (d) to (f). This result coincides with the results of Qin, et al., who demonstrated that hydrogen−induced cracking cannot occur at the interfaces when the sizes of MnS inclusions are less than 3.3μm in austenitic steels since the inclusion−matrix interfaces can not provide sufficient space for crack nucleation[43]. Based on the issues mentioned above, it can be inferred that Al_7Cu_2Fe intermetallic particles with an average diameter of 4.6μm[44] mainly act as hydrogen trapping sites instead of hydrogen−induced fracture origins in the High Fe Si−high H_2 specimen. Although intermetallic particles are undesirable because they are generally detrimental to the mechanical and damage tolerant properties of structural materials[45], it is reasonable to assume that the resistance to hydrogen embrittlement can be improved through changing the content, chemical compositions, size and shape of intermetallic particles in Al-Zn-Mg-Cu aluminum alloys.

3.2 Hydrogen repartitioning among various trap sites during deformation

As mentioned in Section 3.1, the existence of intermetallic particles affects the hydrogen-induced strain localization behavior and related internal hydrogen redistribution during deformation. Compared to the hydrogen trapping behavior at the unloading state, internal hydrogen atoms are likely accumulated in the strain localization region due to the heterogeneous initiation of dislocations and vacancies during deformation, leading to a localized repartitioning of internal hydrogen among various trap sites and resulting in hydrogen-induced premature fracture in the strain localization region during deformation.

Bhuiyan, et al. revealed that interstitial lattices, dislocations, vacancies, intermetallic particles, precipitates, grain boundaries and hydrogen micro pores are the main hydrogen trap sites in Al-Zn-Mg-Cu aluminum alloys[17]. According to Oriani, et al., hydrogen trapped at each trap site is calculated by assuming that hydrogen trapped at normal interstitial lattices and at other trap sites is in a thermal equilibrium as follows[41]:

$$\frac{\theta_T}{1-\theta_T} = \theta_L \exp\left(\frac{E_b}{RT}\right) \quad (1)$$

where E_b is the trap binding energy (the trap binding energy of each trap site is summarized in Table. 3 [25-27,46,47]), R is the gas constant (8.31 J mole^{-1}K^{-1}), T is the absolute temperature, θ_L is the occupancy of the interstitial sites and θ_T is the occupancy of other trap sites. The total hydrogen concentration, C_H^T, which represents the sum of the hydrogen stored in the normal interstitial lattice and all the trap sites is expressed as:

$$C_H^T = \theta_L N_L + \sum \theta_{Ti} N_{Ti} + C_{pore} \quad (2)$$

where N_L and N_{Ti} are the trap densities in the normal interstitial lattice and the ith trap sites other than micro pores in atoms per unit volume, respectively.

Table 3 Binding energy of each hydrogen trap site in Al-Zn-Mg-Cu aluminum alloys

Trap Site	Binding energy E_b(kJ · mol^{-1})	Trap Site	Binding energy E_b(kJ · mol^{-1})
Edge dislocation	9.64[26]	Precipitates	33.77[27]
Screw dislocation	7.72[26]	Intermetallic particles	28.10[46]
Vacancy	28.95[25]	Pore	67.20[47]
Grain boundary	19.30[25]		

At an unloading state, Parker, et al. reported that the dislocation density is 2.5×10^{13} m^{-2} in Al-Zn-Mg-Cu aluminum alloys with the help of the electron back scattered diffraction (EBSD) technique[48]. The average grain size was 20μm[17] as measured using an optical microscope after etching polished samples by means of the Barker's method. The precipitate density was 2.7×10^{23} m^{-3}, which was determined via the transmission electron microscope (TEM) observations operated at 30kV[49]. The volume fraction, number density and spatial distribution of both hydrogen micro pores and intermetallic particles were calculated through the X-ray tomography images.

According to Calister, et al., vacancy concentration is expressed as[50]:

$$C_0 = \exp\left(-\frac{Q_{fl}}{k_B T}\right) \tag{3}$$

where the vacancy formation energy Q_{fl} is taken to be 0.66eV[50] in the absence of hydrogen, k_B is Boltzmann's constant (8.62×10^{-5}eV/atom) and T is the absolute temperature. Both the molecular hydrogen (C_{H_2}) and adsorbed hydrogen ($C_{H_{adsorbed}}$) are trapped within the hydrogen micro pores, and thus, the trapped hydrogen content in the hydrogen micro pore is calculated as[31,51]:

$$C_{pore} = C_{H_{adsorbed}} + C_{H_2} \tag{4}$$

The content of adsorbed hydrogen in the micro pores is expressed as follows:

$$C_{H_{adsorbed}} = \theta_s N_s \tag{5}$$

where θ_s is the occupancy of adsorbed hydrogen at the surface of the micro pores and N_s is the trap density of adsorbed hydrogen at the surface of the micro pores. The gas pressure of a hydrogen micro pore is in thermal equilibrium with the surface energy of the aluminum matrix[31], based on Eq. (6):

$$P = 4\gamma/d \tag{6}$$

where P is the internal gas pressure, d is the diameter of a pore and γ is the surface energy of aluminum. The surface energy of aluminum is reduced by the adsorbed hydrogen trapped at the surface of a hydrogen micro pore and is expressed as:

$$\gamma = \gamma_0 - [E_S + RT\ln(\theta_L)]\frac{N_s \theta_s}{N_A A} \tag{7}$$

where the original surface energy is $\gamma_0 = 1$J/m^2[51]. $E_S = 67.2$kJ·mol^{-1}[47] is the hydrogen trap binding energy of adsorbed hydrogen at the surface of a micro pore and N_A is the Avogadro's constant. Based on the ideal gas law, the content of molecular hydrogen within a hydrogen micro pore is given by the following:

$$C_{H_2} = 2N_A \frac{4\gamma V}{dRT} \tag{8}$$

where V is the volume of a hydrogen micro pore.

Huang, et al. reported that the total dislocation density under loading is calculated as[52]:

$$\rho = \rho_{GND} + \rho_{SSD} \tag{9}$$

where ρ_{SSD} is the statistically stored dislocation (SSD) density and ρ_{GND} is the geometrically necessary dislocation (GND) density. SSD tends to be accumulated by trapping other dislocations in a random way while GND is required for compatible deformation of various parts of a material[53]. Brinckmann, et al. proposed the calculation for SSD density as follows[54]:

$$\rho_{SSD} = \frac{\sqrt{3}\,\bar{\varepsilon}^p}{bl} \tag{10}$$

where $\bar{\varepsilon}^p$ is the equivalent plastic strain and the Burgers vector is $b = 0.286$ nm. The mean free path of dislocation motion l is the half of a single dislocation slip distance inside a grain[55]. The value of the mean free path is taken as $l = \frac{\sqrt{2}}{2}D = 14\mu m$ [17]. According to Paneda, et al. [56], the GND density is calculated as:

$$\rho_{GND} = \bar{r}\frac{\eta^p}{b} \tag{11}$$

where the Nye-factor is $\bar{r} = 1.9$ in face center cubic (fcc) alloys[56] and η^p is the equivalent plastic strain gradient.

SSD and GND distributions calculated from the equivalent strain (ε_{eq}) mapping, viewed on the y-z (RD-ND) cross-section in both the High Fe Si-high H$_2$ and the Low Fe Si-high H$_2$ specimens under the applied strain, $\Delta\varepsilon_a$ of 4.5% are shown in Fig. 6. In order to reveal the influence of both stain localization and related strain gradient to the distribution of SSD and GND, strain concentration regions (i.e. the strain is approximately two times higher than the applied strain level) are extracted and shown as solid white outline in Fig. 6(b), (c), (e) and (f). The SSD distribution is consistent with the equivalent strain distribution, as shown in Fig. 6(b) and (e). The high GND concentration regions observed in the High Fe Si-high H$_2$ specimen are attributed to the isolated, small regions with a high strain concentration located in the strain localization region, as shown in Fig. 6(c). In contrast, high GND concentration regions are mainly located at the interface of the strain localization region in the Low Fe Si-high H$_2$ specimen, as shown in Fig. 6(f). The average GND density in the Low Fe Si-high H$_2$ specimen is higher compared with that in the High Fe Si-high H$_2$ specimen, which is mainly attributed to a more intense strain localization behavior and a higher strain gradient during deformation.

On the other hand, Clouet, et al. revealed that excess vacancies are formed in the strain localization region due to dislocation plasticity[57]. Militzer, et al. proposed that the vacancy formation rate under loading is expressed as[58]:

$$\frac{dC_v}{dt} = \chi\frac{\Omega_0\sigma}{Q_{f2}}\dot{\varepsilon} + \zeta\frac{C_j\Omega_0}{4b^3}\dot{\varepsilon} - \frac{D_v\rho}{\kappa^2}c_v - \frac{D_v}{L^2}c_v \tag{12}$$

where $\chi = 0.1$ is the dimensionless constant[58], σ is the flow stress, $\Omega_0 = 1.65 \times 10^{-29}$ is the atomic volume, $\dot{\varepsilon}$ is the strain rate, $Q_{f2} = 0.35eV$ is the formation energy of the vacancy in the presence of hydrogen[59], b is Burgers vector, ζ describes the neutralization effect induced by the presence of vacancy emitting and vacancy absorbing jogs, C_j is the concentration of thermal jogs, D_v is the vacancy diffusivity, ρ is the dislocation density, L is the grain size, κ is a parameter representing the distribution of dislocations and c_v is the excess vacancy concentration.

Witzel proposed that the second term is only applied in the temperature range over 0.4 of the melting temperature[60], thus, the second term is not considered in the present research, which was performed at room temperature. In the presence of hydrogen, the majority of vacancies are instantaneously trapped and stabilized by hydrogen after initiation. It is therefore reasonable to assume that the vacancies do not annihilate. The formation rate of vacancies under a high hydrogen concentration is then expressed as:

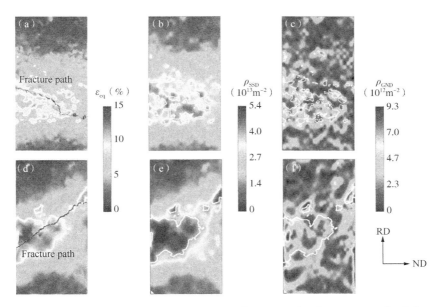

Fig. 6 SSD and GND densities calculated from the equivalent strain (ε_{eq}) mapping under different applied strains, viewed on the y-z (RD-ND) cross-section; (a) equivalent strain map calculated between ε_a of 2.1% and 6.3% in High Fe Si-high H_2, (b) SSD density calculated from (a), (c) GND density calculated from (a); (d) equivalent strain map calculated between ε_a of 2.4 and 6.9% in Low Fe Si-high H_2, (e) SSD density calculated from (d) and (f) GND density calculated from (d). Fracture path is shown as the black line in (a) and (d). It is worth noting that the localization regions that the strain concentration are approximately two times higher than the applied strain levels are extracted and shown as solid white outline in Fig. 6.

$$\frac{dC_v}{dt} = \chi \frac{\Omega_0 \sigma}{Q_f} \dot{\varepsilon} \quad (13)$$

By integrating Eq(13), the vacancy concentration under different applied strain is expressed as:

$$C_v = \chi \frac{\sigma \Omega_0}{Q_f} \varepsilon + C_0 \quad (14)$$

where ε is the true strain and $C_0 = 8.3 \times 10^{-4}$[61] is the initial vacancy concentration.

Comparisons of the hydrogen trapping behavior between the whole specimen at the unloading state and the region ahead of the hydrogen-induced quasi-cleavage crack tip (i.e. a selected region that is 20μm in diameter) at an applied strain, $\Delta\varepsilon_a$ of 4.5% in the Low Fe Si-high H_2 specimen is shown in Fig. 7. Hydrogen atoms are mainly trapped at interstitial lattice, precipitates and hydrogen micro pores due to their high trap densities and high hydrogen trap binding energies at the unloading state. With an increase in applied strain levels, excess vacancies and dislocations are generated in the strain localization region ahead of the crack tip and hydrogen atoms are mainly trapped at vacancies, precipitates and hydrogen micro pores. Of note, the trapped hydrogen content at vacancies at an applied strain of 4.5% was approximately 10^6 times higher compared with that at the unloading state, indicating that hydrogen atoms accumulated and mainly repartitioned to vacancies in the strain localization region during loading, as shown in Fig. 7(a). Although hydrogen induced strain localization was observed in the Low Fe Si-high H_2 specimen, Fig. 7(a) reveals that the repartitioned hydrogen content to dislocations was far lower than the content of hydrogen trapped at other trap sites due to its low hydrogen trap binding energy.

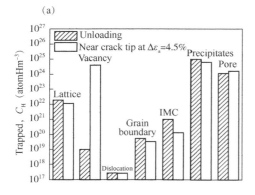 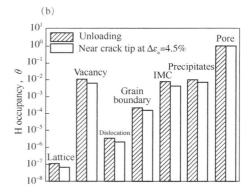

Fig. 7 Comparisons of the hydrogen trapping behavior between the whole specimen at unloading state and a region ahead of the hydrogen-induced quasi-cleavage crack tip an applied strain of 4.5% in Low Fe Si-high H_2 specimen; (a) trapped hydrogen content and (b) hydrogen trapping occupancy

It is assumed that the total hydrogen content C_H^T of the pre-charged specimen does not change during the in-situ tensile test in the present research. According to Eq (1) and (2), hydrogen occupancy of all hydrogen trap sites is affected by both trap densities and trap binding energies. Due to the vacancy production and dislocation multiplication (i.e. the trap densities, N_{Ti}, for both dislocations and vacancies increase during deformation) in the strain localization region, it can be inferred that hydrogen atoms migrate and accumulate in the newly formed hydrogen trap sites, leading to the change in the hydrogen occupancy for different trap sites under loading. Fig. 7(b) reveals that hydrogen atoms are mainly trapped at vacancies, precipitates, intermetallic particles and hydrogen micro pores regardless of the applied strain levels. It is worth noting that hydrogen occupancy of hydrogen micro pore remains almost unchanged before and after deformation, as shown in Fig. 7(b). The relationship between hydrogen occupancy of lattice and other hydrogen trap sites are summarized in Fig. 8. The hydrogen occupancy of vacancy and grain boundary increases with an increase in the hydrogen occupancy of lattice. In contrast, hydrogen occupancy of hydrogen micro pore increases with an increase in hydrogen occupancy of lattice ranges from 10^{-15} to 10^{-11}, and then remains unchanged in the range from 10^{-8} to 10^{-7} in the present research due to its high trap binding energy (i.e. 67.20kJ·mol^{-1}), marked as black dashed lines in Fig. 8. In addition, the trap site occupancy of dislocation is approximately 10^{-6}, which is approximately 10^4 times lower compared with that at precipitates and vacancies. It

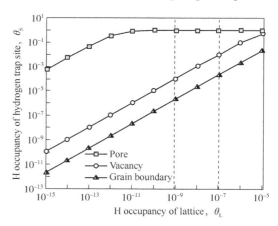

Fig. 8 Relationships between hydrogen occupancy of lattice to the hydrogen occupancy of other hydrogen trap sites

can be inferred that the influence of hydrogen trapped at dislocations on hydrogen-induced quasi-cleavage cracks might be negligible considering the repartitioned hydrogen content and hydrogen trap occupancy ahead of the quasi-cleavage crack tip during deformation.

The spatial concentration mapping of the repartitioned hydrogen at dislocations, vacancies and precipitates calculated from the equivalent strain (ε_{eq}) mapping in both the High Fe Si-high H_2 and Low Fe Si-high H_2 specimens are shown in Fig. 9. Fracture surfaces along the RD-ND virtual cross-sections are also shown and are marked as the solid black lines in Fig. 9. The hydrogen-induced strain localization has a significant influence on the content of hydrogen repartitioned to dislocations, vacancies and precipitates during deformation. For instance, hydrogen atoms are repartitioned to dislocations in the strain localization region during deformation, especially along the entire fracture surface, as shown in Fig. 9 (a) and (d). However, the content of hydrogen repartitioned to dislocations in the strain localization region is approximately 10^7 times lower compared with that repartitioned to vacancies and precipitates, indicating the limited effects of hydrogen concentrated at dislocations to the initiation and propagation of hydrogen-induced quasi-cleavage cracks during deformation.

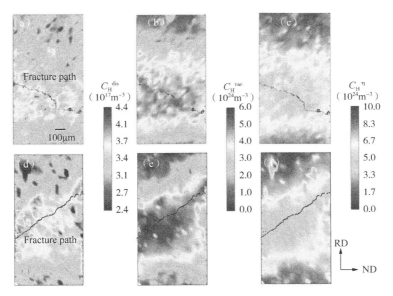

Fig. 9　Hydrogen concentrations at various trap sites including dislocations, vacancies and precipitates that calculated from the equivalent strain (ε_{eq}) mapping at an applied strain, $\Delta\varepsilon_a$ of approximately 4.5%, viewed on the y-z (RD-ND) cross-section; (a) and (d) is the hydrogen concentration at dislocations in High Fe Si-high H_2 and Low Fe Si-high H_2 specimen, (b) and (e) is the hydrogen concentration at vacancies in High Fe Si-high H_2 and Low Fe Si-high H_2 specimen and (c) and (f) is the hydrogen concentration at precipitates in High Fe Si-high H_2 and Low Fe Si-high H_2 specimen. Fracture surface is shown as the black line

Based on the issues mentioned above, a hydrogen embrittlement model considering the thermal equilibrium among various hydrogen trap sites and related in-situ hydrogen repartitioning in the strain localization region was proposed, as shown in Fig. 10. Due to the vacancy production and dislocation multiplication in the strain localization region, hydrogen atoms migrate, accumulate and repartition among various trap sites in the strain localization during deformation. It is worth noting that the content of repartitioned hydrogen to dislocations is far lower than that to vacancies and precipitates, indicating that the repartitioning of hydrogen is affected by both the hydrogen trap density and the hydrogen trap binding energy of each hydrogen trap site. In addition, Yamaguchi, et al. studied the effects of multiple hydrogen concentrations along grain boundaries on the hydrogen-induced intergranular fracture in Al-Zn-Mg-Cu aluminum alloys by analyzing the cohesion strength of grain boundaries with different grain boundary energies [i.e. $\Sigma 3(111)$ GB, $\Sigma 3(112)$ GB and $\Sigma 5(012)$ GB, etc.] in the presence of hydrogen using first-principles simulations[25]. It is concluded that the cohesion strength of grain boundaries decreases with an increase in the content of hydrogen trapped at the grain boundaries. The critical content of hydrogen atoms concentrated at grain boundaries for the occurrence of hydrogen-induced intergranular crack is approximately 2.8×10^{22} atoms H/m^3, which is approximately 3×10^5 times higher than the content of hydrogen trapped at grain boundaries of Al-Zn-Mg-Cu aluminum alloys at the unloading state (i.e. 9.9×10^{16} atoms H/m^3)[17]. This result indicates that a highly localized repartitioning of hydrogen due to hydrogen-induced strain localization during deformation is necessary for the occurrence of hydrogen-induced cracking.

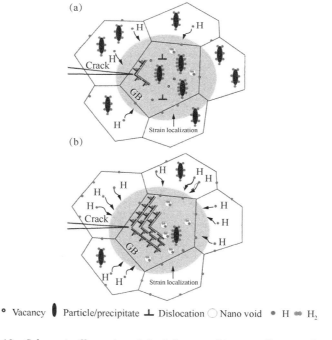

Fig. 10 Schematic illustration of the influence of intermetallic particles to the hydrogen partitioning and related crack propagation in the strain localization region, (a) high intermetallic particle content material and (b) low intermetallic particle content material

Although the trapped hydrogen content at precipitates slightly decreases with an increase in the applied strain levels, it can be seen that hydrogen atoms are still mainly repartitioned to intermetallic particles, hydrogen micro pores, vacancy and precipitates in the strain localization region at an applied strain of 4.5% in Al-Zn-Mg-Cu aluminum alloys, as shown as Fig. 7. In the previous research, it is revealed that intermetallic particles and hydrogen micro pores are relatively macroscopic features that are micrometer in size (i.e. approximately 2.0μm and 3.0μm for hydrogen micro pores and intermetallic particles in low Fe and Si content specimen, respectively[44]). It can be inferred that the nearest neighboring distance between intermetallic particles or pores are micrometer in size, indicating that continuous, river-shaped quasi-cleavage cracks are unlikely to be formed due to either the coalescence of hydrogen micro pores or the decohesion of intermetallic particles. In terms of vacancies, Nagumo, et al. revealed hydrogen trapped at vacancies stabilizes vacancies and promotes the agglomeration of vacancies into nano voids[14]. Su, et al. observed hydrogen – induced nano voids distribute uniformly in the strain localization region at an applied strain of 5.5%, revealing that the repartitioning of hydrogen at nano voids increases the resistance to hydrogen embrittlement of Al-Zn-Mg-Cu aluminum alloys[16]. In addition, neither growth nor coalescence of hydrogen-induced nano voids are observed even at high applied strain levels (i.e. localized strain, e_l, of 24%) during deformation, indicating that nano voids do not result in the initiation and propagation of hydrogen-induced quasi-cleavage cracks in Al-Zn-Mg-Cu aluminum alloys, as observed by Gao, et al. with the help of synchrotron X-ray nanotomography[47]. In terms of precipitates, Wei, et al. studied the hydrogen trapping behavior of TiC precipitates with different interface coherencies in both quenched and tempered 0.05C-0.20Ti-2.00Ni steels, reporting that the trap activation energy (E_a) for hydrogen to semicoherent TiC precipitates and incoherent TiC precipitates is 55.8kJ/mol and 85.0kJ/mol, respectively[62]. Takahashi, et al. revealed that hydrogen is mainly trapped at the carbon vacancy on the (001) broad interface between the coherent V_4C_3 precipitates and ferrite matrix in VC precipitation strengthening steels, as observed by atom probe tomography (APT) technique[63]. In addition, Nagao, et al. studied the effects of nano sized (Ti, Mo) C precipitates on the hydrogen trapping and hydrogen-induced fracture behavior in lath martensitic steels[64-65]. It is proposed that hydrogen was trapped at the interface between nano sized (Ti, Mo) C precipitates and the matrix, which promoted the decohesion of precipitates and resulted in the initiation and propagation of hydrogen-induced quasi-cleavage cracks along the {110} lath boundaries during plastic deformation in lath martensite steels. However, neither hydrogen-induced precipitate decohesion nor the propagation of hydrogen-induced quasi-cleavage cracks along the (Ti, Mo) C precipitate/matrix interface was observed in their research. In addition, the hydrogen trap binding energy of high-angle grain boundaries (e.g. 57.4kJ/mol) is much higher than that of lath boundaries (e.g. 25.7kJ/mol) and precipitates (e.g. 30.5kJ/mol) in their research, indicating that hydrogen should be mainly concentrated at grain boundaries instead of at lath boundaries and precipitates due to its high hydrogen trap binding energy[66]. In terms of hydrogen trapping behavior among various trap sites, the mechanism for why the hydrogen-induced fracture process is dominated by the initiation and propagation of quasi-cleavage cracks along the precipitates/matrix interface and lath boundaries instead of by intergranular cracking along grain boundaries remains unclear. Tsuru, et al. studied the hydrogen trapping behavior at $MgZn_2$ precipitates, including both interstitial sites inside the precipitates and at the interface between $MgZn_2$ precipitates and the matrix in Al-Zn-Mg-Cu aluminum alloys using first-principles simulations[27]. It is proposed that hydrogen atoms are mainly trapped at the $[0001]_{MgZn_2}//\{111\}_{Al}$ interface, and they reported that the hydrogen trap binding energy of the most favorable site at the interface is approximately 0.35eV/H. In addition, it is reported that the cohesion strength of the interface between the precipitates and matrix decreases with an increase in the content of hydrogen trapped at the interface. It is worth noting that the amount of hydrogen for the occurrence of precipitates decohesion is 18.9 atoms H/nm^2, which is approximately 2.5 times higher than the content of hydrogen concentrated at the precipitates (i.e. 7.3 atoms H/nm^2) as calculated through hydrogen repartitioning analysis during plastic deformation in the present research. This difference is mainly because the hydrogen repartitioning analysis only considers the generalized hydrogen behavior for repartitioning to precipitates and the practical hydrogen concentration at each individual precipitate, and in particular, the influence of the size and shape of the precipitates to the localized hydrogen accumulation was not considered in the present research. For individual, plate-shaped precipitate located on the $\{111\}_{Al}$ planes[48], it can be inferred that hydrogen is mainly trapped at the angular corners with the lowest curvature due to the hydrostatic tension concentration during deformation. With an increase in applied strain levels, hydrogen atoms continue to migrate and accumulated at the angular corners of precipitates, leading to a localized decohesion when the content of repartitioned hydrogen exceeds the critical value of 18.9 atoms H/nm^2. As a result, it is reasonable to assume that a hydrogen-induced decohesion of precipitates along the $\{111\}_{Al}$ planes at low applied strain levels during deformation provides both possible fracture initiation sites and a propagation path for hydrogen-induced quasi-cleavage cracks in Al-Zn-Mg-Cu aluminum alloys.

4 Conclusions

Hydrogen-induced premature fracture and the related hydrogen embrittlement behavior in high Zn (10 mass% Zn) Al-Zn-Mg-Cu aluminum alloys with different intermetallic particle contents were studied in the present research. The following conclusions were obtained:

(1) Intermetallic particles are one of the hydrogen trap sites in Al-Zn-Mg-Cu aluminum alloys. The areal fraction of hydrogen-induced quasi-cleavage cracks decreases with an increase in the content of intermetallic particles, indicating an enhancement in the resistance to hydrogen embrittlement due to the existence of intermetallic particles.

(2) Quantitative assessment of localized hydrogen repartitioning behavior during deformation was performed to understand the hydrogen embrittlement behavior in Al-Zn-Mg-Cu aluminum alloys. The hydrogen repartitioning behavior is strongly dependent on pre-existing intermetallic particles, vacancies and precipitates instead of dislocations due to their high hydrogen trap densities and high hydrogen trap binding energies.

(3) Since the concentration of hydrogen trapped at dislocations is extremely limited, it can be assumed thata high hydrogen concentration at the interface between the precipitates and the

aluminum matrix due to the repartitioning of hydrogen results in the decohesion of precipitates along $\{111\}_{Al}$ planes in the strain localization region during deformation, where quasi-cleavage cracking might originate.

Acknowledgment

The authors gratefully acknowledge the support from the New Energy and Industrial Technology Development Organization (part of the Technological Development of Innovative New Structural Materials, Project HAJJ262715) and the Industry-Academia Collaborative R&D Program "Heterogeneous Structure Control" from the Japan Science and Technology Agency, JST (JPMJSK1412). The synchrotron radiation experiments were performed through proposal No. 2016A1199 and 2016B1081 with the approval of JASRI.

References

[1] DEJ. Talbot. Inter Metall Rev. (20)1975: 166-84.
[2] S. Lynch. Corros. Rev. (30)2012: 105-123
[3] H. Su, M. S. Bhuiyan, H. Toda, K. Uesugi, A. Takeuchi, Y. Watanabe, Scripta. Mater. 135 (2017): 19-23.
[4] R. N. Parkins, Corros. Sci. 20 (1980): 147-166.
[5] G. A. Young, J. R. Scully, Metall. Mater. Trans. A. 33 (2002): 1297-1297.
[6] G. A. Young, J. R. Scully, Acta Mater. 46 (1998): 6337-6344.
[7] T. Izumi, G. Itoh, Mater. Trans. 52 (2011): 130-134.
[8] M. L. Martin, B. P. Somerday, R. O. Ritchie, P. Sofronis, I. M. Robertson, Acta Mater. 60 (2012): 2739-2745.
[9] Z. Zhang, G. Obasis, R. Morana, M. Preuss, Acta Mater. 113 (2016): 272-283.
[10] W. W Gerberich, R. A Oriani, M. Li, X. Chen, T. Foecke. Philos Mag A. 63(1991): 363-76.
[11] H. K Birnbaum, P. Sofronis. Mater Sci Eng A. 176(1994): 191-202.
[12] I. M. Robertson: Eng. Fract. Mech. 68(2001): 671-92.
[13] M. L. Martin, I. M. Robertson, P. Sofronis, Acta Mater. 59 (2011): 3680-3687.
[14] M. Nagumo. Mater Sci Technol. 20(2004): 940-50.
[15] T. Neeraj, R. Srinivasan, J. Li, Acta Mater. 60 (2012): 5160-5171.
[16] H. Su, H. Toda, R. Masunaga, et al, Acta Mater. 159(2018): 332-343.
[17] M. S. Bhuiyan, H. Toda, Z. Peng, S. Hang, K. Horikawa, K. Uesugi, et al., Mater. Sci. Eng. A. 655 (2016): 221-228.
[18] G. M. Pressouyre, Acta Metall. 28 (1980): 895-911.
[19] J. Albrecht, I. M. Bernstein, A. W. Thompson, Metall. Trans. A. 13 (1982): 811-820.
[20] Y. Liang, D. C. Ahn, P. Sofronis, R. H. Dodds, D. Bammann, Mech. Mater. 40 (2008): 115-132.
[21] Y. Lee, R. P. Gangloff, Metall. Mater. Trans. A, 38 (2007): 2174-2190.
[22] E. Martínez-Pañeda, C. Niordson, R. P. Gangloff, Acta Mater. 117(2016): 321-332.
[23] M. S. Bhuiyan, Y. Tada, H. Toda, H. Su, K. Uesugi, A. Takeuchi, et al., Int. J. Fract. (2016): 1-17.
[24] M. Kobayashi, H. Toda, Y. Kawai, T. Ohgaki, K. Uesugi, D. S. Wilkinson, et al., Acta Mater. 56 (2008): 2167-2181.
[25] M. Yamaguchi, K.-I. Ebihara, M. Itakura, T. Tsuru, K. Matsuda, H. Toda, Comput. Mater. Sci. 156 (2019): 368-375.

[26] M. Itakura, M. Yamaguchi, K. I. Ebihara, T. Tsuru, K. Matsuda, H. Toda, going to be submitted.

[27] T. Tsuru, M. Yamaguchi, K. Ebihara, M. Itakura, Y. Shiihara, K. Matsuda, et al., Comput. Mater. Sci. 148 (2018): 301–306.

[28] M. Imabayashi, K. Tomita, J. Jpn. Inst. Light. Metals. 22(2014): 73–81.

[29] H. Toda, T. Kamiko, Y. Tanabe, M. Kobayashi, D. J. Leclere, K. Uesugi, et al., Acta Mater. 107 (2016): 310–324.

[30] H. Toda, A. Takijiri, M. Azuma, S. Yabu, K. Hayashi, D. Seo, et al., Acta Mater. 126 (2017): 401–412.

[31] H. Toda, T. Hidaka, M. Kobayashi, K. Uesugi, A. Takeuchi, K. Horikawa, Acta Mater. 57 (2009): 2277–2290.

[32] H. Toda, H. Oogo, K. Horikawa, K. Uesugi, A. Takeuchi, Y. Suzuki, et al., Metall. Mater. Trans. A 45 (2013): 765–776.

[33] P. J. Ferreira, I. M. Robertson, H. K. Birnbaum, Acta Mater. 47 (1999): 2991–2998.

[34] Y. Liang, P. Sofronis and N. Aravas: Acta Mater. 51(2003): 2717–2730.

[35] S. C. Wang, M. J. Starink, Int. Mater. Rev. 50 (2005): 193–215.

[36] G. Sha, A. Cerezo, Acta Mater. 52 (2004): 4503–4516.

[37] A. Deschamps, Y. Bréchet, Scr. Mater. 39 (1998): 1517–1522.

[38] J. Yamabe, T. Awane, Y. Murakami, Int. J. Hydrogen Energy. 42 (2017): 24560–24568.

[39] D. M. Li, R. P. Gangloff, J. R. Scully. Metall Mater Trans A. 35(2004): 849–862

[40] J. Albrecht, A. W. Thompson, and I. M. Bernstein, Metall. Trans. A, 10 (1979): 1759–66.

[41] M. L. Martin, B. P. Somerday, R. O. Ritchie, P. Sofronis, I. M. Robertson, Acta Mater. 60 (2012): 2739–2745.

[42] R. A. Oriani, Acta Metall. 18 (1970): 147–157.

[43] W. Qin, J. A. Szpunar, Philos. Mag. 97 (2017): 3296–3316.

[44] H. Su, T. Yoshimura, H. Toda, S. Bhuiyan, K. Uesugi, A. Takeuchi, et al., Metall. Mater. Trans. A. 47 (2016): 6077–6089.

[45] Q. G. Wang, C. H. Caceres, J. R. Griffiths, Metall. Mater. Trans. A. 34 (2003): 2901–2912.

[46] H. G. Lee, J. Y. Lee, Acta Metall. 32 (1984): 131–136

[47] H. Gao, H. Su, K. Shimizu, C. Kadokawa, H. Toda, Y. Terada, et al., Mater. Trans. 59 (2018): 1532–1535.

[48] C. G. Parker, D. P. Field, Springer, Cham(2012): 383–386.

[49] A. Bendo, K. Matsuda, S. Lee, K. Nishimura, N. Nunomura, H. Toda, et al., J. Mater. Sci. 53 (2018): 4598–4611.

[50] W. D. Callister Jr., D. G. Rethvoisch, Chapter 4: Imperfections in Solids in "Ma‐terials Science and Engineering: An Introduction, 8th Edition", John Wiley & Sons, Inc. (2010): 92–95.

[51] R. Kirchheim, B. Somerday, P. Sofronis, Acta Mater. 99 (2015): 87–98.

[52] Y. Huang, S. Qu, K. C. Hwang, M. Li, H. Gao, Int. J. Plast. 20 (2004): 753–782.

[53] M. F Ashby, Phil Mag. 21(1970): 399–424.

[54] S. Brinckmann, T. Siegmund, Y. Huang, Int. J. Plast. 22 (2006): 1784–1797.

[55] T. Ohashi, M. Kawamukai, H. Zbib, Int. J. Plast. 23 (2007): 897–914.

[56] E. Martinez-Paneda, C. F. Niordson, R. P. Gangloff, Acta Mater. 117 (2016): 321–332.

[57] E. Clouet, Acta Mater. 54 (2006): 3543–3552.

[58] M. Militzer, W. P. Sun, J. J. Jonas, Acta Metall. Mater. 42 (1994): 133–141.

[59] T. Enomoto, R. Matsumoto, S. Taketomi, N. Miyazaki, Zair. Soc. Mater. Sci. Japan. 59 (2010): 596–603.

[60] W. Witzel, Z. Metallkd. 64 (1973): 585-589.

[61] K. Carling, G. Wahnström, T. R. Mattsson, A. E. Mattsson, N. Sandberg, G. Grimvall, Phys. Rev. Lett. 85 (2000): 3862-3865.

[62] F. G. Wei, K. Tsuzaki, Metall. Mater. Trans. A. 37A (2006): 331-353.

[63] J. Takahashi, K. Kawakami, Y. Kobayashi. Acta Mater. 153(2018): 193-204.

[64] A. Nagao, M. L. Martin, M. Dadfarnia, P. Sofronis, I. M. Robertson, Acta Mater. 74 (2014): 244-254.

[65] A. Nagao, C. D. Smith, M. Dadfarnia, P. Sofronis, I. M. Robertson, Acta Mater. 60 (2012): 5182-5189.

[66] A. Nagao, M. Dadfarnia, B. P. Somerday, P. Sofronis, R. O. Ritchie, J. Mech. Phys. Solids. 112 (2018): 403-430.

本论文原发表于《Acta Materialia》2019 年第 176 卷。

Failure Analysis of Casing Dropping in Shale Oil Well during Large Scale Volume Fracturing

Wang Hang[1,2]　Zhao Wenlong[3]　Shu Zhenhui[3]
Zhao Qiang[3]　Han Lihong[1,2]

(1. Tubular Goods Research Institute of CNPC; 2. State Key Laboratory of Performance and Structural Safety for Petroleum Tubular Goods and Equipment Materials; 3. Xinjiang Oil Field Company of CNPC, Kelamayi 834000)

Abstract: Shale oil and gas are regarded as typical unconventional reservoir, their recovery often impedes by casing damage. Failure analysis is performed for casing dropping in shale oil well via optical microscope(OM), scanning electron microscope(SEM), full size test and finite element modeling. Morphology observation reveals that thread remains intact under as-received condition after make up and break out. Meanwhile, galling and improper make-up position occur for slipping thread. Full size test exhibits that strength of connection declines from 783.4Kips to 697.1Kips after 9th stage fracturing, with loss of 11.0%. Finite element modeling demonstrates that contact stress reaches to high values at two ends and low values at middle part in the pin and box threads under complex stress. Casing dropping can be attributed to insufficient strength of connection due to improper make-up position, which accelerates by fracturing operation. Finally, dynamic damage effect is suggested to introduce in design of casing string in shale oil well.

Keywords: Casing damage; Galling; Connection strength; Fracturing operation; Shale oil

1 Introduction

Shale oil and gas belong to strategic energy, revolution of shale gas has succeeded in North America[1,2] using horizontal well and multi-stage fracturing. Currently, output gets into rapid growth in China, such as shale gas in South-West area, shale oil in Xinjiang area[3-5]. Because of rigorous work condition, such as high pressure, large displacement and multi-stage fracturing, geo-stress field alters notably around borehole. As a result, casing damage occurs frequently[6-9].

Corresponding author: Wang Hang, wanghang008@cnpc.com.cn; Shu Zhenhui, shuzh@petrochina.com.cn; Han Lihong, hanlihong@cnpc.com.cn.

Statistical data exhibits that there are 11 horizontal wells in total 24 ones and 23 stages in total 596 ones for casing deformation in Xinjiang area of China[5]. The same problem exists in North America, 32 wells in total 62 ones in shale gas in Marcellus oil field of USA. In addition, 28 wells in Utica shale reservoir in Quebec [10], 11 wells in total 14 ones, with 19 deformation points in Duvernay shale reservoir of Canada [11].

Although many works have been done on casing damage by far[12-17], little focuses on thread slipping during multi-stage fracturing operation[18-20]. The objective of this paper is to discover the failure mechanism of casing dropping in shale oil well using large scale volume fracturing.

2 Engineering background

An abnormal fluctuation occurs in operation curve during 10th stage fracturing. Operation parameters include displacement of $10m^3$/min and pump pressure of 70.0MPa. Casing damage is identified after tripping at 885~890m, as shown in Fig. 1. The structure of shale oil well is vertical-like borehole, three-open configuration, well depth of 3793.9m and cementing up to 3255.0m. The steel grade of casing is P110, specification of (139.7×10.54)mm and API LC thread. Field fracturing operates in the form of fine layered and large scale volume injection.

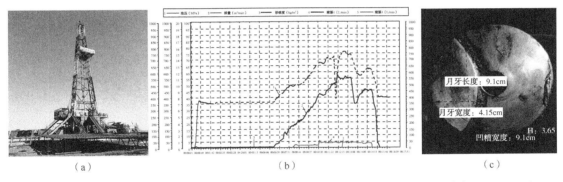

Fig. 1 Failure of casing dropping: (a) field site, (b) 10th stage fracturing curveand, (c) casing stamping

3 Property of materials

Chemical composition and mechanical property meet the requirement of standard of API Specification 5CT for dropping casing, including chemical content tensile strength and impact toughness, as shown in Table 1-3.

Table 1 Chemical composition of dropping casing with steel grade of 110ksi

Element	C	Si	Mn	P	S	Cr	Mo	Ni	Nb	V	Ti	Cu
Body	0.27	0.26	1.32	0.014	0.012	0.13	0.0054	0.01	0.001	0.0038	0.0017	0.011
Coupling	0.26	0.26	0.96	0.017	0.0055	0.96	0.14	0.013	0.0016	0.007	0.003	0.009
API 5CT	—	—	—	≤0.03	≤0.03	—	—	—	—	—	—	—

Table 2 Tensile property of dropping casing with steel gradeof 110ksi

Specimen	Size(mm)	Steel grade (ksi)	Tensile strength (MPa/MPa)	Yield strength (MPa)	Elongation (%)
Body	φ6.25×65.0	110	934.0	857.6	21.7
Coupling	φ6.25×65.0	110	1030.7	959.0	19.3
API 5CT		110	≥862.0	758.0~965.0	≥15.0

Table 3 Impact toughness of dropping casing with steel grade of 110ksi

Specimen	Size(mm)	Notch shape	Temper(℃)	Steel grade(ksi)	Impact energy (J)	Shear fracture ratio(%)
Body	10×10×55	V	0	110	96.7	100
API 5CT				110	≥41.0	—
Coupling	10×10×55	V	0	110	103.3	100
API 5CT				110	≥26.0	—

4 Thread detection and Non-destructive examination(NDE)

The parameters of pin and box threads are listed in Table 4 and Table 5. These results also meet the requirement of standard, based on API Specification 5B. NDE results reveal that there are a lot of deformation traces in the middle part of slipping thread. Meanwhile, a number of pits appear in the local region of thread without slipping, which is salvaged from downhole, as shown in Fig. 2.

Table 4 Parameters of pin thread as-received condition

Type	Close distance (mm)	Taper (mm/m)	Screw pitch deviation(mm/in)	Thread height deviation(mm)	L_4(mm)	D (mm)
Test	1.085	64.0	-0.01	+0.01	88.80	141.00
API 5B	-3.09~+3.27	59.9~67.7	-0.08~+0.08	-0.10~+0.05	82.55~88.91	139.00~141.10

Table 5 Parameters of coupling thread as-received condition

Type	Close distance (mm)	Taper (mm/m)	Screw pitch deviation(mm/in)	Thread height deviation(mm)	Q(mm)	q(mm)
Test	10.2	65.0	0	0	142.56	12.88
API 5B	6.31~12.67	59.9~67.7	-0.08~+0.08	-0.10~+0.05	142.0~143.66	12.70~13.49

Fig. 2　Non-destructive examination using magnetic powder: (a) slipping thread, (b) no slipping thread

5　Morphology observation

The typical morphology is characterized by galling for slipping thread, as shown by dotted box in Fig. 3(a). Based on macroscopic feature, these regions can be divided into three parts, i. e., initiation segment from 1st to 8th thread, intermediate segment from 9th to 18th thread and tail segment from 19th to 26th thread. On a close examination, indentation from clamps is observed in outside surface of coupling neighbor to field-end, as shown by arrow in Fig. 3(b).

Fig. 3　The macroscopic morphology of dropping casing: (a) pin thread, (b) corresponding coupling

Specimens are cut along axial direction and examined using optical microscopy. Metallographic observations reveal that deformation traces occur in the top of thread in initiation segment, as shown by arrow in Fig. 4(a). Meanwhile, some of these threads collapse evidently in the intermediate segment, as shown in Fig. 4(b). However, deformation trace does not even appears in the tail segment, as shown in Fig. 4(c). These evidences indicate that make-up position is not in place for slipping thread, its make-up position only reaches to the intermediate segment. SEM observations reveal that there are obvious friction scratches in the initiation segment, as shown in Fig. 5. It is seemed that these scratches are produced by interference contact between threads during make-up operation. However, this type of scratch does not presents in the tail segment, as shown in Fig. 6. These SEM results verify that make－up position is actually not in place for slipping thread. Chemical composition is analyzed for surface residue using energy disperse spectrum(EDS), which consists of K, Ca, Na, Mn, S, Cl, O, and so on, as shown in Fig. 7.

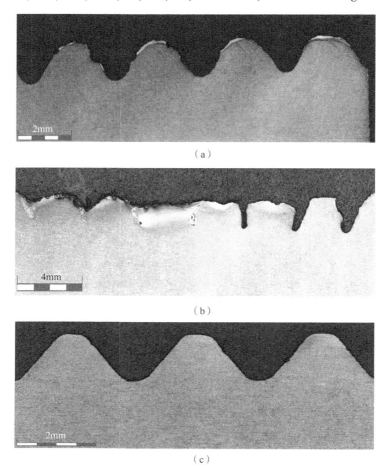

Fig. 4　The morphology of slipping thread indifferent regions:
(a)initiation(b)intermediate and(c)tail part

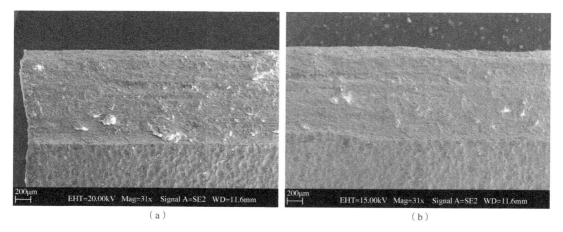

Fig. 5　The morphology feature of guide surface in the 3rd thread: (a)left-side and(b)middle region

Fig. 6 The morphology feature of guide surface in the 21th thread:
(a) middle region and (b) left-side one

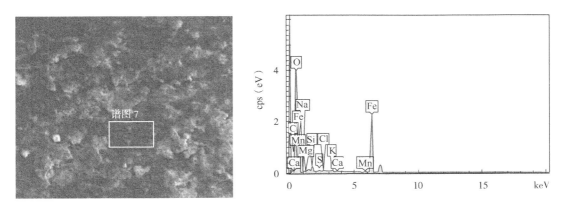

Fig. 7 EDS analysis of the residue in guide surface of the 3rd thread

6 Full scale test

According to standard of API Specification 5B, make up and break out tests are performed to evaluate galling resistance property of thread under as-received condition. The macroscopic morphology characteristics indicate that thread remains intact after three times make-up and two times break-out, as shown in Fig. 8. The morphology of thread is examined after 9th stage fracturing in downhole, as shown in Fig. 9. The box thread remains intact basically, except a little of mechanical knocking in local region. While damage appears in the form of pit and deformation for the pin thread. Tensile to failure tests are employed to evaluate strength of connection of casing thread with different conditions. The experimental results show that strength of connection is 783.4kips for thread under as-received condition, i.e. 3484.5kN, as shown in Fig. 10. After 9th stage fracturing, this strength is 697.1kips, i.e. 3100.7kN, as shown in Fig. 11. These results demonstrate that operation of multi-stage fracturing reduces strength of connection remarkably for thread with proper make-up position.

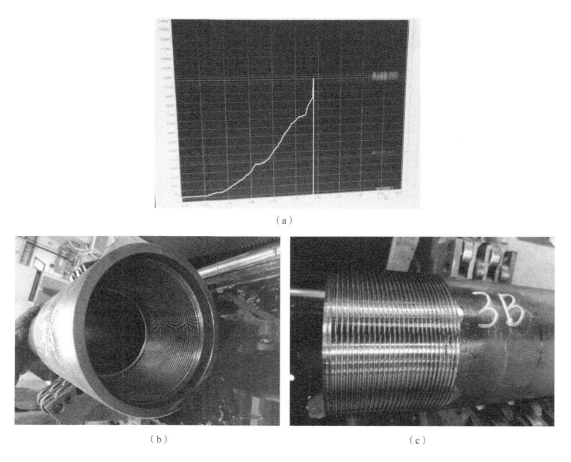

Fig. 8 The torque vs time curve(a) and morphology of box(b) and (c) pin threads after galling resistance test

Fig. 9 The morphology of thread without slipping after 9th stage fracturing: (a) box and (b) pin threads

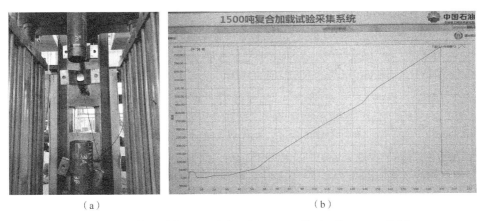

Fig. 10 Tensile to failure test(a) and correspondingloading curve(b) for casing thread as-received condition

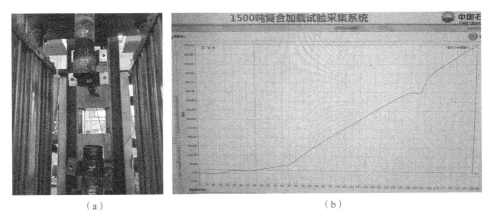

Fig. 11 Tensile to failure test(a) and corresponding loading curve(b) for casing after 9th stage fracturing

7 Finite element analysis

Stress distribution is analyzed for thread type with APILC using finite element modeling. The parameters of material include steel grade of P110, specification of $\phi(139.7\times10.54)$mm, elastic module of 210GPa, poisson ratio of 0.3, density of 7850kg/m^3, and yield strength of 857MPa. Simplification and assumption are as follows[21]: (1) using structure with axial symmetry and ignoring the factor of lead angle, (2) coefficient of friction is supposed to be 0.02, (3) two end of model is elongated with size of 1/3 thread length to eliminate the boundary effect.

Finite element model is showed in Fig. 12, as well assubdivision mesh and local magnification, of which unit number is 9434 with node number of 20935 for casing after subdivision[22]. Meanwhile, this unit number is 9775 with node number of 21614 for coupling. As for make-up simulation, connection torque is applied by means of altering the magnitude of interference between the pin and box threads. As for API LC with 8 thread, its magnitude of interference is 0.099mm along axial direction after machine tight with 1 cycle. Displacement constraint is applied for the top end of coupling in FE model. The load includes axial tension, internal pressure, and other stress.

Finite element analysis reveals that contact stress has high values at two ends and low values at middle part in the pin and box threads under make-up torque. As a whole, this stress level is higher in the pin thread than that in the box one, as shown in Fig. 13. Moreover, this stress distribution is analyzed under make-up torque, internal pressure and tensile stress, as shown in Fig. 14. Based on FE modeling, stress level increases obviously for coupling under internal pressure of 50MPa and three cycle machine tight, the maximum of contact stress is 580MPa in small-end interference thread of coupling. In addition, the maximum of Von Mises stress is 624MPa under 100MPa tension stress[20]. It is believed that contact compression stress is proportional with loading area of thread.

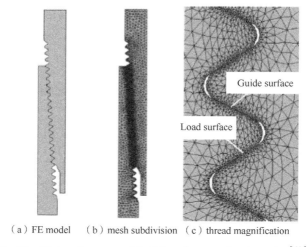

(a) FE model   (b) mesh subdivision   (c) thread magnification

Fig. 12   Finite element model of thread connection in casing[20]

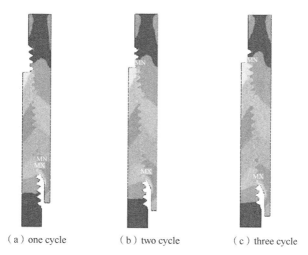

(a) one cycle   (b) two cycle   (c) three cycle

Fig. 13   Von-Mises equivalence stress in thread connection during make up[20]

Meanwhile, strength of connection is a function of loading area of thread. It means that contact stress has an important effect on strength of connection of casing thread. Therefore, strength of connection is much higher for thread with proper make-up position because of much larger loading area in the region of two ends.

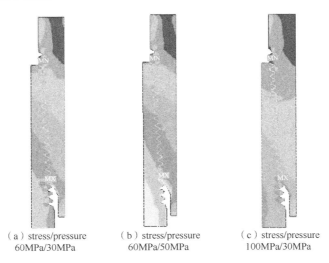

(a) stress/pressure 60MPa/30MPa

(b) stress/pressure 60MPa/50MPa

(c) stress/pressure 100MPa/30MPa

Fig. 14 Von-Mises stress of thread connection after 3 cycle machine tight under complex loading[20]

8 Discussion of results

Chemical content and mechanical property meet the requirement of standard for dropping casing, based on API specification 5CT. Therefore, material factor is not responsible for thread slipping in shale oil well during large scale volume fracturing.

Strength of connection is one of key parameters in design of casing string and maintaining structural integrity of casing string, which is associated with three aspects: galling resistance property, make-up position and fracturing operation. Make-up and break-out tests confirm that this thread type has a fine galling resistance property. Metallographic and SEM observation reveal that make-up position is not in place for slipping thread. On the basis of FEA, contact stress is high values at two ends and low values at middle part in the pin and box threads under complex load [20]. These FE modeling results indicate that strength of connection is higher for thread with proper make-up position due to larger loading area of thread. In addition, full size test exhibits that strength of connection decreases remarkably after 9th stage fracturing, with loss of 11.0%. Consequently, cause of failure is insufficient strength of connection for thread slipping.

Traditional design method of casing string involves in static loading for convention oil and gas wells, which includes hanging load and hole-wall friction. Multi-stage fracturing is used in shale oil well during recovery, casing undergoes tension, internal pressure and collapse in the region of horizontal well without cementing. In addition, there are much higher stress in casing thread due to stress concentration. In this case, plastic deformation would occur when stress reaches to yield strength. Therefore, service behavior can be characterized by low cycle fatigue during multi-stage fracturing, which leads to reduction in strength of connection continuously. This process could be also represent with dynamic damage effect. Based on analysis, it is speculated that degradation of property of material is dominant mechanism for dynamic damage effect.

The similar feature is also documented in thermal well[23]. Plastic deformation happens for thermal casing when injection steam with high temperature, which induces low cycle fatigue during multi-cycle operation of injection and recovery. As a result, plastic strain cumulates continuously and displays dynamic damage effect during this process. This dynamic damage effect finally induces casing damage when cumulative strain exceeds strain capacity in thermal well. Therefore, it can be seen that dynamic damage effect could be explained in terms of degradation of property of material.

9 Conclusions

(1) Chemical composition and mechanical property meet the requirement of standard of API Specification 5CT for dropping casing, including tensile strength and impact toughness, then material factor is not responsible for thread slipping.

(2) Casing dropping can be attributed to insufficient strength of connection due to improper make-up position, which induces loss of loading area of thread. In addition, reduction of strength of connection is accelerated by multi-stage fracturing in field operation.

(3) Dynamic damage effect has an important effect on reduction of strength of connection, this effect is induced by degradation of property of material. As a result, it is suggested to introduce dynamic damage effect in design of casing string in shale oil and gas well.

Acknowledgment

The authors gratefully acknowledge the financial support of CNPC science and technology development project(No. 2019B-4013).

References

[1] E. K. George, Thirty years of shale gas fracturing: What have we leaning? [C]//SPE Annual Technical Conference and Exhibition. Florence: Society of Petroleum Engineering, 2010.

[2] L. Wang, J. L. Ma, F. R. Su, et al. Shale gas factory fracturing technology in North America[J]. Drilling & Production Technology, 35(2012)48-50.

[3] C. J. Xue. Technical advance and development proposals of shale gas fracturing[J]. Petroleum Drilling Techniques, 39(2011)24-29.

[4] C. B. Yin, D. S. Ye, G. B. Duan, et al. Research about and application of autonomous staged fracturing technique series for horizontal well stimulation of shale gas reservoirs in the Sichuan Basin[J]. Natural Gas Industry, 34(2014)67-71.

[5] H. Liu, L. C. Kuang, G. X. Li, F. Wang, X. Jin, J. P. Tao, S. W. Meng. Considerations and suggestions on optimizing completion methods of continental shale oil in China[J]. Acta Petrolei Sinica, 41(2020)489-496.

[6] H. L. Zhang, Z. W. Chen, L. Shi, et al. Mechanism of how fluid passage formed and application in Sichuan shale gas casing deformation analysis[J]. Drilling & Production Technology, 41(2018)8-11.

[7] Q. Dai. Analysis of production casing damage during testing and completion of shale gas well[J]. Drilling & Production Technology, 38(2015)22-25.

[8] Y. Xi, J. Li, G. H. Liu, et al. Overview of casing deformation in multistage fracturing of shale gas horizontal wells[J]. Special Oil and Gas Reservoir, 26(2019)1-6.

[9] Z. L Tian, L. Shi, L. Qiao. Research of and countermeasure for wellbore integrity of shale gas horizontal well[J]. Natural Gas Industry, 35(2015)70-76.

[10] X. L. Guo, J. Li, G. H. Liu, et al. Research on casing deformation for shale gas wells based on focal mechanism[J]. Fault-Block Oil and Gas Field, 25(2018)665-669.

[11] Z. H. Lian, H. Yu, T. J. Lin, J. H. Guo. A study on casing deformation failure during multi-stage hydraulic fracturing for the stimulated reservoir volume of horizontal shale well[J]. Journal of Natural Gas Science and Engineering, 23(2015)538-546.

[12] Pablo. Cirimello, Jose. Luis Otegui, Alberto. Aguirre, Guillermo. Carfi. Undetected non-conformities in material processing led to a failure in a casing hanger during pre-fracture operation[J]. Engineering Failure Analysis, 104(2019)203-215.

[13] Y. Li, W. Liu, W. Yan, J. G. Deng, H. T. Li. Mechanism of casing failure during hydraulic fracturing: lessons learned from a tight-oil reservoir in china[J]. Engineering Failure Analysis, 98(2019)58-71.

[14] K. H. Deng, W. Y. Liu, T. G. Xia, D. Z. Zeng, M. Li, Y. H. Lin. Experimental study the collapse failure mechanism of cemented casing non-uniform load[J]. Engineering Failure Analysis, 73(2017)1-10.

[15] Pablo. G. Cirimello, Jose L. Otegui, Guillermo Carfi, Walter. Morris. Failure and integrity analysis of casing used for oil well drilling[J]. Engineering Failure Analysis, 75(2017)1-14.

[16] K. H. Deng, Y. H. Lin, W. Y. Liu, H. Li, D. Z. Zeng, Y. X. Sun. Experimental investigation of the failure mechanism of P110SS casing under opposed line load[J]. Engineering Failure Analysis, 65(2016)65-73.

[17] C. A. Cheatham, C. F. Acosta, D. P. Hess. Tests and analysis of secondary locking features in threaded inserts[J]. Engineering Failure Analysis, 16(2009)39-57.

[18] Satoshi Izumi, Takashi Yokoyama, Atsushi Iwasaki, Shinsuke Sakai. Three-dimensional finite element analysis of tightening and loosening mechanism of threaded fastener[J]. Engineering Failure Analysis, 12(2005)604-615.

[19] M. Y. Zhang, D. F. Zeng, L. T. Lu, Y. B. Zhang, J. Wang, J. M. Xu. Finite element modeling and experimental validation of bolt loosening due to thread wear under transverse cyclic loading[J]. Engineering Failure Analysis, 104(2019)341-353.

[20] G. J. Yu, X. Q. Chen, A. Q. Duan. Finite element analysis of casing thread connection under impact load[J]. China Petroleum Machinery, 45(2017)14-20.

[21] Z. G. Wang, Y. Zhang. Finite element analysis for stress in tubing connection with API LC under machine tight and tension[J]. Steel Pipe, 30(2001)20-25.

[22] J. T. Xi, G. Nie, X. S. Mei. Finite element analysis of contact of connection property for casing thread connection[J]. Journal of Xi'an Jiao Tong University, 33(1999)63-66.

[23] L. H. Han, H. Wang, J. J. Wang, B. Xie, Z. H. Tian, X. R. Wu. Strain-based casing string for cyclic steam stimulation well[J]. SPE PRODUCTION & OPERATIONS, 33(2018)409-418.

本论文原发表于《Engineering Failure Analysis》2020年第118卷。

A Comprehensive Analysis on the Longitudinal Fracture in the Tool Joints of Drill Pipes

Wang Xinhu   Li Fangpo   Liu Yonggang   Feng Yaorong   Zhu Lijuan

(Tubular Goods Research Institute of China National Petroleum Corporation; State Key Laboratory for Performance and Structural Safety of Petroleum Tubular Goods and Equipment Materials)

Abstract: The Longitudinal fractures or splits in tool joint box of drillpipe often occur, because the numbers of deep, directional, extended and horizontal oil wells are increasing. The mechanism of frictional heat check cracking of drill pipe tool joints was discussed in the standard API RP 7G. However, authors have identified that heat check cracking was just one of the cracking initiation mechanism, not crack propagation mechanism. This paper has reviewed 21 cases of this kind of failure analyzed by authors from year 2000 to 2015. Fracture surfaces and mechanical properties have been examined in this paper. It was found that there were various other causes of crack initiation, in addition to frictional heat check cracking, such as tong tooth bite marks and friction damage in internal threads. Such cracks propagated mostly via stress corrosion cracking (SCC) mechanism, although two cases was brittle cracking due to poor material toughness. The stress corrosion mechanism was related to hydrogen sulphide (H_2S) or ionic sulphur (S^{2-}), which came from the degradation of applied thread greases or drilling fluid ingredients. Although heat check cracks and tooth bite marks as crack initiation are common-place, the failure can be prevented through prevention of crack propagation. Failure data showed that although improved transverse material absorbed energy avoided longitudinal brittle cracking of the tool joint, stress corrosion cracking still occurred. The statistical analysis results showed that material hardness was related to the longitudinal cracking of drillpipe tool joints. It was demonstrated that if material hardness was restricted to less than HB310, the crack did not propagate. It was suggested that the material hardness of tool joints box should be revised to HB285~HB310, in addition that the transverse material Charpy absorbed energy of the tool joint should be specified.

Key words: drill pipe; tool joint; crack; fracture; prevention

1 Introduction

Drill pipe is one kind of important tools for drilling in oil and gas field. The drill pipe's failure

---

Corresponding author: Wang Xinhu, Tel.: + 86 29 81887667; Fax: + 86 29 81887661; E-mail: wangxinhu002@ cnpc. com. cn.

often occurs in the oil and gas field because drill pipe bears continuously changeable tension load, bend load, torsion load, impact load, internal pressure and catastrophic downhole corrosion. Although the wash out caused by fatigue is most popular type of drill pipe failure[1-3], the longitudinal fracture or split of the tool joint box is also one of main failure type of drill pipe[3,4]. The standard API RP 7G[5] explains that friction heating and drilling liquid quenching frequently resulted in the cracking through the tool joint box of drill pipe, and controlling hole angle and lateral force is able to minimize or eliminate the longitudinal fracture of the joint box tool. However, the lateral force on the drill pipe is inevitable during drilling operation. Sometimes, although the lateral force was below the load capacity of the tool joints, the split of tool joints box still occurred. There are growing interests in preventing this kind of failure, and one effective way is to enhance the transversal Charpy absorbed energy of the tool joint box. Xinhu Wang proposed the transversal Charpy absorbed energy requirement[6], which had been accepted by Chinese standard SY/T 5561[7]. However, the longitudinal fracture of tool joints still occurred occasionally. Therefore, this paper has analyzed 21 longitudinal cracking failures of the tool joint box in detail to find the failure mechanism, and prevention measures.

2 Analysis methods and materials

The 21 failure cases in this paper come from the cases analyzed by us since year 2000. Because our research institute (Tubular Goods Research Institute of China National Petroleum Corporation) is the Oil Tubular Goods Failure Analysis Center of Chinese Petroleum Society, many failure analyses were finished every year. The failure analysis reports were submitted to the clients, and kept in the data base. There are 21 longitudinal cracking failure cases of the tool joint box in this data base. All information such as the fracture analysis results, the metallographic analysis results, the material Charpy absorbed energy testing results, hardness testing results, and failure happen time in this paper were from the failure reports in this data base.

Tool joint material is mainly 35CrMnMo steel in China. Table 1 shows the steel chemical composition.

Table 1 The chemical composition of 35CrMnMo steel (Weight%)

Elements	C	Si	Cr	Mn	Mo	P	S
Content	0.32~0.38	0.15~0.35	0.90~1.20	0.85~1.00	0.28~0.35	≤0.015	≤0.008

3 The crack initiation

Among the 21 failures, there were three kind of cracking initiation, which caused the split of tool joint box. The first one is the friction heat cracks on outside surface of tool joint box. The second one is mechanical damage on outside surface of tool joint box. In addition, the third one is the inside thread teeth damaged.

It was realized that the friction heat cracking was main cause which results in longitudinal cracking through tool joint box in the standard API RP 7G[5]. A typical appearance of the longitudinal heat checking through tool joint is shown in Figure 1. The tool joint was worn gleamed, and there were many friction heat cracks on the outside surface. Drillpipe tool joints bore high lateral force and large friction load due to the serious friction in deep well and directional well. In addition, the friction energy could be transformed into heat energy. Once the heated tool joints suffer fast cooling process in drilling liquid, brittle martensite phase would form on the surface of the tool joints. Cracks in the martensite structure could be initiated during thermal cycles in service.

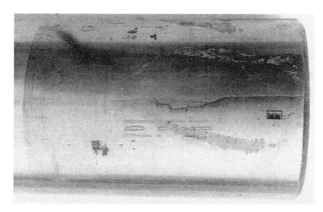

Fig. 1 The friction heat crack appearance of the box tool joint

The mechanical damage on the outer surface is another cracking initiation that results in longitudinal fracture through the tool joint box of drillpipe. Among the 21 longitudinal fracture cases analyzed by the authors, five failures were induced by the friction heat cracking, and six failures were induced by the mechanical damage on outside surface. The tong tooth bite marks was one type of the mechanical damage. Figure 2(a) shows many bite marks and a long longitudinal crack. Figure 2(b) shows another tool joint box, which had many tool bite marks, and fractured at one of bite marks. Individual drill pipe must be connected into drill string by tong in order to drilling well. Tong is the tool of tightening the screws of tool joint (making up tool joint), the clamping part of the tong have many teeth. When holding and making up or breakout the tool joint by a set tong, its teeth will bite the outer surface of tool joint, then the bite marks formed on outer surface of the tool joint box. Due to the effect of stress concentration, cracks initiated from the location of bite marks.

(a)Tooth bite marks and longitudinal crack on tool joint box   (b)Tooth bite marks and fracture at one of bite marks

Fig. 2 The fracture appearance of the tool joints and the tong tooth bite marks on its outer surface

Among the 21 failure cases analyzed by the authors from 2000 to 2015, 10 failure cases were induced by the cracks initiated from inside thread teeth damaged. As the example displayed in Figure 3, the chevron marks on the fracture point back to the thread teeth (the arrows point to the crack origin in Figure 3(b)) and the crack extends toward outer surface of tool joint. Figure 3(c) is the fracture of the tooth pointed by arrows in Figure 3(b), the tooth was crack initiation location, and its crest was damaged. Figure 3(d) and Figure 3(e) are fractographic pictures of the crack origin and its neighboring location in Figure 3(b) and Figure 3(c), cracked white martensite on the surface was fracture initiation. A large crack propagated through the wall thickness as displayed in Figure 3(e), the crack is parallel with the fracture in Figure 3(b) and Figure 3(c). Because friction between pin and box while making up and breaking out of the tool joint, heating and cooling happened at the surface of threads, and a layer of martensite formed in the surface material of threads. Plenty of cracks formed in brittle martensite as displayed in Figure 3(d). Some cracks go into the material structure of wall thickness and propagating to outer surface of tool joint because maximum local stresses is near the root of thread teeth as displayed in Figure 3(e).

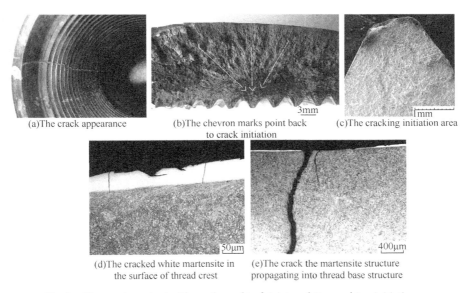

Fig. 3 The crack at the inside surface of tool joint and its cracking initiation

4 Failure Mechanism

The friction between the tool joints of drill pipe and the wall of well hole is inevitable, especially in the directional wells, extended wells and horizontal wells. The scratches and manufacturing defects, and tong tooth bit marks at the outer surface of the tool joints are the potential initiation of cracks. The thread teeth are a potential initiation area of cracks due to friction-induced martensite. Although it is hard to avoid these crack initiation such as heat check cracking, the split failure of tool joints can be eliminated if the further propagation of the crack can be stopped.

The mechanism of the friction heat cracking in the tool joint of drill pipe was discussed in the standard API RP 7G. Authors realized that heat checking was just one of cracking initiation mechanism, but not the mechanism of cracks propagation. Chevron marks were always observed on the fracture surface as displayed in Figure 3 among the 21 failure cases, and these fracture mechanism was brittle fracture. It was well known that higher fracture toughness could prevent the brittle cracking. In order to improve tool joints' reliability, the longitudinal material Charpy absorb energy requirement of tool joint had be put forward in API SPEC 5DP[8]. However, tool joint's longitudinal mechanical property is different from its transversal mechanical property due to steel rolling. Longitudinal Charpy sample's fracture surface is perpendicular to longitudinal axis of tool joint while Transversal Charpy sample's fracture surface is parallel with longitudinal axis of tool joint, higher longitudinal toughness has little effect on preventing the longitudinal cracking of tool joint box. Because longitudinal fracture also is mainly parallel with longitudinal axis of tool joint, higher transversal Charpy absorbed energy is more helpful for preventing longitudinal cracking of the tool joint box. Effective way is to enhance the transversal Charpy absorb energy of the tool joint box. Xinhu Wang put forward the transversal toughness criterion[6], which had been accepted by Chinese petroleum standard of SY/T 5561[7]. The SY/T 5561 is the only standard that introduced the transversal Charpy absorb energy requirement in the world at present. The Charpy absorb energy of the 21 failure tool joint box was shown in Figure 4, and the serial number of the tool joints was in sequence with the failure occurrence date. The size of Charpy impact absorbed energy testing samples were 10mm×10mm×55mm. It was indicated that the transversal toughness of the tool joints has been improved significantly since the year 2008.

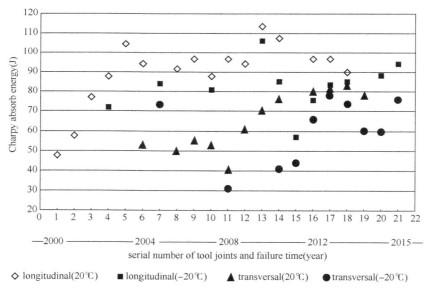

Fig. 4 The Charpy absorbed energy of 21 cracked tool joint box

As shown in Figure 4, the longitudinal Charpy absorbed energy of the No. 1 tool joint box was only 48J. The fracture of the No. 1 was caused by material brittleness. Unfortunately, the transversal Charpy energy was not tested for first five tool joints. Although longitudinal toughness criterion of the tool joint was introduced into the API SPEC 5DP, the transversal Charpy absorbed energy of tool joints box was still quite low. For example, the transversal Charpy absorbed energy of the No. 11 tool joint was only 40J while it's longitudinal Charpy absorbed energy was 90J. Therefore, the transversal Charpy absorbed energy requirement was specified in SY/T 5561 in that the requirement of minimum average absorbed energy for a set of three transverse specimens was 60J at −20℃, and minimum absorbed energy for one of three transverse specimens was 50J at −20℃. The transversal Charpy absorbed energy of the box was greatly improved since then. This is an effective measurement preventing from the brittleness cracking of the tool joint box. Since then, the fracture numbers of tool joint box decreased obviously.

However, the longitudinal cracking through tool joint box still occurred occasionally, even when the transversal material Charpy absorbed energy of tool joints exceeded 70J at −20℃ (Figure 4). Therefore, it was impossible to eliminate the fracture of tool joints only by improving the material toughness.

Among the 21 failure cases, the fracture features of most tool joints were observed by scanning electron microscopy (SEM) with energy dispersive X-ray spectrometer (EDX). It indicated that most fractures were intergranular, the typical feature was showed in Figure 5. This was in accordance with metallographic analysis results, typical intergranular cracks were showed in Figure 6. The transversal Charpy absorbed energy of the tool joints was high enough to prevent it from brittle cracking except for No. 1 and No. 11 tool joints. The transversal Charpy absorbed energy of some tool joints even exceeded 70J. Therefore, it was realized that the stress corrosion crack was the main factor, which induced cracking in the tool joint box with high transversal Charpy absorbed energy. Some tool joints of drill pipe suffered catastrophic corrosion, and it is difficult to distinguish the cracking features.

Fig. 5 The longitudinal fracture appearance of the tool joint

The corrosion products on the fracture surface of tool joints were analysed by EDS. The elemental sulfur was detected on cracking initiation area of the each fracture sample, such as crack

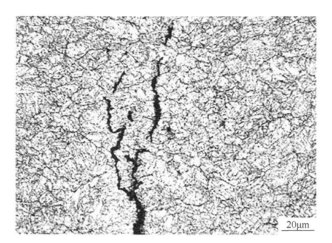

Fig. 6 Optical micrograph of stress corrosion cracks in the tool joint

in bite mark, crack in thread roots, heat cracks. Elemental sulfur in the corrosion products usually comes from the tensile stress[9]. The complicate and rigorous service conditions promote the development of high-strength steel drill pipe. However, the tool joint material which yield strength is 120ksi is one kind of sulfide stress corrosion cracking sensitive material[9-11]. Moreover, the stress corrosion of high-strength steels in the H_2S environment was an old problem in the oil field and results in big economic loss[9]. Thus, it was significant to decrease the stress corrosion cracking susceptibility of tool joints by optimizing material quality.

Many investigations have been conducted on improving the stress corrosion crack resistance of tool joint. The stress corrosion crack resistance of the tool joints depends on the chemical composition, microstructure and mechanical properties of the materials. There are three measures to improve the stress corrosion resistance of tool joints[9]: (1) Reduce the impurity elements contents, especially the sulfur element content. The metallurgy technology has been significantly improved in most of the steel mills in China, the steel purity has reached the advanced world level. The chemical composition and impurity contents are conform to the standard requirements. (2) Maintain the homogeneous distribution of the tempered sorbite, which is the optimum microstructure. The grain sizes of the steel have exceeded a magnitude 8 in most steel mills in China. However, the control of composition segregation and banded structure is an old problem, which is a main gap between the iron and steel industry in China and abroad. (3) Improve the material toughness of the tool joints. The drillpipe failure assessment demonstrates that the brittle split can be avoided when the transverse Charpy absorbed energy of the material is over 60J. The transversal Charpy absorbed energy for the tool joint box has already come up to 60J at the current levels.

In fact, it is a tough work to evaluate the sulfide stress corrosion cracking resistance of material from the above three aspects in the oil field. Therefore, the hardness was put forward to be a comprehensive index to measure the sulfide stress corrosion cracking resistance of material. The lower is the hardness, the better is the sulfide stress corrosion cracking resistance of the steel material[9]. Generally, the steel material which hardness is lower than HRC22 does not suffer sulfide stress corrosion cracking[9-11]. The hardness of the tool joints is HB285 ~ 341 (about HRC29. 8 ~

33.3) in API SPEC 5DP and SY/T5561, that is much higher than HRC22. Therefore, the tool joint is prone to suffer stress corrosion crack. It is impossible to decrease the hardness of the tool joints to HRC22. However, higher stress corrosion crack resistance can be obtained by comprehensive materials quality optimization even when the hardness of the tool joint is above 22HRC.

The hardness of sixteen cracked box tool joints was shown in Figure 7. All of them are above HB 310, and five of them are above HB 340. Few tool joints with the hardness lower than HB 310 suffered stress corrosion cracking. Therefore, HB310 is probably critical hardness for the tool joints of drill pipe with good stress corrosion cracking resistance. Authors suggest that the hardness of tool joints is below HB310 should be the demand of good stress corrosion cracking resistance, and the hardness of tool joints should be HB285~HB310.

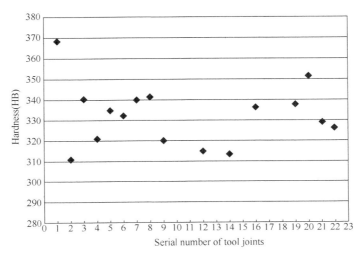

Fig. 7 Hardness of the longitudinal cracked tool joints

5 Conclusions

API RP 7G realizes that the cause of longitudinal fracture in the tool joints box of drillpipe is friction heat checking. This paper analyzed 21 longitudinal fractures in tool joints box in detail. The results show that the friction heat check cracking, tong tooth bite marks, and friction damage thread teeth are the potential cracking initiation of tool joints box. Heat check cracking is just one of the crack initiation mechanisms, not the mechanism of the crack propagation. The propagation mechanism of crack is mainly stress corrosion cracking, and sulfide is one of the mainly corrosive mediums. This paper suggests that the material hardness below HB310 is good measurement for preventing the split through the tool joints box of drill pipe. It is suggested that the material hardness of drillpipe tool joints should be revised to HB285~HB310 in the API SPEC 5DP and SY/T5561, in addition that the transverse material Charpy absorbed energy of the tool joint box should be required.

Acknowledgement

National Science and Technology Projects of China sponsor this work under Grant No. 2011ZX05021-002.

References

[1] Lu S L, Feng Y R, Luo F Q, et al. Failure analysis of IEU drill pipe wash out [J]. International Journal of Fatigue, vol. 27 (2005): 1360-1365.

[2] Li F, Liu Y G, Wang X H, et al. Failure analysis of ϕ127mm IEU G105 drill pipe wash out [J]. Engineering Failure Analysis, vol. 18 (2011): 1867-1872.

[3] Sajad Mohammad Zamani, Sayed Ali Hassanzadeh-Tabrizi, Hassan Sharifi. Failure analysis of drill pipe: A review [J]. Engineering Failure Analysis, vol. 59 (2016): 605-623.

[4] Liu W, Liu Y, Chen W, et al. Longitudinal crack failure analysis of box of S135 tool joint in ultra-deep well [J]. Engineering Failure Analysis, vol. 48 (2015): 283-296.

[5] API RP 7G[S]. Drill stem design and operating limits. 1998.

[6] Wang X H, Xue J J, Gao R, et al. The cracking mechanism and material index of drill pipes [J]. Natural Gas Industry. vol. 27 (2007): 69-71. (In Chinese)

[7] SY/T 5561[S]. Friction Welding Drill Pipe (Chinese standard), 2008.

[8] API SPEC 5DP. Specification for drill pipe. 2009.

[9] Feng Y R, Li H L. Hydrogen-Induced Stress Corrosion of Drill Stem Elements [J]. Corrosion Science and Protection Technology, vol. 12 (2000): 57-59 (In Chinese)

[10] S. M. C. Souza, E. H. de S. Cavalcanti. Failure Analysis in Heavy-Weight Drillpipes during Drilling Operation of Off-Shore Oil Well[C]. Corrosion, 2003: 03524.

[11] Roberto Villalba, Yurmuary Díaz, Jesús Rafael Peñalver, Bad Practices in Drilling Operation Promoting Corrosion Failures Under High Stress[C]. Corrosion, 2005: 05116.

本文原发表于《Engineering Failure Analysis》2017 年第 79 期。

Effects of Chloride Concentration on CO_2 Corrosion of Novel 3Cr2Al Steel in Simulated Oil and Gas Well Environments

Tan Chengtong[1]   Xu Xiuqing[2]   Xu Lining[1]   Yin Chengxian[2]   Qiao Lijie[1]

(1. Corrosion and Protection Center, Key Laboratory for Environmental Fracture (MOE), Institute for Advanced Materials and Technology, University of Science and Technology Beijing; 2. State Key Laboratory of Performance and Structural Safety for Petroleum Tubular Goods and Equipment Materials, CNPC Tubular Goods Research Institute)

Abstract: The corrosion behavior of novel 3Cr2Al steel in simulated oil and gas well environments was investigated using scanning electron microscopy, electron diffraction spectroscopy, potentiodynamic polarization, and potentiostatic polarization. The results revealed that the corrosion rate of 3Cr2Al steel increased as the Cl^- concentration increased. The thickness of the corrosion product film decreased as the Cl^- concentration increased. The increase in Cl^- concentration caused a decrease in pH value, subsequently causing an increase in the solubility of $Cr(OH)_3$ and $Al(OH)_3$ and hindering the deposition of corrosion product, eventually yielding a thinner corrosion film.

Keywords: 3Cr2Al steel; Chloride concentration; CO_2 Corrosion; Corrosion product film; pH value

1 Introduction

In the oil and gas industry, CO_2 corrosion has caused accidents and economic losses[1,2]. CO_2 corrosion is a complex process, which is affected by many factors[3]. When operation temperature increases, CO_2 corrosion rate changes[4,5], as does the chemical composition of the corrosion product film[6]. When CO_2 partial pressure and solution velocity increase[7-10], the CO_2 corrosion rate increases[11-13]. Moreover, when the Cr content in steel substrate increases to 3%, the Cr element may precipitate in the corrosion product film in the form of $Cr(OH)_3$[14]. The Cr enrichment in the corrosion product film can enhance resistance to CO_2 corrosion and suppress pitting corrosion.[15] In oil and gas wells, the occurrence of CO_2 is associated with the formation water produced from reservoirs. Formation water usually contains

---

Corresponding author: Xu Xiuqing, xuxiuqing@cnpc.com.cn; Xu Lining, xulining@ustb.edu.cn.

dissolved salts at a high concentration. Therefore, it is crucial to ascertain whether dissolved salt affects CO_2 corrosion.

Hausler et al.[16,17] demonstrated that Cl^- had no noteworthy effect on CO_2 corrosion. However, Eliyan[18] revealed that corrosion reactions exhibited a special trend with chloride concentration, as they increased and later decreased due to the thickening of corrosion products. Liu[19] found that for carbon steel, Cl^- had no effect on the composition of the corrosion product. Cl^- could destroy the corrosion product film and change its morphology at a CO_2 partial pressure of 20 bar and temperature of 100℃. Ma et al.[20] studied the corrosion of low carbon steel in atmospheric environments of different chloride content and found that a high amount of Cl^- could accelerate the production of β-FeOOH. Schmitt et al[21,22]. demonstrated that increasing Cl^- content could suppress CO_2 corrosion by decreasing the solubility of CO_2 in the solution at room temperature. Liu et al.[23] found that a threshold value of chloride existed for Cr modified steel in a saturated $Ca(OH)_3$ solution. The threshold value of chloride increased with the Cr content in the substrate.

Wang et al.[24] studied Cr and Al alloy steel in a 3.5wt% NaCl solution and found that when an Al element was added to a low-Cr steel, the CO_2 corrosion resistance was strengthened. However, how the Cl^- affects the CO_2 corrosion of Cr and Al alloy steel is not clearly understood. In the present study, 3Cr2Al steel was manufactured and exposed to three Cl^- concentration solutions (0.07, 0.3, and 0.5mol/L). To provide new insights, we demonstrated the effect of Cl^- saturated with CO_2 on the corrosion mechanism of Cr and Al alloy steel by means of electrochemical techniques and an immersion test.

2 Experimental

2.1 Materials and solution

The material used was 3Cr2Al steel; its chemical composition (wt%) is presented in Table 1. The microstructure of the 3Cr2Al steel was ferrite and pearlite, as shown in Figure 1. The specimens for electrochemical measurements were $10 \times 10 \times 3mm^3$ in size. The specimens for weight loss test were arc shaped specimens of 1/8 circle, which were machined with dimensions of 87mm in diameter and 10mm in width (as shown in Figure 2). Prior to each experiment, all the specimens were subsequently ground with 360, 800, and 1200 grit silicon carbide paper, rinsed with deionized water and degreased in acetone. The composition of each solution is shown in Table 2. The Cl^- concentration of solution 3# was 0.5mol/L, which simulated oil field formation water. By changing the content of NaCl on the basis of solution 3#, the Cl^- concentrations of solutions 1# and 2# decreased (0.07 and 0.3mol/L, respectively). Prior to the test, the solutions were first deaerated by pure CO_2 for at least 8 h. During the test, the CO_2 gas was continuously bubbled into the solutions.

Table 1 Chemical composition of 3Cr2Al steel (wt%)

C	Si	Mn	Cr	Mo	Al	Nb	Fe
0.08	0.20	0.55	3.00	0.15	2.00	0.05	Bal.

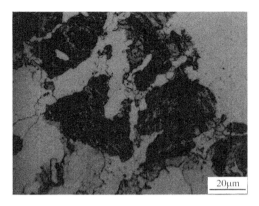

Fig. 1 Microstructure of 3Cr2Al steel. (Color figure can be viewed at wileyonlinelibrary.com)

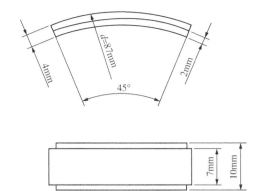

Fig. 2 The size of arc-shaped specimen for weight loss test

Table 2 Cl⁻ concentration in solutions for immersion test and electrochemical measurements

Composition (10^{-3} mol/L)	NaCl	$CaCl_2$	KCl	$MgCl \cdot 6H_2O$	$NaHCO_3$	Na_2SO_4
1# (0.07mol/L Cl⁻)	0	24.8	8.6	9.5	6.2	1.4
2# (0.3mol/L Cl⁻)	222.2	24.8	8.6	9.5	6.2	1.4
3# (0.5mol/L Cl⁻)	432.8	24.8	8.6	9.5	6.2	1.4

2.2 Weight loss tests

Weight loss test was conducted in a 5-L autoclave to investigate the corrosion rate of 3Cr2Al steel in three different Cl⁻ concentration and CO_2-containing environ-ments. The schematic diagram of high-temperature and high-pressure autoclave is shown in Figure 3. Prior to conducting weight loss test, the original weight of each specimen was measured using an analytical balance with an accuracy of 10^{-4} g. After 20 and 144h of immersion, the corroded specimens extracted from the autoclave were immediately rinsed with deionized water. The corrosion products were removed according to ASTM G1-03 standard, then rinsed, dried, and reweighed to obtain the final weight of the specimen. The corrosion rate was obtained by the following equation.

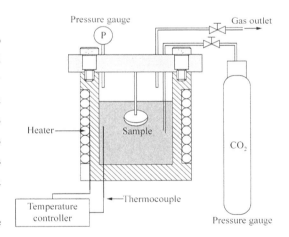

Fig. 3 The schematic diagram of high-temperature and high-pressure autoclave. (Color figure can be viewed at wileyonlinelibrary.com)

$$V_c = \frac{876000(W_0 - W_1)}{t\rho A} \quad (1)$$

Where V_c is the average corrosion rate, mm/a; W_0 and W_1 are the original and final weight of specimen, g, respectively; t represents the immersion time, h; ρ is steel density, g/cm³; and A

is exposed surface area in cm^2.

All immersion tests were performed at 80℃ under stagnant conditions. The weight loss test was conducted in three Cl$^-$ concentration (0.07, 0.3, and 0.5mol/L) solutions, with a CO$_2$ partial pressure of 0.8MPa.

2.3 Morphology observation and composition analysis

The cross-sectional morphology and composition of each corrosion product film were investigated using scanning electron microscopy (SEM) and electron diffraction spec-troscopy (EDS). The cross-sectional morphology was observed using the quadrant back scattering detector technique.

The morphology and energy dispersive X-ray spectros-copy (EDS) analysis of the corrosion product filmwere investigated using a JSM-6510A SEM and a JED-2300 EDS.

2.4 Electrochemical measurements

All electrochemical measurements were performed in a1-L glass cell with a traditional three-electrode system using a Gamry Reference 600 electrochemical workstation. The 3Cr2Al steel was used as working electrode (WE), a platinum sheet as counter electrode (CE), and a saturated calomel electrode (SCE) as reference electrode (RE). In the potentiodynamic polarization tests, three fresh specimens were potentiodynamic polarized in the above three solutions from an initial potential of 1000mV (SCE) to the final potentials of -100mV (SCE), at the scan rate of 0.5mV/s.

Five potentiostatic polarization tests were performed in the solution with a Cl$^-$ concentration of 0.07mol/L. Five individual fresh specimens were potentiostatically polarized for 2500s at the potentials of -500, -480, -430, -420, and -370mV (vs. SCE), respectively.

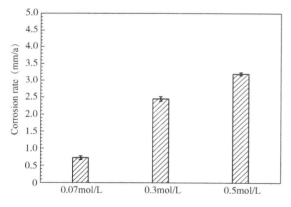

Fig. 4 Corrosion rates of 3Cr2Al steel in solutions with Cl$^-$ concentration of 0.07, 0.3, and 0.5mol/L(80℃, 0.8MPa CO$_2$, 144h)

3 Results

3.1 Corrosion rate

The average corrosion rates (test period of 144h) of 3Cr2Al steel in the three Cl$^-$ concentration solutions (0.07, 0.3, and 0.5mol/L) are shown in Figure 4. The corrosion rate in the 0.5mol/L solution was 4.4 times higher than that in the 0.07mol/L solution. With the increase in Cl$^-$ concentration in the solutions, the average corrosion rate increased notably.

3.2 Macroscopic morphologies of corrosion scales

The macroscopic morphologies of corrosion scales and after scales removal for 3Cr2Al steel are shown in Figure 5. As exhibited in Figure 5, the corrosion scales in the three Cl$^-$ concentration solutions were uniform. After the removal of scale, the 3Cr2Al steel exhibited uniform corrosion and did not show mesa corrosion morphology.

Fig. 5 Macroscopic morphology of corrosion scales and after scales removal in three Cl⁻ concentration solutions, (a and b) 0.07mol/L, (c and d) 0.3mol/L, (e and f) 0.5mol/L. (80℃, 0.8MPa CO_2, 144h) (Color figure can be viewed at wileyonlinelibrary.com)

3.3 Microscopic morphologies of corrosion scales

The microscopic morphologies of the corrosion scales (cross section) for the 3Cr2Al steel are shown in Figure 6. With the increase in Cl⁻ concentration, the average thickness of the scale decreased. The thickness in the solution with a Cl⁻ concentration of 0.07mol/L was approximately 2.6 times that in the 0.5mol/L Cl⁻ concentration solution.

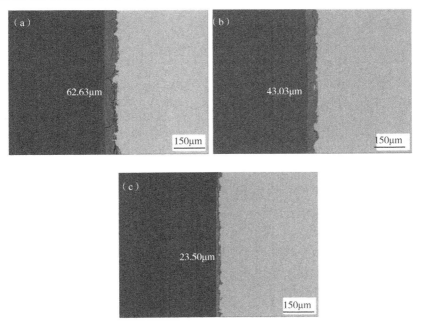

Fig. 6 Cross sectional morphology of corrosion scale in three Cl⁻ concentration solutions (a) 0.07mol/L, (b) 0.3mol/L, (c) 0.5mol/L. (80℃, 0.8MPa CO_2, 144h) (Color figure can be viewed at wileyonlinelibrary.com)

Point electron diffraction spectroscopy (EDS) analysis results based on Figure 6 (as shown by the red point) are compared in Table 3. According to the literature,[25-27] Cr enriches the corrosion product film of low alloy steel containing Cr. As shown in Table 3, Cr enriched the corrosion product film of the 3Cr2Al steel. Al also enriched the corrosion product film, and the Al content in the corrosion product film was 8~9 times higher than that in the steel substrate. With the increase in Cl^- concentration in the solution, the Al content of the corrosion scale decreased. Zhu et al.[28] used the relative content of Cr/Fe ratio to describe the degree of Cr enrichment in the corrosion product film. In the current study, because Ca is present in the solution and Ca can form $Ca_xFe_{1-x}CO_3$ with Fe,[29] we used a Al/(Fe+Ca) ratio to describe the enrichment of Al in the film. As shown in Table 3, the Al/(Fe+Ca) ratio was approximately two, and it monotonously decreased with the increase in Cl^- concentra-tion in the solution. However, the Cr content and Cr/(Fe+Ca) ratio in the film did not decrease monotonically with the increase in the Cl^- concentration, suggesting that the mechanism by which Cl^- affects the Al element may be different from that by which it affects the Cr element. Notably, the (Cr+Al)/(Fe+Ca) ratio monotonously de-creased as the concentration of Cl^- in the solution increased.

Table 3 Point EDS analyze of scale on 3Cr2Al steel (based on scales shown in Figure 6)

Cl^- concentration (mol/L)	Al content (%)	Cr content (%)	Fe content (%)	Ca content (%)	O content (%)	Cl content (%)	Fe+Ca ratio	Al/(Fe+Ca) ratio	Cr/(Fe+Ca) ratio	(Cr+Al)/(Fe+Ca) ratio
0.07	19.01	22.53	2.81	4.11	51.34	0.21	6.92	2.75	3.26	6.01
0.3	18.04	20.16	4.16	4.58	52.23	0.83	8.74	2.06	2.30	4.36
0.5	16.60	24.87	3.65	6.35	48.07	0.47	10.00	1.67	2.49	4.16

To examine the distribution of each element in the corrosion product film, a mapping analysis based on Figures 6a and 6c was conducted. Figures 7 and 8 show the EDS mapping results with Cl^- concentrations of 0.07 and 0.5mol/L, respectively. As shown in Figure 7, the enrichment of Cr and Al in the film is evident, which is consistent with the point EDS results in Table 3. The Fe content in the film was small, which was in sharp contrast with the content in the substrate. This may be the result of Cr and Al enrichment in the film. Ca was present in the film, but the Cl content was small. The distribution of Al/Cr/Fe/Ca/O elements was relatively uniform.

As shown in Figure 8, when the Cl^- concentration was 0.5mol/L, the Fe content in the film increased with respect to 0.07mol/L. In the 0.5mol/L solution, the Al/Cr/Fe/Ca/O element distribution was also uniform. Therefore, the increase in Cl^- concentration did not affect the distribution of each element, but affected their relative content.

The microscopic morphologies of the corrosion scale surfaces in the three Cl^- concentration solutions are presented in Figure 9. Except for NaCl content, the compositions of these three solutions were the same. As shown in Figure 9, the corrosion scales exhibited bilayer structures. The inner layer was a protective Cr-and Al-rich film, which was continuous. The outer layer was mainly composed of scattered $FeCO_3$ grains. At a Cl^- concentration of 0.07mol/L, a large amount of $FeCO_3$ was deposited on the inner layer (Figure 9a), and the inner layer was difficult to be identified from the top. However, when the Cl⁻ concentration increased to 0.3 and 0.5mol/L, the quantity of deposited $FeCO_3$ decreased. Subsequently, the mud-cracking inner layer was visible. The shape of $FeCO_3$ also changed from spherical to flaky, as shown in Figures 9b and 9c.

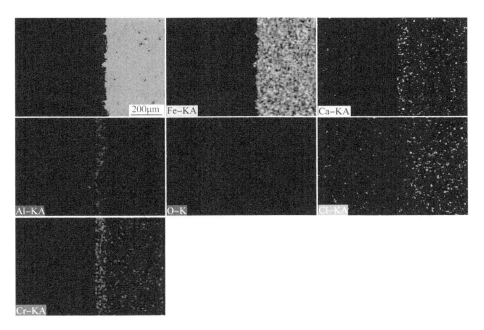

Fig. 7 Elemental map of 3Cr2Al steel cross-section after 144h immersion in 0.07mol/L Cl⁻ concentration solution(Color figure can be viewed at wileyonlinelibrary.com)

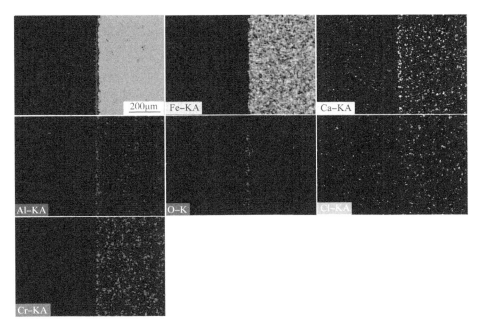

Fig. 8 Elemental map of 3Cr2Al steel cross-section after 144h immersion in 0.5mol/L Cl⁻ concentration solution(Color figure can be viewed at wileyonlinelibrary.com)

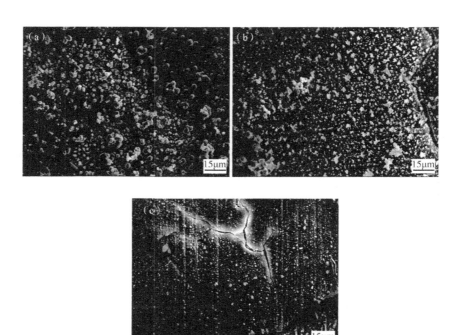

Fig. 9 Microscopic morphology of corrosion product film (a) 0.07mol/L, (b) 0.3mol/L, (c) 0.5mol/L. (80℃, 0.8MPa CO_2, 144h)

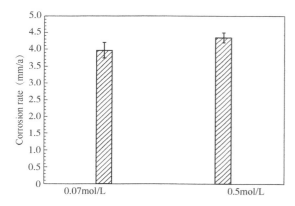

Fig. 10 Corrosion rates of 3Cr2Al steel in solutions with Cl^- concentration of 0.07 and 0.5mol/L(80℃, 0.8MPa CO_2, 20h)

To investigate the growth of the corrosion product film, immersion tests were conducted with a test period of 20h in two Cl^- concentration solutions (0.07 and 0.5mol/L). The average corrosion rates are shown in Figure 10. The microscopic morphologies of the corrosion product film (cross section) are shown in Figure 11.

As exhibited in Figure 10, when the immersion timewas 20h, the corrosion rates in the 0.07mol/L solution were close to those in the 0.5mol/L solution. The corrosion rates of the $3Cr_2Al$ steel after immersion for 20h were higher than those observed after immersion for 144h. This was true forboth solutions (0.07 and 0.5mol/L). For the 0.07mol/L solution, the corrosion rate (3.97mm/a) at the immersion time of 20h was considerably higher than that at 144h (0.72mm/a). This is possibly because during the initial stage of corrosion, no protective corrosion product was present on the surface of the 3Cr2Al steel. With the prolongation of the immersion time, the corrosion product gradually formed, and subsequently, the corrosion rates dropped drastically. As shown in Figure 11, after immersion for 20h in the 0.07mol/L solution, a thin layer of corrosion product film formed on the surface of the $3Cr_2Al$ steel. In the 0.5mol/L solution, the corrosion product did not completely cover the surface of the 3Cr2Al steel after the same immersion time. By comparing the corrosion rates at 20 and 144h, it can be speculated that the corrosion product film formed on 3Cr2Al steel surface was protective.

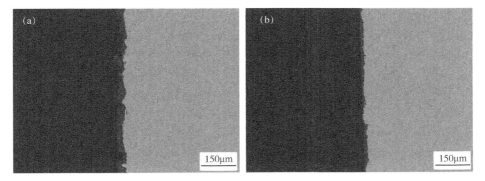

Fig. 11 Cross sectional morphology of corrosion product film in two Cl⁻ concentration solutions (a) 0.07mol/L, (b) 0.5mol/L(80℃, 0.8MPa CO_2, 20h)

3.4 Polarization curves and potentiostatic polarization

We studied the electrochemical properties of the 3Cr2Al steel as bare steel to ascertain whether Cl⁻ affects its corrosion behavior before the formation of the protective film. Potentiodynamic polarization curves of the 3Cr2Al steel in the three solutions are presented in Figure 12. The anode polarization curves changed considerably as the Cl⁻ content increased. In the 0.07mol/L solution, the anode polarization curves underwent a drop in current density, which was a characteristic of pseudopassivation. In the 0.5mol/L solution, the polarization curve represented an activated state. Moreover, when the Cl⁻ concentration changed from 0.07 to 0.3mol/L, the pseudopassivation character weakened.

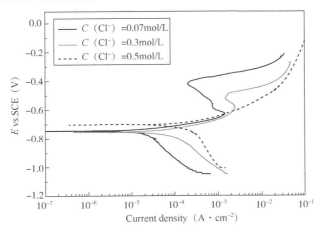

Fig. 12 Polarization curves of 3Cr2Al steel. (80℃, CO_2-saturated brine)(Color figure can be viewed at wileyonlinelibrary.com)

Figure 13 displays the current – time curves of the potentiostatic polarization tests in the 0.07mol/L solution. The potentials at which the current density peaked were −629 and −419mV (vs. SCE), as shown in the inset of Figure 13. According to these two peaks, five feature potentials were selected. Five fresh specimens were potentiostatically polarized to the potentials of −500,

−480, −430mV (in the potential region of pseudopassivation), −420mV (close to the peak potential), and −370mV (higher than the peak potential, out of the pseudopassivation range). Subsequently, the current density for each specimen was monitored for 2500s. When a fresh specimen underwent potentiostatic polarization at −500, −480, and −430mV (vs. SCE), the initial current density value was consistent with the potentiodynamic polarization curves, whose value ranking was −500, −480, and −430mV from large to small. Next, when polarization time was prolonged, all the values decreased at a fairly slow rate, which indicated that the pseudopassive film was becoming thicker. Subsequently, the current density started to increase gradually when a fresh specimen was polarized at −420mV (close to the peak potential), which indicated that the protectiveness of the film reached its maximum. When the fresh specimen was polarized at −370mV (vs. SCE) for 2500s, the initial current density was the highest among the five specimens, and when the polarization time was prolonged, the current density did not decrease but rather slightly increased. This might be because the protectiveness of the corrosion product film weakened.

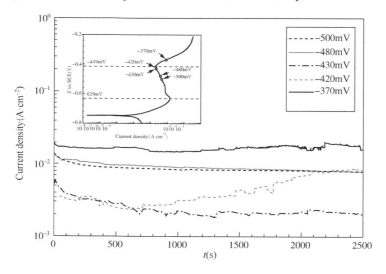

Fig. 13 Current-time curves of potentiostatic polarization of 3Cr2Al steel under $C(Cl^-) = 0.07$mol/L, 80℃, and CO_2-saturated

4 Discussion

By comparing Figures 4 and 10, it can be deduced that the corrosion product film on the 3Cr2Al steel was protective. Therefore, the corrosion rate of the 3Cr2Al steel reduced with the prolongation of the corrosion time. According to Figure 4, in the corrosion environment of 80℃ and 0.8MPa CO_2(144h), the concentration of Cl^- had a noteworthy effect on the corrosion rate of the 3Cr2Al steel; the corrosion rate increased with the increase in Cl^- concentration. However, whether Cl^- affects the corrosion rate directly or indirectly (i.e., by changing other environmental parameters) remains unclear. As shown in Figure 9, with the increase in Cl^- concentration, the amount of $FeCO_3$ deposited on the steel surface decreased, and the crystalline character of $FeCO_3$ also weakened. Some studies[30,31] have illustrated that the development and morphology of $FeCO_3$

crystals were closely related to solution pH. The decrease in solution pH considerably increased the solubility of Fe^{2+}, resulting in slower precipitation and thus a decreased rate of surface coverage. Therefore, the concentration of Cl^- likely affects the pH value of the solution.

We measured the pH value of the three solutions with different Cl^- concentrations. As provided in Figure 14, the pH value decreased with the increase in Cl^- concentration. The pH value of the 0.5mol/L solution was 0.25 lower than that of the 0.07mol/L solution. Therefore, we speculate that one mechanism by which Cl^- concentration affects the corrosion rate of 3Cr2Al steel is by influencing the pH value. Some studies[32,33] have found that the solubility of Cr(OH)$_3$ increases with the decrease in pH (when the pH is lower than 6). Corresponding to the pH value in Figure 14, when the pH was 5.43, the

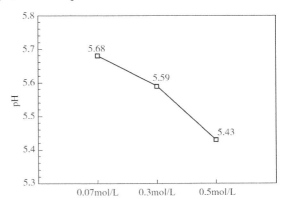

Fig. 14 Measured pH of three Cl^- concentration solutions at 80℃, 1 bar total pressure

solubility of Cr(OH)$_3$ was approximately four times higher than when the pH was 5.68[34]. Therefore, when the Cl^- concentration increased, the pH value decreased, and subsequently, the solubility of Cr(OH)$_3$ increased, thereby causing the deposition of Cr(OH)$_3$ on the 3Cr2Al steel surface to be much harder. Regarding Al[35], the solubility of Al(OH)$_3$ increased with the decrease in pH (when the pH was lower than 5.8). When the pH was 5.43, the solubility of Al(OH)$_3$ was approximately five times higher than when the pH was 5.68. Therefore, when the Cl^- concentration was high (0.5mol/L), the corresponding pH of the solution was low, the depositions of Cr(OH)$_3$ and Al(OH)$_3$ were harder, and eventually, the corrosion product film became thinner (as shown in Figure 6).

The corrosion product film of the 3Cr2Al steel had high protectiveness. When the Cr and Al content in the film were similar, the thicker the film was, the higher the protectiveness was, which subsequently led to a lower corrosion rate. In summary, by influencing the solution pH, the Cl^- concentra-tion changed the thickness of the corrosion film and finally affected the corrosion rate.

The relative Cr and Al content in the corrosion product film ([Cr+Al]/[Fe+Ca] ratio) decreased monotonously with the increase in Cl^- concentration (as shown in Table 3), which indicated that by influencing the pH value, Cl^- not only notably affected the film thickness but also slightly affected the Cr and Al enrichment in the film. Aluminum has a particularly strong tendency to hydrolyze in the solution, and at any pH above 3.5, various combinations of aluminum with hydroxide ions occur.[36] In the present study, one of the main anodic reactions of the 3Cr2Al steel was the dissolution of Al, as described by the following[37].

$$Al \longrightarrow Al^{3+} + 3e^- \tag{2}$$

$$Al^{3+} + 3H_2O \longrightarrow Al(OH)_3 + 3H^+ \tag{3}$$

With the hydrolysis of Al^{3+}, aluminum hydroxide precipitated on the surface. When the solution contained Cl^-, the reactions shown in Eqs. (4)–(6) occurred.[38-40]

$$Al^{3+} + H_2O \rightleftharpoons Al(OH)^{2+} + H^+ \tag{4}$$

$$Al(OH)^{2+} + Cl^- \rightleftharpoons Al(OH)Cl^+ \tag{5}$$

$$Al(OH)Cl^+ + H_2O \rightleftharpoons Al(OH)_2Cl + H^+ \tag{6}$$

With the increase in the Cl^- content, the amount of $Al(OH)_3$ deposition decreased. Therefore, Cl^- may also reduce the degree of Al enrichment in the film by directly affecting the deposition of $Al(OH)_3$. Cr did not form soluble substances with Cl^-. Therefore, the deposition of $Cr(OH)_3$ was not directly reduced, and the $Cr/(Fe+Ca)$ ratio did not monotonically decrease with the increase in Cl^- concentration in the solution (as shown in Table 3). In the 0.5mol/L solution, Cr content slightly increased compared with that in the 0.3mol/L solution. This may be due to the decrease in Al content in the film, subsequently causing an increase in the relative Cr content.

As shown in Table 3, with the increase in Cl^- concentration, the $Al/(Fe+Ca)$ ratio monotonously decreased, but the $Cr/(Fe+Ca)$ ratio did not. This indicated that the mechanism by which Cl^- affects the Al element may be different from that by which it affects the Cr element. Regarding Al, the increase in Cl^- exerted two effects: directly converting $Al(OH)_3$ into a soluble substance and reducing environments at the potentials of -500, -480, -430, -420, and -370mV (vs. SCE), respectively. [Color figure can be viewed at wileyonlinelibrary.com] the pH value and then hindering the deposition of $Al(OH)_3$. However, for Cr, only one mechanism (an indirect mechanism) existed. By reducing the pH value, Cl^- indirectly hindered the deposition of $Cr(OH)_3$.

The corrosion of the 3Cr2Al steel can be divided into two stages. During the stage when the corrosion product film formed on the 3Cr2Al steel, Cl^- considerably affected the thickness of the corrosion product film by changing the deposition of $Al(OH)_3$ and $Cr(OH)_3$, subsequently changing the protectiveness of the film. During the stage when the 3Cr2Al steel was bare, the Cl^- concentration increased from 0.07 to 0.5mol/L, and the anode polarization curve changed from pseudopassivation to activation. Therefore, in a higher Cl^- concentration environment, the corrosion of bare steel is accelerated. In summary, the increase in Cl^- concentration can accelerate the corrosion of 3Cr2Al steel during both stages.

5 Conclusions

The corrosion behavior of novel 3Cr2Al steel in simulated oil and gas well environments of 80℃ and 0.8MPa CO_2, with Cl^- concentrations of 0.07, 0.3, and 0.5mol/L was investigated. The corrosion product film formed at 20 and 144h was studied using scanning electron microscopy and EDS. The effect of Cl^- concentration on the corrosion rate and characteristic of corrosion product film was discussed and the mechanism of Cl^- on corrosion behavior of 3Cr2Al steel was revealed.

(1) The 3Cr2Al steel exhibited uniform corrosion in the three test solutions containing Cl^-. The corrosion rate of the 3Cr2Al steel increased as the Cl^- concentration increased. The thickness of the corrosion product film decreased as the Cl^- concentration increased. The corrosion rate at 144h was lower than that at 20h, and the corrosion product film formed on the 3Cr2Al steel was protective.

(2) Al can enrich the corrosion product film of 3Cr2Al steel. Al content in the corrosion product film was 8~9 times higher than that in the steel substrate, and the Al/(Fe+Ca) ratio in the corrosion product film was approximately 2.

(3) The change in Cl^- concentration affects the deposition of $Cr(OH)_3$ and $Al(OH)_3$ on the steel surface. The increase in Cl^- concentration caused a decrease in pH value, which subsequently caused an increase in the solubility of $Cr(OH)_3$ and $Al(OH)_3$ and hindered the deposition of the corrosion product, eventually yielding a thinner corrosion film.

Acknowledgement

This project was supported by the National Nature Science Foundation of China under grant No. 51871025.

References

[1] F. Farelas, M. Galicia, B. Brown, S. Nesic, H. Castaneda, Corros. Sci. 2010, 52, 509.
[2] O. Palumbo, A. Paolone, P. Rispoli, R. Cantelli, G. Cannelli, S. Nesic, B. J. Molinas, P. P. Zonta, Mater. Sci. Eng. A 2009, 521, 343.
[3] S. Nesic, Corros. Sci. 2007, 49, 4308.
[4] K. Videm, A. Dugstad, Mater. Perform. 1989, 4, 46.
[5] B. Mishra, S. Al-Hassan, D. L. Olson, M. M. Salama, Corrosion 1997, 53, 852.
[6] M. H. Ezuber, Mater. Des. 2009, 30, 3420.
[7] C. de Waard, U. Lotz, D. E. Milliams, Corrosion 1991, 47, 976.
[8] C. de Waard, U. Lotz, Corrosion/1993, Paper No. 69, NACE, Houston, TX, 1993.
[9] K. Videm, A. Dugstad, L. Lunde, Corrosion/1994, Paper No. 14, NACE, Houston, TX, 1994.
[10] S. Nesic, J. Postlethwaite, S. Olsen, Corrosion 1996, 52, 280.
[11] S. Nesic, L. Lunde, Corrosion 1994, 50, 717.
[12] C. de Waard, U. Lotz, A. Dugstad, Corrosion/1995, Paper No. 128, NACE, Houston, TX, 1995.
[13] J. C. Cardoso Filho, M. E. Orazem, Corrosion/2001, PaperNo. 01058, NACE, Houston, TX, 2001.
[14] H. B. Wu, L. F. Liu, L. D. Wang, Y. T. Liu, J. Iron Steel Res. Int. 2014, 21, 76.
[15] C. F. Chen, M. X. Lu, D. B. Sun, Z. H. Zhang, W. Chang, Corrosion2005, 61, 594.
[16] R. H. Hausler, Corrosion/1984. NACE, Houston Texas 1984. p. 72.
[17] A. K. Dunlop, H. L. Hassell, P. R. Rhode, Corrosion/1984. NACE, Houston Texas 1984. p. 52.
[18] F. F. Eliyan, A. Alfantazi, Can. J. Chem. Eng. 2015, 93, 1044.
[19] Q. Y. Liu, L. J. Mao, S. W. Zhou, Corros. Sci. 2014, 84, 165.
[20] Y. Ma, Y. Li, F. Wang, Corros. Sci. 2009, 51, 997.
[21] G. Schmitt, Corrosion/1984. NACE, Houston Texas 1984. p. 1.
[22] J. Han, J. W. Carey, J. Zhang, J. Appl. Electrochem. 2011, 41, 741.
[23] M. Liu, X. Cheng, X. Li, Z. Jin, H. Liu, Constr. Build. Mater. 2015, 93, 884.
[24] R. Wang, S. Luo, M. Liu, Y. Xue, Corros. Sci. 2014, 85, 270.
[25] M. H. Liang, Z. Y. Pan, X. Zhang, Adv. Mater. Res. 2013, 740, 608.
[26] L. N. Xu, S. Q. Guo, C. L. Gao, W. Chang, T. H. Chen, M. Lu, Mater. Corros. 2012, 63, 997.
[27] J. B. Sun, W. Liu, W. Chang, Z. H. Zhang, Z. T. Li, T. Yu, M. X. Lu, Acta Metall. Sin. 2009, 45, 84.
[28] J. Y. Zhu, L. N. Xu, Z. C. Feng, G. S. Frankel, M. Lu, W. Chang, Corros. Sci. 2016, 111, 391.
[29] B. Wang, L. N. Xu, G. Z. Liu, M. X. Lu, Corros. Sci. 2018, 136, 210.

[30] D. Burkle, R. De Motte, W. Taleb, A. Kleppe, T. Comyn, S. M. Vargas, A. Neville, R. Barker, Electrochim. Acta 2017, 255, 127.
[31] S. Q. Guo, L. X. Xu, L. Zhang, W. Chang, M. X. Lu, Corros. Sci. 2012, 63, 246.
[32] D. Rai, D. A. Moore, N. J. Hess, L. Rao, S. B. Clark, J. Solution Chem. 2004, 33, 1213.
[33] M. Ueda, A. Ikeda, Corrosion/1996, Paper No. 96013, NACE, Houston, TX, 1996.
[34] N. Papassiopi, K. Vaxevanidou, C. Christou, E. Karagianni, G. S. E. Antipas, J. Hazard. Mater. 2014, 264, 490.
[35] C. E. Roberson, J. D. Hem, Geol. Surv. Water-Supply Paper 1969, 1827-C, 37.
[36] J. D. Hem, C. E. Roberson, Geol. Surv. Water-Supply Paper 1967, 1827-A, 55.
[37] Z. Sun, D. Zhang, B. Yan, D. Kong, Opt. Laser Technol. 2018, 99, 282.
[38] F. S. da Silva, J. Bedoya, S. Dosta, N. Cinca, I. G. Cano, J. M. Guilemany, A. V. Benedetti, Corros. Sci. 2017, 114, 57.
[39] S. Li, H. A. Khan, L. H. Hihara, H. Cong, J. Li, Corros. Sci. 2018, 132, 300.
[40] E. S. M. Sherif, A. A. Almajid, F. H. Latif, H. Junaedi, Int. J. Electrochem. Sci. 2011, 6, 1085.

本论文原发表于《Materials and Corrosion》2019 年第 70 卷。

Mechanical Performance of Casing in In-Situ Combustion Thermal Recovery

Yang Shangyu[1]　Han Lihong[1]　Feng Chun[1]　Wang Hang[1]
Feng Yaorong[1]　Wu Xingru[2]

(1. State Key Laboratory for Performance and Structure Safety of Petroleum Tubular Goods and Equipment Materials, CNPC Tubular Goods Research Institute; 2. University of Oklahoma)

Abstract: In-situ combustion is often used to develop heavy oil as it has multiple advantages over alternative thermal methods. However, the wellbore integrity can be compromised if the casing is not properly designed or manufactured. Based on the fundamentals of heat transfer, a mathematical model with multiphysics is built to study the temperature and pressure distributions surrounding the wellbore during in-situ combustion. In laboratory, real casings and coupons with grades of N80, P110, and 3Cr110 were tested under high temperature to study their mechanical performances in the in-situ combustion recovery. From the modeling and laboratory testing results, we propose the casing design and manufacture protocols for thermal recovery of heavy oil. Field applications in Liaohe oil field, Du-66 Block, shows that the mechanical deterioration of P110 and 3Cr110 is less than that of N80 and the mechanical performance stability of 3Cr110 is better than others when the temperature is less than 600℃. When the temperature is above 600℃ for more than 10 hours, the mechanical performance of all casing grades declines quickly. Applying finite element analysis, we recommend a safety factor of 1.1 for designing casing at the temperatures of 250℃, 480℃ and 485℃. The finding of this study provides fundamentals for the casing design and material selection for in-situ combustion recovery.

Keywords: Casing performance; Heavy oil; In-situ combustion; Casing design; Steam flooding

1　Introduction

Thermal recovery is often used in enhancing oil recovery, especially for heavy oils which account for 53% of total world reserve. The thermal recovery methods include thermal stimulations, steam flooding, in-situ combustion (ISC), and other variants. The in-situ combustion over steam flooding relies on several advantages of ISC thermal methods. First, ISC has features for being more

---
Corresponding author: Wu Xingru, Xingru.Wu@ou.edu.

efficient than alternative thermal methods such steam flooding. The external energy consumed in ISC is mainly used for compressing and injecting air. Literature shows that air ISC requires only about 23%~39% of the fuel needed for steam, and even more fuel efficiency by burning oxygen. In steam injection, significant heat losses to the surrounding formation during injection, and heat losses to the overburden and underburden during flooding. ISC can eliminate heat losses to the surrounding since heat is generated in the reservoir.

ISC has also many limitations for operators to overcome in practice, which probably are the main reasons for this early but less popular thermal recovery. As the fire front moves forward, gasses containing CO_2 or H_2S will breakthrough at the producer. These gasses are corrosive to the casing, and flowing with water makes the problem of corrosion even worst. At the same time, air is injected at a high rate which leads to solid particles being displaced to the producer and exacerbates the erosion of casing. Furthermore, the casing strength of casing degrades in the elevated temperature environment. These harsh conditions post tremendous challenges for casing design for in-situ combustion thermal recovery. Other technical challenges and experiences of field practices of ISC were highlighted in literature.

Improper design of casing can lead to significant casing damage. Wellbore integrity survey in Du-66 Block in Liaohe oil fields show that by the end of 2015. In the production pad of Well Group 92, 28 wells out of 43 wells showed a variety level of casing damage or deformations after they were switched from steam flooding to in-situ combustion. The damage and deformation of casing are mainly located the perforation intervals and regions close to the upper part of perforation zone. The modes of failure are serve deformation and tensile break.

Before 80's most thermal well casing designs were based on Holliday(1969) model which showed that higher strength materials would be needed for thermal wells. However, field practices show that this criteria alone is not sufficient. Lepper(1998) discussed the casing design for thermal fields should account for collapse resistance, tensile stresses, and connection strength for buttress. Hidayat, Irawan, and Abdullah(2016) studied the casing strength degradation in steam stimulation process using a three-dimensional finite element analysis on N80 casing and concluded that the casing capability to resist the pressure lowers as the number of thermal cycles increases. However, the authors did not present any laboratory testing on this and other casings. Li (2013) presented the mechanisms of casing failure for thermal recovery wells and recommend to use thicker casing to prevent casing failure. Chen, Peng, and Yu(2017) analyzed the stresses on casing resulted from formation and cement thermal expansion and concluded that the casing deformation could be caused by the different expansion rates of surrounding materials. Nowinka, Kaiser, and Lepper(2008) proposed strain-based design of tubulars for extreme conditions such as high temperature wells in thermal recovery.

To date, there is no standard procedure adopted by the industry for ISC casing design when the temperature is above 180℃ except for some proprietary procedures from some operators. Recently a strain-based design concept coupled with laboratory tests and finite analysis have gain ground in designing casing. This study focuses on ISC casing design by testing the casing strengths under high temperatures using laboratory experiments and numerical simulation with an aim to provide fundamental guideline in casing material selection.

2 Mathematical Model for In-Situ Combustion

In the process of in-situ combustion recovery, usually air is injected at an injection well and oil is being produced from a producer as shown in Figure 1. If the air injection pressure and temperature at the bottom hole of the injection well are P_w and T_w, and at far distance ($r = r_\infty$) and the temperature and pressure are T_0 and P_0, respectively; we can set up a mathematical model with the following assumptions:

(1) Initial water and oil saturations are constant;
(2) Overburden and underburden formations are impermeable to oil or water;
(3) Heat losses to overburden and underburden are neglected;
(4) Water and oil are immiscible in the reservoir conditions;
(5) Darcy's flow for oil and water;
(6) Other than viscosity, fluid and rock thermal properties are constant with temperature except for viscosity.

The general governing equation for thermal energy balance in a cylindrical coordinates is given by Eqn. 1:

$$\rho c \frac{\partial T}{\partial t} = \frac{1}{r} \frac{\partial}{\partial r}\left(\lambda r \cdot \frac{\partial T}{\partial r}\right) + \frac{1}{r^2} \frac{\partial}{\partial \theta}\left(\lambda \frac{\partial T}{\partial \theta}\right) + \frac{\partial}{\partial z}\left(\lambda \frac{\partial T}{\partial z}\right) + q \tag{1}$$

Where λ is the heat conductivity of rock, W/(m · K); ρ is density, kg/m³; c is the specific heat capacity of rock, [J/(kg · K)], q is heat flux, W/m²。

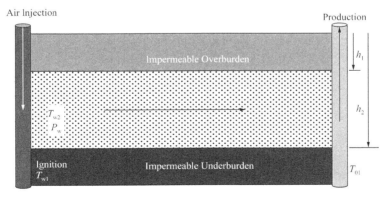

Fig. 1 Model diagram of in-situ combustion from an injection well a production well

For Eqn. 1, since we are mainly interested in highest temperature in the region near by the injection well bore, the heat conduction terms in θ and z directions can be neglected. Therefore, in the rdirection, the Eqn. 1 is written as follows:

$$\rho c \frac{\partial T}{\partial t} = \frac{1}{r} \frac{\partial}{\partial r}\left(\lambda r \cdot \frac{\partial T}{\partial r}\right) + q \tag{2}$$

Other than the combustion of coke in the combustion location, we assume there is no other chemical reactions. For mass balance, for the region nearby well bore, if we assume one-direction flow surrounding the well, the mass balance equations for oil and water in cylindrical coordinates are as follows:

$$\frac{\partial(\phi\rho_o S_o)}{\partial t} = \frac{1}{r}\frac{\partial}{\partial r}\left(\frac{\rho_o k_o k}{\mu_o}\frac{r\partial P}{\partial r}\right) \quad (3)$$

$$\frac{\partial[\phi\rho_w(1-S_o)]}{\partial t} = \frac{1}{r}\frac{\partial}{\partial r}\left(\frac{\rho_w k_w k}{\mu_w}\cdot\frac{r\partial P}{\partial r}\right) \quad (4)$$

Where ρ_o and ρ_w are density of oil and water, respectively, kg/m^3; k is permeability, m^2; P is pressure, Pa; ϕ is porosity; S_O is oil saturation; μ_o and μ_w are oil and water saturations, respectively, Pa·s; k_o, k_w are relative permeability of oil and water, respectively. Given that we are only interested in the near well bore region of the injector, we assume there are oil and liquid water as small gas may exist but in solution form. The generated steam as the result of in-situ solution is also not considered as the steam region can quickly move further into formation and condensate to liquid water.

The above three equations can be solved implicitly as the viscosity is a function of temperature, and the relative permeability is a function of saturations of oil and water. Based on the measurement of relative permeability from the Liaohe Du-66 block, the empirical correlation between saturation and relatives are shown in Eqn. 5.

$$\begin{cases} k_o = 99.95 \cdot e^{[(-[1-S_o])/0.06968]} - 0.01749 \\ k_w = 0.00684 \cdot e^{[(1-S_o)/0.137]} - 0.07708 \end{cases} \quad (5)$$

Matching the experimental data with the Bergman Equation for heavy oil (Bergman and Sutton 2009), The temperature and viscosity relationship for this field is given by Eqn. 6.

$$\log[\log(\mu_o+0.6)] = 9.1138 - 3.5635\log(273.15+T) \quad (6)$$

Where T is temperature is ℃, ad viscosity is in cp. Figure 2 gives the viscosity and temperature relationship based on the equation Eqn. 6. On this figure, oil viscosities are 0.58 mPa·s and 0.68 mPa·s at temperatures of 600℃ and 400℃, respectively. These two temperatures are highlighted because they are the region boundary temperature of in-situ combustion study in this paper. Figure 2 shows that the viscosity decreases dramatically with the increase of reservoir temperature.

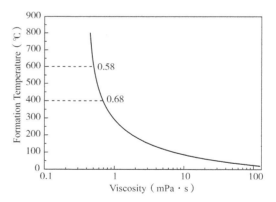

Fig. 2 Viscosity change with the increase of temperature for heavy oil modeling

The boundary conditions for the system is given as:

$$t>0, \ r=r_w, \ T=T_w; \ r=r_\infty, \ T=T_o$$
$$t>0, \ r=r_w, \ P=P_w; \ r=r_\infty, \ P=P_o$$
$$t>0, \ z=h_1, \ \frac{\partial T}{\partial z}=0; \ z=h_2, \ \frac{\partial T}{\partial z}=0$$
$$t>0, \ z=h_1, \ \frac{\partial P}{\partial z}=0; \ z=h_2, \ \frac{\partial P}{\partial z}=0$$

And the initial condition is given by:
$$t=0, \quad r_w \leqslant r \leqslant r_\infty, \quad h_1 \leqslant z \leqslant h_2; \quad S_o = S_{o0}, \quad S_w = S_{w0}$$

Where, T_0 is the initial reservoir temperature, ℃; P_0 is the initial reservoir pressure, MPa; S_{o0} and S_{w0} are initial saturations of oil and water, respectively.

The governing equation can be solved numerically with iterations for given boundary and initial conditions to obtain the pressure and temperature distribution for in-situ combustion thermal recovery.

3 Temperature Distribution in the Region of ISC

Practice shows that the pressure and temperature are main factors driving the in-situ combustion recovery. High temperature lowers the viscosity of the oil dramatically as shown in Figure 2, which leads to the increase of oil mobility. Pressure drives the less viscous oil to the producer. Many classical literature have shown the temperature and pressure profiles for the in-situ combustion, especially on spontaneous combustion and some recent numerical and laboratory studies. The Eqn. 7 is a numerical model to calculate the temperature change in the bottom hole casing with time. The model is derived based on the field data in Liaohe Du-66 block, and Figure 3 gives the model results and actual field data in this block.

$$T_w = 263.52 + \frac{3070}{7.28\sqrt{\pi/2}} e^{\left[-2\left(\frac{t-36.01}{7.28}\right)^2\right]} \quad (7)$$

Where t is time in hour, T_W is the well bore temperature at the bottom hole at time of t with a unit of ℃.

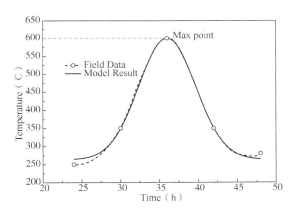

Fig. 3 Matching field ignition temperature and duration using the proposed model. The shape is consistent with the Figure 6 in Burger (1976)

Figure 3 shows that it took about 13 hours for the temperature to increase from the 250℃ to its maximum temperature around 600℃. After the in-situ combustion front moves away from the wellbore, the surrounding of the wellbore gradually cooled down the wellbore.

Temperature distribution is critical in studying ISC performance and casing mechanical properties, especially in the region close to the injection well. The calculated temperature field in the formation are based for the laboratory testing conditions. The temperature impact on oil phase resistance factor, λ_T, is shown in Eqn. 8.

$$\frac{1}{\lambda_T} = 1 + \frac{\left|\partial\left(\frac{\mu_o}{(P-P_0)k_0 k}\right)/\partial P \cdot dP\right|}{\left|\partial\left(\frac{\mu_o}{(P-P_0)k_0 k}\right)/\partial T \cdot dT\right|} \quad (8)$$

Eqn. 8 shows that the temperature of ignition is a critical parameter affecting the oil recovery. Solution of Eqn 8 is plotted in Figure 4. As the fire front moving forward, temperature gradually drops. The best performance is found when the temperature is above 400℃. Therefore, well casing used in ISC has to survival temperatures in the order of 400℃.

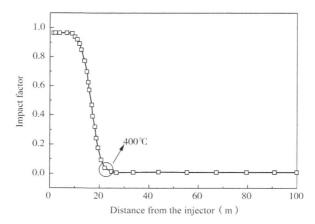

Fig. 4　Temperature impact factor along the distance from the injector

4　Temperature Impact on Casing Mechanics

Many wells are converted to in-situ combustion after steam flooding, and often high case grade is needed for this type of well. For the casing grades of N80, P110, and 3Cr-110, we did a series of laboratory experiments under high temperature conditions to obtain their mechanic properties such as yielding strength, tensile strength, and compressive strength, collapse and burst pressures. The experiments were conducted using both the casing material coupons and real casing to study the actual performance in the field. The testing procedure followed ISO 11960 standard(2010).

4.1　Temperature effect on casing coupons

Figure 5 shows the yielding strength for casing of the three casing material grades from room temperature (20℃) to 700℃. Overall the yielding strength decreases with the increase of temperature. If the temperature is below 350℃, both N80 and P110 meet the API requirement on yielding strength. When the temperature is above 450℃, neither N80 or P110 satisfies the specification of API casing strength. When the temperature is above 550℃, the yield strength declines significantly. At the temperature 700℃, the yielding strengths of N80 and P110 are only 43.4% and 53.6% of that in the room temperature, respectively. Compared with N80 and P110 material, casing of 3Cr-110 has a similar pattern of decrease with the other two casing material grades. When the temperature is above 450℃, the yielding strength of 3Cr-110 decreases slower than the other two. At the temperature of 700℃, the yielding strength is 588 MPa, which is better than N80 and P110 by 27.6% and 17.4%, respectively.

Under the same temperatures, the tensile strengths of N80, P110, and 3Cr-110 casings were tested. Figure 5 demonstrates the testing results. All casing grades have similar magnitude of decreases on tensile strength when the temperature is less than 500℃. The 3Cr-110 Casing performs much better when the temperature is above 700℃ than N80 and P110 casings, and the tensile strength of 3Cr-110 is 12.9% higher than N80, and 8.8% higher than P110 Casing.

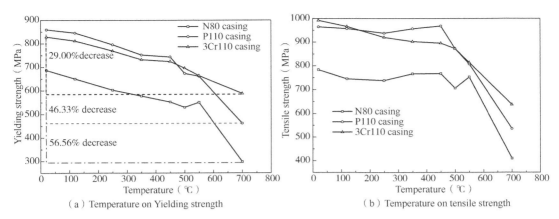

Fig. 5 Temperature impact on the strengths of casing grades of N80, P110, 3Cr110

4.2 Temperature effect on real casing mechanics

Real casings of grades N80, P110, and 3Cr-110 were tested in laboratory under elevated temperatures to check their mechanical properties of collapsing strength, internal pressure strength, and tensile strength, and the testing results are shown in Figure 6. All these mechanical properties declines graduate when the temperature increases up to 550℃. When the temperature is above 550℃, the mechanical performance of casings with grades of P110 and N80 deteriorate severely, while the performance of the casing of 3Cr 110 drops at a much less rate.

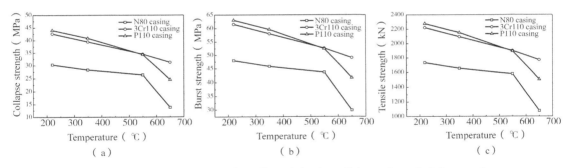

Fig. 6 Temperature effects on the collapse(a), burst(b), and tensile(c) strengths of casing grades of N80, P110, and 3Cr110

4.3 Creep testing on casing coupons under high temperatures

In the in-situ combustion process, casing works under high temperature for several years or even longer. Laboratory tests also shows that the casing mechanical properties are function of temperature and service duration. From above discussion, the casing strength decreases and elasticity increases with the increase of temperature. Under a high temperature, even the casing equivalent stress is much less than the yield strength in this temperature. With the increase of casing service time, the casing gradually elongates till break.

From the working condition for the in-situ combustion, the temperature can be above 600℃ when ignition occurs. To better understand the relationships between the casing strength and temperature and time, three casing grades of N80, P110, and 3Cr-110 were tested in a temperature of 600℃ for 50 hours. Figure 7 show the tensile stress and yield strength change with time. It shows that most changes occur in the first 10 hours, and after it the changes are much gradual.

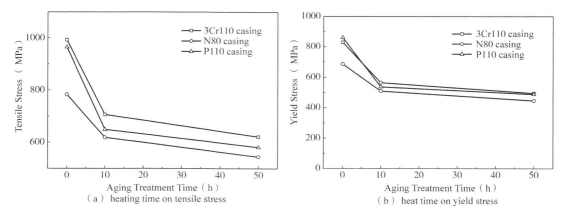

Fig. 7 Aging duration impact on tensile strength of casing grades of N80, P110, 3Cr110

4.4 Creep tests on real casings

Real casing of the three grades are tested in the temperature of 600℃ for 50 hours, and we observed a similar pattern of changes on internal stress, collapsing strength, and tensile strength. The results of measurements for these three casing grades are shown in Figure 8. The elapse time is less than 10 hours, all three properties drop quickly, and after then gradually decline. Overall 3Cr110 casing has a better performance under the same condition than casings of N80 and P110.

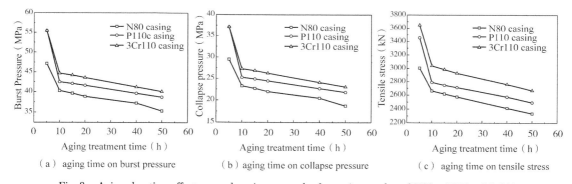

Fig. 8 Aging duration effect on real casing strengths for casing grades of N80, P110, 3Cr110 at the temperature of 600℃

5 Casing Selection and Design for in-situ combustion recovery

If we assume the stress-design theory can be used to select right casing, the design strength for in-situ combustion should satisfy the following criterion:

$$\sigma_T + \sigma_1 = \sigma_{max} \leqslant \frac{\sigma_s}{n} = [\sigma] \tag{9}$$

Where σ_T and σ_1 are thermal stress and mechanical stress, respectively, MPa; σ_{max} is the maximum equivalent stress, MPa; σ_n is maximum yielding stress, MPa; $[\sigma]$ is the allowed stress, MPa; and n is the safety factor.

A coupled model with casing, cement, and formation was setup using Finite element method (FEM) with the above boundary and initial conditions. The model was used to determine the maximum equivalent strengths in the in-situ combustion recovery. Table 1 gives the physical and mechanics parameters that were in the finite element model. These parameters were obtained from the laboratory analysis of samples from Liaohe oil field.

Table 1 FEM calculation physical and mechanics parameters

Parameters	casing	cement	formation
density (kg/m^3)	7850	1830	2720
heat conductivity coefficient[W/(m·℃)]	43.27	0.81	3.44
specific heat[J/(kg·K)]	468.92	879.23	866.67
elasticity modulus (GPa)	194	22	17
poisson ratio	0.26	0.15	0.20
linear expansion coefficient (10^{-6}℃)	11.7	10.3	10.3

If the safety factor of 1.1, is used, the safe temperature upper limits for the tested three casing grades can be determined as shown in Figure 9 for N80, P110, and 3Cr110. The maximum equivalent stress of the casing of grade of N80 is equal to the maximum allowed stress for this casing material at the temperature of 250℃. Therefore, the working temperature for the casing grade of N80 should be less than 250℃. Similarly, the maximum temperature for P110 and 3Cr110 casing grades are 480℃ and 485℃. When these grades of casing are used in the process of in-situ combustion, N80 would prone to fail due to strength. When the temperature is less 550℃, the difference on mechanics properties between P110 and 3Cr110 is negligibly small. Casing grades of P110 and 3Cr110 can be used with some safety margin, and 3Cr110 casing has better mechanic performance than P110 casing.

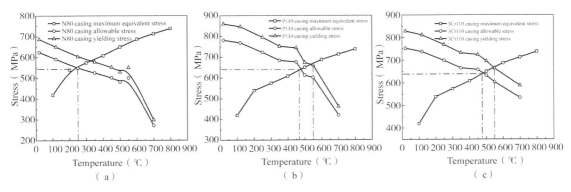

Fig. 9 Determining the maximum allowed working temperature for the grade of
(a) N80 casing, (b) P110 casing, (c) 3Cr110 casing

Table 2 summarizes the observations from the Figure 9 at the temperatures of 250℃, 480℃, and 485℃ for their yielding strength and maximum allowed stress for different casing grades. The table shows that the casing of N80 has a much yielding strength and maximum stress allowed stress than P110 or 3Cr110, and the differences of yielding strength and maximum stress at the tested temperatures are related small. Therefore, if the casing selection is only based on the high yielding strength and maximum allowed stress, the casing of P110 may should be selected because of its price.

Table 2 Casing selection and design results

casing grades	250℃		480℃		485℃	
	yielding strength(MPa)	allowed stress(MPa)	yielding strength(MPa)	allowed stress(MPa)	yielding strength(MPa)	allowed stress(MPa)
N80	605	550	533	485	531	483
P110	797	725	712	647	710	645
3Cr110	772	702	713	648	712	647

6 Conclusions

Many operational challenges limit the applications of in-situ combustion thermal recovery, even it is usually more energy efficient than alternatives. The wellbore integrity and casing stability in the extremely high temperature are major concerns. Given that there is no industry accepted designing protocol for such conditions, this paper proposed a procedure of casing design by integrating finite element analysis of casing performance and laboratory tests on casing coupons and real casings. Under the list assumptions, the temperature field of the in-situ combustion was estimated using a one-dimensional analytical model in the direction of fire flooding front advancing direction. Even the assumptions in the derivation are strict, and there are other factors such as generated gasses and their migration, the actual interested region is the bottom hole temperature and near wellbore region. The bottom hole temperature as a function of time was determined through an empirical correlation validated by real field data. Therefore, we believe that the proposed procedure and conclusions can be widely valid.

The main contributions of this study to the literature exhibits as follows. First, the temperature and pressure profiles as the fire front moves forward are analyzed using numerical models which were tuned with finite field data. This model can be used to other field cases for casing selection and design. Secondly, the laboratory testing results on casing and casing coupons are likely to be the first comprehensive testing under such high temperature. The testing results under these extreme conditions are coupled with finite element modeling to study the variation of mechanical properties of casing, which are useful for setting casing material selection and design. Based on the study presented in the paper, we can draw the following conclusions.

(1) The temperature history at the location of injection well for in-situ combustion was determined numerically and validated by field data from Liaohe oil field.

(2) Mechanical properties of three different casing steel grades of N80, P110, and 3Cr110 are determined experimentally. Testing results shows the mechanical properties drops with the increase of temperature and the increase of testing time. In generally 3Cr110 has better mechanical

performance than the other two materials.

(3) Applying finite element analysis on an integrated model including formation, cement, and well bore, the maximum equivalent stress are determined. Stress-based design protocol is suggested for in-situ combustion. When the working temperature for the casing grade of N80, P110, and 3Cr110 should be less than 250℃, 480℃, and 485℃, respectively.

Acknowledgement

The authors are grateful that China National Oil Corporation(CNPC) to allow publication of this work. This research is supported by Tubular Goods Research Institute of CNPC and State Key Laboratory of Performance and Structural Safety for Petroleum Tubular Goods and Equipment Material of China National Petroleum Corporation(CNPC).

References

[1] Ahmed, Tarek. Reservoir engineering handbook: Elsevier, 2006.
[2] Bergman, David F, Robert P Sutton. "A consistent and accurate dead-oil-viscosity method." SPE reservoir evaluation & engineering, 2009, 12(6): 815-840. doi: https://doi.org/10.2118/110194-PA.
[3] Bird, R Byron, Warren E Stewart, Edwin N Lightfoot. " Transport phenomena. 2002. " JohnWiley & Sons, New York, 2004.
[4] Burger, Jacques G. "Spontaneous ignition in oil reservoirs." Society of Petroleum Engineers Journal, 1976, 16(2): 73-81. doi: https://doi.org/10.2118/5455-PA.
[5] Caruthers, Robert Mack. "High Temperature Oxidation of Crude Oil in Porous Media." Society of Petroleum Engineers(SPE): Richardson, TX SPE-2913-MS, 1970.
[6] Chen, Yong, Xu Peng, Hao Yu. 2017. " Mechanical performance experiments on rock and cement, casing residual stress evaluation in the thermal recovery well based on thermal-structure coupling." Energy Exploration & Exploitation, 2017, 35(5): 591-608. doi: 10.1177/0144598717705048.
[7] Cheng, Qinglin, Yang Liu, Xinyao Xiang. "Enthalpy transfer analysis for oil-water two phase flow in a 1D non-isothermal porous media." Journal of Engineering Thermophysics(Chinese), 2006, 27(1): 20-22.
[8] Han, Lihong, Hang Wang, Jianjun Wang, et al. "Strain Based Design and Field Application of Thermal Well Casing String for Cyclic Steam Stimulation Production." SPE Canada Heavy Oil Technical Conference, Calgary, Alberta, Canada, 7-9 June, 2016.
[9] Hidayat, MIP, S Irawan, Mohamad Zaki Abdullah. "Casing strength degradation in thermal environment of steam injection wells." Journal of Physics: Conference Series, 2016.
[10] Holliday, GH. "Calculation of allowable maximum casing temperature to prevent tension failures in thermal wells. [Preprint]. " Prepr. ASME Petrol. Mech. Eng. Conf.; (United States), 1969.
[11] Jha, Kamal N, Bela Verkoczy. "The role of thermal analysis techniques in the in-situ combustion process." SPE Reservoir Engineering, 1986, 1(4): 329-340. doi: https://doi.org/10.2118/12677-PA.
[12] Lepper, G. B. "Production casing performance in a thermal field." Journal of Canadian Petroleum Technology, 1998, 37(9). doi: http://dx.doi.org/10.2118/98-09-05.
[13] Li, J, SA Mehta, RG Moore, MG Ursenbach, et al. " Oxidation and ignition behaviour of saturated hydrocarbon samples with crude oils using TG/DTG and DTA thermal analysis techniques." Journal of Canadian Petroleum Technology, 2004, 43(7).
[14] Li, Ling Feng. "Research on material selection and sizes optimization of casing material for heavy oil thermal recovery wells." Advanced Materials Research, 2013.

[15] Maharaj, G. "Thermal well casing failure analysis." SPE Latin America/Caribbean Petroleum Engineering Conference, Port-of-Spain, Trinidad, 23–26 April, 1996.

[16] Maruyama, Kazushi, Eiji Tsuru, et al. "An experimental study of casing performance under thermal cycling conditions." SPE drilling engineering, 1990, 5(2): 156–164. doi: https://doi.org/10.2118/18776-PA.

[17] Nesterov, IA, AA Shapiro, et al. "Numerical analysis of a one-dimensional multicomponent model of the in-situ combustion process." Journal of Petroleum Science and Engineering, 2013, 106: 46–61.

[18] Nowinka, Jaroslaw, Daniel Dall Acqua. "New Standard for Evaluating Casing Connections for Thermal-Well Applications." SPE Drilling & Completion, 2011, 26(3): 419–431. doi: https://doi.org/10.2118/119468-PA.

[19] Nowinka, Jaroslaw, Trent Kaiser, et al. "Strain-Based Design of Tubulars for Extreme-Service Wells." SPE Drilling & Completion, 2008, 23(4): 353–360.

[20] Prats, Michael. Thermal Recovery. Edited by Society of Petroleum Engineers, SPE Monograph Series Vol. 7: Society of Petroleum Engineers, 1986.

[21] Ramey, HJ. "In Situ Combustion." 8th World Petroleum Congress, Moscow, USSR, 13–18 June, 1971.

[22] Standardization, International Organization for. ISO 11960: Steel pipes for use as casing or tubing for wells. Geneva, Switzerland, 2010.

[23] Xie, J. "Casing design and analysis for heavy oil wells." First World Heavy Oil Cinference, Beijing, China: 2006-415, 2006.

[24] Xie, J. "A Study of Strain-Based Design Criteria for Thermal Well Casings." SPE Drilling & Completion, 2008, 23(4): 353–360. doi: http://dx.doi.org/10.2118/105717-PA.

[25] Yang, Jian, Xiangfang Li, Zhangxin Chen, et al. "Combined Steam-Air Flooding Studies: Experiments, Numerical Simulation, and Field Test in the Qi-40 Block." Energy & Fuels, 2016, 30(3): 2060–2065.

[26] Zhao, Renbao, Yixiu Chen, et al. "An experimental investigation of the in-situ combustion behavior of Karamay crude oil." Journal of Petroleum Science and Engineering, 2015, 127: 82–92.

本论文原发表于《Journal of Petroleum Science and Engineering》2018 年第 168 卷。

Corrosion Inhibition Performance of 5-(2-Hydroxyethyl)-1, 3, 5-Triazine-2-Thione for 10# Carbon Steel in NH$_4$Cl Solution

Yin Chengxian[1,2]　Ban Xiling[3]　Wang Yuan[1,2]　Zhang Juantao[1,2]
Fan Lei[1,2]　Cai Rui[1,2]　Zhang Junping[3]

(1. State Key Laboratory for Performance and Structure Safety of Petroleum Tubular Goods and Equipment Materials; 2. CNPC Tubular Goods Research Institute; 3. Department of Applied Chemistry, School of Natural and Applied Science, Northwestern Polytechnical University)

Abstract: In this work, 5-(2-hydroxyethyl)-1, 3, 5-triazine-2-thione (HOTAT) was synthesized, and its chemical structure was characterized using FTIR and ^{1}H NMR. The corrosion inhibition performance of HOTAT for 10# carbon steel in 2% NH$_4$Cl solution was studied by weight loss and electrochemical methods. The results showed that HOTAT is a good corrosion inhibitor for 10# carbon steel in 2% NH$_4$Cl solution. The inhibition efficiency of HOTAT increases with increasing inhibitor concentration and decreases with the temperature. The results obtained from potentiodynamic polarization experiments indicate that HOTAT behaves as a mixed-type inhibitor. The adsorption of HOTAT on the 10# carbon steel surface obeys the Langmuir adsorption isotherm, and its adsorption is spontaneous and exothermic.

Keywords: Corrosion inhibitor; Synthesis; Electrochemical; Adsorption isotherm

1 Introduction

Metal materials are often damaged or degraded by the environment during use. Thermodynamic studies have shown that metal corrosion is spontaneous and unavoidable. Corrosion not only leads to the waste of metal resources but also causes corrosion damage to metal structures, which results in large economic losses and catastrophic accidents and depletes noble energy and resources. Of all possible anti-corrosion measures, the use of a corrosion inhibitor is one of the most economic and effective methods for reducing metal corrosion[1-3]. After decades of efforts, great progress has been made in the development and research of corrosion inhibitors, and many types of corrosion inhibitors have been developed[4-8]. Due to environmental protection requirements, high efficiency and low toxicity are driving the development of corrosion inhibitors, and many "green corrosion inhibitors"

Corresponding author: Zhang Junping, zhangjunping@nwpu.edu.cn.

have been developed[9-12] in recent years. Currently, most research focuses on corrosion inhibitors for inorganic acid corrosion[13-15], organic acid corrosion[16-18], salt solution corrosion[19-21] and alkaline corrosion[22-24].

In the oil refinery industry, ammonium chloride (NH_4Cl) corrosion has been reported to be one of the main causes of equipment and piping failures[25-27]. The hydrogenation of hydrocarbons with H_2 is used to saturate olefins and remove impurities[28]. Hydrodenitrogenation is performed according to the following exothermic reaction:

$$R-N+2H_2 \longrightarrow R-H+NH_3(g)$$

In addition, chloride production can occur via reactions such as:

$$R-Cl+H_2 \longrightarrow R-H+HCl(g)$$

In gaseous streams containing NH3 and HCl, acid salts can precipitate according to the following reversible reaction:

$$NH_3(g)+HCl(g) \rightleftharpoons NH_4Cl(s)$$

The solid NH_4Cl can be deposited on the surface of the equipment and cause serious under-deposit corrosion. Water injection processes are often used to dissolve NH_4Cl deposits to minimize under-deposit corrosion. However, NH_4Cl can be hydrolyzed to form a strong acid, which can create an active-passive cell and leads to pitting, crevice corrosion or corrosion cracking. To date, most corrosion inhibitor studies mainly focused on Zn corrosion in NH_4Cl solutions[29,30], which are usually used in batteries. Few investigations of carbon steel inhibitors in NH_4Cl solutions have been reported.

The aim of this study is to synthesize 5-(2-hydroxyethyl)-1,3,5-triazine-2-thione (HOTAT) as a corrosion inhibitor for 10# carbon steel in 2% NH_4Cl solution. The corrosion inhibition action of HOTAT was evaluated by weight loss, polarization and electrochemical impedance spectroscopy (EIS) techniques. Furthermore, the inhibitor adsorption mechanism on the mild steel surface was evaluated by determining the thermodynamic parameters.

2 Experimental

2.1 Instruments and agents

Thiourea, formaldehyde and monoethanolamine were obtained from Sinopharm Chemical Reagent Co., Ltd. (Shanghai, P. R. China). All the chemicals in this study were analytical reagent grade. The instruments used in this work included an AVATAR-360 Fourier transform infrared (FTIR) spectrometer, AVANCE 800 nuclear magnetic resonance spectrometer and JSM 5600LV scanning electron microscope (SEM).

2.2 Synthesis of HOTAT

The synthesis route of HOTAT is shown in Fig. 1.

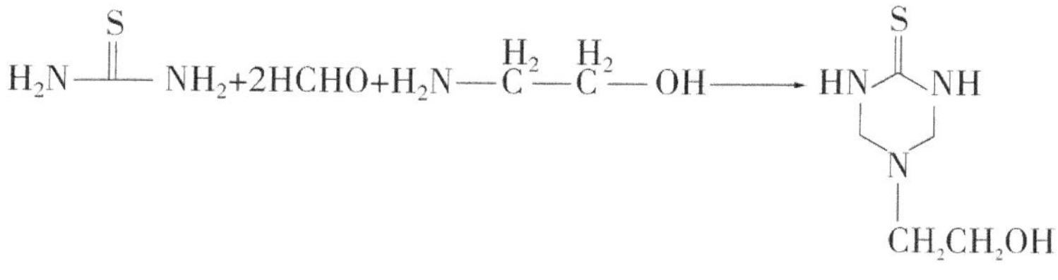

Fig. 1 Synthesis route of HOTAT

Monoethanolamine was added to a mixture of thiourea and 37% aqueous formaldehyde in a molar ratio of 1 : 1 : 2 in a three-necked flask at ambient temperature. Then, the mixture was heated to 70℃ and stirred for two hours. After cooling, it was filtered to give a white solid compound, which was identified as the title compound using FTIR and ^{1}H NMR.

2.3 Weight loss experiments

The specimens used in the weight loss experiments were 10# carbon steel with the following composition in wt.%: C: 0.097, Si: 0.206, Mn: 0.413, P: 0.017, S: 0.007, Ni: 0.004, Cr: 0.019, Mo: 0.002, V: 0.001, Cu: 0.004 and Fe: the balance. The dimensions of the rectangular specimen were 5.0cm × 1.0cm × 0.3cm. The surface of the specimen was polished with silicon carbide paper up to 800 grit, rinsed with distilled water and degreased with acetone before each experiment. The corrosive media were NH_4Cl solutions with different concentrations, which were prepared using NH_4Cl and distilled water. All the experiments were conducted at 50 ~ 80℃ for 6 hours. The specimens were recovered from the solution, and the corrosion product was eliminated using a film-removing solution. Then, the specimens were rinsed with distilled water, degreased with acetone, dried and weighed. The corrosion rate (v) and inhibition efficiency (IE_w) were calculated according to equations (1) and (2)[31-33], respectively.

$$v = \frac{8.76 \times 10^4 \times \Delta m}{\rho \times t \times S} \tag{1}$$

Where v (mm/a) is the corrosion rate of the steel, Δm (g) is the weight loss of the steel, ρ (g/cm^3) is the density of the steel, t (h) is the immersion time, and S (cm^2) is the surface area of the specimen.

$$IE_w = \frac{v_0 - v}{v_0} \times 100 \tag{2}$$

Where v_0 and v are the corrosion rates of steel in the absence and presence of the corrosion inhibitor.

2.4 Electrochemical experiments

Potentiodynamic polarization curves and electrochemical impedance spectra were obtained using a CorrTest instrument (CS350, China). The experiments were performed using a standard three-electrode cell. A steel cylinder inside a Teflon holder served as the working electrode, which had a working area of 1.00cm^2. A graphite rod was used as the counter electrode, and a saturated calomel electrode (SCE).

served as the reference electrode. The polarization curves were recorded from -150 to 200mV vs. the corrosion potential (E_{corr}) with a sweep rate of 0.5mV/s, and the electrochemical impedance spectra were obtained between 100kHz and 10mHz.

The IE_i(%) values were calculated from the potentiodynamic polarization measurements using equation (3)[34,35]:

$$IE_i = \frac{I_{corr} - I'_{corr}}{I_{corr}} \times 100 \tag{3}$$

Where I_{corr} and I'_{corr} are the corrosion currents in the absence and presence of the corrosion inhibitor, respectively.

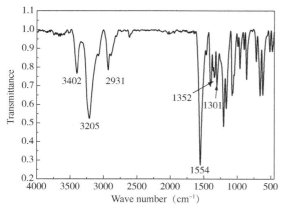

Fig. 2 FTIR spectrum of the product

3 Results and discussion

3.1 FTIR analysis of the synthesized product

Fig. 2 shows the Fourier transform infrared spectrum of the synthesized product.

The peak at 3402 cm^{-1} was attributed to the stretching vibration of –OH. The peak at 3205 cm^{-1} was due to the stretching vibration of –NH–. The peak at 2931 cm^{-1} was attributed to the aliphatic symmetric stretching of CH. The peak at 1554 cm^{-1} was due to the bending vibration of –NH–. The peak at 1352 cm^{-1} corresponded to the stretching vibration of C–N. The peak at 1301 cm^{-1} appeared due to the frequency doubling and combined frequencies of the stretching and deformation vibrations of C=S.

3.2 ^{1}H NMR analysis of the synthesized product

To confirm the structure of the product, ^{1}H NMR spectroscopy was performed. The ^{1}H NMR spectrum of the product exhibited bands at δ = 8.03 ppm (s, 2H, –NH–), δ = 4.59～4.57 ppm (m, 1H, –OH), δ = 4.02 ppm (s, 4H, NH–CH$_2$–N–CH$_2$–NH), δ = 3.51～3.54 ppm (m, 2H, NCH$_2$CH$_2$OH), and δ = 2.61～2.59 ppm (m, 2H, NCH$_2$CH$_2$OH), which confirmed the presence of the expected hydrogen proton (see Fig. 3). The FTIR and 1H NMR results indicated that the synthesized product was HOTAT.

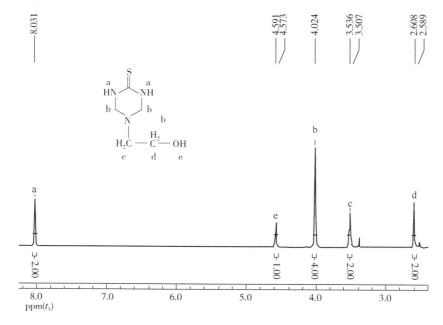

Fig. 3 ^{1}H NMR spectrum of the product

3.3 Weight loss measurements

The corrosion inhibition efficiencies of HOTAT for 10# carbon steel in 2% NH_4Cl solution at different concentrations and temperatures are listed in Table 1.

Table 1 Corrosion inhibition efficiency of HOTAT for 10# carbon steel in NH_4Cl at different concentrations and temperatures

Concentration (mM)	50℃		60℃		70℃		80℃	
	v (mm/a)	IE_w (%)	v (mm/a)	IE_w (%)	v (mm/a)	IE_w (%)	v (mm/a)	IE_w (%)
0	1.0609	—	1.0861	—	1.4268	—	1.8441	—
0.03	0.5474	48	0.6097	44	0.8832	38	1.2743	31
0.06	0.4201	60	0.4855	55	0.7319	49	1.0493	43
0.09	0.3215	70	0.3921	64	0.5978	58	0.8833	52
0.12	0.2429	77	0.3085	72	0.5265	63	0.8096	56

The corrosion rate of 10# carbon steel in the 2% NH_4Cl solution increased with increasing temperature in the absence of the corrosion inhibitor. As shown in Table 1, HOTAT exhibited good inhibition performance for 10# carbon steel in the 2% NH_4Cl solution at various temperatures. The corrosion rate of 10# carbon steel in the presence of the inhibitor decreased with increasing concentration and increased with increasing temperature. Increasing the temperature facilitated the desorption of the corrosion inhibitor molecules from the surface of the carbon steel. As a result, the corrosion rate of carbon steel increased with increasing temperature.

3.4 Potentiodynamic polarization curve

Fig. 4 shows the potentiodynamic polarization behavior of 10# carbon steel in 2% NH_4Cl solution in the absence and presence of different concentrations of the corrosion inhibitor at 50℃. The corrosion potential (E_{corr}), corrosion current density (I_{corr}), and anodic and cathodic Tafel's constant values (b_a and b_c, respectively) were calculated from the polarization plots and are summarized in Table 2.

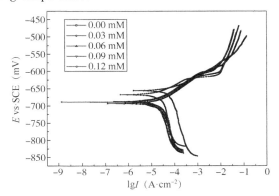

Fig. 4 Potentiodynamic polarization curves for 10# carbon steel in 2% NH_4Cl solution in the absence and presence of different concentrations of the inhibitor at 50℃

As shown in Table 2, the corrosion current density decreased significantly in the presence of the inhibitor, which indicated that HOTAT is an effective inhibitor of the corrosion of 10# steel. The corrosion potential (E_{corr}) decreased slightly in the presence of the inhibitor at all concentrations. The anodic and cathodic Tafel's constant values for 10# steel in 2% NH_4Cl solutions containing the inhibitor varied to some extent. These results indicated that the presence of the inhibitor did not change the corrosion mechanism[36,37]. Both observations suggested that the inhibitor is a mixed-type inhibitor for 10# steel in 2% NH_4Cl solution.

Table 2 Potentiodynamic polarization parameters for 10# carbon steel corrosion in 2% NH_4Cl solution in the absence and presence of different concentrations of the synthesized compound at 50℃

Concentration (mM)	E_{corr} (mV)	I_{corr} (mA·cm^{-2})	$-b_c$ (mV·dec^{-1})	b_a (mV·dec^{-1})	IE_i (%)
0.00	−665	0.6897	143.0	33.7	—
0.03	−667	0.3521	187.3	34.3	49
0.06	−693	0.2295	132.5	42.3	67
0.09	−687	0.2209	106.4	45.9	68
0.12	−688	0.1937	181.3	38.9	72

3.5 Electrochemical impedance spectroscopy measurements

Electrochemical impedance spectroscopy measurements were used to evaluate the influence of HOTAT on the corrosion behavior of 10# carbon steel in 2% NH_4Cl solution. The Nyquist and Bode plots are shown in Figs. 5 and 6, respectively. The semicircles (Nyquist plot) and low-frequency values of the impedance (Bode plot) obtained in the presence of the inhibitor were higher than those obtained in the blank solution, indicating the good inhibitive behavior of HOTAT.

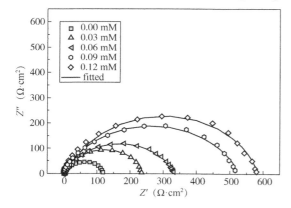

Fig. 5 Nyquist plots for 10# carbon steel immersed in 2% NH_4Cl solution with and without the inhibitor

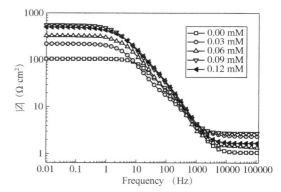

Fig. 6 Bode plots for 10# carbon steel immersed in 2% NH_4Cl solution with and without the inhibitor

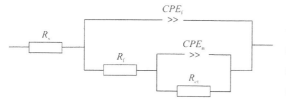

Fig. 7 Equivalent electrical circuit

The equivalent circuit used to fit the electrochemical impedance spectroscopy data is shown in Fig. 7 and is in accordance with other studies[38,39]. The electrochemical impedance spectroscopy parameters determined from the equivalent circuit are shown in Table 3. In Fig. 7 and Table 3, Rs is the solution resistance. The constant phase element representing the double-layer capacitance (C_{dl}) is CPE_n, and R_{ct} is the charge transfer resistance. The parameter CPE_f consists of the film capacitance C_f and the deviation parameter n_1. The inhibitive film resistance is denoted R_f. The values of C_{dl} and C_f were calculated using equations (4) and (5)[40,41], respectively:

$$C_f = Y_0^{\frac{1}{n}} \cdot R_f^{\frac{1-n}{n}} \qquad (4)$$

$$C_{dl} = Y_0^{\frac{1}{n}} \cdot \left(\frac{R_s R_{ct}}{R_s + R_{ct}}\right)^{\frac{1-n}{n}} \qquad (5)$$

According to these results, the R_{ct} and R_f values increased with increasing concentration of the inhibitor. It was proposed that the inhibitor molecules adsorbed on the metal surface, forming a layer that hindered the process of charge transfer[29].

When the inhibitor was added, the C_f values decreased due to the adsorption of the inhibitor molecules on the metal surface[29,42,43]. The decreasing trend in the C_{dl} values indicated that the local dielectric constant decreased and/or the thickness of the electrical double layer increased due to the formation of a protective layer[44].

The corrosion inhibition efficiency (η) values were calculated from the electrochemical impedance spectrum using equation (6)[45,46]:

$$\eta = \frac{R'_{ct} - R_{ct}}{R'_{ct}} \times 100 \qquad (6)$$

Where and R_{ct} are the charge transfer resistances of the solution in the presence and absence of the corrosion inhibitor, respectively.

As shown in Table 3, HOTAT exhibited good inhibition performance on 10# carbon steel in 2% NH_4Cl solution. The corrosion inhibition efficiency (η) increased with increasing concentration of HOTAT.

Table 3 Electrochemical impedance spectroscopy parameters for 10# carbon steel in 2% NH_4Cl solution without and with the addition of the inhibitor at different concentrations

Inhibitor concentration (mM)	R_s ($\Omega \cdot cm^2$)	R_f ($\Omega \cdot cm^2$)	R_{ct} ($\Omega \cdot cm^2$)	C_f ($\mu F \cdot cm^{-2}$)	n_1	C_{dl} ($\mu F \cdot cm^{-2}$)	n_2	η (%)
0.00	1.06	28.99	87.83	86.40	1	92.23	0.8	—
0.03	2.35	32.71	207.73	69.96	0.8	67.77	0.8	58
0.06	1.22	55.78	337.62	44.85	1	65.29	0.7	74
0.09	1.68	67.75	460.02	43.02	1	50.64	0.7	81
0.12	2.76	89.41	492.35	38.47	1	26.95	0.8	82

3.6 SEM analysis

The surface morphologies of 10# carbon steel in 2% NH_4Cl solution at different temperatures are shown in Fig. 8. The surface of the 10# carbon steel was seriously corroded in the 2% NH_4Cl solution in the absence of the inhibitor at different temperatures (Figs. 8a, b, c and d).

The samples retrieved from solutions containing the inhibitor had comparatively smoother surfaces (Figs. 8a', b', c' and d') and were only somewhat degraded. However, the surface was less corroded and more uniform at lower temperatures than at higher temperatures,

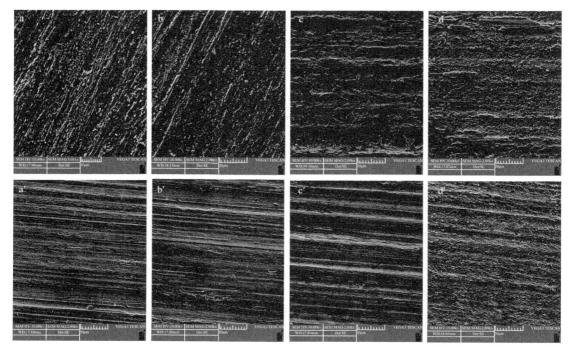

Fig. 8 Surface morphology of 10# carbon steel in 2% NH_4Cl solution at different temperatures
(a-50℃, 0.00 mM; a'-50℃, 0.12 mM; b-60℃, 0.00 mM; b'-60℃, 0.12 mM;
c-70℃, 0.00 mM; c'-70℃, 0.12 mM; d-80℃, 0.00 mM; d'-80℃, 0.12 mM)

demonstrating the superior anti-corrosion performance of the inhibitor at lower temperatures. The micrographs also confirmed the results of the electrochemical and gravimetric analyses.

3.7 Adsorption isotherms

The adsorption isotherm of the corrosion inhibitor describes its adsorption law at a given temperature. It mainly depends on the nature of the corrosion inhibitor itself (polar groups, nonpolar groups, spatial structure) and the metal surface state (non-uniformity). It is generally believed that when the shapes of the anion and anodic polarization curves do not change considerably, the coverage value θ of the corrosion inhibitor on the metal surface is equal to the value of the corrosion inhibition rate.

Fig. 9 shows that the plots of C/θ vs C gave straight lines, suggesting that the adsorption of the inhibitor at the 10# carbon steel/NH_4Cl solution interface obeys the *Langmuir* adsorption isotherm, which is described by the following equation[47-49]:

$$\frac{C_{inh}}{\theta} = \frac{1}{K_{ads}} + C_{inh} \tag{7}$$

Where θ is the surface coverage, C_{inh} is the inhibitor concentration, and K_{ads} is the adsorption equilibrium constant.

The adsorption equilibrium constants (K_{ads}) at different temperatures were estimated from the intercepts of the straight lines in the C_{inh}/θ vs C_{inh} plots (see Fig. 9) and are reported in Table 4. The K_{ads} value decreased with increasing temperature, which was attributed to the increasing desorption of the inhibitor from the metal surface.

The standard free energy of adsorption (ΔG°_{ads}) could be calculated from the adsorption constant (K_{ads}) using equation (8)[50,51]:

$$K = \exp\left(-\frac{\Delta G^\circ}{RT}\right)/55.5 \tag{8}$$

The ΔG°_{ads} values of the inhibitor were negative (Table 4), indicating spontaneous adsorption of the inhibitor on the steel surface. These ΔG°_{ads} values ranged from $-38 \sim -40$ kJ·mol^{-1} (Table 4), which indicated that the inhibitor was adsorbed on the mild steel surface as a consequence of physisorption and chemisorption processes[52].

Table 4 Standard thermodynamic parameters for the adsorption of the synthesized compound on the 10# carbon steel surface in 2% NH$_4$Cl solutions with different inhibitor concentrations at various temperatures

Temperature (°C)	K_{ads}	ΔG°_{ads} (kJ·mol^{-1})	ΔH°_{ads} (kJ·mol^{-1})	ΔS°_{ads} (J·mol^{-1}·K^{-1})
50	29559	−38		63
60	25018	−39	−18	63
70	20488	−40		63
80	16739	−40		63

The adsorption heat (ΔH°_{ads}) was determined using the van't Hoff equation [equation (9)][53] and was obtained from the $\ln K_{ads}$ vs. $1/T$ slopes (Fig. 10). The negative ΔH°_{ads} value revealed that the adsorption of the inhibitor was exothermic[54]:

$$\ln K_{ads} = \left(\frac{-\Delta H^\circ_{ads}}{RT}\right) + \text{constant} \tag{9}$$

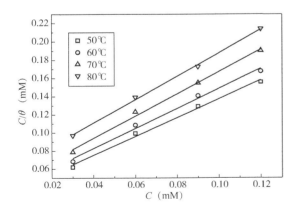

Fig. 9 Langmuir adsorption plots for 10# carbon steel in 2% NH$_4$Cl with different concentrations of the inhibitor at different temperatures

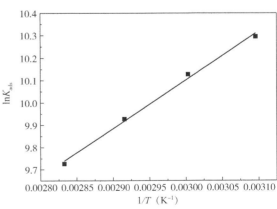

Fig. 10 The relation between $\ln K_{ads}$ and $1/T$ for 10# carbon steel in 2% NH$_4$Cl solution with different concentrations of the inhibitor at different temperatures

The entropy of the inhibitor adsorption (ΔS°_{ads}) was obtained using the following equation[55,56]:

$$\Delta G_{ads}^{o} = \Delta H_{ads}^{o} - T\Delta S_{ads}^{o} \tag{10}$$

The positive ΔS_{ads}^{o} oads values (Table 4) were related to the increase in the inhibitor adsorption disorder. This result indicated that the inhibitor molecules adsorbed on the mild steel surface as water molecules desorbed.

4 Conclusions

HOTAT was synthesized, and its corrosion inhibition performance for 10# carbon steel in 2% NH_4Cl solution was studied by weight loss and electrochemical methods. According to the results, the following conclusions can be drawn:

(1) HOTAT was successfully synthesized, and its chemical structure was characterized by FTIR and ^1H NMR.

(2) The inhibition efficiency of HOTAT on 10# carbon steel in 2% NH_4Cl solution increased with increasing inhibitor concentration. Furthermore, HOTAT exhibited mixed-type inhibitor efficiency.

(3) The mechanism of corrosion inhibition involved the adsorption of the inhibitor on the 10# carbon steel surface as described by the Langmuir adsorption isotherm.

(4) The HOTAT adsorption process was spontaneous and exothermic as revealed by the negative ΔG_{ads}^{o} and ΔH_{ads}^{o} values.

Acknowledgment

This study was funded by the Open Foundation of the State Key Laboratory for Performance and Structure Safety of Petroleum Tubular Goods and Equipment Materials and by the Basic Research and Strategic Reserve Technology Research Fund (project of China National Petroleum Corporation (2018Z-01)).

References

[1] Seyyed Arash Haddadi, Eiman Alibakhshi, Ghasem Bahlakeh, Bahram Ramezanzadeh and Mohammad Mahdavianb, J. Mol. Liq., 284 (2019) 682.
[2] M. Mahdavian, A. R. Tehrani-Bagha, E. Alibakhshi, S. Ashhari, M. J. Palimi, S. Farashi, S. Javadian and F. Ektefa, Corros. Sci., 137 (2018) 62.
[3] S. K. Saha, A. Dutta, P. Ghosh, D. Sukul and P. Banerjee, Phys. Chem. Chem. Phys., 18 (2016) 17898.
[4] Ashish Kumar Singh and M. A. Quraishi, J. Appl. Electrochem., 40 (2010) 1293.
[5] Ivana Jevremović, Marc Singer, Srđan Nešić and Vesna Mišković-Stanković, Corros. Sci., 77 (2013) 265.
[6] Mónica Corrales-Luna, Tu Le Manh and E. M. Arce-Estrada, Int. J. Electrochem. Sci., 14 (2019) 4420.
[7] Adewale Adewuyi, Andrea Göpfert and Thomas Wolff, Ind. Crop. Prod., 52 (2014) 439.
[8] Zhihua Tao, Wei He, Shouxu Wang and Guoyun Zhou, Ind. Eng. Chem. Res., 52 (2013) 17891.
[9] Pavithra M. Krishnegowda, Venkatarangaiah T. Venkatesha, Punith Kumar M. Krishnegowda and Shylesha B. Shivayogiraju, Ind. Eng. Chem. Res., 52 (2013) 722.
[10] A. S. Foudal, M. M. Hegazi and Ali. El-Azaly, Int. J. Electrochem. Sci., 14 (2019) 4668.
[11] Mohammad M. Fares, A. K. Maayta and Mohammad M. Al-Qudah, Corros. Sci., 60 (2012) 112.
[12] Salawu Abdulrahman Asipita, Mohammad Ismail, Muhd Zaimi Abd Majid, Zaiton Abdul Majid, CheSobry

Abdullah and Jahangir Mirzac, J. Clean. Prod., 67 (2014) 139.

[13] V. V. Torres, V. A. Rayol, M. Magalhães, G. M. Viana, L. C. S. Aguiar, S. P. Machado, H. Orofino and E. D'Elia. Corros. Sci., 79 (2014) 108.

[14] Aprael S. Yaro, Anees A. Khadom and Rafal K. Wael, Alex. Eng. J., 52 (2013) 129.

[15] G. Moretti, F. Guidi ang G. Grion, Corros. Sci., 46 (2004) 387.

[16] Salah Abd El Wanees, Mohamed I. Alahmdi, Abla Ahmed Hathoot and Sabry Shaltoot, Chem. Process Eng. Res., 39 (2015) 13.

[17] M. A. Quraishi and H. K. Sharma, J. Appl. Electrochem., 35 (2005) 33.

[18] Farhat Aisha Ansari and M. A. Quraishi. Arab. J. Sci. Eng., 36 (2011) 11.

[19] Sibel Zor, Prot. Met. Phys. Chem. Surf., 50 (2014) 530.

[20] Benchikh A, Aitout R, Makhloufi L, Benhaddad L and Saidani B., Desalination, 249 (2009) 466.

[21] Jasna Halambek, Katarina Berković and Jasna Vorkapić-Furač, Mater: Chem. Phys., 137 (2013) 788.

[22] Layla A. Al Juhaiman, Amal Abu Mustafa and Wafaa K. Mekhamer, Anti – Corros. Method. M., 60 (2013) 28.

[23] El-Deeb Mohamed M., Ads Essam N. and Humaidi Jamal R., Int. J. Electrochem. Sci., 13 (2018) 4123.

[24] Umoren SA, Inam EI, Udoidiong AA, Obot IB, Eduok UM and Kim KW, Chem. Eng. Commun., 202 (2015) 206.

[25] Alvisi PP and Lins VDFC, Eng. Fai. l Anal., 15 (2008) 1035.

[26] Toba K, Ueyama M, Kawano K and Sakai J, Corrosion, 68 (2012) 1049.

[27] Prince Kumar Baranwal and Prasanna Venkatesh Rajaraman, J. Mater. Res. Technol., 8 (2019) 1366.

[28] Paulo Pio Alvisi and Vanessa de Freitas Cunha Lins, Eng. Fail. Anal., 15 (2008) 1035.

[29] Yujie Qiang, Shengtao Zhang, Lei Guo, Shenying Xu, Li Feng, Ime B. Obot and Shijin Chen, J. Clean. Prod., 52 (2017) 17.

[30] M. A. Deyab, J. Power Sources, 280 (2015) 190.

[31] Saviour A. Umoren, Ime B. Obot, A. Madhankumar and Zuhair M. Gasem, Carbohyd. Polym., 124 (2015) 280.

[32] M. R. Noor El-Din and E. A. Khamis, J. Ind. Eng. Chem., 24 (2015) 342.

[33] M. Behpour, S. M. Ghoreishi, N. Mohammadi, N. Soltani and M. Salavati – Niasari, Corros. Sci., 52 (2010) 4046.

[34] Sudhish Kumar Shukla and M. A. Quraishi, Mater. Chem. Phys., 120 (2010) 142.

[35] B. M. Prasanna, B. M. Praveen, Narayana Hebbar and T. V. Venkatesha, Anti – Corros. Method. M., 63 (2016) 47.

[36] A Chetouani, A Aouniti, B Hammouti, N Benchat, T Benhadda and S Kertit, Corros. Sci., 45 (2003) 1675.

[37] Dileep Kumar Yadav, B. Maiti and M. A. Quraishi, Corros. Sci., 52 (2010) 3586.

[38] Francisco Javier Rodríguez-Gomez, Maira Perez Valdelamar, Araceli Espinoza Vazquez, Paulina Del Valle Perezb, Rachel Mata, Alan Miralrio and Miguel Castro, J. Mol. Struct., 1183 (2019) 168.

[39] Ambrish Singh, Yuanhua Lin, Wanying Liu, Shijie Yu, Jie Pan, Chengqiang Ren and Deng Kuanhai, J. Ind. Eng. Chem., 20 (2014) 4276.

[40] Ambrish Singh, Yuanhua Lin, Mumtaz A. Quraishi, Lukman O. Olasunkanmi, Omolola E. Fayemi, Yesudass Sasikumar, Baskar Ramaganthan, Indra Bahadur, Ime B. Obot, Abolanle S. Adekunle, Mwadham M. Kabanda and Eno E Ebenso, Molecules, 20 (2015) 15122.

[41] B. Hirschorn, M. E. Orazem, B. Tribollet, V. Vivier, I. Frateur and M. Musiani, Electrochim. Acta, 55 (2010) 6218.

[42] M. Yadav, T. K. Sarkar and T. Purkait, J. Mol. Liq., 212 (2015) 731.

[43] R. A. Prabhu, T. V. Venkatesha, A. V. Shanbhag, G. M. Kulkarni and R. G. Kalkhambkar, Corros. Sci., 50 (2008) 3356.

[44] E. Alibakhshi, E. Ghasemi and M. Mahdavian, Corros. Sci., 77 (2013) 222.

[45] Mahmoud N. EL-Haddad. Carbohyd, Polymer, 112 (2014) 595.

[46] J. Aljourani, K. Raeissi and M. A. Golozar, Corros. Sci., 51 (2009) 1836.

[47] S. A. Abd El-Maksoud and A. S. Fouda, Mater. Chem. Phys., 93 (2005) 84.

[48] P. Mohan, G. Paruthimal Kalaignan, J. Mater. Sci. Technol., 29 (2013) 1096.

[49] Muzaffer Özcan, Ramazan Solmaz, Gülfeza Kardaş andilyas Dehri, Colloid. Surface. A, 325 (2008) 57.

[50] L. Fragoza-Mar, O. Olivares-Xometl, M. A. Domnguez-Aguilar, E. A. Flores, P. Arellanes-Lozada and F. Jiménez-Cruz, Corros. Sci., 61 (2012) 171.

[51] A. O. Yuce and G. Kardas, Corros. Sci., 58 (2012) 86.

[52] X. Wang, H. Yang and F. Wang, Corros. Sci., 55 (2012) 145.

[53] M. A. Hegazy, E. M. S. Azzam, N. G. Kandil, A. M. Badawi and R. M. Sami, J. Surfactants Deterg., 19 (2016) 861.

[54] H. M. Abd El-Lateef, V. M. Abbasov, L. I. Aliyeva, E. E. Qasimov and I. T. Ismayilov, Mater. Chem. Phys., 142 (2013) 502.

[55] Sh. Pournazari, M. H. Moayed and M. Rahimizadeh, Corros. Sci., 71 (2013) 20.

[56] G. Moretti, F. Guidi and F. Fabris, Corros. Sci., 76 (2013) 206.

本论文原发表于《International Journal of Electrochemical Science》2019年第14卷。

Investigation into the Failure Mechanism of Chromia Scale Thermally Grown on an Austenitic Stainless Steel in Pure Steam

Yuan Juntao [1, 2]   Wang Wen [2]   Zhang Huihui [2, 3]   Zhu Lijuan [1, 2]
Zhu Shenglong [2]   Wang Fuhui [2]

(1. State Key Laboratory of performance and Structural Safety for Petroleum Tubular Goods and Equipment Materials, Tubular Goods Research Institute of CNPC; 2. Key Laboratory for Corrosion and Protection, Institute of Metal Research, Chinese Academy of Sciences; 3. College of Materials Science and Engineering, Xi'an University of Science and Technology)

Abstract   The corrosion behavior of a commercial austenitic stainless steel HR3C was investigated in air and in pure steam at 700℃, and the results are compared. The results show that the chromium concentration in HR3C was sufficient to form protective chromia scale initially in both environments. The oxide scales thermally grown in pure steam, however, were apt to lose their protectiveness by cracking and buckling. A failure mechanism of chromia scale thermally grown in pure steam is proposed by considering influences of hydrogen produced during steam oxidation in terms of molecular hydrogen and hydrogen defects.

Keywords   Stainless steel; SEM; XRD; XPS; High temperature corrosion

1 Introduction

The corrosion resistance of stainless steels is usually attributed to the formation of protective chromia scales. In thermal steam environments such as boiler pipes in fossil-fired power plants and oil pipes in steam-injected oil wells, materials are required to form thin protective and adhesive scales for service safety. The thickening oxides in boiler materials would lead to overheating and premature creep rupture by reducing the thermal conductivity, and exfoliation of these oxides can produce blockage of tubes and erosion damage of valves and turbines [1]. It is necessary, therefore, to study the oxidation mechanism of heat-resistant steels in steam.

In past decades, conventional low-Cr ferritic steels (typically up to 2.25% Cr) and 9–12Cr steels (strictly, ferritic-martensitic steels) [2-7] have been studied extensively, and field experiences indicate that use of these steels is limited to steam temperatures up to 620 ℃ [8]. Austenitic stainless steels with relatively high Cr and Ni, on the other hand, can be used in higher temperatures due to

Corresponding author: Yuan Juntao, yuanjuntaolly@163.com, yuanjt@cnpc.com.cn.

their combination of relatively good creep strength and high temperature oxidation resistance[9]. Extensive studies [10-19] in recent years on the oxidation of austenitic stainless steels (e.g. 304, 310) in wet gases have reported the breakaway oxidation. It seems that the presence of water vapor changes the initial protective Cr-rich oxide scale into a non-protective multilayer scale with a layer of Fe oxides. It is usually considered that the breakaway of protective Cr-rich scale is attributed to the chromium evaporation in the form of $CrO_2(OH)_2$[11,20-25]. However, the chromium evaporation in low-level-oxygen environments (e.g. pure steam) should be negligible because the formation of volatile $CrO_2(OH)_2$ needs oxygen. According to this perspective, chromium evaporation would not be responsible for the breakaway oxidation of Super 304H in pure steam as found in our previous work [26] where mechanical failure (i.e. cracking) of the initially-formed chromia scale was observed. This remains open to debate.

With above considerations in mind, the present work was performed to investigate the protectiveness of thermally formed chromia scale on a commercial austenitic stainless steel HR3C in pure steam, and to determine its failure mechanism. Based on the characterization of microstructure, chemical compositions and oxide phases, the mechanism for the breakaway oxidation in pure steam was discussed.

2 Experimental

Commercial austenitic stainless steel HR3C was studied in the present work and its compositions are listed in Table 1. HR3C tubes as received were cut into coupons with dimensions of $10\times12\times2.5$ mm^3. All coupons were ground to 1000# with SiC papers and cleaned ultrasonically in distilled water and subsequently in ethanol for 15 min.

Table 1 Nominal chemical compositions of austenitic stainless steel HR3C (wt. %)

Element	Fe	C	Mn	Si	Ni	Cr	Nb	N
wt. %	Bal.	0.06	1.20	0.40	20.0	25.0	0.45	0.20

Isothermal corrosion experiments in pure steam under atmospheric pressure were performed in an apparatus as described elsewhere [26]. Flowing steam was generated by pumping ultra-purified water (resistivity $\geqslant 10M\Omega \cdot cm$) containing 8 ppm by weight dissolved oxygen into a pre-heating furnace. Prior to corrosion, the reaction quartz was purged with high-purity argon to prevent the corrosion of samples during the heating process. The test temperature was set at 700 ℃ and the flow rate of inlet water was maintained at 3mL/min. Samples were furnace cooled after the tests to room temperature in high-purity argon. The linear velocity of steam flow roughly calculated from the ideal gas equation was 14.13cm/s. For comparison, isothermal corrosion tests in air were conducted in a muffle furnace at 700℃.

The corroded specimens were characterized carefully in terms of scale microstructure, phase identification and chemical composition. Scale microstructures were investigated by a Scanning Electron Microscopy (SEM). For metallographic cross-section preparation, the corroded specimens were Ni-coated prior to mounting to protect the oxide scales during grinding/polishing and to reveal a clearer contrast between oxide scale and mounting material. Oxide phases were identified by Grazing

Incidence X-ray Diffraction (GIXRD) with an incidence angle of 0.5. The chemical composition of the oxide films by XPS were finished with an ESCALAB250 using Al Kα radiation (hv = 1486.6eV) at a pass energy of 50.0eV. The take off angles of photoelectrons was 45°, with respect to the sample surfaces. The spectrometer was calibrated against the band energy (BE) of the surface carbon contamination C1s at 285.0eV. In order to get the composition-depth profiles by successively removing the oxide surface with argon ion bombardment, a 2keV argon ion sputtering at a target current of $2\mu A/cm^2$ and a pressure of 5.5×10^{-8} mbar was used. The argon ions sputtering rate was approximately 0.2nm/s. To analyze the individual effects of the Ni 2p and Cr 2p core levels, peak decomposition was performed with XPSPEAK4.1 computer software using Gaussian-Lorentzian peak shapes and a Shirley background.

3 Results

The mass gains were collected periodically by weighing specimens. Considering that protective chromia scale could form due to the sufficiently high concentration of chromium in HR3C, mass gains would be small. Fig. 1 shows the mass-gain curves of HR3C in air and in pure steam at 700℃. As expected, mass gains of specimens corroded in both environments are small. The mass gains of specimens corroded in air are around $0.10mg/cm^2$ after exposure from 50h to 300h. For 50h exposure, the mass gain of specimen corroded in pure steam is smaller than that in air. After that, however, the mass gain in pure steam increases remarkably with exposure time and exceeds the mass gain in air for the same exposure duration. After 300h exposure, the mass gain in pure steam reaches $0.20mg/cm^2$.

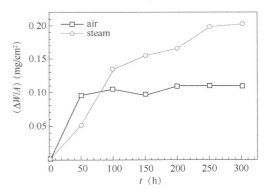

Fig. 1 Mass-gain curves of HR3C steel exposed in two environments at 700℃

Fig. 2 (a) presents the low-magnification surface morphology of oxides thermally formed in air after 300h exposure, where some oxide particles are distributed on the smooth base scale and no cracking and spalling is present. The base scale possesses a high chromium concentration, and the chromium content in the oxide particles is slightly lower than that in the base scale. Fig. 2(b) shows the high-magnification morphology of the oxide particles.

Fig. 3 shows the surface morphologies of oxide scales formed in pure steam, indicating severe cracking and spalling. As shown in Fig. 3(a), there are a large number of pores in the spalled area but no iron-rich oxides on the surface of specimen corroded in pure steam for 50h exposure. This might indicate that the spalling occurred during the cooling process. During the subsequent 50h exposure, the chromia scale could be healed due to the high content of chromium in the steel, and then cracking and spalling could also occur during the subsequent cooling process. With the exposure time extended to 200h, the interfacial chromium concentration was insufficient to heal and sustain the protective chromia scale in the spalled area, and iron oxides formed as shown in Fig. 3(b).

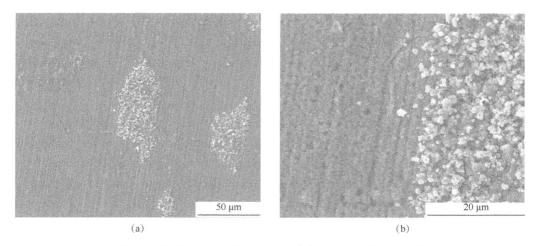

Fig. 2　Surface morphologies under (a) low magnification and (b) high magnification of HR3C steel after 300h exposure in air

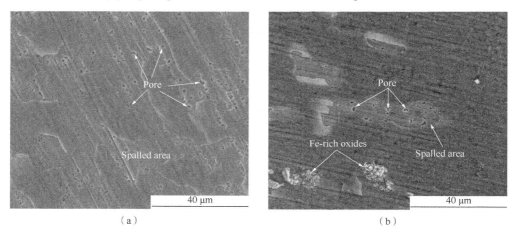

Fig. 3　Surface morphologies of HR3C steel after (a) 50h and (b) 200h exposure in steam

Fig. 4 presents the GIXRD patterns for the surface oxides thermally formed on HR3C steel in

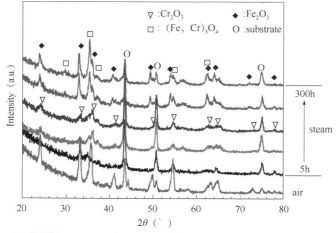

Fig. 4　GIXRD patterns of HR3C steel after exposure in both environments

both environments. In all cases, there are sharp peaks for the HR3C substrates, indicating the surface oxide scales are rather thin. This is in accordance with the small mass gains as shown in Fig. 1. The oxides thermally grown in air mainly consist of chromia. For the specimens corroded in pure steam, the oxides formed after short time exposures (≤100h exposure) consist of Cr_2O_3 and (Fe, $Cr)_3O_4$ and the oxides formed after long time exposures (>100h exposure) consist of (Fe, $Cr)_2O_3$ and (Fe, $Cr)_3O_4$ due to the greater iron content in the oxides.

Considering the small mass gains of HR3C steel in pure steam for short time exposures (i. e. 5h and 50h), the oxide scales were too thin to analyze by SEM technology. In order to analyze the depth distribution of chemical compositions, the argon ion sputtering process was performed. Fig. 5 depicts the XPS depth profiles of Fe, Cr, Ni, and O in the oxides thermally grown in pure steam for 5 h exposure, indicating an outer layer composed of Cr and O, and an inner layer composed of Fe, Cr, Ni, and O. This is inconsistent with the GIXRD result as shown in Fig. 4, where a single Cr_2O_3 scale is indicated. In order to examine the inner layer consisting of Fe, Cr, Ni and O, spectra of Ni 2p and Cr 2p were analyzed. Ni was detected after sputtering 990 s, and the detailed analysis of Ni 2p spectra in Fig. 6 shows the exclusive existence of Ni° (~852. 9eV [27]) for all cases. The detailed analysis of Cr 2p spectra in Fig. 7 shows the exclusive existence of Cr^{3+} (~ 577. 0eV[27]) for spectra taken before sputtering for 990s, and then the coexistence of Cr^{3+} and Cr° (~ 574. 4eV[27]). In this perspective, the outer layer indicated in the XPS depth profiles (Fig. 5) would be a ~198nm thick Cr_2O_3 scale, while the inner layer containing Ni°, Cr^{3+} and Cr° suggests the presence of Cr_2O_3 and metal substrate. This could be explained by followings: (1) Cr_2O_3 is more stable than other oxides such as NiO and Fe_3O_4 beneath an outer Cr_2O_3 layer; (2) the coexistence of Cr° and Cr^{3+} in the inner layer is related to the variation in thickness of chromia scale.

Fig. 8(a) shows the low-magnification cross section morphology of oxides thermally grown in air. Variation of thickness of the oxide scale is evident, in accordance with the observation from the surface morphology. Fig. 8(b) shows the high-magnification morphology of the thicker part which is corresponding to the oxide particles as found on the surface morphology.

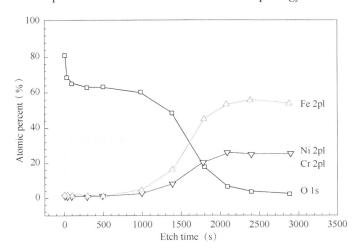

Fig. 5 Composition profiles in the depth of oxides thermally grown in steam for 5h exposure

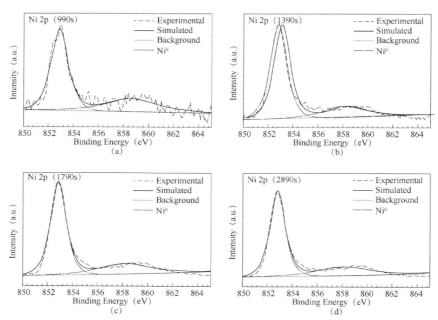

Fig. 6 Detailed XPS spectra of Ni 2p for the oxide scale thermally grown on HR3C in steam for 5h exposure

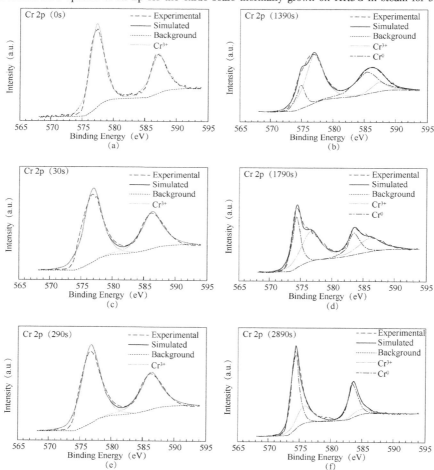

Fig. 7 Detailed XPS spectra of Cr 2p for the oxide scale thermally grown on HR3C in steam for 5h exposure

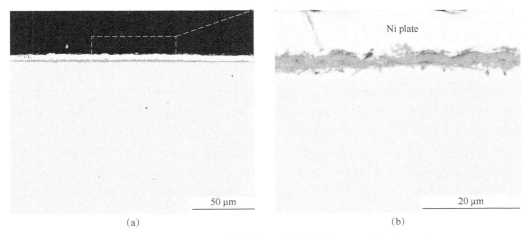

Fig. 8 Cross section morphologies under (a) low magnification and (b) high magnification of HR3C steel after 300h exposure in air

Fig. 9 presents the cross section morphologies of specimens corroded in pure steam for 50h and 200h exposure. Fig. 9(a) shows a thin chromia scale containing cracking and buckling that formed after 50h exposure in pure steam, in accordance with the observation from the surface morphology as shown in Fig. 3(a). Fig. 9(b) shows a thin chromia layer containing corrosion pits, cracking, buckling and healing that formed after 200h exposure in pure steam. The pit corrosion products consist of outer Fe-rich oxides and inner Cr-rich oxides. The pit corrosion indicates the insufficient chromium concentration in the steel to sustain the protective chromia scale, and then results in the rapid formation of less-protective oxides.

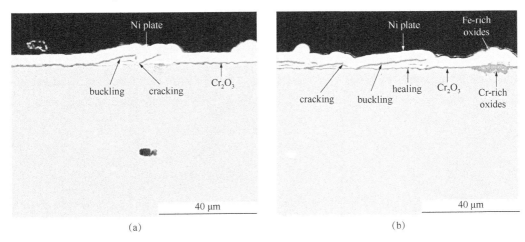

Fig. 9 Cross section morphologies of HR3C steel after (a) 50h and (b) 200h oxidation in steam

4 Discussion

The concentration of chromium in HR3C steel (~25 wt.%Cr) was sufficient to sustain the formation of protective chromia scale in air. In pure steam, protective chromia scale formed after a short time exposure. However, a Cr depletion region formed beneath the oxide scale and significant cracking and buckling occurred during the cooling process. Beneath the chromia scale, a large number of pores were observed in the substrate. As exposure time increased, once the chromium concentration in the steel was insufficient to heal and sustain the protectiveness of chromia scale, iron in the steel became oxidized.

Based on the above observations, it can be speculated that pure steam may cause the breakaway of protective chromia scale by cracking and buckling, and then promote the formation of less protective iron-rich oxides. As mentioned in the Introduction, several mechanisms have been proposed in recent years to interpret the breakaway of chromia scale, such as chromium evaporation[10] and mechanical failure[28]. In a pure flowing steam environment, the evaporation of Cr_2O_3 in terms of $CrO_2(OH)_2$ can be ignored due to the extremely small value of $pH_2O \cdot pO_2^{3/4}$ [20,29]. Shen et al.[28] said that the breakaway of protective chromia might be attributed to the generation of the microcracks and microchannels in the initially-formed Cr_2O_3 scale that allows the inward transport of H_2O molecules, however, they did not find mechanical failure of the initially-formed chromia scale. In our previous work[26] and the present results, we found that the initially-formed chromia scales lost their protectiveness by cracking and buckling as shown in Fig. 3 and Fig. 9. In our point of view, the cracking and buckling of protective chromia scales might result from the large stress in the scale and/or the weak scale/steel interfacial adhesion.

The corrosion reaction of Cr in oxygen can be written as Eq. (1), while the reaction in pure steam can be written as Eq. (2). The difference between these two equations is the production of hydrogen in water. During the process of steam oxidation, the resultant hydrogen may flow into steam or diffuse into the oxides and metal substrate. Several published works employed Secondary Ion Mass Spectrometry (SIMS)[30-31] and Thermal Desorption Spectroscopy (TDS)[32] to study the hydrogen uptake in chromia scales formed in water vapor containing gas, and indicated the highest hydrogen concentration at the metal/oxide interface. Although no currently used techniques can distinguish the value state of hydrogen in the oxides, Norby et al.[33] theoretically discussed the hydrogen defects in oxides in terms of proton (H^+), atomic hydrogen (H), molecular hydrogen (H_2). According to this theory, the resultant hydrogen species may alter the properties of the oxides[34].

$$2Cr + 3/2O_2 = Cr_2O_3 \quad (1)$$

$$2Cr + 3H_2O = Cr_2O_3 + 3H_2 \quad (2)$$

Based on the present results, the breakaway of protective chromia scale formed in pure steam results from the cracking and buckling, which may be related to the hydrogen permeability. The schematic diagram for this mechanism is shown in Fig. 10. During the oxidation process, the growth of chromia scale is mainly supported by the outward diffusion of chromium ions. Because cation outward diffusion usually occurs via a vacancies gradient, vacancies are injected into the metal substrate causing a vacancy supersaturation and eventually a condensation as pores[35], as indicated in Fig. 10(a). The resultant hydrogen may move into the pores at the scale/steel interface which usually act as hydrogen trap[36]. The increasing pressure in the pores would promote the buckling of the protective chromia scale (Fig. 10b). Once the chromia scale cannot relieve the stress by deformation, cracks occur (Fig. 10c). After a short time exposure, the chromium concentration in the steel near the scale/steel interface is sufficient to heal the protective chromia scale. After a longer exposure, more chromium is lost by repeated cracking and healing, and the decreased chromium content in the steel near the outer surface cannot sustain the healing of protective chromia scale. The iron in the steel therefore oxidizes rapidly and forms the nodule/crater structure (Fig. 10d) as found in the case of Super 304H steel[36].

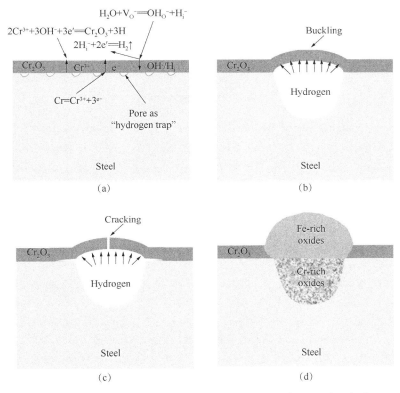

Fig. 10 Schematic diagram for the cracking mechanism of chromia scale formed in high temperature steam

The stress induced by the accumulation of molecular hydrogen in the pores may result in the buckling and cracking of the protective chromia scale. Other possible stresses such as growth and thermal stress of chromia scales are also considered in the scaling process. In the present work, the hydrogen defects [37] may decrease the deformation capability of initially formed protective chromia scale and make the cracking apt to occur.

In summary, three possible causes of cracking of the chromia scales formed in pure steam are: (1) the stress induced by accumulation of molecular hydrogen in the pores, (2) the decreased deformation capability of chromia scale caused by hydrogen defects, and (3) the growth and thermal stresses during the scaling process. Although it is difficult to determine which is the primary contribution to the cracking of chromia scale formed in pure steam, it is likely that the resultant hydrogen during the process of steam oxidation would weaken the protectiveness of chromia scale by causing its mechanical failure.

5 Conclusions

The corrosion behavior of a commercial austenitic stainless steel HR3C was studied in air and in pure steam at 700°C. Based on the above results and discussion, it can be concluded that:

(1) In air, the chromium concentration in HR3C steel is sufficient to form and sustain the protective chromia scale, and therefore HR3C steel shows good corrosion resistance;

(2) In pure steam, the chromium concentration in HR3C steel is also sufficient to form protective chromia scale, but this scale may experience significant cracking and buckling;

(3) The resultant hydrogen during the process of steam oxidation probably weakens the protectiveness of chromia scale by causing its mechanical failure.

Acknowledgement

The authors are grateful for the financial supports from the National Nature Science Foundation of China under Contracts of 51071163 and 51301202, the National Key Basic Research Program of China (973 Program) under grant No. 2012CB625102, and the Special Research Project sponsored by the Education Department of Shaanxi Provincial Government under grant No. 15JK1489.

References

[1] I. G. Wright, R. B. Dooley, A review of the oxidation behaviour of structural alloys in steam, Int. Mater. Rev., 55 (2010) 129–167.

[2] J. Yuan, X. Wu, W. Wang, S. Zhu, F. Wang, Investigation on the enhanced oxidation of ferritic/martensitic steel P92 in pure steam, Materials, 7 (2014) 2772–2783.

[3] L. Liu, Z. G. Yang, C. Zhang, M. Ueda, K. Kawamura, T. Maruyama, Effect of water vapour on the oxidation of Fe – 13Cr – 5Ni martensitic alloy at 973 K, Corros. Sci., 60 (2012) 90–97.

[4] J. Zurek, G. H. Meier, E. Wessel, L. Singheiser, W. J. Quadakkers, Temperature and gas composition dependence of internal oxidation kinetics of an Fe – 10%Cr alloy in water vapour containing environments, Mater. Corros., 62 (2011) 504–513.

[5] D. J. Young, J. Zurek, L. Singheiser, W. J. Quadakkers, Temperature dependence of oxide scale formation on high-Cr ferritic steels in Ar – H_2 – H_2O, Corros. Sci., 53 (2011) 2131–2141.

[6] B. Pujilaksono, T. Jonsson, H. Heidari, M. Halvarsson, J. E. Svensson, G. Johansson, Oxidation of binary FeCr alloys (Fe – 2.25Cr, Fe – 10Cr, Fe – 18Cr and Fe – 25Cr) in O_2 and in $O_2 + H_2O$ environment at 600 °C, Oxid. Met., 75 (2011) 183–207.

[7] A. Aguero, V. Gonzalez, M. Gutiérrez, R. Knodler, R. Muelas, S. Straub, Comparison between field and laboratory steam oxidation testing on aluminide coatings on P92, Mater. Corros., 62 (2011) 561–568.

[8] R. Viswanathan, J. Sarver, J. M. Tanzosh, Boiler materials for ultra-supercritical coal power plants–steamside oxidation, J. Mater. Eng. Perform., 15 (2006) 255–274.

[9] M. P. Brady, K. A. Unocic, M. J. Lance, M. L. Santella, Y. Yamamoto, L. R. Walker, Increasing the Upper temperature oxidation limit of alumina forming austenitic stainless steels in air with water vapor, Oxid. Met., 75 (2011) 337–357.

[10] H. Asteman, J. E. Svensson, L. G. Johansson, Oxidation of 310 steel in H_2O-O_2 mixtures at 600 °C: the effect of water-vapour-enhanced chromium evaporation, Corros. Sci., 44 (2002) 2635–2649.

[11] M. Halvarsson, J. E. Tang, H. Asteman, J. E. Svensson, L. G. Johansson, Microstructural investigation of

the breakdown of the protective oxide scale on 304 steel in the presence of oxygen and water vapour at 600℃, Corros. Sci., 48 (2006) 2014-2035.

[12] F. Liu, J. E. Tang, H. Asteman, J. E. Svensson, L. G. Johansson, M. Halvarsson, Investigation of the evolution of the oxide scale formed on 310 stainless steel oxidized at 600 ℃ in oxygen with 40% water vapour using FIB and TEM, Oxid. Met., 71 (2009) 77-105.

[13] Z. Yue, M. Fu, X. Wang, X. Li, Effect of shot peening on the oxidation resistance of TP304H and HR3C steels in water vapor, Oxid. Met., 77 (2011) 17-26.

[14] M. Calmunger, R. Eriksson, G. Chai, S. Johansson, J. J. Moverare, Surface phase transformation in austenitic stainless steel induced by cyclic oxidation in humidified air, Corros. Sci., 100 (2015) 524-534.

[15] A. Rouaix-Vande Put, K. A. Unocic, M. P. Brady, B. A. Pint, Performance of chromia- and alumina-forming Fe- and Ni-base alloys exposed to metal dusting environments: the effect of water vapor and temperature, Corros. Sci., 92 (2015) 58-68.

[16] T. Dudziak, M. Łukaszewicz, N. Simms, J. R. Nicholls, Steam oxidation of TP347HFG, super 304H and HR3C – analysis of significance of steam flowrate and specimen surface finish, Corros. Eng. Sci. Technol, 50 (2015) 272-282.

[17] Q. Wan, H. Ding, H. D. Liu, R. Y. Wang, Z. G. Li, Y. M. Chen, J. W. Yang, B. Yang, Investigation on short term water vapour oxidation of Super304H, Corros. Eng. Sci. Technol., 50 (2015) 26-33.

[18] G. Rother, J. R. Keiser, M. P. Brady, K. A. Unocic, L. M. Anovitz, K. C. Littrell, R. A. Peascoe-Meisner, M. L. Santella, D. J. Wesolowski, D. R. Cole, Small-angle neutron scattering study of the wet and dry high-temperature oxidation of alumina- and chromia-forming stainless steels, Corros. Sci., 58 (2012) 121-132.

[19] J.-H. Kim, B. K. Kim, D.-I. Kim, P.-P. Choi, D. Raabe, K.-W. Yi, The role of grain boundaries in the initial oxidation behavior of austenitic stainless steel containing alloyed Cu at 700 °C for advanced thermal power plant applications, Corros. Sci., 96 (2015) 52-66.

[20] D. J. Young, B. A. Pint, Chromium volatilization rates from Cr_2O_3 scales into flowing gases containing water vapor, Oxid. Met., 66 (2006) 137-153.

[21] T. Jonsson, B. Pujilaksono, H. Heidari, F. Liu, J. E. Svensson, M. Halvarsson, L. G. Johansson, Oxidation of Fe-10Cr in O_2 and in O_2+H_2O environment at 600℃: a microstructural investigation, Corros. Sci., 75 (2013) 326-336.

[22] X. Peng, J. Yan, Y. Zhou, F. Wang, Effect of grain refinement on the resistance of 304 stainless steel to breakaway oxidation in wet air, Acta Mater., 53 (2005) 5079-5088.

[23] P. J. Meschter, E. J. Opila, N. S. Jacobson, Water Vapor – mediated volatilization of high-temperature materials, Annu. Rev. Mater. Res., 43 (2013) 559-588.

[24] W. Wongpromrat, H. Thaikan, W. Chandra-ambhorn, S. Chandra-ambhorn, chromium vaporisation from AISI 441 stainless steel oidised in humidified oxygen, Oxid. Met., 79 (2013) 529-540.

[25] W. Wongpromrat, G. Berthomé, V. Parry, S. Chandra-ambhorn, W. Chandra-ambhorn, C. Pascal, A. Galerie, Y. Wouters, Reduction of chromium volatilisation from stainless steel interconnector of solid oxide electrochemical devices by controlled preoxidation, Corros. Sci., (2016).

[26] J. Yuan, X. Wu, W. Wang, S. Zhu, F. Wang, The effect of surface finish on the scaling behavior of stainless steel in steam and supercritical water, Oxid. Met., 79 (2013) 541-551.

[27] J. Huang, X. Liu, E.-H. Han, X. Wu, Influence of Zn on oxide films on alloy 690 in borated and lithiated high temperature water, Corros. Sci., 53 (2011) 3254-3261.

[28] J. Shen, L. Zhou, T. Li, High-temperature oxidation of Fe-Cr alloys in wet oxygen, Oxid. Met., 48 (1997) 347-356.

[29] A. Agüero, V. González, M. Gutiérrez, R. Muelas, Oxidation under pure steam: Cr based protective oxides and coatings, Surf. Coat. Technol., 237 (2013) 30-38.

[30] M. Hansel, L. Garcia-Fresnillo, S. L. Tobing, U. Breuer, V. Shemet, Hydrogen uptake and hydrogen profiles in chromia scales formed on Ni25Cr(Mn) in low pO$_2$ test gases at 1000 ℃, Adv. Sci. Technol., 72 (2010) 59-64.

[31] M. P. Brady, M. Fayek, J. R. Keiser, H. M. Meyer Iii, K. L. More, L. M. Anovitz, D. J. Wesolowski, D. R. Cole, Wet oxidation of stainless steels: new insights into hydrogen ingress, Corros. Sci., 53 (2011) 1633-1638.

[32] A. Yamauchi, Y. Yamauchi, Y. Hirohata, T. Hino, K. Kurokawa, TDS measurement of hydrogen released from stainless steel oxidized in H$_2$O-containing atmospheres, Mater. Sci. Forum, 522-523 (2006) 163-170.

[33] T. Norby, M. Wideroe, R. Glockner, Y. Larring, Hydrogen in oxides, Dalton Trans, (2004) 3012-3018.

[34] Z.-Y. Chen, L.-J. Wang, F.-S. Li, K.-C. Chou, Oxidation mechanism of Fe – 16Cr alloy as SOFC interconnect in dry/wet air, J. Alloys Compd., 574 (2013) 437-442.

[35] V. Trindade, H.-J. Christ, U. Krupp, Grain-size effects on the high-temperature oxidation behaviour of chromium steels, Oxid. Met., 73 (2010) 551-563.

[36] A. J. Haq, K. Muzaka, D. P. Dunne, A. Calka, E. V. Pereloma, Effect of microstructure and composition on hydrogen permeation in X70 pipeline steels, Int. J. Hydrogen Energy, 38 (2013) 2544-2556.

[37] S. R. J. Saunders, M. Monteiro, F. Rizzo, The oxidation behaviour of metals and alloys at high temperatures in atmospheres containing water vapour: a review, Prog. Mater Sci., 53 (2008) 775-837.

本论文原发表于《Corrosion Science》2016年第106卷。

NbC-TiN Co-Precipitation Behavior and Mechanical Properties of X90 Pipeline Steels by Critical-Temperature Rolling Process

Zhang Jiming[1] **Huo Chunyong**[2] **Ma Qiurong**[2] **Feng Yaorong**[2]

(1. Institute of Research of Iron and Steel;
2. State Key Laboratory for Performance and Structural Safety of Oil Industry Equipment Materials, Tubular Goods Research Institute of CNPC)

Abstract: The X90 pipeline steel containing Nb and Ti elements was produced using lower (1000℃ rough rolling finished temperature) and higher (1100℃ rough rolling finished temperature) roughing rolling processes (LRRP and HRRP) respectively by thermomechanical controlled processing. Mechanical properties were measured by tensile testing and charpy impact testing, and microstructure was investigated using field emission scanning electron microscopy (FESEM) and field emission transmission electron microscopy (FETEM). LRRP steel has been found to possess similar strength level and almost two times impact toughness compared with HRRP steel. Pancaked austenite grain thickness of LRRP steel was averaged 7.8μm, which was less than a half of grain size of HRRP steel. In addition, precipitates in LRRP steel were mainly NbC-TiN composites, and precipitated intensity was much higher than that of HRRP steel whose precipitates were single NbC or TiN particle. The reason was that rolling deformation at about 1000℃ temperature induced co-precipitation of NbC-TiN composites that promoted epitaxial precipitation of NbC on pre-existing TiN precipitate. High intensity TiN-NbC co-precipitates in LRRP steel promoted adequate Zener drag to retard pancaked austenite growth.

Keywords: Pipeline steel; Mechanical property; Microstructure; Co-precipitation

1 Introduction

High strength microalloy pipeline steels with ultralow carbon and high manganese have been widely applied to the transportation of crude oil and natural gas over a long distance in the past ten years[1,2]. In order to improve the transportation efficiency, outer diameters and transportation pressure of line pipes have been continuously increased. At the same time, high grade pipeline

---

Corresponding author: Zhang Jiming Doctor, Senior Engineer, E-mail: jiming_zhang@126.com.

steels were produced to meet the developing demand of line pipes[3-6]. High grade pipeline steels should possess the combination of high strength and excellent cryogenic toughness according to API 5L standard requirements[7-10]. The refinement of microstructure is an effective way to improve the strength and toughness of pipeline steels by means of thermomechanical controlled processes (TMCP) and addition of niobium and titanium alloying elements[11-15]. The purpose of using TMCP processes, consisting of controlled hot rolling followed by controlled cooling, was to control the transformation temperature and restrain the grain growth at the stage of high temperature after rough and finish rolling. The addition of niobium and titanium elements plays a role of precipitation strengthening and solution strengthening in pipeline steels. Uniform precipitation of nanoparticles has been recognized as one of the most effective methods to increase the tensile strength of steels[16,17]. The degree of strengthening so obtained is highly dependent on the type, number density, size and spatial distribution of the particles, and also the nature of the interaction of the dislocations with the particles. The enhanced mechanical properties of high strength pipeline steels thereby arises from a combination of the refined grain size and the dispersion hardening through the precipitation in matrix.

Added niobium and titanium elements in pipeline steels performed different precipitation behavior according to different manufacturing process. A precipitation temperature of carbonitrides containing Ti approached to the solidus temperature of microalloy steels. However, the precipitation temperature of particles containing Nb was about 1050℃ and substantially lower than that of particles containing Ti[18,19]. The purpose of this work is to investigate effect of precipitation behavior of Nb and Ti particles on mechanical properties of an X90 pipeline steel and optimize the chemical design and rolling process.

2 Experimental material and procedure

The material used in the research is X90 pipeline steel containing Nb and Ti elements, its chemical compositions is listed in Table 1. The steel was melted in 300t converter and poured into ingots having dimensions 300mm. The ingots were soaked at 1200℃ for 2 h and then rolled to 150mm thick plates from 300mm using two different rough rolling processes which were lower rough rolling temperate finished temperatures of 1100℃ (for steel-A) and 1000℃ (for steel-B). Subsequently, the plates were rolled to 19.6mm in a finishing mill at a temperature range of 790~820℃, and followed by accelerated cooling with a rate of 25℃/s.

Samples using for measurement of mechanical properties and microstructure observation were cut from steel plates. Standard tensile specimens with dimensions of ϕ12.5mm×135mm and a gage with a length of 50mm along rolling direction. Tensile tests were carried out with a strain rate of 10^{-3}/s according to ASTM A370 on a MTS 810 servohydraulic machine with an automatic extensometer. Three specimens of each process were tested. Cryogenic impact specimens of Charpy V-Notch(CVN) with a dimension of 10mm×10mm×55mm were machined, and a pre-cut V-notch was perpendicular to rolling direction. Impact specimens were cooled to -10℃ in liquid nitrogen, and testing was performed according to ASTM A370. Five specimens of each process were carried out for impact testing.

Samples for optical microstructure and precipitation observation were polished and etched in 4% nital, and experiments were performed by FEI field emission scanning electron microscopy (FESEM). Specimens for TEM observation were first cut into 300μm thick foils, and then were grinded on SiC abrasive papers to a thickness of 60μm. Lastly the foil was thinned by the twin-jet polishing at 258K, using 5% perchloric acid and 95% ethanol solution as electrolyte. Fine microstructure observation was performed on the FEI F30S field emission transmission electron microscopy(FETEM). At the same time, samples for precipitation analysis were prepared using carbon extraction replicas, and the carbon extraction replicas were examined in FETEM.

Table 1 Chemical compositions of experimental X90 steel (wt%)

Elements	C	Si	Mn	P	S	Mo	Ni	Cu	Nb	Ti	N	Fe
Compositions	0.06	0.25	1.9	0.006	0.002	0.20	0.40	0.25	0.09	0.015	0.005	Bal.

3 Results and discussion

3.1 Mechanical properties

Mechanical properties of X90 steels under different rolling processes were listed in Table 2. Yield strength (YS) and tensile strength (TS) for steel-A were measured as 668MPa and 871MPa, respectively. The values of YS and TS for steel-B were almost the same as that of samples for steel-A, and which were 692MPa and 876MPa, respectively. All tensile properties for samples under two different procedures fulfilled the ASTM A553 standard, and remained within the margins of sufficient strength. However, plasticity and cryogenic toughness of steel-B were more outstanding than that of steel-A. The average uniform elongation and impact toughness at −10℃ temperature reached up to 8% and 499J for steel-B and higher 2% and 238J than that of steel-A, respectively.

Table 2 Mechanical properties of experimental X90 steel

Steels	Tensile strength Rm(MPa)	Yield strength Rt0.5(MPa)	Total elongation (%)	Uniform elongation (%)	CVN (J)
Steel-A	871±11	668±8	23±1	6±1	261±5
Steel-B	876±12	692±9	24±1	8±1	499±8

3.2 Microstructure observation

Fig.1 showed the FESEM micrographs of the hot rolled steel-A and steel-B. Well pancaked prior austenite grain boundaries parallel to rolling direction were observed (Fig.1a and Fig.1b), and the pancaked austenite grain size (thickness) was measured. The pancaked austenite grain thickness of steel-A averaged 18mm(Fig.1a). However, the pancaked austenite grain thickness was less than ten microns and lower than one half of steel-B (Fig.1b). Typical granular bainite and lath-like bainite microstructures were obtained for two steels. A large number of nanoscale precipitates were observed in the matrix. Two categories of different features precipitates in Fig.1c were observed, one category with a regular square shape of about 100 nm and the other category with a spherical shape of less than 40 nm. The energy dispersive spectroscopy(EDS) spectrums of the two categories of nanoscale carbides are presented in Fig.1e and Fig.1f. The EDS analysis revealed that the square precipitates were enriched in Ti, whereas the spherical ones were enriched in Nb. However, the massive number of co-precipitates were founded in Fig. 1d. These co-precipitates in Fig. 1d, regardless of quantity and density, were greatly higher than that of Fig. 1c.

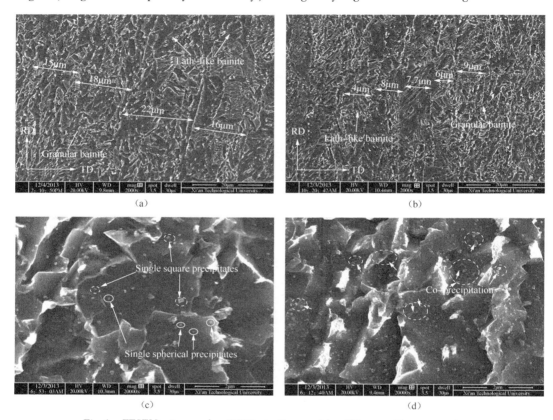

Fig. 1 FESEM micrographs of X90 pipeline steel by different rolling processing,
(a) and (b) for TMCP process and (c) and (d) for CTRP process

3.3 TEM analysis of precipitates

Representative fine microstructure and precipitates of experimental steels observed using TEM were presented in Fig. 2 for steel-A and Fig. 3 for steel-B, respectively. In general, the microstructure in both steels consisted carbide-free bainite(CFB). The CFB was a microstructure made up of packets of parallel plates in morphological packet. The larger plates were also consisted of some small sub-plates. The plate thickness of steel-A was about 4μm for big plates and 2μm for small sub-plates(Fig. 2a and 3a). However, the plate size of steel-B with the sub-plates about 0.2μm thick and the larger plates 0.5~2μm thick was substantially less than that of steel-B. Moreover, high density dislocation substructure was observed in both plates and sub-plates. Fig. 2b-d and Fig. 3b-d presented representative TEM micrographs of precipitates in both steels. The precipitates were of different size and morphology. The precipitates in steel-A were generally divided into two categories, one was regular spherical particles with the range in size from a few nanometers to nearly 100nm, the other was cubic particles with the dimensions of 30~100nm (Fig. 2b and Fig. 2c). The EDS analysis obtained for precipitates identified in Fig. 2b and 2c indicated that spherical particles were NbC and cubic particles were TiN (Fig. 2d). However, precipitates in steel-B were different from steel-A in size and morphology. These precipitates were predominantly co-precipitates including two or three single particles without regular morphologies (Fig. 3b and Fig. 3c). The EDS analysis showed that chemical compositions of co-precipitates were TiN-NbC composites with the average size of about 50nm (Fig. 3d).

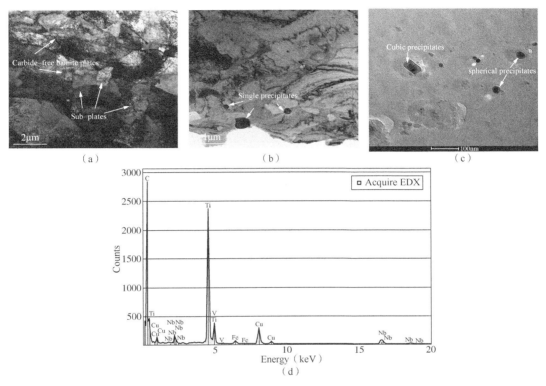

Fig. 2 (a) and (b) Bright field TEM micrographs and (c) precipitates with
(d) EDX analysis in X90 pipeline steel for TMCP process

As it was clearly understood that Ti and Nb elements added during smelting would be precipitated in the form of carbonitrides during cooling and rolling procedure. Turkdogan described a precipitation reaction equation of microalloy elements in steels.

where A and B were constants, $[X][Y]$ was solubility product of X and Y elements in austenite, X indicated microalloy elements, such as Nb, Ti and V etc., Y indicated C and N interstitial atoms. T was kelvin temperature. There existed an equilibrium constant of soluble product at no time, the dissolved Ti and Nb atoms were precipitated as long as the solubility product exceeded this constant. According to the equation 1, calculated precipitated temperature of TiN particles in steel was about 1470℃ close to the temperature of solidus line. The precipitated TiN in grain boundary inhibited austenite grain coarsening at the high temperature. However, the equilibrium temperature for the precipitation of NbC, considerably lower than that of TiN, began to precipitate in the range of 1100℃ to 1050℃. Steel-A was produced by conventional high temperature processing (HTP) that the temperature window of roughing and finishing rolling avoided

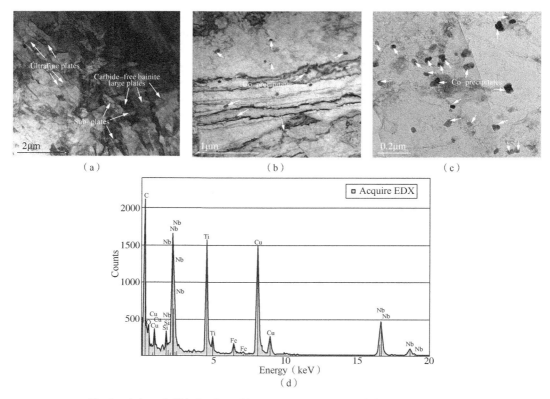

Fig. 3 (a) and (b) Bright field TEM micrographs and (c) precipitates with (d) EDX analysis in X90 pipeline steel for CTRP process

the equilibrium temperature for the onset precipitation of NbC, precipitation behavior of Ti and Nb elements in steel-A was consistent with precipitation law of TiN and NbC single particle with the decrease of temperature. Therefore, precipitates of steel-A displayed regular morphology of TiN and NbC single particles. However, due to roughing temperature was just right belong to temperature window of onset precipitation of NbC, rolling deformation occurred strain induced nucleation of NbC. Pre-existing TiN precipitates were nucleation particles of NbC epitaxial growth and formed TiN-NbC co-precipitates. So a large number of TiN-NbC co-precipitates were observed in steel-B. These Nano-scale TiN-NbC co-precipitates effectively prevented austenite grain coarsening before pancaking and reduced pancaked austenite grain thickness as Zener drag for a given rolling reduction in order to improve impact toughness. Fine microstructure of steel-B was smaller than that of steel-A as a result of nano-scale TiN-NbC precipitation. Ultra-fine TiN-NbC precipitates lowered the brittleness of the matrix that blunted the tip of the propagating crack in ductile fracture. By inducing the formation of ultra-fine TiN-NbC co-precipitate, the impact toughness of steel-B was enhanced.

$$\lg[X][Y] = -\frac{A}{T} + B \qquad (1)$$

4 Conclusions

The main results in the present investigation can be summarized as following:

The critical tempering rolling X90 pipe has the same strength level as the LRRP X90 pipe, however, impact toughness of CTR processing is two times as high as LRRP processing. Results of microstructure observation show that CTR processing X90 pipe possess ultrafine grain size with average 7.8mm of pancaked austenite is less than the LRRP pipe with average 18mm. In addition, the critical rolling promoted strain-induced precipitation of TiN-NbC coprecipitates. Precipitates in the CRT pipe are mainly TiN-NbC, however, precipitates in LRRP pipe are mostly single NbC and TiN particles.

Acknowledgment

This work is financially supported by the National Key Research and Development Program of China (No. 2016YFC0802101).

References

[1] Kong J H, Zhen L, Guo B, et al. Inflence of Mo content on microstructure and mechanical properties of high strength pipeline steel[J]. Mater Design, 2004, 25: 723-728.

[2] Zhang J M, Sun W H, Sun H. Mechanical properties and microstructure of X120 grade high strength pipeline steel[J]. J Iron and Steel Res Int, 2010, 17: 63-67.

[3] Shanmugam S, Ramisetti N K, Misra RDK, et al. Microstructure and high strength – toughness combination of a new 700 MPa Nb-microalloyed pipeline steel[J]. Mater Sci Eng A, 2008, 478: 26-37.

[4] Ghajar R, Mirone G, Arash K. Ductile failure of X100 pipeline steel-Experiments and fractography[J]. Mater Des, 2015, 43: 513-525.

[5] Gräf M K, Hillenbrand H G, Heckmann C J, et al. High-strength large-diameter pipe for long-distance high pressure gas pipelines[J]. Europipe, 2003, 35: 1-9.

[6] Jiming Zhang, Qiang Chi, Lingkang Ji, et al. Microstructure and mechanical properties of twinning M/A islands in a X100 high strength pipeline steel[J]. Materials Science Forum, 2017, 896: 182-189.

[7] Zhang X Y, Gao H L, Zhang X Q, et al. Effect of volume fraction of bainite on microstructure and mechanical properties of X80 pipeline steel with excellent deformability[J]. Mater Sci Eng, 2012, 531 A: 84-90.

[8] Liang P, Li X G, Du C W, et al. Stress corrosion cracking of X80 pipeline steel in simulated alkaline soil solution[J]. Materi Des, 2009, 30: 1712-1717.

[9] Hashemi S H. Comparative study of fracture appearance in crack tip opening angle testing of gas pipeline steels [J]. Mater. Sci. Eng. A, 2012, 558: 702-715.

[10] Li R T, Zuo X R, Hu Y Y, et al. Microstructure and properties of pipeline steel with a ferrite/martensite dual-phase microstructure[J]. Mater Charact, 2011, 62: 801-806.

[11] Zhang J M, Li H, Yang F, et al. Effect of heat treatment process on mechanical property and microstructure of 9%Ni thick steel plate for large lng storage tanks[J]. J Mater Eng Perform, 2013, 22: 3867-3871.

[12] Jiming Zhang, Qiang Chi, Lingkang Ji, et al. Mechanical properties and microstructure of welding joint for high strength-toughness line pipe, Proceedings of the Twenty-sixth (2016) International Ocean and Polar Engineering Conference Rhodes[J]. Greece, 2016: 184-188.

[13] Hashemi S H. Comparative study of fracture appearance in crack tip opening angle testing of gas pipeline steels [J]. Mater. Sci. Eng. A, 2012, 558: 702-715.

[14] J. H. SHIM, Y. W. CHO, S. H. CHUNG, et al. Nucleation of intragranular ferrite at Ti_2O_3 particle in low carbon steel[J]. Acta mater. 1999, 47(9): 2751-2760.

[15] Ji-Ming Zhang, Qiang Chi, Hong-Yuan Chen, et al. Influence of thermal aging on microstructure and

mechanical behavior of X100 high deformability line pipe[C].2015 International Conference on Material Engineering and Mechanical Engineering(MEME2015),2015:524-531.

[16] J. Zhao, W. Hu, X. Wang et al. A novel thermo-mechanical controlled processing for large-thickness microalloyed 560 MPa(X80) pipeline strip under ultra-fast cooling[J]. Mater. Sci. Eng. A, 2016, 673: 373-377.

[17] C. P. Reip, S. Shanmugam, R. D. K. Misra. High strength microalloyed CMn(V – Nb – Ti) and CMn(V – Nb) pipeline steels processed through CSP thin-slab technology: Microstructure, precipitation and mechanical properties[J]. Mater. Sci. Eng. A, 2006, 424: 307-317.

[18] 傅杰,柳得橹. 微合金钢中 TiN 的析出规律研究[J]. 金属学报, 2000, 36(8): 801-804.

[19] 王凤琴,解家英,胡本芙. X70 管线钢中含 Nb 相的析出行为[J]. 北京科技大学学报, 2011, 33(11): 1354-1359.

本论文原发表于《International Journal of Pressure Vessels and Piping》2018 年第 165 卷。

Stress Analysis of Large Crude Oil Storage Tank Subjected to Harmonic Settlement

Zhang Shuxin[1] Liu Xiaolong[2] Luo Jinheng[1] Sun Bingbing[2]
Wu Gang[1] Jiang Jinxu[3] Gao Qi[2] Liu Xiaoben[3]

(1. Tubular Goods Research Institute, China National Petroleum Corporation & State Key Laboratory for Performance and Structure Safety of Petroleum Tubular Goods and Equipment Materials; 2. Petrochina West Pipeline Company; 3. National Engineering Laboratory for Pipeline Safety/MOE Key Laboratory of Petroleum Engineering/Beijing Key Laboratory of Urban Oil and Gas Distribution Technology, China University of Petroleum-Beijing)

Abstract: Differential settlement has a significant effect on the safe operation of tanks. In order to investigate the stress response of crude oil storage tank subjected to harmonic settlement. A numerical simulation model of the steel storage tank was developed in this study. Based on a validated finite element model, parametric analysis was conducted based on the main factors. The results show that the stress at the top of the tank wall is symmetrically distributed along the circumference of the tank. The stress result increases linearly with the increasing harmonic amplitude. The axial stress increases firstly, and then decreases for large wave number. Meanwhile, The stress value increases with the increment of the height-to-radius ratio, and decreases with the increment of the radius-to-thickness ratio. Especially, when the height-to-radius ratio ranging from 1.0 to 1.5 or the ratio r/t is more than 1500, the tendency of stress variation is slighter. This study could be referenced instrength or safety assessment of crude oil storage tank subjected to harmonic settlement.

1 Introduction

With the rapid growth of oil demand, the need of storage devices has increased. As important storage equipment, large scale oil tanks have been widely used in recent decades. Meanwhile, the steel tanks are usually constructed on coastal areas where large differential settlement may easily occur. And the failure behaviour caused by foundation settlement is an urgent problem. In order to ensure the safe operation of storage tanks, investigation on mechanical response of oil storage tank subjected to settlement is important and necessary.

Corresponding author: Liu Xiaoben, xiaobenliu@cup.edu.cn.

In recent years, extensive experimental and numerical research has been conducted to analyze the performance of large scale steel tanks subjected to differential settlement. Gong et al.[1-4] adopted numerical simulation methods to buckling strength of the fixed-roof tank subjected to harmonic settlement. Based on the symmetrical finite element model of the tank shell, Ahmed et al.[5] investigated the effect of tank wall thickness on the buckling mode and the critical buckling settlement displacement. Shi et al.[6,7] studied the effect of wave number, harmonic amplitude and liquid level on the radial deformation of the tank wall. Full-scale model of tank was developed to simulate stress state caused by differential settlement. Li et al.[8] studied the deformation and stress response of large oil storage tanks subjected to foundation settlement. Chen et al.[9,10] investigated the deformation behavior of steel tank wall under uneven foundation settlement, and proposed solution for predicting deformation of steel tank under differential settlement. Shang et al.[11] studied the stress distribution of storage tanks under harmonic settlement, and Fourier series was adopted to represent the measured settlement value. The strength evaluation of the tank wall and bottom was carried out. Yang[12] investigated structural response of thin-walled cylindrical shells under uneven settlement, based on the result of experimental research. He[13] developed the finite element method to obtain the deformation result and stress distribution of steel tanks subjected to uneven settlement.

As it can be seen from there view of previous literature, investigations on the mechanical response of large scale steel tank subjected to differential settlement. In this stage, parametric analysis was conducted based on the main factors that influence the buckling strength of storage tank. While the stress and deformation analysis of tanks is performed by $2D$ finite element method. Numerical simulation results can not accurately reflect the true service status of the storage tank. Meanwhile, influence factors of tank's stress response were not sufficiently investigated. In this study, stress results of large crude oil storage tank was simulated by $3D$ model. The influence of the harmonic amplitude, wave number, height-to-radius ratio and radius-to-thickness ratio on the stress response of tank was studied.

2 Finite element model

2.1 Geometric model

In this study, a full 3D numerical simulation model of crude oil storage tank was developed. The radius and height of the model are $R = 40$m, $h = 22$m, respectively. The variations of wall thickness were considered. The shell and bottom plates of the tank were modeled by S4R. The elements modelling the shell-to-bottom fillet welds and tank foundation are C3D8R. Based on the results of a preliminary mesh sensitivity analysis, the finite element model was divided into 70671 elements as shown in Figure 1.

(a) storage tank model

(b) Section in tank wall

Fig. 1 Finite element model of tank

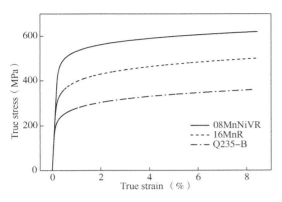

Fig. 2　Stress-strain relationship for tank materials

2.2 Material properties

The elastic modulus, the density and the Poisson ratio of the tank material are 206 GPa, 0.3 and 7850kg/m³, respectively. The yield strength of Q235-B, 12MnNiVR and 16MnR adopted in the model are 235MPa, 490MPa and 345MPa. In order to estimate the mechanical response of steel tank subjected to measured differential settlement. The material nonlinearity was considered in the model. As shown in Figure 2, the well-recognized Ramberg-Osgood model was adopted to regress the true stress-strain curve of the tank materials.

2.3 Boundary conditions

The own-weight of the steel tank was calculated using the gravity constant g = 9.81m/s. Meanwhile, the effect of the hydrostatic pressure should be considered. The liquid pressure value decreases with the increasing height. The effect of the harmonic settlement was investigated by imposing axial displacement load to the circumference of the foundation, while the radial displacement and the circumferential displacement are constrained. The settlement value can be can be indicated as formula 1. The friction coefficient between tank foundation and bottom plate is 0.2 according to verification analysis.

$$u = u_n \cos(n\theta + \varphi_n) \quad (1)$$

Where u_n stands for harmonic amplitude, n represents wave number, φ_n is phase angle.

2.4 FE model verification

The FE model established in this study is validated by a stress result based on field tests[6]. The liquid level of the storage tank is set to 19.76m. The axial stress of the tank wall is plotted against the distance from bottom plate in Figure 3(a). It can be observed from the figure that the trends of the numerical simulation results of axial stress in this study and field test result are basically identical. The maximum of axial stress is mainly located at the shell-to-bottom Fillet

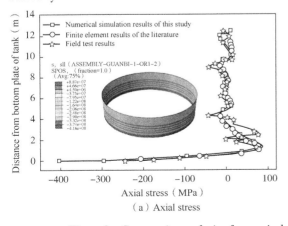

(a) Axial stress

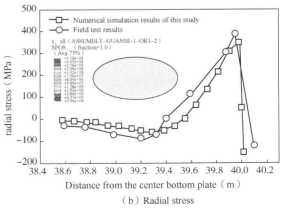

(b) Radial stress

Figure 3　Comparative analysis of numerical simulation results and existing research results

Welds. As can be seen from the figure 3(b), the simulation result of radial stress and test result are highly similar. Comparing the finite element analysis (FEA) results with the existing research results, we can find that the maximum relative error of the model in this paper is 9.15%.

3 Parametricstudy

3.1 Effect of the harmonic amplitude(u)

Harmonic amplitude has a significant effect on the stress response of the tank subjected to harmonic settlement. As the amplitude increases, a large radial displacement, buckling of the shell, and even the failure of the tank may occur. In order to investigate the stress result of tank under differential settlement, the harmonic amplitude varies from 20mm to 100mm, and the Wave number is set as 3. Axial stress and hoop stress are calculated. The stress response result at the top of the tank wall are plotted against the harmonic amplitude in Figure 4. As observed in the figures, the stress is symmetrically distributed along the circumference of the tank. The axial stress and hoop stress presents a similar trend as follow: The stress result increases linearly with the increasing harmonic amplitude.

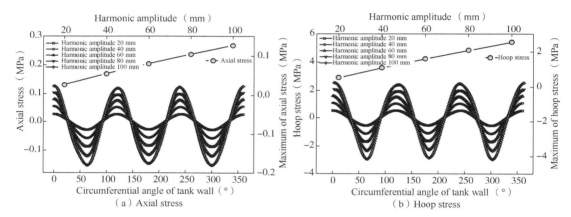

Fig. 4 The stress response results versus circumferential angle of tank wall for various harmonic amplitude

3.2 Effect of the height-to-radius ratio(h/r)

The height-to-radius ratio(h/r) of large crude oil storage tanks varies. As the ratio h/r of tank increases, the stability of the shell will significantly change under the foundation settlement. In order to obtain the effect of the h/r on the stress response of tank wall, the axial stress results at the top of the tank wall with different height-to-radius ratio are displayed together in Figure 5(a). It can be clearly noticed that the axial stress is monotonically increasing. The value of axial stress increases more slightly, when the height-to-radius ratio ranging from 1.0 to 1.5.

3.3 Effect of the radius-to-thickness ratio(r/t)

The wall thickness design of large scale storage tank follows the equal strength design criterion. The value of tank wall thickness in the numerical model varies. Keeping the average wall thickness constant at 18mm, the stress response of tank wall are obtained when radius-to-thickness ratios ranging from 500 to 2000. Figure 5(b) illustrates the effect of the ratio r/t on the axial stress. The plots in the figure show that the axial stress decreases significantly as the radius-to-thickness ratio increases. Meanwhile, as for large ratio (e.g. the ratio r/t is more than 1500), the stress decreases slightly with increasing the ratio. The anti-deformation capacity of tank wall with larger radius-to-thickness ratio will be weaken.

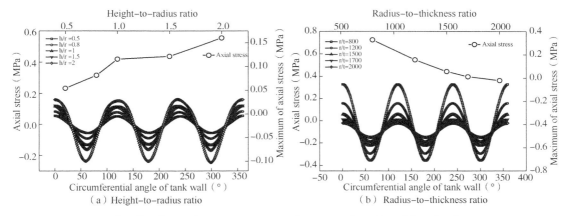

Fig. 5　The axial stress response results versus circumferential angle of tank wall for various tank dimensions

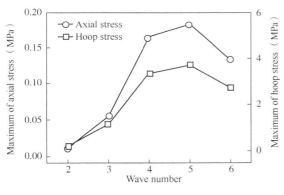

Fig. 6　The stress response results versus wave number

3.4　Effect of the wave number (n)

As the most dangerous settlement type, differential settlement can be simplified as the harmonic forms. The harmonic settlement with wave number varying from 2 to 6 can more accurately reflect the real state of the foundation settlement. In this section, effect of the wave number was investigated by varying n from 2 to 6 while keeping the harmonic amplitude constant at 40mm. Figure 6 present axial stress of tank wall versus wave number. It clearly shows that the wave number has a significant effect on the stress response. The axial stress increases firstly, and then decreases for wave number n=6. This is due to the separation between the tank bottom plate and foundation becomes more obvious, as the wave number increases. The effect of foundation settlement on the tank wall stress is weakened.

4　Conclusions

A comprehensive investigation on stress response of large steel storage tank subjected to harmonic settlement has been performed throughout this paper. A numerical model was established by nonlinear finite element software ABAQUS. Based on the numerical model verified by existing research results, parametric analyses were conducted to derive how the parameters influences the mechanical response of large scale tank. Some remarkable conclusions can be drawn as follows.

(1) The stress value is symmetrically distributed along the circumference direction of the tank. The axial stress and hoop stress presents a similar trend. The stress result increases linearly with the

increasing harmonic settlement amplitude.

(2) The axial stress increases firstly, and then decreases for large wave number(e. g. when the wave number is more than 6). This may be due to the separation between the tank bottom plate and foundation becomes more obvious, as the wave number increases.

(3) When height – to – radius ratio ranges from 0.5 to 2.0. The axial stress increases s monotonically. The axial stress decreases with the increasing radius–to–thickness ratio. As for large ratio(e.g.the ratio r/t is more than 1500), the stress decreases slightly with increasing the ratio.

Acknowledgment

The authors are grateful to the fund support of National Key R&D Program of China (2017YFC0805804).

References

[1] Gong J G, Cao Q S(2017). Buckling strength of cylindrical steel tanks under measured differential settlement: Harmonic components needed for consideration and its effects[J]. Thin-Walled Structures, 119: 345-355.

[2] Zhao Y, Wang Z, Cao Q S, et al. (2013). Buckling behavior of floating-roof steel tanks under measured differential settlement. Thin-Walled structures.

[3] Chen Z P, Fan H G, Cheng, & Jian, et al. (2018). Buckling of cylindrical shells with measured settlement under axial compression. Thin-Walled Structures.

[4] Gong, J, Tao J, Zhao J, Zeng S, & Jin T. (2013). Buckling analysis of open top tanks subjected to harmonic settlement. Thin-Walled Structures, 63(FEB.), 37-43.

[5] Ahmed Shamel Fahmy, Amr Mohamed Khalil(2016). Wall thickness variation effect on tank's shape behaviour under critical harmonic settlement, Alexandria Engineering Journal, Volume 55, Issue 4, Pages 3205-3209, ISSN 1110-0168.

[6] Shi L(2016). Research on the Strength and Stability of Large Crude Oil Tanks[D], China University of Petroleum(Beijing).

[7] Shi L, Shuai J, Xu K, et al. (2014)Assessment of large-scale oil tanks foundation settlement based on FEA model and API 653[J]. China Safety Science Journal, 24(03): 114-119.

[8] Li W, Song W, SongJ L(2016). Deformation and stress analysis of large storage tanks under different settlement modes[J], Spatial structure, 22(04): 78-84.

[9] Chen, Y F, Ma, S, Dong S H. (2020). Deformation of Large Steel Tank under Uneven Foundation Settlement. IOP Conference Series: Earth and Environmental Science.526.012229.10.1088/1755-1315/526/1/012229.

[10] Zhao Y T Lou F Y Chen Y F et al. (2019)Contrastive Analysis on Tank's Behavior Model Under Foundation Settlement[J]. Industrial safety and environmental protection, 04: 30-33.

[11] Shang Guan F Y(2017). Stress Analysis of Large tank under Foundation Settlement[D]. Southwest Petroleum University.

[12] Yang Y. (2011) Experimental research on thin – walled cylindrical shells under uneven settlement[D]. Zhejiang University.

[13] He G F(2012). Deformation and stress analysis of large-scale storage tank under local foundation subsidence [J]. Special Structures, 2012(02)27-31.

本论文原发表于《Earth and Environmental Science》2020 年第 585 卷。

Electrochemical Corrosion Behavior of 15Cr-6Ni-2Mo Stainless Steel with/without Stress under the Coexistence of CO_2 and H_2S

Zhao Xuehui[1,2]   Feng Yaorong[2]   Tang Shawei[3]   Zhang Jianxun[1]

(1. School of Materials Science and Engineering; 2. State Laboratory for performance and Structure Safety of Petroleum Tubular Goods and Equipment Materials, CNPC Tubular Goods Research Institute; 3. School of Materials Science and Engineering, Harbin Institute of Technology)

Abstract: An electrochemical study was performed on 15Cr-6Ni-2Mo stainless steel to evaluate its corrosion behavior. The specimens in the present study were divided into two groups: with and without applied stress. Three conditions were considered and showed significant influence on the corrosion behavior of 15Cr-6Ni-2Mo steel. When the conditions of coexisting CO_2 and H_2S were changed, the passivation state of the material surface had obvious differences. Applied stress decreased the corrosion potential and pitting potential and increased the pitting sensitivity of the materials. The influence of H_2S on the corrosion resistance of materials under stress was more obvious.

Keywords: Stainless steel; Polarization curves; Stress corrosion; Corrosion behavior; Pitting potential

1 Introduction

With the continuously increasing demand for energy and the continuous exploitation of oil-gas fields, downhole tubing corrosion induced by CO_2 is a great threat to integrity of oil-wells, and in most situations, the presence of H_2S and applied stress accelerate corrosion problems[1-4]. Failure accidents caused by corrosion fatigue of an oil well pipe string is a serious environmental threat and also causes great economic loss[5-7]. Thus, the combination of the harsh corrosive environment of an oil well pipe string and the current energy shortage make it is necessary to study and develop more scientific, reasonable, and effective corrosion protection methods[8-10]. However, in terms of economic cost, corrosion resistance, long-term maintenance costs, and service life, each protection method has its advantages and disadvantages. Using a corrosion resistant alloy has proved to be a feasible way to solve the corrosion problem because of its relatively low cost and excellent properties[11,12]. Various corrosion resistant alloys, including low Cr alloy, 13Cr, 22Cr, and nickel-based alloys, have been developed as downhole tubing in different oil field environments[13-19].

Corresponding author: Zhao Xuehui, E-mail: zhaoxuehui@cnpc.com.cn.

However, oil field environments are complicated and distinct, and this limits widespread use of corrosion resistant alloys. At the same time, adaptability between oil field environments and materials is currently a serious problem. In other words, high level material will be overqualified, and low level material will be invalid[20-24]. 15Cr-6Ni-2Mo stainless steel alloy is a new type of martensitic stainless steel that was recently developed in response to the demands of high pressure high temperature oil wells[16]. To date, there have been few works reported on the study of the corrosion behavior of 15Cr-6Ni-2Mo when CO_2 and H_2S were added under different conditions. Meanwhile, there have not been many reports about the corrosion behavior of 15Cr-6Ni-2Mo under stress in an environment of H_2S and CO_2 in the presence of Cl^-. In an oil field environment, CO_2 and H_2S are very aggressive toward the oil well pipe string. In general, CO_2 dissolved in electrolyte can remain in equilibrium with water, and this can be described using equations (1-3)[4,25,26]:

$$CO_2 + H_2O \rightleftharpoons H_2CO_3 \tag{1}$$

$$H_2CO_3 \rightleftharpoons H^+ + HCO_3^- \tag{2}$$

$$HCO_3^- \rightleftharpoons H^+ + CO_3^{2-} \tag{3}$$

H_2S gas can dissolve in water and work as a cathodic depolarizer; this can be expressed using equations (4-7)[17,18]:

$$H_2S \rightleftharpoons H_2S_{ad} \tag{4}$$

$$H_2S_{ad} + e \longrightarrow H_{ad} + HS_{ad}^- \tag{5}$$

$$HS_{ad}^- + H^+ \longrightarrow H_2S \tag{6}$$

$$H_{ad} + H_{ad} \longrightarrow H_2 \tag{7}$$

This paper aims to investigate the electrochemical behavior of 15Cr-6Ni-2Mo with and without stress under coexisting CO_2 and H_2S. Using a combination of potentiodynamic polarization and electrochemical impedance spectroscopy (EIS), the electrochemical behavior of 15Cr-6Ni-2Mo was investigated, and the effects of H_2S and applied stress were evaluated. It is anticipated that this paper can provide insight into the passive film behavior in different test conditions and clarify the relationship between its electrochemical behavior and passivation characteristics.

2 Experimental

2.1 Materials and solutions

Experimental material was selected from high strength 15Cr-6Ni-2Mo stainless steel tubing; according to the manufacturer, the chemical composition (wt%) was: C 0.024, S 0.002, N 0.02, P 0.013, Si 0.28, Cr 15.22, Ni 6.31, Mn 0.18, Mo 2.11, Cu 0.41, and Fe balance. The metallurgic structure of the specimens was tempered martensite (Fig. 1).

The specimensin the present study were divided into two groups: those with and those without applied stress. Rectangular specimens for the electrochemical corrosion test were 57mm×10mm×3mm (width×length×thickness). Before each experiment, all of the specimens were polished with 2000 grit SiC paper to obtain a smooth specimen surface and to minimize errors caused by surface roughness. Finally, the specimens were cleaned with distilled water and ethanol, dried with cool air, and stored in a dry N_2 atmosphere.

Three test conditions were considered in the present study. The first was a saturated CO_2 solution (named as Test 1). In the second condition, H_2S was introduced into the saturated CO_2 solution when the open circuit potential (OCP) was stable (named as Test 2). In the third condition, H_2S and CO_2 were simultaneously introduced into the base solution (named as Test 3). The test solution, which simulated the formation water in an oil field, was made from analytical grade reagents and deionized water, and the Cl^- concentration was 20g/L. An oil bath was used to maintain the temperature of the test solutions. In this test, a certain amount of $Na_2S \cdot 9H_2O$ (0.05g/L) was added to the test solution instead of H_2S.

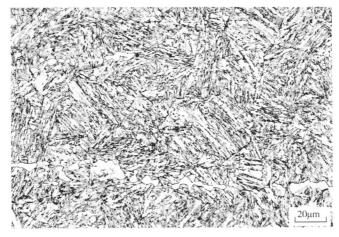

Fig. 1　Microstructure of materials

2.2　Electrochemical tests

Electrochemical measurements were carred out using an electrochemical system with a typical three-electrode electrochemical cell. The working electrode was a rectangular specimen, the counter electrode was two parallel graphite rods, and the reference electrode was silver chloride (Ag/AgCl) with saturated potassium chloride (KCl) solution.

The potentiodynamic polarization curves were recorded over a potential range from −200mV to 800mV versus OCP with a scan rate of 0.3mV s^{-1}. From the polarization curves, OCP, corrosion potential (E_{corr}), passivation current density (I_p), and pitting potential (E_{pit}) were obtained. E_{ocp} is the potential between the working electrode and reference electrode when no electrical current flows. I_p is the passivation current density at which steel is in a stable passive state. E_{pit} is the potential when the corrosion passive film breaks down and leads to a rapid increase in the current density.

A four point bending method was used to load the stress. A diagram of the applied stress is shown in Fig. 2a; the space between the specimen and clamp was insulated by four white Teflon round bars. Electrochemical specimens were welded to a copper wire as the working electrode joint, and the weld points were sealed with epoxy to prevent galvanic corrosion between dissimilar metals. The yield strength (σ_s) of the experimental material at room temperature was 825 MPa, which was provided by the manufacturer, and for experiments, the loading stresses of the specimens were 0% σ_s (no-stress test), 70% σ_s, 80% σ_s, and 90% σ_s.

Also, to reduce the experimental error and to ensure the reproducibility of the experiment, the stress concentration area (region A) of the specimen was exposed to the solution (the exposed area was 2.8 cm^2), and the rest of the area was sealed with a high temperature resistant epoxy resin [Fig. 2(b)]. In these tests, the specimen surface was ground to a smooth finish of $R_a \leqslant 0.2\mu m$. y is just below the proportional limit of the load-strain curve, from which a known elastic stress (σ) can be calculated using the following equation[19]:

$$\sigma = \frac{12Ety}{(3H^2 - 4A^2)} \qquad (8)$$

where E is the elastic modulus, A and H are the distances between the inner and outer supports respectively, and t is the specimen thickness.

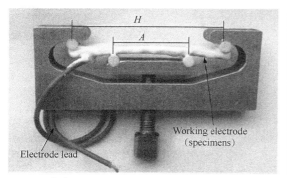
(a) Diagram of applied stress and specimens

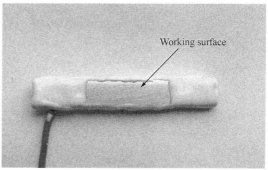
(b) specimen

Fig. 2 Applied stress and specimens

EIS measurements wereperformed at OCP using an alternating current voltage amplitude of 0.005 V. The frequency was varied from 0.01 to 100,000 Hz, and the AC voltage amplitude was 10 mV.

All of the tests were carried out in a circulating oil bath, and atemperature monitoring system was used to control the temperature. All of the curves were tested at least three times to ensure the stability and reproducibility of the tests.

3 Results and Discussion

3.1 OCP measurement

Under the conditions of the CO_2-saturated environment and with added H_2S gas, the pH of the solution system changed obviously with dissolution of the corrosive gas. This change in pH affected the interfacial reaction mechanism of the material and the media, affected electron transfer between the surface of the specimen and the media, and affected formation of the corrosion product film[27-29]. Therefore, when H_2S was added under different test conditions, the corrosion potential of the specimen showed different trends with the dissolution of CO_2/H_2S and with the interfacial reaction.

Time evolution of the OCP for the 15Cr-6Ni-2Mo specimens without stress in three experimental conditions is presented in Fig. 3. Comparison of the test results shows that the obtained OCP values for Test 1 were always more positive, and this indicates that the surface passivation of the specimen was better in the saturated CO_2 environment and that a stable potential was gradually achieved. A more negative potential for the specimens in Test 2 indicate that added H_2S affected the stability of the corrosion film that formed in the early stage.

The OCP of specimens in Test 3 is relatively lower at the beginning, and this indicates that thespecimens were more sensitive to corrosion in the environment with coexisting H_2S and CO_2. The potential increased over time, and this shows that the combination of $H_2S + CO_2$ influenced the kinetics of the anodic reaction and had a significant effect on the passivation of the specimen surface. This difference can be related to the added H_2S. Finally, with formation of the corrosion film, the OCP of the specimens increased gradually and tended to be stable.

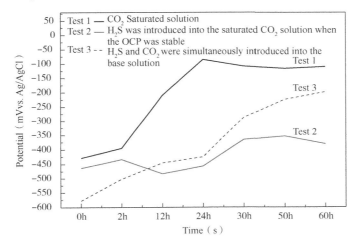

Fig. 3 Changes in the open-circuit potential over time for specimens without stress in different test conditions

3.2 Potentiodynamic polarization behavior

3.2.1 Polarization curves of specimens without stress

Fig. 4 shows the potentiodynamic polarization curves of specimens without stress in the three experimental conditions. A distinct active-passive transition of each specimen was observed in the studied potential range[12]. Nonetheless, there were obvious discrepancies between the polarization behaviors of the different experimental conditions, and the corrosion potential in Test 1 was more positive. When H_2S was added to the experimental solution, the corrosion potential decreased, and the corrosion potential in Test 3 was the most negative. Table 1 lists some of the corresponding electrochemical parameters.

Table 1 Corresponding electrochemical parameters of polarization curves

Conditions	E_{corr}(mV)	I_{corr}(A/cm^2)	E_{pit}(mV)	I_p(A/cm^2)
Test 1	−421	7.5×10^{-7}	15.7	8.9×10^{-6}
Test 2	−496	8.9×10^{-7}	−45	8.5×10^{-5}
Test 3	−605	3.2×10^{-6}	−64	9.4×10^{-6}

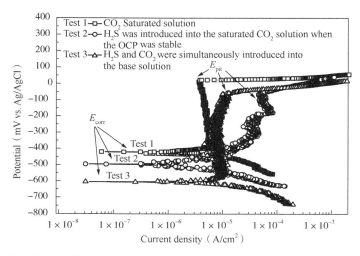

Fig. 4 Potentiodynamic polarization curves of specimens in different CO_2/H_2S conditions

Plateaus of current densities present in curves 1 and 3 indicate distinct passivation-like behavior over a wide range of anodic potentials; the anodic dissolution area was relatively small, and E_{pp} was achieved after a short time. This behavior revealed that the specimen had remarkable passivation behavior under the test conditions, and the passivation film hindered further corrosion of the materials. The value of I_p was relatively stable, and this indicates that the passivation film was dense and completely hindered electron transfer between the material and medium. Thus, when the anodic potential was increased, the corrosion current density remained unchanged. The values of I_p for Tests 1 and 3 were about 8.9×10^{-6} A cm^{-2} and 9.4×10^{-6} A cm^{-2}, respectively. According to the I_p values, the corrosivity of Test solution 1 was relatively small.

Curve 3 is the polarization curve for the combined environment of H_2S and CO_2, and the reaction mechanism was quite different from that of the other condition (curve 1). Absorptivity of HS^- is stronger in acidic solution, and HS^- was prone to adsorb and discharge on the iron electrode. A layer of Fe_xS_y, which was derived from the stoichiometric ratio of ferrous sulfide, was deposited on the surface of the electrode. S^{2-} can form a coordination bond with Fe or Fe^{2+} because of its strong discharging ability. Adhesion of the corrosion product film was improved. As the reaction proceeds, the concentration of S^{2-} decreases, excess Fe^{2+} reacts with CO_3^{2-} and HCO_3^-, and a large amount of carbonate is produced[30,31]. Formation of carbonate was slow, and the size of the product was relatively small. Thus, carbonate filled the pores of the sulfide product, and the surface film became more compact. This film can effectively block electron transfer, and a surface passivation state was obtained.

However, an unstable anodic passivation region was observed for Test 2, and this indicates that there is a dynamic equilibrium reaction between the rupture and self-healing of the passive film. At the beginning of the experiment, the specimens reacted with the CO_2-saturated solution. A carbide corrosion product formed rapidly on the surface of the specimen [equations (9-12)]. When $Na_2S \cdot 9H_2O$ was added to the CO_2-saturated solution, hydrogen sulfide formed rapidly and was involved in the polarization reaction; dissolution of carbide and the formation of new mixed corrosion products formed a complex dynamic equilibrium reaction. Thus, when the anodic current density was increased slightly with an increase in potential, this demonstrated that formation of the passive film only inhibited the corrosion process, whereas local corrosion could not be avoided because of the presence of an electron transport channel in the passive film. When the anodic potential reached −172mV, the new mixed products became stronger; the anodic current density then decreased gradually, and the surface of the specimen was in a passive state. The passivation current density (I_p) was 8.5×10^{-5} A/cm². When the anodic potential reached the critical breakdown potential (E_{pit}), the anodic current density increased immediately.

$$CO_2 + H_2O \longrightarrow H_2CO_3 \tag{9}$$

$$H_2CO_3 + e^- \longrightarrow H^+ + HCO_3^- \tag{10}$$

$$H + H \longrightarrow H_2 \uparrow \tag{11}$$

$$Fe + H_2CO_3 \longrightarrow FeCO_3 + H_2 \tag{12}$$

3.2.2 Effects of loading stress on polarization curves

Not much has been reported on the study of corrosion properties of 15Cr-6Ni-2Mo stainless steel under stress. Under the conditions of high temperature and high pressure CO_2, Zhao et al.[32] studied the effects of different H_2S partial pressures on pitting sensitivity and stress cracking sensitivity of 15Cr-6Ni-2Mo stainless steel at 150℃. Lv et al.[33] studied the effects of an acidic environment and corrosion inhibitor on corrosion potential and corrosion morphology of 15Cr stainless materials at 170℃. Results showed that a high concentration of H_2S decreased the pitting potential and increased the cracking sensitivity of the material, and the corrosion inhibitor effectively improved the corrosion resistance of the material. However, the electrochemical performance of 15Cr-6Ni-2Mo under stress was not further studied.

To study the corrosion performance of the materials under stress, electrochemical corrosion behaviors of the materials under different applied stress conditions were simulated. The corrosion media were saturated CO_2 and the conditions of coexisting $H_2S + CO_2$. The concentration of Cl^- was 20g/L. The experimental temperature was controlled at 90℃.

Corrosion performances of specimens under the combined conditions of stress and corrosion media are shown in Figs. 5 and 6. Table 2 lists some of the corresponding electrochemical parameters.

As seen in Fig. 5, under the stress and test temperature conditions, the values of E_{corr} are −458mV, −500mV, and −475mV. The values of E_{corr} for specimens under applied stress are lower than those for specimens under no stress (−416mV). As observed, with an increase in applied stress, the corrosion potential decreased, and the anode passivation region became increasingly unstable; this indicates that the applied stress enhanced surface activity. The specimen surface remained in a state of dynamic balance between microcorrosion and self-repair.

When the applied stress is greater than or equal to $80\%\sigma_s$, the anodic curve shifted to the right, and the anodic current density obviously increased; this suggests that the corrosion resistance decreased under greater stress. In contrast with the other curves, there is no stable passivation zone in curve 3, and this indicates that the passive film on the specimen surface was not completely formed and that there is still an electron transfer channel between the solution medium and the material, enabling corrosion to continue to occur. Curves 3 and 4 show a similar value of E_{pit} (about -22mV), which is lower than that of the specimens without stress or with lower stress (curves 1 and 2), and this suggests that stress enhanced the pitting susceptibility of materials. When the surface potential reached E_{pit}, the active surface film layer broke down, the corrosion current increased rapidly, and the pitting phenomenon occurred.

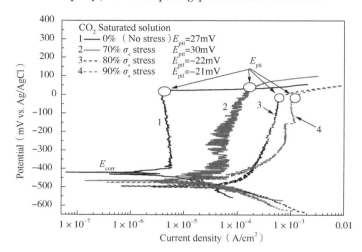

Fig. 5 Potentiodynamic polarization curves of specimens with and without stress in the CO_2 saturated solution system

Table 2 Corresponding electrochemical parameters of polarization curves in CO_2-saturated solution

Applied stress (MPa)	E_{corr} (mV)	I_{corr} (A/cm^2)	E_{pit} (mV)	I_p (A/cm^2)
$0\%\sigma_s$ (no-stress)	-416	7.5×10^{-7}	27	4.5×10^{-6}
$70\% \sigma_s$	-458	5.2×10^{-6}	30	5.8×10^{-5}
$80\% \sigma_s$	-500	7.8×10^{-6}	-22	—
$90\%\sigma_s$	-475	9.5×10^{-6}	-21	7.5×10^{-4}

Polarization curves of specimens with and without applied stress in the combined H_2S+CO_2 condition were obtained (Fig. 6). The values of applied stress were $80\%\sigma_s$ and $90\%\sigma_s$, and the test temperature was 90℃. The corrosion potential of specimens under applied stress decreased from -594mV to -672mV with an increase in the applied stress. At the same time, the polarization curve moved to the right, which indicates that the applied stress also promoted an increase in the corrosion rate.

Compared with the results of the saturated CO_2 condition, the values of E_{corr} for specimens in H_2S+CO_2 conditions all decreased. From thermodynamics analysis, materials are more susceptible to corrosion in the conditions with the combined H_2S+CO_2. In acidic conditions containing H_2S, the cathode reaction was strengthened because of adsorption of H^+ and depolarization effects on the surface; also, the anodic active reaction was accelerated. Moreover, the corrosion medium of H_2S is one of the main factors that lead to stress corrosion cracking.

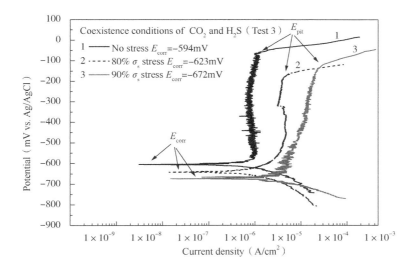

Fig. 6 Potentiodynamic polarization curves of specimens under stress and not under stress in the presence of CO_2 and H_2S

Fig. 7 shows the microscopic morphologies of the surface of each specimen after testing the polarization curves. A small amount of pits are found on the surface of the specimens without applied stress (Fig. 7a), whereas the specimens loaded with $80\%\sigma_s$ and $90\%\sigma_s$ have a large number of tiny pits that are evenly distributed on the surface of the specimens. With an increase in the loading stress, the degree of pitting corrosion increased, and this indicates that stress makes the surface of the material become a relatively high active state and improves the pitting sensitivity.

During potentiodynamic anodic polarization, an increase in the anodic polarization potential caused the passive film on the surface to begin to crack when the potential reached the critical breaking potential. The polarization current increased rapidly, and pitting occurred at this point[34]. In contrast to Fig. 6, curve 1 in Fig. 7 (a) shows that when the polarization potential of the specimen reached E_{pit}, pitting corrosion occurred on the surface of the specimen. When the specimen was loaded with stress, the tensile stress increased the surface activity of the material and accelerated adsorption of Cl^- and OH^- on the surface of the material. Moreover, with an increase in stress, the adsorption rate of Cl^- was accelerated, and the the amount of adsorbed Cl^- increased[35,36].

This is consistent with the results of curves 2 and 3 in Fig. 6. The corrosion current density is relatively larger, and there is no obvious stable passivation region. Curves 2 and 3 also show that the critical breaking potential of the material is relatively lower and that with an increase in the anodic polarization potential, the passivation film was quickly broken down and high density corrosion pits appeared.

3.3 Electrochemical impedance spectroscopy

EIS was measured at the OCP for the specimens both with/without applied stress in different test solutions. The obtained impedance Nyquist spectra are presented in Fig. 8. The specimens all have similar impedance features; the impedance arc has a single incomplete half arc under the condition of no stress, and the radius is relatively larger than that of under stress. When the

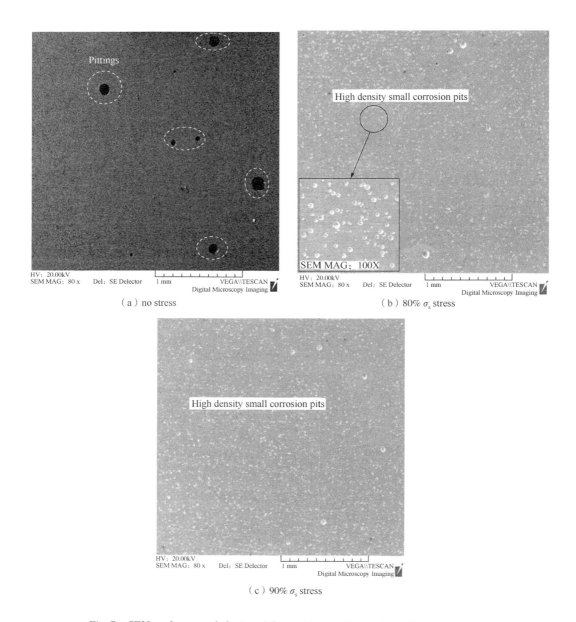

Fig. 7 SEM surface morphologies of the specimens after testing polarization curves

specimens were loaded with different stress (0, 80% σ_s, and 90% σ_s), the impedance radius decreased with an increase in the stress value. This indicates that applied stress affects the activity of the specimen surface, the speed of electron exchange at the metal/solution medium interface is promoted, and corrosion is accelerated. In the combined environment of H_2S+CO_2, the changing amplitude of the impedance arc radius is relatively larger for the stressed state; this indicates that surface corrosion of the material is more sensitive in an H_2S environment, film protection is poor, and charge transfer resistance is relatively small. These observations from the impedance spectra are consistent with the results obtained from the polarization curves.

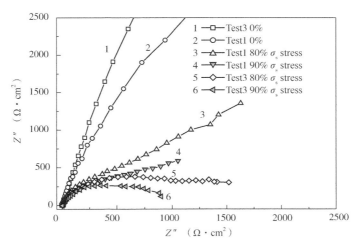

Fig. 8 Electrochemical impedance spectra of the specimens with/without applied stress in saturated CO_2 and H_2S+CO_2 conditions

4 Conclusions

The corrosion behavior of 15Cr-6Ni-2Mo tubing was studied using electrochemistry at 90℃, and the effects of CO_2/H_2S and tensile stress on corrosion behavior were evaluated.

(1) Under the conditions of CO_2-saturated solution and solution of coexisting CO_2+H_2S, the specimens without stress quickly reached a passive state, the passivation zone was relatively wider, and the passivation current density was stable. Although the anodic dissolution zone was relatively wider under the conditions of Test 2, the passivation phenomenon appeared slowly, and passivation current density was relatively larger.

(2) Under the condition of CO_2-saturated solution, with an increase in the applied stress, the passivation state of the material surface was rather unstable. Pitting corrosion potential decreased gradually, but the corrosion potential of the material did not change significantly for states with different amounts of stress.

(3) For the combination of CO_2+H_2S, the corrosion potential and pitting potential decreased with an increase in the applied stress. The polarization curve gradually shifted to the right, and the passive current density increased. Micro-corrosion morphology shows that stress increased the surface activity of the materials and improved the pitting sensitivity of the materials.

(4) The impedance spectrum results show that the impedance radius of the material decreased obviously under stress. Also, the influence of H_2S on the corrosion resistance of the material was more obvious.

Acknowledgement

This research was funded by the China National Petroleum Corporation for Science Research and technology development project (2017D-2307), and the Shaan xi Province Nature Science Foundation of China under Contracts of 2012JQ6014.

References

[1] Shao. Q. Guo, L. N. Xu, L. Zhang, W. Chang, M. X. Lu, Corros. Sci., 110 (2016) 123.

[2] Y. Xiang, C. Li, Z. W. Long, C. Y. Guan, W. Wang, W. Hesitao, Electrochim. Acta., 258 (2017) 909.

[3] G. Hinds, L. Wickström, K. Mingard, A. Turnbull, Corros. Sci., 71 (2013) 43.

[4] G. A. Zhang, Y. Zeng, X. P. Guo, F. Jiang, D. Y. Shi, Z. Y. Chen, Corros. Sci., 65 (2012) 37.

[5] Y. S. Choi, S. Nesic, S. Ling, Electrochim. Acta., 56 (2011) 1752.

[6] A. H. S. Bueno, E. D. Moreir, J. A. C. P. Gomes, Appl. Surf. Sci., 36 (2014) 423.

[7] Y. Liu, L. N. Xu, M. X. Lu, Y. Meng, J. Y. Zhu, L. Zhang, Appl. Surf. Sci., 30 (2014) 768.

[8] Digby D. Macdonald, Adan Sun, Electrochim. Acta., 8 (2006) 1767.

[9] J. Y. Dai, G. Q. Tang, Shandong Chemical Industry, 43 (2014) 58-60.

[10] L. Yang, M. M. Hou, Total Corrosion Control, 28 (2014) 26.

[11] S. Ningshen, M. Sakairi, K. Suzuki, S. Ukai, Corros. Sci., 78 (2014) 322.

[12] E. Oguzie, J. B. Li, Y. Q. Liu, D. M. Chen, Y. Li, K. Yang, F. H. Wang, Electrochim. Acta., 55 (2010) 5028.

[13] B. S. Wang, K. J. Luo, M. C. Zhang, J. X. Dong, The World Steel, 5 (2009) 42.

[14] Q. L. Wu, Z. H. Zhang, X. M. Dong, J. Q. Yang, Corros. Sci., 75 (2013) 400.

[15] Thiago J. Mesquita, E. Chauveau, M. Mantel, N. Bouvier, D. Koschl, Corros. Sci., 81 (2014) 152.

[16] T. J. Mesquita, E. Chauveau, M. Mantel, NKRP. Nogueira, Mater. Chem. Phys., 132 (2012) 967.

[17] A. Pardo, M. C. Merino, A. E. Coy, F. Viejo, R. Arrabal, E. Matykina, Corros. Sci., 50 (2008) 1796.

[18] Y. Xie, Y. Q. Wu, J. Burns, J. S. Zhang, Mater. Charact., 112(2016) 87.

[19] J. Q. Wan, Q. X. Ran, J. Li, Y. L. Xu, X. S. Xiao, H. F. Yu, L. Z. Jiang, Materials & Design, 53 (2014)43.

[20] S. L. Lv, W. W. Song, X. T. Yang, J. X. Peng, Y. Han, G. X. Zhao, X. H. Lv, Z. D. Guo, Y. X. Wen, Corrosion & Protection, 36(2015)76.

[21] S. Fajardo, D. M. Bastidas, M. Criado, J. M. Bastidas, Electrochim. Acta., 129 (2014) 160.

[22] S. D. Zhu, J. F. Wei, R. Cai, Z. Q. Bai, G. S. Zhao, Eng. Fail. Anal., 18 (2011) 2222.

[23] R. Ebara, Procedia Engineering, 2 (2010) 1297.

[24] S. G. Acharyya, A. Khandelwal, V. Kain, A. Kumar, I. Samajdar. Mater. Charact., 72 (2012) 68.

[25] Y. G. Zheng, B. Brown, S. Nesic. Corrosion (Houston, TX, U. S.), 70 (2014) 351.

[26] L. L. Machuca, S. I. Bailey, R. Gubner. Corros. Sci., 64 (2012) 8-16.

[27] X. W. Lei, Y. R. Feng, J. X. Zhang, A. Q. Fu, C. X. Yin, D. Macdonald, Electrochim. Acta., 10 (2016) 640.

[28] Y. Choi, S. Nesic, S. Ling, Electrochim. Acta., 56 (2011) 1752.

[29] C. Plennevaux, J. Kittel, M. Frégonèse, B. Normand, F. Ropital, F. Grosjean, T. Cassagne, Electrochem. Commun., 26 (2013) 17.

[30] G. Verri, K. S. Sorbie, M. A. Singleton, C. Hinrichsen, Q. Wang, F. F. Chang, S. Ramachandran, SPE International Oilfield Scale Conference and Exhibition, Aberdeen, Scotland, UK, 2016.

[31] H. Takabe, K. Kondo, H. Amaya, T. Ohe, Y. Otome, S. Nakatsuka, M. Ueda, CORROSION 2012, ake City, Utah, 2012.

[32] X. H. Zhao, Y. R. Feng, C. X. Yin, Y. Han, Corros. Sci. Prot. Technol., 28(2016)326.

[33] X. H. Lv, F. X. Zhang, X. T. Yang, J. F. Xie, G. X. Zhao, Y. Xue, J. Iron. Steel. Res. Int., 21 (2014) 774.

[34] Y. Sun, J. Hu. Metal Corrosion and Control. Harbin: Harbin Institute of Technology Press, 2003.

[35] C. M. Xu, Y. H. Zhang, G. X. Cheng, W. S. Zhu, Mater. Charact., 59(2008) 245.

[36] J. H. Ding, L. Zhang, M. X. Lu, J. Zhang, Z. B. Wen, W. H. Hao. Appl. Surf. Sci., 15 (2014) 33.

本论文原发表于《International Journal of Electrochemical Science》2018年第13卷。

Fracture Failure Analysis of C110 Oil Tube in a Western Oil Field

Zhu Lixia[1,2]　Kuang Xianren[1,2]　Xiong Maoxian[3]
Xing Xing[3]　Luo Jinheng[1,2]　Xie Junfeng[3]

(1. CNPC Tubular Goods Research Institute; 2. State Key Laboratory for Performance and Petroleum Tubular Goods and Equipment Materials; 3. Tarim Oil Field)

Abstract: There was an oil pipeline fracture found in the unlock process of Z well in western oilfield. The reason to make the oil tube fracture was systematically studied using macroscopic analysis, physical and chemical property test, scanning electron microscope with an energy dispersive X-ray spectrometer and X-ray diffraction analysis in the present study. The results suggested that the requirement of relevant standard for C110 steel grade was satisfied with the chemical component and mechanical property of the oil tube. The perforation of oil tube made H_2S in the fluid medium enter the oil set of connected rings. In addition, the high-pressure gas used in gas lift contained certain H_2S and O_2 so that the tube was fractured and invalid under the combined action of sulfide stress corrosion and oxygen corrosion.

Keywords: Oil tube; Fracture; Stress corrosion; Oxygen corrosion

1 Introduction

Oil tube is one of the most widely used tubing materials in petroleum industry and plays an important role in production. In recent years, with the development of oil and gas fields in harsh geological and environmental conditions, corrosion, bending, surface damage and cracking[1-4] of tube in work are increasing day by day, which causes serious economic losses. In February 2018, during the unlock process of Z well in western oilfield, it was found that the second tubing was fractured at the distance of 1.15m from the master buckle collar, and the specifications of the failed tube is ϕ88.9mm × 7.34mm, and the steel grade was C110. This oil tube began service in 2012. During the oil test, the packer test-retrofit and completion-integrated tubular column was put into the well, and the production zone was activated by gas lift. Then the high pressure gas was obtained. The high pressure gas used in gas lift contained H_2S gas. The high pressure gas used in gas lift contains H_2S gas. During the blowout production of this well, the H_2S concentration at the sampling port was 2100mg/L, and the H_2S concentration in the mouth of the tank was 7~21mg/

Corresponding author: Zhu Lixia, zhulx@cnpc.com.cn.

L. In this paper, the failure process and reasons of the tube were analyzed in combination with the field conditions, in order to provide technical references for the selection and application of the oil tube in oil and gas field.

2 Analysis method

2.1 Macro analysis

The macroscopic morphology of failed tube was shown in Fig. 1. Serious corrosion could be observed on the inner and outer surfaces of the tube. The fracture surface was flat and no obvious plastic deformation was observed after preliminary cleaning of the floating rust on fracture surface, it could be seen that the surface of the fracture was dark gray and could be divided into two areas: shear lip zone(final fracture zone) and flat zone(propagation source area and crack area). Slight radial pattern could be seen in the flat zone and converges to the outer surface. The fracture is stepped with multi-source characteristics.

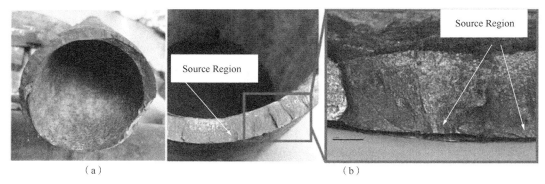

Fig. 1 (a) Macroscopic morphology of fracture.
(b) Multi-source characteristics of fracture source area

2.2 Chemical composition analysis

The samples were taken from the tube body, and chemical composition was analyzed by ARL 4460 direct reading spectrometer. The results were shown in Table 1, and the chemical composition analysis results conformed to the technical requirements.

Table 1 Results of chemical composition analysis ($wt \times 10^{-2}$)

The sample location	C	Si	Mn	P	S	Cr	Mo	Ni	Nb	V	Ti	Cu	B	Al
The tube body	0.28	0.21	0.44	0.0077	0.0010	0.49	0.85	0.030	0.033	0.074	0.0034	0.047	0.0002	0.024
Technical requirements	—	—	—	≤0.015	≤0.005	—	—	—	—	—	—	—	—	—

2.3 Mechanical properties analysis

In the sample of tube body, tensile test, Charpy impact test and hardness test were carried out by UTM5305 material testing machine, PIT302D impact testing machine and RB2002 Rockwell hardness tester, and the results were shown in Table 2. It could be seen from the analysis results that the mechanical properties of the tube conformed to the requirements of the technical requirements.

Table 2 Results of mechanical properties test of the tube

Item	Tensile properties				Impact properties					Hardness
	Diameter ×distance (mm)	Yield strength (MPa)	Tensile strength (MPa)	Elongation (%)	Test temperature (℃)	Impact energy (J)		Shear rate (%)		HRC
						Single	Average	Single	Average	
Results	19.1×50	777	826	23	-10	80 80 78	79	100 100 100	100	24.0 24.7 25.1 25.4
Technical requirements	Tube body 758~828	≥793	≥16	-10	≥44			—		≤30

2.4 Metallographic analysis

Microscopic analysis was carried out on the sample of tube body and fracture. The microstructures of the specimens are fine tempered sorbite with grain size of 11.0 grade. Fig. 2 showed that there were secondary cracks in the fracture, and there were many corrosion pits on both inner and outer surfaces of the sample. The corrosion pits were filled with gray materials, and no abnormal structure was found. The bottom of the corrosion pit on both inner and outer surfaces is sharp, especially on the inner surface. The corrosion pit bottom extended like cracks. From the perspective of corrosion depth, the maximum depth of inner surface corrosion pit and bottom crack was 0.13mm, and the maximum depth of outer surface corrosion pit was 0.24mm. The morphology of outer surface corrosion pit bottom was wider, indicating that the corrosion on the outer surface of tube was more serious than that on the inner surface. The fracture crack analysis of the sample showed that the cracks propagate from the outer surface to the inner surface, and the crack was bifurcated and no abnormal structure was observed around the cracks.

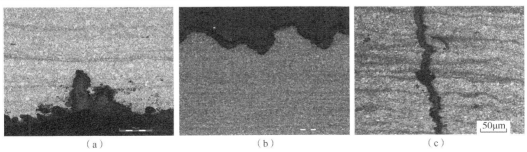

Fig. 2 (a) Microstructure around corrosion pit on inner surface.
(b) Microstructure around corrosion pit on inner surface.
(c) Fracture crack morphology

2.5 SEM and EDS analysis

The fracture source of failed tube was sampled and analyzed by SEM and EDS. The low magnification morphology of the source region was shown in Fig. 3(a). It could be seen that the crack originated from the outer surface, which was consistent with the results of the fracture macroscopic and metallographic analysis. Fracture source zone, crack area and final fracture zone were observed at the low power of fracture. High magnification observation showed that the surface of the fracture was muddy and covered with corrosion products, as shown in Fig. 3(b). After the surface corrosion products of the fracture were removed by mechanical and chemical methods, the morphology of the crack area was observed as shown in Fig. 3(c), and the fracture was characterized by intergranular cracking.

Fig. 3 (a) Low magnification morphology of fracture source zone.
(b) Highmagnification morphology of fracture source zone.
(c) Fracture morphology of crack area

The results of EDS analysis of corrosion products on the outer surface, fracture and bottom of the corrosion pit are shown in Fig. 4. All of corrosion products contained Fe, O, S and other elements. According to the elemental composition of energy spectrum analysis, the corrosion on fracture surface is was mainly composed of oxides and sulfides. After chemical cleaning, the fracture still contained certain S content.

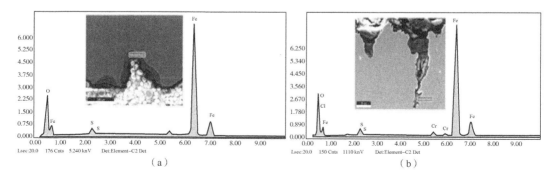

Fig. 4 (a) Energy spectrum analysis of corrosion products on outer surface and bottom of pit.
(b) Energy spectrum analysis of corrosion products in inner surface cracks of corrosion pit Bottom

The corrosion products from the inner and outer wall of the failed tube were stripped off and fully ground into powder samples. XRD analysis method was used to analyze the phase of corrosion products, and the results were shown in Fig. 5. The results showed that the corrosion products of the inner and outer walls were mainly calcium carbonate($CaCO_3$), ferrous sulfide(FeS) and iron oxide (Fe_2O_3). According to the peak phase of corrosion products, the content of calcium carbonate in the inner wall was the highest, followed by ferrous sulfide and iron oxide. The content of ferrous sulfide in the outer wall was the highest, followed by iron oxide and calcium carbonate. The analysis showed that calcium carbonate originates from the scaling of well fluid rather than corrosion products.

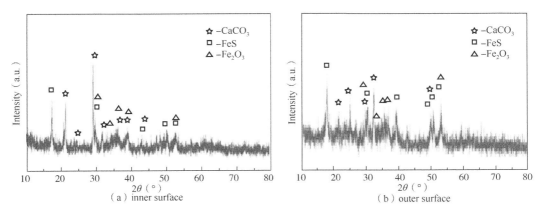

Fig. 5 XRD analysis of corrosion products

3 Results and discussion

The chemical composition analysis and mechanical properties test results of tube could be concluded that the chemical composition, tensile properties, impact toughness and hardness test results of ϕ88.9mm×7.34mm C110 tube used in the well accorded with the requirement of technical requirements; Metallographic analysis showed that the structure of the tube was tempered Sorbite with fine grain size of 11.0 grade. Charpy impact test results showed that the tube has good toughness at low temperature. According to the above test results, the types and causes of tube fracture were analyzed.

3.1 Macroscopic analysis results

Firstly, the macroscopic analysis results of the failed tube showed that the whole tube fracture showed no obvious plastic deformation and showed brittle fracture characteristics. The fracture surface was gray black with obvious radial pattern on the surface, indicating that the fracture originated from the outer surface of the tube. At the same time, the fracture was characterized by step-like and multi-source crack initiation. According to macroscopic characteristics of tube fracture, it was preliminarily judged that tube fracture belonged to H_2S stress corrosion cracking[5-6].

3.2 High magnification morphology of fracture

Secondly, the morphology could be seen from the high magnification of fracture. The fracture surface was covered by corrosion products, showing a muddy pattern. After the surface corrosion the products were removed by mechanical and chemical methods, obvious intergranular fracture characteristics were observed in the local areas. The results of energy spectrum analysis on the fracture surface of tube showed that Fe, O and S were the main elements which had characteristic of oxygen corrosion. The fracture microstructure was characterized by H_2S stress corrosion cracking. In addition, metallographic analysis of fracture showed that there were corrosion pits on both inner and outer surfaces of the tube. In terms of corrosion depth, the corrosion pit depth on the outer surface was greater than that on the inner surface, which indicating that the corrosion on the outer surface of the tube was more serious than that on the inner surface.

3.3 Phase analysis of the inner and outer walls

The phase analysis of the inner and outer walls of the oil tube showed that the corrosion products of the inner and outer walls contained FeS and Fe_2O_3. Therefore, the results of macroscopic and microscopic characteristics analysis, cross section energy spectrum analysis and physical phase analysis of the fracture could be determined that the fracture of the tube was the result of the combined action of H_2S stress corrosion cracking and oxygen corrosion. According to the factors of H_2S stress corrosion cracking and corrosion medium, the failure reasons of the tube were briefly analyzed as follows.

Material of tube: according to literature, one of the controlling factors to prevent H_2S stress corrosion cracking of steel is hardness $\leqslant$ HRC22[7,8]. According to the physical and chemical properties test results of the tube, the Rockwell hardness of the tube was between HRC 23.6 and HRC 25.4, indicating that the material of tube was sensitive to stress corrosion cracking.

Stress: according to the string structure of Z well, the failed tube was the second tube attached to the well tubing, with a total length of 6092.42m. The self-weight of the lower part of the tube produced a large axial tensile load. So, the tube had the stress factor of stress corrosion cracking sensitivity. Combined with the results of metallographic analysis, corrosion occurred on both inner and outer surfaces of the tube, with the maximum corrosion depth of 0.24mm on the outer surface and 0.13mm on the inner surface, equivalent to a maximum wall thickness reduction of 5%, which reduced the stress load that the tube could bear. At the same time, combined with phase analysis results, the corrosion products of the inner and outer surface of the tube are basically the same. But due to the more serious corrosion on the outer surface, the cracking source is located on the outer wall of the tube.

Corrosion medium: the oil and gas composition of the well shows that the Z well is a typical oil and gas well containing H_2S, and with H_2S concentration of 2100mg/L. In the well, tube packer was added at the lower end of the tube at 5920m, and annulus protection fluid was added to the oil set of annulus. The failed tube was the second tube attached to the well tubing, which was located above the packer. Normally, the outer surface of the short tube could not contact the H_2S medium, which meant there should be no risk of stress corrosion cracking. However, the well's operating history showed that the well was perforated through the tube at 3900~3901m and converted to gas lift production. Visibly, although the failed tube is above packer, the tube perforating made the H_2S in the tube internal fluid into the casing and tube annulus, and to the data showed that high pressure gas contained H_2S gas used in the gas lift. It could be seen that the oil casing annulus of this well had content of higher H_2S corrosive medium, indicating that the tube corrosion factors of the stress corrosion cracking sensitivity. In addition, according to the results of phase analysis, the corrosion products of the inner and outer walls of the oil pipe contained not only FeS but also Fe_2O_3. According to the construction conditions on site, the high-pressure gas used in the gas lift contained certain O_2, which caused oxygen corrosion of the oil pipe.

4 Conclusion

The failed tube had a high Rockwell hardness and was in the second tube attached to the well tubing, which itself bear a large tensile stress. In addition, the presence of H_2S corrosion medium and certain O_2 in the annulus lead to H_2S stress corrosion and oxygen corrosion, resulting in fracture.

(1) The chemical composition, tensile properties, impact toughness and hardness test results of the fractured tube conform to the technical requirements. The microstructure of the tube is tempered Sorbite with the grain size of 11.0 grade.

(2) The perforation of the tube makes H_2S in the fluid medium enter the oil casing annulus. In addition, the high-pressure gas used in the gas lift contains certain H_2S and O_2, making the tube fracture failure under the combined action of sulfide stress corrosion and oxygen corrosion, and the source of the fracture is located on the outer wall of the tube.

(3) It is recommended to treat high-pressure gas used in gas lift and to reduce H_2S and O_2 in injected gas as much as possible.

References

[1] Yang Long, Li Helin. Failure analysis prediction & prevention and integrity management[J]. Heat Treatment of Metals, 2011, 36: 15-16.

[2] Long Yan, Li Yan, Ma Lei, et al. The Fracture Reason of Repaired Tubing in a Western Oilfield[J]. Corrosion & Protection, 2018, 39(05): 359-364.

[3] Zhong Bin, Chen Yiqing, Meng Fanlei, et al. Perforation failure analysis of N80 pipe[J]. Corrosion & Protection, 2018, 39(08): 647-650.

[4] Lin Anbang, Peng Bo, Du Jinnan, et al. Corrosion failure analysis of a oil well tubing[J]. Total corrosion control, 2017, 31(10): 42-48.

[5] Li Helin, Li Pingquan, Feng Yaorong. Failure analysis and prevention of oil drill string[M]. Beijing: Petroleum Industry Press, 1999.

[6] Sun Zhi, Jiang Li, Ying Zhanpeng. Failure Analysis-Foundation and Application[M]. Beijing: China machine press, 2009.

[7] Yi Tao, Zhang Qing, Wan Qiang, et al. Stress corrosion of hydrogen sulfide and control of hardness[J]. Journal of Xinjiang petroleum institute, 2002, 14(2): 77-80.

[8] NACE International. NACE MR0175-2000. Sulfide stress cracking resistant metallic materials for oilfield equipment[S].

本论文原发表于《Materials Science Forum》2020年第993卷。

第三篇　专利及标准

授权专利目录

序号	专利名称	专利类型	授权号	发明人
1	一种双金属复合管环焊缝对焊焊接方法	发明专利	201410645571.0	上官丰收；常泽亮；李发根；孟繁印；李广山
2	一种抗硫化物应力开裂的油套管	发明专利	201310028359.5	魏斌；蔡长赟；李发根；韩燕；冯耀荣
3	一种铝合金钻杆管体与钢接头的连接结构	发明专利	201410645836.7	刘永刚；李幸军；冯春；李方坡
4	一种热采井套管材料选用方法	发明专利	201310544749.8	韩礼红；谢斌；王航；曾德平；王建军；田志华；李方坡；路彩虹；冯耀荣；张学鲁；秦长毅；聂明虎
5	一种换热器管束的检测方法	发明专利	201410019925.0	杨锋平；赵新伟；罗金恒；吉玲康；张广利；苗健
6	管道环焊焊缝的全位置自动焊方法	发明专利	201210440045.1	何小东；张雪翠；王月霞；朱丽霞；仝珂；刘荣勤；蔺卫平；宋娟
7	评价溶解态二氧化碳在防腐涂层中扩散及渗透性能的方法	发明专利	201310232373.7	付安庆；白真权；李广山；李为卫；林冠发；苗健；蔡锐；赵雪会
8	一种生产油气管道用弯管的两相感应加热装置及方法	发明专利	201410641070.5	熊庆人；张冠军；李发根；张伟卫；罗金恒；许晓锋
9	低毒曼尼希碱化合物，由其制备的酸化缓蚀剂及其制备方法	发明专利	201410117306.5	王红艳；白真权；魏斌
10	一种超深井用超高强度铝合金钻杆管体及其制造方法	发明专利	201410693499.9	冯春；宋生印；冯耀荣；刘永刚
11	一种H2S-HCl-H2O体系用高温缓蚀剂及其制备方法	发明专利	201410671028.8	张娟涛；李谦定；孙晓玉；王远；尹成先；韩燕；蔡锐；李发根
12	一种定位取样尺	发明专利	201310746769.3	张伟卫；王海涛；李炎华；池强；黄呈帅；齐丽华；李洋；王鹏
13	内螺纹接头中径定位样板及中径测量方法	发明专利	201310253636.2	白小亮；张伟卫；秦长毅；卫尊义；张华；余志；吴健；付咸
14	外螺纹接头中径测量仪、杆，定位样板及中径测量方法	发明专利	201310255353.6X	白小亮；秦长毅；冯娜；艾裕丰；李光峰
15	一种膨胀管膨胀装置	发明专利	201210124196.6	申昭熙；上官丰收；李磊；王建军；刘强；宋生印
16	一种焊钢中M/A岛组织腐蚀的显示方法	发明专利	201210139451.4	仝珂；李鹤；李金凤；宋娟；卫尊义
17	利用单边缺口拉伸试验测量管线钢断裂韧性的方法	发明专利	201310479327.7	李洋；张伟卫；李鹤；吉玲康；池强；霍春勇；冯耀荣
18	用于管道止裂韧性测量的装置及其测量方法	发明专利	201310479358.2	李洋洋；张伟卫；李鹤；王鹏；吉玲康；池强；霍春勇；冯耀荣

续表

序号	专利名称	专利类型	授权号	发明人
19	V150钢级油井管全尺寸实物试验制备方法	发明专利	201310109959.4	王蕊；李东风；韩军；杨鹏；娄苗；张广路；张丹；韩新利
20	稠油蒸汽吞吐热采井用套管的包申格效应腐蚀评测方法	发明专利	201310134329.2	韩礼红；琚诚；王航；曾路平；田志华；李方坡；路彩虹；冯耀荣；张学鲁；綦长毅；裴明虎；潘志勇
21	一种确定管道近中性pH值应力腐蚀开裂敏感区段的方法	发明专利	201310590658.8	赵新伟；张广利；罗金恒；宫少涛
22	地下储气库危害的预测系统和方法	发明专利	201310032090.8	李丽锋；罗金恒；赵新伟；马卫锋；杨锋平；王河
23	确定热采井套管柱总应变的方法	发明专利	201310714246.0	王建军；韩礼红；王航
24	地下储气库完井管柱内压疲劳试验装置及其试验方法	发明专利	201210500560.4	王建军；宋生印
25	一种研究大口径热挤压成型三通壁厚的测试方法	发明专利	201410112212.9	齐丽华；刘迎来；杨放；池强；王鹏；张伟卫；王海涛；李鹤
26	一种油气田用缓蚀剂及其制备方法	发明专利	201210589886.9	魏斌；王红艳；杨发根；李发旺；赵雪会；白真权
27	一种获取一级地区油气输送管试压压力的方法及装置	发明专利	201310028000.8	杨锋平；罗金恒；赵新伟；张广利；李丽锋；张良
28	一种双金属复合管环焊缝焊接方法	发明专利	201310201303.5	李为卫；刘亚旭；许晓锋；方伟；贾昔君；杨扬；徐婷
29	油气输送管道及其制备方法	发明专利	201210242464.4	白小亮；卫尊义；余志；冯娜
30	便携式石油管外螺纹锥度测量仪	发明专利	201110216831.9	白小亮；韩新利；韩礼红；田育洲；艾秋丰；李光峰
31	一种热采井套管柱应变判断方法及其制备方法	发明专利	201310714515.3	王建军；韩航；王鹏；张伟卫
32	一种油田管材用高性能防腐涂料及涂覆方法	发明专利	201210394313.0	蔡锐；尹成先；苗健；张娟涛；路彩虹；付安庆
33	油气输送管道及管接头	发明专利	201210518116.5	张淑慧；戚东茹；蔡雪华；丁楠；冯春
34	一种接箍	发明专利	201110216831.9	申昭熙；李磊；王鹏；宋生印
35	一种接箍	发明专利	201310177929	潘志勇；申昭熙；宋生印
36	一种带环形裂纹的圆棒拉伸试样及其制备方法	发明专利	201310081615.7	熊庆人；马秋荣；张伟卫；刘亚辉；徐欣；孙良；韩新利；高健忠
37	一种油气田用缓蚀剂及其制备方法	发明专利	201210590016.3	魏斌
38	一种管线钢中M/A岛组织的面积含量评定方法	发明专利	201210150104.1	仝河；何小东；朱丽霞；邵晓东；冯耀荣；韩礼红
39	用于提高接头螺纹接头耐磨性的方法	发明专利	201310746042.5	李方坡；刘永刚；韩礼红
40	一种稀土高强韧钻杆接头的制备工艺	发明专利	201210560084.5	徐欣；李方坡；路彩虹；刘永刚

·576·

续表

序号	专利名称	专利类型	授权号	发明人
41	全尺寸非金属管材气体渗透性能的测试装置及其测试方法	发明专利	201310694098.0	李厚木；戚东涛；张学敏；蔡雪华；王熙
42	管线钢平面应变断裂韧性和安全临界壁厚的确定方法	发明专利	201310008460.4	赵新伟；张华；王亚龙；李娜
43	外加轴向载荷的钢管全尺寸四点弯曲试验装置及方法	发明专利	201410238717	聂向晖；王高峰；王耀光；赵金兰；刘迎来；李记科；王长安
44	采用对内衬管拉伸实现内衬双金属复合管复合管生产的方法	发明专利	201310652690.4	李记科；杨红兵；李安强；王长安；吴金辉；聂向晖；李记科；王耀光
45	一种热挤压三通成型工艺	发明专利	201310355865.5	刘迎来；王长安；杨红兵；王耀光；李记科；聂向晖；赵金兰
46	大口径高强度油气输送管用热挤制三通壁厚计算方法	发明专利	201210243066.4	刘迎来；吴宏；齐丽华；王高峰；郭志梅；李云龙；王长安
47	一种用于管材整圆扩径或用于内衬复合管加工的扩径头	发明专利	201110446789.X	李记科；李安强；张永红；杨红兵；李云龙
48	钢管焊缝自动超声水柱耦合监视检测的水柱耦合检测装置设计方法	发明专利	201410240366.6	黄磊；赵新伟；李记科；王长安
49	一种水压试验补偿器及其进行管材水压试验的方法	发明专利	201210154395.1	李记科；杨红兵；吴金辉；王长安；王耀光
50	一种螺旋缝埋弧焊接钢管的焊接生产方法	发明专利	201210254591.6	李记科；杜伟；刘迟；李安强；杨红兵；杨专钊；高建忠；许诗宁
51	一种采用同步扩径工艺生产内衬双金属复合管的加工设备	发明专利	201110429594.4	李安强；李记科；杨红兵；李云龙；王长安；张永红
52	类条形码引伸计系统及其测量应力应变全曲线的方法	发明专利	201210126232.2	杨专钊；李云龙；李记科；王高峰；杜伟；魏亚秋；王耀光
53	四道焊接完成的双金属冶金复合管的焊接工艺及其工艺方法	发明专利	201410225216.8	杨专钊；杨溪；王高峰；惠非；赵哈鸡；冯慧
54	一种DWTT试样人字型缺口的加工设备	发明专利	201110440295.0	李云龙；吴金辉；杨专钊；李记科；高建忠；王长安；杨红兵；许诗宁
55	裸露油气管道最低管壁温度的测量方法及装置	发明专利	201210848627	王海涛；田伟；冯耀荣；吉玲康；刘迎来；张广利
56	稠油热采管材料热蠕变性能预测试验装置及方法	发明专利	201410773837	胡美娟；王鹏；杨放；韩礼红；池强
57	一种能修控制断裂行为的天然气线及其延性断裂控制方法	发明专利	201410643967.1	杨坤；李鹤；霍春勇；吉玲康；李炎华；马秋荣

· 577 ·

续表

序号	专利名称	专利类型	授权号	发明人
58	一种测试管材抗结垢结蜡性能的方法及装置	发明专利	201410610 7680	李厚朴；杨永利；张学敏；王守泽；戚东涛；韩飞
59	一种金属焊接结构的真实应力应变曲线的测量方法	发明专利	201410671 0292	李洋；杨放；张鹤；张伟卫；张继明；李玲康；霍春勇；马秋荣
60	一种油田用玻璃钢管外保温层结构	发明专利	201510383 2911	丁楠；戚东涛；毛学强；李厚朴；常泽亮；齐国权
61	一种抗硫化物应力开裂试验用pH值自动控制装置及其配制方法	发明专利	201510317 288X	吕乃欣；付安庆；尹成先；上官丰收；李发根；徐秀清
62	一种钻杆接头及其制备方法	发明专利	201310571 8095	刘永刚；李方坡
63	适用于直连型套管和油管用螺纹连接头内压气密检测装置	发明专利	201410618 663	李磊；申昭熙；王鹏
64	一种输气管道完整性评价周期的计算方法及系统	发明专利	201410115 4384	王河；罗金恒；张广利；诺贵宁；李丽锋；张良
65	铝合金钻杆接头与管体的连接结构	发明专利	201410738 4750	刘永刚；李方坡；石舸；冯春
66	一种用于管子管体检测的方法及其设备	发明专利	201310746 124X	刘琰；姚欢；余伟不；严密林；杨锋平；贾鹏军
67	一种580MPa级铝合金钻杆用管体及其制造方法	发明专利	201410855 5732	冯春；刘永刚；韩礼红；冯耀荣
68	一种地下储气库注采管柱设计方法	发明专利	201410713 5391	王建军
69	一种输送管适用性测试装置及测试判断方法	发明专利	201410641 5412	李炎华；吉玲康；罗金恒；熊庆人；杨放；胡美娟；张继明
70	一种高强韧160钢级钻杆材料及其制备方法	发明专利	201310590 8371	徐欣；王鹏；王新虎；李方坡；宋生印
71	一种氨基化石墨烯/高密度聚乙烯纳米复合膜及其制备方法	发明专利	201410635 150X	李厚朴；张学敏；戚东涛；任鹏刚；蔡雪华；张华
72	热采井用实体膨胀套管及膨胀预测方法	发明专利	201410220 0479	李德君；刘强；白强；吉玲康；宋生印；吕能；贾鹏军
73	一种管道腐蚀速度预测方法及装置	发明专利	201210548 324X	王鹏；陈茂达；宋生印；王振；胡美娟；申昭熙；冯耀荣；贾春君
74	一种天然气管道止裂预测的方法	发明专利	201410768 2529	杨锋平；李洋；罗金恒；张良；吉玲康；王河；李丽锋
75	一种油管套管螺纹接头密封结构的优化设计方法	发明专利	201410841 909X	王建军；朱强；李方坡；王建东
76	一种480MPa级铝合金油管用铝合金及其管材制造方法	发明专利	201510142 8497	冯春；王鹏；朱丽娟；韩礼红；冯耀荣
77	一种H级表面渗铝改性抽油杆用钢及其杆体制造方法	发明专利	201510885 9662	冯春；韩礼红

续表

序号	专利名称	专利类型	授权号	发明人
78	一种钢套筒柔性止裂器	发明专利	201510867551.2	李鹤;胡美娟;池强;闵希华;李炎华;王鹏;王海涛;黄呈帅;霍春勇;杨放
79	处理金属材料异常拉伸试验曲线获得准确屈服强度的方法	发明专利	201410822741.8	许晓锋;李为卫;刘亚旭;徐婷;吕华;李晨;王海涛
80	一种管道止裂预测用临界单位面积损伤应变的确定方法	发明专利	201510350668.3	杨锋平;罗金恒;霍春勇;马卫锋;李丽锋;王珂;张良;张奕
81	一种柔性复合管快速施工装置及施工方法	发明专利	201511034015.0	齐国权;袁晓山;戚东涛;邹应勇;魏斌;毛学强;李厚补
82	一种避免DWTT异常断口的试样及其制备方法	发明专利	201510531840.5	何小东;梁明华
83	钢连接接头及使用其连接钢的方法	发明专利	201610187147.5	王长安;蔡绪明;赵新伟;蔡彬;邓波;余国民;刘文友;王志鹏;黄伟
84	一种可调式非金属管材最小弯曲半径测试装置及方法	发明专利	201510388930.3	李厚补;羊东明;魏斌;朱原原;张志宏;张学敏
85	一种冶金复合管内覆层晶间腐蚀试样的制备装置及方法	发明专利	201410771222.19	仝珂;何小东;张华;宋娟;蔡克;瞿婷婷
86	一种全尺寸非金属管径测试方法及系统	发明专利	201510310108.5	李厚补;张学敏;戚东涛;羊东明;魏斌;朱原原
87	一种使用测量模型管平头或喇叭嘴的工具尺寸的测量方法	发明专利	201410459553	杨东升;杨专利;吴金辉;黄磊;邓波
88	大膨胀率下膨胀管用快速上扣螺纹结构及加工制备方法	发明专利	2.01511E+12	刘强;宋生印;李德君;吕能;田峰;王远;汪蓬劲
89	大膨胀率下实体膨胀管用高密封性螺纹接头	发明专利	201510906690.1	刘强;宋生印;白强;田峰;王远;武刚;李德君
90	一种表面渗铝改性P110级油套管用钢及其管材制造方法	发明专利	201510548184.75	冯春;韩礼红;冯耀荣
91	一种非金属管寿命预测方法	发明专利	201410721347.5	齐国权;李循迹;池强;魏斌;常泽亮;李厚补
92	一种大体积钢中非金属夹杂物尺寸的统计计算方法	发明专利	201511025126.5	张继明;杨坤;李炎华;霍春勇;王海涛;封辉;胡美娟
93	酸性环境环氧树脂用管线钢埋弧焊焊丝	发明专利	201510587864.2	胡美娟;池强;李炎华;王鹏;王海涛;杨放;池强
94	耐氯离子腐蚀型铝合金及其油气管用铝合金及其管材制造方法	发明专利	201511018640.6	冯春
95	一种复合材料增强管道及其制备方法	发明专利	201610294228.5	张冬娜;戚东涛;丁楠;邵晓东;江汛;卞思辰;於秋霞
96	一种盐穴地下储气库溶腔体积收缩率的预测方法	发明专利	201410309280.4	罗金恒;刘新荣;李丽锋;王军保;蔡克;郭建强

· 579 ·

续表

序号	专利名称	专利类型	授权号	发明人
97	一种620MPa级铝合金钻杆用管体及其制造方法	发明专利	201410853240.6	冯春；李方坡；王新虎；朱丽娟
98	一种薄壁内覆耐蚀合金复合管的晶间腐蚀试验方法	发明专利	201410003399.9	朱丽霞；何小东；朱娟；仝河；张华
99	一种凸台角焊缝复合材料修复补强方法	发明专利	201610597208.8	胡美娟；李鹤；罗金恒；邵春明；池强；杨坤；张伟卫
100	一种碳纤维复合材料止裂器及其制作安装方法	发明专利	201510981219.9	池强；李鹤；马卫锋；胡美娟；张继明；张伟卫；黄呈帅；王鹏；齐丽华
101	采用DWTT实验判材料在快速变形时塑性指标的方法	发明专利	201510990642.2	李记科；杨红兵；何小东；王高维；裴向晖；李富强；马飞
102	一种高磷化学镀Ni-P-PTFE管水镀层及其制备方法	发明专利	201510792508.4	徐秀清；赵雪会；付安庆；朱丽娟；杨锋平
103	一种测量金属材料元素硫应力裂纹敏感性的试验方法	发明专利	201610262771.7	袁军涛；杨力能；杨荫；宋成立；李磊；秦长毅；宋文文；李京川
104	一种油田压裂弯头缺陷检测装置及检验方法	发明专利	201510886278.8	罗华权；龙岩；蔡锐；袁军涛；李岩；李磊；张雪零；谢俊峰；王华
105	用于评价油气管道在高流速下冲刷腐蚀的试验装置及方法	发明专利	201610245352.2	付安庆；尹成先；袁军涛；冯春；曹峰；杨力能；张雪零；李京川
106	一种油气田酸化用长效缓蚀剂及其配制方法	发明专利	201510634242.0	吕乃欣；朱丽娟；冯春；王鹏；刘永刚；徐欣；韩礼红；王新虎
107	一种油井管屈曲服役的高温腐蚀试验装置	发明专利	201510763429.0	罗华权；曹峰；杨力能；解学东；李京川
108	一种管道全长直度的测量装置及测量方法	发明专利	201611248300.7	刘迎来；王高峰；裴向晖；赵新伟；王耀光；王长安；李亮；许彦；丰振军
109	一种高钢级感应加热弯管双温温控制方法	发明专利	201610635963.8	李记科；吕能；白露；贾鹏军；曹峰
110	快速接头及快速接头与非金属复合管的连接方法	发明专利	201610503810.8	闫凯；赵新伟；王长安；杨冬钊；卫栋；李呈坤；卫红兵；王新林；田新新
111	油套管质量水平评价方法	发明专利	201610428580.3	王远；张娟涛；田峰；刘强；罗金恒；武刚；白强
112	适用于油气田含H_2S/CO_2腐蚀环境输送管线的缓蚀剂及其制备方法	发明专利	201610009435.1	闫凯；杨党利；卫栋；赵新伟；田新新；王冬林
113	一种使用油套管特殊螺纹加工刀刃加工螺纹的方法	发明专利	201610383305.4	马卫锋；卫栋；蒋承君；费凡；董绍华；杨锋平；王河；李丽超；张良；赵新伟
114	天然气场站压力设备常用材质焊缝缺陷制作及检测方法	发明专利	201510982025.0	韩燕；尹成先；林凯；杨鹏；马庆伟；吕庆欣
115	一种复合加载应力腐蚀试验装置及方法	发明专利	201510874287.5	潘志勇；杨荫；李军军；龙岩；娄琦；宋成立；余志
116	一种油井管特殊螺纹接头	发明专利	201611229820.3	

· 580 ·

续表

序号	专利名称	专利类型	授权号	发明人
117	用于板材电磁超声自动检测的对比试块及其设计方法	发明专利	201610402629.8	黄磊;赵新伟;王长安;杨专钊;吴金辉;常永刚;王增辉
118	基于应变的高速气流摩阻系数计算方法	发明专利	201611227895.8	王建军;杨尚谕;薛承文;李方坡;王鹏
119	一种用机械研磨技术在小尺寸金属管内壁制备纳米Ni-P镀层的装置和方法	发明专利	201611008985.8	袁军涛;付安庆;李磊;龙岩;朱丽娟;李龙根
120	一种模拟管道爆破断口形貌的测试装置及测试方法	发明专利	201611130160.3	何小东;张庶鑫;全洞;梁明华;蔺卫平;宋广三;张雪琴
121	一种评估天然气管爆炸危害的方法	发明专利	201710079096.9	杨坤;沙勇;杨明;张伟卫;李鹤;胡美娟
122	一种复合材料增强金属管道设计压力计算方法	发明专利	201610868067.6	张冬娜;戚东涛;邵晓东;丁楠
123	一种石油套管圆螺纹牙型角度的判定方法	发明专利	201710993344.0	闫凯;杨红兵;卫栋;韩华刚;田新新
124	基于测井大数据驱动的在役油套管缺陷自动判定方法	发明专利	201610214090.3	王鹏;胡美娟;韩礼红;冯耀荣
125	一种实体膨胀管带载荷弯曲膨胀试验装置	发明专利	201710021069.6	刘强;宋生印;白强;汪蓬勃;贾鹏军
126	一种实体膨胀管带载荷弯曲膨胀试验方法	发明专利	201710020054.8	刘强;宋生印;冯耀荣;李德君;吕能;武刚
127	Connection structure between pipe body and steel joint of aluminum alloy drill pipe	发明专利	US10196863	Liu Yonggang; Li Xiaojun; Feng Chun; Li Fangpo
128	Connection structure between pipe body and joint of aluminum alloy drill pipe	发明专利	US10184596	Liu Yonggang; Li Fangpo; Shi Qi; Feng Chun
129	一种油气管道安全评估的方法和装置	发明专利	CN105320994B	张良;罗金恒;张广利;杨锋平;罗柴生
130	一种钻杆剩余寿命的预测方法	发明专利	201611228899.8	李方坡
131	确定高速气流摩阻系数的试验装置和测试方法	发明专利	201611227891X	王建军;杨尚谕;薛承文;李方坡;付太森
132	一种添加纳米稀土氧化物制备基于应变设计的热采套管材料的方法	发明专利	201710943370.2	王航;韩礼红;王建军;杨尚谕;田涛;朱丽娟;路彩虹;蒋龙
133	一种石油管螺纹连接结构	发明专利	201710392652	潘志勇;罗金恒;王建军;卢攀辉;邝献任;李丽锋;宋成立
134	一种抗腐蚀性特殊螺纹接头	发明专利	201710528648	潘志勇;罗金恒;卢攀辉;杨鹏;李丽锋;李孝军;丛深
135	防止湿气集输管道内腐蚀的管道铺设方法	发明专利	201810135170.9	吕能;王远;刘强;李记科;张鸿博
136	一种埋地管道保护板及其安装方法	发明专利	201810785018.5	齐国权;马秋荣;戚东涛;魏斌;丁楠;李厚朴;张冬娜;邵晓东;蔡雪华

· 581 ·

续表

序号	专利名称	专利类型	授权号	发明人
137	一种便携式测井油管内壁缺陷测量仪	发明专利	201810307878.8	余志
138	一种基于R语言的金属拉伸试验验证方法	发明专利	201710720520.3	张席鑫；李娜；梁明华；张华佳；任继承
139	评价测井钢丝抗应力腐蚀开裂性能的模拟试验装置及方法	发明专利	201710583006.X	龙岩；王远；袁军涛；潘志勇；丁晗；林凯
140	一种高温高压气井油管气密封螺纹分析评价方法	发明专利	201810321078.1	王建东；张华礼；汪传磊；周渡；田涛
141	改进的斜Y型坡口焊裂纹敏感性试验试件及其制造方法	发明专利	201611208203.5	李为卫；池强；吉玲康；宫少涛；贾君；许晓锋
142	地下储气库储存介质地层泄漏半径及体积预测方法	发明专利	201611248228.0	李丽锋；罗金恒；赵新伟；王建军；杨锋平；马卫锋；张良
143	一种环境友好型高温酸化缓蚀剂及其制备方法	发明专利	201810849999.5	吕乃欣；王博林；付安庆
144	一种KD级表面渗铝改性油井油管用钢及其杆体制造方法	发明专利	201810468139.7	冯春；路彩虹；朱丽娟
145	一种有机酸体系曼尼希碱类酸化缓蚀剂及其制备方法	发明专利	201811106806.3	张娟涛；李猜述；王福普；张小龙；胡建国；马磊；宋文文；熊茂昌
146	一种B型套筒的加工方法	发明专利	201910107315.9	李记科；王鸿峰；杨锋平；李高莲；赵新伟；刘迎来
147	一种管道内长距离爬行检测机器人延伸装置及延伸方法	发明专利	201810312305.4	刘琰；李为卫；赵新伟；张鸿博；李先明；常泽亮；冯泉
148	一种盐穴地下储气库井口水合物堵塞风险测定方法	发明专利	201611200345.7	李丽锋；罗金恒；赵新伟；任国其；张皓
149	一种屈服强度大于1138MPa的钻杆用钢管及其制造方法	发明专利	201810300984.3	李方坡；王建军；冯耀荣；韩礼红
150	一种海洋环境和油气介质协同作用下腐蚀模拟系统	发明专利	201710322170.5	李发根；徐秀清；尹成先；冼国栋；蔡锐；张娟涛
151	一种三通管件弧度仿形绘制装置	发明专利	201711329714.7	范炜；高建忠；韩新利；房猛；张杰；张华佳
152	一种套管抗剪切性能评价方法	发明专利	201810002310.5	杨尚谕；韩礼红；王鹏；王建东；冯春；王航；亚旭
153	一种超高强度抗钻杆耐腐蚀钢管及其制造方法	发明专利	201810300981.X	李方坡；尹成先；王建军；冯耀荣；韩礼红
154	一种抗CO_2腐蚀集输管线缓蚀剂	发明专利	201811149893.0	尹成先；范磊；张娟涛；王远；付安庆

续表

序号	专利名称	专利类型	授权号	发明人
155	一种油井管全尺寸旋转弯曲疲劳试验装置及方法	发明专利	201710676044X	王蕊；李东风；韩军；杨鹏；张小佳；张乐
156	一种基于双相组织的海洋用高应变焊接钢管及其制备方法	发明专利	CN201910110156.8	何小东；霍春勇；马秋荣；吉玲康；李为卫；池强
157	油井管接头的密封结构的设计与制造方法、密封方法	发明专利	CN201810250603.5	王新虎；王建东；王鹏；冯春；韩礼红
158	一种螺纹接箍	发明专利	CN201811604981.5	潘志勇；韩礼红；冯春；杨尚谕；田涛；朱丽娟；蒋龙；韩军；朱丽霞
159	一种测试油套环空保护液长效性的试验装置和方法	发明专利	CN201711039303.4	王建军；李方坡；韩礼红；孙建华；张娟涛；陈俊
160	一种检测低合金结构中残余奥氏体分布和含量的方法	发明专利	2018110518104	李鹤；诸萍；杨坤；邹斌；伍奕；杨明；尚臣
161	一种无损检测方法	发明专利	2017108614664	姚欢；刘琰；冯挺；来建刚；曹峰；赵淼；苏杭
162	一种对非金属管集输线服役伸长变形在线监测的方法	发明专利	2019106831020	刘强；汪鹏勃；祝国川；白强；丁楠；杨扬；李厚朴；苏杭；宋生印；李计科；张鸿博
163	一种用于集输管线监测的多元智能传感器嵌入及保护方法	发明专利	2019106831247	刘强；白强；祝国川；汪鹏勃；丁楠；李厚朴；杨专钊；宋生印；张鸿博；张月燕
164	基于超声波射流的水力压裂裂缝监测系统及方法	发明专利	2017107365486	杨尚谕；韩礼红；王鹏；王建军
165	一种套管非均匀外挤能力评价方法	发明专利	201810025967	韩礼红；杨尚谕；王鹏；王建东；王航；路彩虹；冯耀荣；王建军；刘亚旭
166	一种用于柔性复合管拉伸性能测试的方法及试验装置	发明专利	2018108499374	丁楠；李厚朴；胡建国；齐国权；邵晓东；刘强；戚东涛
167	一种油田地面集输管线用环保型耐氧缓蚀剂及其制备方法	发明专利	2018108500102	吕乃欣；李鹤；胡美娟；郭刚；庞永利；马勇；王青
168	一种便携式钢管耐腐蚀性能测试装置	实用新型	2016205664797	李炎华；池强；封辉；何小东；王海涛；张继明
169	一种弹簧引伸计悬臂定位装置及弹簧引伸计	实用新型	2015211303281	张睿鑫；何小东；南卫平；李莞；梁明华；李娜；任继承；张华佳
170	钢管连接器	实用新型	2016202501790	王长安；蔡彬；赵新伟；蔡据明；余国民；邓波；赵晗君；黄伟；张超
171	输气钢管全尺寸气体爆破试验用聚能切割装置	实用新型	2015210476211	吉玲康；胡美娟；李鹤；闫美娟；刘剑；黄忠胜；刘希华；张怀卫；杨坤；池强；张继明
172	一种大直径、高钢级管筒柔性止裂器	实用新型	2015211409445	李鹤；胡美娟；池强；霍春勇；刘剑；黄忠胜；马秋荣；张继明

续表

序号	专利名称	专利类型	授权号	发明人
173	一种输气钢管全尺寸气体爆破试验断裂速度测量装置	实用新型	201521132734	李鹤；张伟卫；胡美娟；霍春勇；杨坤；王海涛；吉玲康；李炎华；陈宏远；齐丽华
174	一种碳纤维复合材料止裂器	实用新型	201521089624.1	池强；李鹤；马卫锋；胡美娟；张继明；杨坤；张伟卫；黄呈帅；王鹏；齐丽华
175	一种涂层湿膜厚度测试装置	实用新型	201520845898.2	蔡克；陈志昕；张翔；马小芳；王珂
176	一种提高ERW焊管腐蚀准确性的实验室测试装置	实用新型	201520932448.7	徐秀清；尹成先；张淑慧
177	钢套筒柔性止裂器	实用新型	201520817747	李鹤；胡美娟；池强；闵希华；李炎华；王鹏；王海涛；黄呈帅；霍春勇；杨放
178	一种复合加载应力腐蚀试验装置	实用新型	201520987265.5	韩燕；尹成先；冯春；马庆伟；吕乃欣
179	一种油井管屈曲腐蚀和冲蚀试验装置	实用新型	201520893185.3	朱丽娜；冯耀；王鹏；刘永刚；徐欣；韩礼红；王新虎
180	一种大口径三通弯曲试验装置	实用新型	CN205785795U	刘迎来；许彦；王高峰；聂向晖；齐丽华
181	一种全尺寸钢管弯曲压缩变形试验装置	实用新型	CN205786120U	王鹏；池强；杨放；陈宏远；王海涛；齐丽华
182	一种中频感应加热弯管机械转臂固定装置	实用新型	CN205763150U	王鹏；池强；齐丽华；吴健；张华佳；张庶鑫
183	一种耐蚀合金复合材料剪切强度测试夹具	实用新型	CN205786089U	梁明华；李京川；王鹏；聂新伟；杨光华；王冬林
184	一种油套管特殊螺纹加工刀片	实用新型	CN205764285U	闫凯；杨红兵；卫俅；赵勇涛；李磊；李岩；宋文义；谐俊峰
185	用于评价管道在高流速下冲刷腐蚀的试验装置	实用新型	201620331415.1	付安庆；龙岩；袁军涛；米成立；田新新；王华
186	一种高标准块V型槽校准装置	实用新型	201621457700.4	白小亮；张华；秦长毅；冯娜；艾光峰
187	一种螺纹尾测量仪校准装置	实用新型	201621456913.5	白小亮；秦长毅；冯娜；艾裕丰；吉楠
188	一种钢管全尺寸水压爆破实验的安全防护系统	实用新型	201621491577.8	李东风；王蕊；娄琦；余志；张益铭；张小佳
189	一种齿高标准块V型槽校准装置	实用新型	201621457700.4	白小亮；张华；卫尊义；李光峰
190	一种螺纹锟测量仪校准装置	实用新型	201621456913.5	白小亮；秦长毅；冯娜；艾裕丰；吉楠
191	一种利用机械研磨技术在小尺寸金属管内壁制备纳米Ni-P镀层的装置	实用新型	201621241672.2	袁军涛；付安庆；龙岩；李磊；胡美娟；霍春勇；朱丽娟；李发根
192	一种基于玻璃纤维复合材料对油气输送管道的止裂结构	实用新型	201621209088.9	李鹤；张伟卫；陈宏远；王鹏；杨坤；胡美娟；马秋荣；李炎华；封辉；杨放
193	一种管道深度扫水用复合式泡沫清管器	实用新型	201620427563.3	李炎华；陈宏远；王鹏；张伟卫；李炎华；齐丽华；胡美娟

· 584 ·

续表

序号	专利名称	专利类型	授权号	发明人
194	一种评价钻杆内涂层耐蚀性能的装置	实用新型	201620852451	朱丽娟；刘永刚；冯春；徐欣；袁军涛；杨尚谕
195	油套管内外螺纹综合检验的辅助工具	实用新型	201620938772	田新新；杨红兵；卫栋；同凯；韩华刚；王冬林；雷俊云
196	组合式三点弯曲试验压头装置	实用新型	201620105684	聂向晖；李亮；刘迎来；王高峰；许彦；丰振军
197	一种凸台角焊缝复合材料修补强结构	实用新型	201620871059	胡美娟；李鹤；罗金恒；邵春明；池强；杨坤；张伟卫
198	一种油气井用铝合金套管结构	实用新型	201620902253	冯春；韩礼红；王鹏
199	一种用于高钢级弯管感应加热双温根制的感应线圈	实用新型	201620847910	刘迎来；王高峰；聂向晖；赵新伟；王耀光；王长安；李亮；许彦；丰振军
200	一种连接非金属复合管的快速接头	实用新型	201620768245	李记科；白强；吕能；贾鹏飞；武刚；曹峰
201	一种带密封扣压接头的连续纤维增强热塑复合管	实用新型	2017214213 91X	李厚朴；戚东涛；吴河山；秦鹏
202	一种带密封扣压接头的连续纤维增强热塑性塑料管	实用新型	201721417 4597	李厚朴；戚东涛；吴河山；秦鹏
203	一种内衬抗硫非金属高钢级管	实用新型	201720920 9859	张淑慧；齐丽荣；徐秀清；马秋荣；官少涛；张翔
204	一种带密封自愈和填裂纹纹陶瓷内衬复合钢管结构	实用新型	201721293 24861	李厚朴；刘琰；羊东明；张学敏；朱原原；威东涛
205	一种管道泄漏磁检测系统	实用新型	20172121 71787	姚欢；冯耘；王鹏峰；丰振军；曹峰；蒋东君；徐生东
206	一种冶金复合材料剪切强度试验装置	实用新型	20172117 16851	聂向晖；王鹏；丰振军；刘迎来；李亮；许彦
207	电磁超声换能器及用电磁超声换能器自动检测钢板的设备	实用新型	20172117 17144	黄磊；赵新伟；张鸿博；王长安；杨专利；吴金辉；余国民
208	一种枪式内锥规校准夹持装置	实用新型	201720856 7616	吉娜；白小亮；冯娜；艾裕峰；李光亮
209	一种用于腐蚀溶液的除氧装置	实用新型	2017208006 6414	袁军涛；张慧慧；付安庆；李发根；吕乃欣；李磊
210	一种用于抽油杆柱状拉正器轴向拉伸的辅助装置	实用新型	2017205728 300	张丰佳；李娜；崔巍；梁明平；兰卫平；张华
211	一种防无牙痕扭用金刚砂布落砂装置	实用新型	2017204005 817	李宇军；林凯；潘志勇；袁军涛；刘文红；邝献任
212	一种快速准确定隔热油管隔热性能的试验装置	实用新型	2017204942 928	李磊；宋成立；袁军涛；杨尚谕；林鹏；付安庆；韩礼红；丁晗
213	一种铝合金套管气密封特殊螺纹接头	实用新型	2017219078 539	王鹏；冯春；杨尚谕；王长安；韩礼红
214	ERW钢管焊缝电磁超声自动检测的对比试块	实用新型	2017218805 343	黄磊；张鸿博；聂向晖；冯春；常永刚
215	一种圆棒试样的磨抛装置	实用新型	2017217352 628	路彩虹；朱丽娟；蒋龙；冯春；瞿婷婷
216	一种复合加载试验机用宽板拉伸试验夹具	实用新型	2017218050 983	李东风；王怒；张益铭；陈宏远；杨鹏；韩军

续表

序号	专利名称	专利类型	授权号	发明人
217	一种复合管结合强度测试装置	实用新型	201721901 0428	梁明华；李娜；蔺卫平；李厚朴
218	一种用于评价高温高压下溶解氧在防腐涂层中渗透性的装置	实用新型	201721782 8013	朱丽娟；郭长永；黄新业；王微；冯春；蒋龙；路彩虹
219	一种评价低含水率高产气/井油管高温腐蚀性能的试验装置	实用新型	201721795 3102	朱丽娟；冯春；王建军；王航；李方坡
220	一种钢管端部直度测量仪	实用新型	201820405 6245	韩华刚；祝光辉；杨红兵；卫栋；闫凯；田新新
221	一种便携式涂层冲击强度测试装置	实用新型	201820610 5513	蔡克；张翔；陈志昕；马卫革；王河
222	一种用于高温高压金属腐蚀试验中试片的固定装置	实用新型	201820684 5325	宋成立；付安庆；周飞；刘绍东；吕乃欣；武刚
223	一种钢管轮廓仪的定位装置	实用新型	201820306 7990	吉楠；白小亮；冯娜；李尊义
224	一种一体式带凸台的压力传感器	实用新型	201820287 6485	胡美娟；李鹤；李为卫；吉玲康；封辉；陈宏远；王鹏；张继明
225	大膨胀率下支体膨胀管用快速上扣接头	实用新型	201621019 935.0	刘强
226	大膨胀率下支体膨胀管用高密封特殊螺纹接头	实用新型	201621010 8090.3	刘强
227	双金属复合管环焊缝超声波检测用对比试块及其设计方法	实用新型	201810392 716.9	裴向晖
228	一种气体保护硫覆装置	实用新型	201821465 8689	韩燕；苗健；薛岗；郭刚；张志浩；吕乃欣
229	一种钛合金油管用抗粘扣气密封螺纹接头	实用新型	201821167 0458	刘强；宋生印；白强；汪鹏勃；田峰；张鸿博；杨专钊
230	一种金属试磨样支持装置	实用新型	201821215 3406	马庆伟；韩燕；吕乃欣；付安庆
231	一种高温实时监测溶解氧浓度的腐蚀评价装置	实用新型	201821230 6789	吕乃欣；马秋荣；尹成先；付安先；刘文红；马勇；庞永莉
232	一种腐蚀试验塑料除氧预饱和加药装置	实用新型	201821231 5542	吕乃欣；张志浩；成杰；姚茶；罗慧娟；郭刚
233	一种增强热塑性塑料复合管弯曲试验系统	实用新型	201821032 8660	邵晓东；李春雨；张冬娜；李厚朴；魏斌
234	一种测量外径极值及椭圆度的智能游标卡尺	实用新型	201820401 6835	韩华刚；杨红兵；卫栋；闫凯；田新新
235	一种管材水压试验系统	实用新型	201820091 9861	赵金兰；王长安；杨专钊；王记科；李高峰；吴金辉；蔡彬
236	一种油气输送用钢板电磁超声自动检测系统	实用新型	201821175 160X	黄磊；张鸿博；李记科；王长安；吴金辉
237	双金属复合管环焊缝超声波检测用对比试块	实用新型	201820618 7496	聂向晖；张鸿博；刘琰；魏斌；威东涛；李为卫；方艳；冯泉；丰振军；李亮；许彦；刘迎来
238	一种防腐蚀耐高压复合管接头	实用新型	201820766 0881	邵晓东；威东涛；刘琰；魏斌；张冬娜；李厚朴；李发根；丁楠；齐国权

续表

序号	专利名称	专利类型	授权号	发明人
239	一种纤维增强热塑性塑料复合连续管的连接接头	实用新型	201820694608	邵晓东；戚东娜；张冬娜；李厚朴；李发根
240	一种用于测量高分子涂层温差腐蚀的装置	实用新型	201820728573X	来维亚；尹成先；徐秀清
241	一种用于评价不锈钢化学成分对缝隙腐蚀影响的实验装置	实用新型	201820729398.6	来维亚；尹成先；徐秀清
242	一种评价钢管表面状态对开裂行为影响的试验装置	实用新型	201821475542.4	龙岩；王磊；付安庆；白真权；潘志勇；罗金恒；袁军涛
243	一种弯管弯曲角的测量装置	实用新型	201920468767.5	李富强；王长安；吴金辉；王冬林；蔡雷；延世强
244	一种非金属复合管连接接头	实用新型	201821899566.2	邵晓东；戚东娜；李厚朴；魏斌；张冬娜；齐国权；丁楠；李发根
245	钢接头用钛合金钻杆及钻杆柱	实用新型	201821714371.6	冯春；蒋龙；王鹏；徐欣；李睿哲；曹亚琼
246	一种套管非均匀外挤模拟试验装置	实用新型	201822244791.9	杨尚谕；韩礼红；王建军；王航；路彩虹
247	一种甲烷浓度检测装置	实用新型	202020134530.6	李鹤；胡美娟；封辉；邹斌；谢萍；杨明；尚臣
248	一种温度和振动耦合作用试验装置	实用新型	202020134527.4	张良；霍春勇；马卫笑；任国庆；杨锋平
249	一种内衬PVC钢管螺纹接头的螺纹结构	实用新型	201922265812X	杨力能；罗华权；李京川
250	一种油套管全尺寸弯曲加载试验系统	实用新型	201922430107.0	张益铭；韩军；余忠；张乐；王恣
251	一种套管孔眼冲蚀试验系统	实用新型	201922284026.4	龙岩；王鹏；谢俊峰；付安庆；赵俊锋；邝献任；吉楠；潘志勇
252	一种管线环焊缝宏观金相拍摄分析装置	实用新型	201922294024.3	仝河；樊治海；瞿婷婷；白小亮；何小东；陈志昕
253	油气管线B型套筒焊缝角超声波检测用对比试块	实用新型	201922022010.22	裴向晖；徐斌；张鸿博；刘琰；许彦；丰振军；李亮
254	一种高膨胀率膨胀管外螺纹套螺纹接头	实用新型	201920996004.8	武刚；李德若；刘强；白强；吕能；杨钊
255	一种双层组合套管	实用新型	201921131922.0	刘永刚；杨尚谕；韩礼红；王亚龙；李方坡；李磊
256	一种带环焊缝的双层组合套管	实用新型	201921132838.0	刘永刚；杨尚谕；韩礼红；王恣；路彩虹；冯耀荣
257	一种页岩气井用套管气密封螺纹接头	实用新型	201921251042.7	王鹏；冯耀荣；韩礼红；吉楠；罗金恒；邝献任
258	用于多规格管线监测的可变径多元传感器安装与保护单元及装置	实用新型	201921200081.4	刘强；汪鹏勃；祝国川；白强；丁楠；李厚朴；吴金辉；米生印；王长安；杨专钊
259	一种集输管线监测用多元传感器安装与保护单元装置	实用新型	201921199993.4	刘强；汪鹏勃；祝国川；白强；丁楠；李厚朴；米生印；杨专钊；张鸿博
260	一种非金属集输管轴向变形在线监测装置	实用新型	201921199952.5	刘强；汪鹏勃；祝国川；白强；丁楠；李厚朴；杨杨；米生印；王长安；杨专钊

续表

序号	专利名称	专利类型	授权号	发明人
261	一种圆螺纹牙型角度检测装置	实用新型	20192128373477X	闫凯；王长安；韩华刚；姚欢；苏杭
262	一种等内径合金钛钻杆	实用新型	201921350627.4	刘永刚；冯春；李方坡；路彩虹
263	一种用于测量双金属复合管点蚀性能的装置	实用新型	201921679039.5	冯泉；张明益；李亚军；孟波；崔兰德；常泽亮；赵志勇；宋文文；江代勇；黄强；李发根；陈庆国；袁梓钧；曹雯
264	用于在役油电磁超声自动检测的对比试块	实用新型	201921766890.1	黄磊；张鸿博；刘迎来；杨科；苏杭
265	一种无缝钢管多通道非接触自动超声检测系统	实用新型	CN201922106956.0	黄磊；刘迎来；郭雨童；张鸿博；杨科
266	一种可旋转式原位疲劳拉伸试验用夹具系统	实用新型	CN201921697307.6	路彩虹
267	一种电塔增材制造的三通管件	实用新型	CN211661083U	胡美娟；吉玲康；卓毅；马秋荣；卢迪；李利军
268	一种钢制管道环焊缝斜接缺陷修复补强套筒	实用新型	CN211661447U	武刚；贾海东；朱丽霞；徐春燕；李丽锋；米成立
269	一种保护埋地管道的阻挡板布置结构	实用新型	CN211667396U	齐国权；马秋荣；戚东涛；李厚朴；丁楠
270	一种油套管全尺寸复合加载实验的防护装置	实用新型	CN211668924U	张乐；韩军；余志；张益铭；杨鹏

发布标准目录

序号	标准代号	标准名称
1	ISO 11961：2018	石油天然气工业 钢制钻杆
2	API Spec 5L：2018	管线钢管规范
3	ISO 19345—1：2019	管道完整性管理规范 第1部分：陆上管道全寿命周期管理
4	ISO 19345—2：2019	管道完整性管理规范 第2部分：海上管道全寿命周期管理
5	ISO 20074：2019	油气管道地质灾害风险管理
6	ISO 11960：2020	石油天然气工业 油气井套管或油管用钢管
7	GB/T 21267—2017	石油天然气工业 套管及油管螺纹连接试验程序
8	GB/T 19830—2017	石油天然气工业 油气井套管或油管用钢管
9	GB/T 20659—2017	石油天然气工业 铝合金钻杆
10	GB/T 9253.2—2017	石油天然气工业 套管、油管和管线管螺纹的加工、测量和检验
11	GB/T 34907—2017	稠油蒸汽热采井套管技术条件与适用性评价方法
12	GB/T 34903.1—2017	石油、石化与天然气工业 与油气开采相关介质接触的非金属材料 第1部分：热塑性塑料
13	GB/T 35185—2017	石油天然气工业用复合材料增强管线钢管
14	GB/T 34903.2—2017	石油、石化与天然气工业 与油气开采相关介质接触的非金属材料 第2部分：弹性体
15	GB/T 9711—2017	石油天然气工业 管线输送系统用钢管
16	GB/T 35072—2018	石油天然气工业用耐腐蚀合金复合管件
17	GB/T 35067—2018	石油天然气工业用耐腐蚀合金复合弯管
18	GB/T 37262—2018	石油天然气工业 铝合金钻杆螺纹连接测量
19	GB/T 37265—2018	石油天然气工业 含铝合金钻杆的钻柱设计及操作极限
20	GB/T 37701—2019	石油天然气工业用内覆或衬里耐腐蚀合金复合钢管
21	GB/T 39096—2020	石油天然气工业 油气井油管用铝合金管
22	SY/T 6858.5—2016	油井管无损检测方法 第5部分：超声测厚
23	SY/T 6858.6—2016	油井管无损检测方法 第6部分：非铁磁体螺纹渗透检测
24	SY/T 6417—2016	套管、油管和钻杆使用性能
25	SY/T 6896.3—2016	石油天然气工业特种管材技术规范 第3部分：钛合金油管
26	SY/T 7042—2016	基于应变设计地区油气管道用直缝埋弧焊钢管
27	SY/T 7043—2016	石油天然气工业用高压玻璃钢油管
28	SY/T 6662.7—2016	石油天然气工业用非金属复合管 第7部分：热塑性塑料内衬玻璃钢复合管
29	SY/T 5699—2016	提升短节
30	SY/T 6763—2016	石油管材购方代表驻厂监造规范
31	SY/T 7318.1—2016	油气输送管特殊性能试验方法 第1部分：宽板拉伸试验
32	SY/T 7318.2—2016	油气输送管特殊性能试验方法 第2部分：单边缺口拉伸试验
33	SY/T 6662.8—2016	石油天然气工业用非金属复合管 第8部分：陶瓷内衬管及管件
34	SY/T 5539.2—2016	石油管产品质量评价方法 第2部分：油气输送管

续表

序号	标准代号	标准名称
35	SY/T 7316—2016	油气输送钢管用板材电磁超声自动检测
36	SY/T 7317—2016	海底管线用直缝埋弧焊钢管焊缝自动超声波检测
37	SY/T 7370—2017	地下储气库注采管柱选用与设计推荐做法
38	SY/T 7318.3—2017	油气输送管特殊性能试验方法 第3部分：全尺寸弯曲试验
39	SY/T 6423.7—2017	石油天然气工业 钢管无损检测方法 第7部分：无缝和焊接铁磁性钢管表面缺欠的磁粉检测
40	SY/T 6423.8—2017	石油天然气工业 钢管无损检测方法 第8部分：无缝和焊接（埋弧焊除外）钢管纵向和/或横向缺欠的全周自动超声检测
41	SY/T 6268—2017	油井管选用推荐作法
42	SY/T 6601—2017	耐腐蚀合金管线管
43	SY/T 6476—2017	管线钢管落锤撕裂试验方法
44	SY/T 6478—2017	油管和套管表面镀层技术条件
45	SY/T 6477—2017	含缺陷油气管道剩余强度评价方法
46	SY/T 7369—2017	纤维增强塑料管在油田环境中相容性试验方法
47	SY/T 0510—2017	钢制对焊管件规范
48	SY/T 7409—2018	酸性油气井钻柱安全评价方法
49	SY/T 7414—2018	套管和油管接头扭矩—位置控制方法
50	SY/T 6896.4—2018	石油天然气工业特种管材技术规范 第4部分：钛合金套管
51	SY/T 7318.4—2018	油气输送管特殊性能试验方法 第4部分：全尺寸气体爆破试验
52	SY/T 7415—2018	油气集输管道内衬用聚烯烃管
53	SY/T 6267—2018	高压玻璃纤维管线管
54	SY/T 6794—2018	可盘绕式增强塑料管线管
55	SY/T 6623—2018	内覆或衬里耐腐蚀合金复合钢管
56	SY/T 6897—2018	钻具螺纹上卸扣试验评价方法
57	SY/T 5988—2018	油管和套管转换接头
58	SY/T 5038—2018	普通流体输送管道用直缝高频焊钢管
59	SY/T 7418—2018	抽油杆螺纹量规校准方法
60	SY/T 6951—2019	实体膨胀管
61	SY/T 5989—2019	直缝电阻焊套管
62	SY/T 6530—2019	非腐蚀性气体输送用管线管内涂层
63	SY/T 7456—2019	油气井套管柱结构与强度可靠性评价方法
64	SY/T 7457—2019	石油、石化和天然气工业 油气生产系统的材料选择和腐蚀控制
65	SY/T 5198—2020	钻具螺纹脂
66	SY/T 6662.2—2020	石油天然气工业用非金属复合管 第2部分：柔性复合高压输送管
67	SY/T 6859—2020	油气输送管道风险评价导则
68	SY/T 7495—2020	连续油管的维护与检测
69	SY/T 7496—2020	套管磨损试验方法

第四篇
"十三五"科技大事记

2016 年

▶ 1月23日至24日，管研院国家质检中心/CNAS检测与校准实验室（西安本部）质量体系顺利通过国家认可委员会复评审。

▶ 3月16日，石油管材及装备材料服役行为与结构安全国家重点实验室建设与运行实施方案顺利通过由陕西省科技厅组织的论证。

▶ 3月30日，石油工业专用螺纹量规计量站荣获中国石油集团"优秀计量机构"称号。

▶ 3月31日，全国石油天然气标准化技术委员会暨石油工业标准化技术委员会2016年年会在北京召开。

▶ 4月7日，管研院石油工业专用螺纹量规计量站顺利通过计量建标考核，取得由陕西省质量技术监督局颁发的2项计量标准考核证书。

▶ 4月26日，由管研院和石油管材及装备材料服役行为与结构安全国家重点实验室承办的集团公司科技项目引智计划"复杂工况油气井管柱服役安全技术交流会"在西安成功召开。来自国内外相关技术人员共计60余人参加了此次交流会。

▶ 4月27日至28日，由管研院、美国石油学会（API）、中国石油学会石油管材专业委员会、石油管材及装备材料服役行为与结构安全国家重点实验室联合举办的"油气井管柱与管材国际会议"在西安召开。来自10余个国家的油田、钢厂、管厂、科研院所共240余名代表参加了会议。

▶ 5月4日，管研院石油管材国家质检中心/石油管检测实验室顺利通过中国合格评定国家认可委员会组织的复评审，并取得国家认监委、认可委颁发的石油管材国家质检中心资质认定证书及石油管检测实验室认可证书。

▶ 5月6日，管研院被陕西省能源化工地质工会认定为"厂务公开职代会四星级单位"。

▶ 5月20日，全国油气标委会石油专用管材分技术委员会和油标委石油管材专业标准化技术委员会2016年年会在昆明市召开。来自油田用户、管材生产企业、设计单位、科研院所共计38家单位的近50余名代表参加了会议。

▶ 5月26日，陕西省科学技术厅公布了2016年陕西省创新人才推进计划入选名单。管研院2人入选中青年科技创新领军人才、4人入选青年科技新星，管研院入选创新人才培养示范基地。

▶ 6月15日至17日，管研院承办的中俄石油装备和材料标准化及合格评定会议在西安召开，中国石油天然气集团有限公司和俄企双方就各自国家和企业标准化政策进行了交流。

▶ 7月12日至14日，由国家石油管材质量监督检验中心举办的"石油管材力学性能检测方法研讨会"在西安召开，来自全国各地20余家钢厂、管厂、设备厂家共30余位技术人员参加了会议。

▶ 7月19日，管研院"CGC-1型气密封特殊螺纹套管""TG460型高强度铝合金钻杆""XGCE1型经济型气密封螺纹接头"等3项科技成果被认定为集团公司2015年度自主创新重要产品。

▶ 8月15日至16日，美国Microalloyed Steel Institute的J. Malcolm Gray博士以及美国原Battelle Institute的Brian Leis博士等专家一行到管研院交流访问。双方就全尺寸实验研究、

管道土壤回填、止裂预测技术等内容展开讨论。

▶8月25日，集团公司总经理、党组副书记章建华一行到石油管工程技术研究院调研指导工作。

▶8月25日至26日，石油管材专业标准化技术委员会在西安召开了石油管材国家、行业标准专家审查会和复审会，对128项国家和行业标准进行了复审，并对10项行业标准草案进行了审查。各油田、管道公司、科研院所和生产厂家，共计31家单位的70名委员、专家和代表参加了本次会议。

▶8月31日，管研院隆盛公司获得中国设备监理协会颁发的甲级设备监理单位证书。

▶9月27日，管研院召开科技创新大会暨建院35周年纪念大会。陕西省原副省长曾慎达、集团公司在陕相关企业领导、延长石油集团公司、陕西燃气集团公司、相关科研院所领导、专家，以及管研院全体员工、离退休员工、特邀嘉宾等共200余人参加了会议。

▶9月30日，管研院申报的SY/T 6857.1—2012《石油天然气工业特殊环境用油井管 第1部分：碳钢和低合金钢油管和套管选用推荐做法》、SY/T 6857.2—2012《石油天然气工业特殊环境用油井管 第2部分：酸性油气田用钻杆》获得2016中国标准创新贡献奖三等奖。

▶9月，在中国石油和石油化工设备工业协会第八届二次理事(扩大)会上，《关于成立石油管及装备材料专业委员会的提案》获得一致同意，并予以批准。

▶9月，《创新之路——石油管工程技术研究院35年发展史》编纂工作顺利完成并印刷成册。该书从多个方面追溯和记录了三十五年的发展历程以及取得的辉煌成就，客观地反映该历史进程中发生的重要事件、取得的重要成果和涌现出的关键人物。

▶10月，管研院首次自主评审机械专业高级技术职称，并首次采用网络平台进行职称评审工作。

▶11月15日，管研院三环公司获得中国特种设备检验协会审批并颁发的特种设备无损检测机构B级资质。

▶11月16日，"石油管材及装备材料服役行为与结构安全国家重点实验室管道断裂控制试验场"在新疆哈密正式挂牌。

▶11月16日，由管研院和西部管道公司联合开展的13.3MPa，ϕ1422mm X80螺旋焊管全尺寸气体爆破试验获得圆满成功。爆破试验的圆满完成对于填补国内全尺寸实物爆破试验技术空白，提高我国油气输送管道科技研发实力具有十分深远的意义。

▶11月18日，由管研院主办的《石油钻采装备金属材料手册》出版发布会在北京顺利召开。

▶11月25日，中国管线研究组织(简称CPRO)在西安召开了2016年冬季会议。会议由中国石油集团石油管工程技术研究院承办，来自21个会员单位的代表共计80余人参加会议。

▶12月1日，管研院隆盛公司顺利通过北京市2016年第一批高新技术企业认定。

2017年

▶3月20日，第十七届中国国际石油石化技术装备展览会(CIPPE)在北京中国国际展览中心开幕。管研院作为国内石油行业在石油管工程技术领域的核心科研机构参展

▶ 3月21日，管研院成功攻克了适用于-45℃极端环境的感应加热弯管母管成分设计技术及其配套热加工技术。

▶ 4月12日，管研院牵头组织的中国腐蚀与防护学会油气田及管道腐蚀与安全专业委员会成立大会在西安召开。

▶ 4月12日，由管研院发起的中国油气井管材与管柱技术创新联盟（COSTA）在西安完成了首次筹备研讨会议。

▶ 4月13日至14日，由管研院、美国石油学会（API）、石油管材与装备服役行为与结构安全国家重点实验室联合举办的"石油管及装备材料国际会议（TEC2017）"在西安召开。

▶ 4月21日至23日，管研院特种设备型式试验机构DGX、DSX、DYX、DJX四项压力管道元件项目顺利通过中国特种设备检验协会核准复评审。

▶ 4月27日，管研院组织的全国油气标委会石油专用管材分技术委员会暨油标委石油管材专业标准化技术委员会2017年年会在武汉召开。

▶ 5月9日至16日，由管研院和中国石油学会联合举办的2017年第一期全国石油管材螺纹检测人员资格鉴定与认证培训班在西安圆满完成培训计划，对35家参训企业的67名学员进行了一、二级资格鉴定与认证。

▶ 5月10日，由管研院主办的《石油管材与仪器》期刊通过了国家学术期刊认证机构的认定评审。

▶ 5月25日至26日，管研院牵头的中国管线研究组织在上海召开了2017年春季会议。

▶ 6月17日至23日，中国船级社认证公司对工程技术服务公司（原隆盛公司）QHSE、ES管理体系分别进行了再认证和监督评审，审核组同意推荐QHSE管理体系再注册及ES管理体系保持证书。

▶ 7月5日，管研院举办了"李鹤林院士学术思想和科学精神研讨会"。

▶ 7月5日至7日，由管研院国家石油管材质量监督检验中心举办的"新版API RP 5C5专题"研讨会在西安召开。

▶ 7月9日至12日，由石油管材及装备材料服役行为与结构安全国家重点实验室等单位联合承办的"石油管材及装备材料服役行为与结构安全学术研讨会"在银川顺利举行。

▶ 9月6日，石油管材专业标准化技术委员会组织召开了非金属及复合管重点技术标准宣贯会。

▶ 9月6日至8日，由管研院和中国石油学会石油管材专业委员会联合主办的美国石油学会（API）授权开展的API石油管材标准培训班在山东威海举行。

▶ 9月13日至15日，管研院HSE管理体系通过北京中油健康安全环境认证中心监督审核。

▶ 10月12日至13日，石油管材国家和行业标准专家审查会在西安顺利召开。

▶ 10月17日至24日，由管研院和中国石油学会联合举办的2017年第二期全国石油管材螺纹检测人员资格鉴定与认证培训班在西安圆满完成培训计划。

▶ 11月3日，管研院工程技术服务公司自主研制的高性能实体膨胀管完成首次现场下井补贴试验，取得圆满成功。

▶ 11月9至10日，由管研院和国家重点实验室联合举办的"中国油气井管材及管柱技术创新战略联盟成立大会暨全国套损防治技术研讨会"在西安成功举办。

▶ 11月23日至24日，中国管线研究组织在南京成功召开了2017年春冬季会议。

► 12月1日，中国工程院原副院长、钢铁研究总院院长、国家新材料产业发展专家咨询委员会主任干勇院士来院交流指导。

2018年

► 1月4日，管研院"油气管道与管柱联合研究中心"通过陕西省国际科技合作基地认定。

► 1月5日，管研院通过陕西省科技厅组织的高新技术企业资格认定。

► 1月18日至19日，由管研院工程技术服务公司组织召开了油气管道环焊缝检测评价与典型缺陷处理交流会。来自中国石油西部管道公司、中国石油西气东输公司、中国石油西南管道公司、北京天然气管道、中国海油上海分公司、管道公司和工程技术服务公司的领导和技术人员，等共40多名代表参加了会议。

► 2月1日至2日，由管研院国家石油管材质量监督中心暨石油工业螺纹量规计量站（简称计量站）举办的新版API 5B技术研讨会在成都举行，来自国内主要油气田、油套管生产企业共22家单位的49名技术人员参加了本次研讨会。

► 3月15日，管研院主办的2018远东无损检测新技术论坛石油管及装备专场筹备会在西安召开。

► 3月27日，第十八届中国国际石油石化技术设备展览会（CIPPE）在北京国展（新馆）隆重开幕。管研院与集团公司装备制造企业组团，共同向世界展示了中国石油装备的技术实力和水平。

► 4月7日，管研院与西安交通大学共同签署了"材料服役安全工程学"博士学科点共建的协议。

► 4月13日至15日，管研院质检中心/CNAS检测与校准实验室（西安本部和秦皇岛实验室）的质量体系顺利通过国家认可委评审专家复评审/扩项评审。

► 4月20日，API第46版API SPEC 5L《管线管规范》，正式发布，其中的附录N"用于需要纵向塑性变形能力PSL2钢管的订货"由管研院提案并参与制定。

► 5月11日，全国油气标委会石油专用管材分技术委员会和油标委石油管材专业标准化技术委员会2018年年会在管研院召开。

► 5月17日，管研院石油工业专用螺纹量规计量站（简称螺纹计量站）在通过计量建标考核，顺利取得由国家质量监督检验检疫总局颁发的石油螺纹单项参数检查仪校准装置计量标准考核证书。

► 5月17日，管研院喜获中国石油集团"2017年度质量先进企业"荣誉称号。

► 5月28日，管研院李鹤林院士和全国"百名科学家、百名基层科技工作者"代表付安庆博士参加了两院院士大会开幕式。

► 5月29日，管研院参加陕西省"一带一路"科技创新创业暨重点实验室创新合作论坛。

► 6月1日，管研院"油气输送管道服役安全创新团队"入选陕西省"三秦学者创新团队"支持计划。

► 6月6日上午，"中国材料与试验团体标准委员会石油石化工程及装备材料领域委员会（简称CSTM/FC58）"成立大会暨学术报告会在中国石油集团石油管工程技术研究院召开。

▶ 6月15日，管研院作为发起单位当选陕西省重点实验室创新联盟副理事长单位。石油管材及装备材料服役行为与结构安全国家重点实验室冯耀荣主任当选为副理事长。

▶ 6月20日，管研院牵头申报的2018年度国家重点研发计划"高应变海洋管线管研制"项目成功通过评审。这是管研院成立三十多年来首次牵头承担国家项目，具有非常重要的意义。

▶ 7月12日至15日，石油管材及装备材料服役行为与结构安全学术研讨会在福建厦门国际会展中心召开。

▶ 7月18日，管研院在呼和浩特组织召开了油井管重点技术标准宣贯会。来自油田管具公司、工程技术、物资供应等用户单位、管材制造企业及科研院所等近20家单位的40余名代表参加了会议。

▶ 8月8日，石油管工程重点实验室顺利通过集团公司第二轮运行评估。

▶ 8月16日至18日，管研院和塔里木油田公司在库尔勒联合举办"油气田管道完整性管理技术培训会"。

▶ 9月6日，韩国天然气公司研究院 Lee Seong Min 院长一行来管研院进行技术交流访问。

▶ 10月16日，管研院与美国西南研究院签署合作协议。

▶ 10月17日，俄罗斯天然气公司楚涅夫斯基·安德烈·雅拉斯拉沃维奇总工艺师一行来管研院进行技术交流访问。

▶ 10月18日，由管研院承办的2018年中国油气计量论坛在西安成功举行。

▶ 10月18日，由管研院、巴西矿冶公司（CBMM）、中信金属有限公司共同发起的"国际焊接研究中心（IWTC）"在西安正式成立。

▶ 10月18日，陕西省石油管材及装备材料服役行为与结构安全重点实验室顺利通过陕西省科技厅组织的验收。

▶ 10月26日，《石油管材与仪器》期刊编委会在石油管工程技术研究院召开，来自高校、油气田公司、石油管材与仪器生产企业和科研院所的近五十位编委和嘉宾参加了会议。

▶ 10月30日，由管研院和国家重点实验室举办的集团公司科技项目引智计划"石油管完整性设计及应用技术研习会"在西安成功召开。

▶ 11月8日，管研院在秦皇岛实验室成功进行12MPa内压下 $\phi1219mm\times26.4mm$ X80钢管全尺寸弯曲试验，这是目前国内进行的最大规格的钢管全尺寸弯曲试验。

▶ 11月13日，管研院出席工信部实验室座谈会并荣获"年度优秀行业重点产品质量状况分析报告"奖。

▶ 11月25日，管研院检验机构顺利通过国家认可委员会现场复评审。

▶ 12月17日，管研院主办的《石油管材与仪器》期刊获第五届陕西省科技期刊优秀期刊称号。

2019 年

▶ 1月4日，管研院成功实施-30℃ $\phi1422mm$ X80钢管全尺寸低温部分气体爆破试验，该试验温度为目前国内开展试验的最低温度，标志着我国在低温试验技术上的一次突破，为中俄东线等低温环境用管道的安全服役提供了支撑，填补了国内空白。

▶ 1月7日，管研院三环公司通过特种设备无损检测机构TOFD资质增项审核。

▶ 1月17日至18日，由国家石油管材质量监督中心暨石油工业螺纹量规计量站（简称计量站）举办的第16版API Spec 5B《套管、油管和管线管螺纹的加工、测量和检验》技术研讨会在成都举行。

▶ 1月31日，管研院成功完成了首次低温环境宽板拉伸试验，此项试验技术的开发，使管研院宽板拉伸试验能力得到了进一步扩充。

▶ 3月14日，石油管材及装备材料服役行为与结构安全国家重点实验室平台建设及试验新技术开发项目顺利通过中国石油集团组织的验收。

▶ 3月29日，管研院牵头承担的国家重点研发计划"高应变海洋管线管研制"项目总体方案通过专家审查。

▶ 4月28日，管研院"石油管材及装备材料服役行为与结构安全国家重点实验室"获"陕西青年五四奖章集体"荣誉称号。

▶ 5月9日，管研院"石油管材及装备材料服役行为与结构安全国家重点实验室"获集团公司青年文明号集体荣誉称号。

▶ 5月11日，国内首次全尺寸管道应力腐蚀疲劳加速试验在管研院秦皇岛实验室正式启动。

▶ 5月22日至24日，由管研院和石油管材及装备材料服役行为与结构安全国家重点实验室主办的"石油管及装备材料国际会议（TEC2019）"在西安隆重召开。来自世界各地行业内专家和技术人员共计450余人参加大会。

▶ 5月23日上午，依托管研院建立的NACE新疆分会正式成立。

▶ 7月1日，管研院荣获"中国石油天然气集团有限公司2018年度质量安全环保节能先进企业"称号。

▶ 7月9日，管研院喜获"沣东杯"陕西省科技工作者创新创业大赛银奖。

▶ 7月9日，管研院研制的智慧管道监测系统在长庆油田首试成功。

▶ 7月10日至14日，由管研院和石油管材及装备材料服役行为与结构安全国家重点实验室主办的石油管材及装备材料服役行为与结构安全学术研讨会在成都召开。

▶ 8月27日，依托管研院建立的中国石油和石油化工设备工业协会石油管及装备材料专业委员会正式成立。

▶ 9月7日，管研院油气田及管道腐蚀与防护团队获"中国腐蚀与防护学会四十年贡献奖——优秀科技团队"荣誉称号。

▶ 10月25日，国际标准化组织（ISO）投票通过，由管研院提出的1项ISO国际标准提案《石油天然气工业陶瓷内衬油管》成功立项。这是我国首次主导制定的油井管领域国际标准，标志着我国陶瓷内衬油管技术在国际上走向引领。

▶ 10月27日，管研院国家质检中心顺利通过扩项、变更评审。

▶ 11月11日，管研院高级技术专家罗金恒入选2019年国家百千万人才工程，并被授予"有突出贡献中青年专家"荣誉称号。

▶ 11月7日，管研院举办的钛/铝轻质合金钻杆的研发与应用技术学术交流会在西安顺利召开。

▶ 11月20日，管研院"石油管材及装备材料服役行为与结构安全国家重点实验室"主任冯耀荣荣获"科技领军建功人才"，青年科技人员王建军、王蕊荣获"青年科技立业英才"。

▶ 12月3日，管研院作为主要承担单位形成的"第三代大输量天然气管道工程关键技术"攻关成果在中俄东线上成功应用。

▶ 12月2日，管研院联合西安德信成科技有限公司、衡阳华菱钢管有限公司研制的国内首支双金属冶金复合油管热轧成功。

▶ 12月23日，管研院牵头申报的2019年度国家重点研发计划"石油天然气工业非API石油专用管质量基础设施关键技术体系研究"项目获国家科技部正式立项批复。

2020 年

▶ 2月，新冠肺炎疫情发生以来，石油管工程技术研究院为确保员工的健康安全与科研工作两不误，开通远程办公模式助力员工居家办公；搭建WeLink智能工作平台助力疫情防控工作；拓宽电子文献获取渠道；多渠道及时宣传疫情防控信息，充分利用信息化手段为全院做好疫情防控和科研工作提供有力支撑，助力全面打赢疫情防控阻击战。

▶ 3月3日，管研院隆盛公司获得中国设备监理协会颁发的设备监理单位证书，申请扩项的乙级设备监理资质获得授权。

▶ 3月12日，管研院国家质检中心管材实物实验室获得陕西省委科技工委"2019年度省委科技工委系统学雷锋活动示范点"荣誉称号。

▶ 3月26日，国家自然资源部宣布我国海域天然气水合物第二轮试采创造了"产气总量86.14万立方米，日均产气量2.87万立方米"两项新的世界纪录，攻克了深海浅软地层水平井钻采核心技术，实现了从"探索性试采"向"试验性试采"的重大跨越，管研院在本轮试采中为成功使用水平井套管提供了关键技术支撑。

▶ 4月10日，管研院主导修订的ISO 11960：2020《石油天然气工业油气井套管或油管用钢管》(第六版)正式出版发布。

▶ 4月23日，石油管工程技术研究院自主研发的数字化石油螺纹检测系统首次在山东永利精工石油装备股份有限公司成功签约。

▶ 4月23日，国家重点实验室输送管所与国家质检中心合作将宽板拉伸试验与新型的应变测试方法相结合，顺利完成基于DIC的宽板拉伸试验。

▶ 4月27日，石油工业专用螺纹量规计量站所建立的抽油杆螺纹量规校准装置顺利通过国家计量建标考核，取得国家市场监督管理总局颁发的计量标准考核证书。

▶ 4月28日，管研院与东营高新区管委会签订《中国石油集团石油管工程技术研究院与东营高新技术产业开发区产学研合作意向书》。

▶ 5月14日，中国工程院院士、中国机械工业集团党委常委、副总经理陈学东一行来管研院调研。

▶ 5月22日，《石油管材及装备材料服役行为与结构安全国家重点实验室完善建设》项目顺利通过集团公司可行性论证。

▶ 5月27日，石油管材与装备材料服役行为与结构安全国家重点实验室主办的膨胀管技术研讨会在西安召开。

▶ 6月9日，管研院成功中标西部管道公司油气管道环焊缝适用性评价及管材试验项目，中标金额679万，创管道检测评价领域技术服务单体合同新高。

▶ 6月9日，中国石油储气库评估中心揭牌仪式在北京恒毅大厦储气库公司隆重举行。

挂靠管研院的安全评估分中心依托国家重点实验室开展工作。

▶ 6月18日，管研院研发的 X70B 型套筒在中缅 $\phi1016mm$ X80 管道上成功应用。

▶ 7月16日，由管研院牵头研制并现场技术服务的国产 105ksi 级钛合金钻杆，在西部油田顺利完成了一口设计井深为 7100m 的超短半径水平井的三开定向造斜侧钻任务并成功出井。

▶ 7月31日，由管研院牵头联合国内中世钛业企业研制的钛合金钻杆在大庆超短半径水平井成功应用。

▶ 8月20日，管研院"高含硫油气田集输缓蚀剂""抗硫非金属复合连续管""基于应变设计的 80SH 热采套管"3 项产品被列入集团公司 2019 年度自主创新重要产品目录。

▶ 9月9日，由中国石油学会石油管材专业委员会和石油管工程技术研究院联合主办的"第五届中国石油石化腐蚀与防护技术交流大会暨石油石化腐蚀与防护专家论坛"在西安成功召开。来自中国石油、中国石化、中国海油、延长石油，以及国内相关科研院所共计 400 余名代表参会。

▶ 9月25日，管研院与大庆油田采油工程研究院共同组建的"油气井管柱设计与评价联合研究中心"揭牌。

▶ 10月12日，管研院与英国 TWI 通过网络会议开展"管道环焊缝断裂的力学行为"专题研讨会。

▶ 10月16日，第 29 届孙越崎能源科学技术奖颁奖大会在京举行。管研院油井管与管柱研究所副所长冯春获孙越崎青年科技奖。

▶ 10月21日，管研院荣获"中国石油天然气集团有限公司 2019 年度质量安全环保节能先进企业"称号。

▶ 10月21日，管研院主办的 API 油井管标准技术研讨会在西安召开。

▶ 10月27日，管研院主办的期刊《石油管材与仪器》在"第六届陕西省科技期刊评优活动"中被评为"特色期刊"。

▶ 10月28日，"全国石油和化工期刊百强榜发布会"在重庆召开，管研院主办的《石油管材与仪器》入选"石油和化工学术期刊 50 强"。

▶ 11月9日，由中国石油集团石油管工程技术研究院主办的 2020 年石油螺纹计量国际研讨会在成都顺利召开。来自油田、油井管制造企业、国际知名计量设备制造商、科研院所、高校及计量技术研究机构等 30 余家单位 70 余名代表参加了会议。

▶ 11月13日，由管研院和渤海装备大港新世纪合作生产的经济型气密封螺纹套管，在长庆采油六厂安平 35-332 井下井成功。

▶ 11月22日，由中国石油集团石油管工程技术研究院、威海华腾海洋工程技术有限公司召集举办的"2020 远东无损检测新技术论坛——石油管及装备专场"在江苏省南京市恒大酒店国会厅召开。

▶ 11月24日，油气田用非金属管全产业链项目发展规划研讨会在西安召开。来自中国石油各油田公司、炼化公司、化工销售公司、管道生产公司及科研院所等 17 家 49 名专家代表应邀参加了研讨会。

▶ 12月2日，中国石油和石油化工设备工业协会团体标准审查会在深圳召开。来自油田、制造企业、科研院所等单位的 24 名专家和代表参加会议。

▶ 12月15日，石油管工程技术研究院东营分院揭牌暨启动仪式在东营高新区胜利石油科技创新园隆重举行。